Lecture Notes in Artificial Intelligence   13728

Subseries of Lecture Notes in Computer Science

Series Editors

Randy Goebel
University of Alberta, Edmonton, Canada

Wolfgang Wahlster
DFKI, Berlin, Germany

Zhi-Hua Zhou
Nanjing University, Nanjing, China

Founding Editor

Jörg Siekmann
DFKI and Saarland University, Saarbrücken, Germany

More information about this subseries at https://link.springer.com/bookseries/1244

Haris Aziz · Débora Corrêa · Tim French (Eds.)

AI 2022: Advances in Artificial Intelligence

35th Australasian Joint Conference, AI 2022
Perth, WA, Australia, December 5–8, 2022
Proceedings

 Springer

Editors
Haris Aziz
University of New South Wales
Sydney, NSW, Australia

Débora Corrêa
University of Western Australia
Perth, WA, Australia

Tim French
University of Western Australia
Perth, WA, Australia

ISSN 0302-9743   ISSN 1611-3349 (electronic)
Lecture Notes in Artificial Intelligence
ISBN 978-3-031-22694-6   ISBN 978-3-031-22695-3 (eBook)
https://doi.org/10.1007/978-3-031-22695-3

LNCS Sublibrary: SL7 – Artificial Intelligence

This Springer imprint is published by the registered company Springer Nature Switzerland AG
The registered company address is: Gewerbestrasse 11, 6330 Cham, Switzerland

Preface

This volume contains the papers presented at the 35th Australasian Joint Conference on Artificial Intelligence (AI 2022). The conference was held in hybrid mode during December 5–9, 2022, and was hosted by the University of Western Australia in Perth, Australia. This annual conference is one of the longest running conferences in artificial intelligence, with the first conference held in Sydney in 1987. The conference remains the premier event for artificial intelligence in Australasia, offering a forum for researchers and practitioners across all subfields of artificial intelligence to meet and discuss recent advances.

AI 2022 received 90 submissions and each submission was reviewed by at least two Program Committee (PC) members or external reviewers in a double-blind process. After a thorough discussion and rigorous scrutiny by the reviewers, 35 papers were accepted for oral presentations and 21 papers were accepted for poster presentations in the conference. In total, 56 submissions were accepted for publication as full papers in this proceeding. AI 2021 had four keynote talks by the following distinguished scientists:

- Makoto Yokoo, Kyushu University, Japan, on December 6, 2022, speaking on "Market design for constrained matching".
- Timothy Miller, University of Melbourne, on December 7, 2022, speaking on "The state of explainable AI".
- Bob Coeck, Chief Scientist, Quantinium, UK, on December 7, 2022, speaking on "Quantum AI".
- Stela Solar, Data61, Australia, on December 8, 2022, speaking on "The directions of AI in Australasia".

The following are notable aspects of the AI 2022 conference:

- AI 2022 was the first opportunity for a large face to face meeting of the Australasian AI community since the beginning of the COVID-19 pandemic. Face to face meetings and opportunities to meet and share ideas are critical to the research community. This is especially the case in Australasia where large distances make such meetings all the more important.
- The West Australian economy is powered by the resources and agriculture industries, and these industries have made large investments in automation, artificial intelligence, and data science. AI 2022 included a day with a special industry focus in partnership with the University of Western Australia Data Institute. Panel discussions allowed industry and academia to share challenges and research directions.
- The AI 2022 program included three workshops, held on Friday, December 9: The First Australasian Symposium on Artificial Intelligence for the Environment (AI4Environment) organised by Guodong Long, Yanjun Zhang, and Tao Shen; The First Workshop on Toxic Language Detection (TLD) organised by Jean Lee, Henry Weld, Kunze Wand, and Josiah Poon; and Federated Learning in Australasia: Towards

Large-Scale AI Systems with Privacy-Preservation" organised by Jarrod Trevathan, M A Hakim Newton, and Ashfaqur Rahman.

- The AI 2022 program included four special tutorials, held on Monday, December 5: Developments in Fair Resource Allocation presented by Haris Aziz, Xinhang Lu, Mashbat Suzuki, and Toby Walsh; A Practical Guide to Knowledge Graph Construction from Technical Short Text presented by Michael Stewart; Memory-based Reinforcement Learning presented by Hung Le; and Joint Slot Filling and Intent Detection for Natural Language Understanding presented by Henry Weld, Caren Han, Josiah Poon, and Sharon Long.
- The program included a Doctoral Consortium held on Monday, December 5, to mentor and assist postgraduate students developing their research, with mentorship provided by research leaders.

We especially appreciate the work of the members of the Program Committee and the external reviewers for their expertise and tireless effort in assessing the papers within a strict timeline. We are also very grateful to the members of the Organising Committee for their efforts in the preparation, promotion, and organisation of the conference, especially the general chairs, Wei Liu and Abdul Sattar, for coordinating the whole event. We acknowledge the assistance provided by EasyChair for conference management.

Lastly, we thank the National Committee for Artificial Intelligence of the Australian Computer Society, Springer, for the professional service provided by the Lecture Notes in Computer Science editorial and publishing teams, and our conference sponsors: the Commonwealth Scientific and Industrial Research Organisation; the UWA Data Institute; the UWA School of Physics, Mathematics and Computing; the UNSW AI Institute; the Centre for Transforming Maintenance through Data Science; the WA Data Science Innovation Hub; and DUG Technology.

December 2022   Haris Aziz
                                                                                                            Débora Corrêa
                                                                                                            Tim French

Organization

General Chairs

Wei Liu   The University of Western Australia, Australia
Abdul Sattar   Griffith University, Australia

Program Committee Chairs

Haris Aziz   University of New South Wales Sydney, Australia
Tim French   The University of Western Australia, Australia

Proceedings Chair

Débora Corrêa   The University of Western Australia, Australia

Program Committee

Shadi Abpeikar	University of New South Wales Canberra, Australia
Harith Al-Sahaf	Victoria University of Wellington, New Zealand
Zhuoyun Ao	Defence Science and Technology Group, Australia
Peter Baumgartner	CSIRO, Australia
Ying Bi	Victoria University of Wellington, New Zealand
Taotao Cai	Macquarie University, Australia
Weidong Cai	The University of Sydney, Australia
Zehong Cao	University of South Australia, Australia
Gang Chen	Victoria University of Wellington, New Zealand
Stephen Chen	York University, Canada
Hongxu Chen	The University of Queensland, Australia
Sung-Bae Cho	Yonsei University, South Korea
Dave de Jonge	IIIA-CSIC, Spain
Marcus Gallagher	The University of Queensland, Australia
Xiaoying Gao	Victoria University of Wellington, New Zealand
Manolis Gergatsoulis	Ionian University, Greece
Giorgio Gnecco	IMT School for Advanced Studies Lucca, Italy
Guido Governatori	Independent Researcher, Australia
Ning Gu	University of South Australia, Australia
Hans W. Guesgen	Massey University, New Zealand

Yi Guo   Western Sydney University, Australia
Soyeon Han   The University of Sydney, Australia
Tim Hendtlass   Swinburne University, Australia
Melinda Hodkiewicz   The University of Western Australia, Australia
Eun-Jung Holden   The University of Western Australia, Australia
Changqin Huang   SCNU, China
Guangyan Huang   Deakin University, Australia
Huan Huo   University of Technology Sydney, Australia
Mahdi Jalili   RMIT University, Australia
Asanka Nuwanpriya Kekirigoda   Defence Science and Technology Group,
  Mudiyanselage   Australia
Daniel Le Berre   CNRS - Université d'Artois, France
Ickjai Lee   James Cook University, Australia
Andrew Lensen   Victoria University of Wellington, New Zealand
Weihua Li   Auckland University of Technology, New Zealand
Pauline Lin   The University of Melbourne, Australia
Hui Ma   Victoria University of Wellington, New Zealand
Yi Mei   Victoria University of Wellington, New Zealand
Tim Miller   The University of Melbourne, Australia
Richi Nayak   Queensland University of Technology, Australia
M. A. Hakim Newton   The University of Newcastle, Australia
Bach Nguyen   Victoria University of Wellington, New Zealand
Maurice Pagnucco   University of New South Wales, Australia
Laurence Park   Western Sydney University, Australia
Xueping Peng   University of Technology Sydney, Australia
Laurent Perrussel   IRIT - Universite de Toulouse, France
Bernhard Pfahringer   University of Waikato, New Zealand
Mark Reynolds   The University of Western Australia, Australia
Rafal Rzepka   Hokkaido University, Japan
Khaled Saleh   University of Technology Sydney, Australia
Dilini Samarasinghe   The University of New South Wales, Australia
Rolf Schwitter   Macquarie University, Australia
Harisu Abdullahi Shehu   Victoria University of Wellington, Australia
Yanjun Shu   Harbin Institute of Technology, China
Hannes Strass   TU Dresden, Germany
Maolin Tang   Queensland University of Technology, Australia
Jing Teng   North China Electric Power University, China
Markus Wagner   The University of Adelaide, Australia
Xianzhi Wang   University of Technology Sydney, Australia
Bing Wang   University of New South Wales, Australia
Che Wang   Griffith University, Australia
Huang Wang   Southeast University, Australia

Max Ward	The University of Western Australia, Australia
Kevin Wong	Murdoch University, Australia
Brendon J. Woodford	University of Otago, New Zealand
Shuxiang Xu	University of Tasmania, Australia
Miao Xu	University of Queensland, Australia
Jianhua Yang	University of Western Sydney, Australia
Yi Yang	Hefei University of Technology, China
Nayyar Zaidi	Monash University, Australia
Qin Zhang	University of Technology Sydney, Australia
Xuyun Zhang	Macquarie University, Australia
Dongmo Zhang	Western Sydney University, Australia
Mengjie Zhang	Victoria University of Wellington, New Zealand
Fangfang Zhang	Victoria University of Wellington, New Zealand
Dengji Zhao	ShanghaiTech University, China

Advisory Board

Melinda Hodkiewicz	The University of Western Australia, Australia
Guodong Long	University of Technology Sydney, Australia
Tim Miller	Melbourne University, Australia
Mark Reynolds	The University of Western Australia, Australia
Toby Walsh	The University of New South Wales, Australia
Mary-Anne Williams	The University of New South Wales, Australia
Xinghuo Yu	RMIT University, Australia
Chengqi Zhang	University of Technology Sydney, Australia
Mengjie Zhang	Victoria University of Wellington, New Zealand

Additional Reviewers

Hayden Andersen	Ruijun Li
Bin Chen	Long Nguyen
Matthew Damigos	Jingli Shi
Qinglan Fan	Mashbat Suzuki
Abdul Karim	Ziying Zhao
Ruotong Hu	Xianglin Zheng
Dongnan Liu	

Keynote Presentations

Market Design for Constrained Matching

Makoto Yokoo 🆔

Kyushu University, Fukuoka, Japan
yokoo@inf.kyushu-u.ac.jp
http://https://sites.google.com/view/makoto-yokoo/

Abstract. The theory of two-sided matching (e.g., assigning residents to hospitals, students to schools) has been extensively developed, and it has been applied to design clearinghouse mechanisms in various markets in practice, including resident matching programs and school choice programs. As the theory has been applied to increasingly diverse types of environments, however, researchers and practitioners have encountered various forms of distributional constraints. As these features have been precluded from consideration until recently, they pose new challenges for market designers. One example of such distributional constraints is a minimum quota, e.g., school districts may need at least a certain number of students in each school in order for the school to operate. In this talk, I present an overview of research on designing mechanisms that work under distributional constraints.

Keywords: Two-sided matching · Market design · Game theory

Are the Inmates Still Running the Asylum? Explainable AI is Dead, Long Live Explainable AI!

Tim Miller

School of Computing and Information Systems,
The University of Melbourne, Parkville, VIC 3010
tmiller@unimelb.edu.au

Abstract. In this talk, I will discuss why I believe many of the assumptions we have been making about explainable AI for better decision making are misguided, and how we can address this issue. In the past, I have argued that in the explainable AI community, maybe 'the inmates are running the asylum', re-framing Alan Cooper. By this, I mean that experts in artificial intelligence are not well placed to design explainability tools and techniques that are intended for non-expert users, and we should turn to the social sciences and human-computer interaction to mitigate this. I will review discuss theories from philosophy, cognitive science, and social & cognitive psychology that have been influential in explainable AI in the last few years, improving explainable techniques. I'll discuss that, despite this progress, I believe the inmates have been running the asylum all along without us knowing, but that by questioning our assumptions, we can change direction to see improved outcomes.

Keywords: Explainable AI · Human-centred AI · Decision support systems

From Quantum Picturalism to Quantum AI

Bob Coecke 🆔

Quantinuum, Compositional Intelligence Team, Oxford
bob.coecke@quantinuum.com

Abstract. In 2020 our Oxford-based Quantinuum team performed Quantum Natural Language Processing (QNLP) on IBM quantum hardware [1, 2, 3]. Key to having been able to achieve what is conceived as a heavily data-driven task, is the observation that quantum theory and natural language are governed by much of the same compositional structure [4, 5, 6] – a.k.a. tensor structure. Hence our language model is in a sense quantum-native, and we provide an analogy with simulation of quantum systems in terms of algorithmic speed-up. Meanwhile we have made all our software available open-source, and with support [7].

We will also introduce the notion of compositional intelligence, exploiting the fact that the compositional match between natural language and quantum extends to other domains as well, such as spatio-temporal perception [8], we will argue that a new generation of AI can emerge when fully pushing this analogy. The so-called ZX-calculus [9, 10] for quantum theory (and linear algebra more generally) has been proven to be complete, so can be conceived as a full-bodied reasoning system that go hand-in-hand with modern machine learning.

Keywords: QNLP · Quantum picturalism · Compositional intelligence

References

1. Coecke, B., de Felice, G., Meichanetzidis, K., Toumi, A.: Foundations for near-term quantum natural language processing (2020). arXiv:2012.03755
2. Meichanetzidis, K., Toumi, A., de Felice, G., Coecke, B.: Grammar-aware question-answering on quantum computers (2020). arXiv:2012.03756
3. Lorenz, R., Pearson, A., Meichanetzidis, K., Kartsalkis, D., Coecke. B.: QNLP in practice: Running compositional models of meaning on a quantum computer (2021). arXiv:2102.12846
4. Clark, S., Coecke, B., Sadrzadeh, M.: A compositional distributional model of meaning. In: Proceedings of the Second Quantum Interaction Symposium (QI-2008), pp. 133–140 (2008)
5. Coecke, B., Sadrzadeh, M., Clark, S.: Mathematical foundations for a compositional distributional model of meaning. In: van Benthem, J., Moortgat, M., Buszkowski, W. (eds.) A Festschrift for Jim Lambek, vol. 36. Linguistic Analysis, pp. 345–384 (2010). arXiv:1003.4394

6. Clark, S., Coecke, B., Grefenstette, E., Pulman, S., Sadrzadeh, M.: A quantum teleportation inspired algorithm produces sentence meaning from word meaning and grammatical structure. Malays. J. Math. Sci. **8**, 15–25 (2014). arXiv:1305.0556
7. Kartsaklis, D., et al.: lambeq: an efficient high-level Python library for quantum NLP (2021). arXiv:2110.04236
8. Wang-maścianica, V., Coecke, B.: Talking space: Inference from spatial linguistic meanings. J. Cogn. Sci. **22**(3), 421–463, (2021)
9. Coecke, B., Kissinger, A.: Picturing Quantum Processes. A First Course in Quantum Theory and Diagrammatic Reasoning. Cambridge University Press (2017)
10. Coecke, B., Horsman, D., Kissinger, A., Wang, Q.: Kindergarden quantum mechanics graduates... or how i learned to stop gluing LEGO together and love the ZX-calculus. Theor. Comput. Sci. **897**, 1–22 (2022). arXiv:2102.10984

AI: Our Co-pilot in a Complex World

Stela Solar

Abstract. We live in a world of high complexity. From the data jungle and increasing population, to process complexities and some of the most existential challenges facing our society. In this context of grand challenges, we need a grand tool to help us navigate the complexity. AI is one such tool. AI can be our co-pilot in this complex world, helping us lead our lives in accordance with our human values.

Contents

Ethical/Explainable AI

Genetic Algorithms

Knowledge Representation and NLP

Machine Learning

Optimization

Reinforcement Learning

Computer Vision

Automated On-Vehicle Road Defect Data Collection and Detection

Zachary Todd$^{(\boxtimes)}$ and Heyang Li

Department of Mathematics and Statistics, University of Canterbury, Christchurch,
New Zealand 8041
zachary.todd@pg.canterbury.ac.nz, thomas.li@canterbury.ac.nz

Abstract. This paper proposes a pipeline for the automated on-vehicle data collection, filtering, and classification of road surface defects. The proposed pipeline provides a flexible framework that allows for the integration of a variety of systems. The pipelines flexibly allow for various sensors such as camera, 3D camera and lidar; computational resources such as on-vehicle edge computing or cloud computing; data transfer such as 5G or on-site upload; and data storage. The pipeline was tested using an edge computer on board a contracted road sweeping vehicle with an image taken every 10 s with image processing and evaluation occurring between. Post installation, the pipeline required no input from the driver of the sweeper vehicle besides turning on the road sweeper. The data was transferred via WiFi as the road sweeper was pulling up at the end of its shift. During operation around 21k road, defects were identified with over 90% of these images containing road defects.

Keywords: Deep learning · Data collection · Edge computing

1 Introduction

There have been strong improvements in the camera sensing technology, and the area of autonomous vehicles technology [12,18], with improved driver assistance, and a better understanding of the road environments [2,8,10,11,13,14].

However, there is still a significant gap in using those technologies for automatic road defects data collection and detection. Much of the research and development in this area has been using data collected manually, with manual data transfer and filtration steps that are not suitable for the large-scale automation needed to detect the whole road network.

Performing these tasks post-data collection allows for greater flexibility that is only limited by the computational resources available and the limitations of the data itself. Whereas, performing these tasks on-vehicle available decreases the available power resources which in turn decreases the computational potential. However, there are several advantages, with real-time data collection allowing for the filtering of data, as well as periodic or real-time reporting. In this context can in inform when a section of the road has been significantly damaged and needs to be fixed.

© The Author(s), under exclusive license to Springer Nature Switzerland AG 2022
H. Aziz et al. (Eds.): AI 2022, LNAI 13728, pp. 3–14, 2022.
https://doi.org/10.1007/978-3-031-22695-3_1

The paper aims to utilise the potential advantages gained from performing data collection and detection on a collection vehicle while creating a flexible pipeline that allows for the proposed approach to fit a wide variety of needs and applications. The pipeline, in short consists of several modules, these being; image capture, location tracking, image prepossessing, image evaluation, data transfer, and data storage.

Initial validation of the approach is demonstrated by evaluating the image evaluation module and testing its ability to perform detection. This is then followed up using by running a case study utilising this model. The case study consists of a real-world application, in which a road sweeper completes its normal tasks while having the proposed application pipeline installed on the collection vehicle. The desired outcome of the case study is that despite running data collection and detection on-vehicle that the model is able to detect road defects and do so in a reliable manner.

2   Method

This section covers two pipelines used in both training applications. In addition, this section also covers the dataset used to train the image evaluation models and the apparatus the overall apparatus used in the application pipeline (Fig. 1).

2.1   Pipeline

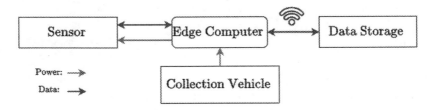

Fig. 1. Overview of the proposed pipeline.

Our approach consists of two pipelines. These are the training and application pipelines. The training pipeline is responsible for training and evaluating the model being used with the application pipeline. The application pipeline is the pipeline used in practices and the case study, with it being is made up of several modules, these being; image capture, location tracking, image prepossessing, image evaluation, data transfer, and data storage.

The image capture module's responsibility is to capture images and to send the captured images to the edge computer for preprocessing. For this, four implementations methods were considered:

(1) Wireless camera such as GoPro and communicate image collected over Bluetooth to edge computer.
(2) Smartphone powered via USB from the vehicle.
(3) Wired camera connected to the vehicle battery and communicating image using power over ethernet (PoE) to edge computer.
(4) Wired camera connected and powered using USB to edge computer.

With implementation (4) being used in the application pipeline. Implementation (1) is the most flexible on its face though it requires the battery of the wireless device to be charged. In addition to this problem, a number of reliability issues with the automatic pairing of a variety of wireless camera systems with the edge computer after they had been turned off, thereby requiring a person to interact with the system to ensure that the camera was charged and paired with the edge computer. Implementation (2) was the cheapest of the four implementations. However, like (1) there are several reliability issues such as image processing slowing down and the unreliable image capture rate, with the likely cause of these issues being due to wear and tear of the smartphone and overheating after extensive use. Implementation (3) though is very similar to implementation (4) because the increased power demand of the PoE made the powering of the system less reliable. Implementation (4) mitigates the reliability issues of the other implementations with there being no overheating, communication issues or major power fluctuations.

For location tracking, a GPS receiver was connected to the edge computer to provide the GPS position of the collection vehicle. This position information is provided with the captured images as metadata so that detected problems can be located.

For computation, a Intel NUC was selected, other options included low-power small GPU systems such as Nvidia Jetson Nanoor Nvidia Jetson Xavier, or the aforementioned smartphone. With the NUC over a GPU system because of cost and availability.

For image prepossessing, the image was cropped to remove the sky and other unrelieved features within the image and resized to the specification of the image evaluation model.

After prepossessing, the image evaluation model evaluates the images, with the responses sent to the data transfer and storage module. The evaluation model is dependent on the type of sensor used to capture the image, the computational resources available and the type of evaluation being performed. As performing the evaluation on-vehicle limits the potentially available power; the available computational resources were also limited.

The data transfer and storage module is responsible for transferring the image and metadata from the edge computer on the vehicle to where the data is stored. There are several methods to transfer the data from the edge computer. The following implementations were considered:

(1) 5G transfer direct from the edge computer to cloud storage.
(2) On board storage and manual retrieval and upload of the data.
(3) On-site WiFi connection with that data being transferred when the collection vehicle returns.

Implementation (1) allows for all of the collected images to be transferred from the edge computer during collection and allows for the possibility of real-time reporting, with the main limiting factor of this implementation being the cost of using 5G infrastructure. Implementation (2) is the simplest from a technology perspective. However, this would result in a lag in detection and reporting time, as well as requiring someone to interact with the edge computer regularly. Implementation 3 provides a compromise between the aforementioned implementations, allowing for a report per shift of the collection vehicle and requiring no manual interaction with the edge computer. However, implementation (3) does limit the number of images that can be transferred due to the bandwidth limitation of transferring data over WiFi while the collection vehicle is leaving and arriving from its station. To mitigate this, only the images (with associated metadata) in which detections occurred are transferred, thereby significantly decreasing the amount of data being transferred (Fig. 2).

2.2 Dataset

The dataset consists of 120k images collected around the South Island, New Zealand, specifically Canterbury, Nelson, Otago and Tasman regions. The dataset consists of all road segments of State-highways 65, 69, 73, 75, 85, and 87 and part of State-highways 1, 6, 7, 8 and 72, with the dataset encompassing over 3000 Km. The annotated dataset is a subset of this, including 19k images taken from these collections. The dataset was labelled with two classes, these being potholes and other defects. During annotations, the images were labelled with an encapsulating polygon. The images were labelled in two passes, with the first labelling the image in a batch of 500 images and then a reviewer to confirm their labels. On average each image took 20s to label with this being just over 30s including the second pass. Around 8% of the labelled images contained other defects and less than 1% contained potholes.

2.3 Apparatus

The training pipeline uses a Nvidia RTX2080 Ti GPU to train the models running with python 3.8, TensorFlow 2.0 and CUDA 10.1.

In the application pipeline, the collection used a road sweeper vehicle that was contracted to sweep roads around several suburbs in Christchurch, New Zealand. To capture the images a Logitech USB web camera was used. For computing, an Intel NUC with an I7 processor was used, with the NUC set to start upon receiving power. The NUC receive power using the road sweeper's auxiliary power outlet. To provide location information for the images, GlobalSat BU-353-S4 USB GPS receiver was used. Running on the NUC was python 3.8

Fig. 2. Dataset examples. The top row consists of example images without defects and the second row contains annotated images with defects.

with TensorFlow 2.0 and Intel Optimization for TensorFlow to run the image evaluation model. For data transfer, the data is transferred over WiFi at the road sweeper station as the road sweeper is departs and arrives at the station.

The main limiting factor of the apparatus is the computing platform. The NUC with an I7 processor [1] was chosen as a compromise between power consumption and computation with lower computation processors such as Pentium and I3 providing similar power requirements though with less computation resources. A potentially better computation platform to use would a low power consumption small GPU systems such as Nvidia Jetson Nano or Nvidia Jetson Xavier. However, due to global supply chain issues, this was not possible at the time.

2.4   Model Training

Two types of models were considered for the image evaluation model; these being image classification and instance segmentation models. Image classification models classify an image into a number of set classes, with each image having an assorted class. Instance segmentation models provide an encapsulating polygon or mask of all of the instances of the relevant objects within an image, as well as classifying these objects. Each input of instance segmenting results in either a set of polygons-class or image-class pairs to communicate what parts of the image belong to each instance object and their assorted classes.

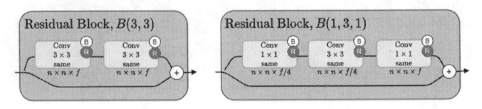

Fig. 3. ResNet Block, With the size of the output volume of each conv layer displayed below each conv layer.

The ResNet [7] convolutional neural network (CNN) was selected as the classification model. ResNet main feature is its residual block (Fig. 3), with the block consisting of two parallel branches. The first is a branch is a series of convolutional (conv) layers either two 3×3 conv layers (B(3, 3)) or 1×1, 3×3 and 1×1 conv layers (B(1, 3, 1)) and the second branch make no changes and passes the input volume froward to be added to the first branch. ResNet starts with a 7 conv layer followed by a maxpool and then followed by fours stages of residue blocks. The number of residual blocks per stage is different for each ResNet variant. After each conv layer there is batch normalisation [9] followed by ReLU activation [17]. After the fourth stage, there are two fully connected layers to finish the network.

Commonly used ResNet variants:

- ResNet18: block: B(3, 3), blocks per stage [2, 2, 2, 2]
- ResNet34: block: B(3, 3), blocks per stage [3, 4, 6, 3]
- ResNet50: block: B(1, 3, 1), blocks per stage [3, 4, 6, 3]
- ResNet101: block: B(1, 3, 1), blocks per stage [3, 4, 23, 3]
- ResNet152: block: B(1, 3, 1), blocks per stage [3, 8, 36, 3]

For the classification experiment, ResNet50 was selected as it is the largest of the ResNet variant while meeting the computation limitation of the edge computer.

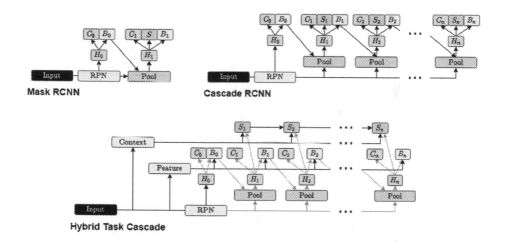

Fig. 4. Instance segmentation method structure.

For instance segmentation, three CNN models were considered; Mask RCNN, Cascade RCNN, and Hybrid Task Cascade (HTC) [3,4,6]. All three methods work similarly by using a backbone network such as ResNet to generate region proposals and the difference in the overall structure is shown in Fig. 4. Mask RCNN then uses a pooling layer to warp the variable size region proposals into a predefined size shape. Finally, these pooled regions are then connected to a head layer that predicts the class, bounding boxes and mask.

Cascade RCNN has two improvements over Mask RCNN these being cascade bounding box (bbox) regression and cascade detection. The cascade bbox regression improves quality by providing a series of regressions with each regressor having a stricter intersection over union (IoU) acceptance threshold. The cascade detection resampling of the positive examples of previous regressors with a higher IoU, as a regressor with an IoU threshold of μ, will produce bboxes with IoU score higher than μ.

HTC interleaves mask and bbox detectors with each bounding box and masks being connected and each aggressor feeding to the next regressor in the cascade. HTC also used a separate spatial context branch for masks instead of using the region proposal network.

For the instance segmentation experiment, HTC with ResNet50 was selected. This is because HTC has better results on large general datasets [15] than Cascade RCNN and Mask RCNN while requiring amount a similar amount of computational resources.

3   Experiments and Results

This section covers the experiments used to evaluate the selected models on the dataset, as well as discusses the case study used to evaluate the application pipeline.

3.1   Model Evaluation

For evaluation, the annotated dataset is split into two subsets with 80% being used to train and 20% being used to test the models. Both the classification and segmentation models use the AutoAugment [5] method to augment the annotate dataset and AdamW optimiser [16] to optimise the gradient descent, and were trained for 300 epochs.

The classification result as shown in Table 1 shows accuracy across all classes of 91.4%. With none of the Pothole images being classified as 'No Defect' and

Table 1. Classification confusion matrix

	No defect	Pothole	Other defect	Total
No Defect	3202	14	297	3513
Pothole	0	32	8	40
Other Defect	15	2	338	355
Total	3217	48	643	3908

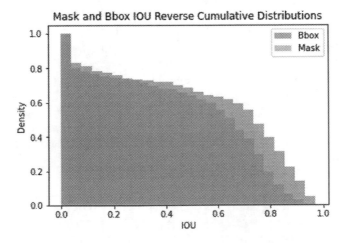

Fig. 5. Mask and Bounding box results

Fig. 6. Detected defect examples, with pothole detection examples in the first two rows and other defects detection examples in the last two rows.

only 15 of the 'Other defect' class being classed as 'No Defect'. These results demonstrated that the model is able to distinguish between images that do and do not contain defects, as well as distinguishing between defects that are and are not potholes (Fig. 5).

The instance segmentation results show the HTC model on the test dataset has a mask and bbox AP@50 (average precision using 0.5 IoU as the acceptance threshold) of 0.69 and 0.71 respectively, and a mask and bbox AP@75 of 0.58 and 0.63 respectively, with there being a sharp drop-off in acceptance with an IoU of 0.6. These results demonstrate that the detection and segmentation are viable in most cases with almost 70% of the images having a greater than 50% mask and bbox IoU.

3.2   Case Study

The computer, camera, and GPS receiver were placed inside and connected to the road sweeper vehicle. The road sweeper ran as normal for a month collecting data and running the application pipeline. The model used to determine if a defect was present was ResNet50. Although the segmentation model performs the classification method has a faster inference time as it does not have to run the associated detection and segmentation components of HTC.

During operation, the only reliability issues experienced were interference between the GPS receiver and a GPS receiver built into the collection vehicle. This caused several of the detected defect images to not have assorted metadata. Although this unreliability can easily be remedied in the case where a collection vehicle has GPS then the image captured time code and the time code of the collection vehicle GPS can be synced to get the location of the captured image. During the month of operation over 21k images were identified with 92.4% containing defects (Fig. 6).

4   Conclusion

The application pipeline demonstrates the viability of automating the detection of road defects in a flexible, accurate and reliable manner. The flexibility is demonstrated in the relatively low complexity of the modules allowing for the accommodation of a variety of sensors, computation platforms and data transformation methods. The accuracy is demonstrated in the results on the training dataset, as well as the presence of road defects within the case study. However, the result that over 90% of the reported images have defects does not guarantee that the result of the dataset has been able to transfer to the context of the case study as there is not a full picture of the images that were determined to not contain a defect. It does provide some confidence that this mapping between the two similar contexts has occurred. Finally, the reliability of the application pipeline is demonstrated by the reliability of the case study, with only minor reliability issues such as the interference in the GPS receiver.

A natural progression from these outcomes would be to increase both the resolution of the problem from the perspective of both the types of road defects being detected, as well as the complexity of reporting being performed by the edge computer. Increasing the class resolution could achieve by training an evaluation model on more diverse classes, for example, adding road defect classes such as cracking, surface defects and surface distress. Increasing the complexity could be achieved by running an instance segmentation model to provide the location and the size of defects within an image, as well as using an onboard GPU to speed up the computation.

References

1. Comparison charts for intel® core™ desktop processor family. https://www.intel.com/content/www/us/en/support/articles/000005505/processors.html
2. Agrawal, R., Chhadva, Y., Addagarla, S., Chaudhari, S.: Road surface classification and subsequent pothole detection using deep learning. In: 2021 2nd International Conference for Emerging Technology (INCET), pp. 1–6. IEEE (2021)
3. Cai, Z., Vasconcelos, N.: Cascade r-cnn: high quality object detection and instance segmentation. IEEE Trans. Pattern Anal. Mach. Intell. **43**, 1483–1498 (2019)
4. Chen, K., et al.: Hybrid task cascade for instance segmentation. In: Proceedings of the IEEE/CVF Conference on Computer Vision and Pattern Recognition, pp. 4974–4983 (2019)
5. Cubuk, E.D., Zoph, B., Mane, D., Vasudevan, V., Le, Q.V.: Autoaugment: learning augmentation policies from data. arXiv preprint arXiv:1805.09501 (2018)
6. He, K., Gkioxari, G., Dollar, P., Girshick, R.: Mask r-CNN. In: 2017 IEEE International Conference on Computer Vision (ICCV). IEEE (2017). https://doi.org/10.1109/iccv.2017.322
7. He, K., Zhang, X., Ren, S., Sun, J.: Deep residual learning for image recognition (2016)
8. Hu, Y., Furukawa, T.: Degenerate near-planar 3d reconstruction from two overlapped images for road defects detection. Sensors **20**(6), 1640 (2020)
9. Ioffe, S., Szegedy, C.: Batch normalization: accelerating deep network training by reducing internal covariate shift. arXiv preprint arXiv:1502.03167 (2015)
10. Jung, J., Bae, S.H.: Real-time road lane detection in urban areas using lidar data. Electronics **7**(11), 276 (2018)
11. Kim, J., Kim, J., Jang, G.J., Lee, M.: Fast learning method for convolutional neural networks using extreme learning machine and its application to lane detection. Neural Netw. **87**, 109–121 (2017)
12. Levinson, J., et al.: Towards fully autonomous driving: systems and algorithms. In: 2011 IEEE Intelligent Vehicles Symposium (IV), pp. 163–168. IEEE (2011)
13. Li, H.T., Todd, Z., Bielski, N.: Equirectangular image data detection, segmentation and classification of varying sized traffic signs: a comparison of deep learning methods (2022)
14. Li, H.T., Todd, Z., Bielski, N., Carroll, F.: 3d lidar point-cloud projection operator and transfer machine learning for effective road surface features detection and segmentation. Visual Comput. **38**(5), 1759–1774 (2022)
15. Lin, T.-Y., et al.: Microsoft COCO: common objects in context. In: Fleet, D., Pajdla, T., Schiele, B., Tuytelaars, T. (eds.) ECCV 2014. LNCS, vol. 8693, pp. 740–755. Springer, Cham (2014). https://doi.org/10.1007/978-3-319-10602-1_48

16. Loshchilov, I., Hutter, F.: Decoupled weight decay regularization. arXiv preprint arXiv:1711.05101 (2017)
17. Nair, V., Hinton, G.E.: Rectified linear units improve restricted boltzmann machines. In: ICML, pp. 807–814 (2010). https://icml.cc/Conferences/2010/papers/432.pdf
18. Yurtsever, E., Lambert, J., Carballo, A., Takeda, K.: A survey of autonomous driving: common practices and emerging technologies. IEEE Access 8, 58443–58469 (2020)

Vision Transformer Based Model for Describing a Set of Images as a Story

Zainy M. Malakan[1,2](✉) (iD), Ghulam Mubashar Hassan[1] (iD), and Ajmal Mian[1] (iD)

[1] The University of Western Australia, Perth Wa 6009, Australia
{ghulam.hassan,ajmal.mian}@uwa.edu.au
[2] Umm Al-Qura University, Makkah 24382, Saudi Arabia
zmmalakan@uqu.edu.sa

Abstract. Visual Story-Telling is the process of forming a multi sentence story from a set of images. Appropriately including visual variation and contextual information captured inside the input images is one of the most challenging aspects of visual storytelling. Consequently, stories developed from a set of images often lack cohesiveness, relevance, and semantic relationship. In this paper, we propose a novel Vision Transformer Based Model for describing a set of images as a story. The proposed method extracts the distinct features of the input images using a Vision Transformer (ViT). Firstly, input images are divided into 16×16 patches and bundled into a linear projection of flattened patches. The transformation from a single image to multiple image patches captures the visual variety of the input visual patterns. These features are used as input to a *Bidirectional-LSTM* which is part of the sequence encoder. This captures the past and future image context of all image patches. Then, an attention mechanism is implemented and used to increase the discriminatory capacity of the data fed into the language model, i.e. a *Mogrifier-LSTM*. The performance of our proposed model is evaluated using the Visual Story-Telling dataset (VIST), and the results show that our model outperforms the current state of the art models.

Keywords: Storytelling · Vision transformer · Image processing

1 Introduction

Visual description or storytelling (VST) seeks to create a sequence of meaningful sentences to narrate a set of images. It has attracted significant interest from the vision to language field. However, compared to image [7,17] and video [18,24,28] captioning, narrative storytelling [14,19,21] has more complex structures and incorporates themes that do not appear explicitly in the given set of images. Moreover, describing a set of images is challenging because it demands algorithms to not only comprehend the semantic information, such as activities and objects in each of the five images along with their relationships, but also demands fluency in the phrases as well as the visually unrepresented notions.

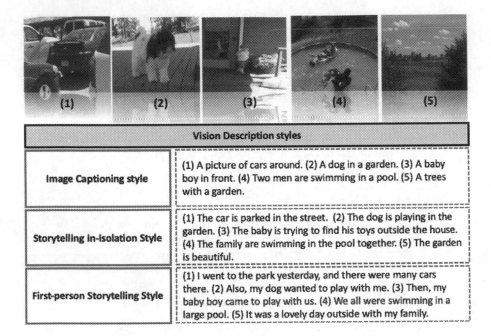

Fig. 1. An example of three vision description techniques includes a single picture caption, story-like caption, and narrative storytelling, which is our aim.

Recent storytelling techniques utilise sequence-to-sequence (seq2seq) models [14,20] to produce narratives based story on a set of images. The key idea behind these approaches is to implement a convolutional neural network (CNN), (*i.e*, *a sequence encoder*), to extract the visual features of the set of images. Then, combining these visual features, a complete set of image representations is obtained. The next step is to input this representational vector into a hierarchical long-short-term memory (LSTM) model to form a sequence of sentences as a story. This approach has dominated this area of research owing to its capacity to generate high-quality and adaptable narratives.

Figure 1 illustrates the technical challenges between single image captioning style, isolation style, and storytelling style for a set of five images. For example, the first sentence of all the three blocks in Fig. 1 annotations show the following: "A picture of cars around.", "The car is parked in the street.", and "I went to the park yesterday, and there were many cars there.". The first description is known as image captioning style which conveys the actual and physical picture information. The second description is known as storytelling in-isolation style which catches the image content as well, but it is not linked to the following sentence. The final description is known as first-person storytelling style which explains more inferences about the image as a story-based sentence and also links to the subsequent sentence.

In order to solve the above challenges and difficulties, we propose a novel methodology that explores the significance of spatial dimension conversion and its efficacy on Vision Transformer (ViT) [6] based model. Our method proceeds by extracting the feature vectors from the given images by dividing them into 16X16 patches and feeding them into a *Bidirectional-LSTM* (*Bi-LSTM*). This models the visual patches as a temporal link among the set of images. By using the *Bidirectional-LSTM*, we represent the temporal link between patches in both forward and backward directions. To preserve the visual-specific context and relevance, we convert the visual features and contextual vectors from *Bi-LSTM* into a shared latent space using a *Mogrifier-LSTM* architecture [22]. During the first layer's gated modulation, the initial gating step scales the input embedding based on the ground truth context, producing a contextualized representation of the input. This combination of multi-view feature extraction and highly context-dependent input information allows the language model to provide more meaningful and contextual descriptions of the input set of images.

The following is a summary of the contributions presented in this paper:

- We propose a novel ViT sequence encoder framework, that utilises multi-view visual information extraction for appropriate narrative based story on the given set of images as input.
- We take into account the context of the past as well as the future and employ an attention mechanism over the contextualized characteristics that have been obtained from Vision Transformer (ViT) to construct semantically rich narratives from a language model.
- We propose to combine *Mogrifier-LSTM* with enriched visual characteristics (patches) and semantic inputs to generate data-driven narratives that are coherent and relevant.
- We demonstrate the utility of our proposed method through multiple evaluation metrics on the largest known Visual Story-Telling dataset (VIST) [12][1]. In addition, we compare the performance of our technique with existing state of the art techniques and show that it outperforms them on various evaluation metrics.

2   Related Works

This section presents a review of literature on different visual captions that directly relate to narrative storytelling techniques, followed by the literature on visual storytelling methods.

2.1   Visual Understanding

Visual understanding algorithms, which include image and video captioning, are the most significant sort of networks utilized to tackle the problem of narrative storytelling. Since it is most relevant to our study, we briefly discuss the recent

---

[1] https://visionandlanguage.net/VIST/.

literature on neural network-based image and video captioning. Typically, these models extract a vector of visual features using a CNN and then transmit this vector to a language model for caption synthesis.

Image Captioning (IC) consists of a single frame (*i.e*, an image) defined by a single phrase. Approaches may be further classified as rule-based methods [2,23] and deep learning-based methods [11,27]. The rule-based approaches apply the traditional methodology of recognizing a restricted number of pre-defined objects, activities and locations in the image, and describing them in natural language using template-based techniques. On the other hand, due to recent advances in deep learning, the vast majority of current methods are dependent on deep learning as well as scientifically advanced techniques such as attention [31], reinforcement learning [29], semantic attributes integration [15], and modeling of subjects and objects [5]. However, none of these algorithms are designed to produce a narrative-based description of a set or collection of images.

Video Captioning (VC) defined as multi-frame description that can explain many frames (*i.e*, a video) in a single statement. VC and storytelling techniques are quite similar as they both utilize an encoder-decoder framework. The encoder is composed of a 2D/3D CNN that extracts visual information from a set of input frames. This information is subsequently converted into normal language phrases using a decoder or a language model based on either a recurrent neural network [4,25] or a transformer network [13,16,33]. Although VC methods can describe multi-frames in a single caption efficiently, it does not generate a story or multi-sentence descriptions for a given set of images.

2.2   Storytelling Methods

Telling a story based on a set of images is an easy task for humans, but an extremely difficult task for machines. Coherent, relevant, and grammatically correct sentences must be generated for a story-based description. For example, Park et al. (2015) [26] illustrated that using bidirectional recurrent neural network (BRNN) is more efficient than a usual recurrent neural network (RNN) because BRNN captures forward and backward image features, which enables the model to interact with the whole story's sentences. Similarly, Sequence-to-Sequence (Seq2Seq) techniques, which utilize CNN+Bi-LSTM [14][2] or CNN+GRU [30] as an encoder and RNN as a decoder enhanced storytelling prediction from a set of images.

In addition, the concept of designing composite rewards as a strategy for storytelling problems was introduced [10][3], which improved the natural flow of the generated story. A novel decoder-encoder framework using *Mogrifier-LSTM* [20] was also proposed to improve the coherence, relevance, and information of

---

[2] https://github.com/tkim-snu/GLACNet.
[3] https://github.com/JunjieHu/ReCo-RL.

the generated story. Recently, the object detection technique using *YOLOv5* [21] is embedded with the encoder to improve the relevance of the story sentences.

Different from previous works, our proposed method derives characteristics from the multiple visual features (i.e., patch features) based on the human-like approach to generate stories. This helps to propose an approach that is both computationally efficient and capable of producing coherent, relevant, and informative stories.

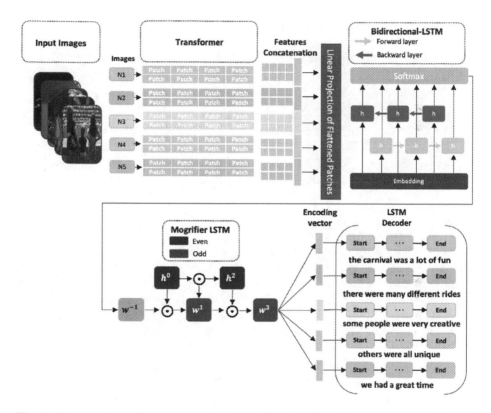

Fig. 2. An overview of our proposed model which consists of a sequence encoder and decoder. The sequence encoder process is implemented by both the Vision Transformer (ViT) and the *Bidirectional-LSTM*. The decoder process is performed by the *Mogrifier-LSTM* as well as the standard LSTM.

3 Proposed Method

Figure 2 presents the overall architecture of our proposed model which comprises of Vision Transformer (ViT), sequence encoder and decoder modules. In the first step, the image features are extracted using ViT, which divides each image into 16×16 patches and encodes them. Then all extracted image patch features are further encoded by *Bidirectional-LSTM* module which extracts the temporal context of the images. The connection between the sequence encoding and image features is captured by the Attention module on two levels: the patch level and the image-set patch level. Finally, the decoder module is responsible for the generation of a sequence of sentences as human story-like by making use of the *Mogrifier-LSTM* architecture. The following discussion delves into the specifics of the aforementioned three modules.

3.1   Vision Transformer (ViT)

A set of I images are fed by the data-loader as $I_s = (I_1, I_2, ..., I_s)$, where

$$I = [I_1, I_2, ..., I_N] \quad s.t. \quad I_N \in \mathbb{R}^{H \times W \times C}, \tag{1}$$

$s \in \{1, 2, 3, 4, 5\}$ which is a set of five images with HxWxC (Height x Width x Channels) shape that presents a unique representation of storytelling from the dataset. To extract image features, we utilized Vision Transformer (ViT) [6] which breaks the given I image into N equal-sized, non-overlapping patches of shape (P, P, C) and linearly maps each patch to a visual representation. We define the extracted features as the combination of patches from the ViT model as follows:

$$I_0 = [I_p^1 E; I_p^2 E; ...; I_p^N E] \quad s.t. \quad E \in \mathbb{R}^{(P^2.C) \times D} \tag{2}$$

where P is the defined parameter as in grid order (left to right, up to down) while C represents the total number of channels. Then we flatten all patches which produces n line feature vectors of shape $(1, P^{2\star}C)$. The patches that have been flattened are multiplied by a trainable embedding tensor of shape $(P^{2\star}C, D)$, which gains the ability to linearly project each flat patch to dimension D. As a result, we produce rich embedded patches of shape $n = (1, D) \in \mathbb{R}^{(1,D)}$.

3.2   Features Encoding

The purpose of visual storytelling is first to comprehend the flow of events occurring in each image and then to produce a consistent narrative similar to how humans narrate a story. As a set of $P = P_1, P_2, ..., P_l$, where P represents the total number of image patches included in I as well as the number of corresponding contexts in each story. In order to represent these relationship features, we utilize a *Bidirectional-LSTM*, which compiles the sequential information of P patches in both *forward* and *backward* direction. Our sequence encoder requires an input of image feature vector $\boldsymbol{f}_i$ at every time step 't' where $i \in \{1, 2, .., 5\}$. Eventually, the sequence encoder part of the model encodes the whole image set, comprising all the image patches and provides contextual information $\boldsymbol{h}_{se} = [\overrightarrow{\boldsymbol{h}_{se}}; \overleftarrow{\boldsymbol{h}_{se}}]$ through the final hidden-state at time step number $t = 5$.

3.3   Story Generation

Since modelling sequential inputs must lead to generating coherent sentences, the solution to the challenge lies in how well the model learns the context. This is particularly problematic for issues that need high levels of coherence and relevance. To solve this, we utilize the standard LSTM [9], which forms the current hidden state denoted by $h^{\langle t \rangle}$, based on the previous hidden state, represented by h_{prev}, and refreshes its memory state $c^{\langle t \rangle}$. Further, standard LSTM utilizes input gates $\boldsymbol{\Gamma}_i$, forget gates $\boldsymbol{\Gamma}_f$, and output gates $\boldsymbol{\Gamma}_o$ which are determined as follows:

$$\boldsymbol{\Gamma}_f^{\langle t \rangle} = \sigma(\boldsymbol{M}_f[\boldsymbol{h}_{prev}, \boldsymbol{w}_t] + \boldsymbol{B}_f), \tag{3}$$

$$\boldsymbol{\Gamma}_i^{\langle t \rangle} = \sigma(\boldsymbol{M}_i[\boldsymbol{h}_{prev}, \boldsymbol{w}_t] + \boldsymbol{B}_i), \tag{4}$$

$$\tilde{\boldsymbol{c}}^{\langle t \rangle} = \tanh(\boldsymbol{M}_c[\boldsymbol{h}_{prev}, \boldsymbol{w}_t] + \boldsymbol{B}_c), \tag{5}$$

$$\boldsymbol{c}^{\langle t \rangle} = \boldsymbol{\Gamma}_f^{\langle t \rangle} \odot \boldsymbol{c}^{\langle t-1 \rangle} + \boldsymbol{\Gamma}_i^{\langle t \rangle} \odot \tilde{\boldsymbol{c}}^{\langle t \rangle}, \tag{6}$$

$$\boldsymbol{\Gamma}_o^{\langle t \rangle} = \sigma(\boldsymbol{M}_o[\boldsymbol{h}_{prev}, \boldsymbol{w}_t] + \boldsymbol{B}_o), \tag{7}$$

$$\boldsymbol{h}^{\langle t \rangle} = \boldsymbol{\Gamma}_o^{\langle t \rangle} \odot tanh(\boldsymbol{c}^{\langle t \rangle}) \tag{8}$$

where w is the word vector embedded in the input at time step 't' (for simplicity, we eliminate t), $\boldsymbol{M}_*$ represents the transformation matrix that is learned at each state, $\boldsymbol{B}_*$ are the biases, σ shows the logistic sigmoid function, and $\odot$ is the product of the vectors' Hadamard transform. In our generation module, the attention vector $\boldsymbol{\zeta}_i$ from the sequence encoder output is used to set up the LSTM hidden state h.

Furthermore, we boost the standard LSTM functionality to generate more cohesive and relevant story-like sentences by integrating a *Mogrifier-LSTM* [22]. The two inputs, w and h_{prev}, modulate each other in an odd and even fashion before being sent into the standard LSTM. In order to accomplish this goal, the *Mogrifier-LSTM* instead scales the columns of each of its weight matrices throughout $\boldsymbol{M}_*$ via *Mogrifier-LSTM* gated modulation. In formal terms, w is gated based on the previous step h_{prev} as gated input. A similar approach of gating prior time step output is used with the previous gated input.

Following the completion of five rounds of mutual gating, as recommended by Malakan *et al.* [21], the most highly indexed versions of w and h_{prev} are subsequently fed into the standard LSTM in the order shown in Fig. 2. Therefore, it may also be stated as: *mogrification* $(w, c_{prev}, h_{prev}) = LSTM(w^{\uparrow}, c_{prev}, h_{prev}^{\uparrow})$ where $w^{\uparrow}$ and $h_{prev}^{\uparrow}$ are the most significant possible indexed for the LSTM inputs w^i and h_{prev}^i respectively. Mathematically,

$$w^i = 2\sigma(\boldsymbol{M}_{xh}^i h_{prev}^{i-1}) \odot w^{i-2}, \text{ for odd} i \in [1, 2, ..., r], \tag{9}$$

$$h_{prev}^i = 2\sigma(\boldsymbol{M}_{hx}^i w^{i-1}) \odot h_{prev}^{i-2}, \text{for even } i \in [1, 2, ..., r], \tag{10}$$

where Hadamard product is $\odot$, which $\boldsymbol{w}^{-1} = \boldsymbol{w}$, $\boldsymbol{h}^0_{prev} = \boldsymbol{h}_{prev} = \boldsymbol{\zeta}_i$ and r represents the total number of mogrification rounds which is a mogrifier hyper-parameter. In addition, the default standard LSTM configuration, with r set to 0, operates without gated mogrification at the input stage. The use of matrix multiplication with a constant of 2 ensures that the resulting transformations of the matrices $\boldsymbol{M}^i_{xh}$ and $\boldsymbol{M}^i_{hx}$ are close to the identity matrix.

3.4   Data Pre-processing and Model Training

A vocab size of 6464 was extracted from the Visual Story-Telling dataset (VIST) with a minimum word count threshold of 8. In addition, we used the size of 256 as a dimension of word embedding vectors. Then, all VIST images are resized to 224×224 pixels from the original size and used as an input to the pre-trained Vision Transformer (ViT). For the training parameters, the Adam optimizer was used, and the learning rate was set at 0.001, while the weight decay was established at 1e-5. Also, a teacher-forcing strategy was utilized in our proposed model to help the model train faster.

All of these settings were calibrated on our NVIDIA GPU, which has 12 GB of memory. For the maximum possible usage of available memory, the batch size was set to 8 during training. This ensured that we obtained the most out of the memory that was available to us. It's worth mentioning that greater GPU memory provides increased batch sizes, which assist the model to train faster. The model has been successfully trained for a total of 83 epochs utilizing about 390K steps. Finally, each epoch of our model was saved locally on our computer. Then, the optimum performance of the model was carefully chosen from epoch 59 since, after epoch 59, the model began to overfit the data and the loss error began to increase, resulting in decreased model accuracy.

4   Experiments and Results

First, we introduce the Visual Story-Telling dataset (VIST) used to evaluate our proposed model. Next, we discuss the results of our proposed model and compare them to other state of the art models. Finally, we give detailed analysis of a few cases in terms of the generated stories and the scores.

4.1   Dataset

Visual Story-Telling dataset (VIST) [12][4] is the only publicly accessible dataset that we are aware for storytelling problems. It comprises 210,819 distinct images that can be found in 10,117 different albums on Flickr and is arranged in sets of five different images. Two types of stories accompany each set of images. One is called Description In Isolation (DII) and includes individual image descriptions that can be useful for research in image captioning. The second one is called

---

[4] https://visionandlanguage.net/VIST/.

Story In Sequence (SIS) which is more relevant to storytelling problems and comprises a whole paragraph in precisely five sentences representing a story. In all dataset statements, it is essential to note that the names of the individuals are adjusted by "[male and female]", places by "[location]", and organizations by "[organization]".

4.2   Performance Comparison

Automatic evaluation metrics are the most common technique for estimating the effectiveness of the automatically generated story. Therefore, we validate our proposed model using automatic evaluation metrics, which also allows us to compare it to the current state-of-the-art methods. Table 1 displays the most recent frameworks used in storytelling challenges. These frameworks were published since 2018 and obtained promising results on the VIST dataset. We compare our proposed model using multiple evaluation metrics, which are: BLEU-1, BLEU-2, BLEU-3, BLEU-4, CIDEr, METEOR, and ROUGE-L. The script for computing the evaluation measures was released by [10][5].

From the experiments, we observe that our model performs better than the state-of-the-art models on all of the given evaluation measures, except for the BLEU-1, BLEU-4, and CIDEr. Table 1 presents the results for all of the mentioned models sorted by the year in which they were released. Overall, our proposed model outperforms the compared models by 0.3 points in BLUE-2, 1.1 points in BLUE-3, 0.7 points in ROUGE-L and 0.2 points in METEOR.

4.3   Storytelling Example Analysis

Automatic metrics are not a perfect reflection of the accuracy of the stories. Therefore, we conducted an in-depth analysis of the stories produced by our proposed model and the ground truth. In addition, we compared our stories with stories produced by the recently proposed CAMT 2021 model [20].

Figure 3 illustrates two different stories from a set of five images from our proposed model, followed by the stories that were Generated using CAMT. The highlighted text in green shows parts that are highly relevant to the story, while the highlighted text in yellow indicates less relevant or general information that is not obvious from the images. We also report the automatic evaluation metrics below each story in Fig. 3.

Text Generation Analysis: Both selected models have shown a persuasive example of a narrative that is representative of how humans write a story, and both of these examples are captivating. In contrast to CAMT model, our proposed model is able to extract more useful information from each input image. For instance, the 3rd sentence in the first scenario shows more relevance to the story, i.e. "they were so happy to be married," as compared to CAMT model

---

[5] https://github.com/JunjieHu/ReCo-RL.

Table 1. A comparison of our proposed model with the recently published methods on the Visual Story-Telling dataset (VIST). Quantitative results were obtained using seven different automated measures of evaluation. "–" indicates that the authors of the corresponding study did not publish the results. The higher scores represent higher accuracy and the results in bold represent the best scores.

Model	B-1	B-2	B-3	B-4	CIDEr	ROUGE-L	METEOR
AREL 2018 [32]	0.536	0.315	0.173	0.099	0.038	0.286	0.352
GLACNet 2018 [14]	0.56	0.321	0.171	0.091	0.041	0.264	0.306
HCBNet 2019 [1]	0.59	0.348	0.191	0.105	0.051	0.274	0.34
HCBNet(w/o prev. sent. attention) [1]	0.59	0.338	0.180	0.097	0.057	0.271	0.332
HCBNet(w/o description attention) [1]	0.58	0.345	0.194	0.108	0.043	0.271	0.337
HCBNet(VGG) 2019 [1]	0.59	0.34	0.186	0.104	0.051	0.269	0.334
ReCo-RL 2020 [10]	–	–	–	0.124	0.086	0.299	0.339
BLEU-RL 2020 [10]	–	–	–	0.144	0.067	0.301	0.352
VS with MPJA 2021 [8]	0.601	0.325	0.133	0.082	0.042	0.303	0.344
CAMT 2021 [20]	0.64	0.361	0.201	**0.184**	0.042	0.303	0.335
Rand+RNN 2021 [3]	–	–	0.133	0.061	0.022	0.272	0.311
SAES Encoder-Decoder OD 2021 [21]	0.64	0.363	0.196	0.106	0.051	0.294	0.330
SAES Encoder-Decoder OD & Noun 2021 [21]	0.63	0.357	0.195	0.109	0.048	0.299	0.331
SAES Encoder OD 2021 [21]	**0.65**	0.372	0.204	0.12	**0.054**	0.303	0.335
Our Proposed Model	0.63	**0.375**	**0.215**	0.123	0.044	**0.310**	**0.354**

which predicted a less relevant sentence "they were very excited." In addition, we noticed that the third and the last sentences in the second examples, which are "They were all smiling" and "Everyone was happy to be at the event."; the two sentences do not relate to the image itself in any manner, and the information they provide seems to be generic and applicable to many images. On the other hand, our proposed model generated a story that was more logically consistent with the story and relevant to the images.

Generated Story Scores: It is essential to demonstrate the model's performance in contrast with traditional automated evaluation metrics. Each set of images comes with a total of five different stories that were written by real people (*i.e.*, ground truth), as mentioned in Sect. 4.1. One of these stories was extracted randomly and removed from the collection. Next, we compared the story generated by our proposed model, CAMT model and the removed ground truth story (we name this Human Generated Story or HGS) with the rest of the four stories in the collection of VIST dataset. In BLEU-1, we found that our proposed model obtains the highest score on the second example, with almost 0.17 points more than HGS and 0.65 points more than the CAMT model; in BLEU-2, our generated story obtains over 0.14 points more than HGS and CAMT model; in BLEU-3, our model obtains 0.9 points more than the HGS and CAMT model in the first example, but the CAMT model receives 0.2 points more than ours and 0.16 points over the HGS in the second example; in BLEU-4, CAMT obtains almost 0.77 points more than our model and the HGS in the first example, while

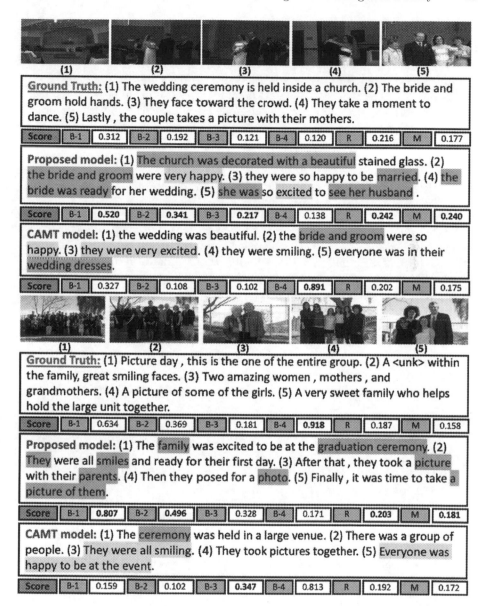

Fig. 3. Examples of our generated stories in comparison to the CAMT model [20] and ground truth. Text highlighted in green indicates high relevance to the image/story, while text highlighted in yellow means that it is not highly relevant but instead contains general information. BLEU-1 (B-1), BLEU-2 (B-2), BLEU-3 (B-3), BLEU-4 (B-4), ROUGE (R), and METEOR (M) scores are shown below each story, with bold scores indicating the highest value.

the HGS receives 0.74 points more than our model and 0.10 points over the CAMT model in the second example; in ROUGE-L, we reported that our model obtains almost 0.2 points more than both the HGS and the CAMT model in both examples; and in METEOR, we find that our model obtains almost 0.7 points more than both the HGS and the CAMT model in both examples. On the other hand, the performance of our proposed model is sufficiently high across practically all of the automated evaluation metrics, with the exception of BLEU-4, as is shown through Fig. 3.

5   Conclusion

This article presented a novel storytelling approach for describing a set of images in a coherent manner. Our proposed framework is robust, which consists of a sequence encoder that receives multi-view image patches from Vision Transformer (ViT) as an input to a *Bidirectional-LSTM*, and a decoder with a standard LSTM enhanced by *Mogrifier-LSTM* that has five rounds of mogrifircation. Furthermore, we utilize an attention mechanism that enables our model to capture a specific significant context in response to a particular visual area while still keeping the more significant story context in mind. We found that our proposed model performs better on most of the automatic evaluation metrics than current state-of-the-art approaches except for BLEU-1, BLEU-4 and CIDEr scores. Additionally, we presented a comprehensive analysis of multiple examples, which indicated that our generated stories are more relevant and coherent.

Acknowledgments. This research received complete funding from the Australian Government, which was sponsored through the Australian Research Council (DP190102443).

References

1. Nahian, M.S.A., Tasrin, T., Gandhi, S., Gaines, R., Harrison, B.: A hierarchical approach for visual storytelling using image description. In: Cardona-Rivera, R.E., Sullivan, A., Young, R.M. (eds.) ICIDS 2019. LNCS, vol. 11869, pp. 304–317. Springer, Cham (2019). https://doi.org/10.1007/978-3-030-33894-7_30
2. do Carmo Nogueira, T., Vinhal, C.D.N., da Cruz Júnior, G., Ullmann, M.R.D.: Reference-based model using multimodal gated recurrent units for image captioning. Multimedia Tools Appl. **79**(41), 30615–30635 (2020)
3. Chen, H., Huang, Y., Takamura, H., Nakayama, H.: Commonsense knowledge aware concept selection for diverse and informative visual storytelling. arXiv preprint arXiv:2102.02963 (2021)
4. Cho, K., Van Merriënboer, B., Bahdanau, D., Bengio, Y.: On the properties of neural machine translation: encoder-decoder approaches. arXiv preprint arXiv:1409.1259 (2014)
5. Ding, S., Qu, S., Xi, Y., Sangaiah, A.K., Wan, S.: Image caption generation with high-level image features. Pattern Recogn. Lett. **123**, 89–95 (2019)
6. Dosovitskiy, A., et al.: An image is worth 16×16 words: transformers for image recognition at scale (2020). https://doi.org/10.48550/ARXIV.2010.11929

7. Fei, Z., Yan, X., Wang, S., Tian, Q.: Deecap: dynamic early exiting for efficient image captioning. In: Proceedings of the IEEE/CVF Conference on Computer Vision and Pattern Recognition, pp. 12216–12226 (2022)
8. Guo, Y., Wu, H., Zhang, X.: Steganographic visual story with mutual-perceived joint attention. EURASIP J. Image Video Process. **2021**(1), 1–14 (2021). https://doi.org/10.1186/s13640-020-00543-1
9. Hochreiter, S., Schmidhuber, J.: Long short-term memory. Neural Comput. **9**(8), 1735–1780 (1997)
10. Hu, J., Cheng, Y., Gan, Z., Liu, J., Gao, J., Neubig, G.: What makes a good story? designing composite rewards for visual storytelling. In: AAAI, pp. 7969–7976 (2020)
11. Huang, L., Wang, W., Chen, J., Wei, X.Y.: Attention on attention for image captioning. In: Proceedings of the IEEE International Conference on Computer Vision, pp. 4634–4643 (2019)
12. Huang, T.H., et al.: Visual storytelling. In: Proceedings of the 2016 Conference of the North American Chapter of the Association for Computational Linguistics: Human Language Technologies, pp. 1233–1239 (2016)
13. Jiang, W., Zhou, W., Hu, H.: Double-stream position learning transformer network for image captioning. IEEE Trans. Circ. Syst. Video Technol. **32**, 7706–7718 (2022)
14. Kim, T., Heo, M.O., Son, S., Park, K.W., Zhang, B.T.: Glac net: glocal attention cascading networks for multi-image cued story generation. arXiv preprint arXiv:1805.10973 (2018)
15. Li, G., Zhu, L., Liu, P., Yang, Y.: Entangled transformer for image captioning. In: Proceedings of the IEEE International Conference on Computer Vision, pp. 8928–8937 (2019)
16. Li, L., Gao, X., Deng, J., Tu, Y., Zha, Z.J., Huang, Q.: Long short-term relation transformer with global gating for video captioning. IEEE Trans. Image Process. **31**, 2726–2738 (2022)
17. Li, Y., Pan, Y., Yao, T., Mei, T.: Comprehending and ordering semantics for image captioning. In: 2022 IEEE/CVF Conference on Computer Vision and Pattern Recognition (CVPR), pp. 17969–17978 (2022). https://doi.org/10.1109/CVPR52688.2022.01746
18. Lin, K., et al.: Swinbert: end-to-end transformers with sparse attention for video captioning. In: Proceedings of the IEEE/CVF Conference on Computer Vision and Pattern Recognition, pp. 17949–17958 (2022)
19. Liu, Y., Fu, J., Mei, T., Chen, C.W.: Let your photos talk: generating narrative paragraph for photo stream via bidirectional attention recurrent neural networks. In: Proceedings of the AAAI Conference on Artificial Intelligence, vol. 31, no. 1 (2017). https://doi.org/10.1609/aaai.v31i1.10760, https://ojs.aaai.org/index.php/AAAI/article/view/10760
20. Malakan., Z., Aafaq., N., Hassan., G., Mian., A.: Contextualise, attend, modulate and tell: visual storytelling. In: Proceedings of the 16th International Joint Conference on Computer Vision, Imaging and Computer Graphics Theory and Applications, vol. 5: VISAPP, pp. 196–205. INSTICC, SciTePress (2021).https://doi.org/10.5220/0010314301960205
21. Malakan, Z.M., Hassan, G.M., Jalwana, M.A.A.K., Aafaq, N., Mian, A.: Semantic attribute enriched storytelling from a sequence of images. In: 2021 Digital Image Computing: Techniques and Applications (DICTA), pp. 1–8 (2021). https://doi.org/10.1109/DICTA52665.2021.9647213
22. Melis, G., Kočiský, T., Blunsom, P.: Mogrifier LSTM. In: International Conference on Learning Representations (2020). https://openreview.net/forum?id=SJe5P6EYvS

23. Mogadala, A., Shen, X., Klakow, D.: Integrating image captioning with rule-based entity masking. arXiv preprint arXiv:2007.11690 (2020)
24. Pan, B., et al.: Spatio-temporal graph for video captioning with knowledge distillation. In: Proceedings of the IEEE/CVF Conference on Computer Vision and Pattern Recognition, pp. 10870–10879 (2020)
25. Pang, S., Chen, Z., Yin, F.: Video super-resolution using a hierarchical recurrent multireceptive-field integration network. Digital Signal Process. **122**, 103352 (2022)
26. Park, C.C., Kim, G.: Expressing an image stream with a sequence of natural sentences. Adv. Neural Inf. Process. Syst. **28**, 1–9 (2015)
27. Phukan, B.B., Panda, A.R.: An efficient technique for image captioning using deep neural network. arXiv preprint arXiv:2009.02565 (2020)
28. Seo, P.H., Nagrani, A., Arnab, A., Schmid, C.: End-to-end generative pretraining for multimodal video captioning. In: Proceedings of the IEEE/CVF Conference on Computer Vision and Pattern Recognition, pp. 17959–17968 (2022)
29. Shen, X., Liu, B., Zhou, Y., Zhao, J., Liu, M.: Remote sensing image captioning via variational autoencoder and reinforcement learning. Knowl.-Based Syst. **203**, 105920 (2020)
30. Wang, J., Fu, J., Tang, J., Li, Z., Mei, T.: Show, reward and tell: automatic generation of narrative paragraph from photo stream by adversarial training. In: The AAAI Conference on Artificial Intelligence (AAAI) (2018)
31. Wang, J., Wang, W., Wang, L., Wang, Z., Feng, D.D., Tan, T.: Learning visual relationship and context-aware attention for image captioning. Pattern Recogn. **98**, 107075 (2020)
32. Wang, X., Chen, W., Wang, Y.F., Wang, W.Y.: No metrics are perfect: adversarial reward learning for visual storytelling. arXiv preprint arXiv:1804.09160 (2018)
33. Zhou, L., Zhou, Y., Corso, J.J., Socher, R., Xiong, C.: End-to-end dense video captioning with masked transformer. In: Proceedings of the IEEE Conference on Computer Vision and Pattern Recognition, pp. 8739–8748 (2018)

Diverse Audio-to-Video GAN using Multiscale Image Fusion

Nuha Aldausari(✉), Arcot Sowmya, Nadine Marcus, and Gelareh Mohammadi

School of Computer Science and Engineering, University of New South Wales,
Sydney, Australia
{n.aldausari,a.sowmya,nadinem,g.mohammadi}@unsw.edu.au

Abstract. Generative adversarial networks have attained synthesised results that are not distinguishable from real examples in domains such as image, audio, text and video. While state-of-the-art image models synthesise images with high and diverse quality in many domains, video synthesis is more challenging and suffers from poor generalisation; moreover, the generated videos are not diverse, especially if the network is trained on a limited dataset. In such cases, the model overfits the training examples and performs poorly at inference time. Dataset collection, in general, is a tedious task, and it is even more challenging for video data due to its size and accessibility. Also, creating a video in the first place requires more time and effort. In this paper, we expand a previously collected video dataset with a supporting image dataset. Then, we apply a multiscale fusion method on multiple conditioned images to facilitate diverse video sample generation. We combine the multiscale fusion model with an audio extractor; then, the encoded features are input to a video decoder to generate videos synchronised with the audio signals. We compare our multiscale fusion model with other image fusion models on the Flowers, VGGFace and Animal Faces datasets. We also compare the overall architecture with other audio-to-video models. Both experiments show the effectiveness of our model over others, based on different evaluation metrics such as FID, FVD and LPIPS.

Keywords: Audio-to-Video GAN · Diverse video synthesis model · Image fusion

1 Introduction

Generative adversarial networks (GAN) [12] have been flourishing in the synthesis domain across multiple sectors such as medicine [38,39] and art [37] and modalities such as images, audio and text and video. Video generation, however, continues to suffer from many limitations. Video generation is a challenging task in general as models deal with the multi-modal nature of video models [5]. Video acquisition is also challenging due to the size and scarcity of video datasets. In addition, when constructing a video dataset from scratch, recording a video requires more time and effort than its counterpart in an image dataset. Training a video generation model on a limited dataset may result in a discriminator

© The Author(s), under exclusive license to Springer Nature Switzerland AG 2022
H. Aziz et al. (Eds.): AI 2022, LNAI 13728, pp. 29–42, 2022.
https://doi.org/10.1007/978-3-031-22695-3_3

overfitting problem. Video generation models are also complex with high data capacity, therefore they tend to memorise the limited training examples so that the back-propagation feedback becomes worthless [13,30]. The trained model, in this case, does not generalise on real-world examples.

There are multiple ways to overcome the overfitting problem in GAN. Transfer learning [33,33,41,46] could be used to overcome the memorising problem. A model could be trained on one dataset; then, the trained model could be trained further and fine-tuned on another dataset of limited size. Augmentation is a popular method that is used in StyleGAN2 ADA [31] to overcome memorisation of training examples. Different types of augmentation such as colour transformations, rotation and scaling are used in the training examples without leaking the augmentation to the generated examples. In this work, we propose a different augmentation approach that first expands the video dataset with an image dataset. For each video sample, there are multiple corresponding images that share semantic similarity. Therefore, each video may be augmented by multiple supporting images. Then, we use a fusion method to fuse these images and generate diverse videos. Image fusion was first introduced in the image generation field. Early methods such as GMN [10] and matchingGAN [6] have limited capabilities and can only generate simple scenes. F2GAN [8] can generate more realistic scenes by using a multilevel fused attention model. However, this model might generate images with unwanted artifacts if the input images are not spatially aligned. Gu et al. [9] overcame this limitation by fusing on a small scale. In this work, we use a fusion technique similar to Gu et al. while taking into account the fact that objects may be of any size. We first compute the multilevel features during the encoding phase. Then, we replace a feature map in one image with other unified resized feature maps in the other images according to the similarity map. Comparing a feature map in an image with multi-scale features in the other images helps in finding better candidates for replacement and results in better image generation.

Previously reported image models [6,8–10] are in the image generation field. To overcome the overfitting problem in the video realm, multiple methods such as augmentation [36,40] have been used. To the best of our knowledge, this paper is the first attempt to improve a fusion method [9] in the image domain and apply it to the audio-to-video domain. In order to apply the fusion method, we first augment our video dataset by adding an image dataset since it is easier to acquire images. We use the collected images to help our model learn from similar features and that results in higher quality and more diverse generations.

1.1   Contributions

In this work, we have made the following contributions:

1. We propose a novel model for audio-to-video synthesis that improves and combines a fusion model from the image generation realm with an audio-to-video synthesis model.
2. We expand the Phonics dataset with an image dataset to enrich the video dataset with image samples, as collecting more video samples is not feasible.

3. We compare our synthesis model with image and video models and across multiple image and video datasets, using several evaluation metrics.

2 Related Work

This section reviews related works on generative adversarial networks, audio-to-video generation and image fusion models.

2.1 Generative Adversarial Networks (GAN)

The usefulness of GAN [12] has been demonstrated in a variety of fields, including picture synthesis [3,22], editing [22,44], denoising [1,43] and retargeting [7,42]. Early GAN models generated content based on noise vectors such as MoCo-GAN [34], G3an [35] and ImaGINator [32]. Later models can generate content given a condition such as image, video, audio, text or key-points. A GAN model consists of two networks that compete with each other to improve the generated results. In most cases, the training GAN continues until convergence, where the generated samples are of sufficient quality and diversity. However, when the training examples are limited, the model tends to overfit the training examples. In other words, since generative models have high capacity, they seem to memorise the examples that are not represented sufficiently in the training dataset, and always generate the same examples. Using fusion models is one way to overcome this limitation in a GAN model, as will be discussed in Subsect. 2.3.

2.2 Video Generation

Video generation is a complex task for several reasons. First, a video is made up of multiple static images. Besides the spatial complexity, these models need to generate videos with smooth temporal trajectories. In addition, when these models are conditioned on audio, they must generate motion-audio synchronisation. This work focuses on the audio-to-video task. The state-of-the-art audio-to-video models rely on supervised signals such as 3D mesh [14–17] or landmarks [18–20,23,24]. These supporting signals help the model learn the changes in motion along the time dimension. However, annotating videos with landmarks or 3D meshes requires time and effort. Also, it is not possible to annotate some datasets such as fireworks or ocean waves where that content has irregular shapes. Given that these models depend on supervised signals, they are not covered in this paper. The focus of this work instead is on pixel-level audio-to-video generation models [21,29,45]. However, the existing models do not account well for the diversity of the generated examples. To address this issue, our proposed model adopts a fusion method so that the generated frames have higher quality as the model learns from the supporting images, as will be presented in Sect. 3.

2.3 Image Fusion Methods

Image fusion models are used to combine images from one category at the feature level and decode that to an image from the same category. Generative Matching Network (GMN) [10] is one of the first attempts at fusion-based image generation. GMN consists of a matching network and a Variational AutoEncoder (VAE). The matching network is responsible for projecting the noise vector and the conditional images in a unified space and then calculating the similarity score which is used to define the interpolation coefficient, which is proportional to the information from each image to be fused to construct the generated image. Because GMN uses VAE architecture, the generative capabilities are limited to simple tasks as the model is implemented on a simple character dataset named Omniglot [11]. MatchingGAN [6] replaces the VAE model in GMN with a GAN to take advantage of GAN capabilities. MatchingGAN can produce more realistic images, but the generated quality can be degraded for complex natural scenes. F2GAN [8] uses multilevel local attention fusion modules with skip connections. Each image participates in the generated image based on a random coefficient. This method might not perform well when the input images are misaligned [9]. Recently, Gu et al. [9] introduced a different fusion methodology in LoFGAN. They go through three steps, namely selection, matching and replacing. The first step is to select local positions in one of the images and then find the best match in the other images in the second step. In the last step, the best matches are replaced by the original positions. The fused image is decoded to generate the actual image.

Previous audio-to-video models [21,45] were conditioned on the input frame to generate a video with frames quite similar to the input frame, but after adjusting the objects according to the temporal features. This will result in videos with low diversity. To produce more diversity in the generated videos, we employ a fusion model to augment each video with similar images. To the best of our knowledge, we are among the first to apply fusion images to expand a video dataset and provide more diversity and higher quality in the generated videos.

3 Proposed Method

3.1 Audio-to-Video GAN Framework

The framework proposed in this work aims to generate diverse videos given a few conditional images from the same category and an audio signal. The overall architecture of the framework is illustrated in Fig. 1. The process begins by encoding the first frame and multiple conditional images from the same category. The image encoder produces features with different spatial sizes. After that, we use a multiscale fusion model to combine the images at the feature level and to gain more diversity, as described in Sect. 3.2. Meanwhile, the corresponding audio signal for the first frame is transformed to a log Mel-spectrogram, and then the signal is encoded using GRU units. Next, the fused image features, encoded audio features and the class (see Sect. 4.2) are concatenated in the channel

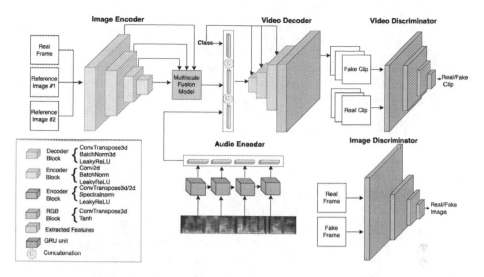

Fig. 1. The overall architecture of the proposed framework that consists of a generator and two level discriminators. The generator has an image encoder, multiscale fusion model, audio encoder and video decoder.

dimension. On the decoder side, the concatenated features are input to the first layer and the class features are input in each decoding stage after reshaping the class feature to the same dimension as the input. The class of a video is encoded using one hot encoding to label a sample from a list of categories as described in Sect. 4.2 There are two discriminators, one each for the image and video levels, to determine whether the generated examples are realistic. While both image and video discriminators evaluate the spatial aspect of a sample, the latter also evaluates the temporal aspect of a video.

3.2 Multiscale Fusion Module

Objects in an image could be of any size. The proposed architecture deals with this issue by seeking similarities in a multiscale manner. To be specific, the model takes features of different sizes and then uses interpolation to resize them. After obtaining all the feature maps of the same size, we look for the top similarity between the base image and the references of all sizes according to cosine similarity metrics. The last step is to replace the feature in the base image with similar features in the reference images [9].

3.3 Loss Function

We trained our model using three loss functions. First, we use adversarial loss that consists of video adversarial loss and image adversarial loss as in Eq. 1 and 2 respectively.

$$\mathcal{L}_V(DV, G) = E_{V \sim p_{data}}[log DV(V)] + E_{I,c,a \sim p_{data}}[log(1 - DV(G(I, c, a)))] \quad (1)$$

$$\mathcal{L}_I(DI, G) = E_{v_i \sim p_{data}}[log DI(v_i)] + E_{I,c,a \sim p_{data}}[log(1 - DI(x)] \qquad (2)$$

In our GAN architecture, a video discriminator DV evaluates the realism of the generated video $G(I, c, a)$ and real video $V = v_1, ..., v_k$, where k is the number of frames. In contrast, the image discriminator DI is used to verify if the input is a frame from a real video (v_i) or is a fake one $x \in G(I, c, a)$. The generator G has multiple inputs for different components. The audio a is input to the audio encoder. The initial frame $f \in v_1, ..., v_k$ and the supporting images $S = s_1, ..., s_n$ are concatenated to form I to be encoded in the image encoder. Then, the class label c along with encoded audio a and image features I are input to the decoder (See Fig. 1).

Generative models use reconstruction loss to enhance the quality of the generated content. However, different variations of the reconstruction loss have been proposed with the introduction of fusion models [8,9]. Weighted reconstruction loss [8] is one of the variations, but this loss might result in implausible artifacts due to comparison with the fused image that has misaligned objects. Local fused reconstruction loss [9] produces higher quality results since it re-applies the fusion model and projects the results at the pixel-level, to compare it with generated images. We chose to apply Local Fused reconstruction loss [9] by using $\mathcal{L}_1$, as in Eq. 3, to compare the fused images $Fusion(I)$ with real frame f.

$$\mathcal{L}_{rec} = \|f - Fusion(I)\|_1 \qquad (3)$$

The overall architecture is trained using the following loss function:

$$\mathcal{L} = \lambda_I L_I + \lambda_V L_V + \lambda_{rec} L_{rec} \qquad (4)$$

4   Experiments

We compared our model with image and video models, as our model is built upon a fusion approach that was first implemented in image models. We compared our model with image fusion models such as GMN [10], MatchingGAN [6], F2GAN [8] and LoFGAN [9] across multiple datasets such as Flowers [27], Animal Faces [28] and VGGFace [25]. Also, another set of comparisons were made in the video form and across multiple datasets such as Phonics audio-video dataset [21,45] and VidTIMIT audio-video dataset [26]. We compared our model with a audio-to-video model [21]. Since audio-to-video pixel-level models are limited, we compared our model with a modified version of unconditional video models such as MoCoGAN [34], G3an [35] and ImaGINator [32].

4.1   Implementation Details

In **all experiments**, the encoder has five 2D convolutional blocks. Each convolutional layer is followed by Leaky-ReLU activation and batch normalisation. The encoder produces features with spatial sizes 32 * 32 and 16 * 16 besides the bottleneck features that are used in the fusion model of size 8 * 8. We use the Adam optimizer with a learning rate of 0.0002.

In the **video experiments**, there are three conditional images: a frame from a video, and two images from the supporting image dataset. The created videos have 32 frames in total. The frames' spatial dimension is 64 × 64. Similar to the video dimension, there are 32 audio segments in total, and each audio segment is 64 × 20 in size. The video decoder consists of 5 layers, and each layer has ConvTranspose3d, BatchNorm3d and LeakyReLU. We used Tanh in the last layer to replace LeakyReLU.

In the **image experiments**, there are three conditional images. We replaced the video decoder with an image decoder. The decoder consists of 5 main layers. Each layer has an up-sampling sub-layer, convolution 2D sub-layer and leaky-ReLU activation. As in the video decoder, we apply Tanh in the last layer.

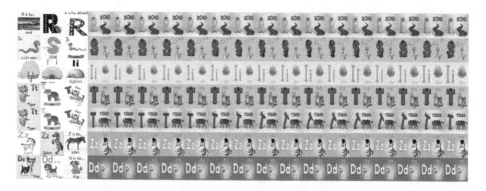

Fig. 2. Samples from our video and image datasets. The first three columns are from the image dataset while the rest of the columns represent frames of a video.

4.2   Datasets

Phonics Dataset. This dataset consists of two parts: the video dataset and the image dataset. The Phonics video dataset is publicly available [47] with 1570 samples. It is an audio-visual video dataset, which has a song about an alphabet letter and a video of an object whose name begins with that letter. The video is an animation of the letter and object that are moving or transforming in different ways. This dataset has limited samples if used to train a high-capacity GAN network directly. However, collecting more video samples requires time and effort because of the pre-processing involved [21]. Moreover, there are limited videos available in the wild. For these reasons, we have expanded the Phonics video dataset by collecting the Phonics image dataset. For each object in a video, we collected three image samples, see Fig. 2 illustrating a video and the three supporting image samples. The total number of image samples is 1533, and the total number of object classes is 511, while the number of letter classes is 26. For example, in the first row in Fig. 2, the object class is "road" while the letter class is "R". Collecting an image dataset does not require the pre-processing steps and helps in increasing the quality and diversity of the generated videos.

The image dataset construction procedure starts by searching for the following phrase: "x is for y" where x is the letter and y is the object that starts with the letter x. Then, three images are downloaded and the class letter and object are recorded for each sample. In addition, the corresponding letter and object classes are recorded for each video sample in a CSV file. Thus, we can relate each video sample with the images that share the same letter and object.

Other Evaluation Datasets. Besides using Phonics dataset in the video experiments, we also trained our model on the VidTIMIT Audio-Video dataset [26]. The videos feature 43 persons saying 10 brief sentences and were taken from the head region. We used the person's identity as a class label. There are 430 videos in total. The model was conditioned on three images of the same person. Because the data was gathered in a laboratory environment, there is not much variation in the conditional images.

For the image experiments, we used Flowers dataset [27], Animal Faces dataset [28] and VGGFace dataset [25]. The Flowers dataset is divided into 102 categories, where each category is limited to 40 images. The Animal Faces dataset contains 149 categories, each with 100 images. The VGGFace dataset has 2354 categories. Each category has 100 images.

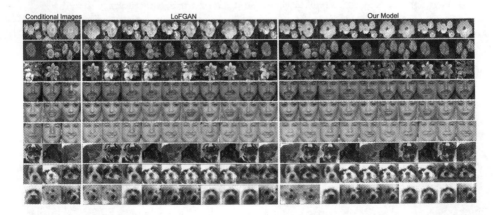

Fig. 3. Images produced from LoFGAN and our proposed network and across multiple datasets such as Flowers, Animal Faces and VGGFace. The first three columns are the conditional images. The remaining columns are the results of inputting the same set of images in multiple iterations.

Fig. 4. The generated videos from our model and audio-ImaGINator. Our model is conditioned on three images (the first three columns) while audio-ImaGINator inputs one image (the first column). The remaining columns are the generated frames sampled with time-step=2

Table 1. Utilising FID and LPIPS scores for quantitative analysis, our model performs better in terms of quality and diversity of the images produced. The symbol (*) means that the results are quoted from F2GAN [8].

	Flowers		Animal faces		VGGFace	
	FID↓	LPIPS↑	FID↓	LPIPS↑	FID↓	LPIPS↑
GMN*	200.11	0.0743	220.45	0.0868	136.21	0.0902
MatchingGAN*	143.35	0.1627	148.52	0.1514	118.62	0.1695
F2GAN*	120.48	0.2172	117.74	0.1831	109.16	0.2125
LoFGAN	79.33	0.3862	112.81	0.4964	20.31	0.2869
Our Model	**60.17**	**0.4021**	**110.32**	**0.5022**	**19.43**	**0.2993**

Table 2. Using FID and FVD scores for quantitative analysis, our model performs better in terms of quality and diversity of the generated videos trained on Phonics dataset.

	MoCoGAN	phonicsGAN	G3an	ImaGINator	Our model
FID	59.81	46.82	36.93	35.23	**31.22**
FVD	1248.73	961.85	936.99	377.25	**323.45**

4.3   Qualitative and Quantitative Evaluations

We compared our model with other image fusion models since the fusion app-roach was first introduced in the image domain. A comparison between LoF-GAN [9] and our network using the image decoder is shown in Fig. 3. Our model shows better quality across three datasets. We noticed when we deal with a dataset that has multiscale objects, such as flowers, our model surpassed other models as shown in the first three rows in Fig. 3. We also used different evaluation metrics such as Fréchet Inception Distance (FID) [4] and Learned Perceptual Image Patch Similarity (LPIPS) [2] to compare our model with image fusion models such as GMN [10], MatchingGAN [6], F2GAN [8] and LoFGAN [9]. FID uses pre-trained inception-v3 to compare the distribution of real and fake images, while LPIPS uses VGG, Alexnet or SqueezeNet to measure the perceptual sim-ilarity between two images by comparing the activation of a certain convolution layer. While a lower FID implies better sample quality, a higher LPIPS suggests higher sample quality. In Table 1 the FID and LPIPS scores are shown. As visual inspection suggests, the improvement is greater in the Flowers dataset than in others. As mentioned earlier, the Flower dataset has flowers and leafs in different sizes. As our network is designed to deal with this fact, it performers better than other models in such datasets.

We also performed another set of experiments for two video datasets, and Fig. 4 shows a comparison between our model and audio-ImaGINator. We chose to compare our model with audio-ImaGINator since it has the second-best per-formance in phonics and VidTIMIT datasets [45]. We did not compare the pro-posed model with another reported in the literature [45] since the latter model uses temporal augmentation, which is out of the scope of this work. We found that the supporting images did help in learning the spatial features and gener-ating higher quality objects. In some cases, we can see the effect of fusion in the generated videos, as shown in Fig. 4 first row. However, most cases produce results similar to the base image, since the variation in the conditional images for the VidTIMIT dataset is not as large as in the other datasets such as flow-ers. We used FID and Fréchet Video Distant (FVD) in the video experiments as quantitative measures. FVD evaluates the temporal dimension of a video besides the spatial dimension. Our model surpasses others in the spatial and temporal dimension as shown by FID and FVD score in Table 2.

Fig. 5. Ablation studies: a comparison between our video model w/ and w/o the multi-scale fusion component trained on the Phonics dataset and sampled with time-step=2.

Conditional Images

Fig. 6. Ablation studies: a comparison between our image model w/ and w/o the multiscale fusion component trained on the Flowers dataset.

4.4 Ablation Study

We performed several ablation studies to evaluate the effectiveness of multiple components. First, we removed the fusion model and trained the same architecture by inputting the initial frame only instead of fusing multiple images. We noticed that if we input only the initial frame, the videos produced have lower quality, as shown in Fig. 5. This is because the decoder only decodes the image features from one source, while with the fusion model the decoder produces an image from multiple images, thereby enhancing the quality.

We also performed the same experiment for our image model. We removed the fusion model, and input a single image to the model without fusion. The same decoder and discriminator in our original experiment was used in this ablation study. We found that every time we input the same image to our model without fusion, the same result was produced. For example, in Fig. 6 first and third rows, we input the same image to the model without fusion 10 times, and the output is almost the same. However, in Fig. 6 second and fourth rows, each time we input the same set of images to our model with fusion, we can see diverse generations.

5 Conclusion

In this work, we attempt to address the diversity of audio-to-video generation results by using a fusion method from the image generation domain. We propose a multiscale image fusion model that aims to overcome the problem of limited video datasets by expanding the dataset with images to enrich the spatial aspect of a video. By fusing multiple images, we can generate more diverse videos. We compared our model with image fusion models and video models, and the results support the effectiveness of the proposed framework.

Acknowledgment. The first author is supported by a scholarship from Princess Nourah bint Abdulrahman University, KSA.

References

1. Yang, Q., et al.: Low-dose CT image denoising using a generative adversarial network with Wasserstein distance and perceptual loss. IEEE Trans. Med. Imaging **37**, 1348–1357 (2018)
2. Zhang, R., Isola, P., Efros, A., Shechtman, E., Wang, O.: The unreasonable effectiveness of deep features as a perceptual metric. In: Proceedings of the IEEE Conference on Computer Vision and Pattern Recognition, pp. 586–595 (2018)
3. Wang, L., Chen, W., Yang, W., Bi, F., Yu, F.: A state-of-the-art review on image synthesis with generative adversarial networks. IEEE Access **8**, 63514–63537 (2020)
4. Heusel, M., Ramsauer, H., Unterthiner, T., Nessler, B., Hochreiter, S.: Gans trained by a two time-scale update rule converge to a local nash equilibrium. Adv. Neural Inf. Process. Syst. **30** (2017)
5. Aldausari, N., Sowmya, A., Marcus, N., Mohammadi, G.: Video generative adversarial networks: a review. ACM Comput. Surv. (CSUR). **55**, 1–25 (2022)
6. Hong, Y., Niu, L., Zhang, J., Zhang, L.: Matchinggan: matching-based few-shot image generation. In: 2020 IEEE International Conference on Multimedia And Expo (ICME), pp. 1–6 (2020)
7. Lee, J., Ramanan, D., Girdhar, R.: Metapix: few-shot video retargeting. ArXiv Preprint ArXiv:1910.04742 (2019)
8. Hong, Y., Niu, L., Zhang, J., Zhao, W., Fu, C., Zhang, L.: F2gan: fusing-and-filling gan for few-shot image generation. In: Proceedings of the 28th ACM International Conference on Multimedia, pp. 2535–2543 (2020)
9. Gu, Z., Li, W., Huo, J., Wang, L., Gao, Y.: Lofgan: fusing local representations for few-shot image generation. In: Proceedings of the IEEE/CVF International Conference on Computer Vision, pp. 8463–8471 (2021)
10. Bartunov, S., Vetrov, D.: Few-shot generative modelling with generative matching networks. In: International Conference on Artificial Intelligence and Statistics, pp. 670–678 (2018)
11. Lake, B., Salakhutdinov, R., Tenenbaum, J.: Human-level concept learning through probabilistic program induction. Science **350**, 1332–1338 (2015)
12. Goodfellow, I., et al.: Generative adversarial nets. Adv. Neural Inf. Process. Syst. **27** (2014)
13. Zhang, D., Khoreva, A.: Improving GAN training by progressive augmentation, PA-GAN (2018)
14. Ji, X., et al.: Audio-driven emotional video portraits. In: Proceedings of the IEEE/CVF Conference on Computer Vision and Pattern Recognition, pp. 14080–14089 (2021)
15. Chen, L., et al.: Talking-head generation with rhythmic head motion. In: Vedaldi, A., Bischof, H., Brox, T., Frahm, J.-M. (eds.) ECCV 2020. LNCS, vol. 12354, pp. 35–51. Springer, Cham (2020). https://doi.org/10.1007/978-3-030-58545-7_3
16. Song, L., Wu, W., Qian, C., He, R., Loy, C. Everybody's talkin: let me talk as you want. ArXiv Preprint ArXiv:2001.05201 (2020)
17. Lahiri, A., Kwatra, V., Frueh, C., Lewis, J., Bregler, C. LipSync3D: data-efficient learning of personalized 3D talking faces from video using pose and lighting normalization. In: Proceedings of the IEEE/CVF Conference on Computer Vision and Pattern Recognition, pp. 2755–2764 (2021)
18. Zhou, Y., Han, X., Shechtman, E., Echevarria, J., Kalogerakis, E., Li, D.: MakeltTalk: speaker-aware talking-head animation. ACM Trans. Graph. (TOG). **39**, 1–15 (2020)

19. Das, D., Biswas, S., Sinha, S., Bhowmick, B.: Speech-driven facial animation using cascaded GANs for learning of motion and texture. In: Vedaldi, A., Bischof, H., Brox, T., Frahm, J.-M. (eds.) ECCV 2020. LNCS, vol. 12375, pp. 408–424. Springer, Cham (2020). https://doi.org/10.1007/978-3-030-58577-8_25
20. Chen, L., Maddox, R., Duan, Z., Xu, C.: Hierarchical cross-modal talking face generation with dynamic pixel-wise loss. In: Proceedings of the IEEE/CVF Conference on Computer Vision and Pattern Recognition, pp. 7832–7841 (2019)
21. Aldausari, N., Sowmya, A., Marcus, N., Mohammadi, G.: PhonicsGAN: synthesizing graphical videos from phonics songs. In: Farkaš, I., Masulli, P., Otte, S., Wermter, S. (eds.) ICANN 2021. LNCS, vol. 12892, pp. 599–610. Springer, Cham (2021). https://doi.org/10.1007/978-3-030-86340-1_48
22. Wu, X., Xu, K., Hall, P.: A survey of image synthesis and editing with generative adversarial networks. Tsinghua Sci. Technol. **22**, 660–674 (2017)
23. Zhou, H., Liu, Y., Liu, Z., Luo, P., Wang, X.: Talking face generation by adversarially disentangled audio-visual representation. In: Proceedings of the AAAI Conference on Artificial Intelligence, vol. 33, pp. 9299–9306 (2019)
24. Mittal, G., Wang, B.: Animating face using disentangled audio representations. In: Proceedings of the IEEE/CVF Winter Conference on Applications of Computer Vision, pp. 3290–3298 (2020)
25. Cao, Q., Shen, L., Xie, W., Parkhi, O., Zisserman, A.: Vggface2: a dataset for recognising faces across pose and age. In: 2018 13th IEEE International Conference On Automatic Face & Gesture Recognition (FG 2018), pp. 67–74 (2018)
26. Sanderson, C., Lovell, B.C.: Multi-region probabilistic histograms for robust and scalable identity inference. In: Tistarelli, M., Nixon, M.S. (eds.) ICB 2009. LNCS, vol. 5558, pp. 199–208. Springer, Heidelberg (2009). https://doi.org/10.1007/978-3-642-01793-3_21
27. Nilsback, M., Zisserman, A.: Automated flower classification over a large number of classes. In: 2008 Sixth Indian Conference On Computer Vision, Graphics Image Processing, pp. 722–729 (2008)
28. Liu, M., et al.: Few-shot unsupervised image-to-image translation. In: Proceedings of the IEEE/CVF International Conference on Computer Vision, pp. 10551–10560 (2019)
29. Tsuchiya, Y., Itazuri, T.: Others generating video from single image and sound. In: CVPR Workshops, pp. 17–20 (2019)
30. Arjovsky, M., Bottou, L.: Towards principled methods for training generative adversarial networks. ArXiv Preprint ArXiv:1701.04862 (2017)
31. Karras, T., Aittala, M., Hellsten, J., Laine, S., Lehtinen, J., Aila, T.: Training generative adversarial networks with limited data. Adv. Neural Inf. Process. Syst. **33**, 12104–12114 (2020)
32. Wang, Y., Bilinski, P., Bremond, F., Dantcheva, A.: Imaginator: conditional spatio-temporal gan for video generation. In: Proceedings of the IEEE/CVF Winter Conference on Applications of Computer Vision, pp. 1160–1169 (2020)
33. Wang, Y., Gonzalez-Garcia, A., Berga, D., Herranz, L., Khan, F., Weijer, J. Minegan: effective knowledge transfer from gans to target domains with few images. In: Proceedings of the IEEE/CVF Conference on Computer Vision and Pattern Recognition, pp. 9332–9341 (2020)
34. Tulyakov, S., Liu, M., et al.: Mocogan: decomposing motion and content for video generation. In: Proceedings of the IEEE Conference on Computer Vision and Pattern Recognition, pp. 1526–1535 (2018)

35. Wang, Y., Bilinski, P., Bremond, F., Dantcheva, A.: G3AN: disentangling appearance and motion for video generation. In: Proceedings of the IEEE/CVF Conference on Computer Vision and Pattern Recognition, pp. 5264–5273 (2020)
36. Babaeizadeh, M., et al.: FitVid: overfitting in pixel-level video prediction. ArXiv Preprint ArXiv:2106.13195 (2021)
37. Shahriar, S.: GAN computers generate arts? a survey on visual arts, music, and literary text generation using generative adversarial network. Displays, 102237 (2022)
38. Yi, X., Walia, E., Babyn, P.: Generative adversarial network in medical imaging: a review. Med. Image Anal. **58**, 101552 (2019)
39. Sorin, V., Barash, Y., Konen, E., Klang, E.: Creating artificial images for radiology applications using generative adversarial networks (GANs)-a systematic review. Acad. Radiol. **27**, 1175–1185 (2020)
40. Logacheva, E., Suvorov, R., Khomenko, O., Mashikhin, A., Lempitsky, V.: DeepLandscape: adversarial modeling of landscape videos. In: Vedaldi, A., Bischof, H., Brox, T., Frahm, J.-M. (eds.) ECCV 2020. LNCS, vol. 12368, pp. 256–272. Springer, Cham (2020). https://doi.org/10.1007/978-3-030-58592-1_16
41. Noguchi, A., Harada, T.: Image generation from small datasets via batch statistics adaptation. In: Proceedings of the IEEE/CVF International Conference on Computer Vision, pp. 2750–2758 (2019)
42. Bansal, A., Ma, S., Ramanan, D., Sheikh, Y.: Recycle-gan: Unsupervised video retargeting. In: Proceedings of the European Conference on Computer Vision (ECCV), pp. 119–135 (2018)
43. Zhong, Y., Liu, L., Zhao, D., Li, H.: A generative adversarial network for image denoising. Multimedia Tools Appl. **79**, 16517–16529 (2020)
44. Jo, Y., Park, J.: Sc-fegan: face editing generative adversarial network with user's sketch and color. In: Proceedings of the IEEE/CVF International Conference on Computer Vision, pp. 1745–1753 (2019)
45. Aldausari, N., Sowmya, A., Marcus, N., Mohammadi, G.: Cascaded siamese self-supervised audio to video GAN. In: Proceedings of the IEEE/CVF Conference on Computer Vision and Pattern Recognition, pp. 4691–4700 (2022)
46. Robb, E., Chu, W., Kumar, A., Huang, J.: Few-shot adaptation of generative adversarial networks. ArXiv Preprint ArXiv:2010.11943. (2020)
47. Phonics Dataset. github.com/NuhaAldausari/Cascaded-Siamese-Selfsupervised-Audio-to-Video-GAN. Accessed 2 Oct 2022

Zero-shot Personality Perception From Facial Images

Peter Zhuowei Gan$^{(\boxtimes)}$ (ID), Arcot Sowmya (ID), and Gelareh Mohammadi (ID)

University of New South Wales, Sydney, Australia
{zhuowei.gan,a.sowmya,g.mohammadi}@unsw.edu.au

Abstract. Personality perception is an important process that affects our behaviours towards others, with applications across many domains. Automatic personality perception (APP) tools can help create more natural interactions between humans and machines, and better understand human-human interactions. However, collecting personality assessments is a costly and tedious task. This paper presents a new method for zero-shot facial image personality perception tasks. Harnessing the latent psychometric layer of CLIP (Contrastive Language-Image Pre-training), the proposed PsyCLIP is the first zero-shot personality perception model achieving competitive results, compared to state-of-the-art supervised models. With PsyCLIP, we establish the existence of latent psychometric information in CLIP and demonstrate its use in the domain of personality computing. For evaluation, we compiled a new personality dataset consisting of 41800 facial images of various individuals labelled with their corresponding perceived Myers Briggs Type Indicator (MBTI) types. PsyCLIP achieved statistically significant results ($p < 0.01$) in predicting all four Myers Briggs dimensions without requiring any training dataset.

Keywords: Personality · Personality perception · Personality computing · Data-driven approach · Computational modeling · Transfer learning

1 Introduction

When seeing a person's face for the first time, we instinctively form an impression of their personality [31]. While we may be taught at a young age to "not judge a book by its cover", psychology studies have shown that this innate first perception yields considerable accuracy [6,21,30,43]. Indeed, face-reading skills are critical in our daily lives. They are employed by salespersons to assess prospective clients, film directors to choose the optimal actor, and even while considering which stranger to ask for directions [36]. Regardless of its accuracy, personality perception influences our behaviour towards others [39].

It is a major research area to determine if a computer may gain this capability. Personality perception [42] is the automatic perception of a subject's personality based on their audio, visual, or other features. Rather than attempting

H. Aziz et al. (Eds.): AI 2022, LNAI 13728, pp. 43–56, 2022.
https://doi.org/10.1007/978-3-031-22695-3_4

to recognise an individual's true personality as in personality recognition tasks, automated personality perception seeks to predict the perceived personality or how an individual's personality is perceived by others.

Automatic personality perception (APP) [42] may be a critical first step in developing a more affable conversational agents [34] that are perceived in certain ways. In theoretical psychology, APP can help better understand social interactions and group dynamics [20] In clinical psychology, it can serve as a coaching system to assist socially challenged individuals such as those diagnosed with autism spectrum disorder (ASD) and social anxiety disorder (SAD), in understanding how their behaviours affect others' perception, thus helping them in coping with social norms. Furthermore, the ability to reliably infer personality impressions from facial images enables us to employ it as a discriminator in an adversarial network [15] for creating face images based on certain personality attributes [12].

Currently, we can attain a high degree of object detection accuracy in some domains [22], partially owing to the vast number of training datasets available [46], such as the ImageNet [10] dataset with 14 million images. This is not the case, however, in areas such as personality computing [18]. The state-of-the-art model [19] of facial image personality perception has a training set of 28,230 images. In contrast, even the MNIST dataset [11], which is often used in beginner deep learning courses, has 60,000 images.

The size of datasets used by cutting-edge deep learning models is ever increasing, such as GPT-3 [8] with 410 billion training tokens. However, we have yet to replicate the success of large datasets in personality computing [42]. Creating an annotated dataset of psychometrics (i.e. personality measures) is far more complex and costly, since it is considerably more difficult to label psychological features [18].

This highlights the value and attraction of zero-shot classification in personality perception tasks [42]. The introduction of contrastive language-image pre-training (CLIP) [33] made zero-shot personality perception promising: Trained on 400 million image-caption pairs, CLIP has been shown to grasp abstract classification labels [4]. We hypothesise that there is a potential way for CLIP to comprehend cues meant to elicit personality attributes and hence be used in personality perception tasks, by accessing a latent psychometric layer within the CLIP model.

The goal of our study is to utilise CLIP to build a zero-shot model of personality perception from unlabelled images by harnessing latent psychometric information from the CLIP pre-trained model. PsyCLIP (Psychometric-CLIP) adopt the CLIP's text/image encoder structure. As CLIP was pre-trained on image-caption pairings, we translate each psychometric label into CLIP-style text prompts (i.e., image captions). To find the optimal text-prompt, we first generate a list of candidate prompts using GPT-3's text-davinci-002 text completion engine [1]. We then proceed to eliminate biased prompts that favour a particular personality trait and select the prompt that results in the highest accuracy. To evaluate the performance of PsyCLIP, we have created a large dataset of

41800 facial images, labelled with Myers Briggs [29] personality types. The result from PsyCLIP is encouraging: We achieved statistically significant results (p < 0.01) in all personality dimensions, which are comparable to those obtained by the state-of-the-art supervised model [19]. With PsyCLIP, we make the following contributions:

- Establish the existence of a latent psychometric layer in CLIP, and demonstrate how it can be harnessed in the domain of personality computing.
- Provide a new personality dataset consisting of 41800 facial images of various individuals labelled with their corresponding perceived MBTI personality.
- Introduce a novel approach in handling zero-shot personality perception tasks that produces results comparable to those of a state-of-the-art supervised model, without the need for any training sets.

PsyCLIP is significant because it provides a reasonable base model for computational social scientists, potentially capable of perceiving *any* psychological attribute [37]. It may serve as a playground for rapidly testing psychological theories and sparking new psychological discoveries.

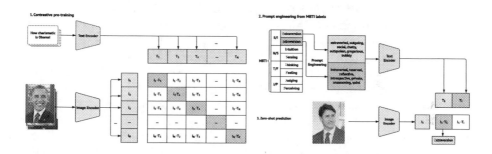

Fig. 1. Summary of our approach. We perform prompt engineering (i.e. translate MBTI subscale traits into CLIP-style text prompt) for each MBTI subscale and encode them using CLIP text encoders. We then assess the classification results of 16000 evaluation samples by encoding them with CLIP image encoders.

2   Related Work

2.1   CLIP

CLIP (Contrastive Language-Picture Pre-training) [33] combines image and text encoding to anticipate appropriate image-text pairing of training instances. Then, for zero-shot object classification, the classification labels are translated to captions such as "a photograph of an extroverted person," and CLIP predicts the caption class that most closely matches the provided photograph. Although CLIP is zero-shot, it outperforms some state-of-the-art supervised models.

However, CLIP has been mostly used for standard object classification tasks, and there is a dearth of research on CLIP's performance with psychological labels. Evidence suggests that CLIP might comprehend abstract prompts, as seen in BigSleep [23] and DeepDaze [24] CLIP functions as a discriminator in these projects, combining with a BigGAN [7] or SirenNetwork [38] to generate abstract artworks from arbitrary inputs.

This leads to the hypothesis that there might be hidden information about personality measures within the pre-trained CLIP model. This work explores the effectiveness of CLIP in personality perception tasks and intends to spark discussion on using pre-trained models in personality computing.

2.2   Personality Measures

Two psychometric instruments stand out among contemporary personality models: the Big Five [35] and the Myers-Briggs Type Indicator (MBTI) [29]. The Big Five (the five-factor usually assessed using NEO Personality Inventory) is more prominent in academia, whereas the MBTI is more prevalent in the consulting and training industries [14]. Big Five model describes each person's personality across five dimensions: Extraversion, Openness, Agreeableness, Conscientiousness, and Neuroticism that are revealed from semantic analysis of personality descriptors. However, MBTI indicates preferences in how people perceive the world and make decisions [29] with four categories: Extraversion-introversion, intuiting-sensing, thinking-feeling, and judging-perceiving. Studies show that there is a strong correlation [14] between the MBTI and the four dimensions of the Big Five, as shown in Table 1: The Big Five Extraversion is highly correlated with the MBTI Extraversion/Introversion (E-I) dimension; the Big Five Openness is highly correlated with the MBTI Intuition/Sensing (N-S) dimension; and the Big Five Agreeableness is only associated with MBTI thinking; The Big Five conscientiousness is associated with both the thinking-feeling (T-F) and judging-perceiving (J-P) dimensions; Neuroticism as measured by the NEO-PI is unrelated to any MBTI subscale score.

The primary distinction between MBTI and Big Five is that MBTI employs a binary classification system (e.g., either extrovert or introvert), whereas Big Five employs a linear scale (e.g., a number associated with each dimension) [9]. As a result, MBTI naturally lends itself to classification tasks, whereas Big Five dimensions lend themselves to regression. Therefore, MBTI was a more natural choice for evaluating PsyCLIP's performance, as CLIP was designed as a classification model.

Another reason we choose MBTI is that the large dataset we gathered was in MBTI. In practice, it is easier to collect big datasets of personality perceptions using MBTI. We posit that APP on dichotomy-based datasets (such as the MBTI) could be a necessary prelude to APP on scaling-based datasets (such as Big Five).

Table 1. The table shows how MBTI correlates to big five spectrums. [14]

MBTI	Big Five
E-I	Extraversion
N-S	Openness
T-F	Agreeableness, Conscientiousness
J-P	Conscientiousness

2.3   Personality Perception

Personality perception [42] is the automatic perception of a subject's personality based on their audio, visual, or other features [26]. Recent personality perception work includes textual personality perception [13,25,44], audio personality perception [27,41,45], visual perception from videos [5,16] and multimodal perception [28].

2.4   Facial Image Perception

In the field of personality perception, there are fewer studies simply on the basis of visual images. This can be ascribed in part to the difficulty inherent in gathering sufficiently large image datasets for personality computation [18, 19] described a supervised model based on ResNet and multi-layer Perceptron. It uses a person's face image to predict their Big Five traits. It was trained using 28,230 face images of 11,202 subjects. Although the connection between predicted and true scores is modest, it can correctly predict the relative standing of two randomly picked persons on a personality dimension in 58 % of situations (as against the 50 % expected by chance).

In the current study, instead of using conventional supervised models, we explores the possibility of Zero-Shot classificaton through large pretrained models like CLIP.

3   Method

To achieve zero-shot personality perception, We first translate MBTI subscale traits into CLIP-style text prompt for each MBTI subscale and encode them using CLIP text encoders. This is the primary distinction between PsyCLIP and CLIP: rather than utilising classification labels directly, as is customary in CLIP, we design psychological classification labels into text prompts that capture the relevant features for each category. We then assess the classification results of 16000 evaluation samples by encoding them with CLIP image encoders. This section describes the dataset we used and all steps to perform a zero-shot personality perception.

3.1   Dataset

Although the proposed method does not require any training, it needs a dataset to evaluate its performance. We have built a dataset from the largest online MBTI database [3]. The website contains 51800 profiles of famous people and characters. Each profile consists of a profile image of size 256×256. The profile has been scored by a number of voters for their perceived personality type. The personality type of each profile is determined by the perceived personality type with the highest vote. In post-processing, we took the top 1000 most voted non-fictional profiles for each 16 MBTI personality types (e.g., INPT type for Introvert, Intuition, Perceiving, Thinking) , resulting in a final sample size of 16000. The minimum number of votes per profile is 6, maximum is 5049, and average is 87.

3.2   CLIP Encoders

As explained earlier CLIP assigns each input image to the encoded text prompt that results in the highest similarity. In PsyCLIP we introduce prompts that are pertinent to each personality type. The prompts are engineered using GPT-3 as explained in the next section. The text and picture encoders in PsyCLIP were built using the ViT32 CLIP model,which has shown to achieve the best performance [33]. We retain the encoders in their current state in order to test CLIP's baseline performance against the MBTI evaluation dataset and to ascertain their potential for discriminating psychological features.

As seen in Fig. 1, we evaluated the effectiveness of PsyCLIP across the four MBTI dimensions. We apply a prompt engineering technique, detailed in the next section, to determine the ideal prompt that best describes each Myers Briggs subscale feature. For instance, we discovered that the prompt that best captures the extroversion attribute is "extraverted, outgoing, sociable, talkative, outspoken, gregarious, effervescent." We next repeated the prompt engineering procedure for each class of the four categories, resulting in a total of eight subscale feature sets.

3.3   Prompt Engineering

Prompt Generation. We produce prompts for each dimension using Generative Pre-trained Transformer 3 (GPT-3) [8]. GPT-3 is the state-of-the-art text generation model, trained on 499 billion tokens. We hypothesise that GPT-3 might assist us in converting psychological labels to CLIP-style instructions. We employed a temperature of 0.7 and the text-davinci-002 text completion model [1], and use the text completeion engine to complete the following: "list a series of adjectives that describes MBTI extroversion." This results in a list of potential candidates that capture the psychological qualities, as shown in Table 2. We then evaluated the performance of the generated prompts against a small test set of 100 samples per personality for prompt selection.

Table 2. Sample prompts generated by GPT-3 text-davinci-002 engine with the input "list a series of adjectives that describes MBTI Extraversion/Introversion/Thinking/Feeling".

Trait	Generated prompts from GPT3
Extraversion	Extraverted, Outgoing, Social, Chatty, Outspoken, Gregarious, Bubbly
Introversion	Introverted, Reserved, Reflective, Introspective, Private, Unassuming, Quiet
Thinking	Analytical, Logical, Rational, Objective, Introspective, Thoughtful
Feeling	Empathetic, Compassionate, Sympathetic, Cooperative, Caring

Prompt Selection. After generating prompts, we proceed to finding the optimal prompts. We begin by determining the accuracy of each prompt candidate's categorisation in a test pool of 100 randomly chosen candidates for each personality type. The samples are randomly chosen amongst the set of 35800 profiles that are not in the evaluation set. The findings are then utilised for eliminating biased prompts. Biased prompts are prompts that result in skewed results in favour of a certain sub-scale. For instance, if we use the raw GPT-3-generated prompt "analytical, logical, rational, objective, introspective, thoughtful" for the MBTI Thinking type and the raw GPT-3-generated prompt "empathetic, compassionate, sympathetic, cooperative, caring" for the MBTI Feeling type, the result would heavily favour the thinking type. While it is 90.3% accurate in identifying the thinking attribute of an INTP (introverted, intuiting, thinking, perceiving), it is only 20.0 percent accurate in identifying the feeling trait of an INFP (introverted, intuiting, feeling, perceiving). In this case, although the average accuracy of INPs is 55.6% in this scenario, the result cannot be considered statistically significant. As a consequence, we reject prompts that result in skewed outcomes and retain only those that result in above-expectation accuracy in both sub-scales. After rejecting biased prompts, we then find the prompts that would result in highest overall perception accuracy.

4   Results

The prediction accuracy for each MBTI category is shown in Tables 3, 4, 5 and 6. Predictions for each category conditioned on other categories are also reported to help better understand the model behaviour. Overall, PsyCLIP performed above the 50% chance level in all four categories and is statistically significant at $p < 0.01$ on the 16000-person sample size. This corroborates the hypothesis that CLIP do contain latent psychometric information.

4.1   Comparison to Similar Models

In aggregate, the average accuracy is 56.95%. There is a dearth of research on MBTI-based face personality perception with which to make direct comparisons. However, we may still make comparisons to models based on Big Five [19]. In

Table 3. Accuracy score for Thinking/Feeling classification. Result is significant at p < 0.001 against the baseline of 50% (chance level). For example, the first entry means the model has 68.8% accuracy in classifying INTPs as Thinking amongst 1000 INTP samples.

	T(%)	F(%)	Overall(%)
INP	68.8	53.8	57.1
INJ	68.4	46.0	61.3
ENP	66.1	55.1	60.6
ENJ	57.1	54.0	55.6
ISJ	63.8	52.7	58.2
ISP	65.6	47.7	56.7
ESJ	54.3	68.1	61.2
ESP	54.4	60.5	57.2
Overall			58.5

Table 4. Accuracy score for Judging/Perceiving classification. Result is significant at p < 0.001 against the baseline of 50% (chance level). For example, the first entry means the model has 51.6% accuracy in classifying INTJs as Judging amongst 1000 INTJ samples.

	J(%)	P(%)	Overall(%)
INT	51.6	60.9	56.3
ENT	53.9	64.4	59.2
ESF	31.7	72.7	52.0
ISF	43.2	70.9	57.1
EST	49.7	66.5	58.1
IST	54.3	60.3	57.3
ENF	38.0	74.2	56.1
INF	46.4	65.8	56.1
Overall			56.5

58% of situations (as opposed to the 50% anticipated by chance), the supervised model [19] can correctly predict the relative standing of two randomly picked persons on a personality dimension, which could be used as a reference point for comparison. Unfortunately, the dataset used in [19] is not publicly available, making direct comparisons of PsyCLIP's performance difficult. We plan to make the evaluation dataset for PsyCLIP publicly accessible, so that other researchers can test their model and we can make a direct comparison to PsyCLIP performance.

Table 5. Accuracy score for introversion/extraversion classification. Result is significant at p < 0.001 against the baseline of 50% (chance level). For example, the first entry means the model has 55.1% accuracy in classifying INTJs as Introverted amongst 1000 INTJ samples.

	I(%)	E(%)	Overall(%)
NTJ	55.1	44.9	50.0
NTP	54.3	63.3	57.5
SFJ	29.4	82.3	55.3
SFP	42.5	61.4	51.8
STP	35.4	68.3	51.9
STJ	65.6	47.7	55.6
NFJ	54.3	68.1	64.9
NFP	54.4	60.5	67.8
Overall			56.9

Table 6. Accuracy score for Sensing/Intuiting classification. Result is significant at p < 0.001 against the baseline of 50% (chance level). For example, the first entry means the model has 65.8% accuracy in classifying INTJs as Intuitive amongst 1000 INTJ samples.

	N(%)	S(%)	Overall(%)
ITJ	65.8	41.3	53.6
ETJ	60.0	49.1	54.6
ITP	58.8	61.5	60.2
ETP	55.4	59.9	57.7
IFP	49.4	61.6	55.5
EFP	45.1	63.2	54.2
EFJ	57.3	52.5	54.9
IFJ	62.0	50.4	56.2
Overall			55.9

4.2   Comparison Amongst Predictors in Each Dimension

The top performing dimension is the thinking/feeling subscale 2, where Psy-CLIP attained an overall accuracy level of 58%. With a prediction accuracy of 55.9%, the intuiting/sensing dimension is the lowest predictor. Interestingly this is consistent with the supervised model [19], which attained the highest accuracy scores for Big Five conscientiousness and the lowest accuracy score for openness. (For inter-scale relationships, see Table 1.) According to their work [19], the attributes most associated with cooperation (conscientiousness and agreeableness) should be more easily represented in the human face from an evolutionary standpoint. Our results add to the evidence supporting this theory.

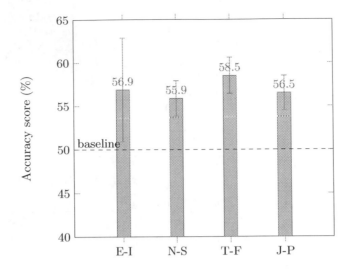

Fig. 2. Percentage accuracy of PsyCLIP with respect to each MBTI dimension. The classification accuracy exceeds the 50% prediction baseline in all personality dimensions.

4.3   Significance

The significance of the result can be interpreted in three ways:

- **It is competitive to SOTA model, without any training set.** PsyCLIP is competitive to the state-of-the-art model out of the box without any fine-tuning, hence it has much potential when datasets are fed into it.
- **It is highly generalisable.** The model is not conditioned on a particular set of psychometric prompts nor designed specifically for MBTI. This means the model has a potential to be a good base model for **any** image-based psychometric classification task.
- **It is statistically significant as a proof of concept.** It proves the existence of a psychometric layer within contrastive language-image pretraining models. We hope it can inspire more affective computing research utilising large pretrained models.

5   Ethical Impact

5.1   Societal Value

On the positive side, automatic personality perception is of significant societal value:

- In affective computing (AC), automatic personality perception is a necessary step in creating a social AI. To be social, an AI must understand how human perceive one another. The perception can be used to create a social avatar, or to build social conversational agents that are perceived in certain way, among other things.

- In clinical psychology, it can serve as a coaching system to assist socially challenged individuals such as those diagnosed with autism spectrum disorder (ASD) and social anxiety disorder (SAD), in understanding how their behaviours affect others' perception, thus helping them in coping with social norms.
- In theoretical psychology, computational models of such complex perception processes could potentially provide new insights or evidences into psychology theories. For example, as elaborated in Result section, our paper provided a data point to theory that the attributes most associated with cooperation (conscientiousness and agreeableness) should be more easily represented in the human face from an evolutionary standpoint [19]

5.2   Potential Misuses

The abuse of personality computing and its repercussions have been graphically described in several fictions [17] involving a dystopian society in which people are mercilessly evaluated and classified by a computer system.

Beyond fiction, there have been reports [32,40] of HR departments use AI to analyse a candidate's personality based on their web footprint. It would be devastating if PsyCLIP or a similar technology were utilised in this manner to assess a person's personality based on their appearances.

PsyCLIP was not designed for such purposes. One argument is that since PsyCLIP was trained for perception rather than recognition, it is only capable of predicting an applicant's perceived personality and hence has little use for candidate screening.

However, as individual researchers, we have little influence over whether a third party will recognise the delicate distinction between personality detection and apparent personality perception, or how third parties would use such technology.

Does this, however, imply that we should never do research on computer modelling of the human psychological traits? Is this to indicate that AIs are meant to be heartless machines forbidden the knowledge of human emotions, personality, or psychology? One may argue that if research into automated personality perception is halted, we will never be able to build a social AI [2].

We call upon the community to come together and come up with ethical frameworks and regulations on the usage of personality computing technologies, especially in sensitive areas such as recruitment, user profiling and surveillance.

We are also concerned about the potential biases in the dataset. Although by theory [29] all sixteen personalities are equal in value and none are preferable to another, the dataset is labelled by humans, who could be typing a person based on their racial or cultural stereotype. We attempted to mitigate this issue by only evaluating data points that received at least ten votes. Additionally, we would make the dataset and model available to the public upon publication, as we believe that increased transparency and openness are critical in identifying and combating such biases.

6 Conclusion

Based on our experiments, we provide new evidence on the correlation between personality and the facial image. With a sample size of 16000, the findings are statistically significant at $p < 0.01$ and consistently better than the baseline across all four dimensions.

The effectiveness of CLIP in personality perception, along with its zero-shot nature, offers up new possibilities for personality computing applications. It is a complement to conventional supervised models and opens a new direction in study of personality perception phenomenon. With some improvements, computational psychologists now can have a simple model that can be used to predict *any* perceived personality attributes and use it to better understand the semantic associations between words/phrases and personality types.

One area for future study is to investigate how consensus among personality voting influences PsyCLIP's performance. There may be a difference between forecast outcomes for persons with a high personality voting consensus and those with a low voting consensus. Another possibility is to broaden the scope of PsyCLIP's examination beyond MBTI to include additional psychological qualities. Finally, the concept of fine-tuning the PsyCLIP model against a certain personality scale is intriguing and worth exploring.

References

1. Engines - openai api. https://beta.openai.com/docs/engines/gpt-3. Accessed 22 Apr 2022
2. The ethics of artificial intelligence: Issues and initiatives: Think tank: European parliament (2020). https://europarl.europa.eu/thinktank/en/document/EPRS_STU634452
3. Personality database. www.personality-database.com/vote (2022). Accessed 06 Jan 2022
4. Ali, S., Parikh, D.: Telling creative stories using generative visual aids. arXiv preprint arXiv:2110.14810 (2021)
5. Biel, J.I., Teijeiro-Mosquera, L., Gatica-Perez, D.: Facetube: predicting personality from facial expressions of emotion in online conversational video. In: Proceedings of the 14th ACM International Conference on Multimodal Interaction, pp. 53–56 (2012)
6. Borkenau, P., Brecke, S., Möttig, C., Paelecke, M.: Extraversion is accurately perceived after a 50-ms exposure to a face. J. Res. Pers. J. Res. Pers. **43**(4), 703–706 (2009). https://doi.org/10.1016/j.jrp.2009.03.007
7. Brock, A., Donahue, J., Simonyan, K.: Large scale gan training for high fidelity natural image synthesis (2019)
8. Brown, T.B., et al.: Language Models are Few-Shot Learners. arXiv (2020). arxiv.org/2005.14165v4
9. Celli, F., Lepri, B.: Is big five better than mbti? a personality computing challenge using twitter data. In: CLiC-it (2018)
10. Deng, J., et al.: ImageNet: a large-scale hierarchical image database. In: CVPR 2009, pp. 248–255 (2009)

11. Deng, L.: The mnist database of handwritten digit images for machine learning research. IEEE Sign. Process. Mag. **29**(6), 141–142 (2012)
12. Durupinar, F.: Personality-Driven Gaze Animation with Conditional Generative Adversarial Networks. arXiv (2020). arxiv.org/2012.02224v1
13. Farnadi, Get al.: Computational personality recognition in social media. User Model. User-Adapt. Interact. **26**(2–3), 109–142 (2016)
14. Furnham, A.: The of big five versus the big four: the relationship between the myers-briggs type indicator (mbti) and neo-pi five factor model personality. Pers. Individ. Differ **21**(2), 303–307 (1996)
15. Goodfellow, I., et al.: Generative adversarial nets. Adv. Neural Inf. Process. Syst. **27** (2014)
16. Gürpınar, F., Kaya, H., Salah, A.A.: Combining deep facial and ambient features for first impression estimation. In: Hua, G., Jégou, H. (eds.) ECCV 2016. LNCS, vol. 9915, pp. 372–385. Springer, Cham (2016). https://doi.org/10.1007/978-3-319-49409-8_30
17. Hermann, I.: Artificial intelligence in fiction: between narratives and metaphors. AI Soc. 1–11 (2021)
18. Junior, J.C.S.J., et al.: First impressions: a survey on vision-based apparent personality trait analysis. IEEE Trans. Affect. Comput. 1 (2019). https://doi.org/10.1109/TAFFC.2019.2930058
19. Kachur, A., Osin, E., Davydov, D., Shutilov, K., Novokshonov, A.: Assessing the big five personality traits using real-life static facial images. Sci. Rep. **10**(8487), 1–11 (2020). https://doi.org/10.1038/s41598-020-65358-6
20. Kenny, D.A.: Person: a general model of interpersonal perception. Pers. Soc. Psychol. Rev. **8**(3), 265–280 (2004)
21. Kramer, R.S., King, J.E., Ward, R.: Identifying personality from the static, nonexpressive face in humans and chimpanzees: evidence of a shared system for signaling personality. Evol. Hum. Behav. **32**(3), 179–185 (2011)
22. Liu, L., Ouyang, W., Wang, X., Fieguth, P., Chen, J., Liu, X., Pietikäinen, M.: Deep learning for generic object detection: a survey. Int. J. Comput. Vis. **128**(2), 261–318 (2019). https://doi.org/10.1007/s11263-019-01247-4
23. lucidrain: Github - lucidrains/big-sleep. http://github.com/lucidrains/big-sleep (2021). Accessed 06 Jan 2022
24. lucidrain: Github - lucidrains/deep-daze. http://github.com/lucidrains/deep-daze (2021). Accessed 06 Jan 2022
25. Majumder, N., Poria, S., Gelbukh, A., Cambria, E.: Deep learning-based document modeling for personality detection from text. IEEE Intell. Syst. **32**(2), 74–79 (2017)
26. Mohammadi, G., Vinciarelli, A.: Automatic personality perception: prediction of trait attribution based on prosodic features. IEEE Trans. Affect. Comput. **3**(3), 273–284 (2012)
27. Mohammadi, G., Vinciarelli, A.: Automatic personality perception: prediction of trait attribution based on prosodic features extended abstract. In: 2015 International Conference on Affective Computing and Intelligent Interaction (ACII), pp. 484–490. IEEE (2015)
28. Mohammadi, G., Vuilleumier, P.: A multi-componential approach to emotion recognition and the effect of personality. IEEE Trans. Affect. Comput. (2020)
29. Myers, I.B.: The myers-briggs type indicator: Manual (1962) (1962)
30. Naumann, L.P., Vazire, S., Rentfrow, P.J., Gosling, S.D.: Personality judgments based on physical appearance. Pers. Soc. Psychol. Bull. **35**(12), 1661–1671 (2009). https://doi.org/10.1177/0146167209346309

31. Oosterhof, N.N., Todorov, A.: The functional basis of face evaluation. Proc. Natl. Acad. Sci. **105**(32), 11087–11092 (2008)
32. Patil, S.M., Singh, R., Patil, P., Pathare, N.: Personality prediction using digital footprints. In: 2021 5th International Conference on Intelligent Computing and Control Systems (ICICCS), pp. 1736–1742. IEEE (2021)
33. Radford, A., et al.: Learning transferable visual models from natural language supervision. arXiv preprint arXiv:2103.00020 (2021)
34. Ruane, E., Farrell, S., Ventresque, A.: User Perception of Text-Based Chatbot Personality. In: Følstad, A., Araujo, T., Papadopoulos, S., Law, E.L.-C., Luger, E., Goodwin, M., Brandtzaeg, P.B. (eds.) CONVERSATIONS 2020. LNCS, vol. 12604, pp. 32–47. Springer, Cham (2021). https://doi.org/10.1007/978-3-030-68288-0_3
35. Salgado, J.F.: The big five personality dimensions and counterproductive behaviors. Int. J. Select. Assess. **10**(1–2), 117–125 (2002)
36. Shevlin, M., Walker, S., Davies, M., Banyard, P., Lewis, C.A.: Can you judge a book by its cover? evidence of self-stranger agreement on personality at zero acquaintance. Pergamon-Elsevier (2003). http://irep.ntu.ac.uk/id/eprint/16819
37. Sijtsma, K.: Introduction to the measurement of psychological attributes. Measurement **44**(7), 1209–1219 (2011)
38. Ruane, E., Farrell, S., Ventresque, A.: User perception of text-based chatbot personality. In: Følstad, A., et al. (eds.) CONVERSATIONS 2020. LNCS, vol. 12604, pp. 32–47. Springer, Cham (2021). https://doi.org/10.1007/978-3-030-68288-0_3
39. Uleman, J.S., Adil Saribay, S., Gonzalez, C.M.: Spontaneous inferences, implicit impressions, and implicit theories. Annu. Rev. Psychol. **59**, 329–360 (2008)
40. Upadhyay, A.K., Khandelwal, K.: Applying artificial intelligence: implications for recruitment. Strat. HR Rev. (2018)
41. Valente, F., Kim, S., Motlicek, P.: Annotation and recognition of personality traits in spoken conversations from the AMI meetings corpus. In: Thirteenth Annual Conference of the International Speech Communication Association (2012)
42. Vinciarelli, A., Mohammadi, G.: A Survey of Personality Computing. IEEE Trans. Affect. Comput. **5**(3) (2014). DOI: https://doi.org/10.1109/TAFFC.2014.2330816
43. Walker, M., Vetter, T.: Changing the personality of a face: perceived big two and big five personality factors modeled in real photographs. J. Pers. Soc. Psychol. **110**(4), 609–624 (2016). https://doi.org/10.1037/pspp0000064
44. Yu, J., Markov, K.: Deep learning based personality recognition from facebook status updates. In: 2017 IEEE 8th International Conference on Awareness Science and Technology (iCAST), pp. 383–387. IEEE, Taichung (2017). https://doi.org/10.1109/ICAwST.2017.8256484, https://ieeexplore.ieee.org/document/8256484/
45. Yu, M., Gilmartin, E., Litman, D.: Identifying personality traits using overlap dynamics in multiparty dialogue. arXiv preprint arXiv:1909.00876 (2019)
46. Zou, Z., Shi, Z., Guo, Y., Ye, J.: Object Detection in 20 Years: A Survey. ResearchGate (2019). www.researchgate.net/publication/333077580

Multi-view Based Clustering of 3D LiDAR Point Clouds for Intelligent Vehicles

Haoxiang Jie[1], Zuotao Ning[1]([✉]), Qixi Zhao[1], Wei Liu[1,2], Jun Hu[1], and Jian Gao[1]

[1] Neusoft Reach Automotive Technology Company, Shenyang, China
ningzt@reachauto.com
[2] School of Computer Science and Engineering, Northeastern University, Shenyang, China

Abstract. 3D point clustering is important for the LiDAR perception system involved applications in tracking, 3D detection, etc. With the development of high-resolution LiDAR, each LiDAR frame perceives richer detail information of the surrounding environment but highly enlarges the point data volume, which brings a challenge for clustering algorithms to precisely segment the point cloud while running with a real-time processing speed. To meet this challenge, we innovate a multi-view (bird's eye view and front view) based clustering method, named MVC. The method contains two stages. In the first stage, we propose a density image based algorithm, PG-DBSCAN, to segment the point cloud in bird's eye view (BEV), which derives the preliminary division with fairly low computation resources. Then in the second stage, a front view (FV) clustering process is integrated to refine the under-segmented clusters. Our method takes both the speed and precision advantages of BEV and FV clustering, and this coarse-to-fine architecture reasonably allocates the computation resources and shows a real-time outstanding clustering performance. We evaluate the MVC algorithm both on the publicly available dataset with 64-line LiDAR and our own dataset with 128-line LiDAR. Compared with other clustering methods, MVC is able to derive more accurate clustering results. Specifically, toward the 128-line LiDAR with large data volume, our method shows an outperforming running speed, which perfectly fits on the LiDAR perception tasks.

Keywords: Point Cloud Segmentation · High Resolution LiDAR · PG-DBSCAN

1 Introduction

In the LiDAR perception system, Deep-Learning based 3D detection modules are widely used to provide important evidences for the free driving space prediction. However, sometimes such kind of modules may perform miss detection or incorrect detection when meeting untrained rare scenes, and further cause wrong

H. Aziz et al. (Eds.): AI 2022, LNAI 13728, pp. 57–70, 2022.
https://doi.org/10.1007/978-3-031-22695-3_5

drivable area prediction. That could be dangerous. For solving this problems, engineers and researchers have been developing 3D object clustering methods using as the back-up plan that is able to perceive the obstacle locations when detection modules make wrong judgments.

With the development of laser sensors and electronic chips, the resolution of LiDARs is designed higher and higher. For example, the LiDAR "Ruby" from Robosence company[1] is assembled with 128 lines and 0.2 °C horizontal resolution. Such LiDAR is able to reflect over 2.3 million 3D points in each frame and provides much richer 3D information of the surrounding environment compared with the lower resolution LiDARs. But the denser the point clouds are, the larger the data volume would be. So the high resolution LiDAR raises more requirements for the processing speed while maintaining the segmentation accuracy.

However, facing the dense LiDAR point cloud, the traditional clustering methods, such as DBSCAN [6], Mean Shift [4], pose difficulties of high computational complexity and fail in real-time processing. The point cloud clustering methods based on the range maps, such as [13] and [21], may be able to meet the real-time running requirement, but would bring serious over-segmentation issues when the LiDAR resolution goes up.

So, in this paper, in order to decrease the computational complexity and achieve a satisfactory clustering performance, we propose a multi-view based 3D point cloud clustering algorithm (MVC). This method is inspired by the real-world spatial distribution of objects on the streets, that in the bird's-eye view (BEV), the majority of objects in the driving scenes are naturally separated. Thus we project the point cloud in BEV and design a preliminary clustering stage. In order to improve the processing speed, we down-sample the points in BEV with polar grid maps. Meanwhile, we modify the traditional DBSCAN [6] method and reduce the computational complexity. However, the BEV clustering module cannot segment the objects located at the same place in BEV but different places on the vertical direction, such as a billboard and the car below. Thus, for improving the clustering accuracy, we introduce front-view (FV) refining clustering stage to solve the vertical under-segmentation problem.

The main contributions of this paper are the following items.

- We innovate a new point cloud clustering method combining the BEV and FV, which utilizes the point cloud geographical features to accurately segment the 3D obstacles.
- We raise a PG-DBSCAN [6] algorithm which highly reduces the computational complexity for this 3D point clustering task.
- We compare our method with 4 effective traditional point cloud clustering methods on both semanticKITTI dataset [1] and a self-collected 128-line LiDAR dataset.

---

[1] https://www.robosense.ai/en/rslidar/RS-Ruby.

Fig. 1. A demonstration of MVC clustering result.

2 Related Work

In this section, through analysing existing technologies, we divide clustering algorithms into two categories: free-trained methods and Deep-learning based methods.

2.1 Free-trained Point Cloud Clustering Methods

Clustering Method Based on Voxel/Grid Map. Most of Voxel/Grid map based algorithms need minimal computational overhead and perform fast processing speed through down-sampling the point cloud. So this kind of clustering algorithms is widely utilized in the field of robots and unmanned vehicles due to their real-time requirement. For example, in the 2007 DARPA Challenge [19,20], many teams chose this kind of methods to separate objects from the ground. [5] created hybrid elevation maps to extract the non-ground objects, then down-sampled the non-ground point cloud by voxel cells and cluster the points according to the voxel connectivity. Similarly, [9] utilized the BEV grid map and 3D voxel to down-sample the non-ground points, then combined the connectivity of the grids and the height difference between the voxels to further cluster the point cloud [15] proposed the curved voxel clustering method considering the difference of horizontal and vertical angular resolutions for LiDAR. However, the segmentation accuracy of these methods highly depends on the size of grids or voxels, and some of the spatial information of point clouds is lost due to the point down-sampling.

Clustering Method Based on Range Image. The point cloud clustering methods based on the range image also attracted the interest of many researchers [2] projected 3D point clouds into range images. They performed a N4-searching by BFS algorithm and clustered the point cloud following a given angle threshold. On the basis of [2,12] further increased the constraints of distance and reflection intensity difference between adjacent points and reduced the over-segmentation rate. [22] proposed the Scan Line Run (SLR) clustering method based on the range image. [8] combined density and connectivity information of the range image to achieve real-time clustering performance.

Clustering Method Based on Graph Model. Applying graph theory to achieve point cloud clustering is also a research direction. For example, [11] proposed a clustering method based on Radially Bounded Nearest neighbours (RBNN) graphs. They represented the 3D laser point clouds as directed graphs, then cluster the LiDAR point clouds based on a given threshold [14] generated point cloud undirected graphs based on the scanning characteristics of the mechanical rotating LiDAR. The point cloud is then separated according to the local convexity criterion which is calculated based on the normal vector of the nodes. Similarly, [3] considered the hardware parameters of Velodyne HDL-64 LiDAR when creating the undirected graphs. In order to balance the accuracy and running speed of the algorithm, they used a 4-connected region growing method to cluster 3D point clouds.

2.2 Deep-learning Based Point Cloud Clustering Methods

Pointnet [16] firstly generated Deeplearning network for solving the point cloud classification and clustering problem. They used multilayer perceptrons followed by max-pooling to extract the point feature and resulted a decent performance. A novel work, [7] proposed a proposal-free point cloud clustering method by a simplified framework with a Deeplearing-based solution. The method does not rely on any post-process, and is able to reach a good performance [18] presented a top-down Deep-learing based LiDAR segmentation architecture with a MASK R-CNN instance head. The method also formulated a pseudo labeling framework to enhance the clustering performance by training the network on unlabelled dateset. [10] used cylinder convolution extract grid-level features for each LiDAR frame and proposed Dynamic Shifting for complex point distributions then raised Consensus-driven Fusion to finally derive instance preditions.

3   The Proposed Method

Our work mainly focuses on the non-ground targets clustering. So a relevant ground segmentation method [14] is utilized for data preprocessing. The pipeline of MVC is briefly demonstrated in Fig. 2, which follows a coarse-to-fine architecture. It works with two stages: BEV coarse-segmentation and FV fine-segmentation.

The details of MVC are described in the following subsections.

3.1   Preliminary Clustering Based on BEV Projection

The demonstration of the BEV clustering procedure is shown in Fig. 3.

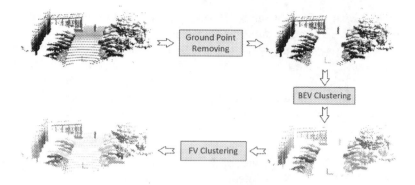

Fig. 2. The pipeline of MVC.

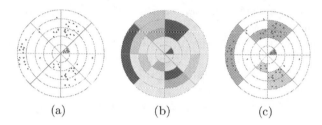

(a)   (b)   (c)

Fig. 3. (a) shows the BEV grids and the projected 2D points. We take the grids as pixels and count the point number in each grid as pixel values, then generate the density image as (b). The brightness of the color represents the density value. Using the modified DBSCAN method, pixels are clustered as shown in (c) and the points located in the grid with same color share the same cluster label.

Firstly, we project the object point cloud onto $x - o - y$ plane in Polor Coordinate $\{\rho, \theta\}$ by

$$\rho = \sqrt{x^2 + y^2} \tag{1}$$

$$\theta = i \times res \tag{2}$$

where i is the horizontal index of the point anti-clockwise counted from the positive direction of the x axis, and res is the horizontal angle between two adjacent laser beams in the same scanning line. Clearly, the ρ is equal to the range value of each target point.

Secondly, the grids of the BEV map are generated with a manually selected angle unit size θ_{thres} and range unit size r_{thres}. By using these grids we generate density image and count the point number in each gird as the density value dv. The reason why we choose Polar Coordinate in the BEV grid map is inspired by a related work CVC [15] that in such Coordinate the grid area expands with the range value increasing, which perfectly fits the near-dense-far-sparse geometrical characteristic of the LiDAR points.

Based on the traditional DBSCAN [6], we propose a density-based clustering method Polar-Grid-DBSCAN(PG-DBSCAN).

Compared with the traditional DBSCAN [6], instead of going through each point and calculating the surrounding data density for clustering, we go through each of the pixel in the density image to segment the point cloud. Firstly the pixels with density value dv lower than 4 are marked as noise pixels. For the other pixels, we start from a random pixel as target and search its 8 neighbour pixels. If the neighbour pixels are not noise pixels, these neighbours are marked by the same label with the target. By recurrence, the whole density image is segmented to different areas, and the points located in each area share the same label.

Our PG-DBSCAN greatly accelerates the clustering speed compared with the traditional DBSCAN [6]. While dealing with large-scale point cloud data, as the area query operation of the traditional DBSCAN [6] is calculated based on the Euclidean distance, the average running time complexity is $O(\log(n))$, so the average computational complexity of traditional DBSCAN [6] is $O(n \times \log(n))$, where n is the number of the points. However, PG-DBSCAN finishes the region query by inquiring the 8-neighbour of each pixel in the density image, so the computational complexity of one point becomes to $O(1)$, and the average computational complexity of one frame is reduced to $O(n)$, where n is the pixel number of the density image.

After the operations mentioned above, we derive the preliminary segmentation result of the non-ground targets.

3.2 Refining Based on Range Image

Fig. 4. After BEV preliminary segmentation, we project each clustered point cloud to a range image separately. The pixel brightness represents the range value of the points. Pure black pixels means no point or the point out of the region.

After BEV segmentation, for each cluster, we calculate the height difference ΔH between the highest point and lowest point. Only when ΔH is higher than 2 meters, we consider the cluster may be under-segmented that requires fine FV clustering.

For FV segmentation, the preliminary cluster is projected into range images [13], as Fig. 4.

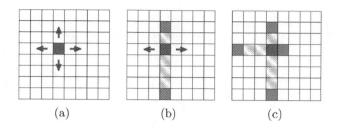

(a)   (b)   (c)

Fig. 5. The process of the modified N-4 neighbour searching. (a) indicates the searching directions. (b) and (c) are the searching process. We go through the pixels following the arrow directions and find the first non-zero-pixel specified by solid green blocks. The shadow green blocks represent the zeros pixels that are ignored.

Here we introduce a modified N-4 neighbor searching, and define neighbors of the target pixels as the adjacent pixels in the same row or column with the non-zero pixel values shown as Fig. 5

With this neighbour searching method, we cluster the pixels by judging if the height difference Δh and range difference Δr between the target and neighbour pixels are within a given threshold γ. We consider the points to belong to the same cluster on the condition that Δh and Δr meet Eq. (4) and Eq. (5). Here r is directly derived from the pixel value and h is calculated by:

$$h = r \times tan(\alpha) \tag{3}$$

$$\Delta r < \gamma \tag{4}$$

$$\Delta h < \gamma \tag{5}$$

where α is the vertical LiDAR beam angle which can be found from the LiDAR product specification. Through recursive searching method, all pixels in the range image are fine clustered, and the process is illustrated in Algorithm 1.

Algorithm 1: Refining based on Range Image

Input: An image in FV as *img* and its cluster as *cluster*
Output: *cluster*

foreach p_i *in img* **do**
    if $p_i.flag! = is\_visited$ **then**
        RecursiveClustering(p_i,cluster)
        Update(cluster)
    else
        L continue
  return *cluster*

Function RecursiveClustering(p_i, c_i)

Input: A pixel of the image in FV as p_i and its initialised cluster as c_i
Output: c_i

RecursiveClustering(p_i, c_i)
$Neighbour_i \leftarrow$ FindNeighbour(p_i)
foreach n_i in $Neighbour_i$ **do**
   | **if** $is\_same\_cluster(n_i,p_i)$ **then**
   | | $Update(c_i)$
   | | $n.flag \leftarrow is\_visited$
   | | $RecursiveClustering(n_i, c_i)$
   | **else**
   | L continue

---

4   Experiment

In this section, we test the proposed MVC algorithm and provide the experiment setup and evaluation metrics. We report the comparisons with different clustering methods on both SemanticKITTI dataset and our own dataset(NRS). Also, we carry ablation study for better understanding the advantages of the clustering processes of our method in the two views(BEV, FV).

We conduct experiments on a desktop with an Intel Xeon(R) CPU E3-1231 v3 @ 3.40 GHz × 8, 32 Gb RAM.

4.1   Experiment on SemanticKITTI

A Related work, [23] evaluated 4 different clustering methods on the SemanticKITTI dataset using Panoptic Quality(PQ) as evaluation metrics. For comparing the clustering performance of MVC with those 4 methods, we apply the same clustering process and the evaluation metrics. The result is shown in Table 1.

Table 1. Comparison between our method and the methods reported in [23] on SemanticKITTI dataset.

Methods	Settings	PQ
Euclidean cluster	$d_{th} = 0.5$ m	56.9
Supervoxel cluster	$w_c, w_s, w_n = 0.0, 1.0, 0.0$	52.8 52.7
Supervoxel cluster	$w_c, w_s, w_n = 0.0, 1.0, 0.5$	
Depth cluster	$\theta = 10°$	55.2
Scan-line run	$th_{run}, th_{merge} = 0.5, 1.0$	57.2
Ours	$\theta_{thres}, r_{thres}, \gamma = 2, 0.5, 0.6$	**58.8**

Performance Evaluation. Following [23], all the clustering methods work as a post-process step after a semantic segmentation method, [24]. The experiment setting of the upper four methods remain same as in [23].

The experiment result shows that our method outperforms in the comparison group. It is worth mentioning that in this experiment, we abandon the ground point removing process, since the semantic segmentation process has already removed the ground point. Moreover, this pre-process also removes other background points, such as trees, which consequently deletes almost all the objects with large vertical size. However, our FV clustering processing happens only when the clusters from BEV segmentation are higher than 2 m. Thus, the FV clustering process seldom works in this experiment, but MVC still derives the best performance among all the methods.

4.2   Experiment on Self-Recorded Dataset

Importantly, for meeting engineering design requirements and feeding the needs from customers, it is necessary to test MVC method on our own dataset(NRS). NRS dataset is collected with RS-Ruby 128-line LiDAR sensor in the company NEUSOFT REACHAUTO[2] including the scenarios on the campus roads at Neusoft headquarters and the street of Shenyang city(China). However, because of the different labeling method between SemanticKITTI and NRS datasets, we have to change the evaluation metric from PQ to the method reported in [13,17].

We adopt the over-segmentation, under-segmentation and precision as criteria to evaluate the proposed algorithm. Further, we introduce four states of clustering results to quantify the clustering performance: Precision (P), True Positive (TP), Over-segmentation-rate (OSR) and Under-segmentation-rate (USR).

- TP is the number of clustered objects that are successfully segmented.
- OS is the total number of over-segmentation clusters.
- US is the total number of under-segmentation clusters.

Using the above-mentioned states, the following three metrics are formulated as:

$$OSR = 1 - \frac{TP}{TP + OS}, \tag{6}$$

$$USR = 1 - \frac{TP}{TP + US}, \tag{7}$$

$$P = \frac{TP}{TP + OS + US} \tag{8}$$

According to the related works, most of the previous algorithms are tested and validated via 64-beam LiDARs or even fewer. To the best of our knowledge, this paper is the first attempt to evaluate a clustering algorithm using a 128-line LiDAR.

---

[2] https://www.reachauto.com/.

Table 2. Experimental settings of different methods

Euclidean cluster	$d_{th} = 0.5m$
Supervoxel cluster	$w_c, w_s, w_n = 0.0, 1.0, 0.0$
Depth cluster	$\theta = 5°$
Scan-line run	$th_{run}, th_{merge} = 0.3, 0.5$
Ours	$\theta_{thres}, r_{thres}, \gamma - 1, 0.2, 0.5$

Performance Evaluation. Considering the point on NRS is denser than that on SemanticKITTI, we delicately adjust the coefficients of the methods for better performance, as shown in Table 2.

We separate NRS dataset into 3 scenario types: Easy, Medium and Hard. Easy type only consists of some sparse road participates without the vertical structure either, Fig. 6(a); Medium type has a dense road participates distribution, but there are not many vertical structures in this type of point clouds, Fig. 6(b); Hard type point cloud has crowded road participates and also objects with vertical structures such as trees and cars below, Fig. 6(c).

Since the NRS dataset does not have semantic segmentation labels, we cannot train a good segmentation network for pre-processing. So we choose a free-trained ground removing method to pre-process the point cloud.

Table 3 reports the comparison results as well as the processing speed of the group. In the Easy scenarios, all the methods have similar precision rate because of the dense point clouds but sparse objects. In the medium and hard scenarios, the precision gap between ours and the other methods becomes larger. Euclidean Cluster, Supervoxel Cluster and Scan-line Run suffer from under-segmentation caused by the background points (trees, traffic lights and etc.). Depth Cluster shows higher over-segmentation rate because smaller objects near the LiDAR would block the laser beams and truncate the big objects, which cannot be handled by this angle based clustering method. While being beneficial from the view combination in MVC, the performance of our method experiences a slight going down but still remain a decent precision rate.

Besides, we also report the clustering speed by frame-per-second (FPS) of all the methods in these three scenarios without taking the pre-processing stage of ground points removal into account. As the amount of points becomes larger, our method stably runs in a high speed at about 10 ms per frame, which satisfies the real-time requirement.

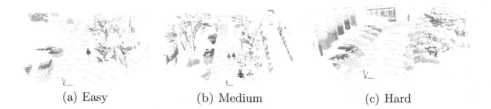

(a) Easy   (b) Medium   (c) Hard

Fig. 6. A demonstration of the three scenario types in NRS dataset.

Table 3. Segmentation results on NRS dataset

Scenarios	Algs	OS	US	TP	OSR	USR	P	FPS
Easy	Ours	34	44	1088	0.030	0.039	**0.933**	**110**
	Euclidean cluster	15	108	965	**0.015**	0.100	0.887	73
	Supervoxel cluster	18	143	927	0.019	0.134	0.852	24
	Depth cluster	109	31	948	0.103	**0.032**	0.871	70
	Scan-line run	75	35	978	0.071	0.035	0.899	54
Medium	Ours	104	217	2908	0.035	0.069	**0.900**	**100**
	Euclidean cluster	85	322	2822	0.029	0.102	0.874	67
	Supervoxel cluster	73	419	2737	**0.026**	0.133	0.848	16
	Depth cluster	346	90	2793	0.110	**0.031**	0.865	63
	Scan-line run	114	263	2852	0.084	0.038	0.883	54
Hard	Ours	190	407	2922	0.061	**0.122**	**0.830**	**98**
	Euclidean cluster	242	811	2276	0.096	0.263	0.684	63
	Supervoxel cluster	105	1126	2098	**0.048**	0.349	0.630	12
	Depth cluster	597	485	2248	0.210	0.177	0.675	60
	Scan-line run	383	391	2555	0.130	0.133	0.767	54

4.3 Ablation Study

The ablation study mainly focuses on assessing the clustering process in different views. We evaluate the clustering performance and running speed in this experiment and report the statistics in Table 4.

Table 4. The ablation study of different modules in MVC

Group	OS	US	TP	OSR	USR	P	FPS
1. BEV(PG-DBSCAN)+FV	104	217	2908	0.035	0.069	**0.900**	100
2. BEV(PG-DBSCAN)	63	1052	2014	**0.028**	0.332	0.655	**250**
3. BEV(DBSCAN [6])+FV	97	236	2886	0.032	0.075	0.897	2
4. FV	328	81	2820	0.104	**0.028**	0.873	18

FV Clustering Module. As shown in Table 4, we compare the performance of FV clustering removed MVC (Group 2) with the original MVC (Group 1). Without the FV clustering, the number of over-segmentation vehicles slightly goes down and OSR remains steady, however, the amount of under-segmentation rate increases sharply. Because, in some scenes, background points may combine different target objects together, as show in Fig 7. Without FV clustering process, the points from these objects are clustered into the same cloud and cause under-segmentation.

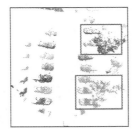

Fig. 7. Left image shows only the BEV clustering result, and the right image is the result of whole MVC algorithm

BEV Clustering Module. In this part, we analyse the importance of BEV clustering process. Comparing Group 1 with Group 4 in Tab. 4, we conclude that PG-DBSCAN efficiently accelerates the clustering process and slightly improve clustering precision.

Also, from Group 3 and 1, we can see that, with nearly same clustering precision, the running speed of MVC with traditional DBSCAN [6] is 50 times slower than that with PG-DBSCAN.

5   Conclusion

In this paper, a multi-view based clustering method is proposed for the 3D point cloud. The algorithm adopts the coarse-to-fine architecture. First, the non-ground point cloud is projected to the BEV density image and down-sampled. We propose PG-DBSCAN based on the traditional DBSCAN [6] for the preliminary segmentation. Then we further separate the under-segmented clusters on vertical direction based on range images. We compare our method with 4 traditional clustering algorithms on both SemanticKITTI and NRS dataset. The experiment results show the real-time performance, stability and accuracy of the MVC algorithm, and prove that this method is suitable for clustering the dense point clouds in various driving scenes.

References

1. Behley, J., et al.: Semantickitti: a dataset for semantic scene understanding of lidar sequences. In: Proceedings of the IEEE/CVF International Conference on Computer Vision, pp. 9297–9307 (2019)
2. Bogoslavskyi, I., Stachniss, C.: Fast range image-based segmentation of sparse 3D laser scans for online operation. In: 2016 IEEE/RSJ International Conference on Intelligent Robots and Systems (IROS), pp. 163–169. IEEE (2016)
3. Burger, P., Wuensche, H.J.: Fast multi-pass 3D point segmentation based on a structured mesh graph for ground vehicles. In: 2018 IEEE Intelligent Vehicles Symposium (IV), pp. 2150–2156. IEEE (2018)
4. Cheng, Y.: Mean shift, mode seeking, and clustering. IEEE Trans. Pattern Anal. Mach. Intell. **17**(8), 790–799 (1995)
5. Douillard, B., et al.: Hybrid elevation maps: 3D surface models for segmentation. In: 2010 IEEE/RSJ International Conference on Intelligent Robots and Systems, pp. 1532–1538. IEEE (2010)
6. Ester, M., Kriegel, H.P., Sander, J., Xu, X., et al.: A density-based algorithm for discovering clusters in large spatial databases with noise. In: KDD, vol. 96, pp. 226–231 (1996)
7. Gasperini, S., Mahani, M.A.N., Marcos-Ramiro, A., Navab, N., Tombari, F.: Panoster: end-to-end panoptic segmentation of lidar point clouds. IEEE Robot. Autom. Lett. **6**(2), 3216–3223 (2021)
8. Hasecke, F., Hahn, L., Kummert, A.: Fast lidar clustering by density and connectivity. arXiv e-prints pp. arXiv-2003 (2020)
9. Himmelsbach, M., Hundelshausen, F.V., Wuensche, H.J.: Fast segmentation of 3D point clouds for ground vehicles. In: 2010 IEEE Intelligent Vehicles Symposium, pp. 560–565. IEEE (2010)
10. Hong, F., Zhou, H., Zhu, X., Li, H., Liu, Z.: Lidar-based panoptic segmentation via dynamic shifting network. In: Proceedings of the IEEE/CVF Conference on Computer Vision and Pattern Recognition, pp. 13090–13099 (2021)
11. Klasing, K., Wollherr, D., Buss, M.: A clustering method for efficient segmentation of 3D laser data. In: 2008 IEEE International Conference on Robotics and Automation, pp. 4043–4048. IEEE (2008)
12. Li, M., Yin, D.: A fast segmentation method of sparse point clouds. In: 2017 29th Chinese Control And Decision Conference (CCDC), pp. 3561–3565. IEEE (2017)
13. Li, Y., Le Bihan, C., Pourtau, T., Ristorcelli, T.: Insclustering: instantly clustering lidar range measures for autonomous vehicle. In: 2020 IEEE 23rd International Conference on Intelligent Transportation Systems (ITSC), pp. 1–6. IEEE (2020)
14. Moosmann, F., Pink, O., Stiller, C.: Segmentation of 3D lidar data in non-flat urban environments using a local convexity criterion. In: 2009 IEEE Intelligent Vehicles Symposium, pp. 215–220. IEEE (2009)
15. Park, S., Wang, S., Lim, H., Kang, U.: Curved-voxel clustering for accurate segmentation of 3D lidar point clouds with real-time performance. In: 2019 IEEE/RSJ International Conference on Intelligent Robots and Systems (IROS), pp. 6459–6464. IEEE (2019)
16. Qi, C.R., Su, H., Mo, K., Guibas, L.J.: Pointnet: deep learning on point sets for 3D classification and segmentation. In: Proceedings of the IEEE Conference on Computer Vision and Pattern Recognition, pp. 652–660 (2017)
17. Shin, M.O., Oh, G.M., Kim, S.W., Seo, S.W.: Real-time and accurate segmentation of 3-D point clouds based on gaussian process regression. IEEE Trans. Intell. Transp. Syst. **18**(12), 3363–3377 (2017)

18. Sirohi, K., Mohan, R., Büscher, D., Burgard, W., Valada, A.: Efficientlps: efficient lidar panoptic segmentation. IEEE Trans. Robot. (2021)
19. Thrun, S., Montemerlo, M., Aron, A.: Probabilistic terrain analysis for high-speed desert driving. In: Robotics: Science and Systems, pp. 16–19 (2006)
20. Urmson, C., et al.: High speed navigation of unrehearsed terrain: Red team technology for grand challenge 2004. Robotics Institute, Carnegie Mellon University, Pittsburgh, PA, Tech. Rep. CMU-RI-04-37 1 (2004)
21. Wen, M., Cho, S., Chae, J., Sung, Y., Cho, K.: Range image-based density-based spatial clustering of application with noise clustering method of three-dimensional point clouds. Int. J. Adv. Robot. Syst. **15**(2), 1729881418762302 (2018)
22. Zermas, D., Izzat, I., Papanikolopoulos, N.: Fast segmentation of 3D point clouds: a paradigm on lidar data for autonomous vehicle applications. In: 2017 IEEE International Conference on Robotics and Automation (ICRA), pp. 5067–5073. IEEE (2017)
23. Zhao, Y., Zhang, X., Huang, X.: A technical survey and evaluation of traditional point cloud clustering methods for lidar panoptic segmentation. In: Proceedings of the IEEE/CVF International Conference on Computer Vision, pp. 2464–2473 (2021)
24. Zhou, H., et al.: Cylinder3d: An effective 3D framework for driving-scene lidar semantic segmentation. arXiv preprint arXiv:2008.01550 (2020)

Deep Learning

FDGATII: Fast Dynamic Graph Attention with Initial Residual and Identity

Gayan K. Kulatilleke(✉), Marius Portmann, Ryan Ko, and Shekhar S. Chandra

School of Information Technology and Electrical Engineering, University of
Queensland, Queensland, Australia
{ryan.ko,shekhar.chandra}@uq.edu.au

Abstract. Despite their recent popularity, deep and efficient Graph
Neural Networks remain a major challenge due to (a) over-smoothing,
(b) noisy neighbours (heterophily), and (c) the suspended animation
problem. Inspired by the attention mechanism's ability to focus on selective information, and prior work on feature preserving mechanisms, we
propose FDGATII, a dynamic deep-capable model that addresses all
these challenges *simultaneously* and efficiently. Specifically, by combining Initial Residuals and Identity with the more expressive dynamic self-
attention, FDGATII effectively handles noise in heterophilic graphs and
is capable of depths over 32 with no over-smoothing, overcoming two
main limitations of many prior GNN techniques. By using edge-lists,
FDGTII avoids computationally intensive matrix operations, is parallelizable and does not require knowing the graph structure upfront.
Experiments on 7 standard datasets show that FDGATII outperforms
the GAT and GCN based benchmarks in accuracy and performance on
fully supervised tasks. We obtain State-of-the-art (SOTA) on the highly
heterophilic Chameleon and Cornell datasets with 1 layer, and come only
0.1% short of Cora SOTA with zero graph pre processing. https://github.
com/gayanku/FDGATII

Keywords: Dynamic attention · Heterophily · Over-smoothing

1 Introduction

Recently, research on graphs has been receiving increased attention due to the
great expressive power and pervasiveness of graph structured data [29]. Many
interesting irregular domain tasks such as 3D meshes, social networks, telecom-
munication networks and biological networks involve data that are not repre-
sentable in grid-like structures [25]. As a unique non-Euclidean data structure for
machine learning, graphs can be used to represent diverse feature rich domains.

A Graph Neural Network (GNN) generalizes deep neural networks (DNNs)
from regular structures to irregular graph data. GNNs perform neighbourhood
structure aggregation and node feature transformation to map nodes to low dimensional embeddings [15,17], mostly differing in how aggregation and combination

© The Author(s), under exclusive license to Springer Nature Switzerland AG 2022
H. Aziz et al. (Eds.): AI 2022, LNAI 13728, pp. 73–86, 2022.
https://doi.org/10.1007/978-3-031-22695-3_6

is performed [4]: Graph Convolutional Network (GCN) [13] uses convolution [16]; Graph Attention Network (GAT) [25] uses attention; GraphSage [8] uses max pooling. Downstream tasks such as node classification, clustering, and link prediction [8,22] use these aggregated low dimensional vectors [28].

Most graphs require the interaction between nodes that are not directly connected, i.e., higher-order information which is achieved by stacking GNN layers [2]. However, stacking layers degrades the performance [5,20] due to over-smoothing: node representations become indistinguishable with increasing number of layers [6,13,26]. Further, GNNs in general are not able to handle long-range information due to over-squashing: information from the exponentially growing receptive field being compressed into fixed-length node vectors [2] due to its unfocused aggregation mechanism. Finally, deeper models stop responding to training due to the suspended animation problem [26], i.e. depth is a problem [6].

To avoid these problems, several works combine deep propagation with shallow neural networks; SGC [26] used the K-th power of the adjacency matrix to capture higher-order information; H2GCN [29] aggregates higher-order information at each round. However, this form of linear combination of neighbour features at each layer looses the powerful expression ability of deep nonlinear architectures, essentially making them shallow models [5].

In another attempt to address the problem and incorporate deeper layers, JKNet [27] used dense skip connections, DropEdge [23] randomly removed graph edges and GCNII [5] added a portion of Initial residual and Identity. GCNII showed remarkable results for up to 64 layers and is the SOTA (Table 2) in Cora, a homophilic benchmark dataset. However, all these are spectral approaches based on the Laplacian eigenbasis and requires the whole graph structure [25]. The normalization used is computationally expensive and not scalable.

Furthermore, due to naive uniform aggregation of the neighbourhood, most of these models, including GCNII, are more suitable for homophilic datasets, where nodes linked to each other are more likely to belong in the same class, i.e., neighbourhoods with low noise. In practice, real-world graphs are also often noisy with connections between unrelated nodes [12], resulting in poor performance in current GNNs. As many popular GNN models implicitly assume homophily, results may be biased, unfair or erroneous [19]. This can result in a 'filter bubble' phenomenon in a recommendation system (reinforcing existing beliefs/views, and downplaying the opposite ones), or making minority groups less visible in social networks [29]. As a result, despite GCNIIs SOTA in homophilic datasets (Cora), its accuracy in heterophilic datasets (Texas, Wisconsin) is relatively poor [29].

On the other hand, [24] showed that self-attention is sufficient for achieving SOTA performance. GAT [25] generalizes attention for graphs using attention-based neighbourhood aggregation. Importantly, GAT improves on simple averaging [13] and max pooling [8] by allowing every node to compute a weighted average of its neighbours [4], which is a form of selective aggregation. The generalization ability of the attention mechanism helps GNNs generalize to larger and more noisy graphs [14]. By determining individual attention on each neighbour, GAT ignores irrelevant neighbours and focuses on those that are relevant [2].

Surprisingly, yet, GATs heterophilic performance is poor (Table 2).

A refinement, GATv2 [4], uses a more expressive dynamic attention, where the ranking of attended nodes is better conditioned on the query node by replacing the supposedly monotonic GAT attention function with a universal approximator attention function that is strictly more expressive. However, GAT or GATv2 *alone, in its current form* cannot handle heterophilic data due to the still present essentially local aggregation operation [17].

In Table 2, under heterophily, only H2GCN outperforms a Multilayer Perceptron (MLP) of 1 layer which uses only node features and no structural information. Furthermore, most GNN models use simple graph convolution based aggregation schemes [8,13], leading to filter incompleteness. While this can be solved by using a more complex graph kernel [1], currently, even attention-based models perform poorly given heterophilic data, despite the ability to focus on the most "relevant" content.

Thus, it remains an open problem to design efficient GNN models that effectively handle (a) over-smoothing, (b) suspended animation and (c) heterophily/noise simultaneously. As observed by [5], it is even unclear whether the network depth is a resource or a burden when designing new GNNs. Motivated by these limitations, we propose a generalizable, efficient, and parallelizable attention based deep-capable model that addresses aforementioned challenges simultaneously. Our main contributions are:

- We introduce a novel deep-capable GNN model, FDGATII, successfully combining strengths of GCN and GAT worlds by using dynamic attention supplemented with Initial residual and Identity, capable of handling the major graph challenges: over-smoothing, noisy neighbours (heterophily) and suspended animation *simultaneously*. To the best of our knowledge, this is the first time a graph attentional model has demonstrated depths of up to 32, a limitation of many prior GNN techniques, attention based or otherwise, and show that dynamic attention is better suited for heterophilic datasets, if used with modifications.
- FDGATII is computationally efficient. It does not require an adjacency matrix as input nor its subsequent, expensive matrix operations or normalizations. Further, its attention layers can be parallelized across edges while feature computation can be parallelized across all nodes.
- FDGATII has the same complexity as SOTA GCN models, but uses significantly fewer layers to achieve comparable or better results, yielding a superior efficiency-to-accuracy ratio across homophilic and heterophilic datasets.

Extensive experiments on 7 benchmarks show that FDGATII outperforms GAT and GCN based benchmarks in accuracy as well as on accuracy vs efficiency, on fully supervised tasks. FDGATII achieves SOTA accuracy results on Chameleon and Cornell datasets, beating H2GCN, a model specifically designed for heterophily. There is zero graph pre processing. FDGATII consumes over a magnitude less computational resources and is only –0.1% below SOTA for Cora, placing a close second. By not assuming homophily, FDGATII minimises its potential negative effects: bias, unfairness and potential for filter bubbles. FDGATII is also capable of inductive learning. Table 1 has a full feature comparison.

Table 1. Feature comparison: GAT, GCN, GCNII and FDGATII. *Cham & Cornell

Feature	GAT	GCN	GCNII	FDGATII
No graph pre processing (ex: normalisation)	Yes	No	No	Yes
Does not require knowing graph structure upfront	Yes	No	No	Yes
Processing can be parallelized	Yes	No	No	Yes
Free from oversmoothing	No	No	Yes	Yes
Free from suspended animation	No	No	Yes	Yes
Heterophilic performance	Poor	Poor	Good	SOTA*
Capable of deep architectures (layers > 8)	No	No	Yes	Yes
Dynamic attention	No	No	No	Yes
Inductive learning	Yes	No	Yes	Yes
# layers for best Cora accuracy	2	2	64	2

2   Related Work

2.1   Notation

$G = (V, E)$ is an undirected graph with n nodes $v_j \in V$ and m edges $(v_i, v_j) \in E$. $\bar{G} = (V, \bar{E})$ is its self-looped graph. A is the adjacency matrix, D the degree matrix of G. Adjacency matrix and degree matrix of $\bar{G}$ is $\bar{A} = A + I$ and $\bar{D} = D + I$. The symmetric positive semi definite *normalized graph Laplacian matrix* is given by $L = I_n - D^{-1/2}AD^{-1/2}$ with eigen-decomposition $U\Lambda U^T$. Λ is its diagonal eigenvalue matrix, $U \in R^{n \times n}$ is the unitary eigenvector matrix.

2.2   Convolution and GCN

Given signal x and filter $g_\gamma(\Lambda) = diag(\gamma)$ the graph convolution operation is $g_\gamma(L) * x = Ug_\gamma(\Lambda)U^T x$ where $\gamma \in R^n$ is the vector of spectral filter coefficients. $g_\gamma(\Lambda)$ can be approximated by a truncated expansion of a Kth order Chebyshev polynomial [9], where $\theta \in \mathbf{R}^{K+1}$ corresponds to a vector of polynomial coefficients:

$$\mathbf{U}g_\theta(\Lambda)\mathbf{U}^T\mathbf{x} \approx \mathbf{U}\left(\sum_{l=0}^{K}\theta_l\mathbf{\Lambda}^l\right)\mathbf{U}^T\mathbf{x} = \left(\sum_{l=0}^{K}\theta_l\mathbf{L}^l\right)\mathbf{x} \tag{1}$$

GCN [13] simplifies graph convolution by *fixing* $K = 1, \theta_0 = 2\theta$ and $\theta_1 = -\theta$ to get $g_\theta * x = \theta(I + D^{-1/2}AD^{-1/2})x$ and uses a normalized adjacency matrix, $\bar{P} = \bar{D}^{-1/2}\bar{A}\bar{D}^{-1/2} = (D+I_n)^{-1/2}(A+I_n)(D+I_n)^{-1/2}$. Each GCN layer (Eq. 2) contains a nonlinear activation function σ, typically ReLU.

$$\mathbf{H}^{l+1} = \sigma\left(\bar{\mathbf{P}}\mathbf{H}^l\mathbf{W}^l\right) \tag{2}$$

However, node embeddings are aggregated recursively layer by layer. Embeddings in the final layer requires all previous embeddings, resulting in high memory cost. GCN gradient update in the full-batch training scheme needs storing all intermediate embeddings, limiting scalability. As the learned filters depend on the Laplacian eigenbasis, which depends on the entire graph structure, a model trained on a graph cannot be directly applied to a different graph structure [25].

2.3   GCNII

GCNII [5] extends the fixed coefficient GCN to a deep model by expressing the K order polynomial filter as *arbitrary* coefficients using Initial residual and Identity (II). Essentially, GCNII 1) combines the preprocessed (normalized) representation $\bar{\mathbf{P}}\mathbf{H}^l$ with an initial residual connection from the first layer $\mathbf{H}^0$; and 2) adds an identity $\mathbf{I}_n$ to the l-th weight matrix $\mathbf{W}^l$. By using a connection to the initial residual $\mathbf{H}^0$, GCNII ensures that the final representation of each node retains at least a α_l fraction from the input layer.

However, as GCNII combines neighbour embeddings by uniformly averaging, its heterophilic performance is relatively poor. GCNs preserve structure over features, regardless of the graph's heterophilic nature, resulting in original node features being destroyed [11]. Further, [20] showed that GCNs tend to fail when graphs are dense and do not always improve with more layers. Alternatively, a selective aggregation of the neighbourhood allows focusing on relevant nodes [29].

2.4   Attention Mechanism and GAT

The DP (dot-product) attention mechanism (Equation 3) [18,24] has been widely used in GNNs [12,28]. Different from DP, GAT [25] uses concatenation followed by a 1-layer feed-forward network parameterized by $\mathbf{a}$ (Eq. 4).

$$e\left(\mathbf{h}_i, \mathbf{h}_j\right) = \text{LeakyReLU}((\mathbf{W}\mathbf{h}_i)^T \cdot \mathbf{W}\mathbf{h}_j) \tag{3}$$

$$e\left(\mathbf{h}_i, \mathbf{h}_j\right) = \text{LeakyReLU}\left(\mathbf{a}^T \cdot [\mathbf{W}\mathbf{h}_i \parallel \mathbf{W}\mathbf{h}_j]\right) \tag{4}$$

In contrast to GCN, which weighs all neighbours $j \in \mathcal{N}_i$ with equal importance, GAT computes a learned weighted average of the representations of $\mathcal{N}_i$ using attention. Compared to GCN, assigning different weights for neighbours can mitigate noise and achieve better results [28] while being more robust in the presence of noisy "irrelevant" neighbours [2].

3   Proposed Architecture

Our proposed design (Fig. 1) is built upon a local embedding step that extracts local node embeddings from feature vectors using GATv2. To extend GATv2 to handle heterophilic and noisy data, we borrow two techniques from GCNII [5] and H2GCN [29] with modifications, namely residual connection and identity.

However, the theoretical foundation of our model, which is grounded in the spatial domain, is completely different from GCNII which is spectral. We do not require edge values; only the presence or absence of an edge: i.e. a simple list of edges. Using only the edge-list as [25], with self-loops as [10,13], we avoid computationally intensive matrix operations such as inversions or eigen-decompositions and the need to know the graph structure upfront. Experiments show our design is efficient, robust and generalizes well to homophilic and heterophilic datasets alike.

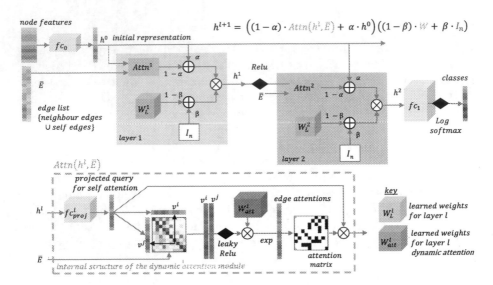

Fig. 1. FDGATII uses dynamic attention to combine relevant neighbours via edge-lists, an $\alpha\%$ of initial representation h^0 projected via fc_0 and a $\beta\%$ of Identity I_n at each layer. Attention module concatenates source (row) and destination (column) features of each edge, projects via W_H^n, applies a non-linearity (leaky-relu) and an $exp()$ to obtain the edgewise attentions before reshaping to a matrix suitable for softmax with the query. After multiple layers, an fc_1 projection and log softmax provides the node classification.

Typically, GNN models follow an iterative learning approach:

$$\mathbf{h}_i^{l+1} = \text{COMBINE}\Big(\mathbf{h}_i^l, \text{AGG}(\{\mathbf{h}_j^l : j \in N_i\})\Big),$$

$$\mathbf{h}_i^0 = \mathbf{X}_i, \text{and } y_i = \arg\max\{\text{softmax}(\mathbf{h}_i^K)\mathbf{W}\}$$

where, AGG is a permutation invariant aggregation operator and COMBINE is a learnable function. By adding self-nodes, we amalgamate COMBINE and AGG to simplify the process and apply a more expressive attention operator ATTN to both tasks simultaneously, defined by:

$$\mathbf{h}_i^{l+1} = \text{ATTN}(\{\mathbf{h}_j^l : j \in N_i \cup i\})$$

3.1 Initial Residual and Identity (II)

We incorporate initial representation $\mathbf{H}^0$ and identity $\mathbf{I}_n$, in α_l and β_l fractions, with edge-list $\bar{\mathbf{E}}$ to formally define the $(l+1)$-th layer of FDGATII as:

$$\mathbf{H}^{l+1} = \sigma\left[\Big((1-\alpha_l)\,\text{ATTN}(\bar{\mathbf{E}}, \mathbf{H}^l) + \alpha_l \mathbf{H}^0\Big) \cdot \Big((1-\beta_l)\,\mathbf{I}_n + \beta_l \mathbf{W}^l\Big)\right] \qquad (5)$$

According to [10], identity mapping of the form $\mathbf{H}^{l+1} = \mathbf{H}^l(\mathbf{W}^l + \mathbf{I}_n)$, as in Eq. 5, satisfies the following properties: 1) the optimal weight matrices $\mathbf{W}^l$ have small norms; 2) the only critical point is the global minimum. The first property allows us to put strong regularization on $\mathbf{W}^l$ to avoid over-fitting, while the latter is desirable in semi-supervised tasks where training data is limited.

Next, it is theoretically proven [20] that a K-layer GNN's convergence rate depends on s^K, where s is the maximum singular value of the weight matrices $\mathbf{W}^l, l = 0, \ldots, K - 1$. By replacing $\mathbf{W}^l$ with $(1 - \beta_l)\mathbf{I}_n + \beta_l \mathbf{W}^l$ and regularizing $\mathbf{W}^l$, resulting singular values of $(1 - \beta_l)\mathbf{I}_n + \beta_l \mathbf{W}^l$ stay closer to 1, which implies that s^K is large, and the information loss is relieved.

3.2   Selection of Proper Attention

It has been shown that GAT is better at learning label-agreement between a target node and its neighbors than DP attention [12]. Variance of GAT depends only on the norm of features, while the DP variance depends on the variance of the input's dot-product and the expectation of the square of the input's dot-product. As a result, with more layers, more features of i and j correlating resulting in a larger dot-product and the subsequent softmax normalization which increases the larger values further, DP is only able to attend to a small set of neighbours.

3.3   Dynamic Attention (GATv2)

According to [4], the main problem in the standard GAT scoring function (Eq. 4) is that the learned layers $\mathbf{W}$ and $\mathbf{a}$ are applied consecutively, and thus can be collapsed into a single linear layer. GATv2 replaces the linear approximator with a universal approximator (Eq. 6) and has been shown to perform better on noisy data [4]. Further, theoretically, DP is strictly weaker than GATv2. We use this form of dynamic attention for our aggregation function.

Specifically, a scoring function $e : R^d \times R^d \to R$ computes a score for every edge (j, i), which indicates the importance of the features of the neighbour j to the node i:

$$e(\mathbf{h}_i, \mathbf{h}_j) = \mathbf{a}^T \cdot \text{LeakyReLU}(\mathbf{W}[\mathbf{h}_i \| \mathbf{h}_j]), \qquad (6)$$

where attention scores $\mathbf{a} \in R^{2d'}$ and weights $\mathbf{W} \in R^{d' \times d}$ are learned. $\|$ denotes vector concatenation. We capture the graph structure using edges, computing $e_{i,j}$ for *all* $j \in N_i$ neighbourhood of node i. Attention scores are normalized across all connected sparse neighbours $j \in \mathcal{N}_i$ using softmax.

$$\alpha_{ij} = softmax_j(e(\mathbf{h}_i, \mathbf{h}_j)) = \frac{exp(e(\mathbf{h}_i, \mathbf{h}_j))}{\sum_{j' \in \mathcal{N}_i} exp\left(e\left(\mathbf{h}_i, \mathbf{h}_{j'}\right)\right)} \qquad (7)$$

Finally, we compute the weighted average of the transformed features of the neighbour nodes (followed by a nonlinearity σ) as the new representation of i, using the normalized attention coefficients:

$$\text{ATTN}_i' = \sigma\left(\sum_{j \in \mathcal{N}_i} \alpha_{ij} \mathbf{W} \mathbf{h}_j\right) \qquad (8)$$

In addition to Eq. 5, following [5], we also propose FDGATII* with dual weight matrices for smoother representation, defined as:

$$\mathbf{H}^{l+1} = \sigma \left[(1 - \alpha_l)\text{ATTN}(\bar{\mathbf{E}}, \mathbf{H}^l) \left((1 - \beta_l)\mathbf{I}_n + \beta_l \mathbf{W}_1^l \right) \right.$$
$$\left. + \alpha_l \mathbf{H}^0 \left((1 - \beta_l)\mathbf{I}_n + \beta_l \mathbf{W}_2^l \right) \right] \tag{9}$$

GCNII [5] uses β_l is to ensure the decay of the weight matrix adaptively increases with more layers. While FDGATII typically achieves best accuracy early with a few layers, we still adopt the same mechanism, $\beta_l = log\left(\frac{\lambda}{l} + 1\right) \approx \frac{\lambda}{l}$, where λ is a hyperparameter, for robustness at high depth. Following [27], we add skip connections in the form of initial representations H^0 as in [5].

FDGATII differs from existing models with respect to its use of a modified attention mechanism. Notably, we demonstrate competitive performance of GATv2+II with only a few layers in non-homophilous networks. Using edge-lists avoids computationally intensive matrix operations. Table 1 summarizes how FDGATII accumulates all benefits from GCN and GAT worlds with none of the drawbacks.

3.4   Datasets and Experiments

Homophily is the fraction of edges which connect two nodes of the same label [17]. A higher value (1) indicates strong homophily; a lower value (0) indicates strong heterophily.

We evaluate FDGATII against SOTA GNNs on benchmark graph datasets for fully supervised classification. Following [5,21], we use 7 datasets (Table 5). Cora, Citeseer and Pubmed are homophilic citation networks where nodes correspond to documents, and edges correspond to citations. The remaining four are heterophilic datasets of web networks, where nodes and edges represent web pages and hyperlinks, respectively. Node feature vectors are bag-of-word representations of the document. Following [5,21] we use the same data splits, 60:20:20 nodes for training:validation:testing, learning rate = 0.01, hidden units = 64 and measure the average performance on the 10 splits for each dataset.

We choose GCNII [5] as our performance and accuracy benchmark as it is (a) more current; (b) most similar to our work in the use of initial representation and identity; (c) actively attempts to solve over smoothing (d) is the SOTA in Cora (a prominent dataset for GNN model comparison) and *most* importantly (d) it is a deep-capable model. We also compare with H2GCN [29] which is the SOTA for Cornel, Texas and Wisconsin; highly heterophelic datasets, but note that H2GCN is a shallow model.

For training and inference time measurements we perform GPU warm-up and synchronization prior to measurements. We take the average time for 1000 inferences to lower any possibility of errors and to be more reflective of real-world use of models. We ignore pre processing times, but point out, unlike the benchmarks, FDGATII has *no* expensive full graph eigen operations or normalizations.

Table 2. Mean classification accuracy of full-supervised node classification. (a) reported by [5], (b) reported by [29], (c) best results running GCNII (official author implementation) and H2GCN (public pytorch repo: github.com/GitEventhandler/ H2GCN-PyTorch) on data splits of [5], (d): our FDGATII, with same splits. Best is bold and second underlined. # of layers in parenthesis.

Dataset	Cora	Cite.	Pumb.	Cham.	Corn.	Texa.	Wisc.
Hormophily %	0.81	0.74	0.8	0.23	0.30	0.11	0.21
MLP[b]	74.75(1)	72.41(1)	86.65(1)	46.36(1)	81.08(1)	81.89(1)	85.29(1)
GCN[n]	85.77	73.68	88.13	28.18	52.70	52.16	45.88
GAT[a]	86.37	74.32	87.62	42.93	54.32	58.38	49.41
Geom-GCN-I[a]	85.19	**77.99**	90.05	60.31	56.76	57.58	58.24
GraphSAGE[b]	86.90	76.04	88.45	58.73	81.18	82.43	75.95
MixHop[b]	87.61	76.26	85.31	60.50	75.88	77.84	75.88
H2GCN-1[b]	86.92	77.07	89.40	57.11	<u>82.16</u>	**84.86**	**86.67**
APPNP[a]	87.87	76.53	89.40	54.3	73.51	65.41	69.02
JKNet[a]	85.25(16)	75.85(8)	88.94(64)	60.07(32)	57.30(4)	56.49(32)	48.82(8)
JKNet(Drop)[a]	87.46(16)	75.96(8)	89.45(64)	62.08(32)	61.08(4)	57.30(32)	50.59(8)
Incep(Drop)[a]	86.86(8)	76.83(8)	89.18(4)	61.71(8)	61.62(16)	57.84(8)	50.20(8)
GCNII[a]	**88.49(64)**	77.08(64)	89.57(64)	60.61(8)	74.86(16)	69.46(32)	74.12(16)
GCNII*[a]	88.01(64)	<u>77.13(64)</u>	90.30(64)	<u>62.48(8)</u>	76.49(16)	77.84(32)	81.57(16)
H2GCN-1[c]	77.3038	74.5220	87.5887	49.0351	73.7838	78.9189	79.0196
GCNII[c]	88.2696	76.9325	90.3499	63.7500	77.2973	78.3784	79.8039
FDGATII[d]	<u>88.3903(2)</u>	76.3082(1)	**90.5502(2)**	**66.1184(1)**	**84.3243(1)**	<u>83.7838(1)</u>	<u>86.0784(1)</u>

Table 3. Inductive learning - F1 (micro) on PPI. (1): Results from [5]. (2): Our results with identical settings and Eq. 5. Note, we do not require any data pre processing

Method	PPI(reported)[1]	Method	PPI(our tests)[2]
GraphSAGE	61.2	FDGATII (2 layers)	98.51
GAT	97.3	FDGATII (3 layers)	98.91
JKNet	97.6	FDGATII (4 layers)	99.18
GeniePath	98.5	FDGATII (5 layers)	99.17
Cluster-GCN	99.36	FDGATII (6 layers)	99.24
GCNII (9 layers)	99.53	GCNII (9 layers)	99.52
GCNII* (9 layers)	**99.56**	GCNII* (9 layers)	99.53

4 Results and Discussion

4.1 Fully Supervised Node Classification

Table 2 reports the mean classification accuracy. We reuse the metrics already reported by [5] and [29]. We observe that FDGATII demonstrates SOTA results on heterophilic datasets while still being competitive on the homophilic datasets. Further, FDGATII exhibits significant accuracy increases over its attention based predecessor, GAT. This result suggests that dynamic attention with initial residuals and identity improves the predictive power whilst keeping the layer count (and hence the model parameters and computational requirements) low.

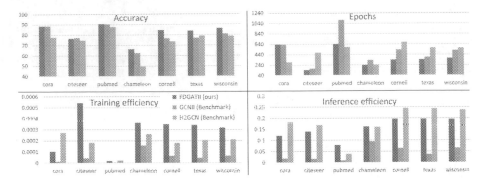

Fig. 2. Accuracy, epochs, training and inference time comparison. For variants, we use the lowest average time taken to run all 10 standard splits. Efficiency = 1/time. Original GCNII is in pytorch. Original H2GCN is in TF. A public pytorch H2GCN is used to eliminate any framework effects. Tested on Google colab with GPU.

Table 4. Ablation study w/0 II and w/0 dynamic attention. * Eq. 5, ** Eq. 9. Hyperparameter settings from [5]. $L1$ and $L2$ are 1 and 2 layers, respectively.

Metric	Cora	Cite.	Pumb.	Cham.	Corn.	Texa.	Wisc.
Without II, L1	86.90	**75.65**	87.01	**65.18**	65.95	62.16	54.51
Without II, L2	86.74	74.45	86.19	49.78	58.92	57.30	51.76
With II*, L1	87.06	75.07	89.96	61.34	76.76	70.00	81.96
With II*, L2	**87.79**	74.88	**90.35**	47.82	79.19	79.73	83.53
With II*, L2 w/o dynamic attention	87.53	75.01	90.34	46.00	79.73	81.90	82.54
With II**, L1	84.91	75.28	89.48	49.12	80.27	78.65	84.12
With II**, L2	86.52	75.14	90.12	44.34	**80.81**	**82.16**	84.90
With II**, L2 w/o dynamic attention	85.98	74.48	90.03	43.99	80.80	80.54	**85.49**

4.2 Inductive Learning

We use the PPI dataset and follow [8] using 20:2:2 graphs for train:validation:test. For settings, we follow [5]: 2048 hidden units, learning rate 0.001. Similar to [5, 25], we add a skip connection from layer l to $l+1$. Table 3 reports the F1 (micro) scores. Results show that FDGATII is capable of competitive inductive learning.

4.3 Ablation Study

In this section, along with Table 4, we consider the effect of various design strategies. Our 1 or 2-layer models, without Initial residual and Identity (II), is theoretically equivalent to GAT(static attention)/GATv2(dynamic attention). The ablation study indicates that the addition of II *together with dynamic attention* results in improvements on the heterophilic dataset performance. This result suggests that both II and dynamic attention techniques are needed to solve the problem of over-smoothing and data heterophily. Figure 4 also confirms GAT/GATv2 cannot handle heterophily or depth unaided, while FDGATII shows significant and consistently better results.

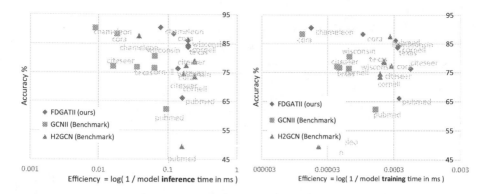

Fig. 3. Efficiency vs accuracy, on GPU with warm-up. **left:** average inference time for 1000 iterations. **right:** average training efficiency for 10 iterations. Efficiency = log(1/time). Top-right is better.

4.4 Performance and Efficiency

Figure 3 summarizes the high accuracy-to-computational-time-efficiency ratio of FDGATII clearly indicating its superior performance mix. The proposed architecture performs consistently better across noisy and diverse datasets with comparable or better accuracy (Table 2) while exhibiting superiority in training and inference times, specifically 12x faster training speeds and up to 9x faster inference speeds over our chosen deep-capable SOTA benchmark, GCNII [5]. FDGATII is 3x faster than H2GCN [29] on Citeseer. Our dynamic attention achieves higher expressive power with fewer layers paying selective attention to nodes, while the II supplements self-node features in highly heterophilic datasets.

By using edge-lists, FDGATII avoids computationally intensive eigen decompositions and matrix operations as well as the need to know the graph structure upfront. Also, output feature computation can be parallelized across nodes while the attention computation can be parallelized across all edges. While FDGATII has the same time complexity of GCNII, by using significantly fewer layers (Table 2 and Table 5), it achieves comparable or better results with superior efficiency-to-accuracy ratios. Note, in Fig. 2, the graph pre processing (inversion, normalization) times for benchmarks were not taken into account due to focus on model training and inference. FDGATII has zero graph pre processing.

4.5 Suspended Animation and Over Smoothing

Responding to training indicates absence of suspended animation [26], while effectively handling higher receptive fields indicates robustness to over-smoothing [6]. Figure 4 shows FDGATII's performance for 3 selected datasets under increasing layer depth. There is no evidence of performance degradation from suspended animation or over smoothing even at depth of 32. Accuracy is achieved early and sustained over higher depths. In Cora, the drop is 0.1 for 32 layers. H2GCN reported OOM for depths over 8.

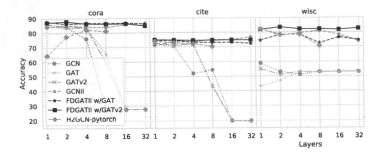

Fig. 4. Accuracy vs layer depth (on Goole Colab with GPU). FDGATII is consistent. H2GCN OOM after 8 layers. Depth and heterophily degrades GAT/GATv2 accuracy.

4.6 Broader Issues Related to Heterophily

Many popular GNN models implicitly assume homophily, producing results that may be biased, unfair or erroneous [29]. This can result in the so-called 'filter bubble' phenomenon in a recommendation system (reinforcing existing beliefs/views, and downplaying the opposite ones), or make minority groups less visible in social networks, creating ethical implications [7]. FDGATII's novel self-attention mechanism, where dynamic attention supplemented with II for feature preservation, reduces the filter bubble phenomenon and its potential negative consequences, ensuring fairness and less bias.

This offers new possibilities for future research into data where 'opposites attract', in which the majority of linked nodes are different, such as social and dating networks (the majority of persons of one gender connect with the opposite gender), chemistry and biology (amino acids bond with dissimilar types in protein structures), e-commerce (sellers with promoters and influencers), and dark web and other cybercrime related activities [29]. In a typical dark web social network, fraudsters are more likely to connect to intermediaries and prospective victims than to other fraudsters. Illicit actors will form ties with other actors who play different roles [3], resulting in heterophilic characteristics.

Table 5. Final model hyperparameters.

Dataset	H%	Clases	Nodes	Edges	Features	α	Dropout	λ	Layers	Varient	WD
Cora	0.81	7	2,708	5,429	1,433	0.3	0.6	0.2	2	Eq 5	1e–4
Citeseer	0.74	6	3,327	4,732	3,703	0.5	0.6	1	1	Eq 9	1e–6
Pubmed	0.80	3	19,717	44,338	500	0.2	0.3	1	2	Eq 5	5e–5
Chameleon	0.23	4	2,277	36,101	2,325	0.1	0.3	0.2	1	Eq 5	5e–4
Cornell	0.30	5	183	295	1,703	0.1	0.5	1	1	Eq 9	5e–4
Texas	0.11	5	183	309	1,703	0.3	0.6	1.5	1	Eq 9	5e–4
Wisconsin	0.21	5	251	499	1,703	0.4	0.3	0.2	1	Eq 9	5e–4
PPI		121	56,944	818,716	50	0.5	0.2	1.0	7	Eq 5	0.0

5 Conclusion

We propose FDGATII, a novel efficient dynamic attention-based model that combines attentional aggregation with dual feature preserving mechanisms based on Initial residual and Identity. FDGATII successfully combines strengths of both GCN and GAT worlds with none of the drawbacks, is inductive, able to handle noise in graphs and achieves depths of upto 32; a first for any attentional model and a limitation of many prior GNN techniques. Extensive experiments on a wide spectrum of benchmark datasets show that FDGATII achieves SOTA or second-best accuracy on benchmark fully supervised tasks. FDGATII has exceptional accuracy and efficiency whilst *simultaneously* addressing over-smoothing, suspended animation and heterophily prevalent in real world datasets.

Acknowledgements. Dedicated to Sugandi.

References

1. Abu-El-Haija, S., et al.: Mixhop: higher-order graph convolutional architectures via sparsified neighborhood mixing. In: International Conference on Machine Learning, pp. 21–29. PMLR (2019)
2. Alon, U., Yahav, E.: On the bottleneck of graph neural networks and its practical implications. In: International Conference on Learning Representations (2020)
3. Bright, D., Koskinen, J., Malm, A.: Illicit network dynamics: the formation and evolution of a drug trafficking network. J. Quant. Criminol. **35**(2), 237–258 (2019)
4. Brody, S., Alon, U., Yahav, E.: How attentive are graph attention networks? In: International Conference on Learning Representations (2021)
5. Chen, M., Wei, Z., Huang, Z., Ding, B., Li, Y.: Simple and deep graph convolutional networks. In: International Conference on Machine Learning, pp. 1725–1735. PMLR (2020)
6. Chien, E., Peng, J., Li, P., Milenkovic, O.: Adaptive universal generalized pagerank graph neural network. In: International Conference on Learning Representations (2020)
7. Chitra, U., Musco, C.: Analyzing the impact of filter bubbles on social network polarization. In: Proceedings of the 13th International Conference on Web Search and Data Mining, pp. 115–123 (2020)
8. Hamilton, W.L., Ying, R., Leskovec, J.: Inductive representation learning on large graphs. In: Proceedings of the 31st International Conference on Neural Information Processing Systems, pp. 1025–1035 (2017)
9. Hammond, D.K., Vandergheynst, P., Gribonval, R.: Wavelets on graphs via spectral graph theory. Appl. Comput. Harmon. Anal. **30**(2), 129–150 (2011)
10. Hardt, M., Ma, T.: Identity matters in deep learning. In: International Conference on Learning Representations (2017)
11. Jin, W., Derr, T., Wang, Y., Ma, Y., Liu, Z., Tang, J.: Node similarity preserving graph convolutional networks. In: Proceedings of the 14th ACM International Conference on Web Search and Data Mining, pp. 148–156 (2021)
12. Kim, D., Oh, A.: How to find your friendly neighborhood: graph attention design with self-supervision. In: International Conference on Learning Representations (2020)

13. Kipf, T.N., Welling, M.: Semi-supervised classification with graph convolutional networks. In: J. International Conference on Learning Representations (ICLR 2017) (2016)
14. Knyazev, B., Taylor, G.W., Amer, M.: Understanding attention and generalization in graph neural networks. Adv. Neural Inf. Process. Syst. **32**, 4202–4212 (2019)
15. Kulatilleke, G.K., Portmann, M., Chandra, S.S.: SCGC: Self-supervised contrastive graph clustering. arXiv preprint arXiv:2204.12656 (2022)
16. LeCun, Y., Bengio, Y., et al.: Convolutional networks for images, speech, and time series. Handb. Brain Theory Neural Netw. **3361**(10), 1995 (1995)
17. Liu, M., Wang, Z., Ji, S.: Non-local graph neural networks. IEEE Trans. Pattern Anal. Mach. Intell. (2021)
18. Luong, T., Pham, H., Manning, C.D.: Effective approaches to attention-based neural machine translation. In: EMNLP (2015)
19. Maurya, S.K., Liu, X., Murata, T.: Simplifying approach to node classification in graph neural networks. J. Comput. Sci. 101695 (2022)
20. Oono, K., Suzuki, T.: Graph neural networks exponentially lose expressive power for node classification. In: International Conference on Learning Representations (2019)
21. Pei, H., Wei, B., Chang, K.C.C., Lei, Y., Yang, B.: Geom-gcn: geometric graph convolutional networks. In: International Conference on Learning Representations, pp. 6519–6528 (2019)
22. Perozzi, B., Al-Rfou, R., Skiena, S.: Deepwalk: online learning of social representations. In: Proceedings of the 20th ACM SIGKDD International Conference on Knowledge Discovery and Data Mining, pp. 701–710 (2014)
23. Rong, Y., Huang, W., Xu, T., Huang, J.: Dropedge: Towards deep graph convolutional networks on node classification. In: International Conference on Learning Representations (2019)
24. Vaswani, A., et al.: Attention is all you need. Adv. Neural Inf. Process. Syst. 5998–6008 (2017)
25. Veličković, P., Cucurull, G., Casanova, A., Romero, A., Liò, P., Bengio, Y.: Graph attention networks. In: International Conference on Learning Representations (2018)
26. Wu, F., Souza, A., Zhang, T., Fifty, C., Yu, T., Weinberger, K.: Simplifying graph convolutional networks. In: International Conference on Machine Learning, pp. 6861–6871. PMLR (2019)
27. Xu, K., Li, C., Tian, Y., Sonobe, T., Kawarabayashi, K.i., Jegelka, S.: Representation learning on graphs with jumping knowledge networks. In: International Conference on Machine Learning, pp. 5453–5462. PMLR (2018)
28. Zhou, J., et al.: Graph neural networks: a review of methods and applications. AI Open **1**, 57–81 (2020)
29. Zhu, J., Yan, Y., Zhao, L., Heimann, M., Akoglu, L., Koutra, D.: Beyond homophily in graph neural networks: Current limitations and effective designs. Adv. Neural Inf. Process. Syst. (2020)

SG-Shuffle: Multi-aspect Shuffle Transformer for Scene Graph Generation

Anh Duc Bui[✉], Soyeon Caren Han, and Josiah Poon

The University of Sydney, Camperdown, Sydney, Australia
abui2208@uni.sydney.edu.au, {caren.han,josiah.poon}@sydney.edu.au

Abstract. Scene Graph Generation (SGG) serves a comprehensive representation of the images for human understanding as well as visual understanding tasks. Due to the long tail bias problem of the object and predicate labels in the available annotated data, the scene graph generated from current methodologies can be biased toward common, non-informative relationship labels. Relationship can sometimes be non-mutually exclusive, which can be described from multiple perspectives like geometrical relationships or semantic relationships, making it even more challenging to predict the most suitable relationship label. In this work, we proposed the SG-Shuffle pipeline for scene graph generation with 3 components: 1) Parallel Transformer Encoder, which learns to predict object relationships in a more exclusive manner by grouping relationship labels into groups of similar purpose; 2) Shuffle Transformer, which learns to select the final relationship labels from the category-specific feature generated in the previous step; and 3) Weighted CE loss, used to alleviate the training bias caused by the imbalanced dataset.

Keywords: Scene graph generation · Long-tailed bias · Unbiased scene graph generation

1 Introduction

Scene Graph Generation (SGG) is a fundamental visual understanding task that aims to encode image structure using the objects in the image as well as the relationships between these objects into a more compact representation with graphs [6]. Such representation allows for a more comprehensive understanding of the visual scene and serves as an intermediate data structure for downstream machine learning tasks between images and text, such as VQA [11] or Text-Image Matching [10]. Significant progress has been made recently in SGG thanks to the advancement of object detection [15]. However, due to the challenges of variation in object-predicate type as well as the extremely long tail bias of objects and predicates, efforts for SGG must be made so that scene graphs can be more effective for other visual understanding tasks. The traditional pipeline of SGG can often be viewed as a design pattern that comprises 2 main parts, with the predicate prediction built on top of the object detector, which generates object feature representation through convolution neural network structure. Most methods

© The Author(s), under exclusive license to Springer Nature Switzerland AG 2022
H. Aziz et al. (Eds.): AI 2022, LNAI 13728, pp. 87–101, 2022.
https://doi.org/10.1007/978-3-031-22695-3_7

focus on leveraging contextual object features in images via a variety of message propagation mechanisms such as LSTM [4,9,18,25] and GNN [7,13,23]. Such methods include the biassed prediction of predicate labels towards the head categories with much lower performance in tail categories. This is a major problem for the intended purpose of scene graphs; head categories frequently have generic meanings but tail categories provide important information that can be used in downstream tasks. Recent research has been conducted toward solving this long tail bias by a number of debiasing methods: data augmentation [5,8], model design [22,24], and bias disentangling [2,17]. These methods focus on making use of predicates' frequency and the hierarchy structure of predicates' correlation with object labels to make their models focus on infrequent predicates. There has been a lack of research into non-correlated predicates, which are used for different purposes but might have a non-mutually exclusive distribution of objects and subjects. We argue that this leads to the problem where the model needs to give attention to the classification between predicate labels of different purposes and semantic features like "above," which is used to describe positional relationships, and "holding," which is used to describe an action. This leads to less focus on differentiating between predicate labels of similar purpose like "above" and "under", which is already challenging due to the long tail biased problem presented in the SGG task. The SGG model can learn the differences between predicates with similar or contrasting semantic correlations and reduce the bias of the tail class towards the head class of different semantic spaces.

To tackle the challenge, we propose the SG-Shuffle architecture that limits the learning of classification between predicate labels in different semantic spaces to improve classification between semantically correlated predicate labels. In order to separate non-correlated predicate labels, we group correlated predicates into four groups: Geometric, Possessive, Semantic, and Misc based on their purpose and super-type following the description in the Neural Motif paper [25]. A stacked transformer encoder is adopted for feature refinement and contextual information encoding of the object feature to generate the category-specific predicate feature with fine-grained information that distinguishes predicate with correlated semantics. A shuffle transformer structure based on Transformer [19] and ShuffleNet [12] is proposed to fuse such fine-grained category-specific features into a more universal feature that can classify between all predicates labels in the dataset. This structure both fuses the fine-grained features generated from the previous step and further propagates contextual information among the scene graphs. We then applied the simple loss weighting strategy at the end of the training process to further handle the long tail bias problem that also exists within the predicate of the same category.

Our contributions are as follows: First, we addressed the SGG issue where uncorrelated labels are classified against each other, which we tackled by categorising correlated labels and learning category-specific predicate features. Second, we also proposed a Shuffle Transformer layer, which is used to fuse features of different focuses to obtain the universal predicate feature for predicate classification as part of our architecture, SG-Shuffle. Third, we evaluated the performance of the proposed SG-Shuffle to demonstrate its effectiveness in the SGG task.

2   Related Works

Scene graphs received an attention in vision and language joint learning research as they can serve as a structural representation of images and have the potential to benefit several downstream vision and language reasoning tasks, such as image generation, image retrieval, visual question answering, and image captioning. Earlier works in scene graph generation involve making better use of visual features. They leverage contextual information for object prediction and predicate prediction using message passing [21], LSTMs [4,9,18,25], and GNN [7,13,23]. Statistics correlation of object and predicate are also used in addition to give the models more information to enhance the results. [25] has used GloVe for implicit statistics correlation, whereas [1] has explicitly used statistical correlation as edges in GNN. While performance was improved, challenges still remain due to the long-tailed data distribution which causes these models to perform poorly on infrequent classes. Recent work has looked at several debiasing methods for unbiased scene graph generations which can mainly categorised into three major types: re-sampling; loss re-weighting; and bias disentanglement from biased results. [8] proposed to oversampling image instances while undersampling common predicates for balanced predicate distribution. [22] and [5] on other hand suggest to use label correlation to realign their training loss while other methods like [7,16,26] propose their additional training loss objectives to reduce the bias problems. Other than re-sampling and loss re-weighting, bias disentanglement is also commonly used, removing bias from biased model result for unbiased scene graph. [17] propose to remove causal inference bias while missing label bias is estimated from label frequency and removed in [2].

One of the challenge in computer vision is the channel sparse connection problem in convolution neural networks for images, where each convolution only operates on a single group of input channels due to the use of group convolutions for reducing model complexity. ShuffleNet [12] was proposed to address the problem by allowing for information exchange between channels of different groups through the use of channel shuffle operations between group convolution layers. Inspired by this, channel shuffling was used in multiple works in a variety of different deep learning research works [3,20] to allow for information flow and strengthen the correlation between components of their model. Based on the success of ShuffleNet, we propose Shuffle Transformer containing the channel shuffling operation for combining multiple category specific predicate features.

3   Methodology

The typical methods of SGG comprise a two-stage process: 1) detecting objects within the images and 2) predicting the relationships between these objects. In the first stage, a standard object detector like Faster RCNN [15] obtains a set of bounding boxes for the set of objects detected in the image. RoIAlign [15] generates the visual feature of these bounding boxes and determines the initial detection of the object label for each of the detected objects. The object bounding boxes, which represent the position of the object in the image; object visual features, which represent the shape, form, and pattern learned by the object detector about the object; and object labels, which represent natural language understanding of the object semantic, are predicted using the input image and used as input for the next step. If ground truth information is used, as is the case of PredCls or SGCls settings, the ground truth information is inserted at the step where the information is intended to be used. In the second stage, the information generated from the object detector is used to predict the predicate between the predicted objects. As Faster RCNN is usually used for object detection, SGG models generally focus on the second stage of the process, which is also aligned with the main focus of our proposed SG-Shuffle. Our proposed model for predicate prediction, in particular, consists of three steps: 1) Four individual transformer sub-models are used to learn the category-specific representation of the objects and predicates; 2) Shuffle Transformer layers are then used to merge and allow information flow between the previous step's output; and 3) Finally, weighted cross-entropy (CE) loss is calculated and used for model optimization as a way to reduce the long tail biased problem.

3.1   Categories

To focus the attention of the model on distinguishing between predicate labels of similar semantic space, we categorize the predicate labels into 4 groups based on their super-type following the description of the Neural Motif [25]: Geometric, Possessive, Semantic, and Misc as shown in Table 1. We limit the need of the model to classify between predicates with different semantic purposes which can often require attention to different aspects of the input, for example: mainly object position for Geometric predicates, object label, and visual feature for Possessive predicates, or a balanced combination of the three for Semantic and Misc predicates. And this, in turn, allows the model to make use of the input aspects selectively to classify between semantically correlated predicates of the same category, which can be challenging for general scene graph models since they are often represented close to each other in the feature space due to semantic similarity, especially with the long tail biased problem of the SGG dataset.

3.2   Parallel Transformer Encoder

For each image, bounding boxes, corresponding object features, and object labels are generated using Faster RCNN [15] as input to our relationship prediction module. In order to incorporate object label information as input, object labels are encoded using GloVe encoding [14]. As these inputs are still non-contextual and are not specifically trained for relationship prediction, as shown in Fig. 1, we made use of the transformer encoder architecture to transverse contextual information as well as refine the feature vectors for relationship prediction. And

Table 1. Predicate categories and predicate labels in each category

Category	Predicate label
Geometric	'above', 'across', 'against', 'along', 'at', 'behind', 'between', 'in front of', 'near', 'on', 'on back of', 'over', 'under', 'in' , 'and'
Possessive	'belonging to', 'has', 'part of', 'wearing', 'attached to', 'of', 'wears', 'with'
Semantic	'to', 'carrying', 'covered in', 'covering', 'eating', 'flying in', 'growing on', 'hanging from', 'holding', 'laying on', 'looking at', 'lying on', 'mounted on', 'painted on', 'parked on', 'playing', 'riding', 'says', 'sitting on', 'standing on', 'using', 'walking in', 'walking on', 'watching'
Misc	'for', 'from', 'made of'

since we needed 4 sub-models to learn the specific details about the relationship categories, the 4 transformer encoders are trained with their own goal of classifying relationships within each of the category groups. For simplicity, we concatenate the three object information streams, including object bounding box, object feature, and object label encoding, as input to our context encoder.

$$Input = W_o[pos(box_i), visual_i, GloVe(label_i)] \tag{1}$$

Our context encoder adopts the out-of-the-box architecture of the transformer encoder as it has been shown to be relatively effective compare to RNN, or CNN in both natural language and computer vision. It is composed of layers of self-attention, feed-forward, and layer normalization stacked.

$$o'_c = LayerNorm(SelfAttention(o_c) + o_c) \tag{2}$$

$$o''_c = LayerNorm(FeedForward(o'_c) + o'_c) \tag{3}$$

where o_c is the input object feature o'_c is the output of the multi-head attention and o''_c is the output of the transformer encoder. The output of the contextual encoder is combined with the bounding boxes and visual features of the unions and intersections through concatenation to predict the category-specific relationships, which calculates a category-specific loss using CE loss.

$$r_c = Softmax(Concat(subject_c, object_c, v_{intersect}, v_{union})) \tag{4}$$

where r_c is the category specific predicate features, v is the visual feature while $subject_c$ and $object_c$ are the output object features of the objects pair.

We applied CE loss to jointly optimise the sub-models in their respective categories. The parameters of the sub-models are trained in parallel and optimized as part of the training process. For each of the sub-models, object pairs with relationships from other categories are not considered in the computation of this loss.

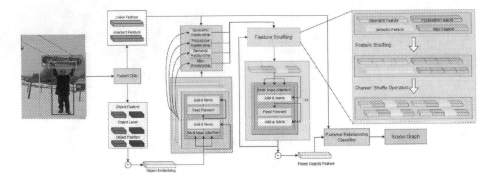

Fig. 1. Overall architecture of SG-Shuffle

3.3   Shuffle Transformer

In the second step of our relationship prediction pipeline, after the sub-model learns the category-specific contextual information for each of the objects in the image, these category-specific contextual object features are merged together in order to classify the actual relationship of the objects pair in the original relationship labels set. Therefore, in this stage, the outputs of the transformer encoder sub-models are used. Furthermore, while the 4 categories used in the previous step are from different semantic domains, they are still correlated as the candidate relationship of the same pair of objects, and hence, information flow between these sub-models is needed. In order for the aforementioned reasons to be incorporated into the model, we need to ensure that all the outputs of the sub-models are relevant in the prediction stage and the correlation between these category-specific object features is taken into account to further improve the prediction result. We proposed making use of the shuffle architecture, which was proposed in ShuffleNet [12] to handle information flow between channels of CNN for computer vision tasks for this purpose. As shown in Fig. 1, this architecture specifically makes use of the channel shuffle operation to allow such information flow. The original ShuffleNet is used with CNN for images and is not directly applicable to our situation, so we replace their convolution layer with a transformer encoder layer with the same architecture as in the previous step, with shuffle layers in between similar to the shuffle net. For SG-Shuffle, since we have 4 category-specific object features of 4 sub-models from the previous stage, for simplicity, the same number of shuffle sub-models are used in our shuffling stage. By using four shuffle sub-models, a quarter of the output features from the previous layer are concatenated to be the input of the next layer.

$$o'_s = SelfAttention([partition_s(o_{(k=1 \longrightarrow 4)})]) \tag{5}$$

$$o_{final} = W_s[o_{(s=1 \longrightarrow 4)}^{final-1}] \tag{6}$$

$$r_{final} = Softmax(Concat(subject_{final}, object_{final}, v_{intersect}, v_{union})) \tag{7}$$

After a few shuffle layers, the output of the shuffle sub models are concatenated and used to predict the predicate of the object pairs using the softmax function.

3.4 Weighted CE Loss

While categorizing the predicate labels into 4 different groups helps alleviate the bias problem to a certain degree, there is still bias between predicates of the same predicate group. To further remedy the long tail bias problem in the SGG, at the end of the training process, we applied a simple re-weighted CE loss to balance the learning process of each predicate label. Traditionally, for classification tasks like predicate prediction in SGG, a network is trained to minimise the CE loss. The predicted probability is obtained by applying the Softmax function to the output of the final layer. This loss penalises errors of predicting each label equally and therefore makes the model skew toward common labels due to the number of instances they have in the dataset. The weighted CE loss is a simple modified version of this CE loss with larger weights for the infrequent labels and lower weights for the frequent labels and penalises error classification accordingly.

$$l(x,y) = L\{l_1, ..., l_N\}^T, l_n = -\sum_{c=1}^{C} w_c log \frac{exp(x_{n,c})}{\sum_{i=1}^{C} exp(x_{n,i})} y_{n,c} \qquad (8)$$

4 Evaluation Setup

Dataset Details. We used the VG150 dataset [25], a subset of the large-scale Visual Genome vision and language dataset. It is the pre-processed split which is specifically used for SGG tasks, with the most frequent 150 object categories and 50 predicates categories. For object and predicate-super categories, we followed [25] criteria to split the predicate classes into 4 super-classes based on their semantic nature. Following the same testing strategy, as [17], we also use the original split with 70% training set and 30% test set, as well as taking 5000 samples from the training set as a validation set for parameter adjustment.

Evaluation Metrics. We use the mean Recall@ K metric. This metric has recently been used in place of regular recall due to the long tail bias problem in the image dataset, which leads to the performance bias in this metric [17]. The evaluation is done by predicting the relationship triplets in 3 settings: Predicate Classification (PredCls): using the image with ground truth object label and bounding box, Scene Graph Classification (SGCls): only ground truth bounding box and Scene Graph Detection (SGDet): using only the ground truth image.

Implementation Details. We use the Faster RCNN as the object detector to focus on the performance of the predicate prediction and stay consistent with previous work. It is pre-trained on ImageNet and fine-tuned on VG150 by [17] with ResNeXt-101-FPN being the backbone for region proposals. For consistency with previous works, the parameters of the object detector were kept frozen during the training and evaluation period. The stacked encoder used as the category sub-models contains 6 layers of transformer encoder with 4 attention heads each. For the weighted CE loss, we applied the inversed square root of predicate frequency as mentioned in [2] as weight for the loss function. We optimised the proposed model using the Adam optimizer with an initial learning rate of 0.001 and the warm-up and decay strategy suggested by [17]. The experiment was conducted on the NVIDIA T4 GPU.

5   Performance Analysis

5.1   Quantitative Evaluation

We compare SG-Shuffle with other SGG methods to demonstrate the ability of the proposed SG-Shuffle architecture to improve upon the feature refinement of objects and relationships in SGG while also displaying that it can be used with debiasing methods for unbiased SGG.

Table 2. Performance evaluation on VG150.

		PredCLS		SGCLS		SGDET	
		mR@50	mR@100	mR@50	mR@100	mR@50	mR@100
Without debiasing	IMP (2017)	9.80%	10.50%	5.80%	6.00%	3.80%	4.80%
	Motif (2018)	13.30%	14.40%	7.10%	7.60%	5.30%	6.10%
	KERN (2019)	17.70%	19.20%	9.40%	10.00%	6.40%	7.30%
	VCTree (2019)	17.90%	19.40%	10.10%	10.80%	6.90%	8.00%
	Our model	**24.39%**	**25.94%**	**13.00%**	**13.90%**	**10.94%**	**12.01%**
With debiasing	Motif + TDE (2020)	25.50%	29.10%	13.10%	14.90%	8.20%	9.80%
	PCPL (2020)	35.20%	37.80%	**18.60%**	**19.60%**	9.50%	11.70%
	Motif + DLFE (2021)	26.90%	28.80%	15.20%	15.90%	11.70%	13.80%
	BGNN (2021)	30.40%	32.90%	14.30%	16.50%	10.70%	12.60%
	Our Model/w Weighted Loss	**35.57%**	**38.67%**	17.96%	19.24%	**13.52%**	**14.91%**

Firstly, we compare SG-Shuffle without weighted CE loss to other biased SGG baselines, including IMP [21], Motif [25], KERN [1], VCTree [18]. These models aims to generate better objects feature representations by traversing context information between objects in the images. We compare SG-Shuffle without weighted CE with these baselines to demonstrate the effectiveness of SG-Shuffle in generating informative feature representation. As shown in the first part of Table 2, the mR@100 of our model is 6.5% higher in the PredCLS setting, 3.1% higher in the SGClS setting, and 4.0% higher in the SGDet setting comparing to the VCTree, which is the best performing model among models without debiasing methods. These models worked on the large label set at once, a challenging task since the distance between the predicate label are not uniform in the feature space. SG-Shuffle was able to gain better performance by learning in-depth features that differentiate predicate with close semantic nature.

Secondly, we compare SG-Shuffle with weighted CE loss with the more recent unbiased SGG models such as TDE [17], PCPL [22], DLFE [2], and BGNN [8], which use debiasing strategies to solve the long-tailed bias problem in the SGG task. We observed that with simple weighted CE loss, SG-Shuffle outperforms baseline models in the PredCls setting and SGDet setting, in which our PredCls score is 0.87% higher than PCPL and 5.77% higher than DLFE in mR@100 score. It only comes slightly lower than only PCPL in SGCls by a minor 0.3%. Among the baseline models, strategy used by PCPL also involve learning a better representation of predicates by modeling the relationship between predicate labels. Compare to our model and PCPL, other models in the baseline are designed to reduce the training bias by removing biased probability or re-sampling, to outperform models without debiasing. But without in-depth learning of predicate representation, their performance is generally lower than the models with this feature like PCPL and our model.

5.2 Hyper Parameter Tuning

While increasing the number of layers is often advantageous in the early layers of deep learning models, at a higher number of layers, it could also lead to diminishing gradient and optimization issues. We conducted hyper-parameter testing with it being tested with a varying number of shuffling layers.

Table 3. Performance with shuffle layers in PredCLS, SGCLS, and SGDET.

# of shuffle layers	PredCLS		SGCLS		SGDET	
	mR@50	mR@100	mR@50	mR@100	mR@50	mR@100
4	30.00%	32.43%	14.72%	15.87%	10.59%	11.90%
5	35.09%	37.68%	**17.96%**	**19.24%**	**13.52%**	**14.91%**
6	**35.57%**	**38.67%**	16.64%	17.81%	11.46%	12.78%
7	32.13%	35.04%	14.89%	16.03%	9.92%	11.23%

In Table 3, we tested the model with 4, 5, 6, and 7 layers of shuffled transformer in all three SGG settings: PredCls, SGCls, and SGDet, and compare the performance using the mR@100 and mR@50 metrics. As shown in the table, the performance of the model increases significantly when the number of shuffle layers goes from 4 to 5 and goes down from 6 to 7. At this depth, challenge in optimization outweighs the performance gain of further layer depth increase. The model performs best with the PredCls setting when using 6 layers of shuffling, while 5-layered models perform best in the SGCls and SGDet settings.

5.3 Alternate Shuffle Layer

In order to learn how the level of connection between the sub-models in the shuffling layer of the model affects the final performance, we also tested a pair-to-pair shuffling layer which is shown in Fig. 2. In this setting, the output features

Table 4. Performance of SG-shuffle with full shuffle layers and pairwise shuffle layers

	PredCLS		SGCLS		SGDET	
	mR@50	mR@100	mR@50	mR@100	mR@50	mR@100
No shuffle	27.53%	29.85%	14.94%	16.32%	6.13%	7.54%
Pair-to-pair shuffle	29.23%	31.62%	13.93%	15.14%	10.89%	12.18%
Full shuffle	**35.57%**	**38.67%**	**16.64%**	**17.81%**	**11.46%**	**12.78%**

from the previous layer are each partitioned into halves and each half is combined with a half from a different sub models and used as input for the current layer. After a few layers of shuffling, the category specific context information is shared in all 4 sub-models pathways. As shown in Table 4, while the pair to pair shuffling procedure does increase performance when compared to the model without any shuffling by 2% in PredCls setting and 5% in SGDet setting, the performance increase was lower than the full channel shuffling. We attribute the higher performance of the full shuffling layer over the pair-to-pair shuffling layer to the direct connection with all 4 sub-model from previous layer that allows it to learn important aspects from the previous layers at a faster rate and give better feature representation for relationship prediction.

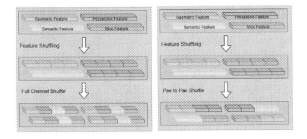

Fig. 2. Full channel shuffling and pair to pair shuffling layer

5.4   Ablation Study

Ablation Testing. We performed an ablation study on our model by removing the shuffling layer or weighted CE, and comparing them with the full model to show the effectiveness of each component. Compare with the model without weighted CE loss, the full model has nearly a 13% increase in performance in the PredCls setting when using the mR@100 metric. Long-tail bias plays a major part in SGG, and with no debiasing methodology, models are generally highly affected by the training bias introduced by the highly imbalanced dataset.

When debiasing is included, the weighted CE loss model also has much lower performance than the full model by around 8% mR@100 in the PredCls setting. This is due to the availability of shuffle layers, which allow information to flow

Table 5. Ablation result of SG-shuffle with mR@100 and mR@50

Shuffle	Weighted CE loss	PredCLS		SGCLS		SGDET	
		mR@50	mR@100	mR@50	mR@100	mR@50	mR@100
✓	–	24.39%	25.94%	13.00%	13.90%	10.94%	12.01%
–	✓	27.53%	29.85%	14.94%	16.32%	6.13%	7.54%
✓	✓	**35.57%**	**38.67%**	**16.64%**	**17.81%**	**11.46%**	**12.78%**

more freely between the sub-models in the full model. Comparing the shuffled only model and the weighted CE loss only model, the weighted CE loss has the advantage of debiasing and has higher performance in PredCls and SGCls but loses out in the SGDet setting in which object position was omitted. This omission leads the model to rely on its own predicted object position for SGG, which is less reliable than ground-truth information and harder to refine without the use of shuffle layer. As shown in Table 5, both of the components of the SG-Shuffle are necessary to achieve higher performance in unbiased SGG.

Categories Breakdown. In Table 6, we look at the effect of the model component with respect to each category of predicate label by comparing the PredCls mR@100 of the models in each of the 4 categories. For the geometric category, the full model has the highest performance, while the weighted CE loss only model has only slightly higher performance than the shuffle only model. Since there are both common and uncommon relationships in this category, so the debiasing advantage of the weighted CE loss in uncommon classes is matched by the shuffle-only model, which performs better in common classes. In the possessive category, which is dominated by the common predicate classes the shuffle-only model has a slightly higher performance than the other two models.

Table 6. Effect on different categories using PredCLS setting with mR@100

Shuffle	Weighted CE loss	Geometric	Possessive	Semantic	Misc	Overall
✓	–	21.27%	**31.33%**	29.71%	4.73%	25.94%
–	✓	22.06%	30.10%	37.16%	9.66%	29.85%
✓	✓	**28.11%**	31.11%	**48.97%**	**29.29%**	**38.67%**

The main difference in the overall mR@100 lies in the semantic category with the largest number of predicates, which are both informative and comparably uncommon in the dataset. There is a large gap between the three models' performances in this category. The shuffle-only model suffered from the long-tailed bias, which affects the uncommon predicate in this category and performs the lowest of the three models. Between the full model and the weighted CE loss only model, the full model has the advantage of the shuffle transformer with more informative feature representation and much higher performance in the semantic category than the other two ablations tested models. Similarly, the mR@100 of the full SG-Shuffle model also has higher performance in the Misc category, which has uncommon predicate labels, than the ablation-tested models.

5.5 Qualitative Analysis and Case Study

We visualize several scene graphs generated in the PredCls setting using the Ablation Tested Models in Fig. 3. We selected 4 samples for the case study based on the objects presented in the images: A person's portrait; B) a person in a large background; C) a building; and D) animal with plants in the background.

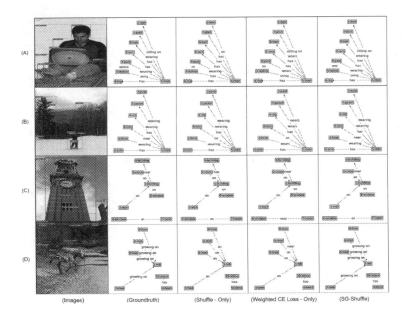

Fig. 3. Sample scene graph generated from ablation tested models. Correct relationship is marked with black arrows and incorrect relationship is marked with red arrows (Color figure online)

Observing samples A and B, between three models, the shuffle only model prefers more common predicates like "on" or "has", but cannot predict less frequent predicates like geometric relationship "near", or semantic relationship "using". The long-tailed bias present in the dataset to heavily affect the prediction of the model. The weighted CE loss only model, on the other hand, favors the infrequent relationship "wear" over the more common but same meaning "wearing". The SG-Shuffle model perform better than the other two models in both common possessive relationships like "has" as well as infrequent semantic relationships like "sitting on" or "using". Similarly, in sample C, the shuffle only model misclassified infrequent relationship "near", the weighted CE loss only does the opposite, misclassified common relationship "on", while the SG-Shuffle correctly classifies both. However, weighted CE loss could not predict the every infrequent semantic relationship as shown in sample D, where "growing on" was misclassified as "on". Comparatively, the full model was able to associate "growing on" with object "tree" thanks to the improved feature representation from jointly learning from category-specific object features.

6 Conclusion

In this paper, we propose the SG-Shuffle model for unbiased SGG by addressing non correlation problem of relationship labels in the existing SGG dataset. We proposed to categorise the set of predicate labels to four category "Geometric", "Possessive", "Semantic", and "Misc" in a divide and conquer approach which is learned as part of the SG-Shuffle SGG pipeline by Transformer Encoder in parallel to provide category-specific context for further SGG. We also propose a shuffle transformer layer, which apply channel wise shuffling operation in combination with the Transformer Encoder architecture to allow information flow between sub-models and merge together the learned category-specific feature representation. We demonstrated the effectiveness of SG-Shuffle in the VG150 dataset in comparison with other state-of-the-art SGG models.

References

1. Chen, T., Yu, W., Chen, R., Lin, L.: Knowledge-embedded routing network for scene graph generation. In: Proceedings of the IEEE/CVF Conference on Computer Vision and Pattern Recognition, pp. 6163–6171 (2019)
2. Chiou, M.J., Ding, H., Yan, H., Wang, C., Zimmermann, R., Feng, J.: Recovering the unbiased scene graphs from the biased ones. In: Proceedings of the 29th ACM International Conference on Multimedia, pp. 1581–1590 (2021)
3. Geng, S., et al.: Dynamic graph representation learning for video dialog via multimodal shuffled transformers. In: Proceedings of the AAAI Conference on Artificial Intelligence, vol. 35, pp. 1415–1423 (2021)
4. Guo, Y., et al.: Relation regularized scene graph generation. IEEE Trans. Cybern. **52**, 5961–5972 (2021)
5. Guo, Y., et al.: From general to specific: informative scene graph generation via balance adjustment. In: Proceedings of the IEEE/CVF International Conference on Computer Vision, pp. 16383–16392 (2021)
6. Johnson, J., et al.: Image retrieval using scene graphs. In: Proceedings of the IEEE Conference on Computer Vision and Pattern Recognition, pp. 3668–3678 (2015)
7. Knyazev, B., de Vries, H., Cangea, C., Taylor, G.W., Courville, A., Belilovsky, E.: Graph density-aware losses for novel compositions in scene graph generation. arXiv preprint arXiv:2005.08230 (2020)
8. Li, R., Zhang, S., Wan, B., He, X.: Bipartite graph network with adaptive message passing for unbiased scene graph generation. In: Proceedings of the IEEE/CVF Conference on Computer Vision and Pattern Recognition, pp. 11109–11119 (2021)

9. Li, Y., Ouyang, W., Zhou, B., Wang, K., Wang, X.: Scene graph generation from objects, phrases and region captions. In: Proceedings of the IEEE International Conference on Computer Vision, pp. 1261–1270 (2017)

10. Long, S., Han, S.C., Wan, X., Poon, J.: Gradual: graph-based dual-modal representation for image-text matching. In: Proceedings of the IEEE/CVF Winter Conference on Applications of Computer Vision, pp. 3459–3468 (2022)

11. Luo, S., Han, S.C., Sun, K., Poon, J.: REXUP: I REason, I EXtract, I UPdate with structured compositional reasoning for visual question answering. In: Yang, H., Pasupa, K., Leung, A.C.-S., Kwok, J.T., Chan, J.H., King, I. (eds.) ICONIP 2020. LNCS, vol. 12532, pp. 520–532. Springer, Cham (2020). https://doi.org/10.1007/978-3-030-63830-6_44

12. Ma, N., Zhang, X., Zheng, H.-T., Sun, J.: ShuffleNet V2: practical guidelines for efficient CNN architecture design. In: Ferrari, V., Hebert, M., Sminchisescu, C., Weiss, Y. (eds.) Computer Vision – ECCV 2018. LNCS, vol. 11218, pp. 122–138. Springer, Cham (2018). https://doi.org/10.1007/978-3-030-01264-9_8

13. Mi, L., Chen, Z.: Hierarchical graph attention network for visual relationship detection. In: Proceedings of the IEEE/CVF Conference on Computer Vision and Pattern Recognition, pp. 13886–13895 (2020)

14. Pennington, J., Socher, R., Manning, C.D.: Glove: global vectors for word representation. In: Proceedings of the 2014 Conference on Empirical Methods in Natural Language Processing (EMNLP), pp. 1532–1543 (2014)

15. Ren, S., He, K., Girshick, R., Sun, J.: Faster R-CNN: towards real-time object detection with region proposal networks. In: Advances in Neural Information Processing Systems 28 (2015)

16. Suhail, M., et al.: Energy-based learning for scene graph generation. In: Proceedings of the IEEE/CVF Conference on Computer Vision and Pattern Recognition, pp. 13936–13945 (2021)

17. Tang, K., Niu, Y., Huang, J., Shi, J., Zhang, H.: Unbiased scene graph generation from biased training. In: Proceedings of the IEEE/CVF Conference on Computer Vision and Pattern Recognition, pp. 3716–3725 (2020)

18. Tang, K., Zhang, H., Wu, B., Luo, W., Liu, W.: Learning to compose dynamic tree structures for visual contexts. In: Proceedings of the IEEE/CVF Conference on Computer Vision and Pattern Recognition, pp. 6619–6628 (2019)

19. Vaswani, A., et al.: Attention is all you need. In: Advances in Neural Information Processing Systems 30 (2017)

20. Wang, Y., Sun, F., Lu, M., Yao, A.: Learning deep multimodal feature representation with asymmetric multi-layer fusion. In: Proceedings of the 28th ACM International Conference on Multimedia, pp. 3902–3910 (2020)

21. Xu, D., Zhu, Y., Choy, C.B., Fei-Fei, L.: Scene graph generation by iterative message passing. In: Proceedings of the IEEE Conference on Computer Vision and Pattern Recognition, pp. 5410–5419 (2017)

22. Yan, S., et al.: PCPL: predicate-correlation perception learning for unbiased scene graph generation. In: Proceedings of the 28th ACM International Conference on Multimedia, pp. 265–273 (2020)

23. Yang, J., Lu, J., Lee, S., Batra, D., Parikh, D.: Graph R-CNN for scene graph generation. In: Ferrari, V., Hebert, M., Sminchisescu, C., Weiss, Y. (eds.) ECCV 2018. LNCS, vol. 11205, pp. 690–706. Springer, Cham (2018). https://doi.org/10.1007/978-3-030-01246-5_41

24. Yu, J., Chai, Y., Wang, Y., Hu, Y., Wu, Q.: CogTree: cognition tree loss for unbiased scene graph generation. arXiv preprint arXiv:2009.07526 (2020)

25. Zellers, R., Yatskar, M., Thomson, S., Choi, Y.: Neural motifs: scene graph parsing with global context. In: Proceedings of the IEEE Conference on Computer Vision and Pattern Recognition, pp. 5831–5840 (2018)
26. Zhang, J., Shih, K.J., Elgammal, A., Tao, A., Catanzaro, B.: Graphical contrastive losses for scene graph parsing. In: Proceedings of the IEEE/CVF Conference on Computer Vision and Pattern Recognition, pp. 11535–11543 (2019)

Explainable Detection of Microplastics Using Transformer Neural Networks

Max Barker[1], Meg Willans[2], Duc-Son Pham[1(✉)], Aneesh Krishna[1], and Mark Hackett[2]

[1] School of Electrical Engineering, Computing and Mathematical Sciences, Curtin University, Bentley, WA 6102, Australia
max.barker1@curtin.edu.au, dspham@ieee.org
[2] School of Molecular and Life Sciences, Curtin University, Bentley, WA 6102, Australia

Abstract. Microplastics are environmental contaminants that put marine and aquatic ecosystems at serious risk. Monitoring microplastics is necessary to understand the level of microplastic pollution in our environment. However, the lack of a standard protocol for quantifying and classifying microplastics causes problems in the reliability and comparability of results. Previous literature has employed deep learning models to classify and quantify microplastic polymers with great success, but the ability of these models to classify microplastics from new domains is unanswered. This paper presents an innovative approach to microplastic classification that employs a deep learning approach using a transformer neural network. Our specific contributions are: (1) A novel way to pre-process FTIR spectral data to dramatically increase classification accuracy. (2) Developed a transformer neural network for classifying microplastic polymer FTIR spectra. With the inclusion of a wider range of data, future deep learning approaches will improve the classification and quantification of microplastic polymers, subsequently reducing the costs and labour involved.

Keywords: Microplastic detection · Transformer · Time-series analysis · Supervised learning · Deep learning

1 Introduction

Microplastics are highly persistent contaminants found in marine and freshwater systems [20]. They are classified as tiny pieces of plastic that form through environmental factors and manufacturing processes [11]. Prolonged environmental exposure fragments larger plastic objects into smaller microplastics over time [11]. Microplastics have adverse effects in the reproduction of certain marine species [1], and have been found within human lung tissue and blood [5,7]. Microplastics are commonly analysed using chemical characterisation methods, such as Fourier-transform infrared spectroscopy (FTIR) and Raman spectroscopy [11]. Spectroscopy methods can be highly effective, but can suffer from

H. Aziz et al. (Eds.): AI 2022, LNAI 13728, pp. 102–115, 2022.
https://doi.org/10.1007/978-3-031-22695-3_8

a lack of precision or costly and time consuming processes [4,11]. Micro-FTIR is a reflectance spectroscopy method which enables high throughput analysis of particles, capable of rapidly recording large quantities of data but result in polymer spectrum's that are hard to distinguish from others [22]. Machine learning techniques could enable micro-FTIR and similar methods to be effective in classifying and analysing microplastics. If a protocol that utilises machine learning and micro-FTIR is developed, microplastic quantification and classification could be conducted in a fraction of the time.

A promising machine learning architecture for the classification of microplastics is the transformer neural network. They are a recent innovation in the field of machine learning [19], and have proven to be effective in both time series forecasting and time series classification tasks [13,24]. This is due to the transformer neural networks ability to capture discrepancies and interactions over long sequences [21]. A successfully modelled transformer network could be capable of classifying microplastics that have been collected using micro-FTIR spectroscopy.

Machine learning methods are becoming more popular in microplastic research, particularly through computer vision and classification [10,16,23,25]. Classification methods have proven extremely effective at microplastic classification, with some methods achieving near-perfect accuracies [25]. Obtaining microplastics from environmental domains is challenging and expensive [16], resulting in the need to manually create microplastics [16,25]. Still unanswered is how a machine learning model trained on the microplastic polymers from one domain performs when classifying polymers from another domain.

The paper is organised as follows. The related background of this application is given in Sect. 2. In Sect. 3, we detail how we adapt the standard transformer to this specific problem and other techniques to process the data. A number of experiments are given in Sect. 4 to demonstrate the effectiveness of our method. Finally, Sect. 5 concludes the paper.

2   Background

2.1   Microplastics

Microplastics pose serious environmental risks. They are classified as either primary or secondary microplastics. Primary microplastics are manufactured for use as abrasives and exfoliates in consumer products. Secondary microplastics result from the fragmentation of larger pieces of plastics through environmental factors, such as solar radiation and ocean waves [12]. The presence of microplastics within our environment is having a significant impact on both marine and terrestrial ecosystems [6]. The aggregation and digestion of microplastics affect the injuries and mortality of aquatic birds, fish, mammals and reptiles [14]. Due to their size and persistence, microplastics have infiltrated human food sources [1] and drinking water [15]. Microplastics have been found within the blood and lung tissue of humans, negatively impacting our health [5,7]. Microplastics can act as a vector for persistent organic pollutants (POPs) which are highly carcinogenic chemical compounds. Contaminated microplastic particles may affect both our ecosystem and food sources [22].

The growing concern over microplastics in our environment has introduced a need for a robust and straightforward method of analysing microplastics. Some of the most common spectroscopic techniques come in the form of FTIR spectroscopy [11] and electron microscopy [16]. To date, a standardized solution for microplastic analysis is yet to be developed, as each method has its strengths and limitations. A standardised method of microplastic analysis would facilitate the comparability, validity and accuracy of future microplastic research [6,22].

2.2   Time Series and Transformers

Transformer neural networks [19] have shown great performance in numerous time series applications. This is largely due to their capabilities of modelling long-range dependencies and interactions in sequential data [21]. Transformers have been successfully applied to various time series tasks. The original transformer architecture was designed as a sequence-to-sequence deep learning model, initially created for use in natural language processing (i.e., language translation). This is achieved through its encoder-decoder architecture. Encoder-decoder transformers have been used in time series forecasting [8,24] and anomaly detection [17], where the input time series sequence results in some output sequence from the transformer model. Classification transformers modify the deep learning model by removing the decoder framework. The transformer neural networks can then be used in the feature representation stage through its encoder architecture and then for classification tasks using the feature representations [9,13]. A multivariate time series classifier proposed by [9] uses a two-tower transformer, where the towers respectively focus on time-step-wise and channel-wise attention. This modification of the transformer architecture achieved state-of-the-art results in numerous multivariate time series. Recent work by [13] achieved greater classification rates on raw optical satellite time series compared to recurrent and convolutional neural networks (CNN).

2.3   Microplastics and Deep Learning

Deep learning methods have proven extremely useful in microplastic quantification and classification. A recent study conducted Shi [16] used electron micrographs in combination with a CNN to achieve a microplastic polymer classification accuracy of 98.33%, displaying a cost-effective and rapid way of analysing microplastics. Yurtsever [23] used a CNN to classify microbeads from distinct cleansers within wastewater samples with accuracies of 89%. CNNs have also been used in conjunction with FTIR spectroscopy. Zhu [25] have developed a CNN that can classify the FTIR spectra of polyethylene and polypropylene polymers with high accuracy (99%). Deep learning methods can be extremely beneficial tools for classifying microplastics. Thus far, much of the research has only tested their deep learning models on one microplastic dataset without further validating performance by using microplastics obtained from another domain or environment. This is largely due to the lack of publicly available datasets and the costs involved in collecting environmental microplastic samples [3].

3 Methodology

3.1 Transformer Model

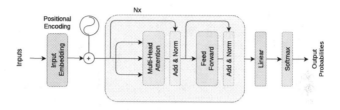

Fig. 1. Modified transformer neural network

The transformer neural network [19] was originally proposed for natural language processing, such as machine translation tasks. Whilst it has very powerful modelling capabilities for sequences, the original model architecture is unsuitable for our desired classification task. Rather than a sequence-to-sequence model, we require a sequence-to-vector model capable of taking an input time series sequence and outputting a binary classification of either polyethylene or polypropylene. The aim is to leverage the modelling power of the transformer neural network in the encoding stage such that we can use the found feature representations to perform classifications. A modified TNN architecture was created, aimed at classifying this type of data (see Fig. 1). Following previous research [9,13], this study's transformer model employs an encoder layer for representation learning. The decoder in the traditional transformer is replaced with a feed-forward neural network for classification instead. This implementation is a modification of a previous transformer model [2] and is designed to classify reflectance micro-FTIR spectra samples. We aim to utilise the transformer's ability to capture long-term dependencies in sequences to classify microplastic polymer identities.

3.2 Dataset Creation

This study used two microplastics datasets: a marine and a standard polymer dataset. Each dataset contains spectra samples of polyethylene (PE) and polypropylene (PP) polymers that have been analysed using reflectance micro-FTIR spectroscopy. This FTIR spectroscopy method enables high throughput and a non-destructive way of analysing microplastics [22]. Each sample in the dataset represents the measurement of the interaction of infrared radiation with

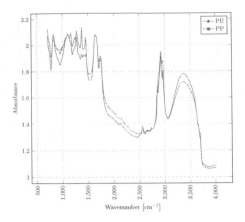

Fig. 2. Average polyethylene and polypropylene spectra.

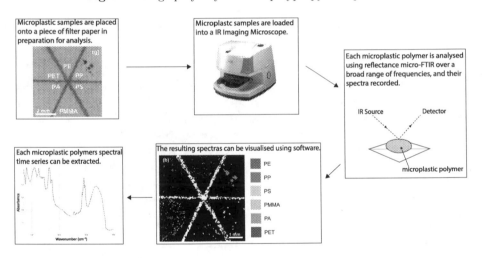

Fig. 3. Workflow of microplastic analysis using micro-FTIR

a polymer (PE or PP) by absorption. Figure 3 displays the process involved with collecting spectra data through micro-FTIR. The marine polymer dataset contains spectra samples of PE and PP polymers sourced from marine saltwater. A total of 512 spectra samples were obtained. The marine polymer dataset was then split into training, validation and testing sets. 70% of the samples were used to train the transformer model, and the remaining 30% were split to form validation and testing sets.

A standard polymer dataset was obtained containing spectra samples of PE and PP polymers. Unlike the marine polymer dataset, the standard polymers were not sourced from the environment but purchased as standard samples. The standard polymer dataset contains 89 PE and 80 PP polymer spectra. The standard polymers is be used to test the trained transformer model to see how well the model generalises to polymers that were collected from another domain. Specifically, the marine polymer dataset is used for model training and testing. The standard dataset evaluates how well the trained model can generalise to unseen data from another domain.

3.3   Pre-processing

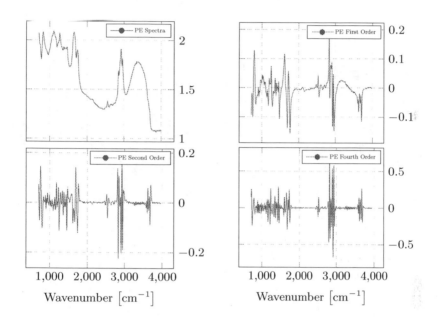

Fig. 4. Plotted marine polyethylene spectra showcasing raw data and first, second and fourth-order differencing.

From the domain knowledge, it is known that different plastics react differently to electro-magnetic frequencies. Furthermore, it is their sensitivity that differentiates them than the absolute response. For that reason, time series differencing was applied to the polymer spectra to extract the relationships shared between the samples for both polymer types. Time series differencing was used because recorded spectra for PP and PE polymers share similar trends as seen in Fig. 2 and high-order analysis would be able to differentiate them more easily. This pre-processing technique aims to assist the transformer during the representation learning stage.

The raw spectra data and higher orders of difference were compared to determine the best pre-processing approach to maximise the transformer model's performance. The first-order difference is calculated by the following:

$$A_{[n]} = X_{[n]} - X_{[n-1]}, \tag{1}$$

where X is an array containing a single spectra sample, and A is an empty array. The element at position $A_{[n]}$ is equal to the difference between the element at $X_{[n]}$ and $X_{[n-1]}$, also known as the first-order difference. Higher order differences are achieved by applying the same formula again to the resulting array A. Figure 4 shows the various effects of orders of differencing on a PE spectra sample. The absorbance of the unprocessed PE spectra fluctuates between a range of 2.0 and 1.1. After applying first-order differencing, the range is restricted between 0.20 and -0.15. Figure 4 further shows this range is further restricted by using higher orders. What can be seen is that the peaks along the wavelength are preserved using this method.

3.4   Data Augmentation

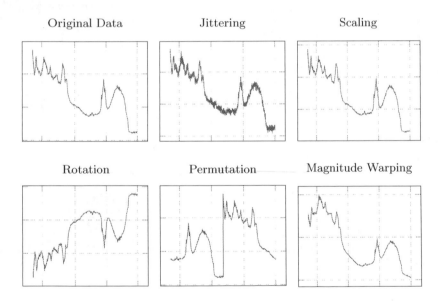

Fig. 5. Effects of augmentation on a single polyethylene spectra.

Due to the minimal training data available, data augmentation was implemented to create new but similar spectral samples. This was done with the aim of improving the model's accuracy and its ability to generalise to new data. A study conducted by [18] tested eight different time series augmentation methods on a data set consisting of wearable sensor data for Parkinson's' Disease monitoring. These methods include jittering, scaling, rotation, permutation, and magnitude warping. Using these augmentation methods, [18] achieved a classification performance improvement of 77.54% to 88.68%. Figure 5 visually represents different augmentation methods' effects on a PP sample. Permutation randomly perturbs the data locations within a set window in the time series. Scaling changes the magnitude of the data by multiplying elements by a random scalar, while jittering simulates additional noise in the time series. Finally, magnitude warping changes the magnitude of each sample by convolving the time series with a smooth curve. Each augmentation method was tested, and their impact on the model's performance in predicting the marine and standard polymers was recorded.

4   Experiments

4.1   Experiment 1: Time Series Difference Selection

Experiment Details. In this experiment, we evaluated the effect of time series differencing on the marine and standard spectra samples on the model's overall performance. The model was be trained with the marine training set and evaluated with the marine testing and standard polymer sets. Varying orders of differencing were tested ten times each, and the resulting accuracies were averaged and compared. The model was trained using 50 epochs per run, a batch size of 64 and a learning rate of 10^{-4}.

Table 1. Model performance with respect to various orders of difference, evaluated by marine set accuracy, loss, and Standard set accuracy.

Diff.	Marine Acc.	Loss	Std. Acc.
Default	70.84 %	0.5298	51.17%
First-order	97.43%	0.081	52.11%
Second-order	98.46 %	0.043	72.35%
Third-order	98.51 %	0.041	72.61%
Fourth-order	**98.71**%	0.035	75.64%
Fifth-order	97.69 %	0.058	77.29%
Sixth-order	97.25%	0.067	**81.52%**

Findings and Discussion. The transformer model's performance on the raw spectra data leaves much to be desired. A test set accuracy of 70.84% and a standard set accuracy of 51% is far too low to be considered useful. The low accuracy may indicate the shortcomings of reflectance-FTIR, where recorded spectra can suffer from distortions due to plastic morphology. The standard set accuracy for the raw data indicates morphological differences between the marine and standard samples. Both datasets contain the same polymers analysed using the same methods, but the model is not generalising well to the standards. Since the marine samples have been collected from the environment, they would have been subject to environmental exposure such as UV radiation [12, 22]. The morphological changes in the microplastic samples due to environmental degradation are significant enough to affect the transformer model's performance. Figure 6 gives us further insight into the impact on the performance by giving a visual representation of the difference in the reflectance spectra for the polymers. Irregular microplastic particle morphology can cause reflectance spectra distortions [22], which may further impact the model's ability to generalise.

Time series differencing has a significant effect on model performance. Table 1 shows a dramatic increase of +26% in testing accuracy after applying first-order differencing, up to a total of 97.43%. Testing accuracy increases until fourth-order differencing, where the test set reaches its peak. The accuracy of the standard set continues to increase throughout the experiments, up to a maximum value of 81.52% using sixth-order differencing. Past this point, both testing and standard set accuracy begin to decrease.

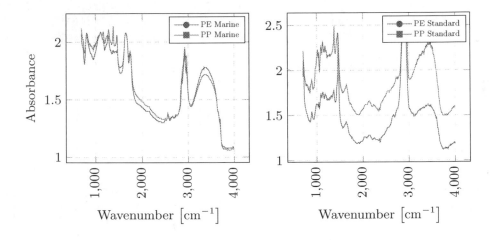

Fig. 6. Average spectra for marine and standard polymers.

4.2   Experiment 2: Data Augmentation

Experiment Details. In this experiment, the aforementioned data augmentation methods were applied to the dataset to observe their impacts on the models performance. Namely, jittering, scaling, permutation and magnitude warping. For each augmentation method, the marine training dataset was increased in size by a factor of two, consisting of the original datasets and the augmented data. Fourth-order differencing was then applied to the augmented dataset. For each augmentation method, the model was trained five times, with the corresponding test and standard set accuracies averaged and recorded. Following Experiment 1, the data augmentation methods were be trained using 50 epochs, a batch size of 64 and a learning rate of 10^{-4}.

Table 2. Model performance on validation, test and standard sets when using data augmentation.

Aug.	Val. Acc.	Marine accuracy	Std. accuracy
No Aug	99.2%	**98.7%**	**75.6%**
Jittering	99.3%	98.5%	72.5%
Scaling	**100%**	**98.7%**	66.7%
Permutation	99.4%	98.5%	60.5%
Mag. Warp	**100%**	**98.7%**	65.7%

Findings and Discussion. The data augmentation methods outlined by [18] did not have any meaningful performance increases on either the marine set or the standard polymer set. Table 2 showcases the outputted accuracies. None of the augmentation methods improved the marine test accuracy, either remaining unchanged or slightly decreasing. All augmentation methods positively influenced the model's validation accuracy and loss. Using augmentation methods, the model was able to reach higher validation accuracies, peaking at 100% for both scaling and magnitude warping. The model performed worse on the standard polymer set using all augmentation methods. The most significant decrease was when using permutation, where a decrease of 12% was observed.

The data augmentation methods could be causing the model to overfit to the marine set, as the augmentation methods are not specifically designed for microplastic spectral data. It could also be the case that key features in the spectral data are not being replicated properly in the augmentation process. Data augmentation's negative impact on the standard polymer dataset accuracy shows that domain-specific knowledge is required to design an appropriate method of spectral data augmentation that can assist with model generalisability.

4.3   Experiment 3: Model Tuning

Experiment Details. This experiment investigated increasing the model's performance through tuning the hyperparameters, aimed at increasing the model's accuracy on the standard polymer dataset. The microplastic spectra were preprocessed using fourth-order differencing.

Findings and Discussion. During the training of the transformer model, the outputted validation loss and accuracy and the performance on the standard set were monitored and recorded. Within the first 10 to 20 epochs, the model's accuracy for both the validation and standard sets increased. This can be observed in Fig. 7, where the accuracies and loss values for both datasets are initially correlated but diverge after 20 epochs of training. Within this 10–20 epoch range, the standard set accuracies average at a value of 88.45%, which is a 7% increase from the best recorded standard accuracy shown in Table 1. During epochs 10–20, the validation accuracy averages at a value of 97%.

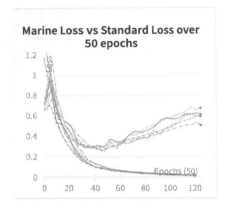
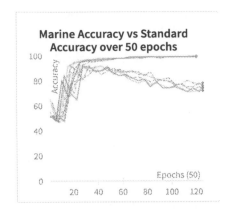

Fig. 7. Relationship between validation and standard set accuracy and loss after 5x training iterations for 50 epochs.

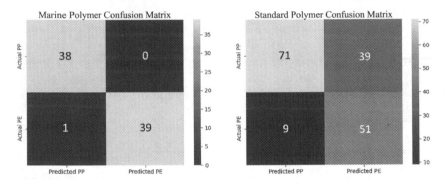

Fig. 8. Confusion matrices for model performance on marine polymer test set and standard polymer set.

The network overfitting could explain the divergent behaviour between the marine and standard polymer datasets. By reducing the number of epochs, we can gain overall accuracy. The performance increase is promising, but future research must consider the performance of deep learning models when classifying microplastics from another domain due to the possibility of overfitting.

By reducing the number of epochs to 20 and adjusting the hyperparameters, a test set accuracy of 98.71% and standard set accuracy of 92.3% were achieved. This final accuracy was achieved with a learning rate of 10^{-4} and a batch size of 128. In this case, the transformer model had two encoder layers both with four heads. Figure 8 displays the resulting confusion matrices. The marine polymer confusion matrix incorrectly classified a PE polymer as a PP polymer, which may result from a polymer that has had more morphological change than the rest considerably. As shown in the standard polymer confusion matrix, the model has a bias toward making PE predictions with 86% of incorrect classifications being PE classifications.

4.4   Experiment 4: Model Performance Comparison

Experiment Details. In this experiment, we evaluated the model's performance by comparing its classification accuracies and F_1-scores against other machine learning models: K-nearest neighbour, Random Forest, Naive Bayes, Linear Discriminant Analysis and Logistic Regression. We have also evaluated the performance against a 1D CNN to compare its performance against similar microplastic classifiers [16,23,25]. Each model was trained with the marine training set and evaluated with the marine and standard testing sets. The resulting accuracies were averaged and recorded for comparison.

Table 3. Comparison between transformer, kNN and Random Forest models on the marine and standard testing sets.

Model	Marine accuracy	Marine F1	Std. accuracy	Std. F1
Transformer	98.7%	98.7%	**92.3%**	**92.0%**
CNN	98.7%	92.6%	91.2%	91.0%
KNN	94.8%	92.6%	88.0%	84.6%
Rand. Forest	97.4%	94.5%	68.0%	44.6%
Naive Bayes	98.7%	98.7%	80.0%	78.9%
LDA	97.4%	97.4%	58.8%	54.2%
Logistic Reg.	94.8%	94.7%	52.9%	19.9%

Findings and Discussion. Table 3 reveals the accuracies of each machine learning model. The best model was the transformer, with a standard accuracy and f1 score of 91.1% and 91.2%, respectively. Each machine learning model performed very well on the test sets in terms of accuracy and F1-score, but most suffered once classifying the standard set. K-nearest neighbour achieves an impressive accuracy of 83.5% on the standard polymer set when nearest neighbour $k = 3$. Similarly, Naive Bayes achieved a score of 80%. Both Naive Bayes and K-nearest neighbour perform similarly to previous iterations of the transformer model. This demonstrates that the transformer model is an appropriate choice over other machine learning models. The 1D CNN performed extremely well on both the standard and marine testing sets, with F1-scores of 98.7% and 91.2% respectively. This performance closely matches that of the transformer architecture and further supports deep learning methods as appropriate tools for analysing microplastics.

5   Conclusions

We have proposed a highly effective transformer model that provides accurate classifications of microplastic polymers. We have shown that time series differencing can dramatically increase performance for both known and unknown data. In addition, we have shown that the morphology of microplastics can influence a machine learning models ability to classify microplastics from other domains. With continually growing datasets, machine learning approaches are expected to classify a broad range of microplastic polymers and dramatically reduce the cost and labour involved in microplastic classification and quantification. Our implementation will be made publicly available at https://github.com/br3nr/microplastic-transformer.

References

1. Barboza, L.G.A., et al.: Microplastics pollution in the marine environment. In: World Seas: An Environmental Evaluation Volume III: Ecological Issues and Environmental Impacts (2018)
2. Cohen, M.: Transformers for time series (2019). https://github.com/maxjcohen/transformer
3. Cowger, W., et al.: Critical review of processing and classification techniques for images and spectra in microplastic research. Appl. Spectrosc. **74**(9), 989–1010 (2020)
4. Harrison, J.P., Ojeda, J.J., Romero-González, M.E.: The applicability of reflectance micro-Fourier-transform infrared spectroscopy for the detection of synthetic microplastics in marine sediments. Sci. Total Environ. **416**, 455–463 (2012)
5. Jenner, L.C., Rotchell, J.M., Bennett, R.T., Cowen, M., Tentzeris, V., Sadofsky, L.R.: Detection of microplastics in human lung tissue using μFTIR spectroscopy. Sci. Total Environ. **831**, 154907 (2022)
6. Lamichhane, G., et al.: Microplastics in environment: global concern, challenges, and controlling measures. Int. J. Environ. Sci. Technol. (2022)
7. Leslie, H.A., Van Velzen, M.J., Brandsma, S.H., Vethaak, A.D., Garcia-Vallejo, J.J., Lamoree, M.H.: Discovery and quantification of plastic particle pollution in human blood. Environ. Int. **163**, 107199 (2022)
8. Li, S., et al.: Enhancing the locality and breaking the memory bottleneck of transformer on time series forecasting. In: Advances in Neural Information Processing Systems, vol. 32 (2019)
9. Liu, M., et al.: Gated transformer networks for multivariate time series classification. arXiv preprint arXiv:2103.14438 (2021)
10. Lorenzo-Navarro, J., Castrillón-Santana, M., Gómez, M., Herrera, A., Marín-Reyes, P.A.: Automatic counting and classification of microplastic particles. In: Proceedings of the ICPRAM (2018)
11. Prata, J.C., da Costa, J.P., Lopes, I., Duarte, A.C., Rocha-Santos, T.: Effects of microplastics on microalgae populations: a critical review. Sci. Total Environ. **665**, 400–405 (2019)
12. Primpke, S., Lorenz, C., Rascher-Friesenhausen, R., Gerdts, G.: An automated approach for microplastics analysis using focal plane array (FPA) FTIR microscopy and image analysis. Anal. Methods **9**(9), 1499–1511 (2017)

13. Rußwurm, M., Körner, M.: Self-attention for raw optical satellite time series classification. ISPRS J. Photogramm. Remote Sens. **169**, 421–435 (2020)
14. Sana, S.S., Dogiparthi, L.K., Gangadhar, L., Chakravorty, A., Abhishek, N.: Effects of microplastics and nanoplastics on marine environment and human health. Environ. Sci. Pollut. Res. **27**(36), 44743–44756 (2020)
15. Schymanski, D., Goldbeck, C., Humpf, H.U., Fürst, P.: Analysis of microplastics in water by micro-Raman spectroscopy: release of plastic particles from different packaging into mineral water. Water Res. **129**, 154–162 (2018)
16. Shi, B., et al.: Automatic quantification and classification of microplastics in scanning electron micrographs via deep learning. Sci. Total Environ. **825**, 153903 (2022)
17. Tuli, S., Casale, G., Jennings, N.R.: TranAD: deep transformer networks for anomaly detection in multivariate time series data. arXiv preprint arXiv:2201.07284 (2022)
18. Um, T.T., et al.: Data augmentation of wearable sensor data for Parkinson's disease monitoring using convolutional neural networks. In: Proceedings of the ACM ICMI, pp. 216–220 (2017)
19. Vaswani, A., et al.: Attention is all you need. In: Proceedings of the NeurIPS, vol. 30 (2017)
20. Wagner, M., et al.: Microplastics in freshwater ecosystems: what we know and what we need to know. Environ. Sci. Eur. **26**(1), 1–9 (2014)
21. Wen, Q., et al.: Transformers in time series: a survey. arXiv preprint arXiv:2202.07125 (2022)
22. Willans, M.: Developing a protocol for microplastic detection using microreflectance FTIR spectroscopy. Honours thesis, Curtin University (2021)
23. Yurtsever, M., Yurtsever, U.: Use of a convolutional neural network for the classification of microbeads in urban wastewater. Chemosphere **216**, 271–280 (2019)
24. Zhou, H., et al.: Informer: beyond efficient transformer for long sequence time-series forecasting. In: Proceedings of the AAAI, vol. 35, pp. 11106–11115 (2021)
25. Zhu, Z., Parker, W., Wong, A.: PlasticNet: deep learning for automatic microplastic recognition via FT-IR spectroscopy. J. Comput. Vis. Imaging Syst. **6**(1), 1–3 (2021)

EDE-NAS: An Eclectic Differential Evolution Approach to Single-Path Neural Architecture Search

Junhao Huang[1]([✉])[iD], Bing Xue[1][iD], Yanan Sun[2][iD], and Mengjie Zhang[1][iD]

[1] School of Engineering and Computer Science, Victoria University of Wellington, PO Box 600, Wellington 6140, New Zealand
{junhao.huang,bing.xue,mengjie.zhang}@ecs.vuw.ac.nz
[2] School of Computer Science, Sichuan University, Chengdu 610065, China
ysun@scu.edu.cn

Abstract. Convolutional neural networks (CNNs) are a very prevalent and powerful deep learning paradigm. In recent years, many neural architecture search (NAS) methods have been developed to automate the design process of CNN architectures, significantly reducing human effort. Among various search techniques, differential evolution (DE), as a popular evolutionary computation algorithm, has advantages of fewer control variables, fast convergence and powerful optimization capability. However, existing DE-based NAS methods simply use conventional search operators, and do not consider the global and local information in the search process well, thus failing to achieve satisfactory results. In this paper, we propose an eclectic DE approach for NAS that can make good use of the search capability of DE. The architectural parameters are encoded into two parts according to their ranges. A discrete mutation operator is proposed to evolve the part that has a small search space, while a versatile mutation operator is devised for the other part with a large search space. The proposed DE algorithm can well balance the global and local search, and yields better overall results than most compared methods with a single-path CNN architecture design based on basic operations on four benchmark image classification datasets.

Keywords: Neural architecture search · Convolutional neural networks · Differential evolution

1 Introduction

Neural architecture search (NAS) [3] has greatly promoted the development of deep learning, as it saves the manpower and labour of manually designing deep neural networks. In particular, various computer vision tasks, especially image classification, have greatly benefited from NAS in designing novel high-performance convolutional neural network (CNN) architectures [17]. Search strategy is one of the most critical components of an NAS algorithm, which has

© The Author(s), under exclusive license to Springer Nature Switzerland AG 2022
H. Aziz et al. (Eds.): AI 2022, LNAI 13728, pp. 116–130, 2022.
https://doi.org/10.1007/978-3-031-22695-3_9

become the focus of NAS research in recent years. The vast majority of existing NAS algorithms are based on reinforcement learning [26], gradient descent [20], and evolutionary computation (EC) [2] to navigate the search space. Since NAS can be considered as a challenging non-differentiable optimization problem, among these search strategies, EC is becoming increasingly popular because of its powerful ability in solving hard problems in a gradient-free fashion [16].

EC is a population-based method, and in EC-based NAS, each CNN architecture is treated as an individual in the population. The architecture search process of EC-based NAS is also the update of the population performed by the EC algorithm. Among various EC algorithms, differential evolution (DE) [24] is a powerful optimizer, which contains few control variables yet is very efficient and robust in dealing with optimization problems over large spaces. During population updating process of DE, new individuals are generated by mutation and crossover operators, where the mutation operation leads to a mutant individual by combining the difference information among multiple individuals in the population, while crossover is a recombination of the mutant individual with the current target individual. In particular, the behavior of the search algorithm in exploring the search space can be purposefully adjusted by different designs of mutation and crossover operators.

DE is rarely involved in NAS, probably due to the early introduction and preconception of other popular EC algorithms in this area, such as genetic algorithm (GA) [18] and particle swarm optimization (PSO) [13]. Wang et al. [29] proposed to use DE to evolve CNN architectures, and showed better performance than the predecessor using PSO [28]. Nevertheless, the conventional random mutation operator used affects the search efficiency since it does not consider the quality of the selected solutions and their correlations with the candidate solution. Hence, the DE operator needs to be improved to better balance the local and global information of the population and thus enhance the search capability.

When designing a CNN architecture, many relevant parameters need to be determined, e.g., layer type, convolution kernel size, stride size, and number of filters for each layer. Basically, layer type, kernel size and stride size have a small search range, while the number of filters for each layer has far more possible values. It is desirable to find an efficient way to encode these architectural parameters with different ranges.

To further exploit the potential of DE-based NAS and improve its search ability, in this work, we propose an eclectic DE approach, dubbed EDE-NAS, to automated CNN architecture design for image classification. We effectively split the architectural parameters that need to be optimized into two categories, namely, one containing those with only a few possible values and one including those with more options. In addition, a discrete mutation operator and a versatile mutation operator concerning elite-based local and global information are devised respectively for these two kinds of architectural parameters. The network architecture is constructed with basic convolutions and pooling operators based on a single-path (no branches) backbone which, though simple, has been proved to be hardware-friendly [5]. Experiments on several benchmark image classification datasets show better results overall compared to those state-of-the-art models. The contributions of this work are:

1. Designing a convolutional architecture search space based on basic standard convolutional and pooling layers, and constructing single-path CNN architectures from the search space. The resulting architecture, despite its simplicity because of the plain topology, is efficient and effective for the target tasks.
2. Encoding the involved four architectural parameters into two parts, 1) *part1*: layer type, kernel size and stride size which have a few options; 2) *part2*: number of filters containing more possible values within a certain range. Such an encoding scheme can handle a large search space and simplify it to facilitate the convergence of the search process.
3. Evolving the CNN architecture in a variable-length manner by DE. A discrete mutation operator is designed to evolve the small search space of *part1* and a versatile mutation operator is devised to evolve the large search space of *part2*. Both operators focus on the elitism of the population which is conducive to continuously improving the evolution quality, and global information which avoids the population being trapped into bad local optima.

2   Background

2.1   Differential Evolution (DE)

DE [24] is a population-based EC algorithm, originally proposed for continuous optimization problems. It produces new individuals by combining the information among multiple individuals in the population. DE is a very effective global optimizer which has few control variables and converges relatively fast. Generally, a DE algorithm procedure is constituted of four main steps: initialization, mutation, crossover, and selection. First, a group of initial solutions/vectors is randomly generated in the encoding space, which then enters the mutation-crossover-selection loop. In each mutation, a mutant vector is generated by performing differential operation among several selected vectors from the population. Taking the classical DE/rand/1/bin operator as an example, the i-th mutant vector in the g-th generation v_i^g is calculated by combining three randomly selected solutions $x_{r_0}^g$, $x_{r_1}^g$ and $x_{r_2}^g$:

$$v_i^g = x_{r_0}^g + F \cdot (x_{r_1}^g - x_{r_2}^g) \tag{1}$$

where $i \neq r_0 \neq r_1 \neq r_2 \in \{1, 2, ..., T\}$ and T denotes the population size; F is a scaling factor controlling the impact of the difference vector. After the mutation, the final trial vector u_i^g is produced by conducting crossover operation on the mutant vector v_i^g given the crossover rate CR. Specifically, the j-th element in u_i^g is set as:

$$u_{i,j}^g = \begin{cases} v_{i,j}^g, & \text{if } \text{rand}() \leq \text{CR or } j = j_{rand} \\ x_{i,j}^g, & \text{otherwise} \end{cases} \tag{2}$$

where rand() is a random number between 0 and 1; To assure there is at least one element comes from the mutant vector, a random position j_{rand} is specified. At the selection stage, the trail vector u_i^g is compared with the current solution x_i^g, and the one with higher fitness value (assuming a maximization problem) will survive in the next population.

$$x_i^{g+1} = \begin{cases} u_i^g, & \text{if } f(u_i^q) \geq f(x_i^q) \\ x_i^g, & \text{otherwise} \end{cases} \tag{3}$$

where $f(\cdot)$ refers to the task-specific fitness function, e.g., classification accuracy.

2.2   EC-Based NAS

NAS is a technology that automates the process of neural network architecture design, which is comprised of three critical components: search space, search strategy, and performance estimation [6]. EC-based NAS algorithms, from the perspective of search strategy, leverage EC algorithm to explore the search space and find the desire architecture. Xie and Yuille [31] proposed to evolve the connections among layers of different stages in CNN by a standard GA. The connection topology is encoded as a fixed-length binary string, and the encoding scheme theoretically can represent many popular network architectures, such as VGGNet [23], ResNet [8], and DenseNet [10]. A variable-length PSO approach [12] is devised to encode single-path CNN architectures for image classification. The individuals are updated in a discrete fashion by a difference vector between the current individual and best-so-far individuals. However, this method does not fully exploit the ability of PSO in local search, and the effectiveness of architecture search is thus limited. In addition to GA and PSO, Wang et al. [29] employed DE to search for single-path CNN architectures encoded by an IP address-based method [28]. A single point crossover operation is introduced as the second crossover in the evolutionary process to enhance local search. Awad et al. [1] also adopted a canonical DE to evolve CNN architecture where the architectural parameters are encoded as real values for continuous evolution, and then mapped back to discretized architecture. Albeit achieving promising performance, these methods still fall short in effective search due to the use of the classic random mutation operator. To address this limitation, in this study, the mutation operation is improved to consider more information in the population.

3   Methodology

3.1   Framework

The overall flowchart of the proposed algorithm is shown in Fig. 1, which basically follows a common EC-based NAS algorithm [16]. First of all, an initial

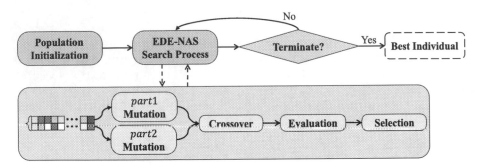

Fig. 1. Framework of EDE-NAS.

population is generated with random configurations for each individual. Then, the population is updated iteratively by the proposed search algorithm. When the stopping criterion is met, the best individual in the latest population is outputted. For a more precise presentation, we also outline the procedure of the proposed algorithm in Algorithm 1. Note that two mutation operators (elaborated in Subsect. 3.3) are conducted on *part1* and *part2* of the encoded individual, respectively. The generation of individuals is carried out one by one, and the new individual takes effect immediately and can be used for the evolution of subsequent individuals.

Algorithm 1: Procedure of EDE-NAS

    Input: Population size T, number of generations G, the differential rate F, and
             the crossover rate CR.
    Output: The best individual.
1 $P^0 \leftarrow$ Initialize and evaluate a population $\{x_i^0\}_{i=1}^T$;
2 **for** $g = 1$ *to* $G - 1$ **do**
3   **for** $i = 1$ *to* T **do**
4         $v_i^{g-1} \leftarrow Mutation(x_i^{g-1}, P^{g-1}, F)$;
5         $u_i^{g-1} \leftarrow Crossover(x_i^{g-1}, v_i^{g-1}, \text{CR})$;
6         $x_i^g \leftarrow Evaluation\&Selection(x_i^{g-1}, u_i^{g-1})$;
7         $P^{g-1} \leftarrow$ Update P^{g-1} by replacing x_i^{g-1} with x_i^g;
8   **end**
9     $P^g \leftarrow$ Updated P^{g-1};
10 **end**
11 $x_{best} \leftarrow$ The best-performing individual from P^{G-1};
12 **Return** x_{best}.

3.2 Search Space and Encoding Scheme

The searched CNN architectures are based on basic standard convolutional (Conv) and pooling layers, with different configurations of kernel size and stride size. We do not introduce additional knowledge regarding primitive operator nor network topology, for mitigating human intervention. The target architecture is constructed based on a single-path backbone, which means that there are no branches in the network and every layer connects to only one preceding layer and one subsequent layer. This leads to a simple CNN architecture overall, which will be demonstrated in the experiments to be sufficiently effective especially in the small and medium-scale image classification tasks.

Many EC-based NAS methods represent a CNN architecture as a vector where each element incorporates multiple architectural parameters through a sophisticated encoding scheme. Despite the fact that such ways are intuitive and straightforward, they may complicate the search space if the number and range of the searchable parameters are large. To alleviate the aforementioned issue, we separate the involved architectural parameters into two parts, i.e., *part1*: layer type, kernel size and stride size, and *part2*: number of filters, which are encoded into two vectors. For the encoding of *part1*, we exhaust the combinations of layer type, kernel size and stride size, and assign each combination a representation code, as illustrated in Table 1.

Table 1. Encoding scheme of one layer in the *part1* of an individual.

Representation code	Layer type	Kernel size	Stride size
0	Conv	1	1
1	Conv	3	1
2	Conv	5	1
3	Conv	3	2
4	avgPool	2	2
5	maxPool	2	2

In this work, three commonly used kernel sizes, namely, $\{1, 3, 5\}$ for the Conv operation, and three different downsampling operations are chosen. Apparently, these combinations are discrete, independent of each other, and result in a relatively small search space. Therefore, we propose a discrete mutation operator to explore such a search space. As for *part2*, a direct encoding is adopted to represent the number of filters of each convolutional layer, where 0 refers to that this convolutional layer has one filter. It is not difficult to see that *part2* could have far more optional values (the range of number of filters per layer is set to [1, 128] in this work) than that of *part1*. A continuous form of mutation operator is devised to navigate such a large search space. Fig. 2 shows an example of the proposed individual encoding method, where the valid (non-white) length of the individual denotes the depth of the decoded network, and each two blocks aligned vertically together represent one layer in the architecture. During the mutation process of *part2*, the updated values of the elements in the encoding vector may fall outside the preset range, e.g., 145 and -26 in Fig. 2. We regard them as invalid values and discard the corresponding layers when decoding the vector into a CNN architecture. This helps the proposed method to realize a variable-length evolution. Based on this encoding scheme, the decoded architecture of the example in Fig. 2 is a six-layer CNN, where the third and fifth layers are maxPool and avgPool, respectively, and the rest are all convolutional layers with a stride size of 1.

part1	1	0	2	5	3	2	4	1
part2	77	36	145	113	-26	109	59	87

Fig. 2. An example of the individual encoding.

Algorithm 2 gives the pseudocode for population initialization, where *part2* and *part1* of an individual are created one by one and then merged together. Particularly, to ensure a high diversity on the depth of searched architectures at the beginning of evolution, a random probability value p_{valid} between 0 and 1 is employed to control the proportion of the valid elements in the individual.

Algorithm 2: Population Initialization

Input: Population size T, individual length l, maximal number of strided/pooling layers max_s, maximal number of filters max_f, *part2* threshold gap thd.

Output: The initial population P_0.

1 $P_0 \leftarrow \emptyset$;
2 **for** $i = 1$ *to* T **do**
3 $\quad$ $p_{valid} \leftarrow$ Randomly generate a value from $[0, 1]$;
4 $\quad$ $part2 \leftarrow$ Generate l integers which are sampled from $[0, max_f - 1]$ with a probability of p_{valid}, and the rest from $[-thd, -1] \cup [max_f, max_f + thd]$;
5 $\quad$ $num_s \leftarrow$ Randomly select an integer from $[0, max_s]$;
6 $\quad$ Randomly generate num_s integers between $[3, 5]$;
7 $\quad$ Randomly generate $l - num_s$ integers between $[0, 2]$;
8 $\quad$ $part1 \leftarrow$ Combine and shuffle the l integers;
9 $\quad$ $P_0 \leftarrow P_0 \cup [part1, part2]$;
10 **end**
11 **Return** P_0.

3.3   Architecture Evolution

As illustrated in Sect. 3.1, the evolutionary process of EDE-NAS includes four major steps, namely, mutation, crossover, evaluation, and selection, which are performed iteratively for the individuals until a termination condition is reached.

Mutation: The mutation operation in DE is to learn from the comparison of multiple solutions in the population. The classical mutation operators, e.g., DE/rand/1/bin as shown in Eq. (1), are lacking in qualitative consideration of the selected individuals and the association with the current candidate individual. In EDE-NAS, we develop a DE/current-to-better/1/bin mutation operator by including the current solution and a solution better than the current one in the process. In this operator, the current solution is co-guided by a better solution as well as random solutions in the population, which well balance the elite-based local search and global search. The discrete form and the continuous form for evolving *part1* and *part2* are respectively formulated as Eqs. (4) and (5), respectively.

$$v_i^g = \begin{cases} x_{r_b}^g \oplus F \cdot (x_{r_1}^g \ominus x_{r_2}^g), & \text{if } \boldsymbol{I}_{x_i^g}^b \neq \emptyset \\ x_{r_i}^g \oplus F \cdot (x_{r_1}^g \ominus x_{r_2}^g), & \text{otherwise} \end{cases} \tag{4}$$

and

$$v_i^g = \begin{cases} x_i^g + F \cdot (x_{r_b}^g - x_i^g) + F \cdot (x_{r_1}^g - x_{r_2}^g), & \text{if } \boldsymbol{I}_{x_i^g}^b \neq \emptyset \\ x_i^g + F \cdot (x_{r_1}^g - x_{r_2}^g), & \text{otherwise} \end{cases} \tag{5}$$

where $A \ominus B$ means removing the elements from A that also appear in B, $C \oplus F \cdot D$ means replacing the elements in C by the elements in D with a probability of F. $\boldsymbol{I}_{x_i^g}^b$ is an individual set containing all the solutions in the population that outperform the current solution x_i^g. In both mutations, $x_{r_b}^g \in \boldsymbol{I}_{x_i^g}^b$ is randomly selected to guide x_i^g toward a potentially better position, and meanwhile, to avoid getting into poor local optima, we also consider global information during mutation by incorporating two random solutions in the population, namely, $x_{r_1}^g$ and $x_{r_2}^g$. if the current solution x_i^g is the best, which means that there is no solution better than x_i^g and thus $\boldsymbol{I}_{x_i^g}^b = \emptyset$, the mutation is reduced to a self-update of x_i^g.

Crossover: Crossover of individuals in EDE-NAS follows the typical operator as in Eq. (2). The two vectors of *part1* and *part2* are treated as a whole, and the crossover are performed based on a preset crossover rate CR and a random position j_{rand}. See Fig. 3 for an example of crossover operation.

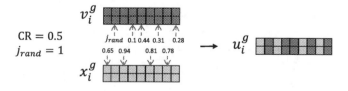

Fig. 3. An example of crossover operation.

Evaluation: The typical evaluation process in NAS is to first train the searched architecture on the training set by stochastic gradient descent (SGD) and then obtain its classification accuracy on the evaluation set. This process is notoriously time-consuming and becomes the most critical factor limiting the algorithm efficiency. In EDE-NAS, we adopt a simple yet widely used acceleration technique, namely, early stopping to improve the search efficiency. Concretely, during the evaluation process, each candidate architecture is trained with a dynamically decayed learning rate in a very small number of epochs. The trained architecture is then directly evaluated on the evaluation set to obtain the classification accuracy, which serves as the fitness value of the corresponding individual.

Selection: After getting the fitness value of the newly generated candidate architecture, selection is conducted by comparing it with that of the current solution. The one with higher fitness value, i.e., classification accuracy on the evaluation set, will be kept to the next generation. Note that the updated individual takes effect immediately and can be used in evolving next individual.

4   Experiment Design

4.1   Datasets and Peer Competitors

To examine the effectiveness of the proposed algorithm, four widely used image classification datasets are selected, i.e., the MNIST Basic (MB) [15], the MNIST with Rotated Digits plus Background Images (MRDBI) [15], Fashion-MNIST [30] and CIFAR-10 [14]. Both MB and MRDBI contain 12,000 training samples and 50,000 test samples of handwritten digits from 0–9. However, the classification task on MRDBI is more challenging than that of MB due to the more complicated background. The Fashion-MNIST dataset has 55,000 training and 10,000 test images of 10 fashion products. Each instance in the above mentioned three datasets is of size $28 \times 28 \times 1$. The CIFAR-10 dataset consists of 50,000 training and 10,000 test color images of 10 objects, such as horse, ship, and truck, and the spatial size of each image is 32×32.

  A number of state-of-the-art methods are chosen for comparison, including the manually designed models as well as NAS methods. Specifically, on MB and MRDBI, CAE-2 [21], PCANet-2 [4], EvoCNN [25], EF-ENAS [22], IPPSO [28], psoCNN [12], DECNN [29], and FPSO [11] are selected as peer competitors; on Fashion-MNIST, VGG16 [23], GoogleNet [27], MobileNet [9], EvoCNN [25], EF-ENAS [22], psoCNN [12], and FPSO [11] are selected as peer competitors; on CIFAR-10, VGG16 [23], ResNet-110 [8], GeNet [31], LS-Evolution [19], and FPSO [11] are selected for comparison.

4.2   Parameter Settings

Most parameter settings in this work follow the community conventions [7] and previous works [11,28,29]. Specifically, the population size T is set to 30, and the number of generations G is 20. The length l of an individual is fixed to 30 with up to 4 strided layers. For each convolutional layer, the maximal number of fiters is 128, and the threshold gap thd is restricted to 30 when evolving $part2$. During the evolutionary process, we set the differential rate F to 0.4 and the crossover rate CR to 0.5. We randomly extract 10% of the training set as the evaluation set, and the test set is never involved during the architecture search process. An SGD optimizer with a cosine annealing learning rate scheduler is employed for early stopping training during the search process. When the search process terminates, the best architecture is fully trained on training set using SGD optimizer with multi-step learning rate decay scheduler, and the accuracy of the trained model on test set is reported. Experiment on each dataset is run 10 times independently, and the best and average results are collected respectively.

5   Experimental Results and Discussions

5.1   Overall Results

In this subsection, the experimental results and discussions of the proposed algorithm on four datasets are provided. First of all, the classification error rates (%) of EDE-NAS and the selected compared methods on MB and MRDBI are shown in Table 2. Then, in Tables 3 and 4, the proposed method is compared with peer competitors in terms of classification performance and number of parameters (#parameters) on Fashion-MNIST and CIFAR-10, respectively.

Table 2 shows the classification performance of EDE-NAS on MB and MRDBI compared with the selected peer competitors. Note that the symbol "–" indicates that the result was not reported in the paper of the compared algorithm. It is shown that the proposed method is superior to most of the compared methods on both datasets, including the handcrafted models (CAE-2 and PCANet-2), the GA-based method (EvoCNN), PSO-based methods (IPPSO and psoCNN), and DE-based method (DECNN). In particular, on MRDBI, EDE-NAS achieves significantly better performance than DECNN with an accuracy advantage of over 25%. When compared with FPSO on the MB dataset, both the best and average performance of the proposed method is slightly worse than that of FPSO. It turns out that FPSO's powerful local search capability helps it find a better architecture on such a simple dataset. On the other hand, on the MRDBI dataset, while the best architecture found by FPSO outperforms that of EDE-NAS, the average result of all searched architectures is worse than that of EDE-NAS.

Table 3 exhibits the classification error rates of the proposed method and the peer competitors on the Fashion-MNIST dataset. Apparently, the best architecture searched by EDE-NAS outperforms almost all the compared methods (except for EF-ENAS) in terms of the classification error rate. In addition, the mean error of the proposed method is very low among the peer competitors with a relatively small number of parameters, demonstrating the effectiveness and stability of EDE-NAS for architecture search on Fashion-MNIST.

With promising results on the previous three small-scale datasets, the proposed method is further conducted on CIFAR-10, to examine its effectiveness on the medium-scale dataset. We report the best result of ten independent runs. As shown in Table 4, the resulting architecture is slightly better than the manually designed models, i.e., VGG16 and ResNet-110, and is very competitive compared with the automatically searched architectures, i.e., LS-Evolution and

Table 2. The classification error rates (%) of EDE-NAS with the peer competitors on MB and MRDBI

Model	Error rate (%)	
	MB	MRDBI
CAE-2	2.48	45.23
PCANet-2	1.4	35.86
EvoCNN (best)	1.18	35.03
EvoCNN (mean)	1.28 (0.15)	37.38 (1.75)
EF-ENAS (best)	–	10.39
EF-ENAS (mean)	–	12.27 (1.69)
IPPSO (best)	1.13	33
IPPSO (mean)	1.21 (0.1)	34.5 (2.96)
psoCNN (best)	–	14.28
psoCNN (mean)	–	20.98
DECNN (best)	1.03	32.85
DECNN (mean)	1.46 (0.11)	37.55 (2.45)
FPSO (best)	0.96	10.17
FPSO (mean)	1.08 (0.06)	11.91 (0.79)
EDE-NAS (best)	0.99	10.97
EDE-NAS (mean)	1.10 (0.06)	11.84 (0.82)

Table 3. The performance of EDE-NAS with the peer competitors on fashion-MNIST

Model	Error rate (%)	#parameters (M)
VGG16	6.5	26
GoogleNet	6.3	23
MobileNet	5	4
EvoCNN (best)	5.47	6.68
EvoCNN (mean)	7.28 (1.69)	6.52
EF-ENAS (best)	4.66	3.31
EF-ENAS (mean)	5.27 (0.36)	4.80
psoCNN (best)	5.53	2.32
psoCNN (mean)	5.90	2.5
FPSO (best)	4.93	0.53
FPSO (mean)	5.22 (0.14)	0.61
EDE-NAS (best)	4.82	0.77
EDE-NAS (mean)	5.16 (0.21)	0.73

Table 4. The performance of EDE-NAS with the peer competitors on CIFAR-10

Model	Error rate (%)	#parameters (M)
VGG16	6.66	20.04
ResNet-110	6.43	1.7
GeNet	7.10	–
LS-Evolution	5.40	5.4
FPSO	6.28	0.70
EDE-NAS	6.04	1.15

FPSO. This also demonstrates that even with a simple building blocks and backbone structure, the proposed approach can still design architectures that perform very well.

5.2 Convergence Analysis

In this subsection, we conduct a convergence analysis of EDE-NAS on the four datasets by visualizing the change of the number of newly generated individuals in each generation, as shown in Fig. 4. Note that the statistics start from the second generation because of the first generation is a random initialization. In EDE-NAS, an offspring individual will survive in the next generation only if it is superior to the current individual. If there is a large proportion of new individuals over the entire population, it means the search algorithm is running actively and finding more and better architectures in the current generation. On the contrary, if only very few new offspring individuals are retained in the population, it reflects that in the current generation, the search algorithm is close to convergence. It is shown in Fig. 4 that on the four datasets, the number of new individuals in the population is high in the early stage and decreases gradually as the search proceeds. In the last generations, new individuals make up a very small percentage of the population, indicating that population update is coming to a halt, in which situation the search algorithm has largely found what it considers the best architecture. We do not exclude that by continuing the search process, the population can find a better architecture through a slow updating process. However, based on this observation and the experimental results, we believe that the current setup is efficient and sufficiently effective.

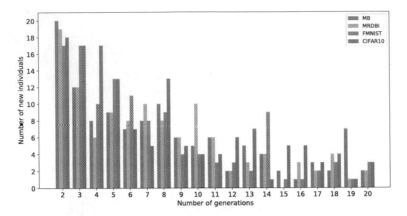

Fig. 4. Convergence analysis.

6   Conclusions

This paper introduced EDE-NAS, an eclectic DE-based approach for single-path CNN architecture design. The involved architectural parameters are effectively split and encoded into two parts based on their range size. Two mutation operators are developed for evolving the two parts, respectively, by utilizing the elitism as well as the random solutions from the population to balance the local and global search. The target network architecture is built on an efficient single-path backbone and evolved in a variable-length way. Experiments on four image classification datasets demonstrate the promising performance of EDE-NAS, which also shows great potential of DE in NAS problem. Since the proposed method simply employs early stopping to speed up the fitness evaluation process which might not be very reliable, more effective and trustworthy acceleration methods are expected to be investigated in the future.

References

1. Awad, N., Mallik, N., Hutter, F.: Differential evolution for neural architecture search. arXiv preprint arXiv:2012.06400 (2020)
2. Bäck, T., Fogel, D.B., Michalewicz, Z.: Handbook of evolutionary computation. Release **97**(1), B1 (1997)
3. Baymurzina, D., Golikov, E., Burtsev, M.: A review of neural architecture search. Neurocomputing **474**, 82–93 (2022). https://doi.org/10.1016/j.neucom.2021.12.014
4. Chan, T.H., Jia, K., Gao, S., Lu, J., Zeng, Z., Ma, Y.: PCANet: a simple deep learning baseline for image classification? IEEE Trans. Image Process. **24**(12), 5017–5032 (2015)
5. Ding, X., Zhang, X., Ma, N., Han, J., Ding, G., Sun, J.: RepVGG: making VGG-style convnets great again. In: IEEE Conference on Computer Vision and Pattern Recognition (CVPR), pp. 13733–13742 (2021)

6. Elsken, T., Metzen, J.H., Hutter, F.: Neural architecture search: a survey. J. Mach. Learn. Res. **20**(1), 1997–2017 (2019)
7. Gämperle, R., Müller, S.D., Koumoutsakos, P.: A parameter study for differential evolution. Adv. Intell. Syst. Fuzzy Syst. Evolut. Comput. **10**(10), 293–298 (2002)
8. He, K., Zhang, X., Ren, S., Sun, J.: Deep residual learning for image recognition. In: IEEE Conference on Computer Vision and Pattern Recognition (CVPR), pp. 770–778 (2016)
9. Howard, A.G., et al.: MobileNets: efficient convolutional neural networks for mobile vision applications. arXiv preprint arXiv:1704.04861 (2017)
10. Huang, G., Liu, Z., van der Maaten, L., Weinberger, K.Q.: Densely connected convolutional networks. In: IEEE Conference on Computer Vision and Pattern Recognition (CVPR), Honolulu, USA, pp. 2261–2269 (2017)
11. Huang, J., Xue, B., Sun, Y., Zhang, M.: A flexible variable-length particle swarm optimization approach to convolutional neural network architecture design. In: IEEE Congress on Evolutionary Computation (CEC), pp. 934–941 (2021). https://doi.org/10.1109/CEC45853.2021.9504716
12. Junior, F.E.F., Yen, G.G.: Particle swarm optimization of deep neural networks architectures for image classification. Swarm Evolut. Comput. **49**, 62–74 (2019)
13. Kennedy, J., Eberhart, R.: Particle swarm optimization. In: International Conference on Neural Networks, vol. 4, pp. 1942–1948 (1995). https://doi.org/10.1109/ICNN.1995.488968
14. Krizhevsky, A., Hinton, G., et al.: Learning multiple layers of features from tiny images. University of Toronto (2009)
15. Larochelle, H., Erhan, D., Courville, A., Bergstra, J., Bengio, Y.: An empirical evaluation of deep architectures on problems with many factors of variation. In: Proceedings of the 24th International Conference on Machine Learning, pp. 473–480 (2007)
16. Liu, Y., et al.: A survey on evolutionary neural architecture search. IEEE Trans. Neural Netw. Learn. Syst. (Early Access), 1–21 (2021). https://doi.org/10.1109/TNNLS.2021.3100554
17. Mi, J.X., Feng, J., Huang, K.Y.: Designing efficient convolutional neural network structure: a survey. Neurocomputing **489**, 139–156 (2022)
18. Mitchell, M.: An Introduction to Genetic Algorithms. MIT Press, Cambridge (1998)
19. Real, E., et al.: Large-scale evolution of image classifiers. In: International Conference on Machine Learning (ICML), pp. 2902–2911 (2017)
20. Ruder, S.: An overview of gradient descent optimization algorithms. arXiv preprint arXiv:1609.04747 (2016)
21. Salah, R., Vincent, P., Muller, X., et al.: Contractive auto-encoders: explicit invariance during feature extraction. In: International Conference on Machine Learning (ICML), pp. 833–840 (2011)
22. Shang, R., Zhu, S., Ren, J., Liu, H., Jiao, L.: Evolutionary neural architecture search based on evaluation correction and functional units. Knowl. Based Syst., 109206 (2022)
23. Simonyan, K., Zisserman, A.: Very deep convolutional networks for large-scale image recognition. In: International Conference on Learning Representations (ICLR) (2015)
24. Storn, R., Price, K.: Differential evolution - a simple and efficient heuristic for global optimization over continuous spaces. J. Glob. Optim. **11**(4), 341–359 (1997)
25. Sun, Y., Xue, B., Zhang, M., Yen, G.G.: Evolving deep convolutional neural networks for image classification. IEEE Trans. Evolut. Comput. **24**(2), 394–407 (2019)

26. Sutton, R.S., Barto, A.G.: Reinforcement Learning: An Introduction. MIT Press, Cambridge (2018)
27. Szegedy, C., et al.: Going deeper with convolutions. In: IEEE Conference on Computer Vision and Pattern Recognition (CVPR), pp. 1–9 (2015)
28. Wang, B., Sun, Y., Xue, B., Zhang, M.: Evolving deep convolutional neural networks by variable-length particle swarm optimization for image classification. In: IEEE Congress on Evolutionary Computation (CEC), pp. 1–8 (2018). https://doi.org/10.1109/CEC.2018.8477735
29. Wang, B., Sun, Y., Xue, B., Zhang, M.: A hybrid differential evolution approach to designing deep convolutional neural networks for image classification. In: Mitrovic, T., Xue, B., Li, X. (eds.) AI 2018. LNCS (LNAI), vol. 11320, pp. 237–250. Springer, Cham (2018). https://doi.org/10.1007/978-3-030-03991-2_24
30. Xiao, H., Rasul, K., Vollgraf, R.: Fashion-MNIST: a novel image dataset for benchmarking machine learning algorithms. arXiv preprint arXiv:1708.07747 (2017)
31. Xie, L., Yuille, A.: Genetic CNN. In: IEEE International Conference on Computer Vision (ICCV), pp. 1379–1388 (2017)

Impact of Mathematical Norms on Convergence of Gradient Descent Algorithms for Deep Neural Networks Learning

Linzhe Cai[1]([✉])[iD], Xinghuo Yu[1][iD], Chaojie Li[2][iD], Andrew Eberhard[1][iD], Lien Thuy Nguyen[1][iD], and Chuong Thai Doan[1][iD]

[1] School of Engineering, RMIT University, Melbourne, VIC 3000, Australia
s3548838@student.rmit.edu.au
[2] School of Electrical Engineering and Telecommunications, University of New South Wales, Sydney, NSW 2052, Australia

Abstract. To improve the performance of gradient descent learning algorithms, the impact of different types of norms is studied for deep neural network training. The performance of different norm types used on both finite-time and fixed-time convergence algorithms are compared. The accuracy of the multiclassification task realized by three typical algorithms using different types of norms is given, and the improvement of Jorge's finite time algorithm with momentum or Nesterov accelerated gradient is also studied. Numerical experiments show that the infinity norm can provide better performance in finite time gradient descent algorithms and give strong robustness under different network structures.

Keywords: Infinity norm · Finite-time convergence · Norms equivalence · Deep neural network

1 Introduction

For a machine learning model, increasing the model complexity can effectively improve the learning ability. For models like neural networks, there are two obvious ways to increase complexity, one is to make the model wider and the other is to make the model deeper [1]. Shallow networks require exponentially increasing the number of units to achieve the same computational results compared with deep networks. Additionally, shallow networks need a good feature extractor that solves the selectivity-invariance dilemma [2], which can be avoided automatically when a deeper structure instead. From the perspective of topology, the transformation of a high-dimensional space by multiple activation functions

---

The authors were supported by the Australian Research Council (ARC) under Discovery Program Grant DP200101197.

makes the multi-classification problem linearly separable [3], thus the study of deep learning attracts more attention.

With the development of gradient descent-based algorithms, stochastic gradient descent (SGD) [4] provides a trade-off between accuracy and speed by modifying the size of the batch, while momentum [5] can help to meet dampen oscillation requirements by considering past velocity when updating. Nesterov accelerated gradient (NAG) [6] can further speed up the process by effectively looking ahead, the gradient of parameters in which with respect to the approximate future position instead of the current one. Other than modifying the direction, Adagrad [7] adapts the learning rate to parameters based on past gradients, reducing the learning rate when approaching the optimum. RMSprop [8] modifies the learning rate through dividing by an exponentially decaying average, solving the dramatically dropping problem. Adam [9] keeps both the adaptive learning rate like RMSprop and the direction adjustment like Momentum. However, most of them can only have asymptotic convergence, which means they cannot complete their learning within a reasonable time.

To solve the problem mentioned above, Recently, a series of algorithms appear to guarantee finite time convergence. Among them, Jorge first provides a kind of finite-time convergent learning algorithm, in particular, gradient flow (continuous gradient descent) through the gradient over the Euclidean distance (L_2 norm) of vectors [10]. After that, Wibisono gives a variant of which by adding a fraction on the Euclidean distance (q rescaled gradient flow) [11]. Besides, Romero and Benosman prove that it is indeed finite-time convergent [12]. Additionally, Garg proposes a fixed-time convergence algorithm that essentially splits the q-RGD into two parts [13]. Although a growing body of research has access to the mathematical norm on convergence, most of them only consider the Euclidean distance (L_2 norm) when rescaling the gradient flow. There is no study focusing on the effect of different types of norms with respect to convergence performance to the best of our knowledge.

This paper aims to study the impact of mathematical norms on the convergence of gradient flow for deep neural networks. Section 2 provides a review of different types of norms, the equivalence of norms, and convergence property. Section 3 gives numerical applications comparing different norms used on specific algorithms, and the potential improvement after involving momentum or NAG methods. Section 4 concludes.

2   Main Results

In this section, we first review the definition of mathematical norms and the most popular used norm types in Sect. 2.1, then give the equivalence of norms as well as the convergence property in Sect. 2.2. The qualitative analysis of different norms based on the expression of algorithms is given in Sect. 2.3, and the related works we used to compare in Sect. 3 are concluded in Sect. 2.4.

2.1 Mathematical Norms

Mathematically, a norm is a function from a vector space to the real numbers describing the distance from the origin, which is an abstract generalization of length [14]. According to the definition, a norm on a vector space $\mathbb{R}^n$ is a real-valued function $\|\cdot\| : \mathbb{R}^n \to \mathbb{R}$ that meets the following properties [15]:

- Triangle Inequality: $\|x + y\| \leq \|x\| + \|y\|$ for all $x, y \in \mathbb{R}^n$.
- Absolute Homogeneity: $\|sx\| = |s| \, \|x\|$ for all $x \in \mathbb{R}^n$ and all scalars s.
- Positive Definiteness: for all $x \in \mathbb{R}^n$, if $\|x\| = 0$, then $x = 0$.

There are some typical types of norms given as follows [16]:

- L_1 norm (Taxicab norm): $\|x\|_1 := \sum_{i=1}^n |x_i|$.
- L_2 norm (Euclidean norm): $\|x\|_2 := \sqrt{x_1^2 + x_2^2 + ... + x_n^2}$.

Both L_1 and L_2 norms are usually used as a regularization term to penalize large weights during logistic regression against the overfitting issue. While L_1 regularization penalizes the sum of the absolute values, L_2 regularization encourages the sum of the square of parameters to be small [17].

- L_p norm $(p \geq 1)$: $\|x\|_p := (\sum_{i=1}^n |x_i|^p)^{1/p}$.

According to [[18], Theorem 3.5.4], L_p is a norm for $1 \leq p < \infty$. However, it will becomes a pseudo-norm for $0 < p < 1$, as it violates the triangle inequality property.

- L_∞ (Infinity Norm): $\|x\|_\infty := \max_i |x_i|$.

The infinity norm is essential for the limit of the L_p norm for $p \to \infty$. According to the expression of the L_p norm, we can figure out that the computation burden increase with the increase of the subscript of the norm symbol. However, after the functional limit operation, the computation of infinity norm as shown in the L_∞ norm only needs to iterate over through vector space once.

2.2 Equivalence of Norms

We recall from [[19], Definition 1.3] that two norms $\|\cdot\|_\alpha$ and $\|\cdot\|_\beta$ on a vector space $\mathbb{R}^n$ are called equivalent if and only if there exist positive real numbers C and D such that for all $x \in \mathbb{R}^n$:

$$C \, \|x\|_\alpha \leq \|x\|_\beta \leq D \, \|x\|_\alpha . \tag{1}$$

A more precise relationship between different norms is obtained through Cauchy-Schwarz inequality and Hoder's inequality: for $p > r > 1$ on $\mathbb{R}^n$ [20], we have

$$\|x\|_p \leq \|x\|_r \leq n^{1/r - 1/p} \, \|x\|_p . \tag{2}$$

In particular,
$$\|x\|_2 \leq \|x\|_1 \leq \sqrt{n}\,\|x\|_2\,,$$
$$\|x\|_\infty \leq \|x\|_2 \leq \sqrt{n}\,\|x\|_\infty\,. \tag{3}$$

According to [[19], Appendix A], the open subset of vector space $\mathbb{R}^n$ defined by equivalent norms are the same, and the convergent sequences and their limits in $\mathbb{R}^n$ defined by equivalent norms are the same. Similar statements are given in [21]: two finite-dimensional linear normed spaces with the same dimension are algebraically isomorphic and topologically homeomorphic. Thus, the convergence property is unchanged no matter what type of norm instead compared with the original algorithm under the Euclidean distance.

2.3 Different Norms Applications

As mentioned in Sect. 1, Jorge [10] and Wibisono [11] proposes finite-time convergence algorithms, Garg [13] provides a fixed-time convergence algorithm, and all of which are gradient flows involving the Euclidean norm (L_2 norm).

Jorge's finite-time convergence algorithm:

$$\frac{dw}{dt} = -\frac{\nabla_w J}{\|\nabla_w J\|_2}. \tag{4}$$

Wibisono's finite-time convergence algorithm:

$$\frac{dw}{dt} = -\zeta \frac{\nabla_w J}{\|\nabla_w J\|_2^{\frac{q-2}{q-1}}}, \tag{5}$$

where $q > 2$.

Garg's fixed-time convergence algorithm:

$$\frac{dw}{dt} = -C_1 \frac{\nabla_w J}{\|\nabla_w J\|_2^{\frac{p_1-2}{p_1-1}}} - C_2 \frac{\nabla_w J}{\|\nabla_w J\|_2^{\frac{p_2-2}{p_2-1}}}, \tag{6}$$

where $p_1 > 2$ and $1 < p_2 < 2$.

As all components in the vector share the same denominator, the relative size among different components in the vector has not changed. Thus, the backpropagation mechanism still works as the core of gradient descent is to figure out which component changes matter more.

According to the expressions of Eqs. (4), (5), and (6), all norms appear in the denominator of gradient flow and only have magnitude but without any direction. Thus, the potential step size when iteration operation will be inversely proportional to the relationship of the magnitude of different norms.

According to Eq. (3), the infinity norm obtains the smallest magnitude among all kinds of norms, which means it provides the largest potential step size after involving gradient flow. We can easily change the norm type in all these algorithms without changing the convergence property considering the convergence property discussed in Sect. 2.2.

2.4 Related Works

This section gives a brief review of the two most popular algorithms, stochastic gradient descent (SGD) [4] and Adam [9], which will be used as the benchmark for the first case study. Two direction adjustment methods, momentum [5] and Nesterov accelerated gradient (NAG) [6], are also introduced, which will be used to analyze the potential improvement for the second case study.

SGD [4] is an iterative approximation method calculated from a randomly selected subset ($50 < n < 256$) achieving faster iterations in trade for a lower convergence rate:

$$\theta = \theta - \eta \cdot \nabla_w J(\theta; x^{i:i+n}; y^{i:i+n}). \tag{7}$$

Momentum [5] accelerates SGD by adding a fraction ($\gamma < 1$) of the previous update vector, which obvious effect dampens oscillation when the gradient in one direction is larger than in others:

$$\begin{aligned} v_t &= \gamma \cdot v_{t-1} + \eta \cdot \nabla_w J, \\ \theta &= \theta - v_t. \end{aligned} \tag{8}$$

Nesterov [6] modifies the momentum one by calculating the gradient with respect to the estimated future position (moved by $\gamma \cdot v_{t-1}$) instead of the current one. The so-called look ahead essential considers the second derivative information of objective function:

$$\begin{aligned} v_t &= \gamma \cdot v_{t-1} + \eta \cdot \nabla_w J(\theta - v_t), \\ \theta &= \theta - v_t. \end{aligned} \tag{9}$$

Adam [9] computes adaptive learning rates, storing an exponentially decaying average (β_2) of past squared gradients and keeping another similar hyperparameter (β_1) on past gradients themselves, which is commonly considered fairly robust to hyperparameter selection:

$$\begin{aligned} m_t &= \beta_1 \cdot m_{t-1} + (1 - \beta_1) \cdot \nabla_w J, \\ v_t &= \beta_2 \cdot v_{t-1} + (1 - \beta_2) \cdot \nabla_w J \odot \nabla_w J. \end{aligned} \tag{10}$$

Additionally, bias correction is considered to offset the shift to the initial value at the beginning of the iteration:

$$\begin{aligned} \hat{m}_t &= \frac{m_t}{1 - \beta_1^t}, \\ \hat{v}_t &= \frac{v_t}{1 - \beta_2^t}, \end{aligned} \tag{11}$$

thus,

$$\theta_{t+1} = \theta_t - \frac{\eta}{\sqrt{\hat{v}_t} + \epsilon} \cdot \hat{m}_t. \tag{12}$$

where ϵ is a smoothing term that avoids singularity.

3   Case Studies

As the Stanford vision and learning lab [22] summarizes that two recommended updates to use for CNN learning in visual recognition are either SGD with Nesterov momentum or Adam. Two case studies with quantitative analysis apply to comparing their performance with different norm-based finite-time algorithms. Section 3.1 compares the accuracy of different norms used in three typical algorithms under ResNet50 architecture, while Sect. 3.2 shows the improvement of Jorge's finite-time involving momentum and Nesterov accelerated gradient (NAG) under a six-layers convolutional network.

3.1   Three Typical Algorithms Using Different Types of Norms

As one of the most popular used image classification databases, CIFAR100 [23] is considered as an example, and the 50-layer ResNet learning framework [24] is introduced to complete the task. We compare the performance of algorithms mentioned in Sect. 2.3 using different types of norms (L_1, L_2, L_3, and L_∞) in each respectively. SGD [4] and Adam [9] are attached as the benchmark. The η and n for SGD in Eq. (7) are fixed at 0.01 and 128, while the η, β_1, β_2, and ϵ for Adam in Eqs. (10), (11), and (12) are 0.001, 0.9, 0.999, and 10^{-8} respectively.

Figure 1 gives the average value of training and testing accuracy of the CIFAR100 database under ResNet50 architecture using three different algorithms within different types of norms. According to Fig. 1a and 1b, Jorge's finite-time algorithm [10] using the L_1 norm instead almost cannot converge under the same step size, while L_2 norm one obtains a reasonable convergent speed. Infinity norm one can obtain the best performance. When focusing on the training result given in Fig. 1a, the performance of SGD and Adam are between L_2 norm one and L_3 norm, while for the testing result given in Fig. 1b, Adam can beat the L_3 norm after 30 epochs. Additionally, the advantage from infinity one to L_3 norm is more obvious in testing accuracy compared with the training one.

Figures 1c and 1d give the accuracy using Wibisono's finite time algorithm [11] within different types of norms, and the q given in Eq. (5) is chosen as 6. As we aim to figure out the improvement from the L_2 norm to the infinity one, only the consistency of the parameter chosen before and after changing the norm types is necessary, while the parameters are not necessarily obtained the best performance. When focusing on the training result given in Fig. 1c, the difference accuracy between L_2 norm one and Adam is relatively smaller compared with Jorge's one Fig. 1a, while the difference accuracy between L_3 norm and infinity one is relatively larger. However, the improvement from L_2 norm to infinity one in Wibisono's finite-time is less significant compared with Jorge's one.

Figures 1e and 1f give the performance of the CIFAR100 database under ResNet50 architecture using Garg's fixed-time algorithm [13] within different types of norms, and the p_1 and p_2 given in Eq. (6) are chosen as 3 and 1.5 respectively. Again, we only maintain the parameter consistency before and after changing the norm types but do not necessarily choose the optimal value. When

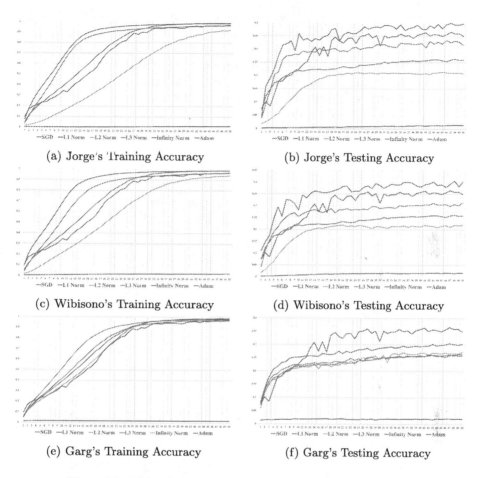

(a) Jorge's Training Accuracy

(b) Jorge's Testing Accuracy

(c) Wibisono's Training Accuracy

(d) Wibisono's Testing Accuracy

(e) Garg's Training Accuracy

(f) Garg's Testing Accuracy

Fig. 1. ResNet50 CIFAR100 performance under different norms

Table 1. Improvement from L_2 norm to infinity one for different algorithms

	Statistics	Jorge	Wibisono	Garg	Jorge-v	Wibisono-v	Garg-v
$L_\infty - L_2$	Max	0.6109	0.5492	0.2441	0.2372	0.2039	0.0562
	Min	0.0596	0.0460	−0.0032	0.0274	0.0920	0.0179
	Median	0.3154	0.2383	0.0269	0.1732	0.1687	0.0338
	Mean	0.3209	0.2615	0.0739	0.1721	0.1636	0.0353
$\frac{L_\infty - L_2}{L_2}$	Max	721.43%	695.65%	51.29%	533.73%	528.45%	39.84%
	Min	6.87%	4.88%	−1.14%	61.81%	61.46%	7.65%
	Median	77.92%	46.96%	4.18%	86.99%	82.24%	15.73%
	Mean	169.30%	119.93%	16.64%	126.54%	108.38%	16.18%

focusing on the training result given in Fig. 1e, the performance of the L_2 norm can almost beat the Adam one, while Fig. 1f shows that the improvement from the L_2 norm to infinity one in Garg's fixed-time algorithm is limited. Specifically, the performance of Garg's fixed-time algorithm under almost all types of norms (excluding the L_1 one) is between SGD and Adam after 20 epochs, which means that the improvement room from L_2 norm to infinity one in Garg's fixed-time algorithm is further compressed.

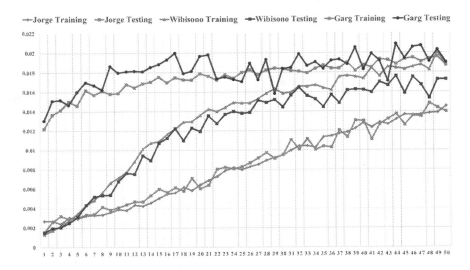

Fig. 2. Accuracy of CIFAR100 database using different algorithms under L_1 norm

According to Fig. 1, L_1 norm-based iteration provides the worst performance for the lowest accuracy for all three algorithms. To indicate they are convergent slower instead of cannot obtain the convergence property, Fig. 2 extracts the accuracy of different algorithms using the L_1 norm from Figure 1. The darker chroma of the same color represents the accuracy difference from training to testing. According to Fig. 2, all three algorithms can converge but with a slower step size, and the difference between the training database and the testing one is relatively small. Although Garg's fixed-time algorithm has a slightly higher original accuracy, it provides the least improvement from the perspective of absolute numerical.

Table 1 gives statistical data on the differences from L_2 norm to infinity one at corresponding iteration times for different algorithms respectively. The absolute values indicate the absolute accuracy improvement from L_2 to infinity norm, while the relative values indicate the absolute differences over corresponding L_2 norm accuracy (percentage improvement). A positive value means that the accuracy of the infinity norm used on the algorithm is higher than Euclidean one, while the negative value implies that the L_2 norm may have higher accuracy at a specific iteration time. The algorithms without -v in the column express the training accuracy, while the algorithms with -v in the column express the validation accuracy (testing datasets).

According to Table 1, the infinity norm used in all algorithms can have an improvement from its original one (L_2 norm). Among them, Garg's fixed-time algorithm provides the smallest improvement, and Jorge's finite time algorithm has slightly more improvement than Wibisono's. Additionally, for a specific algorithm, the improvement in training datasets is always more obvious than which in testing one. The maximum improvement (7 times for Jorge's and Wibisono's training and (7 times for Jorge's and Wibisono's testing) usually appear in the first few iteration times, while the minimum differences (negative for Garg's training accuracy) arise at the latest few steps.

In summary, the effects on different types of norms are obvious for Jorge's finite-time algorithm, and the performance of which using infinity norm can surpass SGD and Adam for training and testing accuracy during the overall process. Although Wibisono's finite-time algorithm with infinity norm also has similar accuracy, the dependency on the parameter chosen weakens its advantage.

3.2   Jorge's Finite-Time Algorithm with Momentum and Nesterov

According to the brief review of gradient descent-based optimization given in Sect. 2.4, there are two mainstreams to refine an algorithm, namely iteration direction (eg. Momentum [5]) and adaptive learning rate (eg. RMSprop [8]). The finite time algorithm is essential one type of rescaled gradient flow, which means the improvement from the perspective of adaptive learning rate is already obtained. Thus, we are interested in whether the direction modification can further improve learning performance.

The second case study focuses on the improvement of Momentum [5] and Nesterov accelerated gradient (NAG) [6] methods used on finite time algorithms with different types of norms. The network structure considered in the case study is a six convolutional layers CNN (filter numbers 32, 32, 64, 64, 128, and 128) with batch normalization and dropout layers attached. To reduce the puzzle caused by the parameter chosen, Jorge's finite-time algorithm is considered as an example.

Figure 3 gives the testing accuracy of the CIFAR100 database using the L_2 norm and L_3 norm with different fractions of momentum (γ in Eq. (8)) or Nesterov (γ in Eq. (9)) accelerate respectively. SGD is still considered as a benchmark. As for the infinity norm, the different performance among different fractions is too small to illustrate, more statistical details will give in Table 3.

According to Fig. 3, the difference between momentum (Figs. 3a, 3c, and 3e) and corresponding Nesterov (Figs. 3b, 3d, and 3f) under the same fraction value is not obvious. Besides, the improvement after involving momentum and Nesterov is outstanding on the L_2 norm-based Jorge's finite-time algorithm as seen in Figs. 3c and 3d. However, the accelerated effect is reduced on the L_3 norm-based one as seen in Figs. 3e and 3f. The enhancement of SGD after adding momentum is between the L_2 norm and the L_3 norm.

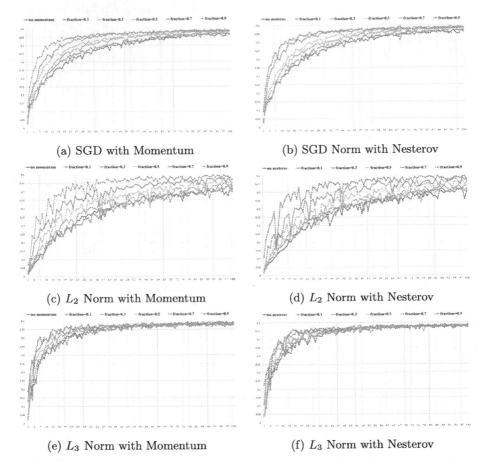

(a) SGD with Momentum   (b) SGD Norm with Nesterov

(c) L_2 Norm with Momentum   (d) L_2 Norm with Nesterov

(e) L_3 Norm with Momentum   (f) L_3 Norm with Nesterov

Fig. 3. CIFAR100 testing accuracy under six convolutional layer structure

Table 2. Improvement of momentum and Nesterov for different types of norms

	Statistic	SGD-M	SGD-N	L_2-M	L_2-N	L_3-M	L_3-N
$f_{0.9} - f_0$	Max	0.2216	0.2085	0.2772	0.2993	0.1634	0.1977
	Min	0.0113	0.0106	0.0567	0.0563	-0.0172	-0.0690
	Median	0.0646	0.0571	0.1349	0.1356	0.0071	0.0134
	Mean	0.0802	0.0773	0.1406	0.1499	0.0270	0.0277
$\frac{f_{0.9}-f_0}{f_0}$	Max	248.24%	154.20%	413.87%	534.09%	540.54%	428.37%
	Min	1.98%	1.86%	11.26%	11.01%	-2.97%	-25.89%
	Median	12.64%	11.48%	31.25%	32.02%	1.27%	2.38%
	Mean	24.71%	23.11%	56.46%	62.38%	12.61%	12.03%

Table 2 concludes the improvement from no momentum to 0.9 fractions (best performance under all subfigures) of both absolute and relative values for SGD, L_2 norm, and L_3 norm respectively. The algorithms with M in the column indicate adding momentum term, and the algorithms with N in the column indicate adding Nesterov term.

When we look at the absolute difference, while the improvement of the L_2 norm is almost double compared with which in SGD, the mean value of SGD is triple compared with the L_3 norm on average and quadruple compared with which median value. As for the relative value, the average improvement of SGD is only double compared with the L_3 norm, but the median value difference between them is five times. Again, the maximum improvement usually appears in the first few iteration times, while the minimum differences arise at the latest few steps. As the extreme value has relatively greater contingency, the statistical significance of which is weakened.

Although SGD, L_2 norm, and L_3 norm obtain the best performance when fractions equal to 0.9, the same fraction gives the worse performance when it comes to the infinity norm, while other fractions almost have no effect on it (as seen in Table 3), which may be caused by the radical acceleration in the same direction, as the increased dimensions almost doubled.

Table 3. Effect of momentum and NAG for infinity norm with different fraction

	Values	Statistic	$f = 0.1$	$f = 0.3$	$f = 0.5$	$f = 0.7$	$f = 0.9$
M	$f_x - f_0$	Mean	0.0042	0.0006	0.0046	−0.0192	−0.0254
		Median	0.0015	0.0003	0.0051	−0.0213	−0.0205
	$\frac{f_x - f_0}{f_0}$	Mean	1.33%	0.51%	0.88%	−3.16%	−5.04%
		Median	0.27%	0.06%	0.91%	−3.72%	−3.57%
N	$f_x - f_0$	Mean	0.0033	0.0005	−0.0017	−0.0006	−0.0143
		Median	0.0051	0.0009	−0.0022	−0.0017	−0.0098
	$\frac{f_x - f_0}{f_0}$	Mean	0.46%	0.11%	−0.22%	0.03%	−3.05%
		Median	0.89%	0.15%	−0.38%	−0.29%	−1.72%

Table 3 concludes the influence of momentum and Nesterov for infinity norm gradient flow under different fractions, where f_x represents the fraction value (γ) in Eqs. (8) and (9).

Among all fraction choices, 0.1 obtain the best performance while the improvement is still limited. The influence is negligible when the fraction is chosen between 0.3 and 0.5. When the fraction comes to 0.7, momentum shows negative effects and even worst when 0.9 is chosen. NAG effectively mitigates the negative effect of momentum by looking ahead effectively. When close to the optimum value, the gradient current time should be smaller than the previous one, and there is reason to believe that it will continue to be smaller, which justifies the radical acceleration deduction.

To have a more intuitive comparison, Fig. 4 concludes the best performance of each norm from Fig. 3 (L_2 norm and L_3 norm with 0.9 fraction Nesterov, SGD with 0.9 fraction momentum, and infinity norm without fraction).

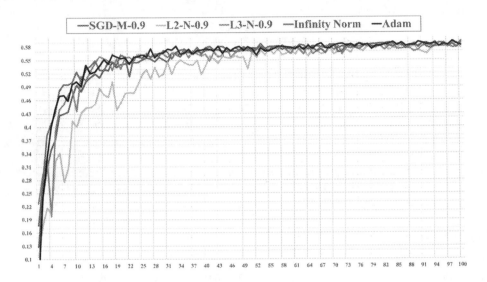

Fig. 4. Highest accuracy of CIFAR100 using different types of norms (Color figure online)

According to Fig. 4, the L_2 norm gives the worst performance (in yellow) while the infinity norm gives the best (in red), while the L_3 norm and the SGD are between the two mentioned above. Although the improvement of the L_2 norm after involving momentum is significant, it cannot surpass the infinity one, which can be imaged as an invisible ceiling existing (infinity norm gradient flow without momentum) no matter what type of norm choose. Thus, the less improvement of the L_3 norm given in Figs. 3e and 3f can be explained as the difference between the L_3 norm and the infinity one being small, thus there is no room for momentum and Nesterov to improve the performance. In other words, the better the performance without momentum or Nesterov acceleration, the less it can be improved through the dampens oscillation methods. Thus, the infinity norm used on Jorge's finite-time algorithm can cover the benefits of Momentum without introducing the updated velocity in the past time, which saves computing costs.

The performance of Adam is also plotted in Fig. 4. Although the accuracy of Adam (in black) and infinity norm gradient flow (in red) is similar, there are no hyperparameters that needed to be adjusted (Jorge's finite-time) related to the infinity norm gradient flow (INGF), and no memory requirement (no momentum or NAG need). Specifically, the average running time of Adam is 13.4% longer than which of the INGF (914.75 s and 806.62 s respectively) under the same GPU model (NVIDIA GeForce RTX 2080 Super with Max-Q Design under CUDA 10.0 support). From that perspective, INGF is superior to Adam which needs not only adaptive learning rates (computational burden) but also history records (memory burden).

4   Conclusion

In this paper, the comparison of different types of norms used in finite-time convergence algorithms is obtained. Qualitative analysis after the equivalence of norms with the help of convergence property verifies the convergence rate. The performance of three typical algorithms using different types of norms is quantitatively analyzed for image classification using the CIFAR100 database under the ResNet50 architecture. Jorge's finite-time algorithm gives the maximum improvement after changing the Euclidean norm to the infinity one. The improvement of Jorge's finite-time algorithm with momentum and Nesterov is studied. Although the better original performance, the less improvement after momentum or Nesterov acceleration involving, infinity norm gradient flow (INGF) without momentum still keeps overwhelming superiority. Although INGF can not always be superior to Adam in accuracy, no hyper-parameters adjustment and no memory requirement of INGF can keep its favorable position in time-consuming compared with Adam. According to the results given in case studies, we have reason to believe that Jorge's finite-time algorithm with infinity norm can provide reliable performance (higher accuracy and less time) for CNN learning tasks, especially visual recognition.

References

1. Goodfellow, I., Bengio, Y., Courville, A.: Deep Learning. MIT Press, Cambridge (2016)
2. LeCun, Y., Bengio, Y., Hinton, G.: Deep learning. Nature **521**(7553), 436–444 (2015)
3. Olah, C.: Neural networks, manifolds, and topology. Blog post (2014)
4. Bottou, L., Bousquet, O.: The tradeoffs of large scale learning. In: Advances in Neural Information Processing Systems 20 (2007)
5. Qian, N.: On the momentum term in gradient descent learning algorithms. Neural Netw. **12**(1), 145–151 (1999)
6. Nesterov, Y.: A method for unconstrained convex minimization problem with the rate of convergence o $(1/k^2)$. In: Doklady an ussr, vol. 269, pp. 543–547 (1983)
7. Duchi, J., Hazan, E., Singer, Y.: Adaptive subgradient methods for online learning and stochastic optimization. J. Mach. Learn. Res. **12**(7) (2011)
8. Tieleman, T., Hinton, G.: Neural networks for machine learning. Technical report (2011). http://www.cs.toronto.edu/tijmen/csc321/slides/lectureslideslec6.pdf
9. Kingma, D.P., Ba, J.: Adam: a method for stochastic optimization. arXiv preprint arXiv:1412.6980 (2014)
10. Cortés, J.: Finite-time convergent gradient flows with applications to network consensus. Automatica **42**(11), 1993–2000 (2006)
11. Wibisono, A., Wilson, A.C., Jordan, M.I.: A variational perspective on accelerated methods in optimization. Proc. Natl. Acad. Sci. **113**(47), E7351–E7358 (2016)

12. Romero, O., Benosman, M.: Finite-time convergence in continuous-time optimization. In: International Conference on Machine Learning, pp. 8200–8209. PMLR (2020)
13. Garg, K., Panagou, D.: Fixed-time stable gradient flows: applications to continuous-time optimization. IEEE Trans. Autom. Control **66**(5), 2002–2015 (2020)
14. Gradshteyn, I.S., Ryzhik, I.M.: Table of integrals, series, and products. Academic Press (2014)
15. Pugh, C.C.: Real Mathematical Analysis, vol. 2011. Springer, Cham (2002). https://doi.org/10.1007/978-0-387-21684-3
16. Weisstein, E.W.: Vector norm (2002). https://mathworld.wolfram.com/
17. Ng, A.Y.: Feature selection, l 1 vs. l 2 regularization, and rotational invariance. In: Proceedings of the Twenty-First International Conference on Machine learning, p. 78 (2004)
18. Wassermann, A.J.: Functional analysis (1999)
19. Conrad, K.: Equivalence of norms. In: Expository Paper, University of Connecticut, Storrs, heruntergeladen von, vol. 17, no. 2018 (2018)
20. Golub, G.H., Van Loan, C.F.: Matrix Computations. JHU Press, Baltimore (2013)
21. Gongqing, Z., Yuanqu, L.: Functional Analysis Lecture Notes. Peaking University Press (1990). (in Chinese)
22. Karpathy, A.: Cs231n convolutional neural networks for visual recognition (2017). cs231n.github.io. Dostopno na. http://cs231n.github.io
23. Krizhevsky, A., Hinton, G., et al.: Learning multiple layers of features from tiny images (2009)
24. He, K., Zhang, X., Ren, S., Sun, J.: Deep residual learning for image recognition. In: Proceedings of the IEEE Conference on Computer Vision and Pattern Recognition, pp. 770–778 (2016)

The Feasibility of Deep Counterfactual Regret Minimisation for Trading Card Games

David Adams[✉]

University of Western Australia, Perth, WA 6009, Australia
david.adams@uwa.edu.au

Abstract. Counterfactual Regret Minimisation (CFR) is the leading technique for approximating Nash Equilibria in imperfect information games. It was an integral part of Libratus, the first AI to beat professionals at Heads-up No-limit Texas-holdem Poker. However, current implementations of CFR rely on a tabular game representation and hand-crafted abstractions to reduce the state space, limiting their ability to scale to larger and more complex games. More recently, techniques such as Deep CFR (DCFR), Variance-Reduction Monte-carlo CFR (VR-MCCFR) and Double Neural CFR (DN-CFR) have been proposed to alleviate CFR's shortcomings by both learning the game state and reducing the overall computation through aggressive sampling. To properly test potential performance improvements, a class of game harder than Poker is required, especially considering current agents are already at superhuman levels. The trading card game Yu-Gi-Oh was selected as its game interactions are highly sophisticated, the overall state space is many orders of magnitude higher than Poker and there are existing simulator implementations. It also introduces the concept of a meta-strategy, where a player strategically chooses a specific set of cards from a large pool to play. Overall, this work seeks to evaluate whether newer CFR methods scale to harder games by comparing the relative performance of existing techniques such as regular CFR and Heuristic agents to the newer DCFR whilst also seeing if these agents can provide automated evaluation of meta-strategies.

Keywords: Artificial intelligence · Machine learning · Extensive-form games

1 Introduction

Attempting to solve problems of increasing complexity is one of the main goals of artificial intelligence (AI) research. Games are often used as a test bed for such research, as they provide a reasonable environment to evaluate but can also be applicable to the real world. Over time different techniques have been created to address different classes of games, starting with simple perfect-information (whole game state is known at all times) deterministic games like Tic-Tac-Toe, to massive imperfect-information (partial unknown game state) extensive form

ⓒ The Author(s), under exclusive license to Springer Nature Switzerland AG 2022
H. Aziz et al. (Eds.): AI 2022, LNAI 13728, pp. 145–160, 2022.
https://doi.org/10.1007/978-3-031-22695-3_11

games like Starcraft. In the case of perfect information games, Monte-Carlo Tree Search (MCTS) [10] and deep neural networks have been used in AIs such as Alpha Zero [21], which surpassed human levels of performance in Chess, Go, and Shogi. In the case of imperfect information games, a technique called Counterfactual Regret minimization (CFR) [6] was used in Libratus [9] to beat top professionals at Heads-up No-limit Texas Holdem Poker. Given that superhuman performance has been achieved at the hardest benchmark for imperfect information games, new harder games are needed to increase benchmarks for existing and future methods. Trading Card Games (TCGs) are a possible direction as, despite having a larger state space, more complex card interactions, and the concept of meta-strategies, they are still feasible to compute in comparison to massive online games like StarCraft or League of Legends, and can still be easily represented as a game tree. This paper will benchmark existing and new game solving methods, such as Deep Counterfactual Regret Minimisation (DCFR) [8], to see if they cope with the demands of more complex games like TCGs and assess whether these methods can evaluate different meta-strategies.

1.1 Foundational Work

In general, games are classified by the following properties:

- Zero-sum: overall reward sums to zero or there is some concept of a winner and loser.
- Information: whether the state is partially or fully known.
- Determinism: whether chance affects the game in any way.
- Sequential: whether actions occur one after another or simultaneously.
- Discrete: whether actions are applied in real time or not.

For simple deterministic perfect-information games with small state spaces, the whole game tree can be evaluated with the classical Minimax [4]. However, for most non-trivial games, an algorithm must decide what part of the game tree to explore. Perhaps the most widely used algorithm is MCTS [10] which was used in DeepMind's AlphaGo [22] to beat the 18-time Go world champion. Instead of hand-crafted game evaluation functions and state selection heuristics that would be required to make Minimax feasible, the algorithm used in AlphaGo used a deep neural network trained with self-play for state evaluation and a MCTS for state selection. Along with its successor AlphaZero [21] AlphaGo serves as the benchmark for perfect-information game playing performance.

Despite this excellent performance in perfect information games there are few real-world scenarios that have perfect information. In fact, most real-world situations, such as **business strategy**, **economic models** or **simple negotiation** can all be modelled as imperfect information games [15]. Whilst extensions can be made to perfect information games to make them imperfect, such as imperfect Chess [19], and simple games like Bridge are used as teaching tools, Poker is the canonical example of an imperfect information game.

Attempting to apply the Minimax algorithm or raw MCTS to Poker will lead to poor results as each game state has uncertainty and it is infeasible to enumerate all combinations. Instead, information sets (**infoset**) are used as a proxy for game state and represent the set of all possible states that could be known with the current information. Whilst MCTS can be modified to accommodate information sets, Regret Matching (or regret minimisation) has been shown to have better convergence and results in practise [24]

Intuitively, the action that you regret not taking the most is the one that should have used. A mathematical representation of regret is the difference between the reward of an action that was taken and the action that could have been taken

$$regret = \mu(\text{possible action}) - \mu(\text{action taken})$$

CFR [26] is an extension to regret matching. It deals with scenarios that have multiple steps and allows an agent to know what the regret of not taking an action is at each step. Instead of calculating the regret for an action, the regret is calculated based on a counterfactual value, which is the value of a state multiplied by the probability of reaching that state.

CFR was used to play the hardest variation of Poker (Heads up no limit Texas-holdem having approximately 10^{161} decision points) and successfully to beat top-level human players [9]. Computing a strategy for this game was obviously infeasible. As such, treating groups of scenarios as strategically identical was required. But, it came at the cost of fixing the implementation to a hand-crafted abstraction and a tabular representation. Overall, this means the original CFR techniques would not generalise well to other games, nor would they scale to extremely large games.

1.2 Current Methods

One of the first methods to deal with both tabular solving and abstraction was Deep CFR [8] (Deep CFR). It performed better than all the previously mentioned approaches and stands as one of the few algorithms that can tackle games whose state or action spaces are too large. It works by using a neural network to approximate, with theoretical convergence, the behaviour of CFR. The neural network architecture used can be seen in Fig. 1.

It is unique compared to the previous methods shown in that it does not calculate and accumulate regrets at each infoset, rather it generalises across similar infosets with the function approximation provided by deep neural networks. Unlike tabular CFR it does not require a hand crafted game abstraction and, as such, learns through self-play.

For each iteration, Deep CFR performs a constant number of partial traversals according to Monte-Carlo CFR [16]. At each infoset it plays its current strategy, which was determined by regret matching the output of the neural network. This neural network takes in information sets as input and has the goal of approximating the regret that tabular CFR would have produced. Like regular CFR, when a terminal node is reached, values are propagated back up the tree.

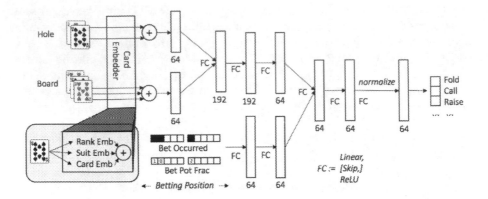

Fig. 1. Neural network architecture of Deep CFR as presented in [8]

These instantaneous regrets are sampled and stored in memory. Then, before the next iteration, a completely new network is trained to minimise the error between the predicted regrets and the samples of regret that have been stored in memory. Once this training is complete, the next iteration can begin.

Despite scaling better than tabular CFR, DCFR is not perfectly scaleable either. The sampling strategies used are simplistic, and introducing more sophisticated methods would likely result in high variance between sampled payoffs. Extensions to Deep CFR such as Variance Reduction Monte-Carlo CFR (VR-MCCFR) [20] and Double Neural CFR (DNCFR) [18] represent the state of the art in solving massive imperfect information games.

Specifically, VR-MCCFR takes the per-iteration estimated value updates of a MCTS and reformulates them as a function of sampled values and state-action baselines whilst still being unbiased. It should be noted that plain Monte-Carlo CFR (also known as chance sampled CFR) was the precursor to this method, but it was only applied to small games and struggled to compete with tabular agents [16]. A visual representation of the difference between MCCFR and VR-MCCFR can be seen in Fig. 2

Fig. 2. The tree traversal of VR-MCCFR as compared to regular MCCFR and normal CFR as presented in [20]

VR-MCCFR was not investigated for implementation in this project due to its inability to be accelerated by GPU compute (unlike DCFR and DNCFR), meaning significantly more computational resources would be required for similar results. Similarly, DNCFR was not investigated due to a lack of open source reference implementations.

1.3   Game Selection

Imperfect information games are a harder class of games than perfect information games due to the number of possible game states growing exponentially due to uncertainty. Trading card games (TCG) are a more difficult class of game because there is not only non-determinism in game state, but also uncertainty in card interactions.Furthermore, instead of being completely turn-based games like Poker, card interactions between players can happen on either player's turn. Specifically, many TCG cards have a logical description of the effect they have on the game state, when this can be applied, and any uncertainty or random conditions that need to be met to carry it out. Being significantly harder than Poker, TCG's could provide an environment to test more powerful imperfect information solving methods. An example of a TCG is Yu-Gi-Oh. From Table 1 not only does Yu-Gi-Oh have a much larger card pool but also a significantly larger number of possible actions.

Table 1. Comparison of Yu-Gi-Oh and Poker

Property	Yu-Gi-Oh	Poker
Move types	20	6
Players	2	2
Multi-interaction	✓	✗
Deck size	40–60	52
Card pool	11,892	52

Even compared to other TGC's Yu-Gi-Oh presents a few unique advantages, such as not having mechanics to re-snuffle the starting hand at the beginning of the game, a limited field size, and generally requires more card interactions overall. This means any agent developed does not have to require a hand evaluation system at the start of the game, can have its state represented efficiently, and can learn common patterns of card interaction more easily.

A Brief Description of Yu-Gi-Oh. A player wins a game of Yu-Gi-Oh by reducing their opponent's life points to zero. Both players start with 8000 and they can be reduced either by attacks from an opponent's *Monster* or card effects. A *Monster* card is one of the three main types of Yu-Gi-Oh cards. It can be placed on the field, termed "summoning" by a player on their turn given certain conditions are met. A simple analogy would be Chess pieces, where a *Monster* is a particular piece and a square on the Chess board is a *Monster* zone on the field. There are also *Spell* and *Trap* cards which cannot directly harm an opponent but do influence the state of the game. These cards are stored in a *Deck* which can be between 40–60 cards of the players' choice.

1.4 Similar Work

Most TCGs have large player bases, and some even have international competitions. Such competitions are usually held in person with physical cards. Additionally, there is a dedicated AI competition in the case of Hearthstone [12], but methods so far have focused on perfect information MCTS. This is also the case for Magic the Gathering [25], where even ensemble MCTS tree search methods showed poor results [11]. This is because MCTS, even with information sets struggles to adequately address the inherent imperfect information nature of the games. Some neural network methods have been attempted but have had poor results [13]. At the time of writing there are no works investigating the application of game solving methods to Yu-Gi-Oh.

1.5 Meta-strategies

The terms Meta-game or Meta-strategy have different interpretations depending on context, but from the perspective of Yu-Gi-Oh the so called "meta" is the specific decks that are the best or most successful. One of the most prudent examples of "meta" is that of the 2013 or "Dragon Ruler" format in Yu-Gi-Oh where 95% of all tournament wins and top positions were taken out by two decks, Dragon Rulers and Spellbooks. Furthermore, the world championships of 2013 were comprised entirely of those two decks [5]. Playing any other deck at the time put a player at a serious disadvantage.

Deciding on a good meta-strategy is a reflection of a player's skill and is not something that is directly addressed by modern card game AIs. Often when playing an AI player their meta-strategy has been pre-determined and does not change, such is the case with AIs provided with community Yu-Gi-Oh simulators and the official Yu-Gi-Oh online games. Because MCTS, CFR and DCFR agents all learn through some form of self-play and attempt to learn the optimal strategy (or policy) given the deck they have, it should be possible to give one of them different decks, train the same agent against itself, and use the results to draw conclusions about the relative performance of those decks.

2 Experimental Design

Being different to almost all games traditionally studied in game theory and AI generally, the implementation of Yu-Gi-Oh for performing experiments will both be distinct and more complicated. There are relatively few digital Yu-Gi-Oh environments, and none that have been used in the context of AI. To play

Yu-Gi-Oh online, players can either purchase the first party app [17] but must unlock cards and cannot play freely with other players. They can go to the website [1] where they are free to choose cards but must still play and perform all the card interactions manually. The final option is a free and open source simulator, EdoPro (formerly YGoPro) which is a C++ game engine that uses Lua scripts to represent card logic. Being originally designed to function as a game server it provides a reasonable API from which game state and available actions can be captured, thus making it a suitable platform to build and train an agent. Being a community built technology means it has the advantage of being regularly updated and remarkably complete in comparison to the latest release of the game. It is also quite fast, supporting Lua scripting for describing card logic.

The overall simulator used for experiments in this project used parts of the core game engine available [3] and a selection of card scripts from [2] as a base. Parts of it were re-written and other parts were added to make tree searching more practical with what was originally a completely linear state machine. On top of this base, a Python abstraction layer was built and linked to the associated algorithms. This Python layer also allowed for parallelisation across different kinds of compute resources. Considering that the successful agents for Poker ran on a supercomputer [9], to be able to achieve any reasonable results, a reduced game was considered. The rule set, card pool and banned card list were all restricted to the original release of the game. Furthermore, every agent used the same pre-constructed deck. This version of the game still captures the complex interactions and vast card pool without making the game overly complicated or too large.

The following agents were implemented:

- **Heuristic agent**
  The EdoPro simulator [3] provides some built-in AIs that are all heuristic agents hard-coded to respond to certain combinations of cards. For example, always attack the weakest monster, always set trap cards in main phase two, and if the opponent's monster is more powerful, set your own monsters to defence position.
- **ISMCTS Agent**
  A simple information set MCTS agent with the UCB [14] tree selection policy
- **Plain CFR agent**
  A custom game abstraction was implemented and the different phases of a Yu-Gi-Oh turn were divided into buckets
- **Deep CFR agent**
  A deep regret matching network trained through partial iterations of MCCFR [16]

For all experiments, a single duel setting was decided upon (where two players play until one wins) as opposed to a match (best two out of 3 duels) to alleviate the need for side decking and to simplify numerical analysis. The agents played 500 duels, and they both played with the same set of cards. They also started with 8000 life-points, had a starting hand of 5 cards, drew one card per turn, and played under no ban list (as was the case for the Original Yu-Gi-oh release). This round of 500 games was repeated three times for each pair of agents to reduce variance further.

Constraints. To save computational resources, for the implementation of all tree searching methods, the following additional constraints were placed on the state of the game to save computational resources:

- What cards were in the graveyard and what order they were in was not recorded.
- The cards that were in both players' extra decks were not recorded.
- Only cards on the field were recorded not the specific placement or ordering and the same for the hands of both players

Multiple instances of each agent were used for training but each referenced the game memory (the game tree in the case of ISMCTS and counterfactual memory for the CFR methods). In the case of ISMCTS random simulations were limited to 100 actions before the evaluation function was applied.

Evaluation Function. Given the sometimes immense length of Yu-Gi-Oh games, waiting until a terminal state in the roll out stage of ISMCTS and CFR would lead to poor performance. As such, the simulated games were cut off and an evaluation function was applied, which is meant to approximate the overall value or result of that state.

The following function was used:

$$v(s) = 1.5 * (cc - oc) + 2 * mf + \frac{cl}{ol}$$

where $cc \in [0, 20]$ is the number of cards the agent controls $oc \in [0, 20]$ is the number of cards the opponent controls $cl \in [0, 8000]$ is the agent's life points, $ol \in [0, 8000]$ is the opponents' life points and $mf \in [0, 5]$ is the number of Monsters the current player has on the field.

This function was chosen based on experience and preliminary testing. It seemed to capture the three main aspects of Yu-Gi-Oh that led to an overall advantage:

- Having more life points than the opponent.
- Having Monsters to attack the opponent with.
- Having more cards than the opponent to play with.

Time constraints limited empirical determination of good values for the weighting of various factors, so crude values were chosen based on experience with the game.

Statistical Significance. Independent sample t-tests were performed to compare two agents in each of the three experiment runs.

Meta-strategy Evaluation. A way of assessing whether CFR methods can evaluate decks at a high level would be to compare them to human evaluations of decks. A common notion among the Yu-Gi-Oh player base is tiers, where if a deck is in tier 1, it is one of the best and is expected to win most major events. If a deck is in tier 3, it is not expected to win much, but is still competitive. In the early history of Yu-Gi-Oh there were no real archetypes (groups of cards that followed a theme and worked well together), so there were only a few popular decks that people played with subtle variations from player to player. Because of this, and the fact that records of top performances during the early 2000s are difficult to find, decks from that time period cannot be used.

The first format to have reasonable records and well defined tiers is that of early 2011. Table 2 shows which decks were selected after compiling popular decks from the Pojo community forums [23] and what tier they are.

Table 2. Decks chosen for Meta-Strategy evaluation and their relative performance

Deck	Tier
Agents	1
Tengu-Plants	1
GraveKeepers	2
Six Samurai	2
Worms	3
Gem-Knights	3

A DCFR agent was trained for each deck. Training consisted of instances of the six agents playing against a random opponent (who was one of the six decks) for a period of four days. They were then placed in a tournament scenario which closely resembles real-life Yu-Gi-Oh tournaments.

3   Results

A summary of the overall win percentages of playing various agents against each other can be seen in Table 3.

3.1   Baseline

Overall the Heuristic Agent beat the random Agent 75% of the games. As was expected, reasonable heuristics crafted by an expert player easily outperformed random play. The difference between the agents was significant; $t(4) = 10.885$, $p < 0.001$. This experiment provided a baseline for the examination of the other agents (Table 4).

Table 3. Overall head to head win percentage of three 500 game matches between different agents.

Deck	Random	Heuristic	ISMCTS	CFR	DCFR
Random	50%	30%	45%	39%	22%
Heuristic	70%	50%	70%	61%	47%
ISMCTS	55%	30%	50%	42%	35%
CFR	61%	39%	65%	50%	37%
DCFR	78%	53%	65%	63%	50%

Table 4. Number of wins, mean and variance of three 500 game matches between the Custom Heuristic Agent and a random Agent

Agent	Round 1	Round 2	Round 3	μ	σ
Random	125	170	130	142	24
Heuristic	375	330	350	352	22

These results are likely due to the fact that heuristics are effective at playing most scenarios with simple decks and most of the actions available do not involve the more complex game mechanics that would require more detailed heuristics.

3.2   Existing Methods

The ISMCTS Agent won 55% of the games when playing the Random Agent (Table 5). The differences between ISMCTS, Random and Heuristic agents were all statistically significant. It fared significantly worse against the Heuristic Agent only winning around 30% of the games played on average. Whilst the ISMCTS Agent was slightly better than random play, it was no match for the expert heuristics, indicating raw MCTS methods are not a good fit for solving Yu-Gi-Oh. The CFR Agent won 61% of the games against the Random Agent (Table 5). However, it was also easily beaten by the Heuristic Agent, only winning 37% of the games. The differences between CFR, Random and Heuristic Agents were all significant. These results indicate that the CFR Agent performs better than the ISMCTS Agent in both random and heuristic scenarios. Therefore, it is still not an ideal fit for Yu-Gi-Oh, especially considering the large training resources required.

Table 5. Number of wins, mean, variance and t score of three 500 game matches between the ISMCTS Agent, CFR Agent, Heuristic Agent and a Random Agent

Match	Round 1	Round 2	Round 3	μ	σ	t ($p < 0.001$)
ISMCTS/Random	275/225	270/230	280/220	275/225	5	12.247
ISMCTS/Heuristic	150/350	141/359	152/348	148/352	6	−42.779
ISMCTS/CFR	175/325	173/357	177/323	175/325	2	13.272
CFR/Random	305/195	306/194	300/200	304/196	3.2	40.894
CFR/Heuristic	200/300	159/341	190/310	183/317	21	−9.9625

3.3 Deep Counterfactual Regret Minimisation Agent

Deep CFR performed the best out of all the agents, winning 53% of games against the Heuristic Agent and 78% against the Random Agent (Fig. 3). The difference between Deep CFR and the Random agent was significant. However, the narrow difference between Deep CFR and the Heuristic Agent was not statistically significant; t (4) = 2.746, $p < 0.052$ (Table 6).

Table 6. Number of wins, mean and variance of three 500 game matches between the Deep CFR Agent, the Heuristic Agent and a Random Agent

Match	Round 1	Round 2	Round 3	μ	σ	t ($p < 0.001$)
DCFR/Random	390/110	387/113	385/115	387/112	2	133.670
DCFR/Heuristic	265/235	250/250	270/230	262/238	10	N/A

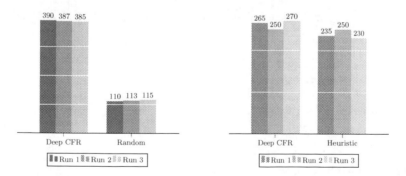

Fig. 3. Visualisation of the number of wins of three 500 game matches between the Deep CFR agent, the Heuristic Agent and a Random Agent

Given this lack of significance, a further test run of three 1000 game matches between the same DCFR and Heuristic agents was run (See Table 7). This run was statistically significant with t (4) = 0.0154, $p < 0.001$ confirming the superior performance of the DCFR agent.

Table 7. Number of wins, mean and variance of three 1000 game matches between the Deep CFR Agent and the Heuristic Agent

Agent	Round 1	Round 2	Round 3	μ	σ
DCFR	565	542	560	555	12.1
Heuristic	435	458	440	444	12.1

3.4   Meta-strategy Evaluation

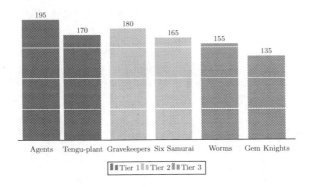

Fig. 4. Number of tournament wins of various decks

The most dominant deck, Agents, won the most games by a reasonable margin (See Fig. 4). However, the tier two deck Gravekeepers came in second place, followed by the other tier one deck, Tengu-Plant. Whilst this slightly matches the trend indicated by Table 2, the tier 1 decks are not as dominant and the tier 3 decks are not completely overpowered.

4   Discussion

4.1   Game Abstractions

The game abstraction used for the CFR agent was both simplistic and small. Looking at some of the Duel replays, it is apparent it hindered the agents' performance in some areas. For example, when the opponent had a powerful Monster like the CFR player would attempt to play a card that only destroys spell cards, thinking it was one that destroyed all cards, leading to a worse position. Furthermore, as the general preprocessing was hand crafted, it is not applicable to other Decks making the agents less general.

Future experiments should incorporate larger game abstractions such as the hierarchical methods outlined in [7] and post-processing methods.

4.2   Agent Results

According to the results against the Heuristic agent, Deep CFR is the most scalable and best performing agent. There are a few factors that were responsible for this success. The first is how it uses the computational resources. Both MCCFR and CFR were limited to using only the CPU for calculations and storage of regret values, whilst the Deep CFR agent was able to make use of 2 GPU's for training its neural networks and some storage, putting it at a significant advantage. This is especially the case when compared to the MCTS agent, which ran out of memory during training and struggled to complete enough iterations to become competitive in terms of win rate. As Yu-Gi-Oh cannot easily be represented as a vector with reasonable memory requirements and hence a matrix game, being able to scale some component of the system to better hardware could be considered a desirable quality in the case of Deep CFR.

In the case of the CFR agent, having a more efficient variant such as CFR+ or Linear CFR could have led to much better results. However, it is likely that both a sampling strategy and advanced abstraction will be required for the best results. These will help alleviate memory issues as the standard CFR agent ran out of memory on multiple occasions and had to cut nodes from the game tree in some cases. Overall, despite being the fastest computationally, CFR does not provide a path to scale to the full game.

A small number of games were played between the Deep CFR Agent and a human player. The Deep CFR Agent was able to win some of these games and made relatively few obvious mistakes, which, whilst expected due to the stochastic nature of Yu-Gi-Oh and the simplified game, is a promising indicator that the agent had achieved a level of human-like play. Future experiments should aim to play more games against humans, perhaps making use of the online facilities provided by EdoPro [3], to ascertain how competitive Deep CFR and agents like it are.

4.3   Meta-strategy Evaluation

Whilst the results are promising for meta-strategy evaluation, there are a few issues that likely led to less than perfect results. The first is training the DCFR agent on much more complicated decks. In the simplified game, the agents hardly ever had to deal with chained effects or extra deck Monsters, yet for the 2011 decks, the usage of extra deck Monsters was critical. This partially explains why the Tengu-Plant deck did not perform as well as expected. Being a strategy that heavily revolves around the extra deck and requires complicated card combinations to be successful, it was probably too much for the Deep CFR agent to learn perfectly. This could also explain why the Gravekeeper deck did better than expected, coming in second place, as its strategy revolves more around controlling the field and playing as few cards as possible. Seeing that the Deep CFR agent prefers more simple strategies, it raises the question of what improvements could be made algorithmically. A possible solution could be focusing on expanding more of the game tree related to the current player's turn as opposed to the opponent's move to encourage exploration of combination moves on a single turn. Despite its shortcomings, the fact that Tengu-Plants came in 3rd place is an indication that Deep CFR can learn complex strategies, and if improvements to the algorithm were made, or if more computational resources were applied, it is likely that it would perform even better.

5   Conclusion

Current CFR methods are almost universally tested using the game of Poker which, whilst providing a complex stochastic imperfect information environment, does not capture complex logic interactions between cards or players. Yu-Gi-Oh, in contrast, is in a class of harder games where card interactions can be stochastic, conditional, or temporally inter-dependent at the same time, and player interaction incredibly situational. These properties make Yu-Gi-Oh a better representation of real world strategic interactions and more apt at addressing modern challenges in AI, such as complex logic, massive state space and hidden information, than Poker. To test Yu-Gi-Oh using CFR, within the bounds of modern computational power, a slightly simplified version of the game that still captures the logic and state space requirements was constructed.

Of the methods tested, Deep CFR and a simple Heuristic Agent performed the best. This indicates that techniques such as MCTS and tabular CFR with custom abstractions are not well suited to address the amount of hidden information Yu-Gi-Oh presents and that even custom abstractions do not capture the relationships between information sets well. Also of concern is that, with the exception of the Heuristic Agent all methods struggled under the CPU, memory, and disk resource limitations of the experimental environment. Future work should look at scaling the experiments to larger computational resources to investigate if MCTS and CFR can perform better under such conditions, especially in the case of sampling variants such as CFR+ and VR-MCCFR.

In the case of using CFR as a way of evaluating meta-strategy, the results are positive but inconclusive. Using Deep CFR as the evaluation system results in similar trends as in real tournament play. Considering that the decks and rules used in those experiments were far more advanced than the simple deck and that each agent had to learn how to play against multiple meta-strategies, the fact that the best deck in the format came out on top is promising. Future work, similarly to the evaluation of CFR methods, should look at what the algorithms do with more training time but also at how different algorithms learn to play different meta-strategies or combinations of them. Furthermore, advanced variants such as Single Deep CFR and Double Neural Deep CFR should be considered for their better computational performance and ability to utilise computational accelerators.

Overall, CFR methods appear to be able to handle the demands of larger and more complicated games. They can produce competitive results when compared to a tuned domain-specific agent by learning similar general behaviours and, if given more resources, would likely outperform them. Furthermore, CFR methods appear to be a promising tool for investigating the construction and evaluation of meta-strategies and, with future research, could lead to intelligent systems that are both able to calculate what resources are required to solve a problem as well as how to best use them when doing so. Such systems do not currently exist for imperfect information contexts, but if they did, they could be revolutionary for business and military strategy, negotiation interactions, and complex planning problems.

References

1. Dueling book. https://www.duelingbook.com/
2. Project ignis card scripts for edopro. https://github.com/ProjectIgnis/CardScripts
3. Project ignis: Edopro. https://github.com/ProjectIgnis/EDOPro
4. Best-first minimax search: Artif. Intell. **84**(1), 299–337 (1996). https://doi.org/10.1016/0004-3702(95)00096-8
5. Akira: Yu-gi-oh! world championship 2013. https://roadoftheking.com/yu-gi-oh-world-championship-2013/
6. Bowling, M., Burch, N., Johanson, M., Tammelin, O.: Heads-up limit Hold'em poker is solved. Science **347**(6218), 145–149 (2015)
7. Brown, N., Ganzfried, S., Sandholm, T.: Hierarchical abstraction, distributed equilibrium computation, and post-processing, with application to a champion no-limit Texas Hold'em agent. In: Workshops at the Twenty-Ninth AAAI Conference on Artificial Intelligence (2015)
8. Brown, N., Lerer, A., Gross, S., Sandholm, T.: Deep counterfactual regret minimization. CoRR abs/1811.00164 (2018). http://arxiv.org/abs/1811.00164
9. Brown, N., Sandholm, T.: Superhuman AI for heads-up no-limit poker: libratus beats top professionals. Science **359**(6374), 418–424 (2018). https://doi.org/10.1126/science.aao1733, https://science.sciencemag.org/content/359/6374/418
10. Browne, C., et al.: A survey of Monte Carlo tree search methods. IEEE Trans. Comput. Intell. AI Games **4**(1), 1–43 (03 2012). https://doi.org/10.1109/TCIAIG.2012.2186810
11. Cowling, P.I., Ward, C.D., Powley, E.J.: Ensemble determinization in Monte Carlo tree search for the imperfect information card game magic: the gathering. IEEE Trans. Comput. Intell. AI Games **4**(4), 241–257 (2012)
12. Dockhorn, A., Mostaghim, S.: Introducing the hearthstone-AI competition. arXiv preprint arXiv:1906.04238 (2019)
13. Grad, L.: Helping AI to play hearthstone using neural networks. In: 2017 Federated Conference on Computer Science and Information Systems (FedCSIS), pp. 131–134 (2017). https://doi.org/10.15439/2017F561
14. James, S., Konidaris, G., Rosman, B.: An analysis of Monte Carlo tree search. In: Proceedings of the AAAI Conference on Artificial Intelligence, vol. 31 (2017)
15. Blar, J., Mutchler, D., Liu, C.: Games with imperfect information (1993)
16. Johanson, M., Bard, N., Lanctot, M., Gibson, R.G., Bowling, M.: Efficient Nash equilibrium approximation through Monte Carlo counterfactual regret minimization. In: AAMAS, pp. 837–846. Citeseer (2012)

17. Konami: Yugioh duel links. https://www.konami.com/yugioh/duel_links/en/
18. Li, H., Hu, K., Zhang, S., Qi, Y., Song, L.: Double neural counterfactual regret minimization. In: International Conference on Learning Representations (2020). https://openreview.net/forum?id=ByedzkrKvH
19. Matros, A.: Lloyd shapley and chess with imperfect information. Games Econ. Behav. **108**, 600–613 (2018). https://doi.org/10.1016/j.geb.2017.12.003, https://www.sciencedirect.com/science/article/pii/S0899825617302221, special Issue in Honor of Lloyd Shapley: Seven Topics in Game Theory
20. Schmid, M., Burch, N., Lanctot, M., Moravcik, M., Kadlec, R., Bowling, M.: Variance reduction in Monte Carlo counterfactual regret minimization (VR-MCCFR) for extensive form games using baselines. In: Proceedings of the AAAI Conference on Artificial Intelligence, vol. 33, pp. 2157–2164 (2019)
21. Silver, D., et al.: Mastering chess and shogi by self-play with a general reinforcement learning algorithm (2017)
22. Silver, D., et al.: Mastering the game of go without human knowledge. Nature **550**(7676), 354–359 (2017)
23. SinL0rtuen: New format's top tiers - let's make a list together ii. https://www.pojo.biz/board/showthread.php?t=991471
24. Syrgkanis, V., Agarwal, A., Luo, H., Schapire, R.E.: Fast convergence of regularized learning in games. arXiv preprint arXiv:1507.00407 (2015)
25. Ward, C.D., Cowling, P.I.: Monte Carlo search applied to card selection in magic: the gathering. In: 2009 IEEE Symposium on Computational Intelligence and Games, pp. 9–16 (2009). https://doi.org/10.1109/CIG.2009.5286501
26. Zinkevich, M., Johanson, M., Bowling, M., Piccione, C.: Regret minimization in games with incomplete information. In: Advances in Neural Information Processing Systems, vol. 20, pp. 1729–1736 (2007)

Are Graph Neural Network Explainers Robust to Graph Noises?

Yiqiao Li[1(✉)], Sunny Verma[1], Shuiqiao Yang[2], Jianlong Zhou[1],
and Fang Chen[1]

[1] University of Technology Sydney, Sydney, Australia
Yiqiao.Li-1@student.uts.edu.au,
{sunny.verma,jianlong.zhou,fang.chen}@uts.edu.au
[2] University of New South Wales, Sydney, Australia
shuiqiao.yang@unsw.edu.au

Abstract. With the rapid deployment of graph neural networks (GNNs) based techniques in a wide range of applications such as link prediction, community detection, and node classification, the explainability of GNNs become an indispensable component for predictive and trustworthy decision making. To achieve this goal, some recent works focus on designing explainable GNN models such as GNNExplainer, PGExplainer, and Gem. These GNN explainers have shown remarkable performance in explaining the predictive results from GNNs. Despite their success, the robustness of these explainers is less explored in terms of vulnerabilities of GNN explainers. Graph perturbations such as adversarial attacks can lead to inaccurate explanations and consequently cause catastrophes. Thus, in this paper, we take the first step and strive to explore the robustness of GNN explainers. To be specific, we first define two adversarial attack scenarios—*aggressive adversary* and *conservative adversary* to contaminate graph structures. We then investigate the impacts of the poisoned graphs on the explainability of three prevalent GNN explainers with three standard evaluation metrics: *Fidelity$^+$*, *Fidelity$^-$*, and *Sparsity*. We conduct experiments on synthetic and real-world datasets and focus on two popular graph mining tasks: node classification and graph classification. Our empirical results suggest that GNN explainers are generally not robust to the adversarial attacks caused by graph structural noises.

Keywords: Graph neural networks · GNN explainers · Adversarial attacks · Robustness

1 Introduction

Generally, a computation graph G can be represented as $G = (V, A, X)$, where V is the node set, $A \in \{0, 1\}$ denotes the adjacency matrix that $A_{ij} = 1$ if there is an edge between node i and node j, otherwise $A_{ij} = 0$, and X indicates the feature matrix of the graph G. It is an ideal data structure for

© The Author(s), under exclusive license to Springer Nature Switzerland AG 2022
H. Aziz et al. (Eds.): AI 2022, LNAI 13728, pp. 161–174, 2022.
https://doi.org/10.1007/978-3-031-22695-3_12

a variety of real-world datasets, such as chemical compounds [3], social circles [21], and road networks [15]. Graph neural networks (GNNs) [5,26,29,33], with the resurgence of deep learning, have become a powerful tool to model these graph datasets and achieved impressive performance. However, a GNN model is typically very complicated and how it makes predictions is unclear; while unboxing the working mechanism of a GNN model is crucial in many practical applications (e.g., criminal associations predicting [24], traffic forecasting [11], and medical diagnosis [1,23]).

Recently, several explainers [19,20,30] have been proposed to tackle the problem of explaining GNN models. These attempts can be categorized into *local* and *global* explainers according to their interpretation scales. In particular, if the method provides an explanation only for a specific instance, it is a *local explainer*. In contrast, if the method explains the whole model, then it is a *global explainer*. Alternatively, GNN explainers can also be classified as either *transductive* or *inductive* explainers based on their capacity to generalize to extra unexplained nodes. We investigate a flurry of recent GNN explainers and decide to use three most representative GNN explainers—GNNExplainer [30], PGExplainer [20], and Gem [19]—in our experiments. GNNExplainer is challenging to be applied into inductive settings as its explanations are limited to a single instance and it merely provides local explanations; while a trained PGExplainer which constructs global explanations and Gem which generates both local and global explanations can be used in inductive scenarios to infer explanations for unexplained instances without the need of retraining the explanation models. Table 1 summarizes the characteristics of these methods.

Table 1. The characteristics of GNN explainers.

	GNNExplainer	PGExplainer	Gem
Interpretation scale	Local explainer	Global explainer	Local & global explainer
Transduction/ Induction	Transductive explainer	Inductive explainer	Inductive explainer
Applications	Node classification Graph classification Link prediction	Node classification Graph classification	Node classification Graph classification

On the other hand, *robustness* is also an important topic in the community of deep learning and has gained significant attention over years. Recently, there are a large number of research studies focusing on the robustness of image classification including adversarial robustness [27] and non-adversarial robustness [10,16]. In addition, researchers start to explore the robustness of GNN models in recent years, having gained several crucial observations and insights [2,34]. Nevertheless, the robustness of GNN explainers is still under exploration. While in real world, graph datasets are never ideal and often contaminated by various nuisance factors such as noises in node features and/or in graph structures. Therefore, one natural question one might ask: *are current GNN explainers robust against these nuisance factors?*

To answer this question, we in this paper take the first step to examine the robustness of GNN explainers. To be specific, we explore two adversary scenarios to contaminate graph datasets:

- *Aggressive adversary.* We introduce noises to graph structures without considering the characteristics of nodes–whether it is an important node or a redundant node. To be more specific, we may pollute any nodes to have edges with others regardless of the impact on the GNN models.
- *Conservative adversary.* In contrast to *aggressive adversary*, we introduce noises to graph datasets in a more cautious way such that we hope the injected noises would not affect the GNN model itself. To achieve this goal, we have to take the characteristics of graph dataset itself into account (e.g., whether the node is an important node or an unimportant node). We then only alter the graph structure by adding edges among unimportant nodes. By doing so, the underlying essential subgraph, which determines the prediction of GNN models, is untouched.

We first use the aforementioned adversary scenarios to contaminate the graph datasets. We then use these generated noisy graph datasets to evaluate the robustness of the GNN explainers. For the baseline, we refer to the performance of the GNN explainers on original (clean) graph datasets. Thus, we track and compare the difference in the performance of GNN explainers between original and polluted graph datasets. Our contributions can be summarized as followings:

- For the sake of comprehensive evaluations, we propose to generate noisy graph data under two scenarios—*aggressive* adversary and *conservative* adversary.
- We empirically investigate the robustness of GNN explainers against these perturbations through two different applications including node classification and graph classification.
- We find that GNN explainers in general are not robust to these perturbations, implying that robustness is another essential factor one should take into account when evaluating GNN explainers.

2 Related Work

2.1 GNNs and the Robustness of GNNs

Graph neural networks (GNNs) have shown their effectiveness and obtained the state-of-the-art performance on many different graph tasks, such as node classification, graph classification, and link prediction. Since graph data widely exist in different real-world applications, such as social networks [25], chemistry [8], and biology [6], GNNs are becoming increasingly important and useful.

Despite their great performance, GNNs share the same drawback as other deep learning models; that is, they are usually treated as black-boxes and lack human-intelligible explanations. Without understanding and verifying the inner working mechanisms, GNNs cannot be fully trusted, which prevents their use in critical applications pertaining to fairness, privacy, and safety [4].

On the other hand, the robustness evaluation for GNNs has received a great deal of attention recently. In recent years, some adversarial attacks and back-door attacks against GNNs are proposed [7,9,28,34]. Specially, in [28], Yang et al. propose a transferable trigger to launch backdoor attack against different GNNs. In [34], authors propose an efficient algorithm NETTACK exploiting incremental computations. They concentrate on adversarial perturbations that target the node's characteristics and the graph structure, therefore taking into account the interdependencies between instances. In addition, they ensure that the perturbations are undetectable by keeping essential data features. Ghorbani et al. [9] demonstrate how to generate adversarial perturbations that produce perceptively indistinguishable inputs that are assigned the same predicted label, yet have very different interpretations. They prove that systematic perturbations can result in drastically different interpretations without modifying the label. Fox et al. [7] investigate that GNNs are not robust to structural noise. They focus on inserting addition of random edges as noise in the node classification without distinguish important and unimportant nodes. On the contrast, we focus on injecting conservative structure noise into unimportant nodes/subgraphs. Overall, in our research, we propose to infuse aggressive and conservative structure noise individually into graph data in order to examine the robustness of GNN explainers.

2.2 GNN Explainers

GNNs incorporate both graph structure and feature information, which results in complex non-linear models, rendering explaining its prediction remain a challenging task. Besides, models explanations could bring a lot of benefits to users (e.g., improving safety and promoting fairness). Thus, some popular works has emerged in recent years focusing on the explanation of GNN models by leveraging the properties of graph features and structures. There are some popular GNN explainers developing explaining strategies based on graph intrinsic structures and features. We will briefly review three different GNN explainers: GNNExplainer, PGExplainer, and Gem.

GNNExplainer [30] is a seminal method in the field of explaining GNN models. It provides local explanations for GNNs by identifying the most relevant features and subgraphs, which are essential in the prediction of a GNN. *PGExplainer* [20] introduces explanations for GNNs with the use of a probabilistic graph. It provides model-level explanations for each instance and possesses strong generalizability. *Gem* [19] is able to provide both local and global explanations and it is also operated in an inductive setting. Thus, it can explain GNN models without retraining. Particularly, it adopts a parameterized graph auto-encoder with Graph Convolutional Network(GCN) [14] layers to generate explanations.

3   Method

In this paper, we examine the robustness of GNN explainers under two adversary scenarios—*aggressive adversary* and *conservative adversary*. In this section, we provide the details of our method. Particularly, we first introduce how we inject noises into graph data and construct noisy graph data (see Sect. 3.1), and we then depict our evaluation flow (see Sect. 3.2).

3.1   Adversary Generation

Without loss of generality, we consider generating aggressive and conservative adversaries in a graph classification task. For a graph $G_i = (V_i, A_i, X_i)$ with label L_i, we have the prediction $f(G_i)$ of a GNN model, and the explanation $E(f(G_i), G_i)$ from a GNN explainer.

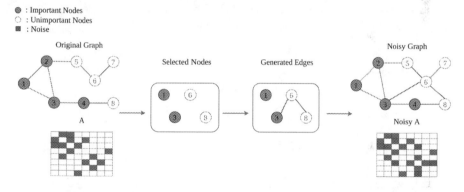

Fig. 1. The instance of generating aggressive structure noise. The orange nodes denote important nodes, while the rest means unimportant nodes in the graph. In this scenario, we do not take the node property into account and we randomly select nodes. (Color figure online)

Aggressive Adversary Generation. The *aggressive adversary* disregards the role of nodes and radically incorporates structure noises into nodes without considering their impacts on the GNN models. For a particular graph G_i, we randomly choose $\varepsilon = \{10\%, 30\%, 50\%, 80\%\}$ nodes from the set V_i, then generate edges among these selected nodes by using random graph generation model with generating edges probability 0.1, meaning that the number of edges is equal to 10% of the number of selected nodes. Figure 1 shows a toy example of *aggressive adversary* generation. After generating aggressive structure noises, we obtain a new noisy graph $\widehat{G}_i = (V_i, \widehat{A}_i, X_i)$ with label L_i, and further obtain the GNN prediction $f(\widehat{G}_i)$ on this new noisy graph as well as its the explanation $E(f(\widehat{G}_i), \widehat{G}_i)$. As we have aggressively changed the structure of the graph, the probability of $f(\widehat{G}_i)$ is expected to be lower, implying that the aggressive structure noises also affect the performance of the GNN models. Furthermore, predictions of GNN model is another input to GNN explainers, which is another factor to influence explanations of GNN explainers.

Conservative Adversary Generation. The *conservative adversary* selectively appends structure noise into unimportant nodes. Particularly, in *conservative adversary*, we build a structure noise which would not alter the prediction of GNN models. For a particular graph G_i, we obtain the unimportant nodes set N_i with the similar ratio of $\varepsilon = \{10\%, 30\%, 50\%, 80\%\}$ we used in the setting of *aggressive adversary*. Then, we use random graph generation model to generate edges among N_i with the generating edges probability 0.1. Similarly, Fig. 2 shows a toy example of *conservative adversary* generation. After developing conservative structure noise, we get a noisy graph $G'_i = (V_i, A'_i, X_i)$ with label L_i. Therefore, we are able to obtain the GNN prediction $f(G'_i)$ and the explanation $E(f(G'_i), G'_i)$. In *conservative adversary*, since the significant subgraph that determines the prediction of GNN models is unmodified, there is a high possibility that $f(G'_i)$ would make the *correct* predictions. Thus, the prediction of GNN as a parameter in GNN explainers inputs keeps stable and unchanged. Therefore, one should expect that the GNN explainers would be more robust against conservative adversary than aggressive adversary.

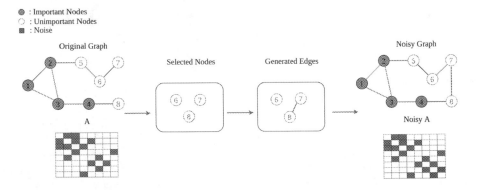

Fig. 2. The instance of generating conservative structure noise. The orange nodes denote important nodes, while the rest are unimportant nodes in the graph. We only select unimportant nodes. (Color figure online)

3.2   Robustness Evaluation Framework

For a GNN model, GNN explainers are used to unveil why the GNN model makes its predictions. Thus, it is intriguing to explore whether these explanations really make sense, especially when the graph data is not clean and polluted by noises, which is often the case in real-world datasets. The contamination can occur in many ways such as during the process of data collection, the defects of sensors, data transmission through network, and many others. In this paper, we insert noises into the original clean graph data to examine whether the explanation of GNN explainers would be affected.

Specifically, in our experiments, we target to investigate the robustness of the GNN explainer to structure noises. We introduce two types of structure noises to graph datasets, of which the detailed information can be found in Sect. 3.1. After obtaining noisy graph dataset, we feed it into a pre-trained GNN that is trained by the original clean graph dataset and get its corresponding predictions. Then a GNN explainer conducts its explanations and we obtain its explanation performance and further conduct comparisons with the explanations on the original graph dataset. The pipeline of our robustness evaluation method is shown in Fig. 3. We further show an example of our experimental flow under the *conservative adversary* in Fig. 4.

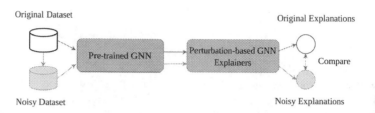

Fig. 3. In this diagram, different lines denote distinct flows. The black lines denote initial flow that generates explanations for the original dataset. The green lines denote flow that generates a noisy graph data from the original graph data as well as its explanations. Finally, we can compare "noisy" explanations with "original" explanations. (Color figure online)

Furthermore, we use accuracy to quantitatively measure the influence of structure noises to the GNN model. We assume that the performance of GNN model would rarely be affected if the prediction accuracy on the noisy graph dataset is roughly the same as the accuracy on the original clean graph dataset. We further assume that if the GNN model itself is not confused by the injected noises, then the GNN explainers would yield similar explanations between original clean graph data and noisy graph data.

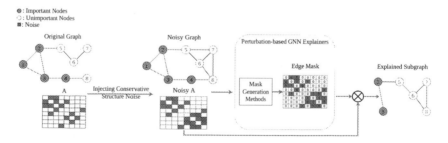

Fig. 4. The instance of generating explanation for noisy graph with *conservative adversary*. The orange nodes denote important nodes, while the rest means unimportant nodes in the graph. The orange nodes and edges are expected to be as an explanation from GNN explainers. However, after injecting structure noise which is highlighted in red colour, the GNN explainers can not get the true important subgraph, which demonstrates that the GNN explainers are not robust to structure noises. (Color figure online)

4   Experiments

In this section, we conduct experiments to inspect the robustness of GNN explainers against structure noises. We first describe the details of the implementation, datasets, and metrics we used in Sect. 4.1. After that, we present and analyze the experimental results for *aggressive adversary scenario* and *conservative adversary scenario* in Sect. 4.2 and Sect. 4.3, respectively.

4.1   Implementation Details, Datasets, and Metrics

Implementation Details. In this paper we choose GCN as the classification classifier. For GNN explainers, we choose GNNExplainer [30], PGExplainer [20], and Gem [19]. In order to obtain the pre-trained GCN models, we split the datasets into percentages of 80/10/10 as the training, validation, and test set, respectively. We follow the experimental settings in Gem [19]. Specifically, we firstly train a three-layer GCN model based on BA-Shapes dataset, Tree-Cycles dataset, and Mutagenicity dataset, respectively. We choose Adam [13] as the optimizer. After that, we utilize the pre-trained GCN models and the explainers to obtain the explanations for both the original clean graph datasets and the noisy graph datasets. Furthermore, by analyzing the experiment settings and results in [19], we note that explainers obtain different levels of accuracy when selecting different top-important edges as explaining edges. Therefore, one should choose an appropriate number of top important edges when evaluating explainers. In our paper, we select top 6 edges for synthetic datasets (BA-Shapes and Tree-Cycles) and top 15 edges for Mutagenicity dataset.

Datasets. We focus on two widely used node classification datasets, including BA-Shapes and Tree-Cycles [18,31], and one graph classification dataset, Mutagenicity [12]. Statistics of these datasets are shown in Table 2. For BA-Shapes and Tree-Cycles datasets the nodes which define a motif structure such as a house or cycle are considered as important nodes. For Mutagenicity datasets, Carbon rings with chemical groups NH_2 or NO_2 are known to be mutagenic. Carbon rings however exist in both mutagen and nonmutagenic graphs, which are not discriminative. Thus, we simply treat carbon rings as the shared base graphs and NH_2, NO_2 as important subgraphs for the mutagen graphs.

Table 2. Dataset information.

	Node classification		Graph classification
	BA-shapes	Tree-cycles	Mutagenicity
# of Graphs	1	1	4,337
# of Edges	4110	1950	266,894
# of Nodes	700	871	131,488
# of Labels	4	2	2

In addition, explainers—GNNExplainer, PGExplainer, and Gem—can obtain higher accuracy when used to explain only important nodes or subgraphs. While in our experiments, we may alter the nodes as well as the subgraph structures, thus we have to explain all nodes or subgraphs (important or unimportant), which may lead to suboptimal accuracy. However, this is not a major issue for us as our goal in this paper is to compare the performance change of GNN explainers on graph datasets before and after adding noises.

Noisy Datasets. Following the noise generation pipeline described in Sect. 3, we inject *aggressive* and *conservative* structure noises into these graph datastes to generate *aggressive* and *conservative* noisy datasets, respectively. For *conservative* structure noisy datasets, we only inject noises into unimportant nodes to minimize the affection of structure noise on GNN prediction. By doing so, we attempt to maintain GNN predictions on *conservative* structure noise datasets.

Metrics. Good metrics should evaluate whether the explanations are faithful to the model. After comparing the characteristic of each quantitative metric [17,32], we chose *Fidelity*$^+$ [31], *Fidelity*$^-$ [31], and *Sparsity* [22] as our evaluation metrics. The *Fidelity*$^+$ metric indicates the difference of predicted probability between the original predictions and the new prediction after removing important input features. In contrast, the metric *Fidelity*$^-$ represents prediction changes by keeping important input features and removing unimportant structures. Besides, *Sparsity* measures the fraction of features selected as important by explanation methods. The *Fidelity*$^+$, *Fidelity*$^-$, and *Sparsity* can be defined as:

$$Fidelity^+ = \frac{1}{N} \sum_{i=1}^{N} (f(\boldsymbol{G}_i)_{y_i} - f(\boldsymbol{G}_i^{1-m_i})_{y_i}), \tag{1}$$

$$Fidelity^- = \frac{1}{N} \sum_{i=1}^{N} (f(\boldsymbol{G}_i)_{y_i} - f(\boldsymbol{G}_i^{m_i})_{y_i}), \tag{2}$$

$$Sparsity = \frac{1}{N} \sum_{i=1}^{N} (1 - \frac{|\boldsymbol{s}_i|}{|\boldsymbol{S}_i|_{total}}), \tag{3}$$

where N is the total number of samples and y_i is the class label. $f(\boldsymbol{G}_i)_{y_i}$ and $f(\boldsymbol{G}_i^{1-m_i})_{y_i}$ are the prediction probabilities of y_i when using the original graph $\boldsymbol{G}_i$ and the occluded graph $\boldsymbol{G}_i^{1-m_i}$, which is gained by occluding important features found by explainers from the original graph. Thus, a *higher Fidelity*$^+$ ($\uparrow$) is desired. $f(\boldsymbol{G}_i^{m_i})_{y_i}$ is the prediction probabilities of y_i when using the explanation graph $\boldsymbol{G}_i^{m_i}$, which is obtained by important structures found by explainable methods. Thus a *lower Fidelity*$^-$ ($\downarrow$) is desired. Furthermore, the $|S_i|_{total}$ represents the total number of features (e.g., nodes, nodes features, or edges) in the original graph model; while $|s_i|$ is the size of important features/nodes found by the explainable methods and it is a subset of $|S_i|$. Note that higher sparsity values indicate that explanations are sparser and likely to capture only the most essential input information. Hence, a *higher Sparsity* ($\uparrow$) is desired.

4.2   Vulnerable to Aggressive Adversary

To measure the robustness of GNN explainers against aggressive structure noises, we estimate the differences in performance of GNN explainers between original and aggressive noisy datasets. We first obtain the explanation performance of each explainers on original clean graph datasets, which serves as our baseline. We then obtain the corresponding explanation performance of each explainers on noisy graph datasets with aggressive adversary. For reference, we also report the GCN accuracy.

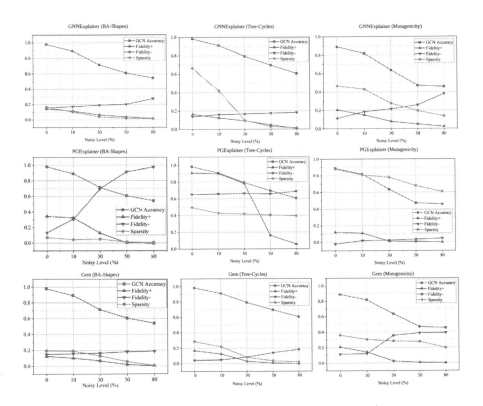

Fig. 5. The results of aggressive adversary in terms of $Fidelity^+$, $Fidelity^-$, and $Sparsity$.

GNN Explainers Are Not Robust to Aggressive Adversary. Figure 5 shows the results of the robustness of GNN explainers against aggressive noise. One can observe that: 1) As the noise level increases, all explanation performance metrics including $Fidelity^+$, $Fidelity^-$, and $Sparsity$ consistently become worse, implying that *aggressive* noises do have negative impacts on the GNN explainers; 2) The accuracy of GCN keeps decreasing as the noise level increases, implying that the aggressively injected noises also affect the performance of GCN itself, which is consistent with the findings in [7,34]; 3) The findings mentioned above are consistent across different datasets and different tasks, suggesting the generality of our findings.

4.3   Vulnerable to Conservative Adversary

Now, we start to explore how conservative adversary affect the GNN explainers. We follow the exact pipeline in Sect. 4.2 expect that we here inject noises in a more cautious way. We believe this conservative adversary would yield negligible impacts on the GCN itself while it may still negatively affect the explanation quality of GNN explainers (see Sect. 3 for more details).

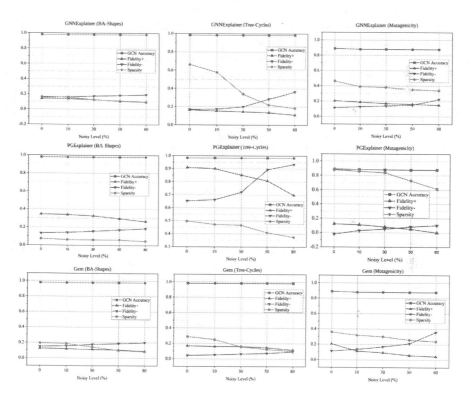

Fig. 6. The results of conservative adversary in terms of $Fidelity^+$, $Fidelity^-$, and $Sparsity$.

GNN Explainers Are Not Robust to Conservative Adversary. Figure 6 shows the experimental results for the setting of conservative adversary. As expected, the accuracy of the GNN is quite stable and does not change much even when the noise level increases, implying that the noises injected in this way do not alter the essential structures of graph datasets. However, in term of $Fidelity^+$, $Fidelity^-$, and $Sparsity$, we see a similar trend as the aggressive adversary (Sect. 4.2) although the impacts here are much benign, which further demonstrates the fragility of GNN explainers to graph noises.

5   Conclusion

In this paper, we attempt to identify the robustness issue of GNN explainers. We propose two types of structure noises—*aggressive adversary* and *conservative adversary*—to construct noisy graphs. We evaluate three recent representative GNN explainers including GNNExplainer, PGExplainer, and Gem, which vary in terms of interpretation scales and generality. We conduct experiments on two different tasks—node classification with BA-Shapes and Tree-Cycles datasets and graph classification with Mutagenicity dataset. Through experiments, we find that the current GNN explainers are fragile to adversarial attacks as the quality of their explanations is significantly decreased across different severity of noises. Our findings suggest that robustness is a practical issue one should take into account when developing and deploying GNN explainers in real-world applications. In our future work, we would develop algorithms and models to improve the robustness of GNN explainers against these adversaries.

References

1. Chen, D., Zhao, H., He, J., Pan, Q., Zhao, W.: An causal XAI diagnostic model for breast cancer based on mammography reports. In: 2021 IEEE International Conference on Bioinformatics and Biomedicine (BIBM), pp. 3341–3349, December 2021. https://doi.org/10.1109/BIBM52615.2021.9669648
2. Dai, H., et al.: Adversarial attack on graph structured data. In: Proceedings of the 35th International Conference on Machine Learning, pp. 1115–1124. PMLR, July 2018
3. Debnath, A.K., Lopez de Compadre, R.L., Debnath, G., Shusterman, A.J., Hansch, C.: Structure-activity relationship of mutagenic aromatic and heteroaromatic nitro compounds. Correlation with molecular orbital energies and hydrophobicity. J. Med. Chem. **34**(2), 786–797 (1991). https://doi.org/10.1021/jm00106a046
4. Doshi-Velez, F., Kim, B.: Towards a rigorous science of interpretable machine learning (2017). https://doi.org/10.48550/ARXIV.1702.08608
5. Duan, W., Xuan, J., Qiao, M., Lu, J.: Learning from the dark: boosting graph convolutional neural networks with diverse negative samples. In: Proceedings of the AAAI Conference on Artificial Intelligence, vol. 36, no. 6, pp. 6550–6558 (2022). https://doi.org/10.1609/aaai.v36i6.20608
6. Fout, A., Byrd, J., Shariat, B., Ben-Hur, A.: Protein interface prediction using graph convolutional networks, pp. 6533–6542, December 2017
7. Fox, J., Rajamanickam, S.: How robust are graph neural networks to structural noise? (2019). https://doi.org/10.48550/ARXIV.1912.10206
8. Fung, V., Zhang, J., Juarez, E., Sumpter, B.G.: Benchmarking graph neural networks for materials chemistry. npj Comput. Mater. **7**(1), 1–8 (2021). https://doi.org/10.1038/s41524-021-00554-0
9. Ghorbani, A., Abid, A., Zou, J.: Interpretation of neural networks is fragile, vol. 33, pp. 3681–3688 (2019). https://doi.org/10.1609/aaai.v33i01.33013681
10. Hendrycks, D., Dietterich, T.: Benchmarking neural network robustness to common corruptions and perturbations. In: Proceedings of the International Conference on Learning Representations (2019)

11. Jiang, W., Luo, J.: Graph neural network for traffic forecasting: a survey. Expert Syst. Appl. **207**, 117921 (2022). https://doi.org/10.1016/j.eswa.2022.117921

12. Kazius, J., McGuire, R., Bursi, R.: Derivation and validation of toxicophores for mutagenicity prediction. J. Med. Chem. **48**(1), 312–320 (2005). https://doi.org/10.1021/jm040835a

13. Kingma, D.P., Ba, L.J.: Amsterdam machine learning lab (IVI, FNWI): adam: a method for stochastic optimization. In: International Conference on Learning Representations (ICLR). arXiv.org (2015)

14. Kipf, T.N., Welling, M.: Semi-supervised classification with graph convolutional networks (2017)

15. Leskovec, J., Lang, K.J., Dasgupta, A., Mahoney, M.W.: Community structure in large networks: natural cluster sizes and the absence of large well-defined clusters. Internet Math. **6**(1), 29–123 (2009). https://doi.org/10.1080/15427951.2009.10129177

16. Li, T., Mehta, R., Qian, Z., Sun, J.: Rethink autoencoders: robust manifold learning. In: ICML Workshop on Uncertainty and Robustness in Deep Learning (2020)

17. Li, Y., Zhou, J., Verma, S., Chen, F.: A survey of explainable graph neural networks: taxonomy and evaluation metrics (2022). https://doi.org/10.48550/ARXIV.2207.12599

18. Lin, C., Sun, G.J., Bulusu, K.C., Dry, J.R., Hernandez, M.: Graph neural networks including sparse interpretability (2020). https://doi.org/10.48550/ARXIV.2007.00119

19. Lin, W., Lan, H., Li, B.: Generative causal explanations for graph neural networks. In: Proceedings of the 38th International Conference on Machine Learning, pp. 6666–6679. PMLR, July 2021

20. Luo, D., et al.: Parameterized explainer for graph neural network. In: Proceedings of the 34th International Conference on Neural Information Processing Systems, NIPS 2020, pp. 19620–19631. Curran Associates Inc., Red Hook, December 2020

21. McAuley, J., Leskovec, J.: Learning to discover social circles in ego networks, pp. 539–547, December 2012

22. Pope, P.E., Kolouri, S., Rostami, M., Martin, C.E., Hoffmann, H.: Explainability methods for graph convolutional neural networks. In: Proceedings of the IEEE/CVF Conference on Computer Vision and Pattern Recognition (CVPR), June 2019

23. Singh, A., Sengupta, S., Lakshminarayanan, V.: Explainable deep learning models in medical image analysis. J. Imaging **6**(6), 52 (2020). https://doi.org/10.3390/jimaging6060052

24. Wang, C., Lin, Z., Yang, X., Sun, J., Yue, M., Shahabi, C.: HAGEN: homophily-aware graph convolutional recurrent network for crime forecasting, vol. 36, pp. 4193–4200, June 2022. https://doi.org/10.1609/aaai.v36i4.20338

25. Wu, Y., Lian, D., Xu, Y., Wu, L., Chen, E.: Graph convolutional networks with Markov random field reasoning for social spammer detection, vol. 34, pp. 1054–1061, April 2020. https://doi.org/10.1609/aaai.v34i01.5455

26. Wu, Z., Pan, S., Chen, F., Long, G., Zhang, C., Yu, P.S.: A comprehensive survey on graph neural networks. IEEE Trans. Neural Netw. Learn. Syst. **32**(1), 4–24 (2021). https://doi.org/10.1109/TNNLS.2020.2978386

27. Xu, H., et al.: Adversarial attacks and defenses in images, graphs and text: a review. Int. J. Autom. Comput., 1–28 (2019). https://doi.org/10.1007/s11633-019-1211-x

28. Yang, S., et al.: Transferable graph backdoor attack. arXiv preprint arXiv:2207.00425 (2022)

29. Yang, S., et al.: Variational co-embedding learning for attributed network clustering. CoRR abs/2104.07295 (2021)
30. Ying, R., Bourgeois, D., You, J., Zitnik, M., Leskovec, J.: GNN explainer: a tool for post-hoc explanation of graph neural networks. CoRR abs/1903.03894 (2019)
31. Yuan, H., Yu, H., Gui, S., Ji, S.: Explainability in graph neural networks: a taxonomic survey. IEEE Trans. Pattern Anal. Mach. Intell., 1–19 (2022). https://doi.org/10.1109/TPAMI.2022.3204236
32. Zhou, J., Gandomi, A.H., Chen, F., Holzinger, A.: Evaluating the quality of machine learning explanations: a survey on methods and metrics. Electronics **10**(5), 593 (2021). https://doi.org/10.3390/electronics10050593
33. Zhou, J., et al.: Graph neural networks: a review of methods and applications. AI Open **1**, 57–81 (2020). https://doi.org/10.1016/j.aiopen.2021.01.001
34. Zügner, D., Akbarnejad, A., Günnemann, S.: Adversarial attacks on neural networks for graph data. In: Proceedings of the 24th ACM SIGKDD International Conference on Knowledge Discovery & Data Mining, KDD 2018, pp. 2847–2856. Association for Computing Machinery, New York, July 2018. https://doi.org/10.1145/3219819.3220078

Ethical/Explainable AI

Towards Explainable AutoML Using Error Decomposition

Caitlin A. Owen[✉][iD], Grant Dick[iD], and Peter A. Whigham[iD]

Department of Information Science, University of Otago, Dunedin, New Zealand
{caitlin.owen,grant.dick,peter.whigham}@otago.ac.nz

Abstract. The important process of choosing between algorithms and their many module choices is difficult, even for experts. Automated machine learning allows users at all skill levels to perform this process. It is currently performed using aggregated total error, which does not indicate whether a stochastic algorithm or module is stable enough to consistently perform better than other candidates. It also does not provide an understanding of how the modules contribute to total error. This paper explores the decomposition of error for the refinement of genetic programming. Automated algorithm refinement is examined through choosing a pool of candidate modules and swapping pairs of modules to reduce the largest component of decomposed error. It is shown that a pool of candidates that are not examined for diversity in targeting different components of error can provide inconsistent module preferences. Manual algorithm refinement is also examined by choosing refinements based on their well-understood behaviour in reducing a particular error component. The results show that an effective process should exploit both the advantages of targeted improvements identified using a manual process and the simplicity of an automated process by choosing a hierarchy of the most important modules for reducing error components.

Keywords: Genetic programming · Automated machine learning · Algorithm refinement · Symbolic regression

1 Introduction

With the successful application of machine learning algorithms to many problem domains [1–3], there is increasing interest from end users who are not experts in machine learning. These applications of machine learning typically involve a wide range of parameters or algorithm module choices [4,5]. This wide range of choices provide a greater opportunity to produce a good predictive model, given that a particular algorithm will not provide consistently better predictive performance than all other candidates [6]. However, even for data scientists, the process of choosing between a large number of algorithms and choosing appropriate parameters/modules is difficult [7,8]. This is the motivation for automated

© The Author(s), under exclusive license to Springer Nature Switzerland AG 2022
H. Aziz et al. (Eds.): AI 2022, LNAI 13728, pp. 177–190, 2022.
https://doi.org/10.1007/978-3-031-22695-3_13

machine learning (AutoML), which involves the data-driven algorithmic selection, composition and parameterisation of machine learning methods, usually in order to minimise prediction error [9,10].

AutoML methods in the literature use only an aggregate measure of prediction error for a continuous response variable (referred to in this paper as total error) to examine combinations of algorithm modules. Performing AutoML using only total error does not provide an understanding of how the modules of the algorithm interact or the role each module plays in reducing total error. Instead, decomposed error will allow an AutoML process to make more accurate and informed decisions, in terms of the compatibility of algorithm module combinations and their parameterisation, by targeting a reduction in the largest component of error.

For a deterministic algorithm, using the same training data for each run produces the same prediction model. These algorithms can be characterised by decomposing error into bias and variance, where the single source of error due to variance is associated with sampling of finite training data [11]. In contrast, a stochastic algorithm involves multiple sources of error due to variance. Therefore, the algorithm can be more fully characterised by splitting error due to variance into error due to external variance (variance due to the training data) and error due to internal variance (variance due to the algorithm itself) using an extended error decomposition [12]. The extended decomposition of error and AutoML are expected to be strongly compatible as they are both empirical processes.

Performing AutoML using "black-box" total error creates ambiguity about which sources of error are being reduced and why a particular combination of algorithm modules has been chosen. In contrast, decomposed error provides an explanation of which error components are being minimised when choosing modules. Therefore, AutoML driven by decomposed error would provide more transparency and an explanation of why a particular combination of modules has been chosen, which are both important for AutoML [13]. This includes an understanding of how the chosen algorithm is appropriate for a given problem. Explainability is important for AutoML because end users need to have confidence and trust in the performance/behaviour of an algorithm [14]; this confidence is gained by understanding why the algorithm modules have been chosen.

In this paper, AutoML driven by an extended decomposition of error is explored by performing algorithm refinement that targets the largest component of error. This process is applied to the refinement of genetic programming (GP) for symbolic regression, although any machine learning algorithm could be refined using this process. Algorithm refinement is first performed using an automated process to determine whether it can be reliable for stochastic algorithms. This method focuses on refining the selection and variation operators of GP, although the method could be used to refine all parts of the GP algorithm. It is shown that error due to internal variance is not sufficiently stable to provide consistent decisions about which module reduces prediction error. This highlights the importance of choosing a diverse set of candidate modules that provide targeted reductions in all components of error, particularly error due to internal variance. To confirm this, algorithm refinement is performed using a manual process. In each iteration, new algorithm components are hand-picked for their well-understood behaviour in terms of reducing the largest component of error, which is shown to successfully reduce total error.

The remainder of this paper is structured as follows: a brief overview of algorithm refinement methods are discussed in Sect. 2; a description of how to decompose error for the refinement of GP is outlined in Sect. 3; results for an automated algorithm refinement process using decomposed error are discussed in Sect. 3.1 and critiqued in Sect. 3.2; results for a manual algorithm refinement process using decomposed error are discussed in Sect. 3.3 and critiqued in Sect. 3.4; finally, conclusions and future work are discussed in Sect. 4.

2   Algorithm Refinement

A number of different methods have been used for AutoML. The combined algorithm selection and hyperparameter optimisation (CASH) problem can be viewed as a "single hierarchical hyperparameter optimisation problem", with the chosen type of algorithm being considered as a hyperparameter [9, p. 847]. Grid search examines all possible combinations of hyperparameters [15]. While this is a simple method, it is computationally expensive and potentially intractable if the number of hyperparameters is large. Random search improves on grid search by not examining the full distributions of hyperparameter values [15]. However, random search is still potentially computationally expensive. Bayesian optimisation, involving a probability surrogate model of objectives, is more computationally efficient and is applicable to any type of objective function. Auto-WEKA [16] involves Bayesian optimisation using tree-based models. Auto-SKLearn [17] also uses Bayesian optimisation, extending Auto-WEKA in order to provide an initial meta-learning step as well as automated ensemble construction. Evolutionary computation has also been used for AutoML. RECIPE (REsilient Classification Pipeline Evolution) uses grammar-based GP, with a grammar representing an algorithm pipeline, i.e., a combination of algorithm modules [18]. RECIPE provides a larger number of algorithm modules than both Auto-SKLearn and Auto-WEKA. TPOT (Tree-Based Pipeline Optimization Tool) also uses a variant of GP to represent algorithm pipelines, allowing parallel processing by using multiple copies of a data set [19].

When applied to regression problems, these AutoML methods involve the same basic process. A portfolio of candidate algorithm modules and parameters is chosen (although the reasons for including the selected module options are not usually explained). During the hyperparameter optimisation process, combinations of these candidates are examined. All of the methods in the literature use total prediction error (for test observations) to guide the improvement of the combination of algorithm modules.

The goal of AutoML, for supervised machine learning, is to minimise total prediction error. However, performing AutoML using total error can lead to a lack of model parsimony and does not focus on an algorithm's sensitivity to noise [20]. Also, AutoML provides no insight as to why a particular combination of algorithm modules should be chosen. Total error does not explain the behaviour of the algorithm, providing a lack of confidence in the chosen algorithm and a discrepancy between AutoML and the growing need for explainable artificial intelligence [21]. It also does not illustrate the stability of the algorithm and therefore the reliability of the AutoML process. Instead, decomposed error should be used in AutoML to examine algorithm behaviour as it is important to provide end users with greater insights into an algorithm's behaviour and therefore enhanced confidence in its performance [22,23]. Therefore, it needs to be explored how decomposed error can be used to guide algorithm refinement.

3   Algorithm Refinement of Genetic Programming Using Decomposed Error

Estimating decomposed prediction error involves splitting up total error into components based on the potential sources of error. In a typical bias-variance error decomposition, a total error measure (e.g., mean squared error) is decomposed into two primary components: a *bias* component quantifies the ability of the method to learn the underlying generating function of a problem, while a *variance* component quantifies the learning method's sensitivity to stochastic effects encountered during the learning process (e.g., the sampling of data) [24]. To enable variance due to the sampling of training data (external variance) as well as variance due to the algorithm (internal variance) to be represented, the standard bias-variance decomposition can be further expanded, as shown in [12]. Error due to internal variance captures the changes in model predictions observed over multiple algorithm runs using the same training data. When further decomposing the error, Tukey's outlier removal is performed in order to provide more stable decomposed error estimates [12]. The error is decomposed using a set of 100 runs (10 training sets and 10 runs per training set). In order to explore how decomposed error can be used to reliably and effectively refine the GP algorithm, this is performed using both automated and manual processes.

3.1   Automated Algorithm Refinement

A desirable option for algorithm refinement is to use an automated process. Starting with traditional GP, alternative modules can be examined, choosing one to minimise the largest component of error (reducing error due to bias, internal variance or external variance). The process can be repeated a specific number of times or until the largest component of error cannot be reduced. The process has been illustrated by focusing on particular key parts of an evolutionary algorithm, rather than examining all possible alternative modules.

The typical framework of an evolutionary algorithm is shown in Fig. 1. A number of design choices present themselves at each block of the diagram, meaning that there are potentially many choices to be made at each module. The automated algorithm refinement process examined in this paper focuses on the

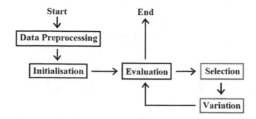

Fig. 1. Modules involved in an evolutionary algorithm process for symbolic regression.

selection and variation (crossover and mutation) parts of the algorithm (shown in red). However, in principle, any part of the algorithm could be examined using this process. The sequential process of targeting the reduction of the largest component of error was performed using these steps:

1. Run GP using an initial configuration of modules.
2. Decompose the error associated with the initial GP configuration.
3. Determine the largest component of error (i.e., error due to bias, internal variance or external variance).
4. Run GP using the current combination of modules except for swapping in each alternative module individually, calculating the decomposed error associated with each combination of modules.
5. Determine which new combination of modules reduces the largest component of error.
6. If the largest component of error cannot be reduced, stop the process and return the current combination of modules. Otherwise, determine the new largest component of error (i.e., repeating Step 3).
7. Repeat Steps 4 to 6 for n time steps (if the process has not already been terminated).

Steps 1 and 4 involve multiple complete runs of GP in order to decompose the error associated with each combination of modules that are examined. Modules from traditional GP, Angle-Driven Geometric Semantic GP ($ADGSGP$) [25] and GP using semantic similarity [26] have been selected to refine GP. The following individual modules were examined:

Selection Operators:

– Tournament selection (TS)
– TS and angle-driven selection (ADS) for crossover [25]
– Double tournament selection (DTS) [27]
– DTS and ADS for crossover

Crossover Operators:

– One point crossover (OPX)

- Perpendicular crossover (PC) [25]
- Semantic similarity-based crossover (SSC) [26]

Mutation Operators:

- Uniform subtree mutation (UM) using full growth
- Random segment mutation (RSM) [25]
- Semantic similarity-based mutation (SSM) based on SSC [26]

The modules of $ADGSGP$ were performed using the implementation of [25]. An initial GP configuration of OPX, UM and TS was used as the starting point for the refinement process, with a maximum of five steps. The refinement of the GP algorithm is explored using a variant of the Keijzer-5 function [28]:

$$f(x, y, z) = \frac{30(x - a)(z - a)}{((x - a) - 10)(y - a)^2} \tag{1}$$

where $a = 10$ for $x, z \in U[9, 11]$ and $y \in U[11, 12]$. A similar adaptation of the Keijzer-5 function is used by [29]. Equation (1) was used to generate 10 training folds of 100 observations and a test fold of 1000 observations. Algorithm refinement driven by decomposed error will generalise to other problems because decomposed error characterises the algorithm for the given problem and therefore will guide the choice of modules for that problem. GP is performed using the Distributed Evolutionary Algorithms in Python (DEAP) framework [30] and the parameters that are not considered for refinement are shown in Table 1, which are typical of those in recent work [31,32].

Table 1. Fixed parameters for GP

Parameter	Value
Population size	100
Number of generations	100
Probability of crossover	0.3
Probability of subtree mutation	0.7
Maximum depth	17
Initial minimum depth	2
Initial maximum depth	6
Minimum depth of subtree mutation	2
Maximum depth of subtree mutation	6
Elitism	Yes (1 individual)
Size of tournament	3
Function set	$\{+, -, \times, \div\}$

The decomposed error values associated with the steps taken to improve the algorithm are shown in Fig. 2. After initially running GP (using OPX, UM and

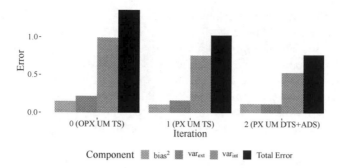

Fig. 2. Decomposed error for data generated by Eq. (1), using the automated algorithm refinement process to change the combination of GP modules implemented.

TS), it was determined that OPX should be swapped with PX in order to reduce the largest component of error (internal variance). Despite this configuration change, internal variance remains the largest error component. In the second iteration, TS was swapped with $DTS + ADS$ in order to continue to reduce error due to internal variance. Subsequent iterations were unable to find combinations of modules that reduced error due to internal variance. Therefore, the process terminated and the mutation operator remained unchanged. However, it is interesting to see that two modules from $ADGSGP$ (PX and ADS for crossover) have been selected using this process. While it is preferable to simultaneously minimise all components of error, the primary goal of each module swap is to reduce the largest component of error, and therefore total error. The process was effective in reducing error primarily due to internal variance but also external variance. However, there was an apparent trade-off between error due to variance and error due to bias. Swapping TS for $DTS + ADS$ provided a reduction in both types of error due to variance but with a slight increase in error due to bias.

3.2 Critique of Automated Algorithm Refinement

The motivations for using this automated algorithm refinement process are clear. First, by automating the process, the only human involvement required is in determining the initial combination of modules examined, the candidate modules and the parameters associated with these modules. Second, the process of changing only one module at each time step allows all candidate modules to be applied without the computational expense of trying all possible combinations. Finally, by choosing a combination of modules that appears to reduce the largest component of error, total error can be reduced (see Fig. 2). However, it needs to be determined whether the estimated decomposed error of GP provides stable enough estimates in order to assess whether the inclusion of a particular module reduces the largest component of error.

For the initial GP configuration of modules $\{OPX, UM, TS\}$ and the subsequent chosen combination of modules $\{PX, UM, TS\}$, the 100 runs ($M = 10$ and $R = 10$) used to estimate the decomposed error were repeated 50 times. The mean decomposed error values (and associated error bars representing one standard deviation) from the 50 repetitions are shown in Fig. 3. The identification of error due to internal variance as the largest component of error for both combinations of modules is consistent across the 50 repetitions. However, the mean internal variance value for the initial combination of $\{OPX, UM, TS\}$ is very similar to that for $\{PX, UM, TS\}$. Also, the error bars have significant overlap, with the inclusion of OPX sometimes providing lower error values compared to PX. While the magnitudes of error due to bias and error due to external variance for OPX and PX are relatively stable across repetitions, this is not the case for error due to internal variance. Across the 50 repetitions, some of them determined that swapping OPX for PX reduces error due to internal variance while other repetitions determined the opposite result (selecting a different module or terminating the process). OPX provided lower error 22 times, larger error 16 times and similar error 12 times (an absolute difference of less than 0.1). As it is not known which error component is targeted by choosing either OPX or PX, it is possible that they both target a reduction in error due to bias and therefore a reduction in error due to variance could not be achieved.

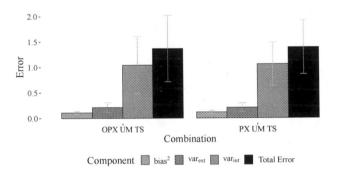

Fig. 3. Mean decomposed error (and associated error bars representing one standard deviation) for data generated by Eq. (1), for 50 repetitions of the first two combinations of GP modules involved in the automated refinement process (see Fig. 2).

These results show that automatic algorithm refinement is associated with difficulties in swapping algorithm modules in order to reduce the largest component of error. In particular, a comparison of modules that exhibit similar decomposed error does not help to substantially reduce the largest component of error. This can also provide inconsistent results when performed for multiple repetitions due to unstable estimates of error due to internal variance. Performing a manual algorithm refinement process might allow for more meaningful comparisons of modules, as we can choose to examine modules with a prior understanding that they are expected or known to target the largest component of error. Such a manual process is investigated in the next section.

3.3 Manual Algorithm Refinement

The motivation for a manual algorithm refinement process is to identify and then target a reduction in the largest component of error using a module known to successfully reduce that component of error. The automated algorithm refinement process used in Sect. 3.1 was able to consistently determine the largest component of error. However, individual runs gave inconsistent results as to whether changing an operator reduced the largest component of error. Therefore, it is plausible that the largest error component can be determined but used within a manual algorithm refinement process as a heuristic for reducing the largest component of error. This was performed by examining a single well-understood adaptation to an algorithm that targets the largest component of error (or is hypothesised to reduce the largest component of error), rather than blindly comparing alternative algorithm module combinations that involve uncertainty and may be too similar in terms of their error reducing behaviour.

Starting with the GP results for the initial combination of $\{OPX, UM, TS\}$ (see Fig. 3), the largest component of error is consistently error due to internal variance. Therefore, an adaptation to the algorithm needs to be applied in order to reduce this component of error. Bagging is well understood to reduce both error due to internal and external variance. However, there is evidence that bootstrapping of the training data is unnecessary when the error due to external variance component is much smaller than the error due to internal variance component [33]. Averaging the predictions from an ensemble of models without bootstrapping will, like bagging, reduce error due to internal variance. Therefore, an ensemble of 25 models (using the operator combination of $\{OPX, UM, TS\}$ and calculating the median value) was used to predict the test observations, with each set of 100 runs ($M = 10$ and $R = 10$) being performed 30 times. The mean and standard deviation of decomposed error for the ensemble algorithm is compared to that of the initial combination of operators in the first two rows of Table 2. An ensemble of models (without bootstrapping) provided a large reduction in error due to internal variance, which is statistically significant ($p < 0.0001$) using the Wilcoxon signed-rank test for the difference between the error due to internal variance components. The additional runs involved in model averaging provided more accurate predictions with a reduction in error due to bias as well as a reduction in error due to external variance.

The new largest error component associated with the operator combination of $\{OPX, UM, TS\}$ in an ensemble is error due to bias. This is consistent across all 30 repetitions. Therefore, a different type of adaptation needs to be applied in order to reduce error due to bias. As both GP with Z-Score standardisation (of explanatory and response variables) and GP with linear scaling have been shown to reduce error due to bias [31,34], GP was performed using a combination of both standardisation and linear scaling (GP_{Z+LS}). The decomposed error of the ensemble algorithm without feature scaling is compared to the ensemble

Table 2. Variants of GP examined using manual algorithm refinement

Variant	$bias^2$	var_{ext}	var_{int}	Total error
OPX UM TS	0.1138	0.2289	1.0826	1.4253
	±0.0005	±0.0951	±0.5985	±0.6959
OPX UM TS Ens	0.0441	0.0263	0.0054	0.0759
	±0.0005	±0.0012	±0.0008	±0.0018
GP_{Z+LS} OPX UM TS Ens	**0.0029**	**0.0025**	**0.0002**	**0.0057**
	±0.0000	**±0.0001**	**±0.0000**	**±0.0001**
GP_{Z+LS} OPX UM TS	0.0031	0.0055	0.0232	0.0318
	±0.0004	±0.0024	±0.0147	±0.0174

algorithm using GP_{Z+LS} in the second and third rows of Table 2 (with the lowest error component values bolded in the third row). Using GP_{Z+LS} provided a large reduction in error due to bias, which is statistically significant ($p < 0.0001$) using the Wilcoxon signed-rank test for the difference between the error due to bias components. It also reduced error due to both external and internal variance (with both differences statistically significant, both $p < 0.0001$). As exhibited when creating an ensemble of models without bootstrapping, all error components were reduced. Therefore, these wrapper methods have reduced total error without exhibiting a trade-off between error due to bias and error due to variance.

The new largest error component associated with an ensemble of models (without bootstrapping) using GP_{Z+LS} was still error due to bias (across all 30 repetitions). However, this component was only slightly larger than error due to external variance, with both components being significantly reduced by performing standardisation and linear scaling. Therefore, this manual refinement process reached an appropriate stopping point for the examined data set.

3.4   Critique of Manual Algorithm Refinement

Estimating which component of error is the largest component gave consistent results over multiple repetitions of the runs required to decompose error. This provides confidence in determining what type of adaptation to the algorithm needs to be made in order to reduce the largest component of error and therefore total error. While this process requires human involvement after performing multiple runs of each adaptation of the algorithm, it allows domain knowledge and targeted decision making to be exploited. It also provides an explainable refinement process by understanding the purpose of selecting particular modules in terms of decomposed error. The two adaptations to the algorithm (see Table 2) were successful in targeting the largest component of error, and therefore total error. However, it needs to be confirmed whether both adaptations were necessary to reduce total error and error due to internal variance in particular.

The final combination of modules before stopping the manual refinement process (GP_{Z+LS} using an ensemble of 25 models) is compared to GP_{Z+LS} without ensembling in the third and fourth rows of Table 2. The results show that a single model provides a statistically significant increase in error due to internal variance ($p < 0.0001$) compared to the ensemble model, using the Wilcoxon signed-rank test for the difference between the error due to internal variance components. This makes it clear that while standardisation and linear scaling provide lower error due to internal variance than the initial set of modules (without standardisation and linear scaling), they do not sufficiently target the component. An ensemble model is needed to target error due to internal variance as standardisation and linear scaling specifically target error due to bias. The standard deviation is also larger for error due to bias and error due to external variance for a single model.

By considering a small set of candidate algorithm adaptations that are known or expected to target the reduction of a specific error component, this manual process requires fewer runs of GP. Instead of performing many runs across many different algorithm adaptations, the focus can be on multiple repetitions of the same algorithm to determine the consistency of both the overall predictive performance and the magnitude of the error components. A small set of candidate algorithms is sufficient if, between them, they capture a reduction in all error components. This algorithm refinement process is not trying to find the algorithm with the best possible predictive performance but instead find an algorithm that provides reasonable performance as well as stable and well-understood behaviour. While only wrapper adaptations to the algorithm have been examined (feature scaling and ensemble models), adaptations internal to the algorithm can also be examined using this manual process. *A set of candidate algorithm adaptations or modules with known or expected behaviour, in terms of targeting a reduction of the largest component of decomposed error*, is a desirable characteristic of an AutoML process. By understanding the behaviour of algorithm adaptations or modules, the set of candidate options can be chosen more carefully in order to provide a diverse range of behaviour, leading to a more effective and explainable reduction of error.

4   Conclusion

This paper introduces the use of decomposed error for performing algorithm refinement. It has been applied to the refinement of GP using both automatic and manual processes. The results for the automatic algorithm refinement process show that comparing algorithm modules with similar decomposed error values makes it difficult to target a reduction of the largest error component. This is particularly the case for algorithms like GP than can exhibit a large and/or unstable error due to internal variance component and therefore can provide inconsistent conclusions about candidate algorithm adaptations. In order to make more meaningful comparisons, the manual refinement process focuses on choosing a candidate algorithm adaptation or module that is known to reduce the largest component of error, determined by manually examining the estimated decomposed error of the algorithm at each step in the process.

For the manual algorithm refinement process, the sequence of algorithm adaptations was successful in reducing the largest component of error, with the type of the largest error component changing throughout the process. Therefore, a set of candidate algorithm adaptations or modules need to provide diversity in reducing different components of decomposed error. Many traditional AutoML processes choose a set of candidate algorithm adaptations or modules without prior examination of their diversity in terms of reducing different components of error. Therefore, choosing candidate modules that coincidentally target a reduction of the same component of error will significantly limit the ability to improve the predictive performance of an algorithm. A greater understanding of how an algorithm module reduces prediction error, and the module's interaction with other modules, can be provided using the extended error decomposition. A more strategically chosen set of candidate algorithm modules, in terms of providing diverse behaviour in reducing different components of error, can then be applied to an automated algorithm refinement process. It is particularly important for the set of candidates to include a module that reduces error due to internal variance. This allows for the prediction error associated with an algorithm to be stabilised, if required, before being able to make reliable further refinements that target other components of error.

Although the manual algorithm refinement process was more successful than the automated process in reducing the largest component of error (and therefore total error), the motivations for automating algorithm refinement are still clear and important. Therefore, mapping the successful elements of the manual algorithm refinement process into an AutoML framework should be explored in future work. This would involve choosing a hierarchy of modules that are most important for reducing a diverse range of error components. This provides efficiency in reducing the search space of hyperparameters while providing confidence in the behaviour of the module combinations. The examination of algorithm refinement in this paper focused on the overall algorithm module structure rather than the tuning of parameters; this should be examined in future work.

Acknowledgment. Thank you to Dr Qi Chen for kindly allowing your ADGSGP code to be used as part of this paper.

References

1. Erickson, B.J., Korfiatis, P., Akkus, Z., Kline, T.L.: Machine learning for medical imaging. Radiographics **37**(2), 505–515 (2017)
2. Tuggener, L., et al.: Automated machine learning in practice: state of the art and recent results. In: 2019 6th Swiss Conference on Data Science (SDS), pp. 31–36. IEEE, New Jersey (2019)
3. Carleo, G.: Machine learning and the physical sciences. Rev. Mod. Phys. **91**(4), 045002 (2019)

4. Mitchell, T.: Machine Learning, ser. McGraw-Hill International Editions. McGraw-Hill, New York (1997). https://books.google.co.nz/books?id=EoYBngEACAAJ
5. Hastie, T., Tibshirani, R., Friedman, J.: The Elements of Statistical Learning: Data Mining, Inference, and Prediction. Springer, Berlin (2009). https://doi.org/10.1007/978-0-387-21606-5
6. Wolpert, D.H., Macready, W.G.: No free lunch theorems for optimization. IEEE Trans. Evol. Comput. **1**(1), 67–82 (1997)
7. Elshawi, R., Maher, M., Sakr, S.: Automated machine learning: state-of-the-art and open challenges, pp. 1–23. CoRR, vol. abs/1906.02287 (2019). http://arxiv.org/abs/1906.02287
8. Olson, R.S., Cava, W.L., Mustahsan, Z., Varik, A., Moore, J.H.: Data-driven advice for applying machine learning to bioinformatics problems. In: Pacific Symposium on Biocomputing 2018: Proceedings of the Pacific Symposium, pp. 192–203. World Scientific, Singapore (2018)
9. Thornton, C., Hutter, F., Hoos, H.H., Leyton-Brown, K.: Auto-weka: combined selection and hyperparameter optimization of classification algorithms. In: Proceedings of the 19th ACM SIGKDD International Conference on Knowledge Discovery and Data Mining, pp. 847–855. ACM, New York (2013)
10. Mohr, F., Wever, M., Hüllermeier, E.: Ml-plan: automated machine learning via hierarchical planning. Mach. Learn. **107**(8), 1495–1515 (2018)
11. James, G., Witten, D., Hastie, T., Tibshirani, R.: An Introduction to Statistical Learning: with Applications in R. STS, vol. 103. Springer, New York (2013). https://doi.org/10.1007/978-1-4614-7138-7
12. Owen, C.A., Dick, G., Whigham, P.A.: Characterising genetic programming error through extended bias and variance decomposition. IEEE Trans. Evol. Comput. **24**(6), 1164–1176 (2020)
13. Drozdal, J., et al.: Trust in automl: exploring information needs for establishing trust in automated machine learning systems. In: Proceedings of the 25th International Conference on Intelligent User Interfaces, pp. 297–307 (2020)
14. Adadi, A., Berrada, M.: Peeking inside the black-box: a survey on explainable artificial intelligence (xai). IEEE Access **6**, 52 138–52 160 (2018)
15. Bergstra, J., Bengio, Y.: Random search for hyper-parameter optimization. J. Mach. Learn. Res. **13**(2), 281–305 (2012)
16. Kotthoff, L., Thornton, C., Hoos, H.H., Hutter, F., Leyton-Brown, K.: Auto-WEKA: automatic model selection and hyperparameter optimization in WEKA. In: Hutter, F., Kotthoff, L., Vanschoren, J. (eds.) Automated Machine Learning. TSSCML, pp. 81–95. Springer, Cham (2019). https://doi.org/10.1007/978-3-030-05318-5_4
17. Feurer, M., Klein, A., Eggensperger, K., Springenberg, J.T., Blum, M., Hutter, F.: Auto-sklearn: efficient and robust automated machine learning. In: Hutter, F., Kotthoff, L., Vanschoren, J. (eds.) Automated Machine Learning. TSSCML, pp. 113–134. Springer, Cham (2019). https://doi.org/10.1007/978-3-030-05318-5_6
18. de Sá, A.G.C., Pinto, W.J.G.S., Oliveira, L.O.V.B., Pappa, G.L.: RECIPE: a grammar-based framework for automatically evolving classification pipelines. In: McDermott, J., Castelli, M., Sekanina, L., Haasdijk, E., García-Sánchez, P. (eds.) EuroGP 2017. LNCS, vol. 10196, pp. 246–261. Springer, Cham (2017). https://doi.org/10.1007/978-3-319-55696-3_16
19. Olson, R.S., Bartley, N., Urbanowicz, R.J., Moore, J.H.: Evaluation of a tree-based pipeline optimization tool for automating data science. In: Proceedings of the Genetic and Evolutionary Computation Conference, ser. GECCO 2016, pp. 485–492. ACM, New York (2016)

20. Brighton, H., Gigerenzer, G.: The bias bias. J. Bus. Res. **68**(8), 1772–1784 (2015)
21. Krawiec, K.: Behavioral Program Synthesis with Genetic Programming, vol. 618. Springer, Cham (2016). https://doi.org/10.1007/978-3-319-27565-9
22. Lipton, Z.C.: The mythos of model interpretability. Commun. ACM **61**(10), 36–43 (2018)
23. Arrieta, A.B., et al.: Explainable artificial intelligence (XAI): concepts, taxonomies, opportunities and challenges toward responsible AI. Inf. Fusion **58**, 82–115 (2020)
24. Bishop, C.M.: Pattern Recognition and Machine Learning (Information Science and Statistics). Springer-Verlag, Secaucus (2006)
25. Chen, Q., Xue, B., Zhang, M.: Improving generalization of genetic programming for symbolic regression with angle-driven geometric semantic operators. IEEE Trans. Evol. Comput. **23**(3), 488–502 (2019)
26. Uy, N.Q., Hoai, N.X., O'Neill, M., McKay, R.I., Galván-López, E.: Semantically-based crossover in genetic programming: application to real-valued symbolic regression. Genetic Program. Evol. Mach. **12**(2), 91–119 (2011)
27. Luke, S., Panait, L.: Fighting bloat with nonparametric parsimony pressure. In: Guervós, J.J.M., Adamidis, P., Beyer, H.-G., Schwefel, H.-P., Fernández-Villacañas, J.-L. (eds.) PPSN 2002. LNCS, vol. 2439, pp. 411–421. Springer, Heidelberg (2002). https://doi.org/10.1007/3-540-45712-7_40
28. Keijzer, M.: Improving symbolic regression with interval arithmetic and linear scaling. In: Ryan, C., Soule, T., Keijzer, M., Tsang, E., Poli, R., Costa, E. (eds.) EuroGP 2003. LNCS, vol. 2610, pp. 70–82. Springer, Heidelberg (2003). https://doi.org/10.1007/3-540-36599-0_7
29. Vladislavleva, E.J., Smits, G.F., Den Hertog, D.: Order of nonlinearity as a complexity measure for models generated by symbolic regression via pareto genetic programming. IEEE Trans. Evol. Comput. **13**(2), 333–349 (2009)
30. Fortin, F.-A., De Rainville, F.-M., Gardner, M.-A.G., Parizeau, M., Gagné, C.: Deap: evolutionary algorithms made easy. J. Mach. Learn. Res. **13**(1), 2171–2175 (2012)
31. Owen, C.A., Dick, G., Whigham, P.A.: Standardisation and data augmentation in genetic programming. IEEE Trans. Evol. Comput. (2022)
32. Dick, G., Owen, C.A., Whigham, P.A.: Evolving bagging ensembles using a spatially-structured niching method. In: Proceedings of the Genetic and Evolutionary Computation Conference, ser. GECCO 2018, pp. 418–425. ACM, New York (2018). http://doi.acm.org/10.1145/3205455.3205642
33. Owen, C.A.: Error decomposition of evolutionary machine learning (Thesis, Doctor of Philosophy). University of Otago (2021). http://hdl.handle.net/10523/12234
34. Owen, C.A., Dick, G., Whigham, P.A.: Feature standardisation in symbolic regression. In: Mitrovic, T., Xue, B., Li, X. (eds.) AI 2018. LNCS (LNAI), vol. 11320, pp. 565–576. Springer, Cham (2018). https://doi.org/10.1007/978-3-030-03991-2_52

Does a Compromise on Fairness Exist in Using AI Models?

Jianlong Zhou[1]([envelope]) [iD], Zhidong Li[1], Chun Xiao[2], and Fang Chen[1]

[1] Data Science Institute, University of Technology Sydney, Sydney, Australia
{jianlong.zhou,zhidong.li,fang.chen}@uts.edu.au
[2] Research Office, University of Technology Sydney, Sydney, Australia
chun.xiao@uts.edu.au

Abstract. Artificial Intelligence (AI) has been increasingly used to assist decision making in different domains. Multiple parties are usually affected by decisions in decision making, e.g. decision-maker and people affected by decisions. While various parties of users may have different responses to decisions regarding ethical concerns such as fairness, it is important to understand whether a compromise on fairness exists in using AI models. This paper takes AI-assisted talent shortlisting as a case study and investigates perception of fairness, trust, and satisfaction with decisions of both recruiters and applicants in AI-informed decision making. The compromises on fairness between decision-maker and people affected by decisions are identified which are then explained by social and psychological theories. The findings can be used to help find compromising points between decision-maker and people affected by decisions so that both parties can reach for a balanced state in decision making.

Keywords: AI ethics · Fairness · Trust · Satisfaction · Compromise

1 Introduction

Artificial Intelligence (AI) and Machine Learning (ML) algorithms have been increasingly used in shaping our everyday lives and activities in different domains especially human related decision making such as allocation of social benefits, hiring, and criminal justice [4,8,11]. As a result, the ethical issues of AI are becoming key concerns in algorithmic decision making. AI algorithms, trained on a large amount of historical data, may not only replicate, but also amplify existing biases or discrimination in historical data [32]. Therefore, fairness has especially been becoming one of actively discussed ethical concerns in AI-informed decision making tasks where multiple parties are usually involved and affected by decisions. Fairness is defined as a global perception of appropriateness – a perception that tends to lie theoretically downstream of justice [9]. In the algorithmic context, fairness means that algorithmic decisions should not create discriminatory or unjust consequences [28]. Examples of bias discrimination are 1) banks evaluating credit risks based on race or gender and not on financial score, and 2) courts judging the recidivism rate of prisoners based on races. Algorithmic fairness is a complicated topic and extensive research has been investigated focusing on fairness definitions (ranging from statistical bias, group fairness, to individual fairness) and unfairness quantification [10,12,22].

Taken the recruiting scenario in the human resources context as an example, AI algorithms are often used to shortlist applicants. The laws such as Australia's Anti-Discrimination Law require that different groups (e.g. male and female) should have equal employment opportunity, which implies that the shortlisting should keep a similar proportion of both male and female candidates for the fairness (equal opportunity for male and female candidates). The AI algorithm is designed and trained to meet such fairness requirement. When the AI algorithm is used to shortlist candidates, female candidates are hurt by the AI algorithm if they are shortlisted with a less proportional number than male candidates. This means that the level of fairness of the AI algorithm is not high enough. In addition, AI model accuracy is another factor that affects user's responses to AI solutions such as user trust [29]. For example, if the AI model accuracy is low, it may affect recruiters' trust because they may not get the most appropriate candidates for a position. However, if the AI model accuracy is very high, the applicants may have questions on the fairness of decisions since fairness usually comes with a trade-off over AI model accuracy [19,23,27]. As it can be seen from this recruiting scenario example, at least two parties are involved in AI-informed decision making: decision-makers (recruiters in this example) and people affected by decisions (applicants in this example). The influence of AI-informed decision making on them and their expectations are different: recruiters prefer high model accuracy to get the most appropriate candidates, while applicants prefer high fairness in recruiting to get equal opportunities. However, decisions usually cannot meet preferences from both parties at the same time so that both parties agree to and are satisfied with decisions.

As a result, important questions are posed on the use of AI:

- Whether people in different roles in AI-informed decision making have different perception of fairness, trust, and satisfaction with decision making?
- Whether there is a compromise on fairness between people in different roles in AI-informed decision making?

In order to answer these questions, this paper takes AI-assisted talent shortlisting as a case study and investigates perception of fairness, trust, and satisfaction with decision making of both recruiters and applicants in AI-informed decision making. Different introduced fairness (refers to the inherent algorithmic fairness) and model performance are introduced and manipulated in AI-informed decision making tasks. The responses in perception of fairness, trust, and satisfaction from recruiters and applicants at each introduced fairness level and model performance are compared to find any differences in responses from recruiters and applicants. Compromises on fairness between decision-maker and people affected by decisions are identified if both parties have the same responses in perception of fairness, trust, or satisfaction under a given introduced fairness level and model performance. A user study has been conducted to answer research questions.

2 Related Work

2.1 Fairness-Accuracy Trade-Off

A large amount of work has shown that fairness usually comes with a trade-off over accuracy. Zliobaite [38] presented a theoretical and empirical analysis of trade-offs between accuracy and fairness. They argued that comparison of non-discriminatory classifiers needs to account for different rates of positive predictions, otherwise conclusions about performance may be misleading in binary classification. Martinez et al. [20] used Pareto frontiers to dynamically re-balance subgroups' risks to minimize performance discrepancies across sensitive groups without causing unnecessary harm. They argue that even in domains where fairness at cost is required, finding a non-unnecessary-harm fairness model is the optimal initial step. Pleiss et al. [23] investigated the tension between minimising unfairness across different population groups while maintaining calibrated predictions. It shows that maintaining cost parity and calibration is desirable yet often difficult in practice. They argue that as long as calibration is required, no lower-error solution can be achieved.

Wang et al. [27] showed that traditional approaches that mainly focus on optimising the Pareto frontier of multi-task accuracy might not perform well on the trade-off between group fairness and accuracy. They proposed a new set of metrics to better capture the multi-dimensional Pareto frontier of fairness-accuracy trade-offs uniquely presented in a multi-task learning setting. Zhao and Gordon [31] theoretically and empirically investigated the problem of quantifying the trade-off between utility and fairness in learning group-invariant representations. They proved a lower bound to characterize the trade-off between fairness and the utility across different population groups.

2.2 Human Responses to AI

Since AI is often used by humans and/or for human-related decision making [26], humans' responses to AI play an important role in AI-informed decision making. This section reviews some of the most investigated human responses to AI including human's perceived fairness (perception of fairness), trust, and satisfaction.

The perception of fairness is a central component of maintaining satisfactory relationships with humans in decision making [1]. The perception of fair treatment on customers is found to be important in driving trustworthiness and engendering trust in the banking context [24].

In AI-informed decision making, algorithmic factors have been studied on how the technical design of an AI system affects people's fairness perceptions. For example, Lee et al. [16] found that people had different variations in the preferences for the three fairness metrics (equality, equity, efficiency) impacted by the decision. Human-related information has also been investigated on their effects on the perception of fairness. For example, education and age have been found affecting both perceptions of algorithmic fairness and people's reasons for the perception of AI fairness [14]. Zhou et al. [37] found that introduced fairness is positively related to perception of fairness.

User trust in AI-informed decision making has been extensively investigated from different perspectives. Zhou et al. [33,36] argued that communicating user trust benefits the evaluation of effectiveness of machine learning approaches. Confidence score, model accuracy and users' experience of system performance have been studied on their effects on user trust [30,34]. Zhou et al. [35] found that the presentation of influences of training data points significantly increased the user trust in predictions, but only for training data points with higher influence values under the high model performance condition.

Theoretical arguments and empirical evidence suggests that satisfaction be among the most important of reactions to the appraisal process [15]. User's satisfaction is another factor that affects the effectiveness of AI-informed decision making. For example, Allam and Mueller [2] found that visual and example-based explanations integrated with rationales had a significantly impact on patient satisfaction in AI diagnostic systems.

These previous work primarily focuses on responses from one party such as decision-maker's response or response of people affected by decisions in AI-informed decision making. However, less attention has been paid to responses from both sides of decision-makers and people affected by decisions in AI-informed decision. This study investigates the responses from both sides in AI-informed decision to find their differences and whether there is a compromise over decisions.

3   Preliminary Knowledge

Fairness is a complex and multi-faceted concept that depends on context and culture [3]. Various mathematical definitions of fairness have been summarised because of various reasons such as different contexts/applications, different stake-holders, impossibility theorems, as well as allocative versus representational harms. It shows that it is impossible to satisfy all definitions of fairness at the same time [3].

In this study, the statistical parity, one of group fairness definitions, is used to represent fairness. The statistical parity suggests that a predictor is fair if the prediction $\hat{Y}$ is independent of the protected attribute Z so that

$$P\left(\hat{Y}|Z\right) = P\left(\hat{Y}\right). \tag{1}$$

It also means that subjects in both protected and unprotected groups have equal probability (P) of being assigned to the positive predicted class. Taken the recruitment as an example, this would imply equal probability for male and female applicants to have positive predicted recruitment:

$$P\left(\hat{Y} = 1|Z = 0\right) = P\left(\hat{Y} = 1|Z = 1\right) \tag{2}$$

where $Z = 0$ represents male applicants and $Z = 1$ represents female applicants. Based on these preliminaries, statistical parity difference (PD) is defined as:

$$PD = \left| P\left(\hat{Y} = 1 | Z = 0\right) - P\left(\hat{Y} = 1 | Z = 1\right) \right| \tag{3}$$

where PD is in the range of $[0, 1]$. $PD = 0$ represents the complete fairness, and $PD = 1$ represents the complete unfairness. This paper manipulates various fairness levels of PD between $[0, 1]$ (called introduced fairness in this paper) to learn how introduced fairness is perceived and affects user responses in algorithmic decision making.

4   Method

4.1   Case Study

A company needs to recruit staff for a position. They posted the position description and a large number of applicants submitted their applications for the position. A machine learning system named Automatic Recruiting Assistant (ARA) is simulated to help to process applications and shortlist applicants for interviewing. ARA is a laboratory simulated candidate assessment tool that is supposed to use historical recruiting data to train a machine learning model and predict whether a candidate will be shortlisted.

In this study, a participant is told to act as either a Recruiter (R) or an Applicant (A) but not both. The participant is then required to conduct tasks and answer questions by giving information on the ARA performance information and shortlisting information of male and female applicants as a role of recruiter or applicant.

4.2   Fairness-Performance Space

In this study, introduced fairness (defined in Eq. 3) and model performance of ML models are manipulated and presented to participants to investigate responses of participants on the perception of fairness, trust, and satisfaction. Therefore, introduced fairness and performance form a 2D space. In this 2D space, each point represents a task condition of introduced fairness and model performance pair (f, p). The values in the dimension of model performance investigated include 70%, 80%, and 90% which correspond to low, middle, and high model performance respectively.

In the fairness dimension of the 2D space, the gender of applicants is used as the protected attribute in the recruitment scenario. The PD is used to measure the fairness and defined as the difference of shortlisted rate by the gender. In this study, fairness is introduced by manipulating PD with its discrete values of $0, 0.1, 0.2, 0.3, \ldots, 0.8, 0.9$, and 1.0, where each PD's discrete value was used as a measure of fairness to define the number of male and female applicants as well as number of male and female applicants shortlisted in each task respectively.

4.3 Task Design

In this study, tasks with different model performance and introduced fairness pair conditions were designed to investigate their effects on user's perception of fairness, trust, and satisfaction in AI-informed decision making. Table 1 shows 11 fairness presentation examples corresponding to different PD values. In this table, "Rate (M)" and "Rate (F)" represent the predicted success rate of male and female applicants respectively, "Male #" and "Female #" represent the number of male and female applicants respectively, and "Listed Male #" and "Listed Female #" represent the number of shortlisted male and female applicants respectively. All together 33 (11 × 3) tasks were designed and conducted by each participant based on eleven (11) fairness presentation examples and three (3) model performance levels (70%, 80%, 90%). Two additional training tasks were also conducted by each participant before the formal tasks. The order of formal tasks was randomized during the experiment to avoid any bias.

Table 1. Examples of fairness presentation in tasks.

Example#	PD	Rate (M)	Rate (F)	Male#	Female#	Listed Male#	Listed Female#
1	0	0.8	0.8	10	10	8	8
2	0.1	0.7	0.8	10	5	7	4
3	0.2	0.6	0.8	5	5	3	4
4	0.3	0.8	0.5	5	10	4	5
5	0.4	0.8	0.4	5	5	4	2
6	0.5	0.7	0.2	10	5	7	1
7	0.6	0.8	0.2	5	5	4	1
8	0.7	0.1	0.8	10	5	1	4
9	0.8	0.9	0.1	10	10	9	1
10	0.9	0.1	1	10	10	1	10
11	1	1	0	5	10	5	0

During the task time, each pair of fairness and model performance is firstly presented to participants with visualisations. Figure 1 shows the screenshot of visualisations in a task conducted in the experiment. The left barchart shows the number of applicants and number of applicants shortlisted by ARA for both males and females, which implies the fairness status in shortlisting for males and females. The right circular chart represents the model accuracy in shortlisting. After reading these information, participants are then asked to agree or reject decisions made by the ARA followed by different survey questions on perception of fairness, trust, and satisfaction in AI-informed decision making.

4.4 Scales of User Responses

Different questionnaires with Likert-type response scales are used in this study to collect responses of perception of fairness, trust, and satisfaction of users. The

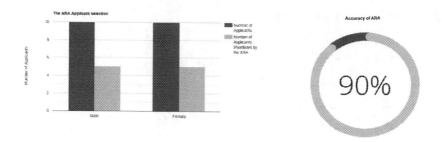

Fig. 1. Visualisation of fairness and model performance (accuracy of ARA).

scale is on a 5-point Likert-type response scale ranging from 1 (strongly disagree) to 5 (strongly agree) for each questionnaire on perception of fairness, trust, and satisfaction respectively.

Trust Scales. Trust is assessed with four items using self-report scales as the following [21].

– I am happy with help provided by the ARA.
– I have confidence in the advice given by the ARA.
– I can depend on the ARA.
– I can trust the ARA to make the correct selection.

Scales of Perception of Fairness. The perception of fairness of participants is assessed with the following two items.

– Overall, female and male applicants are treated fairly by ARA.
– I believe the ARA is a competent performer for both men and women.

Scales of Satisfaction. The satisfaction of participants is assessed with the following item [15,25]: overall, I am satisfied with the recruiting by considering both the performance of ARA and the fairness.

4.5 Experiment Setup

Due to social distancing restrictions and lockdown policies during the COVID-19 pandemic, this experiment was implemented using Python web framework and was deployed on the cloud server online. The deployed application link was then shared with participants to invite them to conduct tasks. In this study, participant responses to tasks were stored in a MySQL database.

4.6   Participants and Data Collection

In this study, 60 participants were recruited to conduct experimental tasks via various means of communications such as emails, text messages and social media posts who were mainly university students and 19 participants were females. Of all participants, 30 participants randomly acted as job applicants and other 30 participants acted as HR recruiters in the experiment.

After each task was displayed on the screen, the participants were asked to answer questions based on the task on perception of fairness, trust, and satisfaction in the AI-informed decision making respectively.

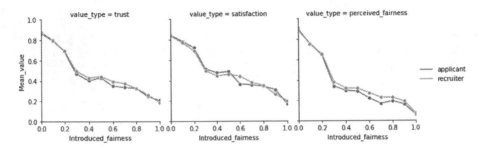

Fig. 2. Overall average responses in trust, satisfaction, and perception of fairness regardless of model performance.

5   Analyses and Results

This section analyses the collected data to answer our questions. We aim to understand whether two parties affected by decisions from AI have the same responses to AI-informed decision making from the perspectives of perception of fairness, satisfaction and trust.

When two parties have similar responses to decisions from AI under a given introduced fairness condition, it shows that they both agree with the effects of the specific introduced fairness on the decision. We can say that there is a compromise between two parties regarding the introduced fairness despite the decision maybe affecting them differently. When two parties show different responses to decisions under a given introduced fairness condition, it implies that there is a disagreement between two parties regarding the introduced fairness. The outcomes of the study can be used to customise user interface or take different measures when there is no compromise.

In order to perform the analyses, we first normalised the collected data of trust, satisfaction, and perception of fairness with respect to each subject to minimise individual differences in rating behavior using the following equation:

$$V_i^N = \frac{V_i - V_i^{min}}{V_i^{max} - V_i^{min}} \tag{4}$$

where V_i and V_i^N are the original rating values and the normalised rating values respectively from the user i, V_i^{min} and V_i^{max} are the minimum and maximum of the ratings of trust, satisfaction, or perception of fairness respectively from the user i in all of his/her tasks.

Figure 2 shows the overall average responses of participants in trust, satisfaction, and perception of fairness (or perceived fairness) regardless of model performance. t-tests were used to compare differences in trust, satisfaction, and perception of fairness between applicants and recruiters at each introduced fairness level. There are no statistically significant differences found in both trust and perception of fairness between applicants and recruiters at each introduced fairness level. However, it was found that recruiters showed statistically significantly higher trust than applicants at the introduced fairness level of 0.6 ($t = 1.9905$, $p < .048$), and no significant differences were found in trust between recruiters and applicants at other introduced fairness levels. The results also show the decreasing trends of trust, satisfaction, and perception of fairness with the increase of PD values on the horizontal axis (the decrease of introduced fairness levels), which is consistent with the previous research [37].

Figure 3 shows the average responses of participants in trust, satisfaction, and perception of fairness per different model performances. t-tests were applied to compare differences in trust, satisfaction, and perception of fairness between applicants and recruiters at each introduced fairness level under different model performances. From Fig. 3, it was found that:

- As it is expected, the recruiters have overall lower satisfaction when performance is low (at the region 1 in Fig. 3), and the applicants have overall lower satisfaction when fairness is low while performance is high. However, we observed the higher satisfaction at the region 1 for applicants even if the fairness is low. If we compare it to the region 2, then we can see the actual value at the region 1 is lower than the region 2. Here we argue that the applicants' satisfaction is higher than recruiters due to the low model performance.
- We observed that there was a significantly higher level of satisfaction from recruiters than applicants at the region 2 ($t = 2.4918$, $p < .0156$). This can be explained that even recruiters thought the fairness was poor, they were still satisfied with ARA.
- We also observed that recruiters showed lower trust under low model performance (at the region 3), this is further affected by fairness. If we compare the region 3 to the region 4, we can see that recruiters trust less at the region 3. We assume that the recruiters may consider fairness-accuracy trade-off here, since we can observe that their trust at the region 4 is higher than applicants, where the fairness is lower.
- We observed that the compromised setting can be achieved. It is obvious that high model performance (90%) and high fairness (close to 0 of introduced fairness) were highly rated and satisfied by both parties (the region 5). And low performance (70%) and low fairness were rated low and less satisfied by both parties (the region 1 and the region 4). The more compromised setting is at the region 7 that both parties had the same satisfaction.

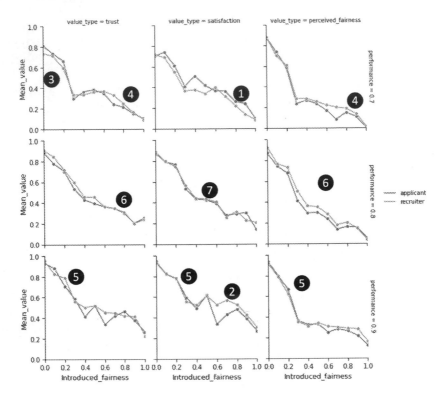

Fig. 3. Average responses of participants in trust, satisfaction, and perception of fairness per different model performances.

– Recruiters showed statistically significantly higher level of perception of fairness than applicants at the introduced fairness level of 0.7 when the model performance is 70% (t = 2.8366, p < .0062). Furthermore, we can see that almost all recruiters rated the perception of fairness higher than applicants when the model performance is 70%. This is maybe because that the recruiters may expect lower fairness to improve the performance given the trade-off between accuracy and fairness. However, recruiters showed statistically significantly lower level of satisfaction than applicants at the introduced level of 0.4 when the model performance is 70% (t = 3.1949, p < .0023). Under each studied model performances, we have not found other significant differences between recruiters and applicants in trust, satisfaction, and perception of fairness at different introduced fairness levels.

6   Discussions

Multiple parties are usually involved in an AI-informed decision making, e.g. decision-maker and people affected by decisions. Different parties may have different responses to a decision from AI-informed decision making. This study took the AI-assisted talent shortlisting as a case study and investigated satisfaction, trust, and perception of fairness of parties (recruiters and applicants) related to decisions respectively. The results showed that compromises on fairness did exist in AI-informed decision making under given model performances and introduced fairness levels.

Fairness heuristic theory [6, 18] suggests that when individuals face uncertain circumstances they rely on impressions of fairness to determine whether to cooperate and enter into exchange relationships with the other party, which suggests that individuals use fairness judgements to form their perceptions of trust. The social exchange theory [5] also argues that fair actions and the treatment by one party generate reciprocation in the form of trust by the other party in the exchange. In the context of talent shortlisting in human resource settings used in this paper, recruiters were unsure about the outcomes from the Automatic Recruiting Assistant when the model performance was low, resulting in the low perception of fairness as shown in the region 4 in the right diagram of the first row in Fig. 3, and therefore also resulting in low trust as shown in the region 4 in the left diagram of the first row in Fig. 3. The similar conclusion was observed for applicants as stated in the previous section.

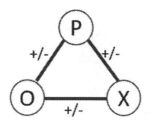

Fig. 4. Heider's POX model.

In the psychology of motivation, *balance theory* proposed by Fritz Heider [13] conceptualizes the cognitive consistency motive as a drive toward psychological balance. It assumes that individuals retain their psychological balance and develop their relationships with others or things within their circumstances. They prefer to maintain a balanced state through a series of cognitive operations that balance out their likes (represented by "+") and dislikes (represented by "-") to create equilibrium. Balance theory is often termed POX theory, representing the balanced/imbalanced state of individuals from the relationships among one person (P), the other person (O), and an attitudinal thing or object (X), as shown in Fig. 4. In this triadic relationship, a balance is achieved when there are three positive (+) links or two negatives (-) with one positive. Balance theory has been used in social psychology to understand various interpersonal relationships such as service quality, customer behaviour understanding [7,17]. Such balance theory can be used to explain the satisfaction of applicants and recruiters across different model performances in the talent shortlisting example conducted in this paper. As shown in Fig. 3, applicants showed an overall higher satisfaction level with AI than recruiters when the model performance is 70%, and vice versa when the model performance is 90%. All these result in "tensions" between applicants and recruiters. To reduce "tensions", this study modulated the model performance to 80%, and recruiters and applicants reached a balanced state (the region 7), where recruiters and applicants compromised and had the similar level of satisfaction.

The findings from this study can be used to help find compromising points between decision-maker and people affected by decisions so that both parties can reach for a balanced state in AI-informed decision making. Such findings also suggest AI developers as well as AI users that different stakeholders can be considered together in AI-informed decision making so that all stakeholders can satisfy with decisions.

7 Conclusion and Future Work

Since multiple parties are usually affected by decisions in AI-informed decision making and they have different responses regarding the fairness, this paper investigated whether there is a compromise on fairness in using AI models by examining user's satisfaction, trust, and perception of fairness in AI-informed decision making. The paper took the AI-assisted talent shortlisting as a case study to compare responses to decisions from recruiters and applicants. The results showed that compromises on fairness did exist in AI-informed decision making under given model performances and introduced fairness levels, which can be used to help find compromising points between decision-maker and people affected by decisions so that both parties can reach a balanced state The future work of this study will focus on the setup of a compromise profile for an AI-informed decision making through investigation of wider model performances such as from 50% to 100% and such profile can be used to guide the use of AI solutions for more effective decision making.

References

1. Aggarwal, P., Larrick, R.P.: When consumers care about being treated fairly: the interaction of relationship norms and fairness norms. J. Consumer Psychol. **22**(1, SI), 114–127 (2012)
2. Alam, L., Mueller, S.: Examining the effect of explanation on satisfaction and trust in AI diagnostic systems. BMC Med. Inform. Decision Making **21**(1), 178 (2021). https://doi.org/10.1186/s12911-021-01542-6
3. Bellamy, R.K.E., et al.: AI fairness 360: an extensible toolkit for detecting, understanding, and mitigating unwanted algorithmic bias. arXiv:1810.01943 [cs] (2018)

4. Berk, R., Heidari, H., Jabbari, S., Kearns, M., Roth, A.: Fairness in criminal justice risk assessments: the state of the art. Sociological Methods & Researchm p. 0049124118782533 (2018)
5. Blau, P.M.: Exchange and Power in Social Life. Wiley, New York, NY (1964)
6. van den Bos, K.: Uncertainty management: the influence of uncertainty salience on reactions to perceived procedural fairness. J. Person. Soc. Psychol. **80**(6), 931–941 (2001)
7. Carson, P.P., Carson, K.D., Knouse, S.B., Roe, C.W.: Balance theory applied to service quality: a focus on the organization, provider, and consumer triad. J. Bus. Psychol. **12**(2), 99–120 (1997). https://doi.org/10.1023/A:1025061816323,
8. Chen, F., Zhou, J.: Humanity Driven AI: Productivity, Well-being, Sustainability and Partnership. Springer, Cham (2022). https://doi.org/10.1007/978-3-030-72188-6
9. Colquitt, J.A., Rodell, J.B.: Measuring justice and fairness. In: Cropanzano, R.S., Ambrose, M.L. (eds.) The Oxford Handbook of Justice in the Workplace, pp. 187–202. Oxford University Press (2015)
10. Corbett-Davies, S., Goel, S.: The measure and mismeasure of fairness: a critical review of fair machine learning. arXiv preprint arXiv:1808.00023 (2018)
11. Feldman, M., Friedler, S.A., Moeller, J., Scheidegger, C., Venkatasubramanian, S.: Certifying and removing disparate impact. In: Proceedings of KDD2015, pp. 259–268 (2015)
12. Glymour, B., Herington, J.: Measuring the biases that matter: the ethical and casual foundations for measures of fairness in algorithms. In: Proceedings of the Conference on Fairness, Accountability, and Transparency, pp. 269–278 (2019)
13. Heider, F.: The Psychology of Interpersonal Relations. Wiley (1958)
14. Helberger, N., Araujo, T., de Vreese, C.H.: Who is the fairest of them all? public attitudes and expectations regarding automated decision-making. Comput. Law Secur. Rev. **39**, 105456 (2020)
15. Jawahar, I.M.: The influence of perceptions of fairness on performance appraisal reactions. J. Labor Res. **28**(4), 735–754 (2007)
16. Lee, M.K., Jain, A., Cha, H.J., Ojha, S., Kusbit, D.: Procedural justice in algorithmic fairness: leveraging transparency and outcome control for fair algorithmic mediation. Proc. ACM Hum. Comput. Interact. **3**, 1–26 (2019)
17. Lin, C.F., Fu, C.S., Chen, Y.T.: Exploring customer perceptions toward different service volumes: an integration of means-end chain and balance theories. Food Qual. Preferen. **73**, 86–96 (2019)
18. Lind, E.: Fairness heuristic theory: justice judgments as pivotal cognitions in organizational relations. In: Advances in Organizational Justice, pp. 56–88. Stanford University Press (2001)
19. Liu, S., Vicente, L.N.: Accuracy and fairness trade-offs in machine learning: a stochastic multi-objective approach. arXiv preprint arXiv:2008.01132 (2020)
20. Martinez, N., Bertran, M., Sapiro, G.: Fairness with minimal harm: a pareto-optimal approach for healthcare. arXiv preprint arXiv:1911.06935 (2019)
21. Merritt, S.M., Heimbaugh, H., LaChapell, J., Lee, D.: I trust it, but i don't know why: effects of implicit attitudes toward automation on trust in an automated system. Hum. Fact. **55**(3), 520–534 (2013)
22. Nabi, R., Shpitser, I.: Fair inference on outcomes. In: Proceedings of the AAAI Conference on Artificial Intelligence, vol. 2018, p. 1931. NIH Public Access (2018)
23. Pleiss, G., Raghavan, M., Wu, F., Kleinberg, J., Weinberger, K.Q.: On fairness and calibration. Adv. Neural Inform. Process. Syst. **30** (2017)

24. Roy, S.K., Devlin, J.F., Sekhon, H.: The impact of fairness on trustworthiness and trust in banking. J. Market. Manage. **31**(9–10), 996–1017 (2015)
25. Sholihin, M.: How does procedural fairness affect performance evaluation system satisfaction? (evidence from a UK police force). Gadjah Mada Int. J. Bus. **15**, 231–247 (2013). https://doi.org/10.22146/gamaijb.5445
26. Starke, C., Baleis, J., Keller, B., Marcinkowski, F.: Fairness perceptions of algorithmic decision-making: a systematic review of the empirical literature (2021)
27. Wang, Y., Wang, X., Beutel, A., Prost, F., Chen, J., Chi, E.H.: Understanding and improving fairness-accuracy trade-offs in multi-task learning. In: Proceedings of the 27th ACM SIGKDD Conference on Knowledge Discovery & Data Mining, pp. 1748–1757 (2021)
28. Yang, K., Stoyanovich, J.: Measuring fairness in ranked outputs. SSDBM 2017 (2017). https://doi.org/10.1145/3085504.3085526
29. Yu, K., Berkovsky, S., Taib, R., Zhou, J., Chen, F.: Do i trust my machine teammate? an investigation from perception to decision. In: Proceedings of the 24th International Conference on Intelligent User Interfaces, pp. 460–468. IUI 2019, ACM (2019)
30. Zhang, Y., Liao, Q.V., Bellamy, R.K.E.: Effect of confidence and explanation on accuracy and trust calibration in AI-assisted decision making. In: Proceedings of the 2020 Conference on Fairness, Accountability, and Transparency, pp. 295–305. FAT* 2020 (2020)
31. Zhao, H., Gordon, G.: Inherent tradeoffs in learning fair representations. Adv. Neural Inform. Process. Syst. **32** (2019)
32. Zhao, J., Wang, T., Yatskar, M., Ordonez, V., Chang, K.W.: Men also like shopping: reducing gender bias amplification using corpus-level constraints. In: Proceedings of the 2017 Conference on Empirical Methods in Natural Language Processing, pp. 2979–2989. Copenhagen, Denmark (2017)
33. Zhou, J., Bridon, C., Chen, F., Khawaji, A., Wang, Y.: Be informed and be involved: effects of uncertainty and correlation on user's confidence in decision making. In: Proceedings of the 33rd Annual ACM Conference Extended Abstracts on Human Factors in Computing Systems, pp. 923–928 (2015)
34. Zhou, J., Chen, F. (eds.): Human and Machine Learning: Visible, Explainable, Trustworthy and Transparent. Springer, Cham (2018)
35. Zhou, J., Hu, H., Li, Z., Yu, K., Chen, F.: Physiological indicators for user trust in machine learning with influence enhanced fact-checking. In: Machine Learning and Knowledge Extraction, pp. 94–113 (2019)
36. Zhou, J., et al.: Measurable decision making with GSR and pupillary analysis for intelligent user interface. ACM Trans. Comput. Hum. Interact. **21**(6), 1–23 (2015)
37. Zhou, J., Verma, S., Mittal, M., Chen, F.: Understanding relations between perception of fairness and trust in algorithmic decision making. In: Proceedings of the International Conference on Behavioral and Social Computing (BESC 2021), pp. 1–5 (2021)
38. Zliobaite, I.: On the relation between accuracy and fairness in binary classification. arXiv preprint arXiv:1505.05723 (2015)

Fairness Aware Swarm-based Machine Learning for Data Streams

Diem Pham[1,2(⊠)] [iD], Binh Tran[1] [iD], Su Nguyen[1] [iD],
and Damminda Alahakoon[1] [iD]

[1] La Trobe University, Victoria, Australia
{thixuandiem.pham,b.tran,p.nguyen4,d.alahakoon}@latrobe.edu.au
[2] Can Tho University, Can Tho, Vietnam

Abstract. Machine learning has been widely applied to extract insights from streaming data. However, ethical issues such as fairness have emerged related to these decision-support systems. Feature engineering methods have shown potential in representing and learning of fairness learning. However, these techniques have not been applied to streaming data. In this paper, we proposed a fairness-aware swarm-based machine learning for streaming data. The novelty of this algorithm is in the utilisation of two swarms, one for classification by building a network of prototypes and one for discrimination mitigating using feature weighting. Experiments with well-known datasets in fairness learning show that the proposed methods can improve fairness while maintaining the classification performance.

Keywords: Fairness · Swarm intelligence · Feature weighting · Data stream

1 Introduction

Machine learning (ML) have recently been adopted in a wide range of applications from finance, education to healthcare [2,23]. ML is applied to analyse and extract insights from data streams have become increasingly popular. Data stream classification is one of important tasks in data stream analytics and has attracted more attention in community due to their practical applications. Since ML models are trained based on data collected from human activities and life styles (e.g. business users, customers), they may inherit some unfairness against underrepresented groups. Discrimination ML methods make decisions toward certain individuals or groups [21]. The issue of discrimination is often related to the use of sensitive or protected attributes such as race, gender, religion in the prediction model. Ethical issues are found in many ML algorithms. For example, the online advertisements on Facebook Ads, Google Ads exclude groups such as families with women and children or disability in housing advertisements [5]. Since these policies/practices can hurt certain groups or individuals, they have become one of the important topics in the ML research community.

H. Aziz et al. (Eds.): AI 2022, LNAI 13728, pp. 205–219, 2022.
https://doi.org/10.1007/978-3-031-22695-3_15

Since discrimination is caused by the use of protected features, a simple solution for fairness would remove the protected features. However, it may result in a poor prediction accuracy due to the potential information loss. In addition, it may not completely remove discrimination due to the existence of admissible features which can work as a proxy for sensitive features to affect on the model's decision. Therefore, feature selection methods [4,6,27] have been proposed as pre-processing methods to select a subset of features that can mitigate the discrimination and maintain the accuracy. Results of these fair FS methods have shown promising in addressing fairness in ML. However, most of them are computationally expensive [4,27], and they mainly focus on static data, so they are not suitable for online learning in which models are continuously updated based on the incoming data streams. In addition, their performance is still limited due to the potential information loss when removing features. Instead of removing features, feature weighting can mitigate information loss by assigning a smaller weight for protected and admissible features and a larger weight for inadmissible features. In this study, a combination of online learning and feature weighting using swarm intelligence to approach fairness for data stream is proposed.

Swarm intelligence (SI) is an important category of evolutionary computation [20] that is inspired by the collective behaviours and social intelligence of animals. The behaviours and interactions of individuals in swarm form the intelligence for learning methods. Individuals find the best solution in their local area and share the information of the current state with others to guide their future search and explore better solutions [26]. SI is a potential approach for learning in dynamic environments because SI methods learn based on interactions between individuals which learn from each other by sharing knowledge to adapt themselves to environments and cooperate to solve a problem. Particle Swarm Optimisation (PSO) is a well-known algorithm which uses a population of particles to search for solutions [14]. PSO has been widely applied to feature weighting methods [25] which assign high weight for informative features and vice versa in different machine learning tasks. Recently, dynamic self-organising swarm method (DSOS) [24] was proposed to incrementally learn a network of prototypes representing data. DSOS can automatically adjust the set of prototypes based on the distribution of incoming data. In this study, we propose a Fairness Aware Swarm-based Machine Learning for Data Streams (FAS Stream). Specifically, FAS Stream system includes a DSOS swarm and a PSO swarm. DSOS is used to learning prototypes for classification while a PSO swarm is used to find feature weights for mitigating discrimination in learning. The major contributions of this study are:

- A new fairness aware online classification method based on two swarms, one for classification task and one for improving fairness.
- A new feature weighting method based on swarm intelligence to reduce discrimination for streaming data.

2   Background

2.1   Machine Learning for Data Streams

Machine learning (ML) has recently been used to explore patterns/knowledge from streaming data. Many online learning algorithms have been proposed for clustering, classification and sequential pattern mining on data streams [22]. However, learning streaming data is still challenging because data arrives continuously and data distribution may change over time (known as concept drifts) [17]. Therefore, to maintain their performance, ML methods are required to evolve and adapt to the changes in learning data stream [22]. Major learning methods which attract more attention from the data stream learning community include ensemble learning, incremental learning and lazy learning. An ensemble is composed of individual learners whose decisions are combined to make a final decision. This approach works based on the assumption that multiple individual learners can cover changes and different distributions in streaming data [16]. Results in Adaptive Random Forest Classification [8] and Accuracy Weighted Ensemble classifier [29] have shown the advantages of this approach. Another approach in data stream learning is incremental learning whose models are adjusted overtime to adapt quickly to the new concepts in streaming data [9]. Hoeffding Tree Classifier [10] and Hoeffding Adaptive Tree Classifier are well known incremental learning methods. Also, lazy methods such as k-Nearest Neighbour (KNN) classifier and regressor are also popular in learning data stream. SAMKNN [19] is a state-of-the-art method in this category. It combines long term memory and short term memory to make decisions in streaming data.

2.2   Fairness Aware Machine Learning

Fairness aware learning methods aim to mitigate discrimination while maintain their prediction performance. Three main approaches for fairness aware machine learning are pre-processing, in-processing and post-processing, which aim to improve fairness by intervening either before, during, or after the model training process. Data transformation to remove discrimination is one of the first pre-processing approaches to fairness. To make the training data distribution fairer among the protected and unprotected groups, these methods change the data class labels (massage data) [12] or re-weight the instances to create a more balance sample [12]. These techniques are lately applied to data stream in Fairness-enhancing interventions (FEI) [11] to modify incoming data chunks before using them for model training. FEI is one of the rare fairness-aware ML methods for data stream that follow the pre-processing approach. The results of these data transformation methods showed that they helped classifiers improve their performance in terms of accuracy and fairness. However, modifying and removing data may lead to meaningless data [15].

Fair feature selection (FS) is another pre-processing approaches that avoid the data transformation limitation. Instead of modify data instances (horizontal approach), feature selection tries to remove sensitive or protected features (vertical approach) to mitigate bias in the trained model. There are two main approaches in fair FS, associational and causal. While associational methods [4] aim to reduce the inequalities in the prediction outcomes between two groups of the sensitive feature, causal methods [27] try to identify and remove the influence of sensitive features on the outcomes. In [4], a fair FS method that follows associational approach is proposed using forward feature selection to find the minimal subset that maximise the objective function of $AUC - weight * discrimination$. Different weights were investigated with three classification methods and six discrimination measures showed that feature selection helps to produce fairer prediction when increasing the weight. Following causal approach, FAIREXP [27] first constructs new features from the original features by recursively applying transformations with the aim of extracting unbiased and useful information for prediction. The constructed features are selected using forward FS to improve accuracy and removed using backward FS for fairness improvement. While the causal approach provides a better understanding of the influence of sensitive features on the outcomes, it usually assume to have the underlying causal structure, which may not always be available, especially in a dynamic environment as data stream. Besides FS and feature construction, feature weighting has also been considered to address fairness. In [3], a system is developed to allow users manually choose feature weights. By showing boundaries that partition the space into regions where the desired fairness constraint is satisfied, the system helps users to obtain greater fairness in their subjective weight selection process. However, this approach can not scale well with high dimensional data. In general, results of fair FS, feature construction, and feature weighting have shown promise; however, they have not been explored in streaming data.

Instead of investigating fairness in the training data, in-processing methods try to modify models in the training process to remove discrimination. They usually inject fairness or discrimination measure to the objective function. Fairness aware strategies have been proposed for tree-based algorithms [1,13], logistic regression and support vector machines [30]. Fair tree-based methods [31,32] are also proposed for streaming data. They are developed based on the Hoeffding Tree algorithm [10] by incorporating fairness gain into information gain to improve both accuracy and fairness.

Although pre-processing and in-processing approaches are popular; they are inapplicable when training models are not allowed to modify. Post-processing techniques are used instead to modify the output predictions for fairness.

In general, most existing studies mainly focus on static data. Research on streaming data has still limited. To the best of our knowledge, feature engineering and feature weighting has not yet been applied to fairness aware online learning.

3   Proposed Method

3.1   Overview

The goal of this research is to develop a fairness-aware classification system for data streams based on swarm intelligence. Figure 1 presents a system overview

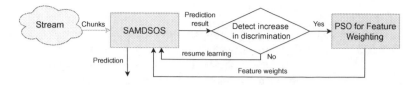

Fig. 1. The overview of FAS Stream.

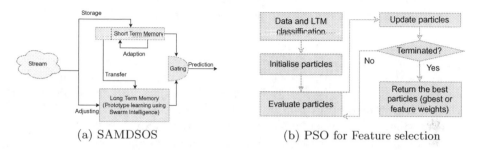

(a) SAMDSOS   (b) PSO for Feature selection

Fig. 2. The two components of FAS Stream.

with two main components, and Fig. 2 provides a closer look at each component. The first component is a dynamic self-organising swarm-based classification with long-short terms memory (SAMDSOS) that can build adaptive classifiers to streaming data incrementally. The second component is a particle swarm optimisation (PSO) feature weighting algorithm. PSO works as a pre-processing step to evolve a weight for each feature to minimise the discrimination of the classification model. It is only triggered when the new coming data chunk has a higher classification discrimination than the average obtained so far. The learnt feature weights are then fed into SAMDSOS for classification. Detail explanation of the two components will be presented in the following subsections.

3.2 Memory Updating Strategy in SAMDSOS

SAMDSOS is a new algorithm extended from the KNN Classifier with Self Adjusting Memory for Heterogeneous Concept Drift (SAM) algorithm [19] which applies long-term memory and short-term memory to address data stream classification problems. The key difference between SAMDSOS and SAM is the incorporation of the dynamic self-organising swarm (DSOS) algorithm [24] to efficiently update and evolve the long-term memory.

SAMDSOS maintains two memories including $\mathcal{M}_S$ to store current concepts and $\mathcal{M}_L$ to store past concepts. While short term memory $\mathcal{M}_S$ records the most recent data points which allow the proposed system to adapt quickly to the changes in data distributions, the key task of the $\mathcal{M}_L$ is to incrementally learn the global data distribution and maintain useful patterns for classification. In SAMDSOS, DSOS is used to construct the long-term memory $\mathcal{M}_L$ by learning a prototype network that capture the non-stationary data distribution from data streams (more details are provided in Sect. 3.3). In the prototype network, each node is a particle whose position will be updated using DSOS. Predictions are made based on the competition between three sub-models including $\mathcal{M}_S$, $\mathcal{M}_L$ and $\mathcal{M}_C$ (combining $\mathcal{M}_S$ and $\mathcal{M}_L$) based on their performance over time.

Algorithm 1 shows the pseudo-code of SAMDSOS. Initially, both $\mathcal{M}_S$ and $\mathcal{M}_L$ is empty. The first data chunk $D_1 = \{(\mathbf{x}_1, y_1), \ldots, (\mathbf{x}_n, y_n)\}$ in the data stream ($\mathbf{x}_i$ and y_i are the feature values and label of example i respectively) will be used to populate $\mathcal{M}_S$ and to create the minimal prototype network of $\mathcal{M}_L$ (the first two particles with positions copied from the two first data points). In the first stage, default feature weights (weights of sensitive features are zero and weights of other features are one) are applied to $\mathcal{M}_L$. Over time, if the size of $\mathcal{M}_S$ reach a pre-defined threshold max_STM_size, the outdated data will be transferred into $\mathcal{M}_L$ (lines 10–15) which will be updated by using DSOS. To avoid conflicts in predictions from $\mathcal{M}_S$ and $\mathcal{M}_L$, a repairing procedure in $\mathcal{M}_L$ is triggered (lines 16–24). With a new data point $(\mathbf{x}_i, y_i)$, SAMDSOS will identify all data points in $\mathcal{M}_S$ with the class y_i and determine the longest distance θ between these points and $\mathbf{x}_i$. The distance θ can be used as an adaptive radius within which data points will belong to the same class. In $\mathcal{M}_L$, the distances between particles and $\mathbf{x}_i$ will be calculated. The particles that have distances less than θ and the label different from y_i will be moved away from $\mathbf{x}_i$. These particles will be moved to their closest neighbour whose label is the same as the particle and out of the threshold radius. The goal of this strategy is to keep similar prototypes close together and far away from inconsistent ones. Furthermore, to ensure that $\mathcal{M}_S$ contains the most current concept, a model adaptation mechanism is applied (lines 25–36). Whenever the concept changes, a bisection method is used to refine $\mathcal{M}_S$ based on interleaved Test-Train errors on the current window data [19]. Discarded data point in $\mathcal{M}_S$ will be transferred to $\mathcal{M}_L$ using DSOS.

3.3   Updating Long Term Memory with DSOS

$\mathcal{M}_L$ is a swarm that includes a set of particles incrementally learning to optimally represent input data based on DSOS algorithm [24]. Each incoming data is considered as a food source that attract particles to move forward to. The two particles become neighbours when they are the two closest ones to a food source. Whenever a food source is in a particle coverage, the particle and its neighbours will share the information to move, consume and gain energy. If the food source is far from particles, a new particle is created. The connection between particles and their neighbours are maintain by repulsive force. If the two particles do not share common food source over time, the repulsive force increase. The connection between two particles is broken when the repulsive force reach a typical threshold. Thank to this mechanism, the swarm avoid crowded groups of particles. When the repulsive force meets a threshold, the connection between two particles will be broken. They are not neighbours do not share food source information any more. Over time, exhausted particles (zero energy) or isolated particles (no connection) will be removed. Due to the page limitation, technical

---

Algorithm 1: SAMDSOS

Input : Data: $D_t = \{(\mathbf{x}_1, y_1), \ldots, (\mathbf{x}_n, y_n)\}$ in which n is the window size, feature weights evolved by PSO

Output: updated $\mathcal{M}_S$ and updated $\mathcal{M}_L$

1 **begin**
2   Initialise Swarm in DSOS: $\mathcal{M}_L = \{\}$;
3   Initialise the short term memory $\mathcal{M}_S = \{\}$;
4   **if** $\mathcal{M}_L$ *is empty* **then**
5   initialise the first two particles in $\mathcal{M}_L$ by using the values of the first two data inputs $\{(\mathbf{x}_1, y_1), (\mathbf{x}_2, y_2)\}$ in D_t ;
6   **end**
7   **for** *each* $\mathbf{x}_i \in D_t$ **do**
8         $\mathcal{M}_S \leftarrow \mathcal{M}_S \bigcup \{\mathbf{x}_i\}$;
9         $m \leftarrow \mathcal{M}_S.size$;
10   **if** $m < max\_STM\_size$ **then**
11          $numshift \leftarrow max\_STM\_size - m$;
12          $moved\_set \leftarrow \{(\mathbf{x}_1, y_1), \ldots, (\mathbf{x}_{numshift}, y_{numshift})\}$;
13          $\mathcal{M}_S \leftarrow$ remove $moved\_set$ from $\mathcal{M}_S$;
14   Use $moved\_set$ and feature weights to update $\mathcal{M}_L$ by using DSOS algorithm ;
15   **end**
16        $\mathcal{X} \leftarrow \mathbf{x}'$ in $\mathcal{M}_S$ with same class as $\mathbf{x}_i$;
17   Find $\mathbf{x}^* \in \mathcal{X}$ with the largest distance from $\mathbf{x}_i$;
18        $\theta \leftarrow distance(\mathbf{x}^*, \mathbf{x}_i)$;
19        $\mathcal{Z} \leftarrow \mathbf{z} \in \mathcal{M}_L$ with distance $(\mathbf{z}, \mathbf{x}_i) < \theta$ and class different from y_i;
20   **for** $\mathbf{z}$ *in* $\mathcal{Z}$ **do**
21          $\mathcal{Z}' \leftarrow \mathbf{z}' \in \mathcal{M}_L$ and same class as $\mathbf{z}$;
22   Choose the closest $\mathbf{z}^* \in \mathcal{Z}'$ with $distance(\mathbf{z}, \mathbf{z}^*) > \theta$;
23   Move $\mathbf{z}$ to $\mathbf{z}^*$ with moving_step equal to $0.5 * distance(\mathbf{z}, \mathbf{z}^*)$
24   **end**
25        $m \leftarrow \mathcal{M}_S.size$;
26   **if** $m > 2 * min\_size$ **then**
27   **for** l *in* $(m, m/2, m/4, \ldots)$, *with* $l > min\_size$ **do**
28           $\mathcal{M}_S^l \leftarrow \{(\mathbf{x}_{m-l+1}, y_{m-l+1}), \ldots, (\mathbf{x}_m, y_m)\}$;
29   **end**
30   Choose l where $\mathcal{M}_S^l$ get smallest error in making prediction. ;
31   **if** $l < m$ **then**
32            $moved\_set \leftarrow \{(\mathbf{x}_0, y_0), \ldots, (\mathbf{x}_{m-l}, y_{m-l})\}$;
33            $\mathcal{M}_S \leftarrow$ remove $moved\_set$ from $\mathcal{M}_S$;
34   Apply DSOS to update $\mathcal{M}_L$ based on $moved\_set$ and feature weights;
35   **end**
36   **end**
37   **end**
38   return updated $\mathcal{M}_S$ and updated $\mathcal{M}_L$
39 **end**

details related to DSOS are not presented in this paper. Interested readers can check the original DSOS for more details [24]. DSOS is used in this paper instead of the long term memory adopted in SAM [19] because DSOS can adapt and grow its swarm based on data rather than replying on pre-defined parameters. This feature will make SAMDSOS more efficient in terms of memory requirements compared to SAM.

SAMDSOS uses KNN algorithm in sub-models to make prediction. Each particle p in $\mathcal{M}_L$ will be assigned a label $Class_p$ based on the distribution of data points hitting the particles in the earlier updating steps, in which $Class_p$ is the majority class of partile p. To catch new trend in data stream, SAMDSOS proposes a weighting strategy which gives newer data a larger weight in voting labels for particles as Eq. (1).

$$W_p^c = \mathbf{N}_p^c * \frac{1}{t} \sum_{i=0}^{i=\mathbf{N}_p^c} t_{p,i}^c \tag{1}$$

where:

- W_p^c: weight of class c in particle p. The class c with the highest weight W_p^c will be assigned to $Class_p$.
- $\mathbf{N}_p^c$: the number of data points with class c hitting the particle p.
- $t_{p,i}^c$: the time when the ith data point with label c hitting the particle p.
- t: arrival time of the last data point. In this paper, we simply use indices of data points in the original data set as time t.

3.4   PSO for Feature Weighting

The goal of PSO is to determine feature weights that minimise discrimination and maintain accuracy in classification. It must be noted that PSO swarm here is independent from the swarm in DSOS discussed earlier. Here, a position of a particle represents the weights of features. The position of particle k is a d-dimension vector $p_k = [p_{k,1}, p_{k,2}, p_{k,3}, ..., p_{k,d}]$ of real numbers corresponding to d features and each weight will range from 0 to 1.

To guide particles moving toward fair solutions, this research proposes a fitness function to evaluate particles. The fitness function shown in Eq. (2) is the weighting function of accuracy and discrimination.

$$Fitness(p_k) = w_1 * acc(p_k) + (1 - w_1) * (1 - disc(p_k)) \tag{2}$$

where $acc(p_k)$ and $disc(p_k)$ are the accuracy and discrimination obtained by applying features with weights for classifying the current window data, respectively. As shown as Eq. (2), a solution with higher accuracy and lower discrimination is the better solution. w_1 allow users to control the trade-off between $acc(p_k)$ and $disc(p_k)$. PSO is triggered when a data chunk arrives.

Any discrimination measure can be used for $disc(p_k)$ in (2). In this method, equal opportunity difference is chosen due to its popularity [7]. It computes the rate prediction difference between unprivileged group and privileged group.

$$Disc(D) = TPR|(D = unprivilegedgroup) - TPR|(D = privileged) \qquad (3)$$

where $Disc(D)$ is discrimination of data D, $TPR|(D = unprivilegedgroup)$ and $TPR|(D = privileged)$ are true positive rate of privileged group and unprivileged group, respectively.

At the first run of PSO, particles are initialised with a random position. The position and velocity of particles are updated by using Eqs. (4) and (5), respectively.

$$p_k^{i+1} = p_k^i + v_k^{i+1} \qquad (4)$$

$$v_k^{i+1} = w * v_k^i + c_1 * r_1 * (pbest_k^i - p_k^i) + c_2 * r_2 * (gbest^i - p_k^i) \qquad (5)$$

where p_k^i and v_k^i are the position and velocity of particle k at iteration i. w is the inertia weight representing the moving momentum of particles. $\boldsymbol{pbest}_k^i$ and $\boldsymbol{gbest}^i$ are the local best position of particle k and the global best position of the swarm respectively. c_1 and c_2 are acceleration. r_1 and r_2 are constant randomised in [0, 1] anew at iteration i.

In this paper, we proposed two PSO algorithms to optimise the fitness function in Eq. (2). The two algorithms PSO_1 and PSO_2 are shown in Algorithm 2 and Algorithm 3 respectively. PSO_1 only uses the traditional updating scheme in PSO to update the local best positions and the global best solutions based on the proposed fitness function. PSO_2 applies a much stricter rule to update the local best and global best positions. In PSO_2, only new positions that provide better accuracy and discrimination will be used to update local best positions. The global best position in PSO_2 is only updated if the particle with the best rank (for both accuracy and discrimination) has a better fitness value. Compared to PSO_1, PSO_2 is more conservative to prevent the SAMDSOS classification system from making big changes in its predictions.

Algorithm 2: PSO_1

Input   : Data window
Output: Best found solution
1 **begin**
2 | Initialise PSO population;
3 | **for** $iter = 1$ to $max\_iterations$ **do**
4 | | **for** $k = 1$ to $Popsize$ **do**
5 | | | Calculate $Fitness(\boldsymbol{p}_k)$ using Eq. (2);
6 | | | Update $\boldsymbol{pbest}_k$ if $Fitness(\boldsymbol{p}_k) > Fitness(\boldsymbol{pbest}_k)$;
7 | | | Update $\boldsymbol{gbest}$ if $Fitness(\boldsymbol{p}_k) > Fitness(\boldsymbol{gbest})$;
8 | | **end**
9 | | Update velocity/position of each particle k using Eqs. (5) and (4);
10 | **end**
11 | Return $\boldsymbol{gbest}$;
12 **end**

Algorithm 3: PSO_2

    Input   : Data window
    Output: Best found solution

1 **begin**
2   Initialise PSO population;
3   **for** $iter = 1$ *to* $max\_iterations$ **do**
4   **for** $k = 1$ *to* $Popsize$ **do**
5   Calculate accuracy $acc(\boldsymbol{p}_k)$ and discrimination $disc(\boldsymbol{p}_k)$;
6   Update $\boldsymbol{pbest}_k$ if $acc(\boldsymbol{p}_k) >= acc(\boldsymbol{pbest}_k)$ and
              $disc(\boldsymbol{p}_k) < disc(\boldsymbol{pbest}_k)$;
7   **end**
8   Rank particles based on acc (descending) and $disc$ (ascending) ;
9   Choose particle $\boldsymbol{p}_r$ with lowest total ranking of acc and $disc$;
10   Update $\boldsymbol{gbest}$ if $Fitness(\boldsymbol{p}_r) > Fitness(\boldsymbol{gbest})$;
11   Update velocity/position of each particle k using Eqs. (5) and (4);
12   **end**
13   Return $\boldsymbol{gbest}$;
14 **end**

---

4 Results and Analysis

This section introduces the datasets, machine learning methods and parameter settings used in our experiments.

4.1 Experiment Setting

To examine the performance of FAS Stream in fairness aware learning from streaming data, this study uses three datasets [18] used commonly in fairness learning (shown in Table 1). For these datasets, we convert categorical features into binary features. Table 1 also shows discrimination (disc) level of the whole dataset calculating by applying Eq. (3) to the entire dataset. Sensitive attribute (S.A) and unprivileged group (U.P) are also shown in Table 1.

Table 1. Datasets used in the experiments

Dataset	N	d	Domain	S.A	U.P	Data disc
Adult (Adult)	32,561	14	Finance	Sex	Female	0.1963
Dutch census (Dutch)	60,420	11	Finance	Sex	Female	0.2951
KDD Census-Income (Census)	284,556	37	Finance	Sex	Female	0.0753

SAMDSOS have been run with $max\_STM\_size = 10\%$ of the number of total instances, $w1 = 0.7$, $Window\_size = 1000$. For PSO_1 and PSO_2 algorithms, the population size is 20, and the maximum iterations are 40. Other PSO parameters

Table 2. Accuracy and discrimination obtained by FAS Stream and other methods

Method	Metric					
	Adult		Dutch		Census	
	Accuracy	Disc	Accuracy	Disc	Accuracy	Disc
FAS_1	0.7923	0.0937	0.7863	0.1964	0.9379	0.0058
FAS_2	0.7896	**0.0859**	0.7907	**0.1947**	0.9377	0.0056
NoFW	0.8067	0.1232	**0.8024**	0.3190	0.9380	0.0054
HeuFW	**0.8068**	0.1142	0.7941	0.2040	0.9379	**0.0053**
KNN	0.8012	0.1336	0.7793	0.2958	0.9377	0.0176
SAM	0.7849	0.0890	0.7654	0.3215	**0.9386**	0.0108

are similar to those adopted in the PSO literature [28]. All the algorithms are implemented in Python.

To examine how feature weighting affects the accuracy and discrimination of FAS Stream, in experiments, we run FAS Stream with different feature selection strategies:

- FAS_1: We run FAS Stream with algorithm PSO_1 for feature weighting.
- FAS_2: We run FAS Stream with algorithm PSO_2 for feature weighting.
- NoFW: We disable feature weighting processing in FAS Stream.
- HeuFW: We set the weight of sensitive features to zero and others to one when running FAS Stream.

4.2 Compare with Other Methods

To show the effectiveness of FAS Stream in building classifiers with a high classification performance and desirable discrimination, this section compares the accuracy obtained by FAS Stream with lazy classifiers. Because PSO is a stochastic algorithm, we execute FAS Stream with PSO feature weighting strategies on these datasets in 30 independent random runs. Table 2 shows the average test accuracy and discrimination (disc) by FAS Stream in 30 runs versus accuracy and discrimination obtained by Lazy learning methods (KNN and SAM). The ideal method is the one with the highest accuracy and lowest discrimination. The results show that although FAS Stream strategies do not reach the highest accuracy, its discrimination is significantly lower than other methods. FAS Stream strategies (first four methods in Table 2) obtain lowest discrimination in the three datasets. The two methods with PSO-based feature weighting (FAS_1 and FAS_2) perform very well in terms of minimising discrimination in the adult and Dutch datasets. FAS_1 and FAS_2 are the second and the third methods on Census, but the difference in discrimination with other FAS Stream strategies is not significant. All above findings are also supported by Wilcoxon statistical tests with a significant level of 0.05.

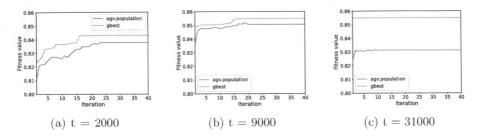

(a) t = 2000   (b) t = 9000   (c) t = 31000

Fig. 3. PSO feature weighting process over time.

The results here demonstrate the effectiveness of FAS Stream (FAS_1 and FAS_2) in weighting relevant features to minimise discrimination in the classification task. The results also shows the limitations of simple/naive strategies such as NoFW and HeuFW in reducing discrimination. Regarding accuracy, HeuFW and NoFW win at Adult and Dutch datasets, respectively while SAM achieves the highest accuracy for the Census dataset. However, the gaps in terms of accuracy between the methods compared in our experiments are very small.

4.3 Further Analysis

Because FAS_2 slightly performs better than others, it is further investigated in this section. In Table 3, metrics including accuracy and discrimination (disc) are used to compare the performance of FAS_2 with different window sizes. In general, there are no significant differences in the performance of FAS_2 with window sizes in the Dutch and Census datasets. For the adult dataset, FAS_2 seems to perform slightly better at minimising discrimination when the window size is 1000. In general, the results suggest that FAS_2 is sensitive to the window size.

Table 3. Influence of window size

Window size	Metric					
	Adult		Dutch		Census	
	Accuracy	Disc	Accuracy	Disc	Accuracy	Disc
200	0.7860	0.0918	0.7892	**0.1825**	0.9367	**0.0054**
500	**0.7992**	0.0930	0.7970	0.1993	0.9370	0.0058
700	0.7812	0.0903	**0.7977**	0.1826	**0.9380**	0.0066
1000	0.7864	**0.0774**	0.7790	0.1862	0.9386	0.0059

We also examine how PSO feature weighting works overtime. As discussed earlier, PSO feature weighting is called to optimise feature weights once discrimination increases. Figure 3 compares fitness values of the best solution (gbest shown in orange) and the average fitness of the whole population (avg.population shown in blue) through 40 iterations at different times t. The first time PSO is executed ($t = 2000$), it needs more than 25 iterations to find the best solution and converge. At $t = 9000$, PSO needs roughly 20 iterations to converge. However, at $t = 31000$, there is no improvement made by PSO. These observations suggest that FAS Stream has successfully and incrementally learned good feature weights for minimising discrimination over time.

5   Conclusion

Handling discrimination is challenging within streaming data. To deal with this challenge, this paper proposes a system made up of two swarms to incrementally build a classifier and reduce discrimination in data. The novelty of this algorithm is the use of feature weighting for discrimination reduction based on swarm intelligence. Experiments on popular datasets in fairness learning show that the proposed method is competitive compared with baseline methods. Further analyses show that FAS Stream has the capability of learning from flexible window sizes and incrementally learning feature weights for mitigating discrimination over time. In future studies, we will try to apply multi-objective optimisation to maximise accuracy and minimise discrimination.

References

1. Aghaei, S., Azizi, M.J., Vayanos, P.: Learning optimal and fair decision trees for non-discriminative decision-making. In: Proceedings of the AAAI Conference on Artificial Intelligence, vol. 33, pp. 1418–1426 (2019)
2. Ara, A., Ara, A.: Case study: integrating IoT, streaming analytics and machine learning to improve intelligent diabetes management system. In: 2017 International Conference on Energy, Communication, Data Analytics and Soft Computing, pp. 3179–3182. IEEE (2018)
3. Asudeh, A., Jagadish, H.V., Stoyanovich, J., Das, G.: Designing fair ranking schemes. In: Proceedings of the 2019 International Conference on Management of Data, pp. 1259–1276. New York, NY, USA (2019)
4. Belitz, C., Jiang, L., Bosch, N.: Automating procedurally fair feature selection in machine learning. In: Proceedings of the 2021 AAAI/ACM Conference on AI, Ethics, and Society, pp. 379–389 (2021)
5. Celis, E., Mehrotra, A., Vishnoi, N.: Toward controlling discrimination in online ad auctions. In: International Conference on Machine Learning, pp. 4456–4465 (2019)
6. Galhotra, S., Shanmugam, K., Sattigeri, P., Varshney, K.R.: Causal feature selection for algorithmic fairness. In: Proceedings of the 2022 International Conference on Management of Data, pp. 276–285 (2022)
7. Garg, P., Villasenor, J., Foggo, V.: Fairness metrics: a comparative analysis. In: 2020 IEEE International Conference on Big Data (Big Data), pp. 3662–3666 (2020)
8. Gomes, H.M., et al.: Adaptive random forests for evolving data stream classification. Mach. Learn. **106**(9), 1469–1495 (2017)
9. Gomes, H.M., Read, J., Bifet, A., Barddal, J.P., Gama, J.: Machine learning for streaming data: state of the art, challenges, and opportunities. ACM SIGKDD Explor. Newslet. **21**(2), 6–22 (2019)

10. Hulten, G., Spencer, L., Domingos, P.: Mining time-changing data streams. In: Proceedings of the Seventh ACM SIGKDD International Conference on Knowledge Discovery and Data Mining, pp. 97–106 (2001)
11. Iosifidis, V., Tran, T.N.H., Ntoutsi, E.: Fairness-enhancing interventions in stream classification. In: International Conference on Database and Expert Systems Applications, pp. 261–276 (2019)
12. Kamiran, F., Calders, T.: Classifying without discriminating. In: 2009 2nd International Conference on Computer, Control and Communication, pp. 1–6 (2009)
13. Kamiran, F., Calders, T., Pechenizkiy, M.: Discrimination aware decision tree learning. In: 2010 IEEE International Conference on Data Mining, pp. 869–874 (2010)
14. Kennedy, J., Eberhart, R.: Particle swarm optimization. In: Proceedings of International Conference on Neural Networks (ICNN), vol. 4, pp. 1942–1948 (1995)
15. Krawczyk, B.: Learning from imbalanced data: open challenges and future directions. Progress Artif. Intell. **5**(4), 221–232 (2016). https://doi.org/10.1007/s13748-016-0094-0
16. Krawczyk, B., Minku, L.L., Gama, J., Stefanowski, J., Woźniak, M.: Ensemble learning for data stream analysis: a survey. Inform. Fus. **37**, 132–156 (2017)
17. Krempl, G., et al.: Open challenges for data stream mining research. ACM SIGKDD Explor. Newslet. **16**(1), 1–10 (2014)
18. Le Quy, T., Roy, A., Iosifidis, V., Zhang, W., Ntoutsi, E.: A survey on datasets for fairness-aware machine learning. Wiley Interdisciplinary Reviews: Data Mining and Knowledge Discovery, p. e1452 (2022)
19. Losing, V., Hammer, B., Wersing, H.: KNN classifier with self adjusting memory for heterogeneous concept drift. In: 2016 IEEE 16th International Conference on Data Mining (ICDM), pp. 291–300 (2017)
20. Mavrovouniotis, M., Li, C., Yang, S.: A survey of swarm intelligence for dynamic optimization: algorithms and applications. Swarm Evolution. Comput. **33**, 1–17 (2017)
21. Mehrabi, N., Morstatter, F., Saxena, N., Lerman, K., Galstyan, A.: A survey on bias and fairness in machine learning. ACM Comput. Surv. **54**(6), 1–35 (2021)
22. Mohamed, M., Arkady, Z., Shonali, K.: Data stream mining. In: Oded, M., Lior, R. (eds.) Data Mining and Knowledge Discovery Handbook, p. 761. Springer, New York (2010). https://doi.org/10.1007/b107408
23. Mulinka, P., Casas, P., Vanerio, J.: Continuous and adaptive learning over big streaming data for network security. In: 2019 IEEE 8th International Conference on Cloud Networking, pp. 1–4 (2019)
24. Nguyen, S., Tran, B., Alahakoon, D.: Dynamic self-organising swarm for unsupervised prototype generation. In: 2020 IEEE Congress on Evolutionary Computation, pp. 1–8 (2020)
25. Niño-Adan, I., Manjarres, D., Landa-Torres, I., Portillo, E.: Feature weighting methods: a review. Expert Syst. Appl. **184**, 115424 (2021)
26. Peška, L., Tashu, T.M., Horváth, T.: Swarm intelligence techniques in recommender systems-a review of recent research. Swarm Evolution. Comput. **48**, 201–219 (2019)
27. Salazar, R., Neutatz, F., Abedjan, Z.: Automated feature engineering for algorithmic fairness. Proc. VLDB Endowment **14**(9), 1694–1702 (2021)
28. Tran, B., Zhang, M., Xue, B.: A PSO based hybrid feature selection algorithm for high-dimensional classification. In: 2016 IEEE Congress on Evolutionary Computation, pp. 3801–3808 (2016)

29. Wang, H., Fan, W., Yu, P.S., Han, J.: Mining concept-drifting data streams using ensemble classifiers. In: Proceedings of the ninth ACM SIGKDD International Conference on Knowledge Discovery and Data Mining, pp. 226–235 (2003)
30. Zafar, M.B., Valera, I., Rogriguez, M.G., Gummadi, K.P.: Fairness constraints: mechanisms for fair classification. In: Artificial Intelligence and Statistics, pp. 962–970 (2017)
31. Zhang, W., Bifet, A.: Feat: a fairness-enhancing and concept-adapting decision tree classifier. In: International Conference on Discovery Science, pp. 175–189 (2020)
32. Zhang, W., Zhang, M., Zhang, J., Liu, Z., Chen, Z., Wang, J., Raff, E., Messina, E.: Flexible and adaptive fairness-aware learning in non-stationary data streams. In: 2020 IEEE 32nd International Conference on Tools with Artificial Intelligence (ICTAI), pp. 399–406 (2020)

Explainable Network Intrusion Detection Using External Memory Models

Jack Hutchison[1], Duc-Son Pham[1]([⊠]), Sie-Teng Soh[1], and Huo-Chong Ling[2]

[1] School of EECMS, Curtin University, Perth, Western Australia, Australia
dspham@ieee.org
[2] Department of Electrical and Computer Engineering,
Curtin University Malaysia, Miri, Sarawak, Malaysia

Abstract. Detecting intrusions on a network through a network intrusion detection system is an important part of most cyber security defences. However, the interest in machine learning techniques, most notably neural networks, to detect anomalous traffic more accurately has led to a rise of these network intrusion detection systems being a black box, opaque to the user with little ability to explain its decisions and robbing the defenders of useful information that could lead them vulnerable to an opportune attacker. This paper makes several contributions to addressing this through augmenting an autoencoder-neural network model with external memory. It first explores the effect of the memory size and the addressing scheme used on F_1-score performance, finding optimal performance plateaus at memory sizes greater than 50, and that addressing schemes to increase the sparsity of the memory usage have negligible effect on performance. In addition, this work has generated several tools to better explain the model. This includes plotting which memory slots are strongly matched with what classes, visually and numerically measuring how much external memory each class takes to be properly encoded, and using the contents of the external memory to not only identify similar previously seen classes, but identify similarity with unseen classes and help gauge how outdated a model may be based on how the results align with domain knowledge. These tools and techniques show promising results in demonstrating the explainability potential of external memory with regards to an intrusion detection system and how they might be applied to help secure networks.

Keywords: Network intrusion detection · Memory-augmented models · Autoencoder · Cyber security · Deep learning

1 Introduction

Detecting cyber crime is becoming more important for businesses and organisations every year. It's estimated to cost Australian businesses up to \$29 billion a year and the threat environment is "worsening" and that the "growth

D.-S. Pham—This work was partly supported by a Curtin Malaysia collaborative research grant.

of cyber crime is outstripping our ability to respond" [1]. A successful breach can leak important information such as customer passwords, compromise the integrity of a business' data, or lead to significantly reduced ability to function if resources are taken offline such as through a ransomware attack. As such, research is constantly being conducted to develop better tools suited to defend against attackers. This work will focus specifically on network intrusion detection systems (NIDS). NIDS detect activity in a variety of ways, which can be summarised into 2 main distinctions, signature-based [9] and anomaly-based [5].

In recent years, deep neural networks in particular have increased in popularity in cyber security [2,4,21]. However, they are usually black-box models, or models where the inner workings cannot be easily explained to the user and the model cannot justify the result clearly. This is detrimental, in particular to the field of cyber security, as without having a working knowledge of the model, potential blind spots in its functionality may go unnoticed by defenders and taken advantage of by hackers. As going back to models such as decision trees, which are highly explainable [12], would lead to a loss of the performance gained through using these models [4], there have been attempts at making neural networks and other black-box models more explainable. These attempts include using the generic LIME [15], which manipulates the input being fed into the model, and measure the effect on the output to provide information such as important features that drive the models' decision making. This was used to explain unsupervised clustering of network traffic analysis in [14], as well as for deep neural networks on supervised datasets [19]. The idea of manipulating inputs to gain insight into the model was also used in [13], which used an adversarial approach to find the minimum change required from the input to change the classification of incorrect samples to their correct class.

To address the shortcomings of deep learning in cyber attack detection, this paper achieves two important goals. The first goal is to not only explore explainability within the specific context of intrusion detection and external memory [7], but to also do so using more recent datasets [16]. The second goal is to explore more characteristics of the external memory. In particular interest is the effect of the size of the memory has on performance, as well as further investigating the hard shrinkage used in memory augmented autoencoder of [6], which is supposed to affect the addressing of the external memory [7] by promoting stronger but fewer matches and thus increasing classification performance. This was only tested on the KDD99 dataset, and has not been visualised using read/write weightings of the external memory to see the actual effect that it has on the makeup of the external memory. This provides a good example to investigate the power of the explainability provided by external memory to explain changes in the model's operations and performance. As such, the chosen model will use the memory augmented autoencoder as a base, and feed the encoded data it generates into a neural network for multi-class classification.

2   Background

External memory was first developed by [7] as part of what was called a Neural
Turing Machine. The premise was that a neural network could be augmented
with external storage (memory), where it could store and retrieve information to
assist it in its task, much the same way that a computer uses its random access
memory (RAM) to expand its capabilities. What makes it different from existing
models, such as LSTM which also has a memory component, is that external
memory can be de-attached from any kind of sequence. It is less about storing
information for the next piece of data, but rather storing information that may
be useful at some later unspecified date.

This model was first tested with simple tasks such as copying sequences, or
a priority sort, but was expanded to handle more complicated tasks in [20] such
as question-answering problems and in [8], including navigating graphs, question
answer problems and the block puzzle. The advantage with external memory is
that the reading and writing of the external memory can be *tracked* and provide
valuable insight into the model, what it is doing and why. This was used by [11]
to create a model that could explain its inference on sequential data on both
a T-maze test, where it must make a correct turn at a T junction based on
information it has seen previously, and the story close test, where it must pick
the best ending to a story based on a series of premises. In addition to tracking,
the contents of the external memory could be used just as easily, as it can be
easily retrieved in a meaningful format due to the external nature of it.

Applications of external memory
to the field of cyber security and
intrusion detection are still *underde-*
veloped. External memory has been
used to classify different types of malware using restricted training data
[18], is briefly mentioned in [6] which
used a memory augmented autoencoder primarily for anomaly detection in images and video, as well

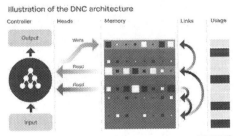

Differentiable neural computers (DNC)

intrusion detection in [3] using a sim-
ple memory network. However, in latter cases, they do not explore using that
external memory to explain their findings or their performance, and evaluation
was limited to the very old KDD99 dataset [17].

3   Methodology

Model Structure. The chosen model consisted of 2 components, the autoencoder and the classification neural network Fig. 1. The autoencoder was based
on the work conducted in [6] and inserts the external memory between the traditional encoder and decoder.

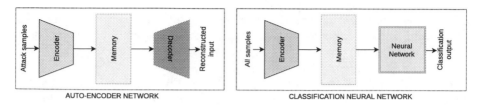

Fig. 1. Autoencoder and classification model structure

3.1 Encoder

The traditional part of the encoder involves having a series of dense layers with the ReLu activation function reduce the dimensionality of the data [6]. This encoded data is then passed to the memory module. This memory module contains the external memory, which is a vector in the shape of $N \times D$. Here, N is the number of memory slots, also referred to as the memory size, and D is the dimensions of the encoded data. In each slot of the external memory, the contents represent a typical pattern of the encoded data found during the training stage.

When a piece of data is passed through the memory module, its similarity to each slot in the memory is calculated using the cosine distance. These similarity scores are stored inside one vector, which is then passed through a softmax operation to generate an attention vector of size N and the contents of which sum to 1. This attention vector represents how strongly each slot in the external memory matches with the input.

Finally the contents of each slot in the external memory is weighted by the matching value in the attention vector, and the weighted contents of the memory are added together to form one final vector of shape D. Thus the memory slots that more strongly match with the input have more of an effect on the final output than those that only weakly match. In this way, the input data is not directly involved with creating the output, but instead serves to generate a query (through the attention vector) to the external memory, with the latter then responsible for creating the output [6].

Hard Shrinkage. This is an optional operation between the creation of the attention weight, and its multiplication with the memory, which takes the form $f_\lambda(x) = 1_{|x| \geq \lambda} x$ where 1 is the indicator function and λ is a sparsity threshold. The goal of this operation is to reduce the number of memory slots the encoder uses to reconstruct the input in training.

Training. The result generated by the external memory is then passed into the decoder, which is a symmetric copy of the Dense layers at the start of the model, creating an output in the same dimensions as the original input [6]. This autoencoder is trained to minimise the difference between the input and the output, using the mean squared error loss function and Adam optimizer [10]. However, it is only trained with data labelled as an attack sample. For the NSL KDD dataset, this is any data not with the label 'normal', and for the 2017/2018 dataset, it is any data not with the label of 'benign'. This is to promote the external memory to generate representations for each type of attack class within, and to further separate out normal/benign traffic from attack traffic by performing well for the attack samples, but poorly for normal/benign traffic and generating a more anomalous output.

Once fully trained, the encoder aspect of the autoencoder, consisting of the dense layers and external memory module, is separated out to be used for the classification component. This component is a neural network consisting of a series of dense layers with ReLU activation, with a final dense layer at the end with a softmax activation, providing the final prediction. This neural network is not trained on the raw input data, but instead the encoded data. Unlike the autoencoder training, this training uses all of the data, including the normal/benign class and is compiled with sparse categorical cross-entropy loss.

Model Shape and Hyperparameters. The number of layers, and the number of units in each dense layer for both the encoding and decoding, and the classification neural network were fine tuned for both the NSL KDD and CIC2017 dataset. This was done by generating potential model shapes based on a list of values, creating and training the model and then evaluating its performance in terms of F_1 score. The CIC2017 results were used for the CIC2018 dataset due to the similarity between the datasets, and the large size of the CIC2018 dataset making it impractical to experiment with many combinations. The chosen model shape and hyperparameters is not guaranteed to be optimal for enough combinations were not tested, however they ensured that a reasonable level of performance could be achieved.

As it is impractical to perform fine-tuning step for every external memory size that was investigated, some preliminary tests were conducted with a makeshift model to find a suitable memory size to conduct all of these fine-tuning experiments with. The chosen value for this memory size was 250, which appeared to show promising results. Memory size and its effects were properly explored after this fine-tuning stage and are detailed in the results section of this paper.

For the NSL dataset, the range of values [110, 90, 70, 50] were used for the number of neurons in the encoding layers, experimenting with 1–3 layers of depth. For the neural network, the neuron counts [60, 45, 30, 15] were used experimenting with 1–2 layers of depth. The optimal model shape was found to be [110, 90] for the autoencoder, and [60, 45] for the neural network.

For the CIC2017 dataset, the range of values [60, 45, 30, 15] were used for the autoencoder, with 1–2 layers of depth and the range of values [45, 30, 15] with a depth of 1–2 layers were used for the neural network. These values were lower than those used for the NSL dataset due to the reduced dimensions of the data. The optimal model shape was found to be [60, 45] for the autoencoder, and [45, 30] for the neural network.

Some of the model's hyperparameters, such as the number of epochs, the learning rate and the batch size, were also fine tuned. The chosen values for these hyperparameters, for both the autoencoder and the neural network, were 150 epochs, a learning rate of 0.0001, and a batch size of 128.

4   Experiments

We use three well-known IDS datasets: NSL-KDD [17], CIC2017 and CIC2018 [16]. Substantial work was needed to prepare the datasets for experiments as per the literature. As the classes are very imbalanced, we mainly measure the performance using F_1 score, with and without weighting by class population.

4.1   Memory

Memory Size. Once the model's general design was finalised, the first hyperparameter of the external memory, the size, was explored. A series of models were trained with various memory sizes for both the NSL KDD dataset and the CIC2017 dataset, and the performance of each model was logged.

NSL-KDD: As the external memory size increases, Fig. 2 highlights the initial poor performance with very low memory and the very quick stagnation and plateau of performance as the number of memory slots reach 100 and over. This is inline with [6], which found that after a certain threshold, the memory size had little to no effect on the performance. Before that threshold, the memory size was insufficient to store enough representations within to perform effectively and performance was compromised. Interestingly, this threshold was much lower for this dataset than it was for the UCSD-Ped2 used in [6], which had performance plateauing at memory sizes 1000 and over. This is likely as a result of the UCSD-Ped2 dataset consisting of video and image data which is comparatively more complex data wise than anomaly detection of network traffic.

When the memory size varies from 3 to 1500, Table 1 shows that the performance gain in the early stages comes from more successfully classifying the rare minority class r2l, with general improvements in the other classes from size 3 to size 15, and what would appear to be slight fluctuations from size 15 onwards.

Table 1. Category-wise F_1 score vs memory size

Memory size	3	15	250	1500
DOS	0.799	0.842	0.861	0.848
Normal	0.798	0.786	0.786	0.798
probe	0.574	0.645	0.619	0.687
R2L	0	0.065	0.327	0.273
U2R	0	0.011	0.012	0.018
macro avg	0.434	0.470	0.521	0.525
Weighted avg	0.670	0.694	0.725	0.728

CIC2017: The general performance profile as shown in Table 2 and Fig. 2, shows a similar pattern to the NSL KDD dataset emerges, albeit with much higher classification performance. It appears that there is a more gradual ascent in low memory performance as the size increases than in the NSL KDD dataset, however it still approaches and reaches a plateau point of performance at memory

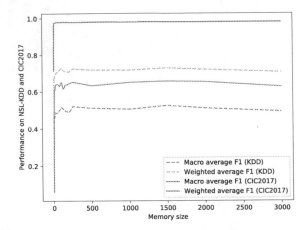

Fig. 2. Graph of memory size and performance on NSL-KDD and CIC2017

sizes greater than 50 where the addition of external memory does not yield any gains at a similar size of external memory.

For individual class performance, one can notice that there exists several attacks, such as the Web Attack variants and Infiltration that universally achieve poor performance. However, for the other classes, once again, significant performance leaps can be found between memory size 3 and 15, and 15 and 250. More notably with this dataset than the NSL KDD dataset is the lack of drastic improvement in the classification of the BENIGN class when compared to attack classes. This is most likely as a result of BENIGN being the majority class, and being omitted from the training helping to more easily differentiate it.

CIC2018: Due to the large size of the 2018 dataset, the full range of memory sizes were not used. Instead, the values 15, 250 and 1500, to represent low, medium and high sizes of external memory were used to gain a snapshot of performance, shown in Table 3. Although the full shape of the curve cannot be seen due to this limited data, it appears to support the findings of the previous 2 datasets, with a relatively significant jump in macro f1-score from 15 to 250, and a relatively minor one between 250 and 1500 that is consistent with random variation found between other models.

The results on 3 datasets suggest that external memory past a certain size offers little to no performance gain in classification. Below this size, performance is affected, increasingly dropping off as memory size tends towards 3. In addition, this threshold value appears lower than the one suggested by [6], highlighting how simpler data, such as network traffic information, can more effectively utilise lower sizes of external memory than more complicated data, such as images and video. The potential explanation behind this is explored later.

Space Addressing. The sparse addressing was tested using the external memory sizes 15, 250 and 1500 to represent low, medium and high sizes of memory respectively. The sparsity threshold to be used was suggested to be between $1/N$

Table 2. Category-wise F_1 score vs Memory Size on CIC2017

Memory size	3	15	250	1500
BENIGN	0.96626	0.98576	0.98737	0.98737
Bot	0.0	0.34069	0.51155	0.51155
DDoS	0.86044	0.98465	0.99096	0.99096
DoS GoldenEye	0.5967	0.90996	0.94308	0.94308
DoS Hulk	0.87957	0.9499	0.9501	0.9501
DoS Slowhttptest	0.72558	0.88077	0.91236	0.91236
DoS slowloris	0.54967	0.92129	0.94969	0.94969
FTP-Patator	0.52839	0.95663	0.97661	0.97661
Heartbleed	0.0	0.38889	0.68889	0.68889
Infiltration	0.0	0.0	0.0	0.0
PortScan	0.83829	0.90817	0.92072	0.92072
SSH-Patator	0.45444	0.92555	0.93237	0.93237
Web Attack Brute Force	0.0	0.0	0.03214	0.03214
Web Attack Sql Injection	0.0	0.0	0.0	0.0
Web Attack XSS	0.0	0.0	0.00509	0.00509
macro avg	0.42662	0.61015	0.6534	0.6534
weighted avg	0.94084	0.97639	0.97915	0.97915

Table 3. Category-wise F_1 score vs Memory Size on CIC2018

Memory size	15	250	1500
Benign	0.993	0.993	0.993
Bot	0.998	0.998	0.999
Brute Force -Web	0.339	0.474	0.445
Brute Force -XSS	0.603	0.606	0.597
DDOS attack-HOIC	0.999	0.999	0.999
DDOS attack-LOIC-UDP	0.831	0.817	0.842
DDoS attacks-LOIC-HTTP	0.993	0.994	0.995
DoS attacks-GoldenEye	0.989	0.994	0.966
DoS attacks-Hulk	0.999	0.999	0.997
DoS attacks-SlowHTTPTest	0.600	0.580	0.597
DoS attacks-Slowloris	0.957	0.967	0.964
FTP-BruteForce	0.782	0.775	0.782
Infilteration	0.010	0.018	0.021
SQL Injection	0.1	0.228	0.306
SSH-Bruteforce	0.999	0.999	0.999
macro avg	0.746	0.763	0.767
weighted avg	0.978	0.978	0.978

and $3/N$ by [6], where N is the size of the external memory. However testing $2/N$ and $3/N$ yielded NaN loss values in practice. As such, $1/N$ was chosen as the value to be used for this experiment.

While the effect of the size of the external memory was consistent with [6], the effects of sparse addressing through hard shrinkage were not. For the NSL KDD dataset, instead of increasing the performance of the model as was suggested, it appeared to have negligible effect, as shown in Table 4 by comparing the No (no hard shrinkage activated) and Yes (hard shrinkage enabled with shrink threshold set to $1/N$) rows. While performance does increase for 250 memory size, the increase is only for the weighted f1 average and not the macro, suggesting that it has performed marginally better for the majority class, and performance decreases for both metrics when comparing memory sizes 15 and 1500.

For the CIC2017 dataset, the reduction in performance in memory size 15 does appear quite significant, dropping around 3% for the weighted, and 16% for the weighted average. However, once again no significant increase of performance is noted with the higher memory sizes. A similar pattern is present for the CIC2018 dataset, although the performance decrease for the memory size of 15 is not quite as significant as for the CIC2017 dataset.

The possible explanation behind this discrepancy between the original paper [6] and this paper is explored in the Explainability experiment.

Table 4. Hard shrinkage comparison

	NSL-KDD		CIC2017		CIC2018	
Size	F_1	Weighted F_1	F_1	Weighted F_1	F_1	Weighted F_1
15 No	0.4646	0.6928	0.6391	0.9758	0.7468	0.9785
15 Yes	0.4559	0.6870	0.4696	0.9488	0.6527	0.9675
250 No	0.5044	0.7093	0.6349	0.9797	0.7632	0.9783
250 Yes	0.4989	0.7161	0.6340	0.9774	0.7716	0.9784
1500 No	0.5143	0.7240	0.6524	0.9778	0.7672	0.9785
1500 Yes	0.5073	0.7123	0.6350	0.9774	0.7605	0.9781

4.2   Explainability

Mapping Memory Activation. The external memory of the autoencoder, representing what patterns and common occurrences in the data it has learnt through the training process, can be used to explain the model's performance and provide insights into each class of attack and the connections between them. The first technique used was examining and graphing which memory slots were strongly activated by each class of attack. In the normal encoding process, the input data is encoded using the initial dense layers. This result is then used to generate a vector the size of the external memory; comparing the encoded input with each slot in memory using the cosine distance, and then passing the final vector through a softmax function. Finally, this vector acts as an attention vector is multiplied with the external memory to generate the output. However, if this final step is left out, the attention vector can be used to indicate which memory slots are strongly matched with the input and which slots are not. When averaged out and normalised for each class, these vectors provide information on the general memory activations, in addition to allowing comparisons between these activations.

Classes that strongly activated similar memory slots could be highly related, matching with the contents of the memory and suggesting that the information learnt in the training process applies to both sets of data. This could be used to explain potential confusion between classes in classification, and common features/patterns that are shared between classes that could be used in further analysis in identifying future types of attacks.

This insight in to the functioning of the model can also be used to explain the findings generated in this paper when investigating the size of the external memory and any hard shrinkage operations on the performance of the model.

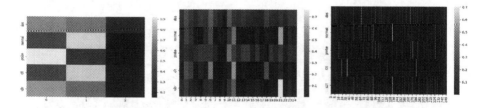

Fig. 3. Plotted memory activations of NSL with size 3, 25 and 250

Size of Memory. It can be hypothesised that the lack of performance with small amounts of external memory can be attributed to not having sufficient slots in memory to represent all the types of attacks. This leads to a homogenisation of output in the model and reduces performance for not only the rarer classes of attacks as it cannot sufficiently reconstruct them, but also the majority classes as their memory encodings are compromised by trying to generalise too much.

When plotting the memory activations of each class in the NSL KDD dataset with memory size of 3, this fact becomes apparent. Looking at Fig. 3, one can see that both dos and u2r, as well as normal and r2l share incredibly similar patterns of activation with each other. It also becomes apparent how difficult it is for the encoder to distinguish between 5 types of classes, using only 3 memory slots due to the limited memory possibilities.

Expanding the memory size to 25, one can see that although there still exists similarities between classes, the difference in memory patterns have become more distinct. For example, memory slot 18 shows clear distinction in values between different classes. This contrast is important in making each class distinctly encoded inside the external memory and ensuring that the encoded representations of each class when later used in the classification module are in the most useful form.

Further expanding memory size to 250, the lack of improvement of in performance when increasing the memory size past a certain point also becomes more apparent. As the memory size increases, the usage of the memory does not increase at the same rate. This leads to higher sparsity in the memory and thus more memory slots not contributing to the output generated. Once this saturation point has been achieved, adding extra memory is only increasing the sparsity of the memory, not expanding any representations found within.

Shrinkage. The same technique, with some modifications, can be applied to investigating whether the hard shrinkage is actually reducing the number of memory slots used by each class and promoting stronger matches with fewer slots. While in extreme cases it can be easily to see the difference in the sparsity of the memory, to be better judge this aspect, another graph can be generated by sorting each row from low to high. While the relation of memory slots between rows is lost, it shows the general shape and tail of the activation for each type of class, and allows comparisons with how much of the memory is used, and how quickly the strength of the match tails off. Figure 4 shows the sorted memory activiation of CIC2017 with two shrinkage thresholds 1 and 0.2 respectively, which clearly illustrates the effect.

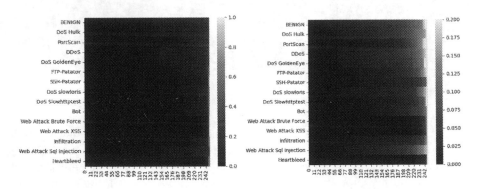

Fig. 4. Sorted memory activations of CIC2017 with max 1.0 and 0.2

When the same process is applied to the model that was trained with hard shrinkage on, the reduction in memory slot usage and stronger matches becomes immediately apparent. This suggests that the hard shrinkage does promote sparsity in the use of the memory as [6] claims for the CIC2017/2018 dataset. However this does not necessarily always translate to performance and might depend on the nature of the data and the model that is in use. Losing too much information in the low memory activations may compromise a model's ability to accurately classify and perform.

Memory Contents. We are interested in examining the memory contents. However, a preliminary plot of the heatmap revealed little connection between memory slots and it had limited insight into explaining anything about the model. The next iteration was to use only some of the memory, namely the memory slots highly activated by each class, excluding memory slots that had little to no effect on the inputted data. While this did improve the readability of the graph by reducing the number of memory slots shown, it did little to increase the insight gathered from the graph.

As such, the contents of the memory had their dimensionality reduced using PCA down to 2 dimensions. The entire contents of the memory was first plotted Fig. 5, and then subsequently only the top 15 activations of each class were plotted to show the most representative memory contents. In this case, the memory slots the attacks activated had quite different contents, shown by the different shape and location of the dots when comparing the two graphs.

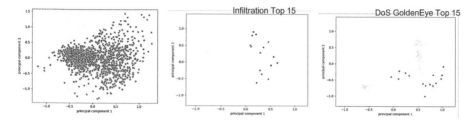

Fig. 5. PCA graph with all memory contents, top 15 infiltration, and top 15 DoS Golden Eye - CIC2017 dataset

Unseen Attacks. Finally, experiments were also done with trying to predict how the external memory would react to an unseen attack that is similar to an existing attack it has been trained on. The reasoning behind this was the fact that one could leverage the external memory to show how up to date the model was with current attacks, and whether or not it would need to be retrained with new data. If the unseen attack activated much of the same memory slots as its similar counterpart than the model could perform just fine. However, if it was something far off, or confused with another, very dissimilar type of attack, it could be used as evidence that the external memory has become outdated and needs to be retrained with newer data. This is incredibly important to know in the field of cyber security, as resources are often constrained and knowing when they need to be spent vs when they can be saved could be essential in the proper operations of a network security system.

This unseen attack was artificially introduced by withholding an attack sample from the training set. For this purpose, the class Web Attack XSS was chosen from the CIC2017 dataset, due to the very similar memory activations it has with Web Attack Brute Force, shown in Fig. 6. The model was trained with only access to the Web Attack Brute Force class, and after training, the memory contents for both classes were graphed using the aforementioned method.

While there exists a slight difference between the graphs, the majority of the points are of the same memory slots, and those that are not exist in the same general area. This shows the adaptability of the encoded memory to handle unseen attacks that are similar to ones that it has trained on before, in addition

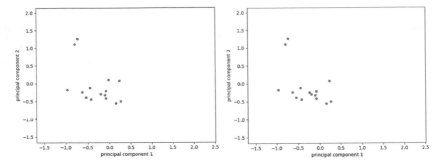

Fig. 6. PCA graph with top 15 Web Attack Brute Force (seen in training) and top 15 Web Attack XSS (unseen in training) - CIC2017 dataset

to highlighting a useful visual metric on how well generalised the model is for unseen attacks that are similar to attacks in the training sample.

5   Conclusion

The comprehensive study has revealed that past a certain threshold, which appeared to be approximately 100, additional memory slots offered little to no extra performance to the model. It also found that the shrinkage function to promote memory sparsity had mixed effects on performance, rather than the positive one found in [6].

Not only was the average match with external memory gathered and plotted as a heatmap, but the contents of the external memory was reduced to 2 dimensions and plotted as a dot graph. These visual tools allow the user to not only identify areas of commonality between classes of attacks through comparison, but to identify when the model may be becoming outdated and in need of fresh data and new training through comparing unseen attacks (simulated through leaving out a class) with existing classes that are expected to be similar.

Similarly, numeric metrics were designed to describe the sparsity of the memory through measuring the maximum value in the averaged activations, as well as the number of memory slots required to reach a threshold of the total match. This provides the reader with a picture of whether the data is strongly matched with few memory slots, or more weakly matched with more memory slots.

References

1. Affairs, H.: Australia's Cyber Security Strategy 2020. Cyber Security, p. 52 (2020)
2. Chawla, A., Lee, B., Jacob, P., Fallon, S.: Bidirectional LSTM autoencoder for sequence based anomaly detection in cyber security. Int. J. Simul. Syst. Sci. Technol. (2019). https://doi.org/10.5013/IJSSST.a.20.05.07
3. Chun, L., Xiaoxian, G., Jing, Z., Wei, W., Hanji, S., Peng, G.: Intrusion detection using end-to-end memory network. In: Proceedings of the ICCIS. pp. 244–249. ACM Press, Wuhan, China (2017)

4. Ferrag, M.A., Maglaras, L., Moschoyiannis, S., Janicke, H.: Deep learning for cyber security intrusion detection: approaches, datasets, and comparative study. J. Inform. Secur. Appl. **50**, 102419 (2020)
5. García-Teodoro, P., Díaz-Verdejo, J., Maciá-Fernández, G., Vázquez, E.: Anomaly-based network intrusion detection: techniques, systems and challenges. Comput. Secur. **28**, 18–28 (2009)
6. Gong, D., et al.: Memorizing Normality to Detect Anomaly: Memory-augmented Deep Autoencoder for Unsupervised Anomaly Detection. arXiv:1904.02639 [cs] (2019)
7. Graves, A., Wayne, G., Danihelka, I.: Neural Turing Machines. arXiv:1410.5401 [cs] (2014)
8. Graves, A., Wayne, G., Reynolds, M.: Hybrid computing using a neural network with dynamic external memory. Nature **538**(7626), 471–476 (2016)
9. Hadi, M.Z.S., Entin, M., Pratiarso, A., Ellysabeth, J.: Intrusion detection system based SNORT using Hierarchical clustering. ISSIT 2011 **1**(1), 85–90 (2011)
10. Kingma, D.P., Ba, J.: Adam: A Method for Stochastic Optimization. arXiv:1412.6980 [cs] (2017)
11. La Rosa, B., Capobianco, R., Nardi, D.: Explainable inference on sequential data via memory-tracking. In: Proceedings of the IJCAI, pp. 2006–2013. Yokohama, Japan (2020)
12. Mahbooba, B., Timilsina, M., Sahal, R., Serrano, M.: Explainable Artificial Intelligence (XAI) to enhance trust management in intrusion detection systems using decision tree model. Complexity **2021**, e6634811 (2021)
13. Marino, D.L., Wickramasinghe, C.S., Manic, M.: An Adversarial Approach for Explainable AI in Intrusion Detection Systems. arXiv:1811.11705 [cs, stat] (2018)
14. Morichetta, A., Casas, P., Mellia, M.: EXPLAIN-IT: towards explainable AI for unsupervised network traffic analysis. In: Proceedings of the Big-DAMA 2019, pp. 22–28. ACM Press, Orlando, FL, USA (2019)
15. Ribeiro, M.T., Singh, S., Guestrin, C.: "Why Should I Trust You?": Explaining the Predictions of Any Classifier. arXiv:1602.04938 [cs, stat] (2016)
16. Sharafaldin, I., Habibi Lashkari, A., Ghorbani, A.A.: Toward generating a new intrusion detection dataset and intrusion traffic characterization. In: Proceedings of the ICISSP, pp. 108–116. Funchal, Madeira, Portugal (2018)
17. Tavallaee, M., Bagheri, E., Lu, W., Ghorbani, A.A.: A detailed analysis of the KDD CUP 99 data set. In: Proceedings of the CISDA, pp. 1–6. IEEE, Ottawa, ON, Canada (2009)
18. Tran, K., Sato, H., Kubo, M.: MANNWARE: a malware classification approach with a few samples using a memory augmented neural network. Information **11**(1) (2020)
19. Wang, M., Zheng, K., Yang, Y., Wang, X.: An explainable machine learning framework for intrusion detection systems. IEEE Access **8**, 73127–73141 (2020)
20. Weston, J., Chopra, S., Bordes, A.: Memory Networks. arXiv:1410.3916 [cs, stat] (2015)
21. Yousefi-Azar, M., Varadharajan, V., Hamey, L., Tupakula, U.: Autoencoder-based feature learning for cyber security applications. In: Proceedings of the IJCNN, pp. 3854–3861. IEEE (2017)

Genetic Algorithms

Handling Different Preferences Between Objectives for Multi-objective Feature Selection in Classification

Ruwang Jiao$^{(\boxtimes)}$ (iD), Bing Xue (iD), and Mengjie Zhang (iD)

School of Engineering and Computer Science, Victoria University of Wellington,
Wellington 6140, New Zealand
{Ruwang.Jiao,Bing.Xue,Mengjie.Zhang}@ecs.vuw.ac.nz

Abstract. Maximizing the classification performance and minimizing the feature subset size are two key objectives in multi-objective feature selection. Most existing works treat these two objectives equally. However, from the perspective of decision-makers, the preferences of these two objectives are different, that is, the classification performance is more important than the number of selected features. Besides, improving the classification performance is also more challenging than reducing the number of selected features. To deal with this issue, this paper proposes a preference-inspired multi-objective evolutionary algorithm, which consists of three major components: 1) a fitness function is proposed to give more preference to the objective of classification performance; 2) based on the analysis of solutions' distribution, an irrelevance learning method is proposed to detect the irrelevant features; 3) a dimensionality reduction method is proposed to remove irrelevant features and further improve the classification performance of feature subsets. By comparing the proposed method with five state-of-the-art multi-objective evolutionary algorithm-based feature selection methods, empirical results on nine classification datasets demonstrate that the proposed method is able to obtain a set of feature subsets with better classification performance.

Keywords: Evolutionary computation and learning · Feature selection · Multi-objective optimization · Classification

1 Introduction

The explosive growth of data in the current era has brought a wealth of information, but also produced a large number of useless data [5]. High-dimensional data poses severe challenges to current learning tasks (such as classification). Having many features can easily lead to the overfitting of the classification model and the degradation of the classification performance. To improve the utility of high-dimensional data, feature selection (FS) plays a crucial role in machine learning and data mining [9,12]. It aims to select the informative features and remove irrelevant, redundant, and noisy features from the original feature set. The benefits of FS in classification are manifold [1]: 1) data compression, reducing the

H. Aziz et al. (Eds.): AI 2022, LNAI 13728, pp. 237–251, 2022.
https://doi.org/10.1007/978-3-031-22695-3_17

data dimensionality when data volume is large and data storage space is limited; 2) performance improvement, boosting both the classification performance and the generalization ability of the classifier; 3) data interpretation and understanding, gaining knowledge about the process of generating data or easily visualize data; and 4) processing acceleration, reducing training time to increase the computational efficiency.

Based on the relationship with classifiers, FS can be categorized into filter, wrapper, and embedded methods [17]. Filter methods independently evaluate each feature by various proxy measures, and then pick out the discriminative features with top rankings. However, they normally neglect the composite effect that arises when selected features are put together. Embedded methods incorporate feature evaluation into models of classifier training, by which a feature evaluator is achieved along with the optimization of a classifier. Among the three methods, wrapper methods are the most time-consuming, but usually exhibit the best classification performance. Wrapper methods first generate candidate feature subsets via a search algorithm, and then evaluate the goodness of these candidate feature subsets based on a classification model. Evolutionary algorithms (EAs) have been recognized as an effective search approach in wrapper methods given their global search ability, and the population-based property can approximate a set of trade-off solutions in a single run [3]. Thus, in this paper, the wrapper method is used to excavate the optima feature subsets for classification with the assistance of an EA.

Maximizing the classification performance and minimizing the feature subset size are two main objectives in FS, which can be viewed as a multi-objective optimization problem [10,17]. Unlike most filter and embedding methods that require the number of selected features to be specified in advance, which is unknown in reality, multi-objective feature selection can automatically obtain a set of optimal feature subsets, which are trade-offs between the number of selected features and classification performance [16]. In the past decade, several multi-objective EAs (MOEAs) have been proposed to optimize these two objectives to obtain a set of trade-off feature subsets [4,16]. Most existing works treat the two objectives in FS as being equally important. However, the preferences of these two objectives are different [6]. It is well known that classification performance is more important than the number of selected features in most cases. In terms of the difficulty of optimizing these two objectives, maximizing the classification performance is also more challenging than minimizing the feature subset size, since the latter can be easily achieved by directly reducing the number of selected features. By contrast, maximizing the classification performance needs to consider the relevance and redundancy of features and deal with the complex interaction issue between features. It is challenging to select informative features from high-dimensional data which generally contains many irrelevant and redundant features [7]. These irrelevant and redundant features often impede classification performance and misdirect classification tasks. Therefore, it is of great significance to remove irrelevant and redundant features and pay more attention to improving classification performance in multi-objective FS.

Goals and Contributions: The overall goal of this paper is to develop an MOEA to find a diverse set of trade-off feature subsets between the classification performance and the feature subset size. More importantly, among the two objectives in multi-objective FS, the designed algorithm aims to find feature subsets that achieve better classification performance. To achieve this goal, this paper proposes a preference-inspired MOEA for multi-objective FS in classification, termed PMOFS. Focusing on how to generate and select feature subsets with better classification performance, a dimensionality reduction operator and a fitness function in PMOFS are proposed to improve the classification performance of feature subsets.

The main contributions of this paper can be summarized as follows:

1. A fitness function is designed during the environmental selection process, which combines the convergence criterion and the classification performance criterion, with the aim of prioritizing the objective of maximizing the classification performance over the objective of minimizing the feature subset size.
2. Based on the analysis of the population distribution, an irrelevance learning method is proposed to detect irrelevant features.
3. A dimensionality reduction technique is proposed, which aims to remove irrelevant features for offspring solutions and improve their classification performance.

The rest of this paper is organized as follows. The details of the proposed PMOFS algorithm are elaborated in Sect. 2. Section 3 presents the experiment design. Section 4 reports the experimental results and some discussions. Finally, Sect. 5 concludes this paper.

2   The Proposed Approach

2.1   Problem Statement

By minimizing the selected feature ratio $f_{ratio}(\boldsymbol{x})$ and the classification error rate $f_{err}(\boldsymbol{x})$, a multi-objective FS problem can be formulated as:

$$\min \quad \boldsymbol{F}(\boldsymbol{x}) = (f_{ratio}(\boldsymbol{x}), f_{err}(\boldsymbol{x})) = (\frac{\sum_{i=1}^{D} x_i}{D}, 1 - \frac{1}{c}\sum_{i=1}^{c}\frac{TP_i}{|S_i|})$$

$$\text{where } \boldsymbol{x} = (x_1, ..., x_D) \in \Omega,$$
$$x_i \in \{0, 1\}, i = 1, \cdots, D, \tag{1}$$

where D is the total number of features. $\boldsymbol{x}$ is the solution (feature subset). $x_i=1$ and $x_i = 0$ mean that the i-th feature is selected and discarded, respectively. The first objective $f_{ratio}(\boldsymbol{x})$ is the selected feature ratio, which is equal to the ratio of the number of selected features and the total number of features. The second objective $f_{err}(\boldsymbol{x})$ is the balanced classification error rate [11], which can be obtained by evaluating a classifier using the selected feature subset $\boldsymbol{x}$. c is the number of classes of a dataset. TP_i represents the number of correctly identified instances in class i, and $|S_i|$ is the number of instances of class i.

2.2   Overall Framework

Figure 1 outlines the overall framework of PMOFS. In initialization, the number of selected features in each solution x (e.g., feature subset) is randomly generated, i.e., $|x| = rand(1, D)$. Then, $|x|$ features are randomly selected and their bits are set to 1, i.e., $x_i=1$, $i \in [1, D]$. In each generation, the canonical genetic operators, i.e., single-point crossover and bit-flip mutation, are used to generate an offspring population $\mathcal{O}$. Subsequently, guided by a learning array $\mathcal{A}$, a dimensionality reduction operator is performed to remove irrelevant features of each solution in $\mathcal{O}$. After evaluating the classification performance of each solution in $\mathcal{O}$, the nondominated sorting operation [2] is performed to divide the union of the parent and offspring populations ($\mathcal{U} = \mathcal{P} \bigcup \mathcal{O}$) into different nondominated front levels. Then, the irrelevance array $\mathcal{A}$ can be learned according to the distribution of the union population $\mathcal{U}$. At last, the environmental selection operator (Algorithm 3) is used to select elite solutions from the union population $\mathcal{U}$ as the next parent population $\mathcal{P}$. Once the stopping criterion is satisfied, i.e., the maximum number of function evaluations, the nondominated feature subsets will be regarded as output.

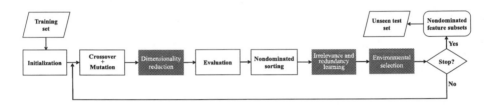

Fig. 1. The framework of PMOFS.

The novelty of PMOFS lies in three operators: irrelevance learning, dimensionality reduction, and environmental selection, which will be described in detail subsequently.

2.3   Irrelevance Learning

Redundant features normally cannot improve the classification performance, but will lead to longer training time, while irrelevant or noisy features can degenerate the classification performance. In evolutionary FS, since the population-based search mechanism can generate a lot of feature subsets. One question may arise: can we learn the irrelevant features through the distribution of the population in the objective space? The answer is yes.

The proposed irrelevance learning is based on the population's distribution. Fig. 2 depicts an illustrative example. In Fig. 2, solution x_1 selects two features, and exhibits the best classification performance (i.e., lowest classification error rate). Solution x_2 selects three features, two of which are selected by both solution x_1 and solution x_2, i.e., the first and fifth features. However, compared with

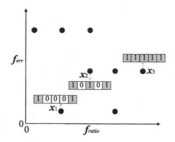

Fig. 2. An example to illustrate the irrelevance analysis.

Algorithm 1: Irrelevance learning

Input: $\mathcal{U}$: the union of the parent and offspring populations $\mathcal{U} = \mathcal{P} \bigcup \mathcal{O}$;
$\mathcal{F}(\boldsymbol{x})$: the nondominated front level for each solution $\boldsymbol{x}$ in $\mathcal{U}$.
Output: the learning array $\mathcal{A}$.

1 $\mathcal{A} = (0, 0, \cdots, 0), \mathcal{A} \in \mathbb{R}^D$;
2 **for** *each solution* $\boldsymbol{x} \in \mathcal{U}$ **do**
3      $\mathcal{S} \leftarrow \{\boldsymbol{x}' | \boldsymbol{x}' \prec \boldsymbol{x} \text{ and } \mathcal{F}(\boldsymbol{x}) - \mathcal{F}(\boldsymbol{x}') \leq 1, \boldsymbol{x}' \in \mathcal{P}\}$;
4   **for** *each* $\boldsymbol{x}' \in \mathcal{S}$ **do**
5   **if** $\boldsymbol{x}' \wedge \boldsymbol{x} == \boldsymbol{x}'$ **then**
6              $\mathcal{A} \leftarrow \mathcal{A} \vee ((\boldsymbol{x}' \vee \boldsymbol{x}) \wedge (\overline{\boldsymbol{x}'}))$;

7 Return $\mathcal{A}$.

$\boldsymbol{x}_1$, $\boldsymbol{x}_2$ selects one more feature (i.e., the third feature) but exhibits a larger classification error rate than that of $\boldsymbol{x}_1$. From this perspective, it can be inferred that the third feature is more likely to be irrelevant which degrades the classification performance. As a result, we can label the third feature as irrelevant and avoid selecting it when generating new feature subsets (solutions). Taking a closer look at Fig. 2, we can find that solution $\boldsymbol{x}_3$ also selects more features (i.e., the second, third, and fourth features) and has worse classification performance than that of $\boldsymbol{x}_1$, but it is hard to deduce that the second, third, and fourth features are all irrelevant, since the number of features selected by $\boldsymbol{x}_1$ and $\boldsymbol{x}_3$ is quite different, so the second, third, and fourth features selected by $\boldsymbol{x}_3$ may contain both relevant and irrelevant ones. In this regard, in this paper, the irrelevance analysis is only performed on two solutions that are in the adjacent nondominated front levels.

The pseudo-code of the irrelevance learning mechanism is presented in Algorithm 1. A learning array $\mathcal{A}$, is used to label the useless features. The i-th bit of $\mathcal{A}$ is equal to 1 ($\mathcal{A}_i = 1$) implies that the i-th feature is likely to be irrelevant. For each solution $\boldsymbol{x}$ in population $\mathcal{U}$, we find solutions $\boldsymbol{x}'$ who can dominate $\boldsymbol{x}$, and $\boldsymbol{x}'$ and $\boldsymbol{x}$ are on adjacent nondominated front levels (line 3 of Algorithm 1). If all features selected by $\boldsymbol{x}'$ are also selected by $\boldsymbol{x}$ (line 5 of Algorithm 1), the irrelevant features can be inferred, which are additional features selected by $\boldsymbol{x}$. Finally, the bits of these irrelevant features are set to 1 for $\mathcal{A}$ (line 6 of Algorithm 1).

2.4　Dimensionality Reduction

For newly generated offspring solutions in $\mathcal{O}$, the purpose of the dimensionality reduction operation is to further reduce the number of their selected features before evaluating them. Algorithm 2 presents the pseudo-code of dimensionality reduction. For each solution in $\mathcal{O}$, we find the locations of its selected features, and these features are also learned as irrelevant by $\mathcal{A}$ (line 3 of Algorithm 2). These selected features are then removed with a random probability to achieve the purpose of removing irrelevance and reducing dimensionality.

Algorithm 2: Dimensionality reduction

Input: $\mathcal{O}$: the offspring population; $\mathcal{A}$: learning array.
Output: the offspring population $\mathcal{O}$.
1 **for** *each* $\boldsymbol{x} \in \mathcal{O}$ **do**
2 　　**for** *j=1,...,D* **do**
3 　　　　**if** $A_j == 1$ *and* $x_j == 1$ **then**
4 　　　　　　**if** *rand()<0.5* **then**
5 　　　　　　　　$x_j \leftarrow 0$;

6 Return $\mathcal{O}$.

2.5　The New Fitness Function for Environmental Selection

In environmental selection, instead of utilizing the nondominated front level and crowding distance to select elite solutions as the next parent population, we define a new fitness function to prioritize the objective of classification error rate. The proposed fitness function consists of two criteria, i.e., the convergence criterion and the classification performance criterion. Specifically, given a feature subset $\boldsymbol{x}$, the convergence criterion can be naturally represented by its nondominated front level, i.e., $\mathcal{F}(\boldsymbol{x})$; and the classification performance criterion is represented by the acute angle between $\boldsymbol{x}$ and the nondominated solution $\boldsymbol{x}^*$ with the minimum classification error rate $(\boldsymbol{x}^* = \arg\min_{\boldsymbol{x} \in \mathcal{F}_1} f_{err}(\boldsymbol{x}))$, i.e., $\theta_{\boldsymbol{x},\boldsymbol{x}^*}$. The smaller the angle $\theta_{\boldsymbol{x},\boldsymbol{x}^*}$, the better the classification performance of $\boldsymbol{x}$. During the optimization process, $\boldsymbol{x}^*$ is constantly updated as the feature subset with a smaller classification error rate is found. In order to balance the convergence criterion and the classification performance criterion, a scalarization approach is proposed as the fitness function:

$$\min fitness(\boldsymbol{x}) = \alpha * \cos\theta_{\boldsymbol{x},\boldsymbol{x}^*} * \min(\mathcal{F}(\boldsymbol{x}) - 1, 1) + (1 - \alpha) * \hat{\mathcal{F}}(\boldsymbol{x}), \qquad (2)$$

where α is the dynamic weight coefficient associated with the current evolutionary generation g, which has the form of [18]:

$$\alpha = (1 + \delta) \exp(-(\frac{g}{\mathcal{G}})^4) - \delta, \tag{3}$$

where $\delta = 1e{-}8$ is a tolerance, $\mathcal{G} = \dfrac{g_{max}}{\sqrt[4]{\ln(\frac{1+\delta}{\delta})}}$, and g_{max} is the maximum number of generations. It is obvious that α value will decrease as the evolutionary generation g increases.

The acute angle between $\boldsymbol{x}$ and $\boldsymbol{x}^*$ in the objective space is calculated as

$$\cos \theta_{\boldsymbol{x},\boldsymbol{x}^*} = \frac{\boldsymbol{F}'(\boldsymbol{x}) \bullet \boldsymbol{F}'(\boldsymbol{x}^*)}{||\boldsymbol{F}'(\boldsymbol{x})|| \times ||\boldsymbol{F}'(\boldsymbol{x}^*)||} \tag{4}$$

where $\boldsymbol{F}'(\boldsymbol{x}) = \boldsymbol{F}(\boldsymbol{x}) - \boldsymbol{z}^* = (f_{err}(\boldsymbol{x}) - z_1^*, f_{ratio}(\boldsymbol{x}) - z_2^*)$ is the normalized objective vector for $\boldsymbol{x}$, and $\boldsymbol{F}'(\boldsymbol{x}) \bullet \boldsymbol{F}'(\boldsymbol{x}^*)$ represents the inner product of two objective vectors $\boldsymbol{F}'(\boldsymbol{x})$ and $\boldsymbol{F}'(\boldsymbol{x}^*)$, and $\boldsymbol{z}^*$ is the ideal point which represents the best objective value found so far. $||.||$ calculates the norm of a direction. It is clear that $\cos \theta_{\boldsymbol{x},\boldsymbol{x}^*}$ is in the range of $[0, 1]$.

$\hat{\mathcal{F}}(\boldsymbol{x})$ in Eq. (2) is the normalized nondominated front level for feature subset $\boldsymbol{x}$ with the form of

$$\hat{\mathcal{F}}(\boldsymbol{x}) = \frac{\mathcal{F}(\boldsymbol{x}) - \mathcal{F}_{min}}{\mathcal{F}_{max} - \mathcal{F}_{min}} \tag{5}$$

Obviously, $\mathcal{F}_{min}$ is equal to 1.

The proposed fitness function in Eq. (2) has the following properties:

Property 1. When $\lim\limits_{g\to 0} \alpha = 1$, and $\lim\limits_{g\to 0} fitness(\boldsymbol{x}) = \cos \theta_{\boldsymbol{x},\boldsymbol{x}^*} * \min(\mathcal{F}(\boldsymbol{x})-1, 1)$, Eq. (2) is mainly determined by the classification performance criterion.

Property 2. When $\lim\limits_{g\to g_{max}} \alpha = 0$, and $\lim\limits_{g\to g_{max}} fitness(\boldsymbol{x}) = \hat{\mathcal{F}}(\boldsymbol{x})$, Eq. (2) gradually degenerates to the original dominance relationship.

Property 3. The nondominated solution $\boldsymbol{x}$ always has the best fitness value, since $fitness(\boldsymbol{x}) = 0$ for $\forall \boldsymbol{x} \in \mathcal{F}_1$.

In the early stage of the search process, α is close to 1, which means that the value of $fitness(\boldsymbol{x})$ is mainly determined by the classification performance criterion (Property 1), thus high selection pressure on the classification error rate is exerted to prioritize the objective of classification performance. In the later stage of the search, with the value of g approaching g_{max}, α is close to 0. There is no much room for classification performance improvement, thus the influence of $fitness(\boldsymbol{x})$ will gradually emphasize the importance of the convergence criterion to generate more well-distributed feature subsets (Property 2). Since nondominated solutions represent the best solution set found so far, they

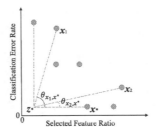

Fig. 3. An example to illustrate how the proposed fitness function selects solutions.

always have the highest priority to be chosen regardless of the changes of the α value (Property 3).

Figure 3 gives an example to illustrate how the proposed fitness function selects solutions. In Fig. 3, solution x_1 is in the second front level ($\mathcal{F}(x_1) = 2$) after nondominated sorting, while x_2 locates in the third front level ($\mathcal{F}(x_2) = 3$). In the traditional environmental selection of NSGA-II, i.e., based on nondominated sorting and crowding distance, the two objectives are treated as equally important, thus x_1 will be chosen prior x_2 since the nondominated front level of x_1 is smaller than that of x_2. However, although x_1 selects a smaller number of features than x_2, its classification error rate is much larger than that of x_2, since x_2 may contain more informative features than that of x_1. Therefore, in the proposed fitness function, x_2 will be preferred to x_1 when α is large. It is worth noting that when x_2 is selected as the solution that forms the next parent population and uses it to generate offspring, although it selects a larger number of features, it contains more relevant features which facilitate the improvement of classification performance. Furthermore, the method proposed in Sect. 2.4 is conducive to removing its irrelevant features for dimensionality reduction.

Algorithm 3: Environmental selection

Input: $\mathcal{U}$: the union of the parent and offspring populations $\mathcal{U} = \mathcal{P} \bigcup \mathcal{O}$;
$\mathcal{F}(x)$: the nondominated front level for each solution x in $\mathcal{U}$;
N: population size.
Output: the parent population $\mathcal{P}$.
1 $\mathcal{P} = \emptyset$;
2 Remove duplicated solutions in the search space from $\mathcal{U}$;
3 $x^* \leftarrow \arg\min_{x \in \mathcal{F}_1} f_{err}(x)$;
4 Calculate the *fitness* value of each solution in $\mathcal{U}$ according to Eq. (2);
5 The solutions in $\mathcal{U}$ are sorted in ascending order according to the *fitness* value, and the top N solutions are added to $\mathcal{P}$;
6 Return $\mathcal{P}$.

Algorithm 3 gives the details of the environmental selection operator. During environmental selection, the duplicated solutions will be removed first, to avoid selecting multiple identical solutions, thus increasing the population diversity. Then, the best N solutions are selected from the union of the parent and offspring populations based on the fitness value defined in Eq. (2).

3 Experiment Design

3.1 Classification Datasets

Table 1. The characteristics of the classification datasets

Dataset	#Features (D)	#Classes	#Instances
Ionosphere	34	2	351
Sonar	60	2	208
Movementlibras	90	15	360
Hillvalley	100	2	1212
Musk1	166	2	476
LSVT	310	2	126
ORL	1024	40	400
DLBCL	5469	2	77
Prostate-GE	5966	2	102

Nine classification datasets collected from the UCI machine learning repository[1] are utilized for testing the performance of PMOFS and its competitors. Table 1 summarizes the characteristics of the datasets. It can be observed from Table 1 that the chosen datasets have different numbers of features, classes, and instances with an expectation to well represent various real-world cases.

3.2 Comparison Algorithms

Five classical and state-of-the-art MOEAs are chosen as competitors of PMOFS: NSGA-II [2], MOEA/D [19], SIOM-NSGA-II [15], PMMOEA [13], and DAEA [14]. NSGA-II and MOEA/D are representatives of Pareto-dominance-based and decomposition-based MOEAs, respectively, while SIOM-NSGA-II, PMMOEA, and DAEA represent state-of-the-art MOEA-based FS methods.

3.3 Parameter Settings

Each classification dataset is randomly divided into a training data subset and a test data subset with the proportions of about 70% and 30% [17], respectively. During the training process, 5-NN with five-fold cross-validation is utilized to calculate the classification error rate on the training data subset to avoid the FS bias [8]. The nondominated feature subsets obtained at the end of the training process are applied to the unseen test set.

---

[1] http://archive.ics.uci.edu/ml.

Each MOEA independently runs 30 times on each dataset. NSGA-II, MOEA/D, and the proposed PMOFS algorithm all employ the single-point crossover and bit-flip mutation operators to generate offspring solutions, where the crossover probability and mutation probability are set to 1.0 and $1/D$, respectively, while SIOM-NSGA-II, PMMOEA, and DAEA adopt their specific genetic operators. The population size for all compared algorithms is set to be the same as the number of features ($N = D$) but bounded by 200 ($N = 200$ if $D > 200$) to avoid high computational costs on high-dimensional datasets [15]. Other parameter settings of the compared algorithms follow their original papers.

3.4   Performance Metrics

The following three performance metrics are used to measure the quality of the final obtained feature subsets:

1. Hypervolume ($\boldsymbol{HV}$) [20]: it measures both the convergence and diversity of nondominated feature subsets. The larger the HV value, the better the performance of a method.
2. Minimal classification error rate ($\boldsymbol{MCER}$) [14]: it is the classification error rate on the test data obtained by the feature subset that has the minimal classification error rate on the training data.
3. The number of selected features ($\boldsymbol{FN}$) [14]: it is the number of selected features by the feature subset with the minimum classification error rate.

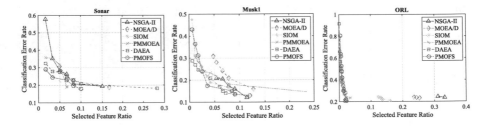

Fig. 4. Distributions of nondominated solutions obtained by each algorithm on test sets in terms of median HV value.

4   Results and Discussions

Performance Comparison in Terms of HV: The comparison results derived from the six compared algorithms in terms of HV are summarized in Table 2. At first glance, the proposed PMOFS method can achieve superior performance on most datasets. Specifically, PMOFS exhibits the best results on six out of nine datasets, and DAEA has the best performance on the rest three datasets, while the rest four algorithms (NSGA-II, MOEA/D, SIOM-NSGA-II, and PMMOEA) do not gain any best result. According to statistical significance tests, the proposed PMOFS method performs significantly better than NSGA-II, MOEA/D, SIOM-NSGA-II, and PMMOEA on eight, eight, three, and three datasets, respectively, and is not significantly worse than its peer algorithms on any dataset.

Figure 4 plots the final nondominated solutions with the median HV obtained by six MOEAs on test sets. As can be seen from the figure, on the small dataset

Table 2. Comparative results (mean and standard deviation) in terms of HV, MCER, and FN metrics, respectively.

Dataset	Algorithm	HV	MCER	FN
Ionosphere	NSGA-II[2]	9.12E-1±1.89E-2≈	6.55E-2±2.16E-2≈	4.07E+0±1.46E+0≈
	MOEA/D[19]	9.04E-1±2.04E-2≈	7.44E-2±2.40E-2≈	3.40E+0±1.30E+0≈
	SIOM-NSGA-II[15]	9.09E-1±2.16E-2≈	6.95E-2±2.51E-2≈	3.87E+0±1.41E+0≈
	PMMOEA[13]	9.05E-1±1.95E-2≈	7.33E-2±2.26E-2≈	3.73E+0±1.60E+0≈
	DAEA[14]	9.16E-1±1.78E-2≈	6.08E-2±2.05E-2≈	3.80E+0±1.21E+0≈
	PMOFS	9.13E-1±1.80E-2	6.43E-2±2.08E-2	3.77E+0±1.38E+0
Sonar	NSGA-II[2]	8.03E-1±1.89E-2+	1.79E-1±4.08E-2≈	1.06E+1±4.32E+0+
	MOEA/D[19]	8.01E-1±2.04E-2+	1.86E-1±3.11E-2≈	7.67E+0±3.46E+0≈
	SIOM-NSGA-II[15]	8.15E-1±2.16E-2≈	1.82E-1±3.27E-2≈	7.30E+0±3.48E+0≈
	PMMOEA[13]	8.12E-1±1.95E-2≈	1.86E-1±3.13E-2≈	6.80E+0±3.43E+0≈
	DAEA[14]	8.16E-1±1.78E-2≈	1.81E-1±2.28E-2≈	9.60E+0±5.14E+0≈
	PMOFS	8.21E-1±1.93E-2	1.75E-1±2.30E-2	7.10E+0±2.68E+0
Movementlibras	NSGA-II[2]	7.32E-1±3.48E-2+	2.39E-1±4.06E-2≈	1.51E+1±6.79E+0≈
	MOEA/D[19]	7.32E-1±2.87E-2+	2.45E-1±3.38E-2≈	1.07E+1±2.84E+0≈
	SIOM-NSGA-II[15]	7.50E-1±3.10E 2≈	2.49E-1±3.74E-2≈	1.09E+1±5.58E+0≈
	PMMOEA[13]	7.50E-1±2.71E-2≈	2.51E-1±3.22E-2≈	1.00E+1±4.16E+0≈
	DAEA[14]	7.54E-1±2.65E-2≈	2.45E-1±3.17E-2≈	1.29E+1±6.60E+0≈
	PMOFS	7.58E-1±2.90E-2	2.39E-1±3.41E-2	1.22E+1±5.29E+0
Hillvalley	NSGA-II[2]	6.05E-1±2.30E-2+	4.03E-1±2.00E-2≈	1.22E+1±5.49E+0+
	MOEA/D[19]	6.15E-1±1.62E-2+	4.12E-1±1.90E-2+	5.53E+0±2.47E+0-
	SIOM-NSGA-II[15]	6.24E-1±1.30E-2≈	4.05E-1±1.51E-2≈	8.37E+0±4.11E+0≈
	PMMOEA[13]	6.13E-1±1.80E-2+	4.19E-1±2.05E-2+	4.63E+0±1.97E+0≈
	DAEA[14]	6.26E-1±1.38E-2≈	4.03E-1±1.57E-2≈	8.83E+0±4.44E+0≈
	PMOFS	6.28E-1±1.05E-2	4.01E-1±1.24E-2	8.53E+0±5.17E+0
Musk1	NSGA-II[2]	8.12E-1±3.06E-2+	1.28E-1±2.92E-2≈	3.24E+1±1.37E+1+
	MOEA/D[19]	8.17E-1±2.83E-2+	1.60E-1±3.14E-2+	1.53E+1±4.65E+0-
	SIOM-NSGA-II[15]	8.62E-1±2.61E-2≈	1.39E-1±3.05E-2≈	2.13E+1±9.34E+0≈
	PMMOEA[13]	8.48E-1±3.05E-2+	1.56E-1±3.58E-2+	1.62E+1±8.59E+0-
	DAEA[14]	8.71E-1±2.30E-2≈	1.27E-1±2.77E-2≈	2.31E+1±8.47E+0≈
	PMOFS	8.64E-1±1.82E-2	1.37E-1±2.11E-2	2.66E+1±1.40E+1
LSVT	NSGA-II[2]	7.65E-1±5.24E-2+	1.69E-1±5.52E-2+	3.71E+1±9.35E+0+
	MOEA/D[19]	7.81E-1±5.05E-2+	1.85E-1±5.15E-2+	2.30E+1±6.74E+0+
	SIOM-NSGA-II[15]	8.58E-1±4.33E-2≈	1.43E-1±4.61E-2≈	1.18E+1±7.28E+0≈
	PMMOEA[13]	8.57E-1±4.33E-2≈	1.53E-1±7.37E-2≈	5.03E+0±2.31E+0-
	DAEA[14]	8.84E-1±3.88E-2≈	1.22E-1±4.31E-2≈	1.22E+1±9.69E+0≈
	PMOFS	8.76E-1±4.47E-2	1.31E-1±5.04E-2	8.40E+0±4.85E+0
ORL	NSGA-II[2]	5.64E-1±2.53E-2+	2.31E-1±3.71E-2+	3.43E+2±2.32E+1+
	MOEA/D[19]	6.03E-1±3.73E-2+	2.45E-1±4.36E-2+	2.61E+2±2.75E+1+
	SIOM-NSGA-II[15]	7.14E-1±2.55E-2+	2.13E-1±2.81E-2≈	1.55E+2±3.33E+1+
	PMMOEA[13]	7.85E-1±3.13E-2+	2.32E-1±3.52E-2+	4.11E+1±2.03E+1-
	DAEA[14]	8.06E-1±2.39E-2≈	2.08E-1±2.68E-2≈	3.89E+1±1.89E+1-
	PMOFS	8.06E-1±2.59E-2	2.05E-1±2.99E-2	7.42E+1±5.77E+1
DLBCL	NSGA-II[2]	5.20E-1±5.64E-2+	1.94E-1±9.97E-2+	2.23E+3±4.33E+1+
	MOEA/D[19]	5.34E-1±5.24E-2+	2.18E-1±7.85E-2+	2.01E+3±6.72E+1+
	SIOM-NSGA-II[15]	6.99E-1±7.16E-2+	1.78E-1±9.41E-2+	1.01E+3±5.26E+1+
	PMMOEA[13]	8.61E-1±9.34E-2≈	1.53E-1±1.03E-1≈	3.63E+0±2.20E+0-
	DAEA[14]	8.76E-1±8.80E-2≈	1.37E-1±9.69E-2≈	2.67E+0±1.12E+0-
	PMOFS	8.95E-1±9.20E-2	1.12E-1±1.01E-1	2.64E+1±3.55E+1
Prostate-GE	NSGA-II[2]	3.15E-1±1.92E-2+	5.46E-1±3.49E-2+	2.47E+3±3.26E+1+
	MOEA/D[19]	3.28E-1±2.19E-2+	5.47E-1±3.75E-2+	2.28E+3±7.26E+1+
	SIOM-NSGA-II[15]	4.20E-1±2.60E-2+	5.45E-1±3.45E-2+	1.11E+3±2.45E+1+
	PMMOEA[13]	5.16E-1±3.68E-2≈	5.33E-1±4.05E-2≈	1.97E+0±1.07E+0-
	DAEA[14]	5.15E-1±3.75E-2≈	5.33E-1±4.13E-2≈	1.93E+0±1.23E+0-
	PMOFS	5.26E-1±2.79E-2	5.20E-1±3.02E-2	2.37E+1±2.05E+1
Friedman's rank	NSGA-II[2]	5.5556	3.3333	6.0000
	MOEA/D[19]	5.1111	5.5556	3.3333
	SIOM-NSGA-II[15]	3.4444	3.8889	3.5556
	PMMOEA[13]	3.7778	4.6667	1.5556
	DAEA[14]	1.7778	2.0000	3.4444
	PMOFS	1.3333	1.5556	3.1111

Sonar, under a similar number of selected features, the solutions obtained by PMOFS tend to have a smaller classification error rate, which can be attributed to the proposed fitness function that focuses more on the classification performance. On the high-dimensional dataset ORL, benefiting from the dimensionality reduction operator, PMOFS can obtain a set of trade-off solutions with very few selected features, and some of them exhibit good classification performance.

Performance Comparison in Terms of Classification Performance: As discussed before, in FS, classification performance is more important than the number of selected features. To verify the effectiveness of the proposed fitness function in emphasizing the classification performance, Table 2 also reports the mean and standard deviation of MCER metric values. Similar to HV results, the MCER results show that PMOFS outperforms the compared methods in most of the datasets, particularly on the high-dimensional datasets. Besides, the Friedman's results show that PMOFS can rank first among the six state-of-the-art methods, which suggests it has the best overall classification performance.

Performance Comparison in Terms of the Number of Selected Features: Intuitively, PMOFS should select a large number of features, since the proposed fitness function gives more preference to the objective of classification performance. Particularly at the early and middle stages of evolution, PMOFS tends to select features with smaller classification error rates but a larger number of selected features, those feature subsets with a smaller number of selected features but higher classification error rates are discarded. However, from Friedman's rank results in Table 2, in terms of the comparison results of the FN metric, we can find that PMOFS can rank second among the six compared methods. The reason is that the proposed dimensionality reduction mechanism can remove a large number of irrelevant features to reduce the size of feature subsets. Meanwhile, it is necessary to reduce the number of selected features while ensuring that the classification performance does not deteriorate. The proposed dimensionality reduction mechanism can achieve this goal because most of the removed features are irrelevant which could degenerate the classification performance.

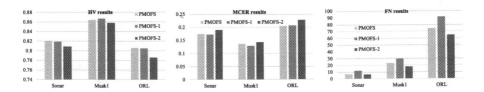

Fig. 5. Comparison results between PMOFS and its variants.

Effectiveness Analysis of the Proposed Irrelevance Learning, Dimensionality Reduction, and Fitness Function: To verify the effectiveness of each major component of PMOFS, we test the performance of PMOFS with its two variants: PMOFS-1 removes the irrelevance learning and dimensionality reduction operators, and PMOFS-2 replaces the fitness function based environmental selection with the nondominated sorting and crowding distance.

Figure 5 plots the results of three methods on the Sonar, Musk1, and ORL datasets. It can be seen that PMOFS and PMOFS-1 have similar HV and MCER results, but the number of selected features obtained by PMOFS is smaller than PMOFS-1. This can be attributed to the fact that the proposed irrelevance learning and dimensionality reduction operators can identify and remove a large number of irrelevant features. Since removing irrelevant features is beneficial to the improvement of classification performance, this can explain why PMOFS selects a smaller number of features but exhibits similar or even better classification performance (e.g., on the Musk1 dataset) with PMOFS-1.

Compared with PMOFS-2, PMOFS can obtain better HV and MCER values, which verifies the effectiveness of the proposed fitness function in prioritizing the objective of classification performance.

5   Conclusions

This paper has proposed a preference-inspired MOEA (PMOFS) for handling the different preferences between objectives for multi-objective FS. In PMOFS, the designed fitness function enables PMOFS to achieve a smaller classification error rate since there is more search pressure on improving the classification performance. In addition, through irrelevance learning, the proposed dimensionality reduction operator is not only conducive to the improvement of classification performance, but also reduces the size of feature subsets. The experimental results verified that PMOFS outperforms the compared state-of-the-art algorithms in terms of HV and classification performance metrics, and is highly competitive in terms of the number of selected features.

Given the promising performance of PMOFS, we envisage the following two directions for future endeavors. First, the idea of incremental learning can be borrowed to learn the irrelevance between features. Second, it is also worth studying to reduce the training time by combining the proposed method with Bayesian statistical learning.

Acknowledgments. This work was supported in part by the Marsden Fund of New Zealand Government under Contracts MFP-VUW1913 and MFP-VUW1914, the Science for Technological Innovation Challenge (SfTI) fund under contract 2019-S7-CRS, MBIE Data Science SSIF Fund under the contract RTVU1914, MBIE Research Program under contract C11X2001 (PFR No. 39186), NZ-SQ Data Science Catalyst Program under contract 2021-37-004A, the University Research Fund at Victoria University of Wellington grant number 223805/3986, and National Natural Science Foundation of China (NSFC) under Grant 61876169.

References

1. Cai, X., Nie, F., Huang, H.: Exact top-k feature selection via $\ell_{2,0}$-norm constraint. In: International Joint Conference on Artificial Intelligence (IJCAI), pp. 1240–1246. Citeseer (2013)
2. Deb, K., Pratap, A., Agarwal, S., Meyarivan, T.: A fast and elitist multiobjective genetic algorithm: NSGA-II. IEEE Trans. Evol. Comput. **6**(2), 182–197 (2002)
3. Eiben, A.E., Smith, J.: From evolutionary computation to the evolution of things. Nature **521**(7553), 476–482 (2015)
4. Feng, C., Qian, C., Tang, K.: Unsupervised feature selection by pareto optimization. In: AAAI Conference on Artificial Intelligence, vol. 33, pp. 3534–3541 (2019)
5. Gui, J., Sun, Z., Ji, S., Tao, D., Tan, T.: Feature selection based on structured sparsity: a comprehensive study. IEEE Trans. Neural Netw. Learn. Syst. **28**(7), 1490–1507 (2016)
6. Jiao, R., Xue, B., Zhang, M.: Solving multi-objective feature selection problems in classification via problem reformulation and duplication handling. IEEE Trans. Evol. Comput. (2022). https://doi.org/10.1109/TEVC.2022.3215745
7. Khurana, A., Verma, O.P.: Optimal feature selection for imbalanced text classification. IEEE Trans. Artif. Intell. (2022). https://doi.org/10.1109/TAI20223144651
8. Kohavi, R., John, G.H.: Wrappers for feature subset selection. Artif. Intell. **97**(1–2), 273–324 (1997)
9. Li, J., et al.: Feature selection: a data perspective. ACM Comput. Surv. **50**(6), 1–45 (2017)
10. Mukhopadhyay, A., Maulik, U., Bandyopadhyay, S., Coello, C.A.C.: A survey of multiobjective evolutionary algorithms for data mining: part i. IEEE Trans. Evol. Comput. **18**(1), 4–19 (2013)
11. Patterson, G., Zhang, M.: Fitness functions in genetic programming for classification with unbalanced data. In: Orgun, A., Thornton, J. (eds.) AI 2007. LNCS (LNAI), vol. 4830, pp. 769–775. Springer, Heidelberg (2007). https://doi.org/10.1007/978-3-540-76928-6_90
12. Telikani, A., Tahmassebi, A., Banzhaf, W., Gandomi, A.H.: Evolutionary machine learning: a survey. ACM Comput. Surv. **54**(8), 1–35 (2021)
13. Tian, Y., Lu, C., Zhang, X., Cheng, F., Jin, Y.: A pattern mining-based evolutionary algorithm for large-scale sparse multiobjective optimization problems. IEEE Trans. Cybern. **52**(7), 6784–6797 (2022)
14. Xu, H., Xue, B., Zhang, M.: A duplication analysis-based evolutionary algorithm for biobjective feature selection. IEEE Trans. Evol. Comput. **25**(2), 205–218 (2020)
15. Xu, H., Xue, B., Zhang, M.: Segmented initialization and offspring modification in evolutionary algorithms for bi-objective feature selection. In: Genetic and Evolutionary Computation Conference (GECCO), pp. 444–452 (2020)
16. Xue, B., Zhang, M., Browne, W.N.: Particle swarm optimization for feature selection in classification: a multi-objective approach. IEEE Trans. Cybern. **43**(6), 1656–1671 (2012)
17. Xue, B., Zhang, M., Browne, W.N., Yao, X.: A survey on evolutionary computation approaches to feature selection. IEEE Trans. Evol. Comput. **20**(4), 606–626 (2015)
18. Zeng, S., Jiao, R., Li, C., Li, X., Alkasassbeh, J.S.: A general framework of dynamic constrained multiobjective evolutionary algorithms for constrained optimization. IEEE Trans. Cybern. **47**(9), 2678–2688 (2017)

19. Zhang, Q., Li, H.: MOEA/D: a multiobjective evolutionary algorithm based on decomposition. IEEE Trans. Evol. Comput. **11**(6), 712–731 (2007)
20. Zitzler, E.: Evolutionary algorithms for multiobjective optimization: methods and applications, vol. 63. Citeseer (1999)

Genetic Algorithm with a Novel Leiden-based Mutation Operator for Community Detection

Anjali de Silva[(✉)] [ID], Aaron Chen[ID], Hui Ma[ID], and Mohammad Nekooei[ID]

Victoria University of Wellington, Wellington, New Zealand
{desilanja,aaron.chen,Hui.Ma,mohammad.nekooei}@ecs.vuw.ac.nz

Abstract. Detecting quality community structures in complex networks is an important and highly active research area. Plenty of methods have been proposed for community detection in recent years. Among them, Genetic Algorithms (GAs) have been widely explored for community detection due to their strong competence at exploring the global discrete search space. However, existing GA algorithms for community detection still face major challenges when handling large and complex networks due to their use of random mutation operators. Whenever any candidate community structure in a GA population is mutated, a mutated node of the network under processing is often associated to a community with loose connections, seriously hurting GA's effectiveness and scalability. To address this issue, a newly designed Leiden-based GA (LGA) with a novel mutation operator based on the Leiden algorithm is proposed in this paper to improve the effectiveness of the mutation operator and the performance of the GA approach. Experiment results clearly show that LGA can achieve highly competitive performance in comparison to several state-of-the-art GA and non-GA community detection algorithms on multiple synthetic and real-world networks.

Keywords: Community detection · Leiden algorithm · Genetic algorithm

1   Introduction

Many complex real-world systems can be modeled as networks [24]. Studying these networks in different fields such as communication, biological and transportation can often lead to deep understandings and scientific breakthroughs [2,7,17]. In recent past years, researchers are increasingly focusing on the *community structures* of complex networks. Identifying the community structure helps to reveal the hidden features of a network that are of significant research and practical values [24].

The current analysis of the community structures focuses on *community detection* with the aim to group densely connected nodes together, meanwhile separating loosely connected nodes into different groups [22]. In recent years,

H. Aziz et al. (Eds.): AI 2022, LNAI 13728, pp. 252–265, 2022.
https://doi.org/10.1007/978-3-031-22695-3_18

many effective community detection algorithms have been proposed, including heuristic based algorithms [18], mathematical optimization algorithms [8,30], evolutionary computation (EC) algorithms [8,22,24], and deep learning-based algorithms [5,10,28].

Among all the existing community detection algorithms, EC algorithms have been shown to be highly effective and scalable [22]. In comparison to other EC algorithms such as ant colony optimization (ACO) [9] and firefly and bat algorithms [18], Genetic Algorithm (GA) is widely explored for community detection due to its strong competence at exploring the global discrete search space with a desirable balance between efficiency and effectiveness without prior domain knowledge and mathematical models.

While showing promise, existing GA approaches for community detection still face major challenges, especially when handling large and complex networks. Particularly, many previously proposed GAs rely on random mutation techniques to evolve new community detection solutions. The mutation operators do not explicitly consider a node's connectivity with all its adjacent communities in a network. Consequently, a node is often associated to a community with loose connections after mutation, seriously hurting a GA's effectiveness on large networks.

In the literature, the Leiden algorithm [30] has been shown to perform reliably well with state-of-the-art performance on many networks. In comparison to GA, Leiden can make more informed decisions when it merges groups of nodes iteratively into increasingly larger communities, driven explicitly by the connection densities among groups. As such we utilize Leiden to enhance the effectiveness of the mutation operator and design a new Leiden-based GA (LGA) in this paper. Meanwhile, as a population-based search strategy, LGA provides diverse starting points for Leiden to carry out its greedy community building process, alleviating Leiden's reliance on the initial node groupings.

The performance of LGA has been evaluated experimentally in this paper and compared to several state-of-the-art algorithms on both real-world and synthetic networks. Our experimental results clearly show that LGA can consistently outperform many existing algorithms on a majority of the benchmark networks evaluated. The key contributions of this paper are listed below:

- We develop a mutation operator based on the Leiden algorithm to generate mutated community structures with significantly high quality. To the best of our knowledge, this is the first time that Leiden has been adopted as a mutation mechanism to boost the performance of GA. Building on Leiden, the mutation operator in LGA can effectively connect densely connected nodes to the same community.
- Building on the Leiden mutation operator, we develop and implement LGA algorithm that can be applied to various networks for community detection.
- Comprehensive experiments have been conducted in this paper on a wide range of benchmark networks. Our experiment results clearly show that LGA can achieve highly competitive performance, in comparison to several state-of-the-art GA and non-GA community detection algorithms.

2   Related Works

During recent years, community detection has received increasing attention in the research community [22]. Existing algorithms for community detection can be categorized as heuristic algorithms [1,18], greedy search algorithms, EC algorithms and machine learning based algorithms [5,10,28]. State-of-the-art learning based algorithms are more suitable for processing high-dimensional networks [14] that are different from the networks considered in our problem formulation in Sect. 3. Since they are also computationally intensive, we will not consider such approaches in this paper.

Among the existing greedy search algorithms [11,26], the Louvain algorithm [4] is an efficient algorithm for identifying community structures in large networks. It gradually merges small communities into larger groups so as to maximize the modularity of the resulting community structure. To further improve Louvain algorithm's effectiveness, Leiden algorithm [30] was proposed in 2019. Similar to Louvain, Leiden is an efficient greedy search algorithm. While merging smaller communities, Leiden also checks the connectivity strength among the nodes in each community, in order to produce high quality community structures. Despite well demonstrated effectiveness, the final performance of Leiden highly depends on the initial community setup as a key algorithm input. Default initial community set up for the Leiden algorithm is that it assigns each node in the network to its own community. However, instead of the default initial community set up, a different initial community set up also can be given as an input to the Leiden algorithm. If the initial community setup is misleading, Leiden may end up with poor community structures.

Besides greedy search, a popular approach is to formulate community detection as modularity maximization problems that can be further solved through various optimization algorithms. Particularly, Barber and Clark showed in [3] that such optimization problems are NP-hard. Therefore, in order to achieve a good balance between effectiveness and efficiency, using EC techniques for community detection has now become a trending research area. A brilliant review of existing EC approaches for community detection can be found in [22].

Among all EC techniques, GA is the most widely studied method for community detection [6,15,16,29]. Most GA approaches rely on purely random mutations to explore the solution space. As a result, the mutated community structures often have poor modularity (see Fig. 2 for an example). To address this issue, the CCGA algorithm introduced in [24], uses the *Clustering Coefficient* (CC) to guide the mutation process. Meanwhile, LSSGA [8] designs a new mutation operator to improve the local structural similarity of each network node.

However, by relying only on the local information of each network node, CCGA and LSSGA cannot guarantee to mutate community structures in the direction of maximizing the global modularity across all communities. In comparison, Leiden can effectively improve the modularity by iteratively merging smaller communities into larger ones, guided by the community structures evolved by GA. Hence, we propose to develop a Leiden-based mutation operator to boost the performance of GA.

3   Problem Formulation

A network SN is a graph $G = (V, E)$, where V is the set of n nodes $V = \{v_1, v_2, ..., v_n\}$ in the network and E is the set of edges, denoted as $E = \{e_{i,j} | e_{i,j} \in V \times V\}$. Similar to many existing works [8,24], this paper considers networks with undirected and unweighted edges.

A *community* in a network refers to a group of nodes that are densely connected among each other and sparsely connected with the nodes in other groups. In this paper, the *community structure* of a network SN is defined as a group of p communities, denoted as $CS = \{C_1, C_2, ..., C_p\}$, $p \geq 1$. Here C_i stands for the i-th community of SN, which is non-overlapping with any other communities, i.e., $\forall i \neq j$, $C_i \cap C_j = \emptyset$. Furthermore, any node of SN must belong to exactly one community such that SN is fully covered by all the communities. The main goal of the community detection problem is to find the optimal community structure CS^*, as defined formally below:

$$CS^* = \arg \max_{CS} Q(CS, SN) \tag{1}$$

In Eq. (1), $Q(CS, SN)$ refers to the modularity of community structure CS with respect to network SN. It is a major and widely adopted quality measure of any community structures and is originally introduced by Girvan and Newman [20]. $Q(CS, SN)$ is formally defined below:

$$Q(CS, SN) = \frac{1}{2m^{SN}} \sum_{ij} (A_{ij}^{SN} - \frac{k_i^{SN} k_j^{SN}}{2m^{SN}}) \delta(i, j, CS) \tag{2}$$

In Eq. (2) $A_{ij}^{SN} = 1$ whenever nodes v_i and v_j are adjacent in network SN. Otherwise, $A_{ij}^{SN} = 0$. m^{SN} is the total number of edges in network SN. k_i^{SN} and k_j^{SN} are the degrees of the nodes i and j in network SN respectively. δ is defined below:

$$\delta(i, j, CS) = \begin{cases} 1 & \text{if nodes } i \text{ and } j \text{ are in the same community of } CS \\ 0 & \text{otherwise} \end{cases} \tag{3}$$

High values of modularity Q in Eq. (2) indicate good community structures [22]. LGA has the aim to maximize Q by evolving the best possible community structures of network SN.

4   Proposed Algorithm

4.1   Overall Algorithm Design

The proposed LGA algorithm follows the general process of conventional GA [20]. The initial population is generated randomly where each chromosome in the population uses the popular Locus based adjacency representation [19]. Based on the community structure it represents, each chromosome is evaluated by Q in Eq. (2) as its fitness. In each generation, LGA adopts the elitism mechanism to pass a certain percentage of chromosomes with the highest fitness directly to the next generation. The rest of the chromosomes of the subsequent generation are created from selected parent chromosomes of the current population using crossover and mutation operators. This iterative process will be performed for multiple generations until the termination criteria is reached. Finally, the fittest chromosome is reported as the best community structure identified by LGA. Since Leiden is an efficient greedy search algorithm, by combining it with GA as in LGA, it produces high quality community structures and accelerates the convergence compared to the other state-of-the-art algorithms. The pseudo-code of LGA is given in Algorithm 1. Each of the main steps of the proposed algorithm is explained in detail below.

4.2   Solution Representation

LGA uses the Locus Based Adjacency (LBA) representation of chromosomes proposed in [19]. For a network with n nodes, a chromosome consists of n genes $g_1, ..., g_n$, with the positions as the indices of nodes. Each gene is assigned with an allele value that spans within a range of $\{1, ..., n\}$. If the allele value of the i-th gene is j, this means that node v_i is connected to node v_j by an edge in the network and the two nodes should be grouped into the same community according to the chromosome. In other words, each gene g_i will choose a neighboring node of node v_i as its allele value. Hence, LBA requires a dedicated decoding process to transform any chromosome into the corresponding community structure. An example network is given in Fig. 1(a) and an example chromosome for this network is illustrated in Fig. 1(b). As shown in the two figures, the allele value is randomly chosen among the neighbors of the each node. For example, node 1 is the allele value of node 0 which is chosen randomly from the list of neighbors $\{1,9\}$ of node 0. Fig. 1(c) illustrates the communities obtained after decoding. Nodes 0 and 1 belong to the same community since the allele value of node 0 is node 1. Similarly node 7 and node 1, node 9 and 7 belong to the same community. Hence $\{0,1,9,7\}$ is a one community which is shown in red in the Fig. 1(c). Similarly it is able to obtain the other two communities given in blue and green in the same figure.

4.3   Population Initialization

In LGA, each chromosome in the initial population is generated by randomly selecting a neighboring node of each node that corresponds to separate genes in the chromosome Sect. 4.2. This approach is widely used in many existing GA methods for community detection [8,24].

4.4   Crossover Operator

The crossover operator generates new offspring chromosomes by recombining different genes from the parent chromosomes. In LGA, following several existing

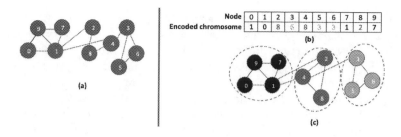

Node	0	1	2	3	4	5	6	7	8	9
Encoded chromosome	1	0	8	5	8	3	3	1	2	7

Fig. 1. An example of the Locus Based Adjacency (LBA) representation.

GA approaches [8,24], we use the uniform crossover operator to generate new offsprings [21]. The gene values to be inherited by the offspring chromosomes are decided based on a random binary vector. If the corresponding value of the binary vector is 1, the offspring inherits the gene value from the first parent, otherwise the gene is inherited from the second parent.

4.5   Mutation Operator

As mentioned in the introduction, random mutation can often lead to poor offspring chromosomes with low modularity. This is because, after performing random mutation on a node in the network, the node is often associated to a community with loose connections in the offspring chromosome. Such an example is illustrated in Fig. 2(a). I_{bm} refers to the chromosome which is to be mutated. $Q(I_{bm})$ is the modularity of I_{bm}. As illustrated in the example, after performing random mutation, the mutated chromosome has lower modularity, since random mutation accidentally separates node 4 and nodes 2 and 8 into different communities despite of the strong connections among them. After performing random-based mutation as illustrated in Fig. 2(a), node 4 has only one connection (edge highlighted in red) within the assigned community, whereas it has two external connections with another community (edges highlighted in purple).

To enhance the effectiveness of the mutation operator, we develop a new mutation operator based on the Leiden community detection algorithm, which has been shown to achieve good performance on many large-scale networks [8]. Leiden accepts an initial community setup in the form of a membership list that assigns each node with an initial community label. Guided by the given community setup, Leiden iteratively merges small communities to form larger communities in the direction of maximizing the modularity. The whole merging process depends on the initial community setup, which is obtained from the evolved chromosome selected for mutation.

In LGA, the chromosome to be mutated is first decoded into the corresponding community structure, according to which the membership list can be established. Subsequently, for all nodes assigned with the same community label, we randomly select a neighboring node for each node. Based on the inter-connections among randomly selected neighbors, we can further break the whole community into smaller sub-communities. Nodes in the same sub-communities are then assigned a new community label. This process ensures that every community in the initial community setup for the Leiden algorithm is reasonably small, facilitating the merging operations to be carried out next by Leiden.

In line with our new mutation operator, any chromosome evolved by GA can be utilized to create the initial community setup that guides the merging activities in Leiden. It alleviates the dependencies of Leiden on any specific community structures, enabling the algorithm to explore all possible formations of communities so as to eventually maximize the modularity. Meanwhile, since Leiden only merges the communities provided by a chromosome when this improves the modularity, it can be guaranteed that the mutated chromosome will never have poor modularity, in comparison to the chromosome before mutation. As illustrated in Fig. 2(b), after performing Leiden-based mutation on chrosome I_{bm}, the mutated chromosome I_{lm} has achieved a higher modularity than I_{bm}. For example in Fig. 2(b) there exists a higher number of internal connections for node 4 (edges highlighted in purple) compared to the external connections (edges highlighted in red). The same observation has also been witnessed in our experiments.

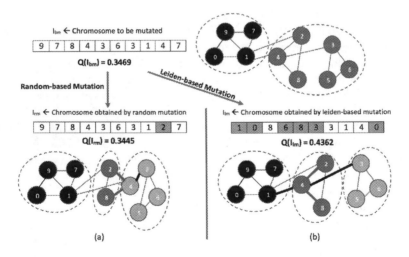

Fig. 2. An example of Random mutation and Leiden-based mutation: (a) the chromosome obtained after the random-based mutation process and (b) the chromosome obtained after the Leiden-based mutation process.

Algorithm 2 presents the pseudo-code for the new Leiden-based mutation operator. Lines 2 and 3 are the key steps of this algorithm. Line 2 decodes chromosome I_{bm} under mutation to the corresponding community structure CS_{bm}. Line 3 breaks every community in CS_{bm} into smaller sub-communities. This is achieved by randomly pairing each node with one of its neighboring node into the same sub-community. Hence, a large community is more likely to be broken

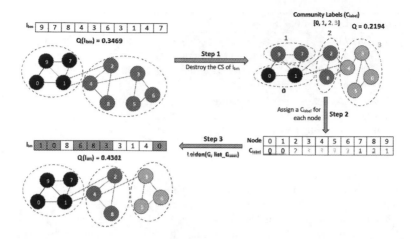

Fig. 3. An example of the key steps of Leiden-based mutation.

into small sub-communities than a small community. This process is illustrated in Step 1 in Fig. 3. Line 4 then generates a list of community labels based on the output of Line 3, as demonstrated in Step 2 in Fig. 3. The list of labels and the network will be given as the inputs to the Leiden algorithm, which will produce the mutated community structure CS_{lm} in Line 5, as illustrated in Step 3 in Fig. 3. Finally CS_{lm} is encoded into the mutated chromosome I_{lm}. The encoding process is described in the next subsection.

4.6   Encoding

To encode any community structure into a chromosome with the LBA representation, LGA first creates a spanning tree over all nodes in the same community. The breadth-first graph search algorithm is further conducted on the spanning tree, starting from a randomly selected root node of the tree, such that every other node will choose its parent node in the tree as its neighbor in the LBA chromosome. On the other hand, the root node will choose one of its child node as its neighbor in the LBA chromosome. The same process is repeated for every community. Eventually, a complete LBA chromosome can be created where each node in the network is associated with a separate gene that captures its selected neighbors.

4.7   Population Update

Upon creating the new population for the next generation, LGA selects 5% of the most fitted chromosomes of the current population. The remaining 95% of the population is filled with the newly evolved offspring chromosomes. Thanks to elitism, the best chromosomes evolved by LGA will never get lost during the evolution process.

Algorithm 1. Leiden-based GA (LGA)

Input: Network SN; Population size N_p; Crossover rate P_c; Mutation rate P_m;
      Elitism ratio P_e; Generation size N_g;
      Population ratio to generate the offsprings P_s
Output: Community structure CS^*

1: Initialize the population with N_p randomly created chromosomes   ▷ Refer
    Subsection 4.3
2: **for** each chromosome I in N_p **do**
3:     Evaluate fitness of chromosome I: $Q(I)$ ▷ Refer Section 3
4: **end for**
5: **for** each generation g in N_g **do**
6:     Pass the best P_e chromosomes to generation $g+1$ ▷ Refer Subsection 4.7
7:     Select P_s parent chromosomes N_{OS}
8:     **for** each chromosome I in N_{OS} **do**
9:         Perform crossover with P_c to obtain offspring I_{bm} ▷ Refer Subsection 4.4
10:        Perform $Leiden\_Mutation(SN, I_{bm}, P_m)$ to obtain I_{lm} ▷ Refer Algorithm 2
11:        Evaluate fitness of chromosome I_{lm}: $Q(I_{lm})$
12:        Update population with I_{lm} for the next generation g+1   ▷ Refer
        Subsection 4.7
13:     **end for**
14: **end for**
15: Return the evolved community structure with the highest modularity

---

Algorithm 2. Leiden-based mutation operator (Refer Subsection 4.5)

1: **function** Leiden_Mutation(SN, I_{bm}, P_m):
2:     Decode I_{bm} to obtain CS_{bm} ▷ Refer Subsection 4.2
3:     Destroy each community of CS_{bm} to obtain a new community structure new_CS
4:     Generate a list of community labels $list\_C_{label}$ based on new_CS
5:     $CS_{lm} = $ Leiden(SN, $list\_C_{label}$)
6:     Encode CS_{lm} to obtain I_{lm} ▷ Refer Subsection 4.6
7:     **return** I_{lm}

---

5 Experiment and Analysis

Real-world networks and synthetic networks are used for the evaluation of LGA. Detailed explanation regarding the benchmark networks are given in Subsect. 5.1. Four different state-of-the-art algorithms were used to compare the performance of LGA, including Louvain [4], Leiden [30], CCGA [24] and LSSGA [8]. They are briefly explained in Subsect. 5.2. We follow strictly the recommended parameter settings of each competing algorithm according to their inventors [24]. The parameter settings relevant to LGA are given in Subsect. 5.3. All the experiments are conducted on MacOS 11.4 with an Intel Core i3 8-core processor and 8GB RAM. Python 3.9 and networkx 2.6.3 have been used for the experiments.

5.1   Benchmark Networks

The experiments of our proposed LGA and the other competitive algorithms have been carried out on real-world and synthetic networks. Descriptions of all benchmark networks are given below.

Real-World Networks. To evaluate the performance of LGA, we used 11 real-world networks, as summarized in Table 1, along with the network type, number of nodes and the number of edges.

Table 1. Description of the real-world and synthetic networks.

Network	Type	Number of nodes	Number of edges
Karate [23]	Social	34	78
Dolphins [23]	Social	62	159
Polbooks [23]	Social	105	441
Football [23]	Social	115	613
Jazz [23]	Collaboration	198	2742
E.coli [27]	Biological	418	519
Email-Eu-core [31]	Communication	1005	25571
Cora [23]	Citation	2708	5429
Facebook [13]	Online social	2888	2981
Citeseer [23]	Citation	3312	4732
Protein [25]	Biological	3724	8748
LFR256	Synthetic	256	911
LFR512	Synthetic	512	1958
LFR1000	Synthetic	1000	4038

Synthetic Networks. The performance of LGA is also evaluated on three synthetic networks based on Lancichinetti-Fortunato-Radicchi (LFR) benchmarks [12]. These networks are summarized in Table 1. The parameter settings for generating the three LFR benchmark networks are given in [24].

5.2   Baseline Algorithms

The performance of LGA is compared to four state-of-the-art algorithms. Out of four algorithms, Louvain [4] and Leiden [30] are greedy search algorithms. CCGA [24] and LSSGA [8] are competing EC algorithms. These four algorithms are specifically designed to maximize the modularity of community structures.

5.3   Parameter Settings

Parameter settings for LGA follow closely with the parameter setting of CCGA [24]. Specifically, population size is 300, generation number is 200, elitism ratio is 0.05, crossover rate is 0.8 and mutation rate is 0.2. The algorithm was run for 30 times independently.

5.4   Results

In this study, modularity in Eq. (2) is used as the performance metric. Table 2 compares the moularity achieved by LGA and other competing approaches introduced in Subsect. 5.2. The highest average modularity over 30 runs achieved on every benchmark network is bolded in the table. Missing modularity is indicated as hyphen (–) since the corresponding results cannot be found in relevant research papers.

According to Table 2, on most networks LGA outperformed the competing algorithms. LGA achieved significant improvement in modularity on Citeseer and Email networks among all algorithms. LGA also achieved competitive modularity on Karate, Football, Facebook and three synthetic networks. Among all the networks, LGA only falls slightly behind LSSGA on the Cora network. Even though LGA does not perform better than LSSGA on Cora, LGA still manages to outperform Leiden. This shows that, by using community structures evolved by GA as the initial community setup for Leiden in our Leiden-based mutation operator, LGA is more effective at handling large complex networks than Leiden.

Table 2. Average modularity over 30 runs, achieved by LGA and competing algorithms.

Network	Louvain [4]	Leiden [30]	CCGA [24]	LSSGA [8]	LGA
Karate	0.4156	**0.4198**	**0.4198**	**0.4198**	**0.4198**
Dolphins	0.5224	0.5256	0.5057	0.5283	**0.5285**
Polbooks	0.5270	0.5269	0.5270	0.5270	**0.5272**
Football	0.6042	**0.6046**	0.5243	**0.6046**	**0.6046**
Jazz	0.4402	0.4449	0.4440	–	**0.4451**
Ecoli	0.7777	0.7806	0.7647	–	**0.7815**
Facebook	**0.8087**	**0.8087**	**0.8087**	–	**0.8087**
Protein	0.7813	0.7863	0.7314	–	**0.7891**
Email	0.4318	0.4331	0.2721	0.4165	**0.4347**
Cora	0.8123	0.8230	0.7646	**0.8257**	0.8247
Citeseer	0.8920	0.8960	0.8147	0.8100	**0.8969**
LFR256	**0.7664**	**0.7664**	0.7143	–	**0.7664**
LFR512	0.8312	**0.8313**	0.7717	–	**0.8313**
LFR1000	**0.8614**	**0.8614**	0.7778	–	**0.8614**

5.5   Further Analysis

Figure 4 presents the convergence curves of LGA, stdGA and CCGA on some real-world networks: Citeseer and Cora. It presents the average curves on maximum fitness values obtained over 30 runs. stdGA refers to the conventional GA that uses the random mutation operator. As evidenced in this figure, LGA converges much faster than other GA algorithms. Specifically, in our experiments, LGA can find its best solution after the 2nd to 10th generation on small networks such as Karate, Dolphins, Polbooks, Football and Ecoli. On large networks such as Citeseer and Cora networks, LGA converges to its best solution after the 30th to 50th generation. In comparison to other GA-based algorithms, LGA requires consistently a much smaller number of generations to find its best solution.

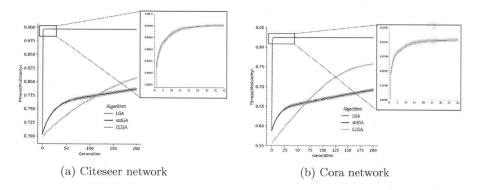

(a) Citeseer network   (b) Cora network

Fig. 4. Convergence curves of LGA, stdGA and CCGA on two real-world networks.

We further conduct experiments to evaluate the performance of using a new Leiden random mutation (LRM) operator that combines both Leiden-based mutation and random mutation. This is achieved through two consecutive steps. In step 1, Leiden-based mutation is applied first to the selected chromosome. In step 2, the mutated chromosome is further updated through random mutation in order to encourage more exploration in GA.

Our experiments show that LRM mutation and Leiden-based mutation can achieve the same best modularity on all networks. However, Leiden-based mutation enables LGA to converge faster. For example, LGA converges after the 31st generation on the Citeseer network. In comparison, GA with LRM converges to the best solution after the 46th generation. Similarly on the Cora network, convergence occurred after the 42nd generation for LGA, while it takes 139 generations for GA with LRM to converge. Hence it does not seem necessary to introduce extra randomness in the mutation process. Using Leiden-based mutation alone is effective at evolving high-quality community structures.

6   Conclusions

In this paper, we proposed a new GA based algorithm, named LGA, with a novel mutation operator based on the Leiden algorithm to improve the performance of community detection. We conducted comprehensive experiments on a wide range of benchmark networks including real-world and synthetic networks. Our experiments demonstrated that the proposed LGA can outperform several state-of-the-art algorithms on most of the networks, in particular large networks such as Citeseer and Email. LGA can also converge much faster than some existing GA-based approaches. In the future, it is interesting to explore the potential combined use of LGA and learning based techniques for community detection in high-dimensional networks.

References

1. Arasteh, M., Alizadeh, S.: A fast divisive community detection algorithm based on edge degree betweenness centrality. Appl. Intell. **49**(2), 689–702 (2019)
2. Barabasi, A.-L., Oltvai, Z.N.: Network biology: understanding the cell's functional organization. Nat. Rev. Genetics **5**(2), 101–113 (2004)
3. Barber, M.J., Clark, J.W.: Detecting network communities by propagating labels under constraints. Phys. Rev. E **80**, 2 (2009)
4. Blondel, V.D., Guillaume, J.-L., Lambiotte, R., Lefebvre, E.: Fast unfolding of communities in large networks. J. Statistic. Mech. Theory Exper. **2008**, 10 (2008)
5. Dhilber, M., Bhavani, S.D.: Community detection in social networks using deep learning. In: Hung, D., Van D'Souza, M. (eds.) ICDCIT 2020. LNCS, vol. 11969, pp. 241–250. Springer, Cham (2020). https://doi.org/10.1007/978-3-030-36987-3_15
6. Ebrahimi, M., Shahmoradi, M.R., Heshmati, Z., Salehi, M.: A novel method for overlapping community detection using multi-objective optimization. Phys. A Statist. Mech. Appl. **505**, 825–835 (2018)
7. Guimera, R., Mossa, S., Turtschi, A., Amaral, L.N.: The worldwide air transportation network: anomalous centrality, community structure, and cities' global roles. In: Proceedings of the National Academy of Sciences, pp. 7794–7799. National Academic Sciences (2005)
8. Guo, X., Su, J., Zhou, H., Liu, C., Cao, J., Li, L.: Community detection based on genetic algorithm using local structural similarity. IEEE Access **7**, 134583–134600 (2019)
9. Hosseini, R., Rezvanian, A.: Antlp: ant-based label propagation algorithm for community detection in social networks. CAAI Trans. Intell. Technol. **5**(1), 34–41 (2020)
10. Jin, D., Ge, M., Li, Z., Lu, W., He, D., Fogelman-Soulie, F.: Using deep learning for community discovery in social networks. In: 2017 IEEE 29th International Conference on Tools with Artificial Intelligence, pp. 160–167. IEEE (2017)
11. Kong, H., Kang, Q., Li, W., Liu, C., Kang, Y., He, H.: A hybrid iterated carousel greedy algorithm for community detection in complex networks. Phys. A Statist. Mech. Appl. **536** (2019)
12. Lancichinetti, A., Fortunato, S., Radicchi, F.: Benchmark graphs for testing community detection algorithms. Phys. Rev. E **78**, 4 (2008)
13. Leskovec, J., Mcauley, J.: Learning to discover social circles in ego networks. Adv. Neural Inform. Process. Syst. **25** (2012)

14. Liu, F., et al.: Deep learning for community detection: progress, challenges and opportunities. arXiv preprint arXiv:2005.08225 (2020)
15. Liu, Z., Sun, Y., Cheng, S., Sun, X., Bian, K., Yao, R.: A node influence based memetic algorithm for community detection in complex networks. In: Pan, L., Cui, Z., Cai, J., Li, L. (eds.) BIC-TA 2021. CCIS, vol. 1565, pp. 217–231. Springer, Singapore (2022). https://doi.org/10.1007/978-981-19-1256-6_16
16. Moradi, M., Parsa, S.: An evolutionary method for community detection using a novel local search strategy. Phys. A Statist. Mech. Appl. **523**, 457–475 (2019)
17. Onnela, J.-P., et al.: Structure and tie strengths in mobile communication networks. In: Proceedings of the National Academy of Sciences, pp. 7332–7336. National Acad Sciences (2007)
18. Osaba, E., Del Ser, J., Camacho, D., Bilbao, M.N., Yang, X.-S.: Community detection in networks using bio-inspired optimization: latest developments, new results and perspectives with a selection of recent meta-heuristics. Appl. Soft Comput. **87** (2020)
19. Park, Y., Song, M., et al.: A genetic algorithm for clustering problems. In: Proceedings of the Third Annual Conference on Genetic Programming, vol. 1998, pp. 568–575. Morgan Kaufmann San Francisco (1998)
20. Pizzuti, C.: GA-Net: a genetic algorithm for community detection in social networks. In: Rudolph, G., Jansen, T., Beume, N., Lucas, S., Poloni, C. (eds.) PPSN 2008. LNCS, vol. 5199, pp. 1081–1090. Springer, Heidelberg (2008). https://doi.org/10.1007/978-3-540-87700-4_107
21. Pizzuti, C.: A multiobjective genetic algorithm to find communities in complex networks. IEEE Trans. Evol. Comput. **16**(3), 418–430 (2011)
22. Pizzuti, C.: Evolutionary computation for community detection in networks: a review. IEEE Trans. Evol. Comput. **22**(3), 464–483 (2017)
23. Rossi, R.A., Ahmed, N.K.: The network data repository with interactive graph analytics and visualization. In: Twenty-Ninth AAAI Conference on Artificial Intelligence (2015)
24. Said, A., Abbasi, R.A., Maqbool, O., Daud, A., Aljohani, N.R.: CC-GA: a clustering coefficient based genetic algorithm for detecting communities in social networks. Appl. Soft Comput. **63**, 59–70 (2018)
25. Salwinski, L., Miller, C.S., Smith, A.J., Pettit, F.K., Bowie, J.U., Eisenberg, D.: The database of interacting proteins: 2004 update. Nucleic Acids Res. **32**, 449–451 (2004)
26. Sanchez-Oro, J., Duarte, A.: Iterated greedy algorithm for performing community detection in social networks. Future Generat. Comput. Syst. **88**, 785–791 (2018)
27. Shen-Orr, S.S., Milo, R., Mangan, S., Alon, U.: Network motifs in the transcriptional regulation network of escherichia coli. Nat. Genet. **31**(1), 64–68 (2002)
28. Su, X., et al.: A comprehensive survey on community detection with deep learning. IEEE Trans. Neural Netw. Learn. Syst. (2022)
29. Sun, X., Sun, Y., Cheng, S., Bian, K., Liu, Z.: Population learning based memetic algorithm for community detection in complex networks. In: Tan, Y., Shi, Y., Zomaya, A., Yan, H., Cai, J. (eds.) DMBD 2021. CCIS, vol. 1454, pp. 275–288. Springer, Singapore (2021). https://doi.org/10.1007/978-981-16-7502-7_29
30. Traag, V.A., Waltman, L., Van Eck, N.J.: From louvain to leiden: guaranteeing well-connected communities. Sci. Reports **9**(1), 1–12 (2019)
31. Yin, H., Benson, A.R., Leskovec, J., Gleich, D.F.: Local higher-order graph clustering. In Proceedings of the 23rd ACM SIGKDD International Conference on Knowledge Discovery and Data Mining, pp. 555–564 (2017)

Evolution Strategies for Sparse Reward Gridworld Environments

Glennn Moy$^{(\boxtimes)}$ and Slava Shekh

Defence Science and Technology Group, Edinburgh, Australia
{glennn.moy,slava.shekh}@defence.gov.au

Abstract. We investigate evolution strategies (ES) - an optimisation technique originally developed in the 1960s – for learning policies that work effectively in gridworld environments with sparse rewards. We combine the evolution strategies algorithm with an intrinsic reward, based on observation counts, to optimise the parameters of a neural network. We find that the resulting approach is able to obtain good scores on a number of MiniGrid environments, despite the challenges of sparse rewards, partial observability and the environments being procedurally generated. These scores are comparable with deep reinforcement learning approaches that optimise neural network parameters using gradient descent. However, evolution strategies has a number of advantages. It is simple to implement and uses relatively few hyperparameters, it has less reliance on specialised hardware and it is highly parallelisable. In combination, these properties make it a promising alternative to reinforcement learning for sparse reward gridworld problems.

Keywords: Evolution strategies · Sparse rewards · Intrinsic rewards

1 Introduction

Over the last few years, there has been significant development in techniques that involve an agent interacting with an environment to learn a policy which maximises some desired reward. The most popular and successful techniques have proven to be deep reinforcement learning (RL) methods. Examples of successful RL applications have been demonstrated across a range of environments, including the ability to learn to play board games, such as Chess [1] and Go [2, 3], computer games of increasing complexity, such as Atari [4], StarCraft [5] and DOTA [6], as well as more general tasks, such as controlling simulated robots in the MuJoCo physics simulator [7].

RL approaches use the Markov Decision Process (MDP) formalism, whereby the system is represented as a 4-tuple of state, action, transition probabilities and reward. The goal of an agent operating in that system is to generate a policy for choosing actions that maximise the long-term reward. A variety of deep RL approaches have been developed for learning such policies. However, a common feature of all deep RL approaches is the use of one or more neural networks to represent the policy (either directly or through a value function) and then updating the weights of these neural networks through an optimisation method, such as stochastic gradient descent, via back propagation.

H. Aziz et al. (Eds.): AI 2022, LNAI 13728, pp. 266–278, 2022.
https://doi.org/10.1007/978-3-031-22695-3_19

While these deep RL methods have been incredibly successful and currently dominate the field, other methods for learning policies are available. One example is black-box optimisation methods, where instead of modelling the agent's interactions with the environment as an MDP, the policy is optimised directly. In this formalism, we are attempting to optimise the value of some arbitrary function that takes a set of numbers (which just happen to describe the policy) as input and returns a real-valued output describing the fitness of the policy input.

One example of a black-box optimisation method is "evolution strategies" (ES). This method was originally developed in the 1960s [8], but gained renewed interested in 2017 when a team from OpenAI [9] demonstrated that ES could successfully learn policies in a range of standard RL environments, such as Atari and MuJoCo. Notably however, evolution strategies failed to produce any non-zero results on two particular Atari games: Montezuma's Revenge and Pit Fall. These two games are difficult to solve due to the presence of extremely sparse rewards – often referred to as "hard exploration problems" [10].

Within the field of RL, a number of techniques have been specifically developed for sparse reward problems. In this work, we explore whether similar techniques could also be used in combination with evolution strategies to improve its performance in the MiniGrid [11] environments. In addition to having sparse rewards, MiniGrid is procedurally generated, with each episode generating a different instance for the agent to explore, limiting the ability of agents to memorise a solution. MiniGrid is also partially observable by default, with only a subset of the environment visible to the agent. This combination of sparse rewards, along with partial observability and a procedurally generated environment provides a different set of challenges for ES agents to navigate, compared with those present in earlier work.

2 Background

We begin with a background of the MiniGrid environments, as well as the RL techniques that have already proven effective on MiniGrid and related sparse reward gridworld environments.

2.1 MiniGrid

The Minimalistic Gridworld Environment (MiniGrid) [11] is actually a collection of gridworld environments, each with different properties and difficulty levels. As introduced above, these environments have a number of characteristics that make them suitable for exploring the performance of evolution strategies on hard exploration problems. First, rewards provided by the environments are extremely sparse, with a non-zero reward typically only provided to the agent once the full task is completed. This can involve many separate steps, such as opening doors, navigating to keys, unlocking doors and ultimately navigating to a goal. In addition, the environments are (by default) partially observable and each instance of an environment is procedurally generated, making memorisation of a fixed policy infeasible.

While a number of MiniGrid environments may be suitable for this work, we focus on one of the hard exploration tasks that is commonly reported in the MiniGrid RL literature, namely KeyCorridor. While we have also explored evolution strategies on other MiniGrid environments, KeyCorridor provides significant challenge to the agent, and is one of the most widely benchmarked and reported results in the MiniGrid RL literature. As such, we focus on KeyCorridor and leave exploration of ES on other environments (in the MiniGrid suite and beyond) for future work.

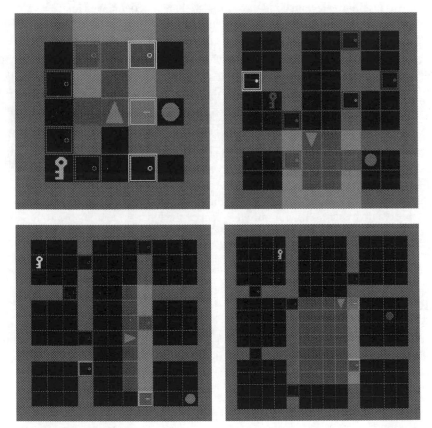

Fig. 1. Examples of the KeyCorridorS3R3 (top-left), KeyCorridorS4R3 (top-right), Key Corridor S5R3 (bottom-left) and Key Corridor S6R3 (bottom-right) environments.

As described in [11], "KeyCorridor is a task where the agent has to pick up an object which is behind a locked door. The key is hidden in another room, and the agent has to explore the environment to find it". KeyCorridor tasks comes in different sizes, each with varying levels of difficulty. In this work, we focus on KeyCorridorS3R3, KeyCorridorS4R3, KeyCorridorS5R3 and KeyCorridorS6R3. Each of these environments is progressively more difficult to solve. While individual instances of these environments are procedurally generated, an example of each environment is shown in Fig. 1.

2.2 Reinforcement Learning

By far the most common approach for learning a policy of an agent interacting with a general environment is through deep RL. One of the most successful forms of deep RL are actor-critic methods. These methods use policy gradient as a conceptual foundation, but include two neural networks – the actor, which learns via a policy gradient approach to decide which action to take, and the critic, which learns a value function to evaluate the action. Three commonly used actor-critic methods are asynchronous advantage actor-critic (A3C) [12], proximal policy optimisation (PPO) [13] and IMPALA [14]. These are popular stand-alone methods, but also form the foundation for the sparse reward approaches discussed in the next section.

While each RL method has different strengths and weaknesses, and varying performance on different problems, most of these techniques perform poorly when applied directly to environments with extremely sparse rewards, such as MiniGrid. For instance, [15] reports that IMPALA achieves a score of zero on the S3R3, S4R3 and S5R3 Key-Corridor environments. Similarly, the PPO implementation provided by the MiniGrid environment gets scores of zero on these three environments, as well as the even simpler S3R2 environment, only demonstrating an ability to learn on the simplest S3R1 environment, as shown in Fig. 2. The runs in Fig. 2 were undertaken using PPO on a machine with one NVIDIA A100 GPU and 54 CPU cores. The S3R2 case failed to perform better than random after 100M steps of training, which took over 8 h on that infrastructure. Other experiments on S3R3 and harder KeyCorridor environments also failed to perform.

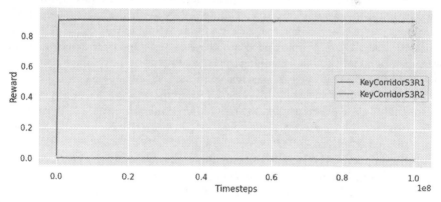

Fig. 2. Performance of PPO on the two simplest KeyCorridor environments, S3R1 and S3R2. Each line represents an average of five runs. Shading shows standard deviation.

2.3 Handling Sparse Rewards

Responding to the challenges facing traditional deep RL methods in handling sparse rewards, a number of techniques have been developed that augment standard RL methods and provide better performance on difficult exploration problems. In general, one of the approaches that has proved successful for handling sparse (or even non-existent) rewards is for the agent to generate its own *intrinsic* reward to guide exploration. These intrinsic rewards have been proposed as analogous to human-centered concepts like intrinsic motivation [16] and curiosity [17], which appear to guide human exploration, especially in children and infants.

Intrinsic rewards in RL typically belong to one of two general classes. Those that encourage the agent to explore novel states that have not previously been explored [18, 19], or those that guide exploration towards states that are in some sense unpredictable or surprising [20–22]. These intrinsic reward mechanisms are combined with standard RL techniques, such as A3C, PPO or IMPALA, to augment the learning process. In recent years, a number of papers exploring various intrinsic reward methods have demonstrated an ability to achieve non-zero scores on increasingly difficult versions of the hard exploration problems presented by MiniGrid. We now give a brief overview of these methods, with a particular focus on the KeyCorridor sparse reward environments. Results for each of the methods introduced here are discussed later in this paper (see Table 1).

In 2016, Count [18] built upon count-based exploration algorithms, extending it to the non-tabular case of playing Atari games from raw pixels. In 2017, Pathak et al. [23] proposed an "Intrinsic Curiosity Module" (ICM), where curiosity is formulated as the "error in the agent's ability to predict the consequences of its own actions". In 2019, random network distillation (RND) [24] introduced a related intrinsic reward concept, where the exploration bonus is based on the error of a neural network predicting features of the observations given by a fixed, randomly initialised neural network – using this as a proxy measure of the familiarity that an agent has with particular states. In 2020, RIDE [25] also built upon ICM, in terms of learning a latent state representation, but in RIDE, the intrinsic reward is based on the difference between the latent representations of two consecutive states, rather than the error of predicting the next state.

In 2021, AMIGO [15] introduced an alternative method of producing intrinsic rewards. AMIGO uses a goal-generating teacher agent that proposes intermediate goals which form an automatically generated and increasingly challenging curriculum for the learning agent, with intrinsic rewards obtained when the learning agent achieves intermediate goals. AMIGO was applied to MiniGrid in a fully observable setting (whereas MiniGrid is normally partially observable). Another novel technique for improving performance on MiniGrid, which was also proposed in 2021, was RAPID [26]. This attempted to improve PPO through a form of imitation learning, where episodes that had previously achieved higher scores are imitated more closely than lower scoring episodes.

Another approach, developed by researchers at Inria and Google Brain, also builds on PPO, but introduces a third agent that they call the adversary, alongside the actor and critic. The adversary attempts to mimic the actor, while the actor attempts to differentiate itself from the adversary's predictions, while still trying to achieve the extrinsic goal of the system. This technique is known as adversarially guided actor-critic (AGAC) [27] and promotes diversity in the actor's exploration, as rewards are given for choices that could not have been predicted by the adversary based on prior exploration. Finally, in another work from 2021, BeBold [28, 29] also builds on PPO but uses visitation counts – in particular, the "regulated difference of inverse visitation counts of consecutive states" [28] – as a criterion for calculating the intrinsic reward. This work also builds on RND to approximate the visitation count, and explicitly encourages exploration at the bounds of previously explored regions.

2.4 Evolution Strategies

RL methods for solving MiniGrid environments have evolved rapidly over the last few years, building on earlier techniques, as well as popular RL approaches, such as A3C, IMPALA and PPO. Common to all of these algorithms is the use of a neural network to represent (directly or indirectly) the policy, and the use of back propagation to update the neural network weights through successive layers. In this paper, however, we explore a different class of algorithm for learning policies on sparse reward MiniGrid environments – evolution strategies.

Evolution strategies is a family of stochastic algorithms for global optimisation, loosely inspired by the biological theory of evolution. It is a member of the broader class of evolutionary algorithms and was initially developed by Ingo Rechenberg and Hans-Paul Schwefel in the 1960s [8]. At its heart, evolution strategies involves a collection of randomly generated candidate solutions being evaluated based on a fitness function, and the resulting evaluations then being used to create a new generation of slightly improved candidate solutions. This process is iterative and the quality of candidate solutions improves over time with respect to the fitness function.

In 2017, OpenAI demonstrated the potential for evolution strategies to be used as "a scalable alternative to reinforcement learning" [9]. They used a variant of natural evolution strategies [30] to demonstrate the performance of ES on a range of tasks, including the MuJoCo robot control environment and the Atari arcade learning environment. Their approach begins with a randomly initialised parameter vector, which describes the weights and biases of a neural network. This neural network defines a policy by which an agent (the parent) acts in the environment. This set of parent weights are used to generate a population of candidate solutions (children, or child policies), by taking the initial parameter vector and perturbing it with Gaussian noise. The various sets of child parameters are used to instantiate child neural networks which are then evaluated and ranked through interactions with the environment. An updated (parent) parameter vector is then generated, as a weighted sum of the parameters in the child networks, weighted according to the total reward that each solution received. This approach proved highly successful, generating results that were extremely competitive with the best available RL methods on the Atari and MuJoCo benchmarks.

In [9], the authors also articulate a number of potential advantages of ES over traditional RL methods, including the simplicity of the approach, the fact that ES only requires a forward pass over the policy and does not require any backpropagation or value function estimation, and the highly parallisable and comparatively robust performance of the technique. The authors also introduce a number of improvements that allow the algorithm to run more efficiently, especially when utilising a large number of parallel workers. This is important, as their key results, including the ability to solve the MuJoCo Humanoid walking task, use significant compute - 1440 parallel workers on Amazon EC2 [9]. In this work, we build on the ideas of [9], to explore evolution strategies in procedurally generated gridworld environments that contain sparse rewards and partial observability. We investigate the potential to improve the performance of ES by combining it with an intrinsic reward signal.

3 Experimental Results

3.1 Our Approach

In this work, we explore ES using comparatively modest computational resources in contrast to [9]. All of our experiments were performed on a Dell R740xd Ubuntu server with 54 Intel Xeon Gold 6258R cores. We implemented our own version of the ES algorithm for the MiniGrid environment, based on a number of example codebases [31–33]. While our implementation was intended to be as simple as possible, we did include a number of the specific improvements that were present in the OpenAI paper [9] and explored a number of alternative design choices, before settling on our final implementation. These improvements include sending a noise seed (rather than all the parameters of the neural network) through to child processes. This greatly reduces the amount of data that is sent to child processes, leading to more efficient communication, thereby increasing the overall speed of execution for the algorithm. We also used mirrored sampling [9, 34]. In mirrored sampling, each set of random perturbations (which are used to generate child candidates) is used twice, once normally and once as a mirror (negated) version of itself. This approach has been found to improve convergence, compared with non-mirrored sampling, in ES [34].

In all of our experiments, we use evolution strategies to parameterise the weights of a neural network, which forms the policy network for the agent. We did early explorations on a range of potential neural network structures but present results here for a relatively simple neural network structure, consisting of two fully-connected linear layers, the first with 588 neurons and the second with 256, connected via a Rectified Linear unit (ReLU) nonlinearity, and an output layer of 7 neurons for the 7 valid actions available in MiniGrid. Observations of size $7 \times 7 \times 3 \times 4$ (using a frame stack of 4, as in [27, 28], for example) from the MiniGrid environment were fed directly into the neural network.

Parameters for the learning rate and mutation rate (sigma) are set to 0.05 (as suggested in [32]) with some parameter exploration undertaken in order to confirm that these values were suitable (see Sect. 4). The number of child evaluations is set to 100 (corresponding to a population of 50, due to mirrored sampling), chosen to limit wasted processing resources, given the 54 cores available on the machine that we used for our experiments. Exploration of other parameters showed limited effect on the final result, as long as a sufficient number of child evaluations (>30) were performed.

3.2 Evolution Strategies Applied to MiniGrid – No Intrinsic Reward

We undertake experiments to explore the potential of applying ES to MiniGrid environments. In Sect. 2.2, we showed that using PPO on the comparatively simple KeyCorridorS3R2 environment did not learn anything meaningful (Fig. 2). By contrast, our initial results in Fig. 3 demonstrates that even without an intrinsic reward, evolution strategies is able to reach scores above 0.8 on KeyCorridorS3R2. It also achieves good scores (above 0.8) on KeyCorridorS3R3 (using a smaller time-budget than was allocated to the PPO runs). This is a promising result, as ES is achieving good scores on KeyCorridorS3R3, even without the presence of any intrinsic reward, outperforming the baseline RL algorithm results (also with no intrinsic reward) that we present in Sect. 2.2. However, despite the promising results on the KeyCorridorS3R3 environment, ES without intrinsic reward, struggles to obtain a good score on the more challenging KeyCorridorS4R3 environment.

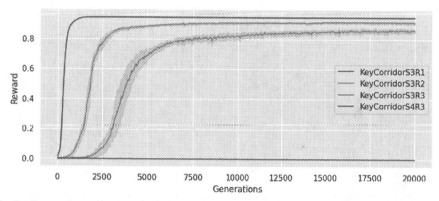

Fig. 3. Comparison of scores obtained by evolution strategies with no intrinsic reward on various KeyCorridor environments. Each line represents an average of five runs. Shading shows standard deviation.

The failure to perform adequately on KeyCorridorS4R3 (Fig. 3), combined with the previous success of intrinsic rewards in RL (Sect. 2.3), motivates the need to explore alternative methods to enhance the base evolution strategies method.

3.3 Evolution Strategies with Intrinsic Reward on MiniGrid

Out of the two broad classes of intrinsic rewards discussed in Sect. 2.3, we use a reward designed to encourage the exploration of novel states which have not previously been seen. Our implementation involves providing a small (scaled) intrinsic reward during the agent's child evaluations whenever a child first encounters a previously unseen observation. For this reward, we use a single frame of observation, as provided to the agent, with no additional processing. This is a simple form of count-based intrinsic reward. At the end of each generation update, this intrinsic reward is switched off and only extrinsic rewards from the game environment are presented.

While there are multiple variations and improvements on this simple reward type in the literature, we use this as an initial basis for exploring intrinsic reward for ES. One advantage of this simple intrinsic reward is the addition of only a single parameter (intrinsic reward strength), which specifies the weighting between the intrinsic and extrinsic rewards. For our initial experiments, we choose an intrinsic reward strength of 0.0001, ensuring that intrinsic rewards will not dominate over any extrinsic reward obtained. We will discuss some alternative intrinsic reward options in Sect. 4.

Figure 4 shows the result of combining ES with an intrinsic reward in a range of KeyCorridor environments of increasing complexity. The results show that the presence of an intrinsic reward significantly improves the performance of the ES method on all of these environments – including the three most difficult environments (KeyCorridorS4R3, KeyCorridorS5R3 and KeyCorridorS6R3) where the base ES method without intrinsic reward failed to perform.

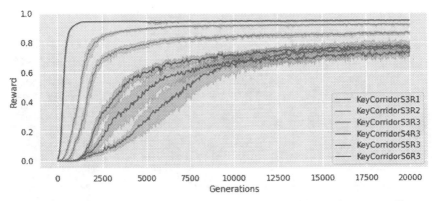

Fig. 4. Performance of ES with intrinsic reward on five KeyCorridor environments of increasing complexity. Each line represents an average of five runs. Shading shows standard deviation.

The final average scores of ES with an intrinsic reward for each of these environments are presented in Table 1, alongside previously reported results for the RL algorithms discussed in Sect. 2.3. For previously reported RL algorithms, where a range of values are reported, this indicates that different papers report different scores for the same technique. In some cases, this may be due to slightly different implementations or parameter choices. In other cases, the authors note that they ran experiments for a shorter duration or modified other settings. For instance, the BeBold/NovelD paper [28, 29] notes that they only evaluated AMIGO for 120M steps, whereas the original AMIGO paper performed evaluations over 500M steps. A hyphen in the table indicates that we are unaware of previously published results for a method/environment combination.

From these results, it is clear that the two most recent RL methods, AGAC and BeBold, remain the current state-of-the-art for KeyCorridor environments. However, evolution strategies with a simple intrinsic reward also performs well on all tested environments and achieves better scores on KeyCorridorS4R3, KeyCorridorS5R3 and Key-CorridorS6R3 than all methods except for AGAC and BeBold. ES with an intrinsic reward is able to consistently reach the goal in these challenging environments, whereas many other recent RL methods cannot. While not equaling the performance of the best RL methods, the ES method performs surprisingly well, given its simplicity. Moreover these results were obtained on a single machine without the use of a graphics processing unit (GPU). The results show that ES can perform well despite the combined challenges presented by sparse rewards, partial observability and the presence of a procedurally generated environment.

Table 1. Reported scores for various techniques on the KeyCorridor environments, alongside our results for ES with an intrinsic reward. Our results are averaged over five runs.

Technique	KC-S3R3	KC-S4R3	KC-S5R3	KC-S6R3
Count (NeurIPS 2016)	0.90	0.00	0.00	–
ICM (ICML 2017)	0.42–0.45	0.00	0.00	0.0
RND (ICLR 2019)	0.89–0.91	0.00–0.23	0.00	0.0
RIDE (ICLR 2020)	0.90–0.91	0.19–0.93	0.00	0.0
AMIGO[1] (ICLR 2021)	0.89–0.93	0.00–0.54	0.00–0.44	0.0
RAPID (ICLR 2021)	0.92	0.47	0.00	–
AGAC (ICLR 2021)	–	0.95	0.93	–
BeBold (NeurIPS 2021)	0.92	0.93	0.94	0.94
ES + intrinsic reward (Ours)	0.86	0.76	0.73	0.75

4 Discussion and Further Work

Evolution strategies provides a promising avenue for future explorations. First, since ES is a fundamentally different approach to RL techniques that have been applied to MiniGrid, it comes with different assumptions and inductive bias tradeoffs. Further exploration of the strengths and weaknesses of this approach is useful in developing a more diverse range of techniques to solve the myriad of different sparse reward problems. Moreover, the form of ES presented in this paper is relatively simple – without the many years of detailed development that has occurred in the space of more traditional sparse reward RL algorithms. Nevertheless, despite that relative simplicity, it manages to achieve scores that only a couple of years ago would have been competitive with the state-of-the-art.

ES also has a number of other properties that make it promising as a practical technique. Firstly, the method has relatively few hyperparameters [9] and we found stable performance across a range of these values. For instance, varying intrinsic reward type (including using simple defaults, like the state-bonus and action-bonus wrappers included in MiniGrid) as well as varying the intrinsic reward strength had little impact on the final result, with many parameter choices leading to learnt policies that reach the goal. We also tested a number of neural network architectures, including the addition of convolutional layers to the network, but found that this did not improve the result on the KeyCorridor environments that we tested. The method is also robust to variations to the observation used as the basis for estimating 'novelty' in the intrinsic reward. For instance, the experiments reported here were based on a single frame of observation – however similar results were obtained using a sequence of the last four frames.

Evolution strategies is also highly parallelisable [9], opening up this technique to relatively cost-effective scaling. The runs presented here use a single machine with 54 cores, but this can be scaled to additional CPU cores relatively easily, opening up the

---

[1] Results for AMIGO were obtained using a fully observable view of the environment.

opportunity for increased child-evaluations and, possibly, better performance. Moreover, as the neural networks are not trained through back propagation and only require forward pass evaluations for inference, large GPU infrastructure is not necessary to take advantage of the method.

There are also many opportunities for improving the method and intrinsic reward demonstrated in this paper. For example, incorporating a more complex intrinsic reward, such as the rewards used in BeBold or AGAC, could lead to further improvements in performance. In addition, the combination of ES and intrinsic reward should be evaluated on additional environments, to understand its performance across a broader range of problems.

5   Conclusion

In this work, we explore evolution strategies, as an alternative to RL, for solving hard exploration gridworld problems. Building on earlier research that showed ES could be a scalable alternative to RL in the MuJoCo and Atari environments [9], we explored the potential for ES within MiniGrid environments. These environments present multiple challenges: they are procedurally generated, they contain sparse rewards, and they are partially observable. ES is able to perform well on small KeyCorridor MiniGrid environments (S1R3, S2R3 and S3R3), outperforming vanilla PPO over a similar timeframe. Nevertheless, ES is unable to solve the more difficult S4R3 environment. However, we find that the combination of ES with an intrinsic reward leads to good results on all KeyCorridor environments, including S4R3, S5R3 and S6R3. In future work, we plan to investigate a number of other intrinsic reward approaches and evaluate this method across a wider range of sparse reward environments, to determine the effectiveness and limits of the method more broadly.

References

1. Silver, D., et al.: Mastering chess, go and shogi by self-play with a general reinforcement learning algorithm. arXiv:1712.01815 (2017)
2. Silver, D., et al.: Mastering the game of go with deep neural networks and tree search. Nature **529**, 484–489 (2016)
3. Silver, D., et al.: Mastering the game of go without human knowledge. Nature **550**, 354–359 (2017)
4. Badia, A.P., et al.: Agent 57: outperforming the Atari human benchmark. In: Proceedings of the 37th International Conference on Machine Learning (2020)
5. Vinyals, O., Babuschkin, I., Silver, D.: Grandmaster level in starcraft II using multi-agent reinforcement learning. Nature **575**, 350–354 (2019)
6. Berner, C., et. al.: Dota 2 with large scale deep reinforcement learning. eprint arXiv:1912.06680 (2019)
7. Todorov, E., Erez, T., Tassa, Y.: MuJoCo: a physics engine for model-based control. In: IEEE/RSJ International Conference on Intelligent Robots and Systems (2012)
8. Beyer, H.-G., Schwefel, H.-P.: Evolution strategies – A comprehensive introduction. Nat. Comput. **1**(1), 3–52 (2002)

9. Salimans, T., Ho, J., Chen, X., Sidor, S., Sutskever, I.: Evolution strategies as a scalable alternative to reinforcement learning. arXiv:1703.03864 (2017)
10. Ecoffet, A.: Montezuma's revenge solved by go-explore, a new algorithm for hard-exploration problems (sets records on pitfall, too), 26 November 2018. https://eng.uber.com/go-explore/. Accessed 20 July 2022
11. Chevalier-Boisvert, M., Willems, L., Pal, S.: Minimalistic gridworld environment for OpenAI gym (2018). https://github.com/Farama-Foundation/gym-minigrid
12. Mnih, V., et al.: Asynchronous methods for deep reinforcement learning. In: International Conference on Machine Learning, pp. 1928–1937 (2016)
13. Schulman, J., Wolski, F., Dhariwal, P., Radford, A., Klimov, O.: Proximal policy optimzation algorithms. arXiv:1707.06347 (2017)
14. Espeholt, L., et al.: IMPALA: scalable distributed deep-RL with importance weighted actor-learner architectures. In: International Conference on Machine Learning, pp. 1407–1416 (2018)
15. Camperon, A., Raileanu, R., Kuttler, H., Tenenbaum, J.B., Rocktaschel, T., Grefenstette, E.: Learning with AMIGo: adversarially motivated intrinsic goals. In: International Conference on Learning Representations (2021)
16. Ryan, R., Deci, E.: Intrinsic and extrinsic motivations: classic definitions and new directions. Contemp. Educ. Psychol. **25**(1), 54–67 (2000). https://doi.org/10.1006/ceps.1999.1020
17. Paul, S.: Curiosity and motivation. The Oxford Handbook of Human Motivation (2012)
18. Bellemare, M., Srinivasan, S., Ostrovski, G., Schaul, T., Saxton, D., Munos, R.: Unifying count-based exploration and intrinsic motivation. In: Advances in Neural Information Processing Systems, pp. 1471–1479 (2016)
19. Lopes, M., Lang, T., Toussaint, M., Oudeyer, P.-Y.: Exploration in model-based reinforcement learning by empirically estimating learning progress. In: NIPS (2012)
20. Houthooft, R., Chen, X., Duan, Y., Schulman, J., Turck, F.d., Abbeel, P.: VIME: variational information maximizing exploration. In: NIPS (2016)
21. Schmidhuber, J.: A possibility for implementing curiosity and boredom in model-building neural controllers. In: From Animals to Animats: Proceedings of the First International Conference on Simulation of Adaptive Behavior (1991)
22. Singh, S., Barto, A.G., Chentanez, N.: Intrinsically motivated reinforcement learning. In: NIPS (2005)
23. Pathak, D., Agrawal, P., Efros, A.A., Darrell, T.: Curiosity-driven exploration by self-supervised prediction. In: Proceedings of the IEEE Conference on Computer Vision and Pattern Recognition Workshops, pp. 16–17 (2017)
24. Burda, Y., Edwards, H., Storkey, A., Klimov, O.: Exploration by random network distillation. In: Proceedings of the 7th International Conference on Learning Representations. ICLR (2019)
25. Raileanu, R., Rocktaschel, T.:RIDE: rewarding impact-driven exploration for procedurally-generated environments. In: International Conference on Learning Representations (2020)
26. Zha, D., Ma, W., Yuan, L., Hu, X., Liu, J.: Rank the episodes: a simple approach for exploration in procedurally generated environments. In: International Conference on Learning Representations (2021)
27. Flet-Berliac, Y., Ferret, J., Pietquin, O., Preux, P., Geist, M.:Adversarially guided actor-critic. In: International Conference on Learning Representations (2021)
28. Zhang, T., et al.: BeBold: exploration beyond the xoundary of explored regions. arXiv:2012.08621
29. Zhang, T., et al.:NovelD: a simple yet effective exploration criterion. In: Advances in Neural Information Processing Systems, vol. 34 (2021)
30. Wierstra, D., Schaul, T., Glasmachers, T., Sun, Y., Peters, J., Schmidhuber, J.: Natural evolution strategies. J. Mach. Learn. Res. **15**(1), 949–980 (2014)

31. Ha, D.:A visual guide to evolution strategies (2017). https://blog.otoro.net/2017/10/29/vis ual-evolution-strategies/
32. Zhou, M.: Evolutionary algorithm (2020). https://github.com/MorvanZhou/Evolutionary-Alg orithm
33. Brownlee, J.:Evolution strategies from scratch in python (2021). https://machinelearningmas tery.com/evolution-strategies-from-scratch-in-python/. Accessed 20 July 2022
34. Brockhoff, D., Auger, A., Hansen, N., Arnold, D.V., Hohm, T.: Mirrored sampling and sequential selection for evolution strategies. In: International Conference on Parallel Problem Solving from Nature, pp. 11–21 (2010)

Niching-Assisted Genetic Programming for Finding Multiple High-Quality Classifiers

Peng Wang[1]([☒])[iD], Bing Xue[1][iD], Jing Liang[2][iD], and Mengjie Zhang[1][iD]

[1] School of Engineering and Computer Science, Victoria University of Wellington, Wellington, New Zealand
{wangpeng,bing.xue,mengjie.zhang}@ecs.vuw.ac.nz
[2] School of Electrical and Information Engineering, Zhengzhou University, Zhengzhou, China
liangjing@zzu.edu.cn

Abstract. Explainable artificial intelligence (XAI) is a recent research focus, aiming to gain trust in machine learning models with clear insights into how the models make certain predictions. Due to its ability to evolve potentially interpretable classifiers, genetic programming (GP) is generally well-suited to XAI. However, many learning algorithms including GP usually learn a single best model. In practice, the best model in terms of training classification accuracy/error rate may not be the most appropriate one from the perspective of a domain expert due to overfitting and limited data. Multiple explicit and high-quality classifiers with the same training performance are therefore needed to increase the chances that the generated models will be considered more reasonable to experts. Therefore, this study designs a niching-assisted GP approach for classification. The results show that the proposed method can significantly increase the classification accuracy on most of the tested datasets. Further analysis shows that the designed algorithm can find different GP programs with the same classification performance, providing good interpretability for classification tasks.

Keywords: Genetic programming · Multiple optimal programs · Classification

1 Introduction

Classification is a major task in data mining, aiming to predict the class labels for unseen data instances. In classification, explainable artificial intelligence (XAI) means that a learning algorithm can provide human-understandable justifications for its output, leading to insights about the inner workings to trust the classifier [20]. Genetic programming (GP), a biological-evolution-inspired technique, is an excellent tool for XAI. As a learning algorithm, GP allows users to use flexible representations such as trees, graphs, and networks with different kinds of operators or functions to represent the model [23]. These enable GP to

© The Author(s), under exclusive license to Springer Nature Switzerland AG 2022
H. Aziz et al. (Eds.): AI 2022, LNAI 13728, pp. 279–293, 2022.
https://doi.org/10.1007/978-3-031-22695-3_20

capture linear and/or non-linear relations offering insights between features and the class labels. From a feature selection point of view, GP is good at embedded-based feature selection, which can simultaneously select a good subset of features and construct a classifier using the selected features [14].

However, most existing machine learning algorithms including GP generate only one learned model for a classification task. Unfortunately, the model might have shortcomings or other issues, such as including a large number of coefficients and/or parameters that make it unlikely to be inherently interpretable [21]. In addition, it might not be the best classifier simply because the data is limited and classifiers with slightly worse classification performance on the training data might be better on the test data. Therefore, it is essential to search for multiple different models that are all well-performing for a classification task. It can provide users with more choices, therefore, the users can pick up classifiers based on their preferences [7,30]. Furthermore, identifying equal informative models has important practical significance [7,26]. In disease diagnosis, for example, Liu et al. [12] found that a classifier with selected features $\{M77836, J02854, T64297\}$ or $\{H06524, H43887, U37019\}$ in the Adeno dataset can achieve the same top classification accuracy (100%). However, the gene/feature, $M77836$ in the first feature subset, is an upregulated protein, while $H06524$ in the second feature subset is a severely down-regulated protein. This shows that different functional modules are likely to separate normal individuals from colon patients.

The existence of multiple optimal classifiers is manifested in GP as programs or trees[1] with different functions and/or features that can achieve the same classification accuracy, which is ignored by many existing studies. As shown in many studies [21,27], using niching techniques is a popular way to find multiple optimal solutions. The key idea of niching techniques is to partition the whole population into several niches. Genetic operators such as mutation and crossover will be performed using the neighbour information, which can be a stabler and more effective way to exchange evolutionary information during training [28]. Motivated by these observations, a niching-assisted GP approach (termed NGP) is proposed. The aim of NGP is to find multiple optimal GP programs for each classification task. To achieve this goal, individuals are preferentially scheduled to perform crossover with their neighbours. Furthermore, an external *Archive* is employed to collect the fittest individuals during the evolutionary training process. A structure score considering the number of selected features, the tree depth, and the number of nodes from a tree is taken as the complexity of the tree. Trees with low structure scores will be preferred.

2   Background

2.1   Genetic Programming

The commonly used representation method in GP is based on the tree-like structure [3]. Each candidate program in GP is represented by a variable length tree

---

[1] Each individual or solution in GP is a classifier or a program with a tree representation. Therefore, this work treats *individual, solution, classifier, tree, program* in GP as the same.

which involves two types of nodes: internal nodes and leaf nodes. The former consist of functions or operators with some arguments, such as $+, -, \times, \div$. The latter are variables or constants that are taken as arguments for the internal nodes. The set of all possible used functions or operators in the internal nodes is named as *function set*. Correspondingly, a term *terminal set* represents the set of all possible used variables or constants. The maximum depth, the longest path from the root to a leaf node, is used to limit the size of a GP tree [8].

The evolutionary process of the standard GP algorithm is introduced as follows. After initialization, all programs in the population are evaluated based on the fitness function. Then, the pair parent trees are selected and then performed genetic operators including elitism, mutation, and crossover. The aim of elitism is to ensure that the best individuals can be preserved during the evolutionary process. Mutation is to maintain the diversity of programs in the population, and crossover in GP aims to combine good building blocks to obtain better programs. When the stopping condition is met, the method will stop and output the best individual.

2.2 Genetic Programming for Feature Analysis

Feature selection approaches are generally grouped into three categories: 1) filter methods, 2) wrapper methods, and 3) embedded methods [27]. Filter methods do not use any learning algorithm to measure the goodness of selected features, relying on the general characteristics of data. Instead, a wrapper method measures the quality of selected features using a learning algorithm. However, a wrapper method usually has high computation cost. In embedded methods, selecting relevant features and training a learning algorithm are done together. Embedded methods can account for the interactions between features and the learning algorithm. In general, an embedded method can achieve comparable classification accuracy to a wrapper method and comparable efficiency to a filter method [14].

GP is popular in designing different kinds of feature selection methods. A GP-based filter feature selection method was proposed in [16]. In [16], a predefined relevance measure function sitting at the root of the tree measures the relevance of its sub-tree to the class labels. Neshatian et al. [17] applied multi-objective GP to classification tasks. However, the obtained trees have very deep depths. Muni et al. [13] designed an embedded feature selection method using multi-tree GP. In [13], if multiple trees output positive values, the class of the tree with the largest weight value will be assigned. Nag et al. [14] claimed that randomly selecting a node in mutation may be too disruptive for a target tree. As a result, each constant of the tree is replaced with another random constant. Each function node is replaced with another random function node. Ahmed et al. [1] designed a GP-based hybrid feature selection method. A primitive feature subset from the original feature set will be obtained according to the search of GP. Then, the selected features are ranked based on the signal-to-noise ratio. Finally, a certain number of top features are selected and passed to classifiers. However, these studies ignore the existence of multiple optimal programs for a feature selection task.

In addition, GP has been widely used for feature extraction, feature construction, regression, and clustering tasks. Nag et al. [15] combined many-objective GP with a support vector machine to extract linearly separable features. Ahmed et al. [2] applied GP to feature construction for biomarker identification. In [24], a semantic-based crossover in GP is proposed for real-valued symbolic regression problems. Lensen et al. [10] applied GP to perform clustering tasks.

2.3   Niching Techniques in Genetic Programming

The essence of the niching technique is to find multiple optima by creating and maintaining different niches. Niching approaches include crowding-based, sharing-based, and speciation-based methods [27]. Lensen et al. [9] combined the speciation technique with GP to perform association rule mining tasks. In [9], the whole population is divided into a number of species, each of which shares a common target feature. Sijben et al. [21] proposed a niching-based multi-objective GP to search for multiple diverse high-quality models for symbolic regression tasks. In each generation, the population is divided into multiple sub-populations by using the distance between solutions in normalized objective space. In [29], the proposed niching GP method separated the population into several clusters based on the tree similarity. In addition, there are other studies applying niching techniques to GP, such as [25] and [6].

Although these niching-based GP methods obtained promising performance, they have not been used to deal with classification tasks, especially for finding multiple optimal classifiers.

3   The Proposed Niching-Assisted GP Method

In this paper, a niching-assisted GP method (NGP) with an external *Archive* is proposed to search for multiple optimal programs.

3.1   Representation of Classifiers and Fitness Function

In NGP, each program is a binary classifier represented by a tree. When a data point is passed through an individual, if the output value is positive, the individual says that the passed data point belongs to one class. Otherwise, it says that the point belongs to the other class. The internal (nonleaf) nodes of these trees are functions. The leaf nodes are features from the data. The aim is to minimize the classification error rate shown in Eq. (1):

$$fitness = \text{Error Rate} = \frac{FP + FN}{TP + TN + FP + FN} \tag{1}$$

where FP, FN, TP, and TN are the false positives, false negatives, true positives, and true negatives, respectively.

Algorithm 1: Overall NGP algorithm

Input: N: population size, G_{max}: maximal generations
Output: *Archive* $\mathcal{A}$
1 **begin**
2   Set $\mathcal{A} = \emptyset$,
3   Initialize population P by ramped half-and-half method,
4   Evaluate individuals in P, and put the fittest individual(s) to $\mathcal{A}$,
5   **while** $g < G_{\max}$ **do**
6   Perform elitism,
7   Get neighborhood matrix $\mathcal{N}$ via Algorithm 2,
8   Perform niche crossover using $\mathcal{N}$,
9   Perform mutation,
10   Evaluate new individuals,
11   Put the fittest offspring(s) to $\mathcal{A}$,
12   Remove solutions with the same selected features in $\mathcal{A}$,
13   **if** $|\mathcal{A}| > N$ **then**
14   Calculate Score values of all solutions in $\mathcal{A}$,
15   Sort solutions in ascending order,
16          $\mathcal{A} = \mathcal{A}(1 : \min(N, |\mathcal{A}|))$
17   **end**
18   **end**
19 **end**

3.2 Overall Algorithm

The overall NGP algorithm is shown in Algorithm 1. NGP is similar to a standard GP search, with the main difference being performing a crossover operator by considering the neighbourhood information of a target individual. In addition, after evaluating the population, the fittest individuals will be stored in the external *Archive* (termed $\mathcal{A}$). When a stop criterion is met, NGP will output all solutions in $\mathcal{A}$.

Niche Crossover Operator: As pointed out by [9], multiple diverse and good programs can be produced by restricting crossover to occurring with neighbours, improving learning efficacy. The introduction of niche in this work simultaneously considers the fitness values of individuals and the number of nodes of individuals. The niche crossover operator (Line 8 of Algorithm 1) randomly selects parents with a proportion of 80% or 20% from the neighbourhood matrix ($\mathcal{N}$) or the global population (P) alternatively. To avoid getting trapped into a local optimum, one individual will be randomly chosen from $\mathcal{N}$ when a randomly generated number is less than 0.8.

Algorithm 2 is to get the neighborhood matrix $\mathcal{N}$. For all individuals in P, their fitness values and the numbers of nodes are obtained and saved into $\mathcal{O}$ and $\mathcal{S}$, respectively. After respectively performing the min-max normalization technique in $\mathcal{O}$ and $\mathcal{S}$, the Euclidean distance of a target individual to other

---

Algorithm 2: Niche Introduction

Input: Population P
Output: Neighborhood matrix $\mathcal{N}$
1 **begin**
2   Set neighborhood size $T = a * |P|$,
3   Store the numbers of nodes of all individuals in P as $\mathcal{S}$,
4   Save the fitness values of all individuals in P as $\mathcal{O}$,
5   Normalize the values in $\mathcal{S}$ and $\mathcal{O}$, respectively,
6   Calculate Euclidean distances among normalized vectors,
7   Generate a $|P| * T$ neighborhood matrix $\mathcal{N}$, and $\mathcal{N}(i)$ consists of T nearest
      solutions to $P(i)$.
8 **end**

---

individuals is calculated (Line 6 of Algorithm 2). Then, T nearest individuals to the target individual will be put into $\mathcal{N}$. It should be noted that the computation cost will be very high if the neighbourhood matrix is calculated in each generation. Therefore, NGP will redefine $\mathcal{N}$ every ten generations. For the neighborhood size T, it is set to $a * |P|$ where $|P|$ means the population size (Line 2 of Algorithm 2). The sensitivity analysis of a will be shown in Sect. 4.2.

Filtering *Archive*: The *Archive* $\mathcal{A}$ preserves the fittest programs from both the historical and current populations. However, some programs may select the same features. Under this situation, only one solution will be randomly chosen from the multiple ones and kept (Line 12 of Algorithm 1). When the size of $\mathcal{A}$ exceeds the predefined capacity (population size), the cleaning steps (Lines 13–17) will be activated. Specifically, each individual in $\mathcal{A}$ will be assigned a structure score since these individuals can include different numbers (Size) of features, have different tree depths (Depth), and contain different numbers of nodes (#Nodes). After getting all Size, Depth, and #Nodes values of all solutions in $\mathcal{A}$, the structure score of each solution in $\mathcal{A}$ is defined as follows:

$$\text{Score}(x) = \frac{\text{Size}^{\circ}(x) + \text{Depth}^{\circ}(x) + \#\text{Nodes}^{\circ}(x)}{3} \tag{2}$$

where Size°, Depth°, and #Nodes° mean the normalized value of Size, Depth, and #Nodes by using the min-max normalization technique, respectively. There is no preference for the three parts, thus Eq. (2) takes $1/3$ as the coefficient.

4 Experiment

4.1 Tested Datasets

To evaluate the potential of the proposed NGP method, a range of real-world binary classification datasets from different domains are used in the experiments. These datasets can be obtained from [5] and are summarised in Table 1, ordered according to the number of features. Some minor data cleaning is done, including

Table 1. The information of datasets

Number	Dataset	# Features	# Classes	# Instances
1	SPECT	22	2	267
2	WBCD	30	2	569
3	Sonar	60	2	208
4	Hillvally	100	2	606
5	Madelon	500	2	4,400
6	QSAR	1,024	2	3,000
7	Leukemia	5,147	2	72
8	DLBCL	7,050	2	77
9	Prostate	10,509	2	102

Table 2. GP parameter settings

Parameter	Value	Parameter	Value
Population size	1024	Generations	50
Crossover rate	0.8	Mutation rate	0.19
Maximal tree depth	4	Elitism rate	0.01
Initialization	Ramped half-and-half	Tournament size	7

removing missing values by deleting whole features or instances as appropriate. Each dataset is randomly divided into two parts: the training set (70% of the data) and the test set (30% of the data). The objective is to minimize the classification error rate on the training set.

4.2   Parameter Setting

On each dataset, 30 runs of the standard GP method from the DEAP package [4] and the proposed NGP method are performed, respectively. In addition, the decision tree (DT) classifier is used for comparison because of its tree-based structure. The parameter settings used in the standard GP method and the proposed NGP method are shown in Table 2. The population size is 1024, and the maximum number of generations is 50 [18]. A small maximum tree depth of four is used to encourage interpretable trees, further reducing the computational cost. The function set includes four arithmetic operators (i.e., $+$, $-$, $\times$, and protected division $\div$). The protected division operator returns zero when dividing by zero. The terminal set consists of all features in a dataset. Additionally, to avoid very large trees on some datasets, the maximum depth of trees in DT is set to be the same as in GP.

For the parameter a in NGP, it controls the size of the niche of an individual. Different niching methods set a to different values, 20% in [28] and 10% in [11]. Therefore, the classification performance of a between 5% and 25% is explored.

Table 3. Average Training Classification accuracy of NGP with different a values

Dataset	$a = 5\%$	$a = 10\%$	$a = 15\%$	$a = 20\%$	$a = 25\%$
SPECT	91.79±0.31	**91.83±0.35**	91.77±0.34	91.74±0.47	91.70±0.30
WBCD	**94.36±1.02**	94.20±1.23	94.25±1.49	94.30±1.49	94.25±0.96
Sonar	80.11±1.6	**81.64±1.88**	80.85±1.87	80.48±2.05	80.85±2.02
Hillvally	**100.0±0.0**	**100.0±0.0**	**100.0±0.0**	**100.0±0.0**	**100.0±0.0**
Madelon	63.65±0.78	**63.76±1.22**	63.73±1.27	63.56±0.7	63.75±1.04
QSAR	92.91±0.37	**92.94±0.32**	92.88±0.29	92.90±0.28	92.87±0.33
Leukemia	98.80±1.60	98.67±1.74	98.53±1.71	98.60±1.28	**99.07±1.61**
DLBCL	96.04±2.64	95.53±2.09	95.72±2.62	**96.35±2.28**	95.47±2.16
Prostate	**92.49±2.61**	91.69±3.48	91.83±3.82	91.88±3.35	91.74±4.08

Here, the average classification accuracy of a at 5%, 10%, 15%, 20%, and 25% on the training sets are reported in Table 3. In Table 3, NGP with $a = 10\%$ achieves the highest classification accuracy on five out of the nine datasets. On the Hillvally dataset, all five methods can obtain the same top classification accuracy, i.e., 100%. Only on one dataset, NGP with $a = 20\%$ or $a = 25\%$ achieves higher classification accuracy than NGP with $a = 10\%$. Although NGP with $a = 5\%$ gets similar training performance on most of the tested datasets to NGP with $a = 10\%$, the latter shows better overall performance. Therefore, the value of a in NGP is set to 10%.

5   Experimental Results

Different from the standard GP method and DT, NGP is to find different GP programs with the same fitness value. Each obtained program from $\mathcal{A}$ in NGP can be considered a classifier consisting of selected features and specific functions. For NGP, the classification accuracy of the solution with the lowest number of the selected features on each test set is shown in Table 4. If multiple solutions have the same number of the selected features, the solution with the lowest number of nodes will be chosen. The classification performance of the standard GP method and DT are also shown in Table 4.

In Table 4, the Wilcoxon test with a significance level of 0.05 is used to judge whether there is a significant difference between algorithms. The signs '↑', '↓', and '≈' mean that the standard GP method or DT is significantly better than, worse than, and has no significant difference from NGP, respectively. The columns W_1 and W_2 show the Wilcoxon comparison results between GP or DT and NGP on subset size and test accuracy, respectively. The more '↓', the better NGP. Finally, the training time of the standard GP and the proposed NGP methods on each dataset is given in the last column in Table 4.

5.1 Test Performance

Classification Accuracy and Subset Size: As shown in Table 4, the proposed NGP method achieves significantly better or similar classification accuracy than DT on eight out of the nine datasets. On the DLBCL dataset, the proposed NGP method obtains more than 10% improvement in accuracy. Only on the Madelon dataset, the DT classifier achieves higher accuracy than NGP but selects more features than GP and NGP. However, on four datasets, DT selects the lowest number of features.

Table 4. Average test results and the training time

Dataset	Method	# Features	W_1	Accuracy	W_2	Time (minutes)
SPECT	DT	**8.3**±0.6	↑	72.47±0.65	↓	
	GP	8.9±1.3	≈	**75.47**±1.52	↑	1.8
	NGP	9.2±1.5		74.90±1.76		2.9
WBCD	DT	5.6±0.5	≈	92.28±0.60	≈	
	GP	6.0±1.4	↓	91.87±1.74	↓	3.7
	NGP	**5.2**±1.1		**92.65**±1.52		4.7
Sonar	DT	9.9±0.9	↓	69.63±3.35	↓	
	GP	8.4±1.7	≈	70.79±6.41	↓	2.2
	NGP	**7.7**±1.8		**72.91**±7.07		3.3
Hillvally	DT	2.9±0.2	↑	99.69±0.09	≈	
	GP	**2.0**±0	≈	99.93±0.15	≈	13.3
	NGP	**2.0**±0		**100**±0		16.7
Madelon	DT	12.6±0.5	↓	**68.15**±0.61	↑	
	GP	**5.1**±2.7	≈	61.12±0.83	≈	103.9
	NGP	5.5±1.5		61.42±0.14		101.9
QSAR	DT	14.0±0	↓	90.04±0.52	↓	
	GP	8.6±1.5	↓	92.33±0.41	≈	243.0
	NGP	**7.4**±1.8		**92.48**±0.51		228.6
Leukemia	DT	**2.0**±0	↑	86.21±3.80	↓	
	GP	5.5±2.1	↓	87.58±9.01	↓	38.8
	NGP	4.7±1.9		**89.85**±6.40		38.4
DLBCL	DT	**3.0**±0	↑	65.83±5.21	↓	
	GP	5.4±1.8	≈	76.39±9.02	≈	56.8
	NGP	5.3±1.7		**76.53**±7.81		53.8
Prostate	DT	**3.0**±0	↑	78.82±5.99	≈	
	GP	6.4±2.6	≈	74.62±8.03	↓	98.1
	NGP	6.3±2.0		**79.79**±9.08		97.3

In Table 4, NGP achieves significantly better test classification accuracy and selects slightly fewer features than the standard GP method on four out of the tested nine datasets. The highest improvement can be seen on the Prostate dataset with more than a 5% increase on the average of the classification accuracy. On the Hillvally dataset, both NGP and GP can achieve the top classification accuracy (100%) using only 2 features on average. Only on the SPECT dataset, the standard GP method achieves slightly higher classification accuracy than NGP. Furthermore, Fig. 1 presents the average convergence curves of the standard GP and NGP methods with generations. Four datasets, WBCD, Madelon, DLBCL, and Prostate datasets, are chosen as representatives. In Fig. 1, a lower fitness value can be obtained from the proposed NGP method on the four datasets.

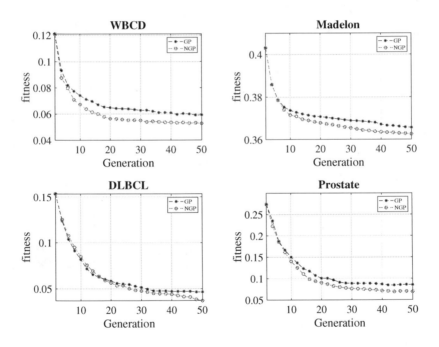

Fig. 1. Convergence curves of the two algorithms during training.

The results indicate that employing neighbourhood information in crossover can help GP to generate fitter individuals and therefore get better classification performance.

Training Time: All experiments use the Mahuika High-Performance Computing (HPC) cluster of the New Zealand eScience Infrastructure (NeSI) [19]. The average computational time (in minutes) of the standard GP and the proposed NGP methods on each dataset is given in the last column in Table 4. In Table 4, both methods spend a relatively short time, even less than 5 min on the SPECT, WBCD, and Sonar datasets. Furthermore, both algorithms can finish one run within 105 minutes on all the tested high-dimensional datasets except for the QSAR dataset. The results show that the proposed NGP method can achieve a much better classification performance and finish the evolutionary training process in a short time.

5.2 Example Programs Evolved by NGP

This section analyzes some example GP programs from the proposed NGP method on the WBCD, Hillvally, and DLBCL datasets. In these examples, multiple GP programs selecting different features can achieve the same classification accuracy. Even, some programs can achieve the same top classification accuracy (i.e., 100%) on both the training and test sets.

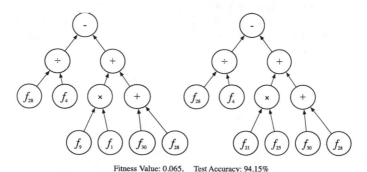

Fitness Value: 0.065,   Test Accuracy: 94.15%

Fig. 2. Two different GP programs with the same classification accuracy on the WBCD dataset.

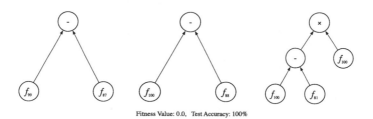

Fitness Value: 0.0, Test Accuracy: 100%

Fig. 3. Three different GP programs with the same top accuracy (100%) on the Hillvally dataset.

As shown in Fig. 2, the two trees, $T_{w1} = f_{28}/f_4 - f_9 * f_1 - f_{30} - f_{28}$ and $T_{w2} = f_{28}/f_4 - f_{21} * f_{25} - f_{30} - f_{28}$, can achieve the same training accuracy of 93.5%. The two trees are equivalent to the following two formulas:

$$T_{w1} = \frac{\text{std concave}}{\text{mean area}} - (\text{mean symmetry}) * (\text{mean radius}) - \text{lar fractal} - \text{std concave}, \quad (3)$$

$$T_{w2} = \frac{\text{std concave}}{\text{mean area}} - (\text{std radius}) * (\text{std smoothness}) - \text{lar fractal} - \text{std concave}, \quad (4)$$

where std and lar mean the standard error and the largest value, respectively. More detailed descriptions of the meanings of these features can be seen in [22]. If $f_{28}/f_4 - f_{30} - f_{28}$, i.e., $\frac{\text{std concave}}{\text{mean area}} - \text{lar fractal} - \text{std concave}$, is considered a new classifier, the training accuracy of 93.0% can be achieved. This indicates that by adding (mean symmetry) * (mean radius) or (std radius) * (std smoothness), the classification accuracy can be slightly improved (from 93.0% to 93.5%). The measures of symmetry in T_{w1} involve several steps. The first step is to find the major axis or the longest chord through the centre of the cell nuclei present in the image. Then, the length difference between lines perpendicular to the major axis to the cell boundary in both directions is measured. Compared with symmetry, smoothness in T_{w2} is easier to collect since it only involves local variation in radius lengths. Therefore, if both T_{w1} and T_{w2} are provided to a user, the user may prefer T_{w2} since it has a lower feature collection cost.

On the Hillvally dataset, the proposed NGP method finds very simple trees but with the same highest classification accuracy. In Fig. 3, the final outputs of the three trees are:

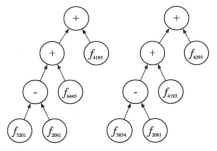

Fitness Value: 0.019,   Test Accuracy: 95.83%

Fig. 4. Two different GP programs with the same classification accuracy on the DLBCL dataset.

$$T_{h1} = f_{99} - f_{87}, \quad (5)$$

$$T_{h2} = f_{100} - f_{88}, \quad (6)$$

$$T_{h3} = f_{100} * (f_{100} - f_{81}). \quad (7)$$

Trees T_{h1} and T_{h2} have the same program structure (same number of nodes and tree depth), but they select different features. Using any of them can perfectly predict the class labels of instances. If all the three GP trees are provided to a user, the user is more likely to use T_{h1} or T_{h2} because of the simpler structure.

In Fig. 4, two different trees shown for the DLBCL dataset have the same both training (98.10%) and test (95.83%) classification accuracies. The two classifiers are the linear combination of different features. Taking either $f_{5201} - f_{2061} +$

$f_{6445} + f_{4193}$ or $f_{3834} - f_{2061} + f_{4193} + f_{4293}$ as a classifier, the same classification performance can be obtained. One interesting point is that there are two common features (f_{4193} and f_{2061}) included. If $f_{4193} - f_{2061}$ is taken as a new classifier, the classification accuracy of 66.67% can be achieved. By adding another two features, the classification accuracy is further improved.

According to these example programs, NGP is capable of automatically evolving multiple different effective GP programs with the same high classification performance.

6   Conclusions

This paper aimed to use GP to evolve multiple optimal classifiers with good interpretability for classification tasks. This goal has been successfully achieved by employing a niching technique and an external *Archive*. During the evolutionary learning process, the fittest individuals were collected into the *Archive*. When performing the crossover operator, the niching information of individuals was considered. The results showed that the proposed NGP algorithm achieved better overall classification accuracy than the standard GP method and the DT classifier. More importantly, the results showed that the proposed NGP method can find different GP programs with the same classification performance, revealing novel insights into the data. It may be possible to support the user in choosing a good model in a powerful and sensible manner if they have options, possibly in conjunction with expert knowledge.

Future work will primarily focus on extending NGP on multiple-class classification tasks and improving NGP further through the use of more advanced concepts to better measure the tree complexity.

References

1. Ahmed, S., Zhang, M., Peng, L.: Enhanced feature selection for biomarker discovery in LC-MS data using gp. In: IEEE Congress on Evolutionary Computation, pp. 584–591 (2013)
2. Ahmed, S., Zhang, M., Peng, L., Xue, B.: Multiple feature construction for effective biomarker identification and classification using genetic programming. In: Annual Conference on Genetic and Evolutionary Computation, pp. 249–256 (2014)
3. Bi, Y., Xue, B., Zhang, M.: Genetic Programming for Image Classification: An Automated approach to Feature Learning, vol. 24. Springer, Heidleberg (2021). https://doi.org/10.1007/978-3-030-65927-1
4. De Rainville, F.M., Fortin, F.A., Gardner, M.A., Parizeau, M., Gagné, C.: Deap: a python framework for evolutionary algorithms. In: Annual Conference Companion on Genetic and Evolutionary Computation, pp. 85–92 (2012)
5. Dua, D., Graff, C.: UCI machine learning repository (2017). http://archive.ics.uci.edu/ml
6. Harada, T., Murano, K., Thawonmas, R.: Proposal of multimodal program optimization benchmark and its application to multimodal genetic programming. In: IEEE Congress on Evolutionary Computation, pp. 1–8. IEEE (2020)

7. Kamyab, S., Eftekhari, M.: Feature selection using multimodal optimization techniques. Neurocomputing **171**, 586–597 (2016)
8. Koza, J.R., Koza, J.R.: Genetic Programming: On the Programming of Computers by Means of Natural Selection, vol. 1. MIT press, Cambridge (1992)
9. Lensen, A.: Mining feature relationships in data. In: Hu, T., Lourenço, N., Medvet, E. (eds.) EuroGP 2021. LNCS, vol. 12691, pp. 247–262. Springer, Cham (2021). https://doi.org/10.1007/978-3-030-72812-0_16
10. Lensen, A., Xue, B., Zhang, M.: Genetic programming for evolving similarity functions for clustering: representations and analysis. Evol. Comput. **28**(4), 531–561 (2020)
11. Li, H., Zhang, Q.: Multiobjective optimization problems with complicated pareto sets, MOEA/D and NSGA-II. IEEE Trans. Evol. Comput. **13**(2), 284–302 (2008)
12. Liu, J., Xu, C., Yang, W., Shu, Y., Zheng, W., Zhou, F.: Multiple similarly effective solutions exist for biomedical feature selection and classification problems. Sci. Rep. **7**(1), 1–10 (2017)
13. Muni, D.P., Pal, N.R., Das, J.: Genetic programming for simultaneous feature selection and classifier design. IEEE Trans. Syst. Man Cybern. Part B (Cybern.) **36**(1), 106–117 (2006)
14. Nag, K., Pal, N.R.: A multiobjective genetic programming-based ensemble for simultaneous feature selection and classification. IEEE Trans. Cybern. **46**(2), 499–510 (2015)
15. Nag, K., Pal, N.R.: Feature extraction and selection for parsimonious classifiers with multiobjective genetic programming. IEEE Trans. Evol. Comput. **24**, 454–466 (2019)
16. Neshatian, K., Zhang, M.: Genetic programming for feature subset ranking in binary classification problems. In: Vanneschi, L., Gustafson, S., Moraglio, A., De Falco, I., Ebner, M. (eds.) EuroGP 2009. LNCS, vol. 5481, pp. 121–132. Springer, Heidelberg (2009). https://doi.org/10.1007/978-3-642-01181-8_11
17. Neshatian, K., Zhang, M.: Pareto front feature selection: using genetic programming to explore feature space. In: Annual Conference on Genetic and Evolutionary Computation, pp. 1027–1034 (2009)
18. Pei, W., Xue, B., Shang, L., Zhang, M.: High-dimensional unbalanced binary classification by genetic programming with multi-criterion fitness evaluation and selection. Evol. Comput. **30**(1), 99–129 (2022)
19. Pletzer, A., Hayek, W., Scott, C., Corrie, B., Rae, G.: How NeSI helps users run better and faster on New Zealand's supercomputing platforms. In: IEEE International Conference on e-Science (e-Science), pp. 465–466 (2017)
20. Poursabzi-Sangdeh, F., Goldstein, D.G., Hofman, J.M., Vaughan, J.W., Wallach, H.: Manipulating and measuring model interpretability. arXiv preprint arXiv:1802.07810 (2018)
21. Sijben, E., Alderliesten, T., Bosman, P.A.: Multi-modal multi-objective model-based genetic programming to find multiple diverse high-quality models. In: Genetic and Evolutionary Computation Conference, pp. 440–448 (2022)
22. Street, W.N., Wolberg, W.H., Mangasarian, O.L.: Nuclear feature extraction for breast tumor diagnosis. In: Biomedical Image Processing and Biomedical Visualization, vol. 1905, pp. 861–870. SPIE (1993)
23. Tran, B., Xue, B., Zhang, M.: Genetic programming for feature construction and selection in classification on high-dimensional data. Memetic Comput. **8**(1), 3–15 (2016)

24. Uy, N.Q., Hien, N.T., Hoai, N.X., O'Neill, M.: Improving the generalisation ability of genetic programming with semantic similarity based crossover. In: Esparcia-Alcázar, A.I., Ekárt, A., Silva, S., Dignum, S., Uyar, A.Ş (eds.) EuroGP 2010. LNCS, vol. 6021, pp. 184–195. Springer, Heidelberg (2010). https://doi.org/10.1007/978-3-642-12148-7_16

25. Vanneschi, L., Tomassini, M., Clergue, M., Collard, P.: Difficulty of unimodal and multimodal landscapes in genetic programming. In: Cantú-Paz, E., et al. (eds.) GECCO 2003. LNCS, vol. 2724, pp. 1788–1799. Springer, Heidelberg (2003). https://doi.org/10.1007/3-540-45110-2_70

26. Wang, P., Xue, B., Liang, J., Zhang, M.: Multiobjective differential evolution for feature selection in classification. IEEE Trans. Cybern. (2021). https://doi.org/10.1109/TCYB.2021.3128540

27. Wang, P., Xue, B., Liang, J., Zhang, M.: Differential evolution based feature selection: a niching-based multi-objective approach. IEEE Trans. Evol. Comput. (2022). https://doi.org/10.1109/TEVC.2022.3168052

28. Xu, H., Xue, B., Zhang, M.: A duplication analysis-based evolutionary algorithm for biobjective feature selection. IEEE Trans. Evol. Comput. 25(2), 205–218 (2020)

29. Yoshida, S., Harada, T., Thawonmas, R.: Multimodal genetic programming by using tree structure similarity clustering. In: International Workshop on Computational Intelligence and Applications, pp. 85–90 (2017)

30. Yue, C., Liang, J., Qu, B., Yu, K., Song, H.: Multimodal multiobjective optimization in feature selection. In: IEEE Congress on Evolutionary Computation, pp. 302–309 (2019)

Evolving Effective Ensembles for Image Classification Using Multi-objective Multi-tree Genetic Programming

Qinglan Fan[✉], Ying Bi, Bing Xue, and Mengjie Zhang

Victoria University of Wellington, PO Box 600, Wellington 6140, New Zealand
{qinglan.fan,ying.bi,bing.xue,mengjie.zhang}@ecs.vuw.ac.nz

Abstract. The high variations across images make image classification a challenging task, where the limited number of training instances further increases the difficulty of achieving good generalization performance. Applying ensemble learning to classification often yields better generalization results on unseen data than using a single classifier. However, for an ensemble to generalize properly, its base learners should be accurate and diverse. Genetic programming (GP) has achieved promising results in image classification. However, existing methods typically employ single-tree representation (i.e., an individual contains a single tree) and are not easy to evolve multiple base learners especially when only limited training data is available. This paper proposes a new ensemble construction method for image classification using multi-objective multi-tree GP (i.e., on individual contains multiple trees). In the new method, a GP individual forms an ensemble, and its multiple trees are base learners that can learn informative features from a relatively small number of training instances. To find effective GP individuals/ensembles, i.e., to make its multiple trees accurate and diverse, the proposed method formulates the ensemble learning problem as a multi-objective task explicitly. Thus, the new objective functions are developed to maximize the diversity and minimize the classification error simultaneously. The proposed method achieves significantly better generalization performance than many competitive methods on four datasets of varying difficulty. Further analysis demonstrates the effectiveness and potentially high interpretability of the constructed ensembles.

Keywords: Image classification · Ensemble learning · Multi-tree genetic programming · Multi-objective

1 Introduction

Image classification aims at predicting class labels of unknown instances/images by building a model based on training instances. It is a fundamental task in computer vision with many applications, including healthcare, the automobile industry, and manufacturing [30]. However, it is not easy to develop an effective

H. Aziz et al. (Eds.): AI 2022, LNAI 13728, pp. 294–307, 2022.
https://doi.org/10.1007/978-3-031-22695-3_21

image classification method due to high image variations, such as illumination, rotation, and scale. Moreover, in some computer vision applications, receiving large annotated datasets is challenging and expensive, increasing the difficulty of obtaining good generalization performance.

Ensemble methods have been applied to various classification tasks, including image classification [3,10]. An ensemble contains multiple base learners and performs a good prediction by combining the results of its base learners via a combination method, e.g., voting and averaging. Much research work has shown that applying ensemble learning for classification can achieve better generalization performance than using a single classification algorithm on unseen data [28]. However, it is challenging to construct an effective ensemble since its base learners are required to be accurate and diverse. In the image classification domain, an accurate base learner is expected to extract informative features from raw images and conduct effective classification simultaneously.

Genetic programming (GP) [13], an evolutionary learning technique, has been successfully applied for image feature learning and classification [2]. It is a population-based search approach based on the Darwinian theory. It typically uses tree-based representation and starts by randomly generating a population of programs/individuals/solutions and further updates these individuals by using evaluation, selection, and genetic operators, i.e., crossover and mutation, until finding the best solution(s). Since a tree-based GP model has potentially good interpretability, a strong global search ability, and flexible representation with variable length, it has achieved promising classification results in image classification [3,8,25]. GP-based methods can learn informative features through multiple levels of transformations, i.e., linear and nonlinear, in a GP individual. The evolved GP individuals can be used for building accurate base learners in an ensemble. However, most existing methods employ a single-tree representation, i.e., an individual consists of a single tree, because it is relatively easy to be implemented. However, using single-tree GP is not easy to build multiple accurate and diverse base learners, especially when only a limited number of training instances are available. With the flexible representation, GP can evolve multiple trees in a single individual, i.e., multi-tree GP (MTGP) [16]. MTGP is a suitable candidate for constructing multiple base learners and forming an ensemble for image classification, but not much work has been done in this direction.

Another crucial issue is that for an ensemble to be more accurate than any of its base learners, the base learners must be diverse, namely, making different errors on the same inputs [28]. However, the objectives of maximizing the diversity and minimizing the classification error (maximizing the classification accuracy) of base learners are often in conflict with each other [5,6]. GP has unique advantages in addressing multi-objective tasks since it can find a set of trade-off (between the diversity and the classification error) solutions in a single run using its population-based search mechanism. In addition, the diversity of the base learners is not easy to measure and control [28].

The overall goal of this paper is to address the above-discussed limitations by evolving effective ensembles using multi-objective multi-tree GP for image classification containing a limited amount of training data. For simplification, the new method is named MMGP, i.e., multi-objective multi-tree GP. Specifically, the goal can be divided into five following objectives:

– Develop a new ensemble construction method for image classification using multi-tree GP that can achieve better generalization performance than many competitive methods;
– Employ an effective tree representation with a relatively simple structure and a few parameters that allows base learners in the MMGP method to learn discriminative and informative features without requiring a large number of training instances;
– Design the new objective functions to minimize classification error and maximize diversity objectives simultaneously, which enables the new method to search for the best ensembles containing accurate and diverse base learners during the evolutionary learning/training process;
– Evaluate the generalization performance of the proposed MMGP method on four image datasets of varying difficulty with a limited number of training instances; and
– Analyze the effectiveness and interpretability of the constructed ensembles using MMGP.

2   Related Work

2.1   GP for Image Classification

With flexible tree representation, many studies have investigated the effectiveness of GP for image classification. Below is a brief summary of these studies.

Atkins et al. [1] used GP for developing a domain-independent image classification method. The GP program evolved by this method had three tiers (3TGP) performing different sub-tasks, such as filtering. Its output was a high-level feature (i.e., a floating-point number) that was used to make the classification decision. This method was comparable to other methods based on domain-specific features. Ruberto et al. [23] presented a genetic program feature learner (GPFL) for image classification. The method could learn a specific high-level feature from raw images at each generation and combine them. On the MNIST dataset, it performed better than the benchmark method, i.e., LeNet-5 [14]. Bi et al. [4] developed an image classification method based on GP with a flexible representation (FGP), which employed multiple image-related functions, including filters and feature extraction methods, as the internal nodes of GP trees for extracting features, and conducted classification using support vector machine (SVM). FGP has shown promising performance on various image classification tasks. Fan et al. [8] proposed a new tree representation to fully utilize the features produced by the different nodes of GP trees. This method could flexibly reuse useful features from internal nodes of a GP tree, automatically choose one classification algorithm, such as SVM or random forest(RF), and perform classification. The method achieved better classification performance compared with many competitive methods.

Existing methods of image classification based on GP have shown promising results. However, these methods usually train a classifier using a GP tree or rely on an external classification algorithm, e.g., SVM, to perform classification [2]. When there is a limited number of training instances, these methods are not easy to generalize well.

2.2   Ensemble Learning for Classification

An ensemble method consists of training a diverse set of base learners and combining their predictions to obtain a good prediction. Thus, it usually achieves better generalization performance than a single model on the test data and has been extensively applied to address various classification problems [28].

Pooja et al. [22] used a linear program boosting classifier to enhance the accuracy of weather predictions. The method employed K nearest neighbor (KNN) as a weak classifier, and its weight was updated depending on the training error. To increase diversity among the base learners, Lu et al. [18] trained the base learner by building the decision group that randomly selected a classifier from three classifiers, i.e., KNN, decision tree (DT), and naïve bayes (NB), repeated three times, and obtained the classification decision by voting. In addition, GP has been used to build ensembles for classification. To improve the classification of text documents, Zhang et al. [27] utilized multiple GP individuals from a population to form an ensemble, which was then able to find a set of similarity functions. Based on majority voting, the similarity functions were more effective than other methods of fusion, such as content-based SVM. Bi et al. [3] proposed an ensemble method by integrating several classification algorithms as the functions of the GP programs. The final class label was returned from the output of the GP program that combined the predictions of multiple classifiers accordingly. The method achieved competitive performance on the image classification datasets. However, it tended to evolve GP programs with large sizes, harming its interpretability.

Existing studies have presented promising results using GP-based ensembles for classification. However, only a few studies have been conducted on building ensembles based on GP for image classification. Moreover, using single-tree GP to construct multiple accurate and diverse base learners is not easy, especially when the number of training instances is limited. This paper aims to address these issues by developing a new ensemble construction method using multi-tree GP for image classification.

3   The Proposed Approach

This section presents in detail the new multi-objective multi-tree GP-based ensemble approach (i.e., MMGP) to image classification. We first provide the overall process of the proposed image classification framework. Second, this section introduces the tree representation to build accurate GP trees/base learners. Finally, a new fitness evaluation strategy with multi-objective is developed.

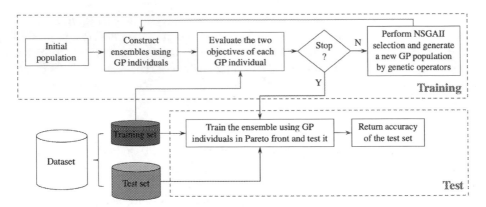

Fig. 1. Flowchart of MMGP. It contains a training process and a test process.

3.1 Overall Algorithm

The goal of this paper is to propose a new ensemble construction approach using multi-objective multi-tree GP to image classification with limited training data. Unlike most existing GP-based methods that usually contain a single tree in each individual, the proposed MMGP approach can evolve multiple trees in each individual. In MMGP, each GP individual forms an ensemble in which multiple trees are base learners and obtains the classification results via majority voting. Furthermore, the evolutionary multi-objective optimization method is employed to search for the best individual/ensemble with accurate and diverse base learners during the evolutionary learning process. MMGP can maximize both objectives of classification accuracy and diversity simultaneously (see Sect. 3.3). Thus, MMGP is expected to achieve better generalization performance than the methods using a single GP tree on unseen data.

Figure 1 shows the flowchart of the proposed image classification framework using MMGP that contains two components, i.e., the evolutionary training/learning process and the test process. In the training process, MMGP takes a training set containing training instances/images and corresponding class labels as the input. Its output is a set of non-dominated solutions/individuals/ensembles that are employed to predict the class labels of the test instances. Firstly, a population of GP individuals with multiple trees is automatically generated via *ramped half-and-half* [13]. To obtain accurate base learners/GP trees, MMGP employs the tree representation proposed in GP-FR [8] (see Sect. 3.2). GP-FR is a state-of-the-art image classification method that achieves competitive performance on various image classification datasets. MMGP constructs ensembles using GP individuals and then evaluates them to obtain the corresponding objective values, i.e., classification accuracy and diversity. Then, GP individuals with higher objective values are chosen using the NSGAII selection method [7]. NSGAII is used in this work since it is a commonly used evolutionary multi-objective optimization (EMO) method, and other EMO methods can also be used here. The selected individuals are used to generate new ones by conducting random-index crossover and mutation [17] (i.e., randomly selecting a tree from the individual to perform crossover or mutation). This process will continue until a predetermined stopping criterion, i.e., the maximum number of generations is reached. Finally, a set of non-dominated GP individuals are returned, all of which are used to construct an ensemble using the training set. The obtained ensemble is used to predict the class labels of test instances via majority voting.

3.2 Multi-tree Representation

MMGP employs the multi-tree GP representation to evolve ensembles for image classification. Figure 2 shows an example GP individual containing three trees. MMGP evolves each tree in one GP individual based on the tree representation in GP-FR [8] to train accurate base learners and then construct effective ensembles using all trees. It is noted that although each tree in the GP individual uses the same representation developed in GP-FR, their sizes/shapes and internal/leaf nodes can be different because of the flexible representation, as shown in Fig. 2. GP-FR is based on strongly typed GP (STGP) [19]. Its tree structure contains multiple layers with different functions, which conduct image filtering, region detection, feature extraction, feature concatenation, and classification, respectively. The GP tree takes a raw image as input, and its output is the predicted class label. GP-FR can automatically learn informative features from raw images by layer-by-layer transformation, select a suitable classification algorithm, such as SVM or logistic regression (LR), and conduct effective classification. Moreover, GP-FR can flexibly reuse features generated by different nodes of the GP tree (see dotted arrows in Fig. 2), thus learning much richer image features and achieving higher classification accuracy than GP-based methods without feature reuse. More details about the tree structure, functions, and terminals of GP-FR can be found in [8].

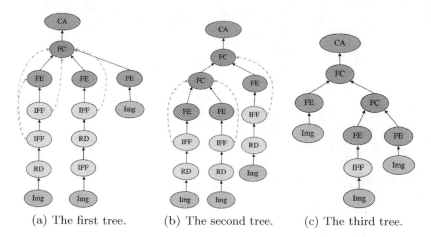

(a) The first tree.   (b) The second tree.   (c) The third tree.

Fig. 2. An example GP individual that can be evolved by MMGP (Img: Input Image; RD: Region Detection Function; IIF: Image Filtering Function; FE: Feature Extraction Function; FC: Feature Concatenation Function; CA: Classification Algorithm Function (e.g., SVM)).

To obtain ensembles that generalize well, base learners are required to be not only accurate but also diverse. MMGP uses the evolutionary multi-objective optimization method to achieve this goal.

3.3   Objective Functions

In the proposed MMGP approach, classification accuracy and diversity are employed as two objectives to measure an individual's fitness.

Objective 1 – Accuracy.

$$acc = \frac{N_{correct}}{N}, \tag{1}$$

where N is the total number of training instances, $N_{correct}$ denotes the number of instances being correctly classified, and acc is the classification accuracy.

Objective 2 – Diversity. MMGP applies pairwise failure crediting (PFC) [5] as a diversity measure, which provides an indication of how different each tree/base learner is from all the other trees/base learners in the GP individual/ensemble.

$$PFC = \frac{1}{T} \sum_{p=1}^{T} \sum_{j=p+1, j \neq p}^{T} \frac{\sum_{i=1}^{N} I(c_i^p, c_i^j)}{Err^p + Err^j}, \tag{2}$$

where

$$I(c_i^p, c_i^j) = \begin{cases} 1, & if\ c_i^p \neq c_i^j, \\ 0, & otherwise \end{cases}$$

In Eq. (2), T is the number of trees in the GP individual, N is the number of training instances, and c_i^p and c_i^j are the predicted class labels of the instance i using the trees p and j, respectively. Indicator function $I(\cdot)$ returns 1 if the predicted class labels between two trees are different for a given training instance, or 0 otherwise. Err^p and Err^j are the number of incorrect predictions for the trees p and j in a GP individual on the training set. Equation (2) will return the value between 0 and 1, where the higher the PFC, the better the diversity.

To avoid overfitting, MMGP uses the stratified K-fold cross-validation method to calculate the accuracy and diversity. The training set is divided into K folds. Each time one fold is used to test the built ensemble/individual, and the remaining folds are used to train base learners/trees and form an ensemble. MMGP assigns the average test accuracy and the average diversity of the K folds as the objective values of the individual. The value of K is set to 3 according to [29].

4   Experiment Design

4.1   Benchmark Datasets

In the experiments, four image classification datasets with limited training data are employed to evaluate the performance of MMGP. They are DSLR [24], web-cam [24], Outex [20], and EYALE [15]. These datasets are representative of various tasks related to image classification, i.e., object classification (DSLR and webcam), texture classification (Outex), and face recognition (EYALE). These tasks with different types of images can comprehensively illustrate the effectiveness of MMGP. Figure 3 presents the example images of the benchmark datasets. Table 1 lists their detailed information. In order to reduce the computational cost, the images are resized and/or converted to gray-scale images.

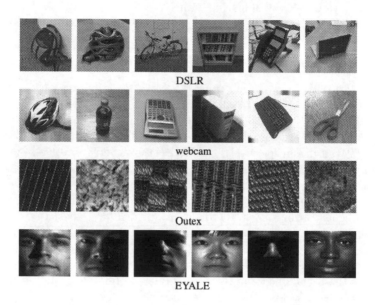

Fig. 3. Example images of the benchmark datasets.

Table 1. The benchmark dataset properties

No	Dataset	Training set	Test set	Image size	#Class
1	DSLR	155 (5)	343	50×50	31
2	Webcam	155 (5)	640	50×50	31
3	Outex	480 (20)	3840	64×64	24
4	EYALE	380 (10)	720	50×45	38

4.2 Benchmark Methods

We employ 11 effective benchmark methods, including ensemble-based methods, CNN-based methods, and GP-based methods, for comparisons to demonstrate the effectiveness of MMGP. Six traditional image classification methods, namely, SIFT+RF, HOG+RF, uniform LBP (uLBP)+RF, SIFT+SVM, HOG+SVM, and uLBP+SVM, which extract SIFT, HOG, and uLBP features respectively and then use RF or SVM for classification. RF is an ensemble classification algorithm. Since the benchmark datasets only contain limited training instances, it is not easy to train a very deep CNN. Three typical shallow CNNs, i.e.,

LeNet-5 [14], CNN-5 [25], and MobileNetV3-Small [11] are used for comparisons in the experiments. Considering that MMGP is a GP-based approach, the comparison methods also include two state-of-the-art GP methods, i.e., IEGP [3] and GP-FR [8]. IEGP is an ensemble method that combines several classification algorithms as the functions of GP programs. MMGP uses the tree representation proposed in GP-FR. Comparing with GP-FR can verify the effectiveness of constructed ensembles in MMGP.

4.3   Parameter Settings

Parameter settings of MMGP are consistent with the commonly used settings in the GP community [2,8]. The maximum number of generations is 50 and the population size is 100. The *ramped half-and-half* method is used to generate initial population [12]. The tree depth is between 4–10 [8]. The mutation and crossover rates are 0.2 and 0.8, respectively. NSGAII is employed to evolve a good set of non-dominated GP individuals/solutions in MMGP. To conduct effective majority voting and save the computational cost of constructing ensembles, MMGP evolves three trees in each GP individual.

The implementation of IEGP, GP-FR, and MMGP is based on the DEAP (distributed evolutionary algorithm in Python) package [9]. The classification algorithms, i.e., SVM, LR, RF, and ERF, used in MMGP and the benchmark methods are implemented based on the scikit-learn package [21]. The parameter settings for these classification algorithms refer to references [26] and [29] due to their effectiveness. The SVM uses a linear kernel with a penalty parameter of 1 [26]. The penalty parameter in LR is also set to 1 [26]. In RF and ERF, there are 500 trees and their maximum depth is 100 [29]. In CNN-5, LeNet-5, and MobileNetV3-Small, the batch size is set to 16 and the epochs is set to 100. On each dataset, all methods have been executed 30 independent times with different random seeds and the results of the 30 runs are reported.

5   Results and Discussions

In this section, we report and discuss the classification results of MMGP and the compared methods on the four benchmark datasets, showing the effectiveness of MMGP for image classification with limited training instances. Table 2 presents the classification accuracy, including the average classification accuracy and the standard deviation (Mean±Std) from 30 runs. To compare MMGP with the benchmark method, we use the Wilcoxon rank-sum test with a 95% confidence interval. The symbols "+" and "−" indicate that MMGP achieves significantly better and worse performance than a particular benchmark method. "=" suggests that MTGPE and the benchmark method obtain similar classification results. From Table 2, MMGP performs better than the benchmark methods, i.e., it obtains 42 "+" and 2 "=".

Comparisons with Traditional Methods: The classification performance of SIFT+SVM, HOG+SVM, and uLBP+SVM on the four benchmark datasets is

Table 2. Classification accuracy(%) of the proposed approach and comparison methods on the DSLR, webcam, Outex, and EYALE datasets

Methods	DSLR	Webcam	Outex	EYALE
SIFT+SVM	51.31 ± 0.00 +	50.00 ± 0.00 +	34.74 ± 0.00 +	66.49 ± 0.00 +
HOG+SVM	36.73 ± 0.00 +	35.00 ± 0.00 +	16.28 ± 0.00 +	65.26 ± 0.00 +
uLBP+SVM	34.40 ± 0.00 +	36.09 ± 0.00 +	86.25 ± 0.00 +	33.46 ± 0.00 +
SIFT+RF	53.52 ± 1.06 +	55.13 ± 1.01 +	63.89 ± 0.37 +	46.42 ± 0.49 +
HOG+RF	48.67 ± 1.03 +	47.98 ± 0.84 +	48.49 ± 0.43 +	54.21 ± 0.48 +
uLBP+RF	41.42 ± 1.15 +	40.49 ± 0.94 +	93.04 ± 0.22 +	30.09 ± 0.42 +
LeNet-5	37.93 ± 2.98 +	36.26 ± 2.88 +	71.20 ± 6.28 +	58.23 ± 3.85 +
CNN-5	42.96 ± 1.68 +	43.86 ± 2.05 +	60.57 ± 3.56 +	73.87 ± 2.17 +
MobileNetV3-Small	52.54 ± 3.30 +	53.63 ± 2.24 +	96.47 ± 1.10 +	77.54 ± 6.52 +
IEGP	57.03 ± 4.72 +	59.86 ± 5.76 +	98.97 ± 0.38 +	92.65 ± 2.27 =
GP-FR	58.78 ± 2.48 +	59.51 ± 2.52 +	99.05 ± 0.42 +	92.42 ± 2.20 =
MMGP	**62.11 ± 0.02**	**63.21 ± 0.02**	**99.45 ± 0.01**	**92.67 ± 0.02**
Overall	**11+**	**11+**	**11+**	**9+, 2=**

not satisfying. SIFT+RF, HOG+RF, and uLBP+RF outperform SIFT+SVM, HOG+SVM, and uLBP+SVM, respectively, on the DSLR, webcam, and Outex datasets, which indicates that ensemble methods often achieve better performance than using a single classifier when only limited training data is available. On the other hand, the proposed MMGP approach performs significantly better than these methods in all comparisons. The results reveal that only using one feature extraction method may not be sufficient for image classification.

Comparisons with CNN-Based Methods: Compared with three CNN-based methods, i.e., LeNet-5, CNN-5, and MobileNetV3-Small, MMGP archives significantly better classification performance on all benchmark datasets. For example, MMGP reaches a 9.57%, 9.58%, 2.98%, and 15.13% increase over the best results obtained by the three CNN-based methods on the four benchmark datasets in terms of mean accuracy. A possible reason for these phenomena is that CNN-based methods usually contain many parameters even for shallow CNNs, and require a large number of training instances to train the model. However, MMGP using a relatively simple tree representation to represent solutions, has a powerful ability to learn informative features and constructs effective ensembles for image classification with limited training instances.

Comparisons with GP-Based Methods: IEGP performs significantly worse than the proposed MMGP approach on DSLR, webcam, and Outex, i.e., decreases by 5.08%, 3.35%, and 0.48% mean accuracy and reaches the similar results to MMGP on EYALE. GP-FR and MMGP employ the same tree representation, but MMGP obtains three "+" and one "=" in four comparisons, illustrating the effectiveness of the constructed ensembles.

In summary, MMGP achieves better generalization performance than the compared methods by constructing effective ensembles. In addition, MMGP obtains a small standard deviation on the four benchmark datasets, indicating its good stability.

6 Further Analysis

6.1 The Effectiveness of the Constructed Ensembles Using MMGP

Since MMGP is a multi-objective method, a set of non-dominated GP individuals/ensembles will be received when the evolutionary training process is completed. To further improve the generalization performance of the new method, we build the ensemble using all returned individuals in the Pareto front on the training set and examine it on the test set. In addition, we report the classification results of the constructed ensemble using the GP individual with the lowest training classification error in the Pareto front for comparisons. The method is named MMGP_LE. Table 3 lists the classification accuracy of the baseline method (i.e., GP-FR), MMGP_LE, and MMGP on the four benchmark datasets. From Table 3, MMGP_LE achieves higher classification accuracy than GP-FR, indicating the effectiveness of the constructed ensembles since they employ the same tree representation. The classification performance of MMGP_LE has been further improved according to the results of MMGP, showing that using all individuals in the Pareto front to build an ensemble is more effective for image classification.

Table 3. Classification accuracy(%) of the constructed ensembles on the DSLR, webcam, Outex, and EYALE datasets

Methods	DSLR	Webcam	Outex	EYALE
GP-FR	58.78 ± 2.48 +	59.51 ± 2.52 +	99.05 ± 0.42 +	92.42 ± 2.20 =
MMGP_LE	60.68 ± 0.02 +	60.65 ± 0.02 +	99.16 ± 0.01 =	92.45 ± 0.02 =
MMGP	**62.11 ± 0.02**	**63.21 ± 0.02**	**99.45 ± 0.01**	**92.67 ± 0.02**

6.2 An Example GP Individual of MMGP

As shown in Figs. 4 and 5, an example GP individual/ensemble evolved on EYALE (i.e., a face recognition dataset) is used in this section to explain how MMGP learns informative features and performs classification, thus demonstrating its effectiveness and potentially high interpretability. The three trees in this GP individual achieve 80.00%, 82.08%, and 88.33% test accuracy, respectively, while it achieves 90.14% test accuracy by considering the results of these three trees. In addition, it is straightforward to know which features are extracted and which classification algorithm is used according to the trees in Fig. 4. For example, from Fig. 5, the texture features extracted from the eye region via the uLBP function and the features extracted from the whole image via the DIF function can effectively distinguish different faces, showing that the evolved GP trees are potentially interpretable.

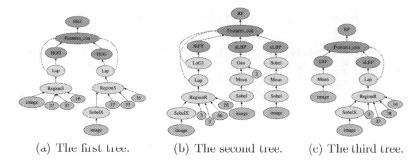

(a) The first tree.   (b) The second tree.   (c) The third tree.

Fig. 4. An example GP individual evolved by MMGP on EYALE.

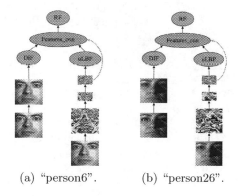

(a) "person6".   (b) "person26".

Fig. 5. The example classification process of the third tree from Fig. 4.

7   Conclusions

The goal of this paper was to develop an effective image classification app-roach based on GP with limited training data. This goal has been successfully achieved by proposing a new ensemble construction method using MMGP. In MMGP, each GP individual formed an ensemble, in which three trees were base leaners. An effective tree representation containing a relatively simple structure and only a few parameters were used to train accurate base learners that could learn informative features from limited training data. To ensure the accuracy and diversity between base learners in an ensemble, MMGP regarded ensemble learning problems as multi-objective optimization problems. As a result, a set of non-dominated GP individuals/ensembles were obtained, which were used to predict the class labels of the test instances. The performance of MMGP has been evaluated on four different image classification datasets containing limited training instances. The experimental results showed that MMGP received signif-icantly higher classification accuracy than those competitive image classification methods in almost all comparisons. Further analysis verified the effectiveness and potentially high interpretability of the constructed ensembles.

This paper investigated the potential of combining multi-tree GP and ensemble learning for effective image classification with limited training data. However, it is computationally more expensive to use multi-tree GP than single-tree GP. In the future, we will use efficient methods, e.g., surrogate models, to further reduce the computational costs of multi-tree GP.

References

1. Atkins, D., Neshatian, K., Zhang, M.: A domain independent genetic programming approach to automatic feature extraction for image classification. In: 2011 IEEE Congress of Evolutionary Computation (CEC), pp. 238–245 (2011)
2. Bi, Y., Xue, B., Zhang, M.: Genetic Programming for Image Classification: An Automated Approach to Feature Learning. Springer, Heidelberg (2021). https://doi.org/10.1007/978-3-030-65927-1
3. Bi, Y., Xue, B., Zhang, M.: Genetic programming with a new representation to automatically learn features and evolve ensembles for image classification. IEEE Trans. Cybern. **51**(4), 1769–1783 (2021)
4. Bi, Y., Xue, B., Zhang, M.: Genetic programming with image-related operators and a flexible program structure for feature learning in image classification. IEEE Trans. Evol. Comput. **25**(1), 87–101 (2021)
5. Chandra, A., Yao, X.: Ensemble learning using multi-objective evolutionary algorithms. J. Math. Model. Algor. **5**(4), 417–445 (2006)
6. Chen, H., Yao, X.: Multiobjective neural network ensembles based on regularized negative correlation learning. IEEE Trans. Knowl. Data Eng. **22**(12), 1738–1751 (2010)
7. Deb, K., Pratap, A., Agarwal, S., Meyarivan, T.: A fast and elitist multiobjective genetic algorithm: NSGAII. IEEE Trans. Evol. Comput. **6**(2), 182–197 (2002)
8. Fan, Q., Bi, Y., Xue, B., Zhang, M.: Genetic programming for image classification: a new program representation with flexible feature reuse. IEEE Trans. Evol. Comput. (2022). https://doi.org/10.1109/TEVC.2022.3169490
9. Fortin, F.A., Rainville, F.M.D., Gardner, M.A., Parizeau, M., Gagné, C.: Deap: evolutionary algorithms made easy. J. Mach. Learn. Res. **13**(70), 2171–2175 (2012)
10. Galar, M., Fernandez, A., Barrenechea, E., Bustince, H., Herrera, F.: A review on ensembles for the class imbalance problem: bagging-, boosting-, and hybrid-based approaches. IEEE Trans. Syst. Man Cybern. Part C (Appl. Rev.) **42**(4), 463–484 (2012)
11. Howard, A., et al.: Searching for mobilenetv3. In: Proceedings of the IEEE/CVF International Conference on Computer Vision, pp. 1314–1324 (2019)
12. Koza, J.R.: Genetic Programming: On the Programming of Computers by Means of Natural Selection. MIT Press, Cambridge (1992)
13. Koza, J.R.: Genetic programming as a means for programming computers by natural selection. Stat. Comput. **4**(2), 87–112 (1994)
14. Lecun, Y., Bottou, L., Bengio, Y., Haffner, P.: Gradient-based learning applied to document recognition. Proc. IEEE **86**(11), 2278–2324 (1998)
15. Lee, K.C., Ho, J., Kriegman, D.J.: Acquiring linear subspaces for face recognition under variable lighting. IEEE Trans. Pattern Anal. Mach. Intell. **27**(5), 684–698 (2005)

16. Lensen, A., Xue, B., Zhang, M.: Generating redundant features with unsupervised multi-tree genetic programming. In: Castelli, M., Sekanina, L., Zhang, M., Cagnoni, S., García-Sánchez, P. (eds.) EuroGP 2018. LNCS, vol. 10781, pp. 84–100. Springer, Cham (2018). https://doi.org/10.1007/978-3-319-77553-1_6

17. Lensen, A., Xue, B., Zhang, M.: Genetic programming for evolving similarity functions for clustering: Representations and analysis. Evol. Comput. **28**(4), 531–561 (2020)

18. Lu, H., Gao, H., Ye, M., Wang, X.: A hybrid ensemble algorithm combining adaboost and genetic algorithm for cancer classification with gene expression data. IEEE/ACM Trans. Comput. Biol. Bioinf. **18**(3), 863–870 (2021)

19. Montana, D.J.: Strongly typed genetic programming. Evol. Comput. **3**(2), 199–230 (1995)

20. Ojala, T., Maenpaa, T., Pietikainen, M., Viertola, J., Kyllonen, J., Huovinen, S.: Outex-new framework for empirical evaluation of texture analysis algorithms. In: Object Recognition Supported by User Interaction for Service Robots, vol. 1, pp. 701–706. IEEE (2002)

21. Pedregosa, F., et al.: Scikit-learn: machine learning in python. J. Mach. Learn. Res. **12**(85), 2825–2830 (2011)

22. Pooja, S., Balan, R.S., Anisha, M., Muthukumaran, M., Jothikumar, R.: Techniques tanimoto correlated feature selection system and hybridization of clustering and boosting ensemble classification of remote sensed big data for weather forecasting. Comput. Commun. **151**, 266–274 (2020)

23. Ruberto, S., Terragni, V., Moore, J.H.: Image feature learning with genetic programming. In: Bäck, T., et al. (eds.) PPSN 2020. LNCS, vol. 12270, pp. 63–78. Springer, Cham (2020). https://doi.org/10.1007/978-3-030-58115-2_5

24. Saenko, K., Kulis, B., Fritz, M., Darrell, T.: Adapting visual category models to new domains. In: Daniilidis, K., Maragos, P., Paragios, N. (eds.) ECCV 2010. LNCS, vol. 6314, pp. 213–226. Springer, Heidelberg (2010). https://doi.org/10.1007/978-3-642-15561-1_16

25. Shao, L., Liu, L., Li, X.: Feature learning for image classification via multiobjective genetic programming. IEEE Trans. Neural Netw. Learn. Syst. **25**(7), 1359–1371 (2014)

26. Young, S., Abdou, T., Bener, A.: Deep super learner: a deep ensemble for classification problems. In: Bagheri, E., Cheung, J.C.K. (eds.) Canadian AI 2018. LNCS (LNAI), vol. 10832, pp. 84–95. Springer, Cham (2018). https://doi.org/10.1007/978-3-319-89656-4_7

27. Zhang, B., et al.: Intelligent gp fusion from multiple sources for text classification. In: Proceedings of the 14th ACM International Conference on Information and Knowledge Management, pp. 477–484 (2005)

28. Zhou, Z.H.: Ensemble Methods: Foundations and Algorithms. CRC Press, Boca Raton (2012)

29. Zhou, Z.H., Feng, J.: Deep forest. Natl. Sci. Rev. **6**(1), 74–86 (2018)

30. Zhu, H., Jin, Y.: Real-time federated evolutionary neural architecture search. IEEE Trans. Evol. Comput. **26**(2), 364–378 (2022)

Knowledge Representation and NLP

QUARRY: A Graph Model for Queryable Association Rules

Michael Stewart[(✉)] [iD]

Department of Computer Science and Software Engineering,
The University of Western Australia, Perth, Australia
michael.stewart@uwa.edu.au
https://nlp-tlp.org

Abstract. Association rule mining is a pivotal technique for knowledge discovery, but often involves time-intensive manual labour when performed on large datasets. In this paper we propose a solution for this problem: QUARRY, a graph model that enables consumable and queryable insights from association rules. In contrast to existing systems which take a list of rules and display them in a purpose-built visualisation, our graph-based model enables association rules to be queried directly via graph queries. Through a case study on maintenance data we show how this model enhances knowledge discovery by eliminating the need for domain experts to trawl through large lists of rules to find useful information. QUARRY, which is designed for compatibility with existing knowledge graphs, provides users with the means to easily search for rules pertaining to specific items as well as roll up and drill down on their searches using the concept hierarchy. Domain experts may also query for association rules based on transaction properties such as costs and dates, enabling critical insights into their data.

Keywords: Association rule mining · Text mining · Information retrieval · Knowledge discovery · Knowledge graphs

1 Introduction

Knowledge discovery is an important task facilitated by artificial intelligence (AI) that supports data-driven decision making. Association rule mining is one of the foundational techniques for knowledge discovery, and is one of the most important and well researched techniques in data mining [1,10].

An *association rule* is an implication that an *antecedent* predicts a *consequent*, for example, "people who buy bread and eggs often also buy milk". The antecedent and consequent are sets of items, known as *itemsets*. Association rules are commonly evaluated by two measures. The *support* is the fraction of transactions that contain both the antecedent and consequent divided by the total number of transactions in the entire dataset. The *confidence* measures the fraction of records containing both the consequent and antecedent divided by the

© The Author(s), under exclusive license to Springer Nature Switzerland AG 2022
H. Aziz et al. (Eds.): AI 2022, LNAI 13728, pp. 311–324, 2022.
https://doi.org/10.1007/978-3-031-22695-3_22

number of transactions containing the consequent. Rules can also be measured by their *lift*, which is the ratio of the confidence of the rule and the expected confidence of the rule.

Once a list of association rules has been mined, the traditional approach to interrogating these rules involves copious manual labour. Data scientists or domain experts typically sort rules by confidence or some other measure (such as support or lift) in order to find "useful" rules that can inform decision making. However, if one is interested in searching for rules pertaining to particular items, one must manually search through the list of all rules to find such rules [9]. This can be prohibitively time-intensive on large datasets with many item types. Furthermore, a list of association rules does not capture the concept hierarchy - for example, a rule stating that `bread` is commonly associated with `milk` would not appear in a search for all rules pertaining to `drinks`, despite `milk` being a subclass of `drink`.

Several researchers have addressed this issue by introducing methods for visualising association rules. The most recent of these techniques typically involve graph visualisations that are generated from the list of association rules [6,9]. While these techniques improve the interpretability of the association rules, they do not inherently change the way the association rules are stored, and thus do not provide any means to *query* the rules outside of the visualisation itself.

Another growing area of research in the knowledge discovery space is the construction of knowledge graphs from unstructured text. This involves the automatic extraction of entities and relationships in unstructured text in order to build a knowledge graph [13,20]. Despite the fact that entities appearing within documents naturally form itemsets, existing association rule mining techniques are not designed with these knowledge graphs in mind. One must therefore write purpose-built scripts to extract item sets from the knowledge graph in order to run association rule mining, and there is no natural cohesion between the mined list of association rules and the original knowledge graph.

In this paper we aim to bridge the gap between association rule mining and graphs by introducing QUARRY, a graph model for storing association rules alongside transactional data. QUARRY enables association rules to be easily queried via graph queries, allowing for the filtering of rules containing specific items that are of interest to domain experts. The concept hierarchy is attached to the items in each rule, allowing for queries over different levels of the hierarchy. Additionally, rules can be queried based on document properties such as costs and dates. The schema is designed to be compatible with an existing knowledge graph and supports further graph-based visualisations.

This paper is structured as follows. We begin by reviewing prior work in association rule mining and graph-based visualisations for association rules in Sect. 2. We then present our model for storing and querying association rules in Sect. 3. In Sect. 4 we present a case study of our model being applied to maintenance work orders in order to demonstrate the effectiveness and utility of our approach. Finally, we conclude this paper and present an outlook to future work in Sect. 5.

2   Related Work

Association rule mining, also known as frequent pattern mining, is a long-standing task in the data mining space [10]. On transactional (tabular) data, it is often performed using the apriori algorithm [2], which identifies frequent patterns in the data and continuously expands the size of these patterns until they can no longer be expanded. Apriori is often contrasted with the FP-Growth algorithm [11], which uses a tree structure to store frequent patterns.

Another means to generate association rules is to build an association graph, then traverse the graph to generate all large itemsets [25]. The main benefit of this is performance, as it reduces the number of database scans necessary when compared to the Apriori algorithm. Tiwari et al. [23] introduce the FP-Growth-Graph algorithm, an extension of the FP-Growth algorithm that uses a graph rather than a tree to store itemsets.

With the growing popularity of graph databases such as social networks [18] and knowledge graphs constructed from unstructured text [13,20], more recent research into association rule mining has looked at mining frequent patterns from graphs directly. Wang et al. [24] propose the concept of graph-pattern association rules, where the antecedent and consequent of each rule are patterns found in a graph rather than itemsets from a transaction. Fan et al. [5] focus on large graphs, proposing a pruning strategy facilitated by a machine learning model in order to improve performance.

A related area of research in the association rule mining space is the effective visualisation of association rules in order to maximise their interpretability. Visualisation tools are often split into several categories: table-based visualisations, matrix-based visualisations, group matrix-based visualisations, and graphical visualisations [6]. Table visualisations include the educational dashboard presented by Garcia et al. [7] and the Aquila tool [21], which allows for the rolling up and drilling down over rules based on the concept hierarchy. Hasler and Karpienko [9] demonstrate a grouped matrix visualisation technique that groups similar rules together, which is available as an R package [8].

Graphical association rule mining visualisations provide a more interactive display of association rules. The visualisations presented by Fernandez-Basso et al. [6] provide a graph view of association rules which supports fuzzy association rules, where the antecedent or consequent are a collection of fuzzy sets [14]. VisAR [22] filters out large subsets of frequent itemsets in order to display a visualisation of notable rules to the user, also allowing the user to specify items of interest. WiFIsViz [16] similarly displays rules in two-dimensional space, grouping itemsets containing similar terms together. More recently, RadialViz [17] presents rules in an orientation-free visualisation, eliminating the impact of the graph's orientation on the legibility of the rules.

Despite the growing body of research on both graph-based techniques for generating association rules, and graph-based visualisations designed to make association rules more interpretable, to the best of our knowledge no research has yet investigated the potential for *storing* association rules in a graph so that they can be *queried* effectively. Systems that generate association rules using an underlying graph, and visualisation systems, both rely on a list of association rules with some underlying purpose-built storage mechanism. Our research thus does not aim to propose another visualisation system, but rather a graph model for storing association rules that supports a wide range of queries over those rules.

3   Model

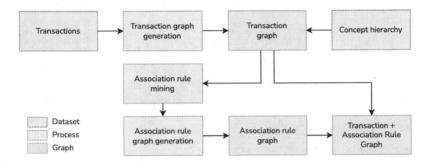

Fig. 1. A block diagram showing how transaction data can be transformed into a queryable transaction + association rule graph (QUARRY).

The goal of QUARRY is to provide a queryable model for association rules. We do not focus on the generation of association rules in this paper; rather, we focus on a way to store association rules after they have been generated such that they can be queried using graph query languages to hasten and improve the process of knowledge discovery.

The overall design of QUARRY is shown in Fig. 1. The input to the model is a list of transactions (sourced from an existing database) and a concept hierarchy. The tabular transaction dataset is first transformed to a "transaction graph", i.e. a graph representation of a list of transactions. If a knowledge graph with the same schema to our transaction graph (such as an entity co-occurrence graph built via a text to knowledge graph pipeline) is already present, this stage can be skipped entirely. Association rules are then mined from this transaction data, and fed into an "association rule graph" which is then merged with the transaction graph. In this section we describe how to build the transaction graph, generate association rules, model those association rules as a graph, and combine the transaction and association rule graphs into a single QUARRY graph that facilitates knowledge discovery.

3.1   Constructing the Transaction Graph

Association rule mining finds patterns appearing in *transactions*, where each transaction t is a row that contains a set of items x. A small example of such

Table 1. An example dataset of transactions.

TID	Itemset
001	Beer, Bread, Eggs, Milk
002	Beer, Bread, Eggs
003	Bread, Eggs, Milk

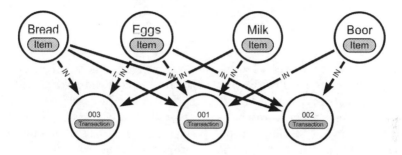

Fig. 2. A translation of Table 1 to a graph schema. Labels have been colour coded.

a dataset with three transactions is shown in Table 1. The first step to building a queryable graph of association rules is to transform this tabular data into a graph. We hereby denote this graph as the "transaction graph" (not to be confused with cryptocurrency transaction graphs), which is a graph representation of a list of transactions.

A graph $G(V, E)$ comprises a set V of vertices and a set E of edges. The edges may be directed or undirected; in our case, QUARRY is a directed graph. A *property graph* allows for each node to have zero or more *properties* (key-value pairs). Nodes may be labelled with one or more *labels*. Edges must always have a start and end node, may have a *type*, and may also contain zero or more properties [3].

Figure 2 depicts a graph schema containing the three itemsets from Table 1. Each transaction is represented as a node with the `Transaction` label. While not shown in this example, the transaction nodes may have properties sourced from the transaction dataset (such as the date of the transaction, the cost, and so on) if required. Each unique item is represented as a node with the `Item` label. An edge is formed between an item x and a transaction t if that x appears in t. These edges are assigned the type `IN`.

3.2   Attaching a Concept Hierarchy

Concept hierarchies are a fundamental concept in data warehousing and are critical in supporting the ability to roll up and drill down on data. An example concept hierarchy for our example items (`beer`, `bread`, `eggs` and `milk`) is shown in Fig. 3.

Fig. 3. An example concept hierarchy.

There are two main ways to represent concept hierarchies in the transaction graph. Firstly, every level of the hierarchy can be represented as a node, with edges formed between each level and its parent in the hierarchy. Alternatively, every level of the concept hierarchy can be stored as labels on the leaf-level item nodes directly. While storing the concept hierarchy as labels allows for simpler queries, it results in the same piece of information being stored multiple times and does not allow for the modelling of multi-level rules in our graph schema.

In light of this, we opt for the former approach. We create nodes for every level of the concept hierarchy, even if those nodes do not appear directly in a transaction. We then create a CHILD_OF relationship between to each node in the transaction graph to its parent node. In our example, relationships are formed from Bread to Food, Eggs to Food, Beer to Drink, and Milk to Drink, and finally from Drink to Item and Food to Item.

3.3   Building the Association Rule Graph

Once the transactions have been transformed into a transaction graph, the next step is to mine the association rules. The algorithm used to mine the rules is not important - one may use apriori [2] over the original list of transactions for its ubiquity and ease of use, or employ a graph-based algorithm [25] directly on the transaction graph for performance. In our implementation of QUARRY we opt to use apriori for simplicity.

Running the apriori algorithm on the rules from Table 1 yields 27 association rules with confidence > 0.6. A selection of five of these rules is shown below:

- Eggs $\longrightarrow$ Bread (Conf: 1.00, Supp: 1.00)
- Beer, Eggs $\longrightarrow$ Bread (Conf: 1.00, Supp: 0.67)
- Beer, Eggs, Milk $\longrightarrow$ Bread (Conf: 1.00, Supp: 0.33)
- Bread, Eggs $\longrightarrow$ Beer (Conf: 0.67, Supp: 0.67)
- Bread $\longrightarrow$ Milk (Conf: 0.67, Supp: 0.67)

We then take these rules and transform them into the association rule graph. Unlike existing work in building graphs from association rules [23,25], QUARRY is not using the graph to speed up the computation of the association rules but is instead modelling rules as a graph so that they be queried using a graph query language. Our schema is thus designed to be easy to write queries

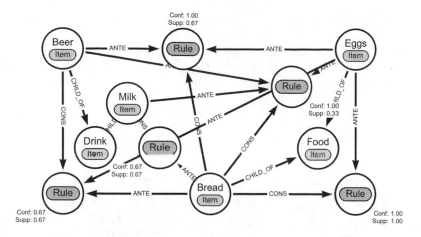

Fig. 4. A translation of five example association rules to the association rule graph schema of QUARRY. ANTE relationships represent antecedents, while CONS relationships represent consequents. We have also added higher-level concepts in the concept hierarchy (the top level is omitted for brevity).

for. An example of the schema of the association rule graph is shown in Fig. 4. Our process for building the association rule graph from a list of transactions is as follows:

1. For each unique Item in the dataset, create a node with the label Item.
2. For each Item with a parent, create a relationship from that item to their parent Item via the CHILD_OF relationship.
3. For each rule, create a node with the label Rule.
4. For each rule, create a relationship of type ANTE between each corresponding Item node in the antecedent (i.e. the left hand side) and the Rule node. Do the same for the consequent (the right hand side), linking each Item to each Rule via a CONS relationship.
5. Assign the Conf (confidence) and Supp (support) properties to the Rule node (optionally also the Lift).

This schema supports multiple level association rule mining [25], which allows for rules to be formed between different levels in the concept hierarchy. For example, Drink could predict Bread with a certain level of confidence. Figure 4 shows nodes for Drink and Food, which are linked to their children via the CHILD_OF relationship. If these items appeared in any rules, they would be linked to those rules the same way as the lower-level items. Pathfinding graph queries can then be used to discover all rules containing items from a certain branch in the hierarchy. This idea is discussed and exemplified in further detail in Sect. 4.

3.4   Combining the Transaction Graph and Association Rule Graph

The final stage of QUARRY combines the transaction graph with the association rule graph. This allows us to query the graph at the item/entity level while also being able to query association rules.

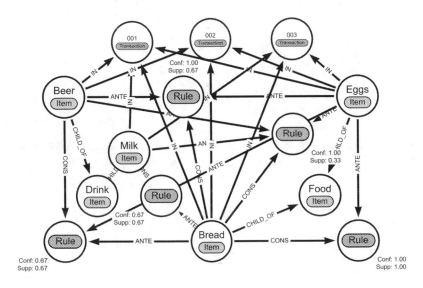

Fig. 5. The final graph, which is a combination of the Transaction and Association graphs.

Connecting the two graphs is relatively straightforward; all that needs to be done is to merge the graphs on the Item nodes. Each Item now has up to four types of outgoing relationships: IN to denote which transactions it is contained in, ANTE to denote in which rules it appears as the antecedent, and CONS to denote in which rules it appears as the consequent. Items that are children of other items also have an outgoing CHILD_OF relationship. The final graph is shown in Fig. 5.

4   Case Study - Maintenance Work Orders

In this section we aim to demonstrate the ability for QUARRY to save vast amounts of time manually searching through association rules by allowing users to query for specific rules of interest. We aim to accomplish this through a case study on maintenance data. Maintenance work orders are an abundant source of rich information capturing work that must be performed on engineering assets such as pumps and compressors [4]. Work orders are of particular interest to reliability engineers who work proactively to improve maintenance strategies.

Our prior work [19] has presented a technical language processing-based model for constructing a knowledge graph from maintenance data. In this case study we now show how to incorporate association rules into this knowledge graph using the method described in Sect. 3. We begin by providing an overview of the dataset, describing our implementation of the association rule mining and graph, and finally demonstrate and discuss a range of queries that are made possible by the QUARRY graph schema.

4.1   Dataset Description

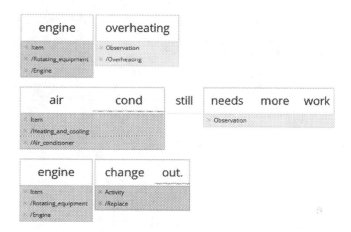

Fig. 6. Three example tagged work orders from the maintenance work order dataset.

The case study dataset[1] consists of 10,000 maintenance work order transactions [12], hereby denoted as "work orders". Each work order contains an average of 4.5 words describing the work that must be completed, for example "replace air conditioner", "fix pump", and so on. The work orders also contain other structured information such as the cost of the work order, the date on which the work was performed, and the type of work order (scheduled or unscheduled). The costs were synthetically generated for demonstration purposes as the original data is anonymised.

The work orders have been automatically labelled using a named entity recognition model in order to identify every entity appearing in the work order. This model was trained using manually annotated data from our research group, the (redacted). Another model was used to automatically extend the singular labels to a maintenance concept hierarchy, so terms such as **engine** are also labelled as Item and Item/Rotating_Equipment. The vast majority of these entity types are sourced from the ISO 15926 asset hierarchy taxonomy [15]. Any entities

---

[1] The dataset and source code of QUARRY is available on GitHub.

that could not be successfully placed into the hierarchy were not included in
the association rule mining. Further details of how the named entity recognition
and hierarchy expansion tasks were performed is available in the corresponding
paper [19].

Figure 6 shows three example work orders that have been tagged using named
entity recognition and the hierarchy expansion model. Itemsets were constructed
from the leaf-level entity labels of each work order in order to build the Transaction Graph. In total, there are 10,000 transactions in the graph, and a total
of 281 entity types, 207 of which are leaf-level and 12 of which are root-level.

4.2   Association Rule Mining and Graph Implementation

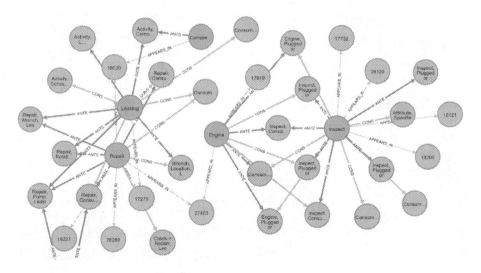

Fig. 7. A subgraph of the combined transaction and association rule graph. Yellow
nodes are transactions, green are entities, and brown are association rules. The values
written on the transaction nodes are their ids. (Color figure online)

Multiple-level association rule mining was performed on these transactions with a
minimum confidence threshold of 0.5 and minimum support threshold of 0.0005,
which yielded a total of 1,086 rules. These rules were then used to construct
the Association Rule Graph, and the two graphs were merged to form the final
QUARRY graph. A subset of the final graph is shown in Fig. 7.

The code for running the association rule mining and generating the graph
is written in Python. The graph itself is stored in Neo4j[2], and the queries in
the following section are written in Cypher. We do not cover Cypher in detail
in this paper, but instead encourage readers to visit the Neo4j Cypher manual[3]
for more information on the syntax.

---

[2] https://neo4j.com.
[3] https://neo4j.com/docs/cypher-manual/current/.

4.3   Querying Association Rules Using QUARRY

QUARRY's graph model allows for the querying of association rules using graph queries. The following queries were written for the maintenance work order database and showcase the types of queries the schema facilitates.

```
MATCH (e1:Entity)-[c:CONS]->
    (r:Rule)<-[a:ANTE]-(e2:Entity)
WHERE r.confidence > 0.8
AND e1.name = "Observation/Leaking"
RETURN r.name, r.confidence, r.support, r.lift
ORDER BY r.lift DESCENDING
```

Listing 1.1. Querying association rules where a certain entity is in the consequent.

The first query, Listing 1.1, aims to find all rules containing the entity Observation/Leaking in the consequent. The WHERE clause ensures that only rules exceeding a confidence value of 0.8 are returned. A selection of three rules yielded by the first query are displayed in Table 2.

Table 2. A selection of rules returned by the query in Listing 1.1. Note the full name of the entity (e.g. "Observation/Leaking") has been shortened to the most granular label (e.g. "Leaking") for brevity.

Rule	Conf.	Supp.	Lift
Repair, Consumable, Wrench $\longrightarrow$ Location, Leaking	1.0	0.0007	21.05
Consumable, Seal, Location $\longrightarrow$ Leaking	1.0	0.0007	10.32
Repair, Consumable, Pump $\longrightarrow$ Leaking	0.89	0.0008	9.03
Consumable, Pipe $\longrightarrow$ Leaking	0.87	0.0013	8.94
Activity, Consumable, Compressor, Location $\longrightarrow$ Leaking	0.83	0.0005	8.60

This query is particularly valuable to engineers as it searches for rules pertaining to a specific failure mode, i.e. leaks. Without using this query, the user must manually search through 1,084 rules to find rules pertaining to leaks. The query returned a total of 57 results, considerably reducing the search time. By adding another conditional to the where statement, such as AND e2.name ="Activity/Repair", the user can drill down on these results to quickly find all association rules where a repair has been performed due to a leak. Virtually any combination of entities appearing in the antecedent and consequent can be included in the search using this method.

```
MATCH (e3:Entity)<-[*0..3]-(e1:Entity)
  -[c:CONS]->(r:Rule)<-[a:ANTE]-(e2:Entity)
WHERE r.confidence > 0.8
AND e1.name = "Observation/Leaking"
RETURN r.name, r.confidence, r.support, r.lift
ORDER BY r.lift DESCENDING
```

Listing 1.2. Querying association rules across multiple levels (up to 3 levels) of the hierarchy.

The second query, Listing 1.2, showcases the ability to use pathfinding to find rules across different levels of the hierarchy. Here, the query is searching for all rules where any type of `Observation` is present in the consequent. The path matching syntax (`[*0..3]`) matches graph patterns up to three levels, thus every entity that is a descendant of the `Observation` class will be matched. In this particular case study it means that engineers can search for rules pertaining to any type of failure mode (i.e. `Observations`), as well as the `Observation` class itself, which is important as the class may appear in multiple-level association rules.

```
MATCH (t:Transaction)<-[:APPEARS_IN]-(e:Entity)-->(r:Rule)
WHERE t.cost > 99000
RETURN DISTINCT r.name, r.confidence, r.support, r.lift
ORDER BY r.lift DESCENDING
```

Listing 1.3. Querying association rules based on transaction properties.

The third query, Listing 1.3, demonstrates the potential to query over the association rules using the properties stored within the transactions. Here we are searching for all rules where any entity in the consequent or antecedent appears in any work order with a total cost greater than $99,000. This query shows the importance of merging the Transaction Graph and Association Rule Graph - without doing so, it would be impossible to query rules based on transaction properties. This particular query is notable in our case study as association rules about items with high replacement costs are generally more important to reliability engineers.

It is possible to also extend this rule to ensure some minimum support threshold for the entities appearing in each transaction. For example, the query could be adjusted to ensure that the entities appearing in the rule appear in at least 10 work orders with a cost greater than $99,000, or that at least some percentage of transactions in which that entity appears has a cost greater than $99,000.

Overall the three queries demonstrate the ability for QUARRYto facilitate knowledge discovery that was not previously possible. Storing association rules in a graph alongside the transactional data provides unprecedented control over the rule mining process, empowering end users to search for and visualise rules based on their domain knowledge. The three queries demonstrated in this paper are only a small sample of the potential queries made possible by the graph schema.

5   Conclusion

In this paper we have presented QUARRY, a graph model for storing association rules alongside transactional data. QUARRY enables a wide range of queries to be performed on association rules, eliminating the need to search through a list to find rules of interest to domain experts. The model transforms AI algorithmic outputs into easy to consume insights for domain experts. We have demonstrated the effectiveness of QUARRY through a case study on maintenance work orders. We have also shown how the schema allows for rules to be queried based on document properties such as cost, dates, and so on, enhancing the process of knowledge discovery. The graph schema is designed such that it can be integrated with an existing knowledge graph.

In future we aim to investigate ways to use the QUARRY transaction graph to support the generation of association rules, rather than running Apriori to build the graph initially. We are also working on a method to run the association rule mining whenever the transaction graph receives new or updated data in order to eliminate the need to run association rule mining over the entire dataset whenever the data changes.

Acknowledgments. This research is supported by the Australian Research Council through the Centre for Transforming Maintenance through Data Science (grant number IC180100030), funded by the Australian Government.

References

1. Agrawal, R., Imieliński, T., Swami, A.: Mining association rules between sets of items in large databases. In: Proceedings of the 1993 ACM SIGMOD International Conference on Management of Data, pp. 207–216 (1993)
2. Agrawal, R., Srikant, R., et al.: Fast algorithms for mining association rules. In: Proceedings 20th International Conference Very Large Data Bases, VLDB, vol. 1215, pp. 487–499. Citeseer (1994)
3. Angles, R.: The property graph database model. In: AMW (2018)
4. Brundage, M.P., Sexton, T., Hodkiewicz, M., Dima, A., Lukens, S.: Technical language processing: unlocking maintenance knowledge. Manuf. Lett. **27**, 42–46 (2021)
5. Fan, W., Fu, W., Jin, R., Lu, P., Tian, C.: Discovering association rules from big graphs. Proc. VLDB Endow. **15**(7), 1479–1492 (2022)
6. Fernandez-Basso, C., Ruiz, M.D., Delgado, M., Martin-Bautista, M.J.: A comparative analysis of tools for visualizing association rules: a proposal for visualising fuzzy association rules. In: 11th Conference of the European Society for Fuzzy Logic and Technology (EUSFLAT 2019), pp. 520–527. Atlantis Press (2019)
7. García, E., Romero, C., Ventura, S., De Castro, C.: A collaborative educational association rule mining tool. Internet High. Educ. **14**(2), 77–88 (2011)
8. Hahsler, M., Chelluboina, S.: Visualizing association rules: introduction to the r-extension package arulesviz. R Proj. Module **6**, 223–238 (2011)
9. Hahsler, M., Karpienko, R.: Visualizing association rules in hierarchical groups. J. Bus. Econ. **87**(3), 317–335 (2017)
10. Han, J., Pei, J., Tong, H.: Data Mining: Concepts and Techniques. Morgan kaufmann, Burlington (2022)

11. Han, J., Pei, J., Yin, Y.: Mining frequent patterns without candidate generation. ACM Sigmod Rec. **29**(2), 1–12 (2000)
12. Ho, M.: A shared reliability database for mobile mining equipment. Ph.D. thesis, University of Western Australia (2015)
13. Kertkeidkachorn, N., Ichise, R.: T2kg: an end-to-end system for creating knowledge graph from unstructured text. In: Workshops at the Thirty-First AAAI Conference on Artificial Intelligence (2017)
14. Kuok, C.M., Fu, A., Wong, M.H.: Mining fuzzy association rules in databases. ACM Sigmod Rec. **27**(1), 41–46 (1998)
15. Leal, D.: ISO 15926 "Life cycle data for process plant": an overview. Oil & Gas Sci. Technol. **60**(4), 629–637 (2005)
16. Leung, C.K.S., Irani, P.P., Carmichael, C.L.: Wifisviz: effective visualization of frequent itemsets. In: 2008 Eighth IEEE International Conference on Data Mining, pp. 875–880. IEEE (2008)
17. Leung, C.K.-S., Jiang, F.: RadialViz: an orientation-free frequent pattern visualizer. In: Tan, P.-N., Chawla, S., Ho, C.K., Bailey, J. (eds.) PAKDD 2012. LNCS (LNAI), vol. 7302, pp. 322–334. Springer, Heidelberg (2012). https://doi.org/10.1007/978-3-642-30220-6_27
18. Nettleton, D.F.: Data mining of social networks represented as graphs. Comput. Sci. Rev. **7**, 1–34 (2013)
19. Stewart, M., Hodkiewicz, M., Liu, W., French, T.: Mwo2kg and echidna: constructing and exploring knowledge graphs from maintenance data. Proc. Inst. Mech. Engineers, Part O: J. Risk Reliabil. (2022)
20. Stewart, M., Liu, W.: Seq2kg: an end-to-end neural model for domain agnostic knowledge graph (not text graph) construction from text. In: Proceedings of the International Conference on Principles of Knowledge Representation and Reasoning, vol. 17, pp. 748–757 (2020)
21. Stewart, M., Liu, W., Cardell-Oliver, R., Griffin, M.: An interactive web-based toolset for knowledge discovery from short text log data. In: Cong, G., Peng, W.-C., Zhang, W.E., Li, C., Sun, A. (eds.) ADMA 2017. LNCS (LNAI), vol. 10604, pp. 853–858. Springer, Cham (2017). https://doi.org/10.1007/978-3-319-69179-4_61
22. Techapichetvanich, K., Datta, A.: VisAR?: a new technique for visualizing mined association rules. In: Li, X., Wang, S., Dong, Z.Y. (eds.) ADMA 2005. LNCS (LNAI), vol. 3584, pp. 88–95. Springer, Heidelberg (2005). https://doi.org/10.1007/11527503_12
23. Tiwari, V., Tiwari, V., Gupta, S., Tiwari, R.: Association rule mining: a graph based approach for mining frequent itemsets. In: 2010 International Conference on Networking and Information Technology, pp. 309–313. IEEE (2010)
24. Wang, X., Xu, Y., Zhan, H.: Extending association rules with graph patterns. Expert Syst. Appl. **141**, 112897 (2020)
25. Yen, S.J., Chen, A.L.P.: A graph-based approach for discovering various types of association rules. IEEE Trans. Knowl. Data Eng. **13**(5), 839–845 (2001)

Using Context-Free Grammar to Generate Synthetic Technical Short Texts

Tyler Bikaun[1]([✉])(iD), Michael Stewart[1](iD), and Melinda Hodkiewicz[2](iD)

[1] Department of Computer Science and Software Engineering, University of Western Australia, Perth 6009, Australia
tyler.bikaun@research.uwa.edu.au
[2] Department of Engineering, University of Western Australia, Perth 6009, Australia

Abstract. Valuable technical information are buried in the under-utilised, user-generated technical texts in engineering domains, such as manufacturing, logistics and maintenance. For maintenance and reliability personnel, the unstructured technical text in maintenance work orders (MWO) hold crucial information about failures and work performed on physical assets. However, the domain-specific language used and scarcity of shared labelled data sets in these contexts present formidable challenges to contemporary natural language processing (NLP) techniques, resulting in inability to achieve performance similar to those in non-engineering domains. In this work, we explore the structure of language in technical short texts by learning a context-free grammar (CFG) through unsupervised grammar induction on industrial MWO texts. We exploit the grammar's generative properties for novel sentence generation and corpus construction and assess its viability for developing synthetic MWO data sets. The results demonstrate a) there exists a grammar in the MWOs, b) the grammar was able to model aspects of the maintenance technical language to produce 12k of synthetic MWO texts 93% as natural and 87% as correct as real texts, and c) the domain-specific language used in technical short text remains challenging to parse due to low data quality and sparsity. Contributions of this work include baseline results for a grammar-based synthetic technical text generation and an appreciation for challenges in assessing the engineering correctness and naturalness of the new synthetic texts.

Keywords: Technical language processing · Maintenance · Language generation

1 Introduction

Large-scale deep neural networks are currently synonymous with state-of-the-art performance in the majority of natural language processing (NLP) tasks. Pre-trained language models take advantage of hundreds of millions of open-domain training examples derived from sources such as the English Wikipedia, Toronto

H. Aziz et al. (Eds.): AI 2022, LNAI 13728, pp. 325–338, 2022.
https://doi.org/10.1007/978-3-031-22695-3_23

BookCorpus, and Common Crawl Corpus to learn powerful representations over sequences of words.

In specialised technical domains such as industrial engineering and asset maintenance, there are only a handful of small-scale data sets currently in existence [3]. This is a result of most engineering and asset-related data being closely guarded by the companies that design, own or operate the assets. The consequence of this data bottleneck is that practitioners are unable to build shared, large-scale, pre-trained industrial language models that domain-specific tasks can benefit from due to insufficient training data. Although there has been initial attempts at fine-tuning open-domain language models on industrial maintenance data [20], it is a worthwhile task to be able to exploit the structure of technical language to produce synthetic data sets.

The structure of language can be modelled through the use of language models such as grammars or neural networks. Though neural networks are powerful, they are often unexplainable black boxes. In contrast, a grammar is more easily interpreted, and thus is valuable for gaining insight into a language. Language that is used in the context of a technical domain is referred to as technical language, and has a particular style [3,6]. Key characteristics of a technical language include a large vocabulary of technical jargon, domain specific terms, acronyms, abbreviations, and erroneous spelling [14]. In addition, the grammatical structure of the texts is often different from natural language where functional words are typically omitted due to length constraints [6]. These characteristics of technical language make it distinct from natural language, resulting in poor performance by contemporary NLP techniques in, for example, the maintenance domain [1,3,6].

In the absence of readily available and large data sets we look to generate synthetic data to build new or augment existing sets. More concretely, we focus on the application of grammar learning to technical language generation. Thus, we ask the questions:

1. What does a grammar learnt from technical language look like?
2. Can we use a learnt grammar to generate synthetic texts that are natural and correct?

2 Background

2.1 What Is a Grammar?

A grammar is a language model. A language can be defined as a set of expressions, which is usually infinite. A language has an alphabet, the characters of which can be combined together to form words. Additionally, the words can be combined together to form sentences. Specific languages have constraints. These take the form of the alphabet allowed, and rules imposed on it. For example, the English language has an alphabet consisting of the 26 characters a ... z (and the capital letters, numbers, and symbols), and it is known which constructions from the alphabet form valid English expressions. "Pump", for example, is a valid English word, but "pmp" is not. Likewise, "the pump is working" is a valid English expression, but "pump is the working" is not. We can say that English is governed by "rules" that define what kinds of expressions are valid. To fluent English speakers, it is quickly decipherable which expressions are valid and which are not, but to quantify these rules is not a trivial task. This is the task of a grammar.

A grammar models a language through a set of *production rules*, of the form *left hand side → right hand side*. Grammar is commonly described using a four-tuple (Σ, V, R, S) [15] as follows:

- Σ is a finite set of characters that constitute the alphabet of the language. As mentioned earlier, the characters combine to form words, referred to as *terminals*.
- V is a finite set of symbols (variables), also known as *non-terminals* (NTs). They represent one or more terminals, and may themselves be represented by other NTs. Capital letters are often used to denote NTs.
- R is a finite set of rules. They are also known as *production rules*. They take the form of X → Y, where X and Y are strings of terminals and non-terminals. Specific types of grammars have additional constraints imposed on the X and Y strings. The rules define the sentences that are valid in the language.
- S is a special NT, the *starting symbol*. This is the symbol that each sentence must start from.

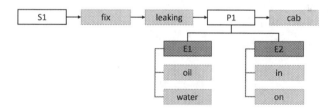

Fig. 1. Example grammatical structure of a technical text.

Consider the example in Fig. 1. Let us define the grammar that could generate sentences such as "fix leaking water in cab" and "fix leaking oil on cab". We need the start variable S, non-terminals, terminals, and rules that combine them all. For this example, these are:

- Σ: {fix, leaking, oil, water, on, in, cab}
- S: S1
- V: {P1, E1, E2}
- T: {{oil, water}, {on, in}}
- R: {P1 →{E1, E2}, E1 →{oil, water}, E2 →{on, in}}

The set of sentences that can be derived from a grammar constitutes the language of the grammar. Given a sentence, we can determine whether the grammar can successfully derive the sentence, and if so, then we say that the grammar can *parse* the sentence (i.e. break down the sentence into smaller units and assign them to meaningful categories). In this way, we can determine if a sentence is part of the language of the grammar, and we can also determine the grammatical structure of sentences that can be parsed.

2.2   Types of Grammars

There is a hierarchy of grammars, first described by Chomsky [2], of which two types are of interest: the context free grammar (CFG), and the context sensitive grammar. A CFG is called such because the left hand side of its rules must be context free - it consists of one NT only. That is, the rules must be of the form A $\rightarrow \gamma$, where A is a NT, and γ is a string of NTs and terminals. A context sensitive grammar has a more relaxed requirement, in that the left hand side of its rules may contain more than one symbol. CFGs have been the most extensively studied grammar in the Chomsky hierarchy, as shown by the survey performed by D'Ulizia et al. [4]. The characteristics of importance when choosing a type of grammar are the power of the grammar in capturing language, and the complexity of parsing a sentence using the grammar. CFGs can capture much of natural language, but not all of it.

Three methods for learning a CFG were considered: ADIOS (Automatic DIstillation Of Structure), eg-GRIDS (based on the GRammar Induction Driven by Simplicity algorithm), and wGCS (weighted grammar-based classification system). All are unsupervised algorithms. ADIOS, developed in 2005, searches for significant patterns and equivalence classes, based on a statistical measure of importance [16]. eg-GRIDS [13] developed in 2004 is based on the GRIDS algorithm developed by [10]. wGCS, a more recent development by [19] in 2020 uses the genetic operators of crossover and mutation. ADIOS iterates through all subpaths in its graph. eg-GRIDS and wGCS are both genetic algorithms. ADIOS uses a statistical measure and greedy approach to update its grammar whereas eg-GRIDS works to minimise a heuristic called minimum description length at each iteration. wGCS uses the weights of its rules based on how many times a particular rule is used when parsing the example sentences as its heuristic to update its grammar. Both ADIOS and eg-GRIDS take only positive examples as input, while wGCS takes in both positive and negative examples.

To the best of our knowledge none of these three grammar-learning algorithms have been tested on technical short texts. ADIOS has an advantage in that it has previously been applied on artificial grammar data [16] and on natural-language corpora such as ATIS [8] and CHILDES [11]. eg-GRIDS has only been tested on the artificial Dyck language with limited sentences generated that have lengths longer ($\approx$20 tokens) than technical short texts in MWOs that are extremely terse ($\approx$5 tokens) [12]. wCGS has, thus far, only been tested on three synthetic context free languages [19]. Two aspects of technical short text influenced our decision to use ADIOS, instead of eg-GRIDS and wGCS. The first is that there are clear patterns in the texts, the presence of these favours the pattern matching approach of ADIOS, the second is that the texts contain a lot of abbreviations. For example, "air conditioner" may be abbreviated to "air cond.", or "air con", or "a/c". These abbreviations are likely to be accommodated by the equivalence classes approach in ADIOS, where an equivalence class (E) is a set of interchangeable nodes within a pattern (P), e.g. P = repair E filter, E = {oil, fuel}.

3   Experiments and Results

In this section, we perform grammar induction using the ADIOS algorithm on a real industrial data set (Table 1). The data set is derived from user-generated texts in the unstructured free-text field of MWOs within the context of heavy mobile mining equipment [9]. The short-text field in MWO's are terse with only 5 words in length on average [1,3,6].

Table 1. Overview and statistics of industrial data set.

	All	Train	Test
Text samples	55k	49.5k	5.5k
Vocabulary size	13.2k	12.4k	3.6k
Hapaxes (token freq. = 1)	7.6k (58%)	7.1k (58%)	2.1k (56%)
Text Lengths	$\mu = 5.0, \sigma = 1.4$		

A detailed discussion of the ADIOS algorithm is beyond the scope of this paper and we refer interested readers to [16] for a thorough explanation. Due to unavailability of the ADIOS codebase, ModifiedADIOS, publicly available at https://github.com/shaobohou/madios, was used for all experiments. After running MADIOS, an evaluation of the induced grammar was performed by considering three main aspects - i) quality of rules produced, ii) generation of novel texts with human evaluation, and iii) ability to parse (*reduce into predefined grammar rules*) unseen technical texts.

3.1   Pre-processing

Before performing grammar induction, the following pre-processing of the technical texts was performed:

- Masking unique equipment identifiers with the special token *itemid*
- Masking manufacturer names with the phrase *manufacture n* where n is enumerated based on the number of unique manufactures,
- Removing special characters except for _, -, &, # and / (these preserve conjunctions, etc.),
- Removing casing, and
- Adding MADIOS specific delimiters (* and #)

3.2   Grammar Induction

To train MADIOS to learn a grammar, the 49.5k training texts were used. Four trials of the MADIOS algorithm were performed with varying parameterisations to explore the affinity of MADIOS to produce patterns and equivalence classes (Table 2). Each grammar took between 5–10 hours to be induced, depending on the parameters used. The *default* parameters were those suggested by MADIOS. Significance test threshold (α) and context size (L) were modified to enhance pattern production and limit over-generalisation in equivalence classes. Multiple grammars were explored to gain insight into the algorithms suitability for grammar induction from technical language texts. The context size was adjusted relative to the mean length of texts in the training corpus (Table 1). An overview of the trials performed is provided in Table 2.

Table 2. Results of parameter tuning (η - divergence threshold, α - significance test threshold, L - context size, ω - coverage, **P** - production patterns, **EC** - equivalence classes, **R** - texts in corpus reduced by grammar).

Trial	η	α	L	ω	P	EC	R
1 *(default)*	0.9	0.01	5	0.65	303	43	39%
2	0.9	0.05	3	0.65	463	329	69%
3	0.9	0.1	4	0.65	766	415	59%
4	0.9	0.05	4	0.65	546	309	52%

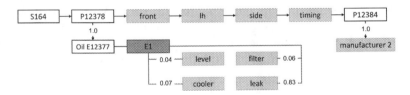

Fig. 2. Simple example of reduced text derived from *Trial 4* grammar with edge weights.

3.3   Synthetic Text Generation

The grammar produced in *Trial 4* was used to generate explainable novel technical texts. This grammar was selected as its production patterns and equivalence classes exhibited high quality groupings and plausible relationships between rules upon human evaluation. To generate novel texts, firstly texts in the corpus reduced by the grammar were identified, consisting of 25.7k texts (52% of the corpus). Secondly, the patterns and equivalence classes in the reduced texts were probablistically filled until terminals were reached by sampling from a distribution weighted with the edge weights produced by the CFG. An example of a reduced text with its weights is shown in Fig. 2. Here, it can be seen that there is an 83% chance of filling the equivalence class *E12377* within the pattern *P12378* with the terminal "leak" and so forth.

To evaluate how well the grammar can produce texts with technical language, human evaluation (using experienced engineers) was performed using a 5-point considering *naturalness* from very unnatural to very natural, and *correctness* from technically incorrect to technically correct, on a Likert scale. The two attributes are described as:

- Naturalness. *Is the text coherent and could it have been plausibly created by a domain expert?*
- Correctness: *Does the text make technical sense e.g. are the casual and hierarchical relationships and/or interactions between entities feasible?*

These attributes are analogous to those used in common domain NLG [7], however they differ to reflect the language generation process in technical settings such as industrial maintenance. For example, the notion of *naturalness* differs from the conventional sense that is typically assessed on syntactical, semantic and grammatical correctness of a generated text. In the case of TLG that is learnt from characteristically noisy source text [3,6], *naturalness* must be relative to the native form of texts, e.g. a level of noise including abbreviations, common misspelling and so forth must be tolerated. Similarly, *correctness* differs due to the necessity in technical domains to precisely represent causal and hierarchical relationships and/or interactions between entities either explicitly or implicitly stated in the text. For example, consider the two example texts that highlight this necessity - "oil blowing" and "change out engine in piston". The former exhibits an incorrect causal relationship as oil cannot blow. Whereas the latter misrepresents the hierarchical relationship between the two items *engine* and *piston*.

Table 3. Overall evaluation for real and novel texts.

Attribute	Real	Novel
Naturalness	4.72/5	4.41/5
Correctness	4.42/5	3.85/5

Using the reduced texts, 12k novel technical texts were generated. To determine whether the grammar can produce natural and correct texts, a set of 100 real texts and 100 novel texts were sampled *i.i.d*, pooled and shuffled before human evaluation. This mixing was performed to establish a fair benchmark to evaluate the novel texts with respect to the attributes of real texts. The results of the evaluation are presented in Table 3. It was found that the novel texts were 93% as natural and 87% as correct as the their real counterparts. A selection of evaluated novel and real texts is provided in Table 4. Moreover, we release the 12k novel texts to the public at https://github.com/nlp-tlp/cfg_technical_short_text.

Table 4. Samples of novel and real texts with corresponding annotator agreement. Highlighted rows indicate novel texts.

	Naturalness	Correctness
Replace steering outer bucket tooth & weloed	1.0	1.0
Under cab cut out	2.0	1.67
Air o-ring on hot and cold	2.33	2.0
Fit out r a r belt with tools	3.33	1.67
Tread depths march	2.67	3.0
Fit temp sender to slew pumps 1&3 u/s	3.33	3.33
Engine alarm on fault	5.0	5.0
Main lube line wont build pressure	5.0	5.0
Movement on bucket frame pins/bushes	5.0	5.0
Crankcase pressure high fault	5.0	5.0
Oil leak behind control valves	5.0	5.0
r/h rear mudflap needs to be replaced	5.0	5.0

Table 5. Examples of unsuccessful (grey) and successful texts parsed by the CFG from *Trial 4*.

Repair hydraulic system leaks	Replace all grease injectors on saddle block
Replace damaged trans filter housing	l/h front strut leaking oil
Replace lh engine ac oil pressure switch	Re torque top cover bolts on rear swing gear case

3.4   Parsing Technical Text

To understand how well the grammars can recognise technical language, the Earley parser [5] was used. The Earley parser is a fast, chart-based, unrestricted context-free language parser implemented with dynamic programming. For each trial, the parser was loaded with the learnt grammar and attempts at parsing unseen texts in the test set was performed. Here, the metric recall was used as defined by [16]. This metric is equal to the proportion of C_{test} that is accepted by the grammar induced from $C_{training}$ where the corpus of synthetic texts C is split into two sets $C_{training}$ and C_{test}. Overall, the recall of the grammars was universally very low ($\approx 1\%$). A set of unsuccessful and successful parsed examples is shown in Table 5.

4   Discussion

In this section, we try to answer the following two questions:

1. What does a grammar learnt from technical language look like?
2. Can we use grammar to generate synthetic texts that are natural and correct?

4.1 A Grammar Learnt from Technical Language

All four grammars had a significant number of starting rules ($\mu = 47.3$k, $\sigma = 1$k), nearly a unique rule for each text in the corpus. The size of starting rules is attributed to a significant sized vocabulary (12.4k unique tokens) and number of hapaxes (58% of vocabulary) derived from the training corpus. A review of the starting rules indicated that many texts were not being reduced by the grammar and were simply representing the input text explicitly. For example, consider the two texts and their corresponding reduced forms - "repair/replace carrier roller plates" $\rightarrow$ "repair P12432 P12541 plates" and "clean after cooler after turbo failure" $\rightarrow$ "clean after cooler after turbo failure". The former was reduced successfully by the grammar whereas the latter was not. Factors such as a large vocabularies and significant numbers of hapaxes are not atypical of technical language used in MWOs [3,6] and could be overcome with retrospective data quality improvement strategies such as lexical normalisation [17,18]. Employing normalisation strategies could enable vocabulary and hapax reduction by normalising tokens to their canonical forms, e.g. {lh, l/h, lefth, ..., l/hand, l/h} $\rightarrow$left hand.

Semantic Grouping. An analysis of the equivalence classes produced by the grammars highlights groupings of tokens that fall under similar semantic concepts. Table 6 exhibits the most common semantic groups captured by the grammars equivalence classes. Taking the equivalence class *E12513* in the relative location semantic class in Table 6 as an example, it is evident that a single category of meaning is captured. For example, "r/h/s" $\rightarrow$"right hand side" and "r/h/f" $\rightarrow$"right hand front" both correspond to a (relative) location word. In addition, eight abbreviations were captured. A subset of the concepts is shown in Table 6. It was found that context size of the grammar had a significant influence over the quantity and quality of groups identified.

Common Lexical Units. Exploration of the production patterns from the grammars highlighted their affinity to capture common lexical units such as bi-grams and tri-grams. This is similar to collocations that find common lexical units based on token co-occurrence frequencies. However, instead of only finding lexical units, the patterns found by ADIOS can nest within equivalence classes and have utility in disambiguation. Table 6 shows examples of common lexical units including those related to *activities* ("change out", "inspect/repair"), *observed states* ("not working"), *events* ("shut down"), *acronyms* ("sos"), and *items* ("torque converter").

Table 6. Overview of conceptual groupings extracted from production patterns and equivalence classes of trial grammars.

Class	Rule(s)
Semantic (relative location)	E12513 →r/h/s |leaking |r/h/f |r/h/r |under |l/h/front |in |from |l/h |rh |lh |l E12555 →access |side |door
Semantic (consumable)	E112466 →oil |water |coolant |fuel |grease |small |air |gas
Semantic (absolute time)	E12528 →600hr |100hr |of |1000hr
Semantic (item)	E12441 →P12387 |pipe |spring |arm |o-ring |valve |P12425 |... |pin
Common Lexical Units	P12381 →['change', 'out'] P12435 →['not', 'working'] P12668 →['shut', 'down']
Abbreviation / Misspellings	E12517 →alternator |... |alternotor |alternater |alernator |altenator |alt/ |alt |ac E12420 →hr |hour

Abbreviations and Common Misspellings. As noted previously, ADIOS is able to identify abbreviations and other variations in word spellings. The equivalence class E12517 in Table 6 demonstrates the extent of lexical variation that ADIOS can capture for the items *air conditioner* and *alternator*.

Parsing Technical Text. Consider the text "repair hyd leak at front of radiator" that was successfully parsed. Figure 3 shows a partial parse tree for this text. This is a complex structure that has a large depth. The starting rule for the text is "S → P950 P1000 radiator", which contains three symbols in the RHS. The text is parsed by being split into three segments, corresponding to the number of segments in the starting rule. However, for the text "pump box temp sensor not working", there was no exact starting rule that matched the structure. One of the closest starting rules was "S → P225 sensor P26", which originated from the text "magic eye sensor not working". The type of sensor was not recognised as an equivalence class in this case, which resulted in a failure to parse.

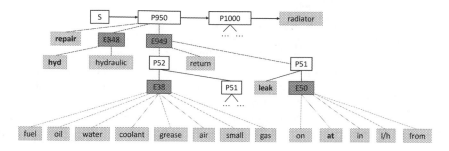

Fig. 3. Partial parse tree for the text "repair hyd leak at front of radiator".

Misspellings are a problem when parsing texts. The text "air in removing drll bit" contains a misspelling ("drll" should be "drill"). If the misspelling had insufficient support while training, then the parse will fail. If a text being parsed contains a new word not present in the training set, the parse will also fail. Technical language contains a lot of jargon, which means that there can be a high percentage of words that only occur a few times [3]. As noted before, in the training set 58% of the words only occur once. In the test set, 717 out of 5500 texts (13%) contained new words. Given that not many new words would be expected in a natural language corpus test set, 13% of the text is a substantial proportion.

4.2   Synthetic Text Generation

Using the grammar from Trial 4, 25.9k texts were generated, 12k which were novel. Human evaluation of 100 novel texts were performed after being randomly mixed with 100 real technical texts. We found that the synthetic texts produced were 93% as natural (representing the conventions of maintainers and engineers in how they communicate) and 87% as technically correct (from an engineering perspective) as their real counterparts. To better understand the performance gap experienced, we inspect the texts under four different conditions, namely those that are 1) natural and correct, 2) natural but not correct, 3) correct but not natural, or 4) neither natural or correct.

By correct, we mean that it is correct statement from an engineering perspective. For example "repair hyd leak" is deemed correct as hydraulic systems do leak. However a statement "replace gas on motor" is not correct as motors do not have gas. By natural, we mean that it is in the style of the MWO's we seek to replicate in terms of syntax and use of abbreviations and jargon.

Of the 100 novel texts evaluated, 24% scored 5/5 on both naturalness and correctness. Examples of these include "blade positioner cylinder has low power when lifting" and "repair crack in left side of swing arm". The reason this text is considered novel is that the trigram "blade position cylinder" was not present in the training set. Although novel, it is also plausible (from an engineering perspective) that a hydraulic component related to the blade on a piece of heavy mobile equipment. The remainder of the text is also technically correct given that hydraulic components can experience low power under lifting operations.

There were 34 novel texts deemed natural but not correct. An example of technically incorrect is "repair hyd mirror brace lhs of cab". Here, the grammatical structure of the technical language is plausible, containing a combination of *activity* ('"repair"), *item* ("hyd mirror brace", "cab", "bolts") , *relative_location* ("lhs") concepts. However, the item "hyd mirror brace" in the former is invalid as it is not an engineering item. Only 6 texts were considered correct and not natural. The text "fit new tyres to pos five and two" is technically correct but the ordering of the positions is unnatural to a human agent who would enumerate in ascending order. Lastly, 36 texts were imperfect in both naturalness and correctness. Poor performers include "replace steering outer bucket tooth & weloed" and "air o-ring on hot and cold". Here, the structure of the phrases are unnatural due to lack of information and/or ordering of lexical units and the interactions and relationships between entities are unconventional and do not represent engineering reality.

In contrast to open-domain NLG evaluation, the notion of "naturalness" and "correctness" has to be adapted to suit the technical text generation process. This meant relaxing constraints on syntactic, semantic and grammatical correctness to match the native style of domain experts. A technical "correctness" score was introduced whereby relationships and/or interactions between entities within the texts were assessed on their engineering feasibility.

5   Conclusion

Motivated by the need to overcome confidentiality challenges and alleviate data scarcity in technical settings, this paper investigated whether a context-free grammar could be learnt from technical language and be used to generate synthetic technical texts. To achieve this, the ADIOS (Automatic DIstillation Of Structure) [16] algorithm was applied to user-generated texts derived from maintenance work order records. Multiple parameterisations of the algorithm were investigated, with the learnt grammars evaluated on their ability to i) produce production rules and equivalence classes, ii) generate natural and correct novel technical texts, and iii) parse (break down into grammar rules) unseen technical texts.

The grammars were found to capture meaningful patterns from technical texts, including conceptual grouping (*activities, relative locations*, etc.), common lexical units, and frequent abbreviations and misspellings. Although, a trade-off between the quality and quantity of patterns produced was observed depending on the parameters used in the ADIOS algorithm. The ability of the grammars to parse unseen technical texts was universally low. This was largely attributed to data quality issues arising from user-generated texts in industrial maintenance, resulting in a significant number of starting rules being created by each grammar.

To demonstrate how context-free grammars could be used to alleviate data scarcity issues, the most promising grammar was used to generate synthetic technical text. A corpus of 12k novel synthetic technical texts were produced, about a quarter the size of the training corpus. Human evaluation of a subset of the texts was performed to assess their naturalness and technical correctness. It was found that the synthetic texts were 93% as natural and 87% as correct as their real counterparts. The contributions of this work are a) the identification of a need for agreed (and improved) measures of naturalness and technical correctness for assessing synthetic texts, b) the challenge with parsing and the opportunity to take the insights from grammar to inform generation of technical texts using state of the art deep learning methods.

Acknowledgements. The team acknowledges the initial work by Laura Peh in her honours thesis to use Adios and generate grammar in 2020. The extension of this work presented in this paper was supported by the Australian Research Council through the Centre for Transforming Maintenance through Data Science (grant number IC180100030), funded by the Australian Government. Additionally, Bikaun acknowledges funding from the Mineral Research Institute of Western Australia.

References

1. Brundage, M.P., Sexton, T., Hodkiewicz, M., Dima, A., Lukens, S.: Technical language processing: unlocking maintenance knowledge. Manuf. Lett. **27**, 42–46 (2021)
2. Chomsky, N.: On certain formal properties of grammars. Inf. Control **2**(2), 137–167 (1959)
3. Dima, A., Lukens, S., Hodkiewicz, M., Sexton, T., Brundage, M.P.: Adapting natural language processing for technical text. Appl. AI Lett. **2**, e33 (2021)
4. D'Ulizia, A., Ferri, F., Grifoni, P.: A survey of grammatical inference methods for natural language learning. Artif. Intell. Rev. **36**(1), 1–27 (2011)
5. Earley, J.: An efficient context-free parsing algorithm. Commun. ACM **13**(2), 94–102 (1970)
6. Gao, Y., Woods, C., Liu, W., French, T., Hodkiewicz, M.: Pipeline for machine reading of unstructured maintenance work order records. In: Proceedings of the 30th European Safety and Reliability Conference/15th Probabilistic Safety Assessment and Management Conference (2020). https://doi.org/10.3850/981-973-0000-00-0
7. Gatt, A., Krahmer, E.: Survey of the state of the art in natural language generation: core tasks, applications and evaluation. J. Artif. Intell. Res. **61**, 65–170 (2018)
8. Hemphill, C.T., Godfrey, J.J., Doddington, G.R.: The ATIS spoken language systems pilot corpus. In: Speech and Natural Language: Proceedings of a Workshop on Speech and Natural Language, pp. 96–101 (1990)
9. Hodkiewicz, M., Ho, M.T.W.: Cleaning historical maintenance work order data for reliability analysis. J. Qual. Maint. Eng. (2016)
10. Langley, P., Stromsten, S.: Learning context-free grammars with a simplicity bias. In: López de Mántaras, R., Plaza, E. (eds.) ECML 2000. LNCS (LNAI), vol. 1810, pp. 220–228. Springer, Heidelberg (2000). https://doi.org/10.1007/3-540-45164-1_23
11. MacWhinney, B., Snow, C.: The child language data exchange system. J. Child Lang. **12**, 271–295 (1985)

12. Petasis, G., Paliouras, G., Karkaletsis, V., Halatsis, C., Spyropoulos, C.D.: e-GRIDS: computationally efficient gramatical inference from positive examples. Grammars **7**, 69–110 (2004)
13. Petasis, G., Paliouras, G., Spyropoulos, C.D., Halatsis, C.: eg-GRIDS: context-free grammatical inference from positive examples using genetic search. In: Paliouras, G., Sakakibara, Y. (eds.) ICGI 2004. LNCS (LNAI), vol. 3264, pp. 223–234. Springer, Heidelberg (2004). https://doi.org/10.1007/978-3-540-30195-0_20
14. Sexton, T., Hodkiewicz, M., Brundage, M.P., Smoker, T.: Benchmarking for keyword extraction methodologies in maintenance work orders. In: PHM Society Conference, vol. 10 (2018)
15. Sipser, M.: Introduction to the Theory of Computation, vol. 27. ACM, New York (1996)
16. Solan, Z., Horn, D., Ruppin, E., Edelman, S.: Unsupervised learning of natural languages. Proc. Natl. Acad. Sci. **102**(33), 11629–11634 (2005)
17. Stewart, M., Liu, W., Cardell-Oliver, R.: Word-level lexical normalisation using context-dependent embeddings. arXiv preprint arXiv:1911.06172 (2019). https://arxiv.org/pdf/1911.06172.pdf
18. Stewart, M., Liu, W., Cardell-Oliver, R., Wang, R.: Short-text lexical normalisation on industrial log data. In: 2018 IEEE International Conference on Big Knowledge (ICBK), pp. 113–122. IEEE (2018)
19. Unold, O., Gabor, M., Wieczorek, W.: Unsupervised statistical learning of context-free grammar (2020)
20. Usuga Cadavid, J.P., Grabot, B., Lamouri, S., Pellerin, R., Fortin, A.: Valuing free-form text data from maintenance logs through transfer learning with Camem-BERT. Enterp. Inf. Syst., 1–29 (2020)

Predicting Marimba Stickings Using Long Short-Term Memory Neural Networks

Jet Kye Chong[1]([✉]) and Débora Corrêa[2,3]

[1] Conservatorium of Music, The University of Western Australia,
Crawley, WA 6009, Australia
jetkye@jetkyechong.com
[2] Department of Computer Science and Software Engineering,
The University of Western Australia, Crawley, WA 6009, Australia
debora.correa@uwa.edu.au
[3] ARC Centre for Transforming Maintenance Through Data Science,
The University of Western Australia, Crawley, WA 6009, Australia

Abstract. In marimba music, 'stickings' are the choices of mallets used to strike each note. Stickings significantly influence both the physical facility and expressive quality of the music performance. Choosing 'good' stickings and evaluating one's stickings are complex choices, often relying vaguely on trial-and-error. Machine learning (ML) approaches, particularly with advances in sequence-to-sequence techniques, have proved suited for similar complex classification problems, motivating their application in our study. We address the sticking problem by developing Long Short-Term Memory (LSTM) models to generate stickings in 4-mallet marimba music trained on exercises from Leigh Howard Stevens' *Method of Movement for Marimba*. Model performance was measured under a range of metrics to account for multiple sticking possibilities, with LSTM models achieving a maximum average micro-accuracy of 97.3%. Finally, we discuss qualitative observations in sticking predictions and limitations of this study and provide direction for further development in this field.

Keywords: Long short-term memory neural network · Marimba sticking model · Marimba sticking dataset · Music performance

1   Introduction

When a musician is technically proficient at their instrument, subtle technical choices facilitate fluid and musical performance, such as in piano fingering or string bowing. These choices service musicality, technical ease and fluency, and are often the result of a blend of tradition, experience, and tedious trial-and-error [27]. In the context of the marimba, a tuned percussion instrument, the analogous challenge is that of 'sticking'. The modern 5-octave concert marimba comprises a set of 61 wooden bars tuned to produce specific pitches when struck. The bars are laid out in a piano arrangement, but their larger size creates an instrument

H. Aziz et al. (Eds.): AI 2022, LNAI 13728, pp. 339–352, 2022.
https://doi.org/10.1007/978-3-031-22695-3_24

around two and a half metres long and one to two wide deep. Marimba players typically play with four mallets (or 'sticks'), with two held in each hand (Fig. 1). The choice of which mallet is used to strike each bar is called a 'sticking'.

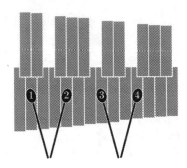

Fig. 1. Four mallets over a section of a marimba enumerated conventionally.

Well-chosen stickings facilitate fluid movements around the instrument and aid the expression of musical phrases, while poor stickings hinder movement and work against expressing musicality. Professional players have proposed general guidelines to sticking, but due to the subjectivity of musicality, different suggestions can be contradictory with no unique and ideal solution. However, advances of machine learning (ML) models for music information retrieval related tasks show precedents for assistance. Deep-learning techniques, including the recurrent neural networks as the Long Short-Term Memory (LSTM) networks, have been used extensively for several musical related tasks (the interested reader is referred to Choi et al. [9] for a tutorial on the field), including, but not limited to, music generation and composition [6,7,12,13,18,19,23], music transcription [37,39], music recommendation [36,42,44], music classification [10,49] and music prediction and modelling [17,24,47].

Current research of marimba stickings is mainly found in path planning algorithms for robotic percussionists. For instance, in Yang et al. [45], human-level musical expressivity of mechanical marimba players is leveraged with musical mechatronics interfaces. In Savery and Weinberg [34], robotic marimba players and melody generation with LSTM neural networks are used for the implementation of a software-based film composer. A comprehensive reference on robotic musicianship and musical path planning is Bretan [5]. In each of these cases, path planning algorithms are implemented to prevent the self-harm of the robot and minimise movement between passages of notes to facilitate music-making. While some similar concerns are shared in marimba sticking techniques, stickings associated with the music itself, and idiomatic stickings used in human performance, are not considered.

The analogues of marimba sticking in other more common instruments, like guitar fretting and piano fingering, have been explored more broadly in previous studies. For the guitar, The Optimum Path Paradigm (OPP) [35] represents a simple and effective approach to minimising movement in musical technique; however, Sayegh acknowledges "more subtle parts of the task [such as musicality and style] might not be possible to capture [with the OPP]." The OPP was further developed in project Robotaba [8] as part of a broader audio-to-tablature algorithm for guitar music transcription. However, neither physical nor musical considerations are meaningfully addressed [4,22]. Other studies have found other approaches minimise the physical movement involved in guitar technique [26,33,46]. Physical limitations remain the primary concern, but other considerations are quantified beyond finger movement and position shifts, such as cognitive factors of reading 'well-written tablature'. Musical and technical factors are explored by Tuohy and Potter [41] who employ numerous genetic algorithm approaches on existing guitar tablature. Another closely related work resembling this study is the TabGen project [25] that tackles sheet-music-to-tablature conversion for guitar. It employs the LSTM to parse music notation and fretting choices as time-series data. Musical issues involved in fretting are also addressed. Subjective evaluation of fretting outputs is conducted alongside quantitative methods, which illuminate approaches for improving conventionally written tablature.

The fingering problem on the piano (also closely analogous to the sticking problem) is the task of finding a suitable fingering given a piano score. A suitable fingering is usually determined utilizing ergonomic, cognitive and music-interpretive constraints [31]. Various approaches have been proposed to tackle this problem, including Hidden Markov Models (HMMs) [29,30,48], dynamic programming and other constraints or cost-based models [1–3,20,31], use of musical rules [40], and a neuro-fuzzy inference system [16].

The similarities between the geometry of the piano and marimba, and the similar concerns of the piano fingering problem with the marimba sticking problem, and a common data-driven approach, motivated us to adopt HMMs as a baseline comparison to ours.

2   The Sticking Problem

Choices of stickings used to play musical passages greatly influence the physical ease or difficulty of performance and the quality of musical interpretation [43]. In a four-mallet marimba performance, the performer holds two mallets in each hand, exerting independent control over each. Following the convention of major marimba texts, this paper refers to the mallets as '1', '2', '3' and '4' from left to right (Fig. 1), represented in sheet music as numbers annotated above or below notes.

While any mallet can be used to strike any individual note, a poorly sequenced set of stickings can cause physical and musical hindrance in practice. Physical hindrances include collisions, strained wrist positions, and tense or repeated rapid arm and wrist movement. These are not always obvious in looking at stickings in sheet music alone. Figure 2a shows a short musical passage with good stickings. The sticking is fairly easily achieved, as the left hand resides on keys away from the body and slightly elevated, leaving space for the right hand to cross under the left and strike the low G on the bar's near edge. Figure 2b shows the same musical passage transposed down two semitones. Now the same sticking no longer works well. Due to the irregular positioning of the bars, now the hands and arms run into one another.

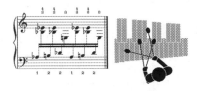

(a) A good sticking for the given music. The left hand glides above the right at all times.

(b) The same passage of music is transposed down two semitones. The same sticking now requires the mallets to interleave and collide.

Fig. 2. The quality of one sticking applied to slightly different music can vary greatly.

Musical hindrances can look different. Figure 3a shows a simple music passage that is annotated with alternating '2' and '3' stickings. Such a sticking minimises movement of the sticks and is physically simple to perform.

(a) An simple sticking for a passage of music minimising movement.

(b) A more physically complex sticking frees the left hand to express the indicated melodic phrase of the lower voice.

Fig. 3. A simple, movement-efficient sticking may lose musical expression.

However, the sticking ignores the indicated musical line represented in the lower voice. A sticking that better represents the musical phrase is shown in Fig. 3b, in which the left hand is free to express and phrase the melodic line while the right hand keeps the non-melodic notes in the background. This more complex sticking lends itself more readily for expressive performance.

Guidelines such as choosing stickings to 'minimise motion' or 'set up a "clean hand"' to play important notes currently assist players [50]. Evidently, these guidelines can contradict each other.

The complexities involved in assigning stickings motivated us to structure the sticking problem as a data-driven classification problem: every note of a musical passage has to be paired with a stick number. The problem can be modelled mathematically, which invites ML approaches to assist with the more complex classification process. This study is the first step in this direction, and results have indicated that our approach may provide practical assistance with the marimba sticking problem.

3   Data Preparation

A generative model for automatic annotation of marimba stickings requires a dataset of sample marimba music with sticking annotations to facilitate the ML process. The data source employed in this investigation is a standard pedagogical text in the marimba literature which assures sticking quality. It contains 590 exercises from the book *Method of Movement for Marimba* by Stevens [38]. All exercises are completely annotated with stickings. The exercises are written in C major but are intended to be transposed into all twelve keys, with alternate stickings provided where required. The exercises consist mainly of pitch and rhythmic information.

The exercises were transcribed as music notation using Musescore notation software [28] as Music XML (`.mxl`) files. Using the `music21` package [14] the exercises were parsed into a Python 3.7.3 environment [32]. Stickings were added within the Python environment as lyrics attached to each note. Stickings for dyads and chords were annotated as lists of sticking numerals. Each exercise was then converted into a list of notes, where each note consisted of a triple: pitch information, rhythmic information and sticking information. Pitch information is stored as a Musical Instrument Digital Interface (MIDI) number. Rhythmic durations are encoded numerically relative to the length of a crotchet—e.g. a crotchet is encoded as 1.0, a quaver as 0.5, semiquaver as 0.25, and so on. Note that stickings are valid at a range of tempi provided by Stevens. The fastest rhythms in the dataset are sextuplets at 132 beats per minute (13.2 notes per second), while the slowest rhythms are quavers at 25 beats per minute (0.83 notes per second). The applicability of stickings at this range of speeds applies to this dataset, but does not necessarily hold for all music (see Sect. 6). The example

$$\mathbf{n} = (\quad \overset{pitch}{60} \quad \overset{duration}{0.5} \quad \overset{sticking}{2} \quad)$$

represents the note C4 for a quaver duration played with stick two, while

$$\mathbf{n} = (\quad \overset{pitch}{(60, 62)} \quad \overset{duration}{1.0} \quad \overset{sticking}{(2, 3)} \quad)$$

represents the dyad C4 and D4 for a crotchet duration played with sticks two and three, respectively.

As numerous stickings can be appropriate for a given musical passage, Stevens often includes alternate stickings within a single exercise. In constructing the dataset, alternate stickings were treated as separate exercises, forming 921 exercises in the Stevens dataset before transposition and further encoding.

3.1 Pitch Encoding

Pitch-Vector Encoding. Pitch-vector encoding (PVE) transforms the MIDI number representation of pitch to a one-hot encoded input vector with 61 components corresponding to the 61 keys of the marimba. The numerically encoded rhythmic duration is added to the input vector as a 62nd component. The corresponding sticking (the output/ground truth) is a one-hot encoded vector with four components. For example,

$$\begin{array}{cccccccc} & C_2 & ... & B_3 & C_4 & C\ _4 & ... & C_7 & duration \\ \mathbf{n}_x = (& 0 & ... & 0 & 1 & 0 & ... & 0 & 0.5 &) \end{array}$$

$$\begin{array}{c} 1\ 2\ 3\ 4 \\ \mathbf{n}_y = (0\ 1\ 0\ 0) \end{array}$$

represents the note C4 for a quaver duration played with stick two. Chords are split into sequences of individual notes, ascending from lowest to highest. The top note of the chord is assigned the rhythmic value of the chord, while all lower notes are assigned a rhythmic value of 0. Exercises are transposed into all possible keys and ranges of the instrument. Duplicate exercises arising in this process are removed. Alternate stickings for different keys are applied. These steps generate 23,286 unique exercises when using PVE.

Interval Transition Encoding. The interval transition encoding (ITE) adapts music encoding assumptions employed by Nakamura et al. for piano fingerings to the marimba sticking problem and uses the musical intervals and physical distances between notes [30]. Chords are first split into sequences of notes as above, and all exercises are transposed into every key.

Pitches are then converted from MIDI number representation to 'lattice encoding'. In this representation, the keys of the marimba are represented geometrically as a two-dimensional 'lattice'. The x dimension runs along the length of the marimba, with each successive natural note representing a step, and y dimension runs along the width of the marimba, which distinguishes natural and accidental notes with a 0 and 1, respectively. A single pitch is represented by a two-dimensional vector (x, y). Intervals between notes are then calculated by taking the element-wise difference between successive pitches (Fig. 4). Note that while in [30] further model assumptions about leaps and symmetry into their methodology, these considerations are less applicable to marimba performance.

After applying ITE to the dataset, duplicate exercises are again removed to avoid data leakage (for example, arpeggios in D major and A major are identical

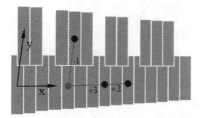

Fig. 4. Lattice encoding of pitch along two axes, representing the geometry of the marimba keyboard. Moving between G , C and E show that the same musical interval may be represented in different ways.

under interval transition encoding) to produce a total of 6,757 unique exercises using this encoding.

Interval Transition Variant for Hidden Markov Models. Hidden Markov Models (HMMs) designed by Nakamura et al. [30] were also trained and tested on the dataset for comparison. We acknowledge that these models were designed for predicting piano fingerings rather than marimba stickings. Still, we include them as a baseline standard for the capability of existing models on a novel but closely related problem. We encode marimba stickings for the HMM algorithms transforming stickings 1, 2, 3 and 4 into analogous piano fingerings –2, –1, 1 and 2 (left-hand index finger and thumb, right-hand thumb and index finger). The exercises are transposed into each key without octave transpositions, and duplicates are removed. The HMMs pre-process the data in a similar method to interval transposition encoding but are geared toward estimating piano fingerings, with the additional assumptions of the equal treatment of large leaps and reflection symmetry between hands.

Sticking Distribution. The distribution of four sticking classes across the exercises is approximately uniform under each of the data encoding methods, mitigating class bias.

4   Machine Learning Methodology

A sequence-to-sequence learning model is required to output sticking sequences for pitch and rhythm sequence inputs. As the prediction of sticking data depends on musical structures that often contain long-term dependencies (e.g. recurring motifs, phrases and patterns), the Long Short-Term Memory (LSTM) neural network is a suitable choice. The LSTM preserves a basic recurrent neural network structure and includes a secondary network state running in parallel that preserves long-term patterns. For simplicity, we omit the LSTM formulations here, which are well-known, and instead focus on how we used the LSTM networks for learning sticking annotation patterns. We refer the reader to Hochreiter and Schmidhuber [21] for formulation details of the LSTM.

LSTM models were built with Keras using the TensorFlow backend [11]. The models in this implementation require inputs to be of constant length. As the length of the exercises in the datasets vary, exercises are pre-padded with the value -99 (the conventional value of 0 cannot be used as $(0, 0, 0)$ is a valid note under interval transition encoding). Consequently, the first layer of the LSTM model is a masking layer which allows the model to ignore all input padding and process the musical data only. The data is fed into an LSTM input layer, and a dense output layer of four units with softmax activation represents a sticking class. For models trained with pitch-vector inputs, the LSTM layer is made bidirectional to allow models to pass input samples forward and backward, allowing classification predictions based on future context. For models trained with interval transition inputs, simply reversing the order of intervals would not represent a reversed exercise. Instead, exercises are were reversed first then re-encoded under ITE. The models were also trained on these inputs, mimicking a bidirectional layer.

We use LSTM models of different sizes (5, 10, 25, 50, 100 and 200 hidden units), trained under the Adam optimiser and categorical accuracy metric with a batch size of 5. Early stopping regularisation is implemented in all models to reduce overfitting.

First, second and third-order HMMs (HMM 1, HMM 2 and HMM 3) using the interval transition encoding variant geared toward piano fingerings are also built and trained to provide a baseline performance level for comparison.

5   Evaluation

Under each method of encoding, twenty per cent of exercises are partitioned for testing. Exercises with identical input vectors but multiple ground truths are grouped to be entirely within the training or testing set to avoid data leakage.

Two approaches are taken to evaluating model performance accounting for multiple ground truths. The naive evaluation is a strict measure where predictions are evaluated against their corresponding ground truth. A more useful evaluation better accounts for multiple ground truths, where the predicted sticking is evaluated against all possible ground truths with the same input, and we take the highest accuracy across all scores. We label this multiple ground truth evaluation 'MGT'.

For example, consider two exercises of length 4 with identical inputs but two different valid ground truth stickings, say $(1, 3, 2, 4)$ and $(1, 2, 3, 4)$. Suppose the model predicts the same sticking $(1, 3, 2, 3)$ for both. The strict accuracy is 50% $(\frac{3+1}{8})$, and the MGT accuracy is 75% $(\frac{3+3}{8})$.

For each model, we evaluate the predictions under strict and MGT approaches on the separated 20% exercises in the test set for each encoding. Results are shown below, with highest scores under each metric highlighted.

Results of LSTM neural network models and HMMs under these evaluation metrics are provided in Table 1. Highest scores of each model under each metric are highlighted.

Table 1. Results showing accuracy of HMM, PVE and ITE models on the test dataset.

Model	Accuracy	
	Strict (%)	MGT (%)
HMM 1	61.9	63.9
HMM 2	69.4	72.2
HMM 3	70.0	72.9
PVE LSTM 5	66.0	83.5
PVE LSTM 10	70.2	89.8
PVE LSTM 25	73.5	92.6
PVE LSTM 50	72.6	91.6
PVE LSTM 100	76.7	97.3
PVE LSTM 200	75.5	95.4
ITE LSTM 5	63.3	79.2
ITE LSTM 10	66.7	84.1
ITE LSTM 25	69.2	87.2
ITE LSTM 50	71.2	89.8
ITE LSTM 100	72.4	92.1
ITE LSTM 200	72.3	92.0

Considering that the baseline HMMs are not optimised for the marimba sticking problem and that the model architecture and training is simpler, these existing models perform reasonably well against small LSTM networks. However, we found that the LSTM performance tends to improve with network size (increasing the complexity of the model) up to 100 units in the hidden layer before performance plateaus or declines. For the ITE, higher performances may be achieved with large LSTMs.

6   Discussion

Our results support the use of LSTM neural networks in predicting marimba stickings. This section discusses qualitative aspects of obtained results, limitations of the current methodology and directions for extending this study.

6.1   Qualitative Analysis of Sticking Predictions

We inspected a sample of sticking predictions made on the test set of the PVE LSTM 100 and ITE LSTM 100 models (the optimal models of each approach).

We first notice basic sticking principles in the model predictions, for example, that stickings generally reflect the use of sticks 1 and 2 in lower notes and sticks 3 and 4 in higher notes.

We also observed that most predicted stickings were playable and efficient, and only occasional predictions were clearly erroneous. Moreover, common to both models was a higher frequency of anomalous predictions near the beginning or end of input samples, as shown in Fig. 5a. These stickings were sometimes unplayable or were sometimes viable alternate stickings that did not match the pattern adopted during the middle of the sample. This behaviour may be a consequence of the influence caused by model learning on samples with multiple ground truths, where the 'correct' sticking is uncertain. Bidirectionality (PVE) and the reversal of inputs (ITE) may have helped training with past and future contexts in the middle of samples. The longer and more repetitive the sample, the higher quality predicted stickings were generally observed.

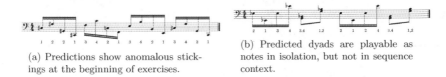

(a) Predictions show anomalous stick-ings at the beginning of exercises.

(b) Predicted dyads are playable as notes in isolation, but not in sequence context.

Fig. 5. Sticking prediction characteristics from ITE LSTM 200 model.

The models sometimes struggled with dyads and chords in context of longer sequences. For example, dyads in Fig. 5b are playable in isolation but are not practical within the context of the sequence, requiring the hands to cross.

The sticking predictions made by the baseline HMM 3 model in the inspected samples appeared similar to the ITE LSTM 200 in terms of playability and con-sistency despite the lower evaluation score. While anomalous stickings were still observed in some short exercises and at the beginning and ends of longer exer-cises, sticking quality was good despite the model being designed for the piano fingering problem (Fig. 6). The main source of error in the HMM 3 model was the occasional prediction of a sticking number '5'—an invalid stick and an artefact of the HMM implementation for piano fingering despite being trained exclusively on marimba exercise inputs. We expect the HMM model's adaptation to directly address marimba stickings may greatly improve its quantitative evaluation score. As such, we see both the ITE LSTM and HMM model approaches being useful in practice for assisting marimbists and percussionists with the sticking problem and suggest these avenues for further development.

6.2   Limitations

Dataset Limitations. While there are advantages in the use of Stevens' *Method of Movement for Marimba* as the dataset for this study described previously, the

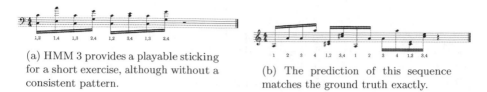

(a) HMM 3 provides a playable sticking for a short exercise, although without a consistent pattern.

(b) The prediction of this sequence matches the ground truth exactly.

Fig. 6. The HMM 3 model exhibits some successful prediction characteristics despite being designed for piano fingering.

dataset is also the key limitation of the methodology. Exercises from the dataset represent typical marimba sticking patterns, but not concert marimba music, which typically contains a greater variety of pitch, rhythm, and patterns. Development of the dataset is challenging with marimba music. The instrument is far less commonly played and studied than popular instruments like the piano and guitar, so music and sticking data is far less available. As a relatively new instrument within the Western music tradition, marimba sheet music generally remains within copyright, and the capacity to share and develop datasets is limited. Data may be more readily obtained by recording the performers live rather than transcribing written marimba music. Such approaches have already been undertaken with piano fingering by Johnson et al., employing camera systems to generate depth maps of pianists' hands while playing [15].

Model Limitations. The models we employ in this study are simplified to predict stickings based on pitch and rhythm only. These models may be unable to capture finer aspects of sticking considerations, which are influenced by other musical variables not modelled in this study such as dynamic, articulation, or expressive markings.

While we model pitch comprehensively, we include rhythmic information only in the context of the musical meter, not in absolute speed terms. While appropriate for pedagogical exercises, in performance practice, the sticking of a passage at a slow tempo will often change compared to when played quickly. Additionally, the model does not account for rolls—rapid successive strikes to sustain a note—which are important and frequently used techniques on the marimba. Accounting for rolls in predicting marimba stickings is a necessary area of development before sticking prediction becomes practical.

Modelling these additional parameters in further work will help to refine sticking predictions for more practical musical applications.

7   Conclusion

In this study, we explore a novel problem of predicting marimba stickings with LSTM neural networks. Leigh Howard Stevens' pedagogical marimba text, *Method of Movement for Marimba* is used as the dataset representing examples of sticking patterns. Our data-driven approach uses Long Short-Term Memory neural networks that are trained on examples extracted from this dataset. We simplify the complexities involved in marimba music by modelling two core variables influencing sticking: pitch and rhythm. The best performances are achieved with LSTM networks that employ interval transition encoding with a maximum multiple ground truth micro-accuracy of 97.3%.

This study has demonstrated an applicability of data-driven machine learning approaches to novel problems in musical instrument technique, which include quantifiable physical constraints as well as qualitative assessments of musicality. The sticking problem of the marimba, a technical issue associated with an instrument that is rapidly growing in popularity, has not previously been analysed using a computational methodology of this nature, and the promising results along side clear paths for development invite further study of the field.

References

1. Al Kasimi, A., Nichols, E., Raphael, C.: Automatic fingering system (afs). In: Poster presentation at ISMIR, London (2005)
2. Al Kasimi, A., Nichols, E., Raphael, C.: A simple algorithm for or automatic generation of polyphonic piano fingerings (2007)
3. Balliauw, M., Herremans, D., Palhazi Cuervo, D., Sörensen, K.: A variable neighborhood search algorithm to generate piano fingerings for polyphonic sheet music. Int. Trans. Oper. Res. **24**(3), 509–535 (2017)
4. Barbancho, A.M., Klapuri, A., Tardon, L.J., Barbancho, I.: Automatic transcription of guitar chords and fingering from audio. IEEE Trans. Audio Speech Lang. Process. **20**(3), 915–921 (2012). https://doi.org/10.1109/TASL.2011.2174227
5. Bretan, P.M.: Towards an embodied musical mind: generative algorithms for robotic musicians. Ph.D. thesis, Georgia Institute of Technology (2017)
6. Briot, J.P., Hadjeres, G., Pachet, F.: Deep Learning Techniques for Music Generation. Springer, Heidelberg (2020). https://doi.org/10.1007/978-3-319-70163-9
7. Briot, J.P., Pachet, F.: Deep learning for music generation: challenges and directions. Neural Comput. Appl. **32**(4), 981–993 (2020)
8. Burlet, G., Fujinaga, I.: Robotaba guitar tablature transcription framework. In: Proceedings of the 14th International Society for Music Information Retrieval Conference, ISMIR, Curitiba, Brazil, pp. 517–522 (2013)
9. Choi, K., Fazekas, G., Cho, K., Sandler, M.: A tutorial on deep learning for music information retrieval. arXiv preprint arXiv:1709.04396 (2017)
10. Choi, K., Fazekas, G., Sandler, M., Cho, K.: Convolutional recurrent neural networks for music classification. In: 2017 IEEE International Conference on Acoustics, Speech and Signal Processing (ICASSP), pp. 2392–2396. IEEE (2017)
11. Chollet, F., et al.: Keras (2015). https://github.com/fchollet/keras
12. Coca, A.E., Corrêa, D.C., Zhao, L.: Computer-aided music composition with LSTM neural network and chaotic inspiration. In: The 2013 International Joint Conference on Neural Networks (IJCNN), pp. 1–7. IEEE (2013)
13. Corrêa, D.C., Levada, A.L., Saito, J.H., Mari, J.F.: Neural network based systems for computer-aided musical composition: supervised x unsupervised learning. In: Proceedings of the 2008 ACM Symposium on Applied Computing, pp. 1738–1742 (2008)

14. Cuthbert, M., Ariza, C., Hogue, B., Oberholtzer, J.W.: music21 (version 5.7.2) [python package] (2006–2021). https://web.mit.edu/music21

15. Johnson, D., Damian, G.T.D.: Detecting hand posture in piano playing using depth data. Comput. Music J. **43**(1), 59–78 (2019). http://muse.jhu.edu/article/746693

16. De Prisco, R., Zaccagnino, G., Zaccagnino, R.: A differential evolution algorithm assisted by ANFIS for music fingering. In: Rutkowski, L., Korytkowski, M., Scherer, R., Tadeusiewicz, R., Zadeh, L.A., Zurada, J.M. (eds.) EC/SIDE -2012. LNCS, vol. 7269, pp. 48–56. Springer, Heidelberg (2012). https://doi.org/10.1007/978-3-642-29353-5_6

17. Eck, D., Lapalme, J.: Learning musical structure directly from sequences of music. University of Montreal, Department of Computer Science, CP 6128, 48 (2008)

18. Eck, D., Schmidhuber, J.: Finding temporal structure in music: blues improvisation with LSTM recurrent networks. In: Proceedings of the 12th IEEE Workshop on Neural Networks for Signal Processing, pp. 747–756. IEEE (2002)

19. Eck, D., Schmidhuber, J.: A first look at music composition using LSTM recurrent neural networks. Istituto Dalle Molle Di Studi Sull Intelligenza Artificiale **103**, 48 (2002)

20. Hart, M., Bosch, R., Tsai, E.: Finding optimal piano fingerings. UMAP J. **21**(2), 167–177 (2000)

21. Hochreiter, S., Schmidhuber, J.: Long short-term memory. Neural Comput. **9**(8), 1735–1780 (1997). https://doi.org/10.1162/neco.1997.9.8.1735

22. Humphrey, E.J., Bello, J.P.: From music audio to chord tablature: teaching deep convolutional networks to play guitar. In: 2014 IEEE International Conference on Acoustics, Speech and Signal Processing (ICASSP), pp. 6974–6978 (2014). https://doi.org/10.1109/ICASSP.2014.6854952

23. Liu, I., Ramakrishnan, B., et al.: Bach in 2014: music composition with recurrent neural network. arXiv preprint arXiv:1412.3191 (2014)

24. Lyu, Q., Wu, Z., Zhu, J.: Polyphonic music modelling with LSTM-RTRBM. In: Proceedings of the 23rd ACM International Conference on Multimedia, pp. 991–994 (2015)

25. Mistler, E.: Generating Guitar Tablatures with Neural Networks. University of Edinburgh, Thesis (2017)

26. Miura, M., Hirota, I., Hama, N., Yanagida, M.: Constructing a system for finger-position determination and tablature generation for playing melodies on guitars. Syst. Comput. Japan **35**(6), 10–19 (2004). https://doi.org/10.1002/scj.10609. https://search.ebscohost.com/login.aspx?direct=true&db=iih&AN=13217635&site=ehost-live

27. Musafia, J.: The art of fingering in piano playing. MCA Music (1971)

28. MuseScore: MuseScore (version 3.1.0.7078) [computer software] (2019). https://musescore.org

29. Nakamura, E., Ono, N., Sagayama, S.: Merged-output hmm for piano fingering of both hands. In: ISMIR, pp. 531–536 (2014)

30. Nakamura, E., Saito, Y., Yoshii, K.: Statistical learning and estimation of piano fingering. Inf. Sci. **517**, 68–85 (2020). https://doi.org/10.1016/j.ins.2019.12.068. http://www.sciencedirect.com/science/article/pii/S0020025519311879

31. Parncutt, R., Sloboda, J.A., Clarke, E.F., Raekallio, M., Desain, P.: An ergonomic model of keyboard fingering for melodic fragments. Music Percept. **14**(4), 341–382 (1997)

32. Python: Python (version 3.7.3) [computer software] (2001–2021). https://python.org

33. Ramos, J.V., Ramos, A.S., Silla, C.N., Sanches, D.S.: An evaluation of different evolutionary approaches applied in the process of automatic transcription of music scores into tablatures. In: 2016 IEEE 28th International Conference on Tools with Artificial Intelligence (ICTAI), pp. 663–669 (2016). https://doi.org/10.1109/ICTAI.2016.0106

34. Savery, R., Weinberg, G.: Shimon the robot film composer and deepscore. In: Proceedings of Computer Simulation of Musical Creativity, p. 5 (2018)

35. Sayegh, S.I.: Fingering for string instruments with the optimum path paradigm. Comput. Music Jou. **13**(3), 76–84 (1989). https://doi.org/10.2307/3680014. http://www.jstor.org.ezproxy.library.uwa.edu.au/stable/3680014

36. Schedl, M.: Deep learning in music recommendation systems. Front. Appl. Math. Stat. **5**, 44 (2019)

37. Sigtia, S., Benetos, E., Dixon, S.: An end-to-end neural network for polyphonic piano music transcription. IEEE/ACM Trans. Audio Speech Lang. Process. **24**(5), 927–939 (2016)

38. Stevens, L.H.: Method of Movement for Marimba: With 590 Exercises. Marimba Productions, Inc., Neptune City (2005)

39. Sturm, B.L., Santos, J.F., Ben-Tal, O., Korshunova, I.: Music transcription modelling and composition using deep learning. arXiv preprint arXiv:1604.08723 (2016)

40. Takegawa, Y., Terada, T., Nishio, S.: Design and implementation of a real-time fingering detection system for piano performance. In: ICMC (2006)

41. Tuohy, D.R., Potter, W.: Guitar tablature creation with neural networks and distributed genetic search. In: Proceedings of the 19th International Conference on Industrial and Engineering Applications of Artificial Intelligence and Expert Systems, IEA-AIE06, Annecy, France (2006)

42. Van Den Oord, A., Dieleman, S., Schrauwen, B.: Deep content-based music recommendation. In: Neural Information Processing Systems Conference (NIPS 2013), vol. 26. Neural Information Processing Systems Foundation (NIPS) (2013)

43. Walter, D.W.: The Performance of Contrapuntal Music on the Marimba and Vibraphone. Ph.D. thesis, Temple University (1984)

44. Wang, X., Wang, Y.: Improving content-based and hybrid music recommendation using deep learning. In: Proceedings of the 22nd ACM international conference on Multimedia, pp. 627–636 (2014)

45. Yang, N., Savery, R., Sankaranarayanan, R., Zahray, L., Weinberg, G.: Mechatronics-driven musical expressivity for robotic percussionists. arXiv preprint arXiv:2007.14850 (2020)

46. Yazawa, K., Itoyama, K., Okuno, H.G.: Automatic transcription of guitar tablature from audio signals in accordance with player's proficiency. In: 2014 IEEE International Conference on Acoustics, Speech and Signal Processing (ICASSP), pp. 3122–3126 (2014). https://doi.org/10.1109/ICASSP.2014.6854175

47. Ycart, A., Benetos, E., et al.: A study on LSTM networks for polyphonic music sequence modelling. In: ISMIR (2017)

48. Yonebayashi, Y., Kameoka, H., Sagayama, S.: Automatic decision of piano fingering based on a hidden markov models. In: IJCAI, vol. 7, pp. 2915–2921 (2007)

49. Yu, Y., Luo, S., Liu, S., Qiao, H., Liu, Y., Feng, L.: Deep attention based music genre classification. Neurocomputing **372**, 84–91 (2020)

50. Zeltsman, N.: Four-Mallet Marimba Playing: A Musical Approach for All Levels. H. Leonard, Milwaukee, WI (2003)

Systematic Monotonicity and Consistency for Adversarial Natural Language Inference

Brahmani Nutakki[(✉)] [iD], Akshay Badola[iD], and Vineet Padmanabhan[iD]

School of Computer and Information Sciences, University of Hyderabad,
Hyderabad, India
brahmani3110@gmail.com, {badola,vineetnair}@uohyd.ac.in

Abstract. Natural Language Inference is a fundamental task required for understanding natural language. With the introduction of large Natural Language Inference (NLI) benchmark datasets such as SNLI and MultiNLI, NLI has seen an uptake in models achieving near-human accuracy. Deeper analyses through adversarial methods performed on these models however have cast doubts on their ability to actually understand the inference process. In this work, we attempt to define a principled way to generate adversarial attacks based on monotonic reasoning and consistency to examine their language understanding abilities. We show that the language models trained for general tasks have a poor understanding of monotonic reasoning. For this purpose, we provide methods to generate an adversarial dataset from any NLI dataset based on monotonicity and consistency principles and conduct extensive experiments to support our hypothesis. Our adversarial datasets preserve these crucial aspects of monotonicity, consistency and semantic similarity and are still able to fool a model finetuned on SNLI 79% of the time while preserving semantic similarity to a much greater extent than previous methods.

1 Introduction

Natural Language Inference (NLI), initially known as Recognizing Textual Entailment, was introduced as a PASCAL Challenge Benchmark task (RTE-1) [17]. The task involves determining if a natural language hypothesis h can be reasonably inferred from the given premise p [15]. Owing to its use as a comparison metric to quantify the semantic inference of models, it is often used as a proxy to gauge a model's ability to understand natural language. Significant advances have been made in the field of NLI, which were further propelled by the advent of huge benchmark datasets such as the Stanford Natural Language Inference Corpus (SNLI) [2] and the Multi-Genre Natural Language Inference Corpus (MNLI) [32].

Language Models and specifically Neural Language Models based on Recurrent Neural Network (RNN) [26] and large Transformers [29] have been a paradigm shift in Natural Language Modeling and have achieved state-of-the-art results in many Natural Language tasks including NLI. However, adversarial

H. Aziz et al. (Eds.): AI 2022, LNAI 13728, pp. 353–366, 2022.
https://doi.org/10.1007/978-3-031-22695-3_25

attacks and stress tests have questioned the actual language understanding ability of these models. The NLI task is particularly amenable to logical inspection and assessment and a model's failures for a given example helps to identify its shortcomings. A very instructive example is [8] which analyzes negation and shows the Language Model's inability to understand it.

In this work, we investigate the role of semantic *monotonicity* and logical *consistency* in the NLI task and introduce a framework for lexical attacks based on them. Monotonicity in this case refers to the semantic relations between generalizations and specializations of a word and inferences which can be drawn from them. By consistency we mean rules of logic; e.g. symmetry transitivity etc. are maintained across the sentences. We transform a given $< premise, hypothesis, label > \equiv (p, h, l)$ triplet in the dataset, by substituting certain words such that the change in label l is deterministic corresponding to the *monotonicity* and *consistency* rules.

For example, consider the sentence pair <*People are marching towards the mountains, The people are going towards the mountains*> $\equiv < p, h >$, with the label $l = entailment$ or e. Replacing *marching* in p with its hypernym *walking* does not change the meaning of p or the label, as it is an upward monotone. Similarly, we can derive rules for label changes for various combinations of substitutions in both p and h which lead to a specific change in label l. We call these substitutions *two-hop* label shifts as they transform both p and h. Our approach differs from prior work which have used brute force or embeddings-based perturbations [11,19] and have focused on transforming only premises. These attacks reveal critical deficiencies in the Language Model's lexical and syntactic understanding. Although we focus on NLI datasets, the methods can be generalized to other language tasks. To the best of our knowledge, this is the first work that uses attacks based on both monotonicity and consistency rules across both the premise and hypothesis.

To sum up, our contributions are:

- We provide a general principled adversarial attack method using our novel two-hop label shift rules.
- We demonstrate the efficacy of our generated datasets on State-of-the-art NLI models, and compare them against existing adversarial text generation frameworks.
- We release the code for the experiments which can be found at https://github.com/nbrahmani/Two-hop-adversarial-attacks

The rest of this paper is organized as follows: Sect. 2 gives an overview of the existing work. Section 3 gives an overview of NLI and Adversarial NLI. Section 4 describes our methodology of the proposed attacks, and Sect. 5 is about the experiments performed and the results obtained. We follow up with discussions in Sect. 6 and conclude in Sect. 7.

2 Related Work

Adversarial methods in Neural Models have been gaining prominence with the success of Image Classification models [21]. With the growing success of Neural Language Models, methods to determine the weaknesses of these models have also gained attention [6, 10]. These methods are usually classified into White-box and Black-box attacks, and the Black-box attacks can be further classified into Score-based, Decision-based, and Transfer-based attacks [16].

White-box attacks have access to the gradient information of the loss function and construct the adversarial instances based on this information. Li et al. [12] use the loss function gradient of each word to find their importance and replace the words with similar words. Ebrahimi et al. [5] attack the model by flipping a character in the sentence that maximizes the model loss. Although these attacks are successful, their methodology is cumbersome.

Black-box attacks, on the other hand, only use the model outputs to generate the adversarial instances. They do not require access to the model's gradient information and are agnostic to the model. For example, Jin et al. [11] use the model's confidence scores to create adversarial perturbations. Zhao et al. [34] use only the final predicted output of the model to generate attacks instead of the confidence scores. A different approach is taken in [30] who train a classifier to mimic the decisions of the model, after which attacks are performed on this model and are then transferred to the original model.

As useful as these attacks are, they are not systematic in nature and introduce random perturbations in the data to craft adversarial examples. While in search of a more principled manner to analyze the adversarial examples in text, research has turned to gauge the model's understanding of logic. It has been observed that language models struggle to understand logic due to its discrete nature. Traylor et al. [28] test whether the models can differentiate between logical symbols such as disjunction ($\vee$), conjunction ($\wedge$) or negation ($\neg$). They find the models largely fail on their newly generated dataset. Meanwhile, the model's ability to infer over conjuncts is probed in [24]. Tarunesh et al. [27] create a huge dataset that tests the models against 17 reasoning tasks, including logical tasks such as Boolean (sentences containing logical and ($\wedge$), or ($\vee$) and their combinations) and quantifier (sentences containing universal ($\forall$) and existential ($\exists$) operators) apart from world knowledge, causality etc.

Richardson et al. [23] and Naik et al. [20] probe the models on various semantic phenomena, including logical aspects such as negation, along with monotonicity-related aspects. Glockner et al. [6] generate perturbations by replacing one word in the premise using lexical knowledge. Similarly, Yanaka et al. [33] have proposed the MED dataset that checks the model's understanding of monotonicity. They synthesize examples based on the monotonicity inference rules using contextual grammar.

Gururangan et al. [7] showed that a simple classification model achieves 67% accuracy on SNLI and 53% on MNLI when only hypotheses are given, thus showing that the models are sensitive to annotation artifacts. Certain words such as negations and gender-neutral terms lead to false predictions by the model.

Poliak et al. [22] tested a hypothesis-only model on ten different datasets and found that the model performed better than most baselines.

Our work follows [33] and [14] in that we use monotonicity and consistency to generate an adversarial dataset from the given dataset[1]. Our approach differs in our use of *two-hop* label shift rules across the premise-hypothesis pair.

3   Adversarial NLI

We discuss NLI first and then Adversarial NLI in detail:

The standard NLI task consists of predicting a label l from a sentence pair of Premise and Hypothesis (p, h). For example, the sentence pair $<A$ *man is riding a horse in a meadow, A person is outside*$>$ has the label *entailment*. Usually we deal with only three labels, *entailment, contradiction, neutral*. For our purposes we'll focus on Neural Language Models, specifically variants of BERT [4] which have achieved state-of-the-art in many NLP tasks. These models transform the sentences into distributed representations and posit them as a classification task.

For NLI, the data is a set of ordered triplets of Premise, Hypothesis and Label: $\mathcal{D} = \{(p, h, l)\}$. The objective is to find a model $\mathcal{M}$ parameterized by weights Θ, such that it predicts the correct label l given (p, h), i.e.:

$$\mathcal{M}_\Theta : (P, H) \to L$$

In this case, the model here is a Neural Language Model which is learned by maximizing the likelihood of Θ over the dataset. That is, the number of predicted labels l_i over the input sentences (p_i, h_i) in the dataset.

$$\mathcal{M}_\Theta = \underset{\Theta}{\mathrm{argmax}}\ \mathcal{L}_\Theta = \underset{\Theta}{\mathrm{argmax}}\ P(l_i | p_i, h_i) \quad \forall (p_i, h_i, l_i) \in \mathcal{D}$$

Adversarial NLI on the other hand can be considered as the process of finding a set of transformations $\mathcal{T} : (S, L) \to (S, L)$ where (S, L) is the set of all $<$*sentence, label*$>$ pairs, such that the *trained model* fails for a given example. Formally:

$$\mathcal{M}(\mathcal{T}(p_i, h_i)) \neq l_i', \quad (p_i, h_i, l_i) \in \mathcal{D}$$

where $\mathcal{T}(p_i, h_i)$ changes either p_i or h_i or both, and l_i is the true label corresponding to the transformation $\mathcal{T}(p_i, h_i)$.

In other words, the goal is to find a method to transform the inputs so that the model's output is not the same as the expected output.

4   Towards Systematic Adversarial NLI

As we mentioned earlier, while approaches for Adversarial NLI exist, they are not systematic in nature. Here, we describe our approach used in determining the transformation $\mathcal{T}$ for Systematic Adversarial NLI.

---

[1] We use both SNLI and MNLI, but in practice, it can be any NLI dataset or the methods can even be adapted for any other language dataset.

Consider a data point $(p, h, l) \in \mathcal{D}$. The transformation $\mathcal{T}$ we propose is based on *two-hop* rules. Recall from Sect. 1 that these are rules which apply to *both* the premise and hypothesis, instead of only the premise. We focus only on single word substitutions using an existing ontology. We choose Wordnet [18] for our purpose, but any other ontology can be used.

Monotonicity and Consistency Rules. Let, $E(p, h)$ denote an entailment, $C(p, h)$ a contradiction and $N(p, h)$ a neutral label for premise-hypothesis pair (p, h). For a sentence $s \in \{p, h\}$, the following rules are applicable:

1. Rules of Consistency [14]:
   - $E(p, h) \wedge E(h, z) \to E(p, z)$
   - $E(p, h) \wedge C(h, z) \to C(p, z)$
   - $N(p, h) \wedge E(h, z) \to \neg C(p, z)$
   - $N(p, h) \wedge C(h, z) \to \neg E(p, z)$
   - $C(p, h) \to C(h, p)$
2. Rules of Equivalence:
   - $s' = W_{Eq}(s) \to E(s, s') \wedge E(s', s)$
   Where W_{Eq} stands for equivalent word substitution.
3. Rules of monotonicity [33]:
   - $s' = W_{ME}(s) \to E(s, s') \wedge N(s', s)$
   - $s' = W_{MN}(s) \to N(s, s') \wedge E(s', s)$
   Where W_{ME}, W_{MN} stand for Monotonically Entailment and Neutral word substitutions, respectively.

Deriving the Label Changes. Using the aforementioned consistency, equivalence and monotonicity based rules, the corresponding changes in label (shifts) for each transformation are deterministic and can be derived. We list here only the effective shift rules for the transformations as the rest of the shift rules do not induce a label change required for an adversarial attack.

We use the following notation for describing the transformations:

- **Single Sentence Transformation**: $\mathcal{T}_M(p, h) : (p', h)$ (or (p, h')) is a transformation $\mathcal{T}$ for a premise-hypothesis pair (p, h) such that only p (or h) is changed to p' (or h') via method M.
- **Dual Sentence Transformation**: $\mathcal{T}_{M,M}$, e.g., $\mathcal{T}_{E,ME}(p, h) : (p', h')$ means that premise p is changed to p' using an equivalent substitution and hypothesis h is changed to h' using a monotonically entailed substitution.

We take $\neg C(p, h)$ and $\neg E(p, h)$ to be $N(p, h)$. Based on a given transformation $\mathcal{T}_M$, we then determine the new label l'. Table 1 lists all the label shift rules.

One issue we faced was that effecting multiple transformations can cause an exponential increase in the number of possible combinations of label changes. To mitigate that, we find the words (which we call markers) which are most representative of the meaning of the word and transform them which we describe in the next Sect. 4. We use a separate model to determine the markers.

Table 1. Table of transformations

$T_E(p,h) : (p',h) \rightarrow E(p,p'), E(p',p)$
$- E(p,h) \rightarrow E(p',h)$
$- C(p,h) \rightarrow C(p',h)$

$T_{ME}(p,h) : (p',h) \rightarrow E(p,p'), N(p',p)$
$- E(p,h) \rightarrow \neg C(p',h)$
$- C(p,h) \rightarrow \neg E(p',h)$

$T_{MN}(p,h) : (p',h) \rightarrow N(p,p'), E(p',p)$
$- E(p,h) \rightarrow E(p',h)$
$- C(p,h) \rightarrow C(p',h)$

$T_E(p,h) : (p,h') \rightarrow E(h,h'), E(h',h)$
$- E(p,h) \rightarrow E(p,h')$
$- C(p,h) \rightarrow C(p,h')$
$- N(p,h) \rightarrow \neg C(p,h')$

$T_{ME}(p,h) : (p,h') \rightarrow E(h,h'), N(h',h)$
$- E(p,h) \rightarrow E(p,h')$
$- N(p,h) \rightarrow \neg C(p,h')$

$T_{MN}(p,h) : (p,h') \rightarrow N(h,h'), E(h',h)$
$- C(p,h) \rightarrow C(p,h')$

$T_{E,E}(p,h) \qquad : \qquad (p',h') \rightarrow$
$E(p,p'), E(p',p), E(h,h'), E(h',h)$
$- E(p,h) \rightarrow E(p',h')$
$- C(p,h) \rightarrow C(p',h')$

$T_{E,ME}(p,h) \qquad : \qquad (p',h') \rightarrow$
$E(p,p'), E(p',p), E(h,h'), N(h',h)$
$- E(p,h) \rightarrow E(p',h')$

$T_{E,MN}(p,h) \qquad : \qquad (p',h') \rightarrow$
$E(p,p'), E(p',p), N(h,h'), E(h',h)$
$- C(p,h) \rightarrow C(p',h')$

$T_{ME,E}(p,h) \qquad : \qquad (p',h') \rightarrow$
$E(p,p'), N(p',p), E(h,h'), E(h',h)$
$- E(p,h) \rightarrow \neg C(p',h')$
$- C(p,h) \rightarrow \neg E(p',h')$

$T_{ME,ME}(p,h) \qquad : \qquad (p',h') \rightarrow$
$E(p,p'), N(p',p), E(h,h'), N(h',h)$
$- E(p,h) \rightarrow \neg C(p',h')$

$T_{ME,MN}(p,h) \qquad : \qquad (p',h') \rightarrow$
$E(p,p'), N(p',p), N(h,h'), E(h',h)$
$- C(p,h) \rightarrow \neg E(p',h')$

$T_{MN,E}(p,h) \qquad : \qquad (p',h') \rightarrow$
$N(p,p'), E(p',p), E(h,h'), E(h',h)$
$- E(p,h) \rightarrow E(p',h')$
$- C(p,h) \rightarrow C(p',h')$

$T_{MN,ME}(p,h) \qquad : \qquad (p',h') \rightarrow$
$N(p,p'), E(p',p), E(h,h'), N(h',h)$
$- E(p,h) \rightarrow E(p',h')$

$T_{MN,MN}(p,h) \qquad : \qquad (p',h') \rightarrow$
$N(p,p'), E(p',p), N(h,h'), E(h',h)$
$- C(p,h) \rightarrow C(p',h')$

Selection of the Markers and Extraction of Sense. Changing all words or a random combination of words would be too computationally intensive and not helpful in generating good adversarial examples. Therefore, based on a transformation T, we select the top 5 most similar words (markers) in the sentence S ($S \in \{P,H\}$). These are selected by comparing the cosine similarities between individual word embeddings and sentence embedding. The word and sentence embeddings are obtained using a pre-trained model.

After that, a word sense disambiguation model is used to obtain the sense of the markers to ensure that the generated examples are semantically similar to original sentences. For this, we use Wordnet sense ids [18]. These transformations and the *two-hop* rules which change only the markers form the basis of our adversarial attacks.

Other methods like TextFooler [11] replace the selected word in the hypothesis from a list of synonyms by comparing the cosine similarities of their embeddings. The attack labels of such perturbations are riddled with errors. The sense of the word can also change due to the replacements. Our attacks are performed by the *two-hop* rules governed by the word-replacement technique and the ground truth and do not suffer from these issues. We also perform sense-based replacement to ensure the sense of the perturbations remains the same.

4.1 Word-Replacement Techniques

After selecting the markers and their sense, the sentences are perturbed using the three word-replacement techniques based on the type of transformation applied. They are 1) Equivalent 2) Monotonic-entailment and 3) Monotonic-neutral. These replacements govern the selected word substitute and the corresponding label. The monotonicity of the word is obtained using a polarity annotator.

- **Equivalent** word replacement is achieved by replacing the marker with one of its synonyms. It always results in an entailment in both directions.
- **Monotonic** replacement substitutes a marker by a general phrase (hypernym) or a specific phrase (hyponym). If the word is upward monotone, replacing it with hypernym results in an inferable sentence (*entailment* label), while replacing it with hyponym results in a neutral sentence. Similarly, replacing a downward monotone word with its hyponym results in an inferable sentence, and a hypernym leads to neutral classification. Corresponding to these rules we define two-word replacement methods: **Monotonic-Entailment** and **Monotonic-Neutral**.

The replacement words obtained are then modified to match the morphology of the original word after which they are filtered based on their grammar score or acceptability score. The model is now asked to classify these transformations along with the labels. Only those input sentence pairs are used whose ground truth is the same as the predicted label; the rest are skipped. If the label predicted for the perturbation differs from the one obtained using the derived rules, the attack is successful, else unsuccessful. The complete Algorithm 1 is given below.

5 Experiments and Results

5.1 Experimental Setup

Before detailing the results of the attacks, we briefly give an overview of the different models and approaches used for individual modules mentioned in Sect. 4.

Algorithm 1. Adversarial Attack using Logical Rules

1: **Input:** T_M, p, h, l, markers $\{m\}$
2: **Output:** Transformed tuple (p', h', l')
3: Select p, h or both based on T_M
4: Treating it as a single sentence s of two clauses, select top 5 words from $s \equiv \{m\}$.
5: **for** $m_i \leftarrow \{m\}$ **do**
6:     Extract the sense and replace the marker according to method M with a word
7: **end for**
8: Remove perturbations where grammar score varies significantly from that of s
9: Query model $\mathcal{M}$ with the perturbed sentence pair (p', h') and check with expected label l'

Selecting Markers and Extracting Sense. For selecting top 5 markers, the embeddings for the premise-hypothesis pair are extracted using a MPNet [25] based sentence encoder which has been fine-tuned on a 1B sentence dataset. This model takes the input sentences and produces word embeddings and sentence embeddings. The top 5 similar words based on cosine similarities between the word and the given sentence embedding are chosen as essential markers. The perturbations are generated by extracting sense from ESCHER [1]. These senses are then used to mine synonyms, hypernyms and hyponyms from Wordnet.

Polarity Annotation and Grammar Score. To get the monotonicity of a marker, we need the monotonic polarity. We follow [9] for polarity annotation. The input sentences are first parsed using a CCG parser and *ccg2mono* proposed in [9] is used to polarize the words as *upward, downward,* or *no polarity.* We then compare the grammar scores of the original and the modified sentences with a BERT model fine-tuned on the COLA dataset [31]. The model gives a probability output of the given sentence being acceptable or not. An absolute difference greater than a threshold between the original and the perturbed sentence is ignored. We found empirically that a threshold value of 0.1 works well.

5.2   Results

Using the models mentioned above, we build our attack pipeline to generate adversarial attacks. We randomly sample 5000 sentence pairs from the train splits of the SNLI [2] and MNLI [32] datasets. We then generate perturbations for all 15 types of transformations, picking a different number of markers each time. Then using the *two-hop* label shift rules, attacks are performed on the model with these perturbations. Example perturbations can be found below:

Example 1. p: Man smokes while sitting on a parked scooter.
h: A man smokes a cigarette while sitting on his scooter.
Marker_p: Man, *Marker_h*: Man
Ground Truth: Neutral, *Predicted Label*: Neutral
Transformations:

1. $\mathcal{T}_E(p, h) : (p, h')$: No perturbations as no valid perturbation exists.
2. $\mathcal{T}_{ME}(p, h) : (p, h')$:
    - H': an *adult* smokes a cigarette while sitting on his scooter.
      Label: Neutral, *Attack Status*: Failed
    - H': a *person* smokes a cigarette while sitting on his scooter.
      Label: Neutral, *Attack Status*: Failed
    - H': a *male* smokes a cigarette while sitting on his scooter.
      Label: Neutral, *Attack Status*: Failed
    - H': an *organism* smokes a cigarette while sitting on his scooter.
      Label: Contradiction, *Attack Status*: Success
3. Remaining Transformations: No perturbations as the label shift rule does not exist for this transformation.

We run the experiments on the BERT base model with both SNLI and MNLI datasets. The results for a different number of markers are given in the tables below.

Table 2. Attack Results on BERT finetuned on SNLI and MNLI

No. of markers	SNLI			MNLI		
	Successful attacks	Failed attacks	Attack accuracy	Successful attacks	Failed attacks	Attack accuracy
1	2181	3086	41.4%	244	4699	4.9%
2	3289	1978	62.4%	466	4477	9.4%
3	3833	1434	72.7%	588	4355	11.8%
4	4095	1172	77.7%	711	4232	14.3%
5	4199	1068	79.7%	763	4180	15.4%

6 Discussion

As seen in Table 2, our attacks achieved an attack accuracy of 79% on the BERT model finetuned on SNLI. This shows that though the model performed well on benchmark datasets, it has a poor understanding of monotonic reasoning and fails at simple lexical monotonic inferences. Meanwhile, BERT finetuned on MNLI has achieved 84.6% accuracy (Attack accuracy being 15.4%) on the adversarial dataset. $BERT_{MNLI}$ being more powerful than $BERT_{SNLI}$, it can be surmised that the model can withstand the attacks better than the latter. From these results, we may assume that the $BERT_{MNLI}$ model has managed to capture simple monotonic inferences. However, keeping in mind the length of the sentences in MNLI it may be that single-word substitutions performed might not be sufficient to validate their monotonic reasoning capacity.

We compare our attack accuracies with adversarial attack methods, namely TextFooler [11] and BERT-Attack [13] as seen in Table 3. We also give a detailed comparative analysis of our model with TextFooler and BertAttack. TextFooler is a state-of-the-art baseline to generate adversarial text. Similar to our methodology, they select markers and replace them to create perturbations. In TextFooler, a marker is selected by sorting the words on their importance ranking and picking the highest word after removing the stop words. Once the marker is selected, its synonyms are extracted for replacement. Synonyms are picked by comparing the cosine similarities of the words in the vocabulary with that of the marker. Parts of speech is ensured to be the same to generate grammatically valid statements. The semantic similarity of the sentences is obtained from the cosine similarity of their embeddings. The attacks are performed by replacing the marker with the best synonym resulting in label preserving perturbations.

Similarly, BertAttack finds vulnerable words by masking each word in the sentence and comparing their logit scores. K replacement words for the vulnerable words are then generated using the BERT model. No additional grammatical or semantic checks are performed as BERT is context aware. Although the accuracies of TextFooler and BertAttack are higher than our attack accuracy, the semantic similarity score for our attacks obtained using Universal Sentence Encoding model [3] is considerably greater as seen in Table 3.

As earlier we also note that the attack labels of the two above methods can be prone to errors due to lack of checking of sense of the word and illegitimate words being introduced into the text. Our method for generating adversarial examples is much more computationally efficient than TextAttack [19]. We give some examples below.

Table 3. Accuracies and semantic similarity of the attacks

Attack	Accuracy on SNLI	Accuracy on MNLI	semantic Similarity
TextFooler	96%	90.4%	0.45
BERT-Attack	92.6%	92.1%	0.40
Ours	79.7%	15.4%	0.87

6.1   Comparison of Examples with TextFooler and BertAttack

The following examples illustrate the issues with the approach followed by TextFooler and BertAttack:

- **Errors in label shifts**: The replacement words considered are not always synonyms, thus leading to incorrect attacks as the perturbations are not label preserving.
    - **TextFooler**- Original: A man in a blue shirt is looking up at a *dog.*
      Perturbation: A man in a blue shirt is looking up at a *canine.*
    - **BertAttack**- Original: A person throwing something for her *dog.*
      Perturbation: A person throwing something for her *puppy.*
      Explanation: The relation between canine and dog is hypernymy, while that between dog and puppy is hyponymy rather than synonymy. The label will therefore be dependent on the monotonicity of the word.
- **Improper Perturbations**
    - Original: There is a little *boy* who *likes* the colour brown.
      Perturbation:
        * **TextFooler**: There is a little boy who *iikes* the colour brown.
        * **Ours**: There is a little *person* who likes the colour brown.
    - Original: Girl *plays nintendo.*
      Perturbation:
        * **BertAttack**: Girl *and facebook.*
        * **Ours**: *Scout* plays nintendo.
      Explanation: Non-existent words or unrelated words.
- **Incorrect Sense** The sense of the replacement word is completely different from the original sense, thus changing the semantics of the sentence. Though parts of speech is considered to ensure the grammaticality of the text, the morphology of the words is not maintained, resulting in sentences with improper grammar.

- Original: The dogs are *running* along the shore to meet their master who just beached his *kayak*.
  Perturbation:
  * **TextFooler**: The dogs are *executed* along the shore to meet their master who just beached his kayak.
  * **Ours**: The dogs are running along the shore to meet their master who just beached his *canoe*.

7 Conclusion

We have proposed a novel approach to generate adversarial datasets from benchmark NLI datasets. These attacks help in assessing a Neural Language Model's understanding of monotonicity reasoning. We evaluate the generated datasets on state-of-the-art NLI models and analyze their performance. We conclude with a comparison with state-of-the-art adversarial attacks and show that our methods produce more semantically similar sentences and do not suffer from lexical errors.

While single word substitutions are easy to incorporate and effective, not all concepts can be encapsulated by a single word. Future work can focus on structural changes with phrase replacement to better test the model's monotonic reasoning ability. Another line of work can be explanation-based attacks that can probe the model's ability to generalize utilizing the context of the sentences. While adversarial analysis illuminates the workings of the model, it remains to be seen if such rules can be incorporated into the models efficiently. So far, while there's work [6] which tries to do so, retraining a model for such a task is computationally expensive while humans can integrate such logical reasoning much more easily. This remains an open area of research.

Acknowledgements. Part of this work was funded by the Institute of Eminence Grant, UoH-IoE-RC3-21-050.

References

1. Barba, E., Pasini, T., Navigli, R.: ESC: Redesigning WSD with extractive sense comprehension. In: Proceedings of the 2021 Conference of the North American Chapter of the Association for Computational Linguistics: Human Language Technologies. ACL, June 2021. https://doi.org/10.18653/v1/2021.naacl-main.371
2. Bowman, S.R., Angeli, G., Potts, C., Manning, C.D.: A large annotated corpus for learning natural language inference. In: Proceedings of the 2015 Conference on Empirical Methods in Natural Language Processing. ACL, September 2015. https://doi.org/10.18653/v1/D15-1075
3. Cer, D., et al.: Universal sentence encoder for English. In: Proceedings of the 2018 Conference on Empirical Methods in Natural Language Processing: System Demonstrations. ACL, November 2018. https://doi.org/10.18653/v1/D18-2029

4. Devlin, J., Chang, M.W., Lee, K., Toutanova, K.: Bert: Pre-training of deep bidirectional transformers for language understanding. In: Proceedings of 2019 Conference of the North American Chapter of the Association for Computational Linguistics NAACL (2019). https://doi.org/10.18653/v1/N19-1423

5. Ebrahimi, J., Rao, A., Lowd, D., Dou, D.: HotFlip: White-box adversarial examples for text classification. In: Proceedings of the 56th Annual Meeting of the Association for Computational Linguistics (Volume 2: Short Papers). ACL, July 2018. https://doi.org/10.18653/v1/P18-2006

6. Glockner, M., Shwartz, V., Goldberg, Y.: Breaking NLI systems with sentences that require simple lexical inferences. In: Proceedings of the 56th Annual Meeting of the Association for Computational Linguistics (Volume 2: Short Papers). ACL, July 2018. https://doi.org/10.18653/v1/P18-2103

7. Gururangan, S., Swayamdipta, S., Levy, O., Schwartz, R., Bowman, S., Smith, N.A.: Annotation artifacts in natural language inference data. In: Proceedings of the 2018 Conference of the North American Chapter of the Association for Computational Linguistics: Human Language Technologies, vol. 2 (Short Papers). ACL, June 2018. https://doi.org/10.18653/v1/N18-2017

8. Hossain, M.M., Kovatchev, V., Dutta, P., Kao, T., Wei, E., Blanco, E.: An analysis of natural language inference benchmarks through the lens of negation. In: EMNLP (2020). https://doi.org/10.18653/v1/2020.emnlp-main.732

9. Hu, H., Moss, L.: Polarity computations in flexible categorial grammar. In: Proceedings of the Seventh Joint Conference on Lexical and Computational Semantics. ACL, June 2018. https://doi.org/10.18653/v1/S18-2015

10. Jia, R., Liang, P.: Adversarial examples for evaluating reading comprehension systems. In: Proceedings of the 2017 Conference on Empirical Methods in Natural Language Processing. ACL, September 2017. https://doi.org/10.18653/v1/D17-1215

11. Jin, D., Jin, Z., Zhou, J.T., Szolovits, P.: Is BERT really robust? A strong baseline for natural language attack on text classification and entailment. In: Proceedings of the AAAI Conference on Artificial Intelligence, vol. 34(05), April 2020. https://doi.org/10.1609/aaai.v34i05.6311

12. Li, J., Ji, S., Du, T., Li, B., Wang, T.: TextBugger: generating adversarial text against real-world applications. In: Proceedings of the Symposium on Networks and Distributed System Security, December 2018. https://doi.org/10.14722/ndss.2019.23138

13. Li, L., Ma, R., Guo, Q., Xue, X., Qiu, X.: BERT-ATTACK: Adversarial attack against BERT using BERT. In: Proceedings of the 2020 Conference on Empirical Methods in Natural Language Processing (EMNLP). ACL, November 2020. https://doi.org/10.18653/v1/2020.emnlp-main.500

14. Li, T., Gupta, V., Mehta, M., Srikumar, V.: A logic-driven framework for consistency of neural models. In: Proceedings of the 2019 Conference on Empirical Methods in Natural Language Processing and the 9th International Joint Conference on Natural Language Processing (EMNLP-IJCNLP). ACL, November 2019. https://doi.org/10.18653/v1/D19-1405

15. MacCartney, B., Manning, C.D.: An extended model of natural logic. In: Proceedings of the Eighth International Conference on Computational Semantics. ACL, January 2009. https://doi.org/10.3115/1693756.1693772, https://aclanthology.org/W09-3714

16. Maheshwary, R., Maheshwary, S., Pudi, V.: Generating natural language attacks in a hard label black box setting (2021). https://doi.org/10.1609/aaai.v35i15.17595

17. Marelli, M., Bentivogli, L., Baroni, M., Bernardi, R., Menini, S., Zamparelli, R.: SemEval-2014 task 1: evaluation of compositional distributional semantic models on full sentences through semantic relatedness and textual entailment. In: Proceedings of the 8th International Workshop on Semantic Evaluation (SemEval 2014). ACL, August 2014. https://doi.org/10.3115/v1/S14-2001
18. Miller, G.A.: WordNet: a lexical database for English. Commun. ACM. **38**(11), 39-41 (1995). https://doi.org/10.1145/219717.219748
19. Morris, J., Lifland, E., Yoo, J.Y., Grigsby, J., Jin, D., Qi, Y.: TextAttack: a framework for adversarial attacks, data augmentation, and adversarial training in NLP. In. Proceedings of the 2020 Conference on Empirical Methods in Natural Language Processing: System Demonstrations. ACL, October 2020. https://doi.org/10.18653/v1/2020.emnlp-demos.16
20. Naik, A., Ravichander, A., Sadeh, N., Rose, C., Neubig, G.: Stress test evaluation for natural language inference. In: Proceedings of the 27th International Conference on Computational Linguistics, pp. 2340–2353. ACL, Santa Fe, New Mexico, USA, August 2018. https://aclanthology.org/C18-1198
21. Nguyen, A., Yosinski, J., Clune, J.: Deep neural networks are easily fooled: high confidence predictions for unrecognizable images. In: 2015 IEEE Conference on Computer Vision and Pattern Recognition (CVPR) (2015). https://doi.org/10.1109/CVPR.2015.7298640
22. Poliak, A., Naradowsky, J., Haldar, A., Rudinger, R., Van Durme, B.: Hypothesis only baselines in natural language inference. In: Proceedings of the Seventh Joint Conference on Lexical and Computational Semantics. ACL, June 2018. https://doi.org/10.18653/v1/S18-2023
23. Richardson, K., Hu, H., Moss, L., Sabharwal, A.: Probing natural language inference models through semantic fragments. Proceedings of the AAAI Conference on Artificial Intelligence, vol. 34(05), April 2020. https://doi.org/10.1609/aaai.v34i05.6397
24. Saha, S., Nie, Y., Bansal, M.: ConjNLI: Natural language inference over conjunctive sentences. In: Proceedings of the 2020 Conference on Empirical Methods in Natural Language Processing (EMNLP). ACL, November 2020. https://doi.org/10.18653/v1/2020.emnlp-main.661
25. Song, K., Tan, X., Qin, T., Lu, J., Liu, T.Y.: MPNet: masked and permuted pre-training for language understanding. In: Larochelle, H., Ranzato, M., Hadsell, R., Balcan, M., Lin, H. (eds.) Advances in Neural Information Processing Systems, vol. 33, pp. 16857–16867 (2020)
26. Sutskever, I., Vinyals, O., Le, Q.V.: Sequence to sequence learning with neural networks. In: Ghahramani, Z., Welling, M., Cortes, C., Lawrence, N., Weinberger, K. (eds.) Advances in Neural Information Processing Systems, vol. 27 (2014)
27. Tarunesh, I., Aditya, S., Choudhury, M.: LoNLI: an extensible framework for testing diverse logical reasoning capabilities for NLI (2021). https://doi.org/10.48550/ARXIV.2112.02333
28. Traylor, A., Feiman, R., Pavlick, E.: AND does not mean OR: using formal languages to study language models' representations. In: Proceedings of the 59th Annual Meeting of the Association for Computational Linguistics and the 11th International Joint Conference on Natural Language Processing, (Volume 2: Short Papers). ACL, August 2021. https://doi.org/10.18653/v1/2021.acl-short.21
29. Vaswani, A., et al.: Attention is all you need. In: Proceedings of the 31st International Conference on Neural Information Processing Systems. NIPS 2017, vol. 30 (2017)

30. Vijayaraghavan, P., Roy, D.: Generating black-box adversarial examples for text classifiers using a deep reinforced model. In: Brefeld, U., Fromont, E., Hotho, A., Knobbe, A., Maathuis, M., Robardet, C. (eds.) ECML PKDD 2019. LNCS (LNAI), vol. 11907, pp. 711–726. Springer, Cham (2020). https://doi.org/10.1007/978-3-030-46147-8_43

31. Warstadt, A., Singh, A., Bowman, S.R.: Neural network acceptability judgments. Trans. Assoc. Comput. Linguist. **7**, 625–641 (2019). https://doi.org/10.1162/tacl_a_00290

32. Williams, A., Nangia, N., Bowman, S.: A broad-coverage challenge corpus for sentence understanding through inference. In: Proceedings of the 2018 Conference of the North American Chapter of the Association for Computational Linguistics: Human Language Technologies, vol. 1 (Long Papers). ACL, June 2018. https://doi.org/10.18653/v1/N18-1101

33. Yanaka, H., et al.: Can neural networks understand monotonicity reasoning? In: Proceedings of the 2019 ACL Workshop BlackboxNLP: Analyzing and Interpreting Neural Networks for NLP. ACL, August 2019. https://doi.org/10.18653/v1/W19-4804

34. Zhao, Z., Dua, D., Singh, S.: Generating natural adversarial examples. In: International Conference on Learning Representations (2018). https://openreview.net/forum?id=H1BLjgZCb

Understanding Document Data Sources Using Ontologies with Referring Expressions

Alexander Borgida[1,2,3], Enrico Franconi[1,2,3], David Toman[1,2,3(✉)], and Grant Weddell[1]

[1] Department of Computer Science, Rutgers University, New Brunswick, USA
[2] KRDB Research Centre for Knowledge and Data,
Free University of Bozen-Bolzano, Bolzano, Italy
[3] Cheriton School of Computer Science, University of Waterloo, Waterloo, Canada
david@uwaterloo.ca

Abstract. We show how JSON documents can be abstracted as concept descriptions in an appropriate Description Logic (DL). This representation allows the use of a DL ontology, which includes naming conventions ("referring expression types (RETs)") for instances of certain primitive concepts, in order to locate (perhaps multiple) subdocuments of the original JSON document capturing information about some particular conceptual entity. Detecting such situations allows for normalizing the JSON document into several separate smaller documents that capture all information about each such conceptual entity. This transformation preserves all the original information present in the input document. The RET assignment enables more refined and normalized capture of documents, and lead to query answers that adhere better to user expectations. We also show how RETs allow checking for a document admissibility condition ensuring that each final subdocument describes a single conceptual entity.

1 Introduction and Motivation

Suppose we have a JSON/mongoDB document, and a Description Logics (DL) ontology attaching semantics to (most) fields in JSON objects (called "keys" in the JSON definition; for the rest of this paper we reserve the word "key" for database-like keys).[1] More precisely, we treat the fields in JSON objects as (functional) roles in an underlying DL, and use the TBox of our ontology to introduce, when appropriate, additional concepts for the subject domain of the document. For example, for the JSON document

```
{ "fname": "John", "lname": "Smith", "age": 25,
  "wife": { "fname" : "Mary", "lname": "Smith" } }
```

---

[1] We assume that the reader is generally familiar with standard DL terminology, such as individuals, roles, TBox, and ABox, as well as JSON documents. For formal definitions of these please see Sect. 2.

© The Author(s), under exclusive license to Springer Nature Switzerland AG 2022
H. Aziz et al. (Eds.): AI 2022, LNAI 13728, pp. 367–380, 2022.
https://doi.org/10.1007/978-3-031-22695-3_26

the TBox might contain subsumptions stating that:

- PERSONs are objects that have *fname* and *lname* fields,
- *fname* and *lname* form a key for PERSONs,
- a *wife* of a PERSON is also a PERSON, and so on.

We are interested in asking (conjunctive) queries over the document, with answers being returned in some "answer language" $\mathcal{L}_{answer}$. We could simply create distinct identifiers for each node in the JSON tree, and use roles (obtained from the fields of JSON objects) to connect them, resulting in an ABox, which in the above case might contain an assertion like (id434,25) : age. In this case, a simple query like q(P) :- age(P,25) would return id434, which is quite meaningless without laboriously tracing back the creation of the ABox; and this answer makes no connection with the knowledge in the ontology. A more desirable answer would use the TBox, and return something like *PERSON with fname= "John" and lname= "Smith"*. This issue, and many others concerning the appropriate choice of description for objects in query answers was first addressed in [2], where the notion of "singular referring expression" was introduced. It stood for a *concept description* that was guaranteed to denote a single individual.

Given the use of a referring expression in $\mathcal{L}_{answer}$ to, among others, avoid meaningless object ids, another paper [13] suggested using a similar approach in some "update language" $\mathcal{L}_{tell}$: replacing the ABox with a "CBox"—a set of concept descriptions that serve as singular referring expressions for objects and are used to assert facts as special concepts the KB knows about. For example, the JSON fragment above could be captured by the concept

$$\exists \text{fname}.\{"\text{John}"\} \sqcap \exists \text{lname}.\{"\text{Smith}"\} \sqcap \exists \text{age}.\{"25"\} \sqcap$$
$$\exists \text{wife}.(\exists \text{fname}.\{"\text{Mary}"\} \sqcap \exists \text{lname}.\{"\text{Smith}"\})$$

Here, nominal concepts play a central role, as do functionality and key constraints for singularity, and a general rule that CBox concepts cannot have empty interpretations.

An algorithm was given in [13] for computing certain answers to conjunctive queries, which returns singular referring expressions appearing in the CBox. This work was carried out in a dialect of the FunDL family of description logics [8]. FunDL and its dialects replace roles with (possibly partial) unary functions called *features*, reify general roles, and always include a so-called *path functional dependency* (PFD) concept constructor to express constraints that are equality generating (including keys). In particular, the dialect used for these purposes in [13] has the property that DL reasoning and query answering are both polynomial-time (in the size of the underlying knowledge base). The above paper also gave, as an example of CBox use, one way to represent a JSON document (in particular, for a document in a MongoDB collection), as a singular concept description in FunDL.

Example 1 ([13]). Consider the case where "person" is the name of a MongoDB JSON collection with a value as given in Fig. 1. The "person" document is captured by a CBox containing the concept description formulated in FunDL and shown in Fig. 2.   □

Intuitively, JSON values are mapped to concepts as follows: (a) primitive values to nominals, (b) (compound) objects to conjunctions of existential restrictions of features corresponding to the field names to the (mapping of) values, and (c) arrays (treated as sets) to multi-valued *roles*, which are then reified using the reserved features dom and ran (see Definition 4). One can also infer that the concept corresponding to the original document, such as the concept in Fig. 2, will have the above-mentioned singularity property by asserting in a FunDL TBox that collection names ("person" in our case) are unique.

```
{ "collection": "person",
  "data" : [
   { "fname": "John", "lname": "Smith", "age": 25,
     "wife": { "fname" : "Mary" },
     "phone": [
        {"colour": "red", "dnum": "212 555-1234"}
      ] } ,
   { "fname": "Mary", "lname": "Jones", "salary": "$150,000 (CAD)",
     "spouse": { "fname": "John" },
     "phone": [
        {"loc": "home", "dnum": "212 555-1234"},
        {"loc": "work", "dnum": "212 666-4567"}
      ] }
  ] }
```

Fig. 1. JSON *PERSON* Document.

$\exists$ collection.{"person"} $\sqcap$
$\exists$ data(
  $\exists$ dom$^-$.$\exists$ ran(
    $\exists$ fname.{"John"} $\sqcap$ $\exists$ lname.{"Smith"} $\sqcap$
    $\exists$ age.{"25"} $\sqcap$ $\exists$ wife.$\exists$ fname.{"Mary"} $\sqcap$
    $\exists$ phone.$\exists$ dom$^-$.$\exists$ ran($\exists$ colour.{"red"} $\sqcap$ $\exists$ dnum.{"212 555-1234"})) $\sqcap$
  $\exists$ dom$^-$.$\exists$ ran(
    $\exists$ fname.{"Mary"} $\sqcap$ $\exists$ lname.{"Jones"} $\sqcap$
    $\exists$ salary.{"$150000CAD"} $\sqcap$ $\exists$ spouse.$\exists$ fname.{"John"} $\sqcap$
    $\exists$ phone.$\exists$ dom$^-$.$\exists$ ran($\exists$ loc.{"home"} $\sqcap$ $\exists$ dnum.{"212 555-1234"}) $\sqcap$
        $\exists$ dom$^-$.$\exists$ ran($\exists$ loc.{"work"} $\sqcap$ $\exists$ dnum.{"212 666-4567"}))))

Fig. 2. FunDL Encoding of the *PERSON* Document.

Since in our case query answers are elements of the CBox, the above translation of a full JSON document into a *single* CBox entry is undesirable because we can only return as answer the individual described by it. In this paper we propose to break up the one CBox entry into several conceptual entities that make sense with respect to the terminology in the TBox, and which correspond to breaking the original document into sub-documents. This *normalization* is achieved by using *referring expression types* (RETs) [3], a part of our ontology that determines *how objects of discourse are referred to* in the knowledge base. In particular, we will show how a combination of a TBox and an RET assignment to the primitive concepts occurring in the TBox enable mapping an initial CBox obtained directly from a MongoDB database, as illustrated above, to an alternative *normalized* CBox, as illustrated below.

Example 2 (JSON normalization). Applying our normalization procedure to the CBox in Example 1 will obtain the following concepts:

DOCUMENT $\sqcap \exists$ collection.{"person"} $\sqcap$
  $\exists$ data($\exists$ dom$^-$.$\exists$ ran($\underline{\text{PERSON} \sqcap \exists \text{fname.\{"John"\}} \sqcap \exists \text{lname.\{"Smith"\}}}$) $\sqcap$
    $\exists$ dom$^-$.$\exists$ ran($\underline{\text{PERSON} \sqcap \exists \text{fname.\{"Mary"\}} \sqcap \exists \text{lname.\{"Jones"\}}}$))

$\underline{\text{PERSON} \sqcap \exists \text{fname.\{"John"\}} \sqcap \exists \text{lname.\{"Smith"\}}} \sqcap$
    $\exists$ age.{"25"} $\sqcap \exists$ wife.$\exists$ fname{"Mary"} $\sqcap$
    $\exists$ phone.$\exists$ dom$^-$.$\exists$ ran($\text{PHONE} \sqcap \exists \underline{\underline{\text{dnum\{"212 555-1234"\}}}}$)
$\underline{\text{PERSON} \sqcap \exists \text{fname.\{"Mary"\}} \sqcap \exists \text{lname.\{"Jones"\}}} \sqcap$
    $\exists$ salary.{"$150000CAD"} $\sqcap \exists$ spouse.$\exists$ fname{"John"} $\sqcap$
    $\exists$ phone.$\exists$ dom$^-$.$\exists$ ran($\text{PHONE} \sqcap \exists \underline{\underline{\text{dnum\{"212 555-1234"\}}}}$) $\sqcap$
      $\exists$ dom$^-$.$\exists$ ran($\text{PHONE} \sqcap \exists \underline{\underline{\text{dnum\{"212 666-4567"\}}}}$))

$\underline{\text{PHONE} \sqcap \exists \text{dnum\{"212 555-1234"\}}} \sqcap \exists$ loc.{"home"} $\sqcap \exists$ colour.{"red"}
$\underline{\text{PHONE} \sqcap \exists \text{dnum\{"212 555-4567"\}}} \sqcap \exists$ loc.{"work"}

In the above, the underlined subconcepts of the CBox concepts serve as *referring expressions* identifying entities, while the remainder of these concepts tells us facts about the entities, using (the dash-underlined) referring expressions when needed, to record facts relating to other identifiable entities. For example, observe how references to phone entities in **phone** facts about persons require only **dnum** facts about phones, but not **loc** or **colour** facts.   □

Our contributions are as follows:

1. We show how a JSON document (or a MongoDB collection) can be abstracted as a concept description, as illustrated in Example 1.
2. We describe the normalization procedure which uses the given TBox and a referring expression type assignment in order to extract additional intuitively reasonable CBox subconcepts, as illustrated in Example 2.
3. We show how a TBox can attach meaning to such concept descriptions (and therefore to the JSON documents), and contrast this with other proposals for assigning a meaning/semantics to JSON documents.
4. We also show how information about the same entity can be consolidated even if it was originally recorded in different parts of the JSON document.
5. Finally, we present a more effective way of diagnosing an admissibility property of a CBox that ensures interpretations of referring expressions are indeed singular.

All of the above tasks are accomplished by relying solely on reasoning about equalities and concept memberships of objects corresponding to values in the input JSON document with respect to the DL ontology (TBox and RET for naming conventions). This sets our approach apart from many other approaches that commonly rely on hand-coded mapping and transformation rules.

The paper is organized as follows: Sect. 2 provides the needed background relating to FunDL and to referring expressions. Section 3 then outlines the main results of this paper relating to the ability to identify subdocuments relating to identifiable entities and subsequent separation of these entities in separate CBox entries/documents. We conclude with a brief overview of related work and with suggestions for follow-on research.

2   Definitions and Background

We now formally define the artifacts introduced in our introductory comments, beginning with a general definition of concept descriptions for members of the FunDL family of DLs with PTIME complexity of logical consequence.[2] Recall that members of this family replace roles with partial functions, and that concept descriptions not only occur in a TBox but also serve as referring expressions in a CBox.

Definition 1 (FunDL Concepts, Referring Expressions, and Knowledge Bases). *Let* F *and* PC *be sets of feature names and primitive concept names, respectively. A* path expression *is defined by the grammar "*Pf ::= f.Pf | id*" for* $f \in$ F. *A concept description is defined by the grammar on the left-hand-side of Fig. 3.*[3]

A subsumption *is an expression of the form* $C \sqsubseteq D$, *where* C *and* D *are parsed by the first six productions in Fig. 3. A* terminology *(TBox)* $\mathcal{T}$ *consists of a finite set of subsumptions. A* concept box *(CBox)* $\mathcal{C}$ *consists of a finite set of concept descriptions parsed by the last six productions in Fig. 3; these are intended to assert the existence of individuals with complex properties. A* knowledge base $\mathcal{K}$ *is a pair* $(\mathcal{T}, \mathcal{C})$.[4]

The semantics *of concept descriptions and path expressions is defined with respect to a structure* $\mathcal{I} = (\triangle^{\mathcal{I}}, \cdot^{\mathcal{I}})$, *where* $\triangle^{\mathcal{I}}$ *is a domain of "objects" and* $\cdot^{\mathcal{I}}$ *an interpretation function that fixes the interpretations of primitive concepts* A *to be subsets of* $\triangle^{\mathcal{I}}$ *and primitive features* f *to be partial functions* $f^{\mathcal{I}} : \triangle^{\mathcal{I}} \to \triangle^{\mathcal{I}}$. *The interpretation is extended in the natural way to path expressions:* $id^{\mathcal{I}} = \lambda x. x$,

---

[2] Some additional conditions must be imposed on PFDs and on conjunctions to guarantee PTIME bounds; see [7,8] for details.

[3] A variety of equality generating dependencies, including keys, can be expressed with the use of a *path functional dependency* (PFD) concept description generated by the second production of this grammar.

[4] In Sect. 3 we also use the standard notion of a FunDL *assertion box* (ABox), a set of assertions of the form "$C(a)$", "$a = b$", and "$f(a) = b$" as defined in [7]. We elaborate on the relationship between knowledge bases that use a CBox and *classical* FunDL knowledge bases, i.e., with an ABox in that section.

$(f.\mathsf{Pf})^{\mathcal{I}} = \mathsf{Pf}^{\mathcal{I}} \circ f^{\mathcal{I}}$; and to complex concept descriptions C or D as indicated on the right-hand-side of Fig. 3.

$$C, D ::= \bot \qquad\qquad\qquad\qquad \emptyset$$

$$\mid\ C : \mathsf{Pf}_1, ..., \mathsf{Pf}_k \to \mathsf{Pf}_0 \quad \{x \mid \forall y.((y \in C^{\mathcal{I}} \wedge (\bigwedge_{i=0}^{k}\{x, y\} \subseteq (\exists \mathsf{Pf}_i.\top)^{\mathcal{I}})$$
$$(\bigwedge_{i=1}^{k} \mathsf{Pf}_i^{\mathcal{I}}(x) = \mathsf{Pf}_i^{\mathcal{I}}(y))) \to (\mathsf{Pf}_0^{\mathcal{I}}(x) = \mathsf{Pf}_0^{\mathcal{I}}(y)))\}$$

$$\mid\ \top \qquad\qquad\qquad\qquad \triangle^{\mathcal{I}}$$

$$\mid\ A \qquad\qquad\qquad\qquad A^{\mathcal{I}} \subseteq \triangle^{\mathcal{I}}$$

$$\mid\ \exists \mathsf{Pf}.C \qquad\qquad\qquad \{x \mid \exists y.(y \in C^{\mathcal{I}} \wedge \mathsf{Pf}^{\mathcal{I}}(x) = y)\}$$

$$\mid\ C \sqcap D \qquad\qquad\qquad C^{\mathcal{I}} \cap D^{\mathcal{I}}$$

$$\mid\ \{a\} \qquad\qquad\qquad \{a^{\mathcal{I}}\}$$

$$\mid\ \exists f^{-1}.C \qquad\qquad\qquad \{f^{\mathcal{I}}(x) \mid x \in C^{\mathcal{I}}\}$$

Fig. 3. SYNTAX AND SEMANTICS OF CONCEPT DESCRIPTIONS.

An interpretation $\mathcal{I}$ satisfies an subsumption $C \sqsubseteq D$ if $C^{\mathcal{I}} \subseteq D^{\mathcal{I}}$, and is a model of a TBox $\mathcal{T}$ if it satisfies all inclusion dependencies in $\mathcal{T}$. $\mathcal{I}$ is a model of a knowledge base $\mathcal{K} = (\mathcal{T}, \mathcal{C})$, written $\mathcal{I} \models \mathcal{K}$, if it satisfies $\mathcal{T}$ and also that $|C^{\mathcal{I}}| > 0$ holds for every $C \in \mathcal{C}$.

Given a TBox $\mathcal{T}$, a concept C is singular *with respect to $\mathcal{T}$ if $|C^{\mathcal{I}}| \leq 1$ for all interpretations $\mathcal{I}$ that are models of $\mathcal{T}$. We call such a concept a* referring expression.

The logical implication problem *asks if $\mathcal{K} \models C \sqsubseteq D$ holds, that is, if $C \sqsubseteq D$ is satisfied in all models of $\mathcal{K}$.* □

Definition 2 (Admissibility and Query Answers). *Let $\mathcal{K} = (\mathcal{T}, \mathcal{C})$ be a FunDL knowledge base and $Q = \{(x_1, \ldots, x_k) \mid \varphi\}$ a conjunctive query. The CBox $\mathcal{C}$ is* admissible *for $\mathcal{T}$ if each $C \in \mathcal{C}$ is a referring expression that is singular with respect to $\mathcal{T}$. (Thus, if $\mathcal{K}$ is consistent and $\mathcal{C}$ is admissible for $\mathcal{T}$, $|C^{\mathcal{I}}| = 1$ for any $C \in \mathcal{C}$ and any interpretation $\mathcal{I}$.)*

A k-tuple of referring expressions $(C_1, \ldots, C_k)$ is a certain answer *to Q in $(\mathcal{T}, \mathcal{C})$ if*

$$\mathcal{K} \models \exists x_1, \ldots, x_k.(\varphi \wedge C_1(x_1) \wedge \ldots \wedge C_k(x_k))$$

for $\{C_1, \ldots, C_k\} \subseteq \mathcal{C}$. □

The second part of our view of an ontology consists of naming conventions for individuals, called *referring expression types* (RETs), which were introduced in [2]. In this earlier work, such types were attached to the free variables of a conjunctive query, and denoted a space of well-formed formulae ψ, with one free variable, over a given FO signature consisting of unary and binary predicates that were eligible to appear in referring expressions for the variable. Such types are essentially patterns of possible ψ. Here, these are patterns of possible descriptions of concept instances, and are now attached to primitive concepts by a user defined *referring expression type assignment* (RTA).[5] They determine a set of possible concept descriptions that are eligible referring expressions for subdocuments. We illustrate this below for our running example, but leave the presentation on how this is accomplished to Sect. 3.

Definition 3 (Referring Expression Types and Assignments). *A refer-ring expression type is defined by the following grammar:*[6]

$$Re ::= A \mid \{?\} \mid \exists \mathsf{Pf}.Re \mid Re \sqcap Re \mid Re \, ; Re$$

A referring expression type assignment (RTA) over a TBox $\mathcal{T}$ is a partial func-tion mapping primitive concepts A occurring in $\mathcal{T}$ to referring expression types RTA(A). We define the language of referring expressions inhabiting Re, $\mathcal{L}(Re)$, as follows:

$$\mathcal{L}(A) = \{A\}$$
$$\mathcal{L}(\{?\}) = \{\{b\} \mid b \text{ is a constant symbol}\}$$
$$\mathcal{L}(\exists \mathsf{Pf}.Re) = \{\exists \mathsf{Pf}.C \mid C \in \mathcal{L}(Re)\}$$
$$\mathcal{L}(Re_1 \sqcap Re_2) = \{C_1 \sqcap C_2 \mid C_1 \in \mathcal{L}(Re_1) \text{ and } C_2 \in \mathcal{L}(Re_2)\}$$
$$\mathcal{L}(Re_1; Re_2)) = \mathcal{L}(Re_1) \cup \mathcal{L}(Re_2)$$

□

Example 3 (CBox normalization). Let CBox $\mathcal{C}$ consist of the single concept description in Fig. 2 obtained from the MongoDB collection given earlier, and let TBox $\mathcal{T}$ consist of the following subsumptions:

$$(\exists \, \mathsf{collection}.\top) \sqcap (\exists \, \mathsf{data}.\top) \sqsubseteq \mathrm{DOCUMENT}$$
$$(\exists \, \mathsf{fname}.\top) \sqcap (\exists \, \mathsf{lname}.\top) \sqsubseteq \mathrm{PERSON}$$
$$\exists \, \mathsf{dnum}.\top \sqsubseteq \mathrm{PHONE}$$

$$\mathrm{DOCUMENT} \sqsubseteq \mathrm{DOCUMENT} : \mathsf{collection} \to id$$
$$\mathrm{PERSON} \sqsubseteq \mathrm{PERSON} : \mathsf{fname}, \mathsf{lname} \to id$$
$$\mathrm{PHONE} \sqsubseteq \mathrm{PHONE} : \mathsf{dnum} \to id$$
$$\mathrm{PERSON} \sqsubseteq \exists \, \mathsf{wife}.\mathrm{PERSON}$$

Consider where our aforementioned normalization procedure, to be detailed in the next section, is given as input: (a) the above knowledge base $\mathcal{K} = (\mathcal{T}, \mathcal{C})$, and (b) the following referring expression type assignment:

$$\mathrm{RTA}(\mathrm{DOCUMENT}) = \mathrm{DOCUMENT} \sqcap \exists \, \mathsf{collection}.\{?\}$$
$$\mathrm{RTA}(\mathrm{PERSON}) = \mathrm{PERSON} \sqcap \exists \, \mathsf{lname}.\{?\} \sqcap \exists \, \mathsf{fname}.\{?\}$$
$$\mathrm{RTA}(\mathrm{PHONE}) = \mathrm{PHONE} \sqcap \exists \, \mathsf{dnum}.\{?\}$$

Our CBox normalization procedure will then replace $\mathcal{C}$ in $\mathcal{K}$ by the CBox in Example 2 of our introduction.   □

---

[5] In [3] it was argued that determining ways to refer to individuals is an integral but distinct step of conceptual modelling/ontology design.

[6] This is a pattern language obtained by abstracting nominals in referring expressions, and by admitting a final production to express *preference* among referring expressions [2].

A function ToC that maps a JSON value to a referring expression in a CBox is straightforward using FunDL concepts, and is detailed next. (Recall that Example 1 in our introduction illustrates an invocation of ToC on a JSON collection.) Some observations and reminders: (a) this mapping assumes *any* JSON value, including an array, will map to some element of an underlying domain; (b) the mapping relies entirely on interpreting field names in field-value pairs comprising JSON objects as feature names; and (c) arrays (treated as sets) are mapped to multi-valued roles reified via the features dom and ran.

Definition 4 (ToConcept). *An arbitrary JSON value is mapped to a CBox referring expression as follows:*

$$\mathsf{ToC}(\texttt{"s"}) \mapsto \{\texttt{"s"}\}$$
$$\mathsf{ToC}(\texttt{null}) \mapsto \top$$
$$\mathsf{ToC}(\{\texttt{"k}_1\texttt{"} : v_1, \ldots, \texttt{"k}_n\texttt{"} : v_n\}) \mapsto \exists \mathsf{k}_1.\,\mathsf{ToC}(v_1) \sqcap \ldots \sqcap \exists \mathsf{k}_n.\,\mathsf{ToC}(v_n)$$
$$\mathsf{ToC}([v_1, \ldots, v_m]) \mapsto \exists\,\mathsf{dom}^-.\exists\,\mathsf{ran}.\,\mathsf{ToC}(v_1) \sqcap$$
$$\ldots \sqcap \exists\,\mathsf{dom}^-.\exists\,\mathsf{ran}.\,\mathsf{ToC}(v_m)$$

where the first case covers all JSON values that are strings, numerics, and Booleans.   □

3   CBox Normalization

Our CBox normalization procedure is based on a pair of normalization rules to be presented in Subsect. 3.2. To enable references to sub-concepts in a concept description required by our formulation of these rules, we define the mapping ToAbox that converts a CBox $\mathcal{C}$ to a standard FunDL ABox $\mathsf{ToAbox}(\mathcal{C})$, a set of assertions of the form "$A(a)$", "$a = b$", and "$f(b) = c$" (with the standard interpretation; for details see [7]).

Definition 5 (ToABox). *Let $\mathcal{C}$ be a CBox. Each concept $C \in \mathcal{C}$ is associated with a fresh constant 'a' and mapped to ABox assertions as follows:*

$$\mathsf{ToAbox}(a : \{b\}) \mapsto \{a = b\}$$
$$\mathsf{ToAbox}(a : A) \mapsto \{A(a)\}, A \ primitive$$
$$\mathsf{ToAbox}(a : \exists f.D) \mapsto \{f(a) = b\} \cup \mathsf{ToAbox}(b : D), b \ fresh$$
$$\mathsf{ToAbox}(a : \exists f^{-1}.D) \mapsto \{f(b) = a\} \cup \mathsf{ToAbox}(b : D), b \ fresh$$
$$\mathsf{ToAbox}(a : D_1 \sqcap \ldots \sqcap D_n) \mapsto \bigcup_{i=1}^n \{a = b_i\} \cup \mathsf{ToAbox}(b_i : D_i), b_i \ fresh.$$

We then define $\mathsf{ToAbox}(\mathcal{C}) = \bigcup_{D_i \in \mathcal{C}} \mathsf{ToAbox}(a_i : D_i)$ for a_i fresh.   □

Intuitively, the $\mathsf{ToAbox}(a : C)$ function converts an input concept $C \in \mathcal{C}$ to a set of ABox assertions by assigning a to its root node and then traversing the syntactic structure of the concept, assigning additional *distinct* constant symbols to subconcepts of C while generating additional ABox assertions. These assertions reflect the meaning of the original concept C. It is easy to see that the models of $(\mathcal{T}, \mathcal{C})$ coincide with the models of $(\mathcal{T}, \mathsf{ToAbox}(\mathcal{C}))$ up to the interpretation of constant symbols introduced by the ToAbox mapping. We can therefore use this connection and the standard FunDL reasoning techniques [7] to reason about entailments with respect to $(\mathcal{T}, \mathcal{C})$. Note also that the ToAbox mapping implicitly associates *subconcepts in* $\mathcal{C}$ with constants. This leads to the following definition.

Definition 6 (Context). *Let* $C \in \mathcal{C}$ *be a concept description. We use the notation* $C[a' : D]$ *to denote a subconcept* D *of the concept* C *where* a' *is the constant symbol assigned to* D *by* ToAbox$(a : C)$. *For the top-level concept we simply use the context* $[a : C]$.

Given a constant symbol a and referring expression type Re, the *projection function* ToRE is intended to generate a referring expression in $\mathcal{L}(Re)$ that identifies a. (Observe the appeal to an auxiliary recursive function in the definition that takes a path function Pf as an argument.)

Definition 7 (Projection on Re **).** *Let* $\mathcal{K} = (\mathcal{T}, \mathsf{ToAbox}(\mathcal{C}))$ *be a consistent knowledge base,* Re *a referring expression type, and 'a' a constant in* ToAbox$(\mathcal{C})$. *We define a* projection function ToRE(a, Re) *to be the result of the following recursive definition of* ToRE(a, Re, id) *on the structure of* Re:

$$\mathsf{ToRE}(a, \mathsf{A}, \mathsf{Pf}) = \mathsf{A} \text{ if } \mathcal{K} \models a : \exists \mathsf{Pf}.\mathsf{A}, \text{ undefined otherwise}$$
$$\mathsf{ToRE}(a, \{?\}, \mathsf{Pf}) = \{b\} \text{ if } \mathcal{K} \models a : \exists \mathsf{Pf}.\{b\} \text{ for some } b, \text{ undefined otherwise}$$
$$\mathsf{ToRE}(a, \exists \mathsf{Pf}'.Re, \mathsf{Pf}) = \exists \mathsf{Pf}'.\mathsf{ToRE}(a, Re, \mathsf{Pf}.\mathsf{Pf}')$$
$$\mathsf{ToRE}(a, Re_1 \sqcap Re_2, \mathsf{Pf}) = \mathsf{ToRE}(a, Re_1, \mathsf{Pf}) \sqcap \mathsf{ToRE}(a, Re_2, \mathsf{Pf}) \text{ if both defined}$$
$$\mathsf{ToRE}(a, Re_1; Re_2, \mathsf{Pf}) = \mathsf{ToRE}(a, Re_1, \mathsf{Pf}) \text{ if defined}, \mathsf{ToRE}(a, Re_2, \mathsf{Pf}) \text{ otherwise.}$$

See earlier work [10,11] for more effective ways of computing the second case by appealing to logical consequence in FunDL knowledge bases. The following is a consequence of this and our previous definitions.

Lemma 1. *For any constant 'a', any* Re, *and any consistent* $\mathcal{K}$, ToRE$(a, Re) \in \mathcal{L}(Re)$ *whenever* ToRE(a, Re) *is defined.*

It will also be useful to appeal to a simplification procedure for referring concepts. The following definition of such a procedure will suffice for illustrative purposes, and clearly preserves concept equivalence.

Definition 8 (Concept Simplification). *We write* Simplify(C) *to denote an exhaustive application of the following to referring expression* C:

1. *If an n-way conjunction contains* $\exists f.C_1$ *and* $\exists f.C_2$, *replace both conjuncts by* $\exists f.C_1 \sqcap C_2$.
2. *If an n-way conjunction contains duplicate conjuncts, remove one of the conjuncts.*   □

3.1   CBox Admissibility

In this subsection, we show how, given a TBox $\mathcal{T}$, one can statically test for admissibility of any CBox obtained by the normalization rules given in Subsect. 3.2 that follows. This is achieved by appeal to a sequence of logical consequence problems for subsumptions expressing functional dependencies with PFDs that are induced by a given RTA assignment. We begin by defining a normalization of an Re that preserves $\mathcal{L}(Re)$.

Definition 9 (Normalized Types). *We write* $\mathsf{Norm}(Re)$ *to refer to an exhaustive application of the following rewrite rules to* Re:

$$Re \sqcap (Re_1; Re_2) \;\mapsto\; Re \sqcap Re_1; Re \sqcap Re_2$$
$$(Re_1; Re_2) \sqcap Re \;\mapsto\; Re_1 \sqcap Re_1; Re_2 \sqcap Re$$
$$\exists \mathsf{Pf}.(Re_1; Re_2) \;\mapsto\; \exists \mathsf{Pf}.Re_1; \exists \mathsf{Pf}.Re_2$$

$\square$

The definition of Norm is an enhanced variant of referring expression type normalization in [2]. The following are consequences: (1) $\mathcal{L}(Re) = \mathcal{L}(\mathsf{Norm}(Re))$, and (2) all *preference operators* (";") are at the top level of $\mathsf{Norm}(Re)$. We call the maximal ";"-free parts of $\mathsf{Norm}(Re)$ *preference-free components*.

To statically test for singularity of referring expressions generated by the ToRE function for a particular referring expression type, we use the following auxiliary definitions:

$$\mathsf{Pfs}(\{?\}) = \{id\} \qquad\qquad \mathsf{Con}(\{?\}) = \top$$
$$\mathsf{Pfs}(A) = \{\,\} \qquad\qquad \mathsf{Con}(A) = A$$
$$\mathsf{Pfs}(\exists \mathsf{Pf}'.Re) = \{\mathsf{Pf}'.\mathsf{Pf} \mid \mathsf{Pf} \in \mathsf{Pfs}(Re)\} \quad \mathsf{Con}(\exists \mathsf{Pf}'.Re) = \exists \mathsf{Pf}'.\mathsf{Con}(Re)$$
$$\mathsf{Pfs}(Re_1 \sqcap Re_1) = \mathsf{Pfs}(Re_1) \cup \mathsf{Pfs}(Re_2) \quad \mathsf{Con}(Re_1 \sqcap Re_1) = \mathsf{Con}(Re_1) \sqcap \mathsf{Con}(Re_2)$$

These functions extract a set of paths leading to nominals and a FunDL concept from the preference-free referring expression type. Altogether, we are now able to formulate the singularity test following the ideas presented in [2], Theorem 20:

Theorem 1. *Let* $\mathcal{T}$ *be a TBox and* Re *a referring expression type. Then all referring expressions in* $\mathcal{L}(Re)$ *are singular if* $\mathcal{T} \models \mathsf{Con}(Re') \sqsubseteq \mathsf{Con}(Re')$: $\mathsf{Pfs}(Re') \to id$ *for every preference-free component* Re' *of* $\mathsf{Norm}(Re)$.

Our static test of admissibility of any CBox generated by our normalization rules then follows by applying the above to any Re in the range of a programmer supplied RTA.

3.2   CBox Normalization Rules

We now have the necessary machinery to present our two rules for normalizing the CBox of a given knowledge base $\mathcal{K} = (\mathcal{T}, \mathcal{C})$ and referring expression type assignment RTA.

Subdocument Extraction. Our first and main rule extracts sub-concepts of a given CBox concept $C \in \mathcal{C}$ as additional separate CBox concepts.

Definition 10 (Subdocument Extraction). *Let* C *be a concept in* $\mathcal{C}$ *that contains a subconcept* D *(i.e.,* $C[a : D]$*) that corresponds to a JSON object and such that* $(\mathcal{T}, \mathsf{ToAbox}(\mathcal{C})) \models A(a)$ *and* $\mathsf{RTA}(A)$ *is defined for a primitive concept* A. *We form a new CBox* $\mathcal{C}'$ *as follows*

$$\mathcal{C}' := \mathcal{C} - \{C[a : D]\} \cup \{C[a : \mathsf{ToRE}(a, \mathsf{RTA}(A))], \mathsf{ToRE}(a, \mathsf{RTA}(A)) \sqcap D\}$$

if ToRE$(a, \mathsf{RTA}(A))$ *is defined.*[7]   □

Intuitively, we replace a single monolithic concept C in $\mathcal{C}$ in which a subconcept D was identified as a representation of an A entity by a modified variant of C in which D has been *replaced* by its *referring expression*. In addition we create a new CBox concept ToRE$(a, \mathsf{RTA}(A)) \sqcap D$ for this recognized entity by extracting the subdocument D. Adding the concept ToRE$(a, \mathsf{RTA}(A))$ to D is needed since the referring expression to the document rooted by 'a' may not be fully contained in D, e.g., due to the use of subsumptions relating to inverses in $\mathcal{T}$.

Note that, because of properties of referring expressions (singularity in particular), we have that the following will hold for any pair of concepts $C[a : \mathsf{ToRE}(a, \mathsf{RTA}(A))]$ and $[b : \mathsf{ToRE}(a, \mathsf{RTA}(A)) \sqcap D]$ introduced by subdocument extraction:

$$(\mathcal{T}, \mathsf{ToAbox}(\mathcal{C}')) \models a = b.$$

Hence, up to renaming of constants, models will always coincide.

The choice of A above is non-deterministic, but does not affect the soundness of the extraction. However, this non-determinism can result in equivalent, but syntactically different referring expressions being used to identify the same subdocument. Although beyond the scope this paper, earlier work in [3] presents additional conditions beyond singularity that can be introduced on RTAs to ensure any choice of A leads to syntactically identical referring expressions.[8] Any RTA satisfying these conditions was called *identity resolving.*

Subdocument Merging. The *Subdocument Extraction* rule can extract multiple subdocuments identified by the same (or logically equivalent) referring expressions from the original document. This arrangement allows information about the same entity to be recorded in several locations in the original document. Moreover, it allows tree-like documents to naturally represent cyclic, graph-like data by simply repeating a particular referring expression in a subdocument.

The above observation leads to the second normalization rule that *collects* such subdocuments in a single CBox entry: it replaces two CBox referring expression entries with a single entry when co-reference is implied.

Definition 11 (Equivalent Subdocument Merge). *Let* $[a : C \sqcap D]$ *and* $[b : C' \sqcap D']$ *be two concepts in* $\mathcal{C}$ *such that* C *and* C' *are referring expressions and* $(\mathcal{T}, \mathsf{ToAbox}(\mathcal{C})) \models a = b$. *We replace* $\mathcal{C}$ *with*

$$\mathcal{C}' := \mathcal{C} - \{[a : C \sqcap D], [b : C' \sqcap D']\} \cup \{[a : C \sqcap \mathsf{Simplify}(D \sqcap D')]\}. \quad □$$

The rule does not assume the programmer supplied RTA is identity resolving. However, it would be sufficient to consider only the CBox entries *with identical*

---

[7] Otherwise we can report a warning about an A entity that cannot be properly identified.

[8] Essentially, this entails ensuring that preference in RTA(A), for any primitive concept A, exhaustively accounts for any primitive concept B for which $(A \sqcap B)^{\mathcal{I}}$ is non-empty for some interpretation $\mathcal{I}$.

referring expressions C and C' if RTA was identity resolving since in this case C would be syntactically identical to C'.

Our main results now follow.

Theorem 2. *Let C be a CBox in a consistent knowledge base $\mathcal{K} = (\mathcal{T}, \mathcal{C})$ and C' a CBox obtained by applying the* Subdocument Extraction *or the* Equivalent Subdocument Merge *rules. Then every model of $\mathcal{K}$ is also a model of $(\mathcal{T}, \mathcal{C}')$ and vice versa (up to renaming of constant symbols assigned by* ToAbox(.)*).*

It is an easy exercise to match constant symbols in $(\mathcal{T}, \mathsf{ToAbox}(\mathcal{C}))$ to those in $(\mathcal{T}, \mathsf{ToAbox}(\mathcal{C}'))$. However, due to equalities stemming from the RET assignment and thus implied by the knowledge base's TBox, this mapping can be many-to-many.

Theorem 3. *Let C be an admissible CBox in a consistent knowledge base $\mathcal{K} = (\mathcal{T}, \mathcal{C})$ and C' a CBox obtained by applying the* Subdocument Extraction *or the* Equivalent Subdocument Merge *rules. Then C' is admissible.*

4   Summary Comments

The main contribution of this paper is showing how JSON-like data sources can be abstracted as concept descriptions in an appropriate DL in a very generic way. This enables their domain-specific semantics to be naturally captured as a TBox in the same logic, which then allows one to draw on mature reasoning services that have been developed for DLs [1,7]. In addition, our approach utilizes *referring expressions* [2] as the means of identifying entities described in such data sources. This is crucial to the proposal's ability to detect multiple subdocuments that provide information about the same entity. Notably, this is achieved completely automatically as the appropriate equalities will be *entailed* by the knowledge base consisting of a domain-specific TBox and the data sources captured as concepts in a CBox.

Our primary technical contribution is our ability to separate entities into distinct documents and to consolidate documents that provide information about the same entity, also by appeal to referring expressions and to entailment in the underlying DL.

4.1   Related Work

There are many papers that take as input a semi-structured document, in JSON or XML (sometimes with a schema) plus an ontology in a DL TBox, and create individual instance descriptions in an ABox; for a survey see [5]. Usually, this is intended for just adding semantics to the document, but reasoning could be used to detect inconsistencies, such as a situation where two properties with disjoint domains apply to the same individual. For example, the use of XPath to navigate XML documents and detect instances of objects belonging to certain OWL classes has been presented in [12]. In particular, their approach and implementation, i.e., their JXML2OWL framework, maps XML documents to existing OWL ontologies via explicit mapping rules that use XPath. Our approach, in contrast, uses a generic mapping of JSON to concept descriptions in a DL and then uses the full power of a DL and reasoning in the DL to capture such mappings and to achieve a variety of other goals, including detection of entities and entity-based equality between subdocuments.

The closest to our work is a virtual OBDA architecture, where data is stored in a JSON MongoDB repository [4]. The architecture extends the basic OBDA architecture by introducing a relational view (an ABox) over MongoDB with respect to a set of type constraints. Rewritten queries by the OBDA framework over the relational view are translated using a fragment of MongoDB aggregate queries. However, as far as we are aware, the ontology, in particular identification constraints such as keys and functional dependencies, is not involved in entity identification issues, nor in separating entities into separate documents.

There is a great deal of work that attempts to synthesize schemata (in various formalisms) from the semi-structured data [5]. However, these approaches are orthogonal to the results in our paper.

4.2 Future Work and Extensions

There are many avenues for future research:

1. *Additional CBox entries generated by reified roles*: in Example 2, we could also have created additional CBox entries for (reified) *phone ownership*, e.g.,

   HAS-PHONE $\sqcap$

   $\exists\, \mathtt{dom.phone}^{-1}(\mathrm{PERSON} \sqcap \exists\, \mathtt{fname}.\{\texttt{"John"}\} \sqcap \exists\, \mathtt{lname}.\{\texttt{"Smith"}\}) \sqcap$
   $\exists\, \mathtt{ran}.(\mathrm{PHONE} \sqcap \exists\, \mathtt{dnum}\{\texttt{"212 555-1234"}\}).$

   This, however, requires generalizing the PFD concept constructor and path descriptions to allow for a limited use of inverse features.

2. *Diagnosis via consistency and pinpointing/data cleaning*: inconsistency of the knowledge base consisting of the domain knowledge $\mathcal{T}$ and the CBox ToC(doc) indicates that either our domain knowledge does not accurately capture the properties of the documents or that the documents themselves contain erroneous data. Also, interactions between a given $\mathcal{T}$ and RTA can lead to additional diagnostic feedback. Axiom pinpointing [9] and data cleaning [6] can help in these situations.

3. *Set-valued properties and referring expressions*: another extension relates to extending the RTAs to allow identification to be based on set-valued properties/values. Such an extension, however, requires extensions to the equality-generating constraints in the underlying DL and the ToRE operation.

References

1. Baader, F., Calvanese, D., McGuinness, D.L., Nardi, D., Patel-Schneider, P.F.: The Description Logic Handbook: Theory, Implementation, and Applications. Cambridge University Press, Cambridge (2003)
2. Borgida, A., Toman, D., Weddell, G.: On referring expressions in query answering over first order knowledge bases. In: Proceedings of Principles of Knowledge Representation and Reasoning, KR 2016, pp. 319–328 (2016)
3. Borgida, A., Toman, D., Weddell, G.: On referring expressions in information systems derived from conceptual modelling. In: Comyn-Wattiau, I., Tanaka, K., Song, I.-Y., Yamamoto, S., Saeki, M. (eds.) ER 2016. LNCS, vol. 9974, pp. 183–197. Springer, Cham (2016). https://doi.org/10.1007/978-3-319-46397-1_14
4. Botoeva, E., Calvanese, D., Cogrel, B., Rezk, M., Xiao, G.: OBDA beyond relational DBs: a study for MongoDB. In: Lenzerini, M., Peñaloza, R., (eds.) Proceedings of the 29th International Workshop on Description Logics, 2016, vol. 1577, CEUR Workshop Proceedings (2016)
5. Hacherouf, M., Bahloul, S.N., Cruz, C.: Transforming XML documents to OWL ontologies: a survey. J. Inf. Sci. **41**(2), 242–259 (2015)
6. Ihab, F., Chu, X.: Data Cleaning. ACM (2019)
7. McIntyre, S., Borgida, A., Toman, D., Weddell, G.E.: On limited conjunctions and partial features in parameter-tractable feature logics. In: The Thirty-Third AAAI Conference on Artificial Intelligence, AAAI 2019, pp. 2995–3002 (2019)
8. McIntyre, S., Toman, D., Weddell, G.E.: FunDL - a family of feature-based description logics, with applications in querying structured data sources. In: Description Logic, Theory Combination, and All That - Essays Dedicated to Franz Baader on the Occasion of His 60th Birthday, pp. 404–430 (2019)
9. Peñaloza, R.: Axiom pinpointing. In: Cota, G., Daquino, M., Pozzato, G.L., (eds.) Applications and Practices in Ontology Design, Extraction, and Reasoning, vol. 49. Studies on the Semantic Web, pp. 162–177. IOS Press (2020)
10. Pound, J., Toman, D., Weddell, G.E., Wu, J.: Query algebra and query optimization for concept assertion retrieval. In: Haarslev, V., Toman, D., Weddell, G.E., (eds.), Proceedings of the 23rd International Workshop on Description Logics (DL 2010), vol. 573, CEUR Workshop Proceedings (2010)
11. Pound, J., Toman, D., Weddell, G.E., Wu, J.: An assertion retrieval algebra for object queries over knowledge bases. In: Walsh, T., (ed.) IJCAI 2011, Proceedings of the 22nd International Joint Conference on Artificial Intelligence, 2011, pp. 1051–1056. IJCAI/AAAI (2011)
12. Rodrigues, T., Rosa, P., Cardoso, J.: Mapping XML to exiting OWL ontologies. In International Conference WWW/Internet, pp. 72–77 (2006)
13. Toman, D., Weddell, G.: Identity resolution in ontology based data access to structured data sources. In: Nayak, A.C., Sharma, A. (eds.) PRICAI 2019. LNCS (LNAI), vol. 11670, pp. 473–485. Springer, Cham (2019). https://doi.org/10.1007/978-3-030-29908-8_38

Tyche: A Library for Probabilistic Reasoning and Belief Modelling in Python

Padraig X. Lamont[✉][iD]

The University of Western Australia, Perth, Australia
paddy.lamont@uwa.edu.au

Abstract. This paper presents Tyche, a Python library to facilitate probabilistic reasoning in uncertain worlds through the construction, querying, and learning of belief models. Tyche uses aleatoric description logic (ADL), which provides computational advantages in its evaluation over other description logics. Tyche belief models can be succinctly created by defining classes of individuals, the probabilistic beliefs about them (concepts), and the probabilistic relationships between them (roles). We also introduce a method of observation propagation to facilitate learning from complex ADL observations. A demonstration of Tyche to predict the author of anonymised messages, and to extract author writing tendencies from anonymised messages, is provided. Tyche has the potential to assist in the development of expert systems, knowledge extraction systems, and agents to play games with incomplete and probabilistic information.

Keywords: Probabilistic reasoning · Learning agents · Software libraries

1 Introduction

Many of the major advances in artificial intelligence in the last decade have embraced the use of data that does not follow hard rules [2,10,11,15]. Systems such as deep CNNs for ImageNet classification, AlphaFold, and GPT-3 have all demonstrated incredible performance over hard rules-based systems for handling data containing a high degree of uncertainty [2,10,11]. However, many of our existing ontological knowledge base systems are designed only to hold facts about our world, with no allowance for uncertainty [1,8,13,16]. Tyche aims to facilitate the creation of knowledge bases with uncertain and probabilistic information (i.e., belief models), through a simple Python API. These belief models can be used to query the probability of truth of logical sentences, and logical observations can be used to learn the probabilities within belief models. The name Tyche was chosen for this library in reference to Tyche the goddess of fortune, from Greek mythology.

H. Aziz et al. (Eds.): AI 2022, LNAI 13728, pp. 381–396, 2022.
https://doi.org/10.1007/978-3-031-22695-3_27

Tyche belief models are structured similarly to knowledge graphs. Knowledge graphs contain many entities, with directed edges between them that represent relationships. For example, you may have a graph with two nodes representing the Earth and the sun. The node representing the Earth may have a relationship to the sun representing that the Earth orbits the sun. The nodes representing the Earth and the sun may also contain knowledge such as their radius or their mass. This representation allows knowledge about entities and their relationships to be flexibly and efficiently represented, and it has been used to create very large knowledge bases such as Wikidata [16]. However, existing knowledge graphs are only built to hold factual information [1,8,13,16]. This restricts their use when information is fundamentally uncertain, such as belief.

Tyche allows the representation of graphs of entities with probabilistic beliefs about them, and probabilistic relationships between them. This extends the data that can be represented within knowledge graphs, by allowing the inclusion of uncertain information. For example, the nodes within a Tyche belief model could represent people, and we could store whether it is believed that each person is friendly, and who we believe their best friend is. However, this information may change each time we interact with each person. Tyche makes use of probability to account for this uncertainty by treating the probabilities of our beliefs about individuals as dice rolls, where the probability represents the chance that a belief will hold each time it is observed. For example, we may believe that 75% of the time that we talk to someone, they will be friendly. This sampling interpretation of probability is unique to aleatoric description logic [7], which forms the basis for the logic that is provided by Tyche. Conversely, other probabilistic logics use probability to represent a degree of belief about the truth of facts [4,5], the state of the world [12], or, in the case of fuzzy logics, the degree of inclusion to a set [3]. The independence of observations in aleatoric description logic leads to a simple recursive interpretation of its semantics, which also gives Tyche computational advantages over other probabilistic logic languages that use logical solvers [4,7,12].

Logical sentences are integral to the querying and updating of Tyche belief models. Aleatoric description logic, which Tyche's logical sentences are based upon, can be used to represent many common constructs of description logics, such as conjunctions ($\wedge$), disjunctions ($\vee$), complement ($\neg$), the universal quantifier ($\forall$), and the existential quantifier ($\exists$). Logical sentences created using these constructs can be used to query probabilities from belief models, or as observations to be used to update belief models. Unlike other probabilistic logics, the logical sentences used to learn a Tyche belief model are not stored in the belief model. Instead, the probabilities within the belief model are updated as observations are made. Other probabilistic logics represent knowledge bases as sets of logical sentences, into which new observations may be added [4,12].

Logical sentences are evaluated in Tyche using belief models to provide the probabilistic beliefs about individuals (termed concepts), and the probabilistic relationships between individuals (termed roles). Tyche belief models are created through the definition and instantiation of Python classes. The classes of individuals define the concepts and roles that are available for a type of individual, and their value may be provided by the fields or methods of the class.

For example, the probabilistic beliefs about an individual may be dynamically calculated from their non-probabilistic properties using other tools such as continuous probability distributions, expert rules, or neural networks. This grants a large degree of flexibility when defining belief models using Tyche. This was a key goal in Tyche's implementation, to facilitate interoperability between Tyche and other systems. The goals of flexibility and interoperability when developing Tyche also guided the choice to implement Tyche in Python, due to its vast ecosystem of available tools for scientific computing [14].

This paper will introduce the use of Tyche to construct belief models, to query them using aleatoric description logic, and to update them based upon observations. An example use of Tyche for knowledge extraction will also be shown, where Tyche is demonstrated as an effective tool to accurately determine the author of anonymised messages.

2   Quantification of Belief: Concepts and Roles

Tyche supports reasoning about belief about both the properties of individuals (termed concepts), and the relationships between individuals (termed roles). The value of concepts are provided as floating-point probability values in the range [0, 1]. The value of roles are provided as probability distributions of potentially related individuals. Tyche currently provides both a mutually exclusive probability distribution for roles, and an independent probability distribution. Mutually exclusive roles only allow a single individual to be related at a time (although, a null-individual may be added to represent "no relation"). Independent roles consider the relation of each individual independently, such that any subset of the potentially related individuals may be related at a time.

Examples of a mutually exclusive role and an independent role are shown in Fig. 1. These roles could be used to represent marriage (where only a single spouse is legally allowed at a time), and friendship (where multiple friends are allowed at a time).

(a) Mutually Exclusive Role.   (b) Independent Role.

Fig. 1. Example roles relating Alice to Bob and/or Jeff.

3   Constructing Belief Models

Tyche supports the creation of belief models as ontological knowledge bases of individuals, the probabilistic beliefs about them (i.e., concepts), and the probabilistic relationships between them (i.e., roles). Python classes may be defined to represent the types of individuals in a belief model, by subclassing `Individual`. The class' fields and methods may then be type-annotated or decorated, respectively, to register them as providing the value of concepts or roles. This allows individuals to be flexibly defined with support for polymorphism of the individual classes. This aims to allow Tyche to be used as a part of other class hierarchies (e.g., database model objects). The code used to define an example individual, `Person`, is shown in Fig. 2.

3.1   Registering Fields as Concepts or Roles

The fields of an `Individual` subclass can be marked as providing the value of a concept or role by type-annotating them with the `TycheConceptField` type, or the `TycheRoleField` type, respectively. This will inform Tyche to access the value of these fields to use as the value for the concepts and roles in your logical queries. The names of the fields will be used as the symbols for the concepts and roles within your queries. The `positive` field and the `conversed_with` field in Fig. 2 are registered as a concept and role, respectively, using this method. These fields could be used to represent whether the person is positive in their messages, and who they are likely to have last sent a message to.

3.2   Registering Methods as Concepts or Roles

The methods of an `Individual` subclass can be marked as providing the value of a concept or role by decorating them with the `@concept()` decorator, or the

```python
class Person(Individual):
    positive: TycheConceptField
    conversed_with: TycheRoleField

    def __init__(self, positive: float, height_cm: NormalDist):
        super().__init__()
        self.positive = positive
        self.conversed_with = ExclusiveRoleDist()
        self.height_cm = height_cm

    @concept(symbol='tall')
    def is_tall(self):
        return self.height_cm > 180
```

Fig. 2. Example definition of a type of individual called Person, with a concept "positive" and a role "conversed_with" defined by fields, and a role "tall" provided by a method.

@role() decorator, respectively. The decorated methods will not be provided with any arguments, and should return the value of the concept or role. The name of the method will be used as the symbol of the concepts and roles by default, although a symbol can also be explicitly provided to the decorators through their symbol parameter. The is_tall method in Fig. 2 is registered as the concept "tall" using this method.

4   Logical Queries and Observations

Logical queries and observations can be used to interact with a Tyche belief model, to query or update its probabilities. Logical queries and observations are created by constructing logical sentence trees within Tyche. Each element within a logical sentence is represented as a node in the tree, with the node's constituent subsentences represented as child nodes. The tree structure of an example logical sentence, $[\rho]((\alpha?\beta : \bot)|\top)$, is shown in Fig. 3. This example sentence represents the expectation that an individual selected from role ρ will have true values sampled for the concepts α and β.

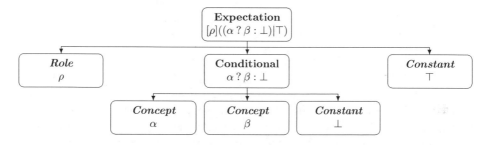

Fig. 3. Tree representation of an example logical sentence. The node class names are shown in bold, and below them are the aleatoric description logic sentences represented by the nodes.

Tyche provides several logical constructs for use in the creation of logical sentence trees, with each represented by their own node class in the language module of Tyche. The available node types, their descriptions, and their aleatoric description logic representations are listed in Table 1. The nodes in a logical sentence tree can be created by first instantiating the node classes needed to represent the leaf nodes of the tree, and then using them as arguments to construct their parent nodes. Tyche also provides shortcut operators for creating conditionals that represent conjunctions (&), disjunctions (¦), and complement (~). For example, the logical sentence from Fig. 3 can be created using the code
Expectation("rho", Concept("alpha") & Concept("beta")).

Table 1. Descriptions of the node types provided by Tyche, along with their aleatoric description logic (ADL) representation.

Node class	Description	ADL
Concept	A sampling of the value of a concept from a belief model	Named variables
Role	A sampling of the value of a role from a belief model	Named variables
Constant	Fixed probabilities such as always (100%), or never (0%)	$\top$ (always), $\perp$ (never)
Conditional	A ternary operator that evaluates to β if α is true, or else evaluates to γ. These operators can be used to construct several common logical operators such as AND, OR, or NOT	$(\alpha\,?\,\beta:\gamma)$
Expectation	A marginalisation operator that evaluates to the chance that α is true, given that an individual in the role ρ was sampled for which β was true. This ignores the null-individual, and is vacuously true if the role contains no individuals for which β could be true	$[\rho](\alpha|\beta)$
Exists	An operator that evaluates to the chance that a role has a related individual (i.e., a relation that is not the null-individual)	N/A

5 Aleatoric Description Logic

Logical sentences within Tyche have been built following the work of aleatoric description logic [7]. Aleatoric description logic (ADL) provides a generalisation of standard description logics to extend true/false information to the closed interval $[0, 1]$, with the extended values representing probabilities of truth [7]. The base syntax of ADL consists of the constant *always* ($\top$), the constant *never* ($\perp$), atomic concepts (named variables), the ternary operator $((\alpha\,?\,\beta:\gamma))$, and the marginalisation operator $([\rho](\alpha|\beta))$. The function of these syntax elements of ADL are described in Table 1. These base syntax elements provide a versatile basis to construct common syntactical elements from other description logics, as shown in Table 2.

Table 2. List of common abbreviations used in description logics, and their equivalent aleatoric description logic (ADL) sentences [7].

Name	Abbreviation	Equivalent ADL
Conjunction	$\alpha \wedge \beta$	$(\alpha\,?\,\beta:\perp)$
Disjunction	$\alpha \vee \beta$	$(\alpha\,?\,\top:\beta)$
Complement	$\neg\alpha$	$(\alpha\,?\,\perp:\top)$
Implication	$\alpha \Rightarrow \beta$	$(\alpha\,?\,\beta:\top)$
Expectation	$E_\rho\alpha$	$[\rho](\alpha\,|\,\top)$
Existential	$\exists\rho.\alpha$	$\neg[\rho](\perp\,|\,\alpha)$

An important feature of ADL is that each occurrence of a concept or role within an aleatoric description logic sentence is treated as an independent sampling of the value of that concept or role [7]. *The probabilities of concepts and roles do not represent the chance that their underlying value is true, but instead the chance that they will be true when sampled.* For example, if the probability of α is not 0 or 100%, then the probability of $(\neg\alpha) \wedge \alpha$ is not 0%. This allows the probability of truth of ADL sentences to be evaluated recursively, without a complicated solver. This is an important feature of ADL, as it facilitates its efficiency, and allows Tyche to model problems that contain sampling more effectively than other probabilistic logics.

6 Evaluation of Logical Queries

The probability of truth of logical queries may be evaluated about an individual in a belief model. The result represents the chance that the sentence will be true when sampled about that individual. For example, you may evaluate the probability that Bob ran into a friend that was happy and relaxed by evaluating the query $[\rho]((\alpha?\beta : \bot)|\top)$. The logical sentence tree of this observation was shown in Fig. 3. To evaluate this query for Bob, you would pass the logical sentence tree representing this query to the *eval* method of the object representing Bob (e.g., `bob.eval(sentence)`). Tyche will then retrieve the values of ρ, α and β from the belief model, and use them to calculate the probability of Bob running into a friend that is happy and relaxed.

This process of value retrieval and calculation is performed in a single recursive pass through the logical sentence tree. The probability of the root node in the logical sentence tree will be calculated by first evaluating the probability of its child nodes. The child nodes will then recursively evaluate their child nodes, and use the results to calculate their own probability. This process allows the efficient recursive calculation of the probability of truth of logical sentences in a single pass of the tree. This is possible due to the sampling nature of aleatoric description logic, as each appearance of a concept or role represents an independent sampling of them [7]. The evaluation procedures for each logical construct provided by Tyche follow the semantics of Modal Aleatoric Calculus [6].

7 Learning Through Observation

Tyche provides mechanisms to learn the value of concepts and roles from logical observations that include them. The observations take the form of logical sentences, and can be supplied to individuals using their *observe* method. Tyche will use Bayes' rule to determine how each term in the observation has influenced the truth of the entire sentence. The values of the concepts and roles used within the sentence may then be learnt using one of the learning strategies supplied by Tyche. These learning strategies must be provided through a method decorator. The code used to define an example individual named Student with learning registered for its "good_grades" concept is shown in Fig. 4.

```
class Student(Person):
    def __init__(self, good_grades: float):
        super().__init__(0.33, NormalDist(175, 6.5))
        self._good_grades = good_grades

    @concept()
    def good_grades(self):
        return self._good_grades

    @good_grades.learning_func(DirectConceptLearningStrategy())
    def set_good_grades(self, good_grades: float):
        self._good_grades = good_grades
```

Fig. 4. A type of individual called Student with a concept "good_grades" that is learnt using a direct learning strategy. Student also inherits the concepts and roles of Person from Fig. 2.

7.1   Calculation of the Influence of Terms in Observations

The influence of each term in logical observations is calculated recursively, through the propagation of two parameters: likelihood and learning rate. The likelihood parameter represents the chance that the current term in the observation is true, and the learning rate quantifies the percentage impact of the current term on the truth of the entire sentence. These two parameters allow the influence of each term in an observation to be calculated through a simple algorithm that recurses through the observation's logical sentence tree. This is important, as it means that learning strategies only need to deal with simple observations that are directly related to the concept or role being learnt (e.g., is_sunny, or $[friend](is\_tired|\top)$). The learning strategies do not need to deal with the interactions between terms in complex observations.

Calculation of the Likelihood Parameter. The likelihood that each term in an observation is true (t) can be calculated using a method built upon the application of Bayes' rule. We apply Bayes' rule to the event that either the parent node of the term was true (p) with chance α_{parent}, or else the parent node was false ($\neg p$) [9]. The term's likelihood, α_{term}, can then be calculated as in Eq. 1. The likelihood parameter, α, is given a value of 1 for the root node of the observation.

$$\alpha_{term} = \alpha_{parent} \cdot P(t \mid p) + (1 - \alpha_{parent}) \cdot P(t \mid \neg p) \tag{1}$$

$$= \alpha_{parent} \cdot \frac{P(p \mid t) \cdot P(t)}{P(p)} + (1 - \alpha_{parent}) \cdot \frac{P(\neg p \mid t) \cdot P(t)}{P(\neg p)} \tag{2}$$

The value of $P(t)$ and $P(p)$ may be calculated directly from the current belief model. The value of $P(p|t)$ may be calculated by replacing the term in the observation with $\top$ (i.e., a constant 100%), and evaluating the resulting sentence in the current belief model.

Calculation of the Learning Rate Parameter. The learning rate of each node in an observation tree can be calculated using the learning rate of its parent, and the difference in truth of its parent if the term were true $(P(p|t))$, and if it were false $(P(p|\neg t))$. The learning rate parameter, r, is given a value of 1 for the root node of the observation. The learning rate, r_{term}, of each child node may then be calculated following Eq. 3.

$$r_{term} = r_{parent} \cdot abs(P(p|t) - P(p|\neg t)) \tag{3}$$

This calculates the influence that the term had on the truth of its parent, and multiplies that by the parent's influence, to determine the influence of the term on the original observation. However, when an expectation over a mutually exclusive role is reached, then we no longer perform this calculation. Instead, the chance that each individual within the role was selected is used to moderate the learning rate that is propagated to each individual within the role. For an expectation $\lceil \rho \rceil (a|b)$, the chance that each individual, x, was selected within the role, ρ, is given by Eq. 4.

$P(x \text{ selected from } \rho \text{ for } obs) =$

$$\frac{P(x \text{ selected from } \rho) \cdot P_x(b) \cdot (\alpha \cdot P_x(a) + (1-\alpha) \cdot (1 - P_x(a)))}{\sum_{y \in \rho} P(y \text{ selected from } \rho) \cdot P_y(b) \cdot (\alpha \cdot P_y(a) + (1-\alpha) \cdot (1 - P_y(a)))} \tag{4}$$

This chance can then be used as a multiplier with the learning rate of the expectation node, r, to calculate the learning rate to propagate to each related individual.

7.2   Learning Strategies

Learning strategies update the value of concepts and roles based upon observations of their use. They do not receive the entire observation, but instead only receive the sub-sentence of the observation that is relevant to them, along with the influence parameters of that sub-sentence. Each inclusion of a `Concept` node in an observation will be passed to the concept's learning strategy, if one is registered. Similarly, each inclusion of an `Expectation` node will be passed to its role's learning strategy, if one is registered. The learning strategies may then use the provided sub-sentence and the propagated influence parameters to update the belief model. This structure provides a simple method for writing learning strategies that do not need to consider the structure of the observation.

Tyche currently provides two learning strategies for concepts, and two learning strategies for roles. The Bayes' rule learning strategy (class `BayesRuleLearningStrategy`) updates the distribution of roles by updating the probability of each individual in a role to the conditional probability of selecting that individual, given the observation. The direct concept learning strategy (class `DirectConceptLearningStrategy`) updates the values of concepts to the weighted sum of the current value of the concept and the likelihood parameter, weighted by the learning rate. The statistical learning strategies (classes `StatisticalConceptLearningStrategy` and `StatisticalRoleLearningStrategy`) maintain a running mean of the concept likelihoods or role distribution weights, which is used as the new value of the concepts or roles.

8   Demonstration: Anonymised Messages

A simple demonstration of Tyche was developed to demonstrate its prediction and learning capabilities for handling anonymised messages where the recipient of each message is known, but the author of the messages is unknown. Tyche may be used to accurately predict the author of these anonymised messages based upon the writing style of the messages. The writing style of messages is represented using three properties: *uses emoji* (E), *capitalises first word* (C) and *is positive* (P). The tendencies for each author to make these stylistic choices are stored as the probability values of each author's concepts in the belief model. Additionally, mutually exclusive roles between the authors are used to represent the chance of each author receiving messages from each other author.

In the evaluation of Tyche's performance for inference, the writing tendencies of each author are considered to be known. However, when the writing tendencies of authors are unknown, Tyche may also be used to learn the author's writing tendencies, without knowing the messages that each author sent. Additionally, Tyche's learning may also be used to learn the probability distribution for who each author is most likely to have received messages from. The only information that Tyche is provided for this learning is the properties of each message, and the assumption that each individual would not receive messages from themselves.

An implementation of this example was also written using ProbLog, to compare it to Tyche. ProbLog is an established tool for probabilistic logic reasoning [4], and as such it is a likely choice for an alternative library to solve this example. However, ProbLog is not built based upon a sampling logic, as Tyche is. Therefore, as will be shown, it takes much longer to run these examples with ProbLog, and ProbLog's lfi learning is not able to learn the belief model from observations.

8.1   Ground-Truth Belief Model

The anonymous messages belief model is created with three authors: Bob, Alice and Jeff. Each author has their own writing tendencies for whether they will use emoji (E), capitalise the first word (C), or be positive (P). The authors also each have their own probabilities for receiving messages from each other author. The values of these writing tendencies and relationships are shown in Fig. 5a. These values are used for the inference and learning demonstrations to randomly sample messages received by each author. An unbiased belief model is also presented in Fig. 5b, which is used as a baseline model with limited prior knowledge for use in learning.

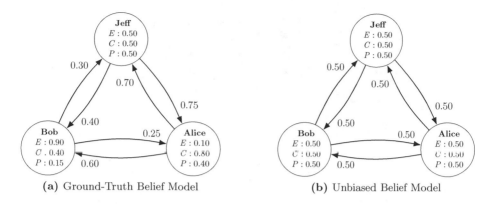

(a) Ground-Truth Belief Model   **(b)** Unbiased Belief Model

Fig. 5. The ground-truth belief model of the anonymous messages example **(a)**, and the unbiased belief model used as a starting point for learning **(b)**.

8.2   Predicting the Author of Sets of Messages

The author of sets of messages is predicted using Tyche by evaluating the chance that each author wrote the messages, based upon the ground-truth belief model. A logical sentence is created that represents the properties of the received messages (e.g., $\neg uses\_emoji \wedge \neg capitalises\_first\_word \wedge is\_positive$). The chance that each possible author wrote this set of messages is then evaluated using the ground-truth belief model, and multiplied by the chance that the recipient received a message from each of the possible authors. The author with the highest chance of having written and sent the set of messages is then selected as the predicted author. To evaluate Tyche's performance on this task, we used the ground-truth belief model to sample 10,000 random sets of messages received by each author, with varying numbers of messages per set from 1 to 10. The accuracy of Tyche at predicting the author of these random sets of messages is recorded in Table 3, and shown in Fig. 6.

Table 3. Author prediction accuracy (%) for each individual in the anonymous messages example, with varying number of messages.

	Number of messages									
	1	2	3	4	5	6	7	8	9	10
Bob	74.8	81.9	85.2	89.2	91.1	92.5	94.5	95.4	96.3	97.1
Alice	76.9	84.5	88.2	91.0	92.8	94.2	95.3	96.6	97.6	98.0
Jeff	90.4	95.5	98.0	99.0	99.5	99.8	99.9	100.0	100.0	100.0

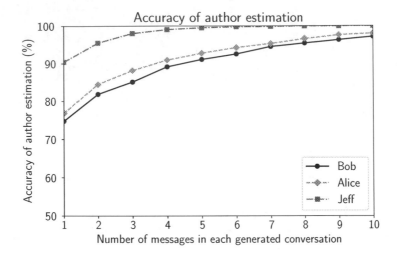

Fig. 6. The accuracy of author estimation as the number of messages in each set of messages is increased, for the three individuals in the anonymous messages example.

These results demonstrate that Tyche can achieve accurate results for author inference. However, the accuracy of the author prediction is not equivalent for all authors. The author prediction for messages received by Jeff is much more accurate than the author prediction for messages received by Alice or Bob. This is due to the fact the writing tendencies of Jeff are similar to Bob and Alice, while the writing tendencies of Alice and Bob are quite dissimilar. Therefore, it is easier to distinguish messages written by Bob and Alice, than to distinguish messages written by Jeff and each other author.

The alternative implementation of this example that uses ProbLog also achieved identical results. This is as expected, as both Tyche and ProbLog calculate the probability of a conjunction over a set of independent variables in the same way. However, the Tyche implementation was much faster than the ProbLog implementation. Author inference using Tyche had a mean duration of 0.5 milliseconds per prediction, whereas inference using ProbLog had a mean duration of 94.5 milliseconds (189x slower).

8.3   Learning the Writing Tendencies of Authors

The ground-truth belief model, shown in Fig. 5a, can be learnt using Tyche based upon observations of messages received by each author in the belief model. The belief model that is being learnt is initialised as the unbiased belief model shown in Fig. 5b. The observations are generated from the ground-truth belief model by first randomly sampling the author that received the set of messages, then sampling who sent them the set of messages, and finally sampling the properties of the messages that they sent. These observations are then passed to the `observe` method of the author that received the messages. An example observation for a set of two messages received by an author is shown in Eq. 5.

$$[received\_message]((\neg E \wedge \neg C \wedge P) \wedge (\neg E \wedge \neg C \wedge \neg P)) \tag{5}$$

To demonstrate the learning of belief models in Tyche, 10 trials were performed, with 5000 total observations per trial, and 2 to 4 messages per observation. At the beginning of each trial, the belief model is reset to the unbiased belief model. The statistical concept learning strategy was used to learn the writing tendencies for each author, with hyper-parameters $decay\_rate = 0.95$ and $decay\_rate\_for\_decay\_rate = 0.95$. The statistical role learning strategy was used to learn the probability distribution of who each author was more likely to receive messages from, with hyper-parameters $decay\_rate = 0.85$ and $decay\_rate\_for\_decay\_rate = 0.9995$. The ProbLog implementation of this example was created through the construction of one large ProbLog program with each observation added as a variable. Each of the observation variables were then passed as evidence to the lfi learning routine. The mean values from the belief models learnt in each trial using the Tyche and ProbLog implementations are shown in Fig. 7. The values are coloured in green if they fall within 0.015 (1.5%) of their true value, amber if they fall within 0.05 (5%) of their true value, and red otherwise.

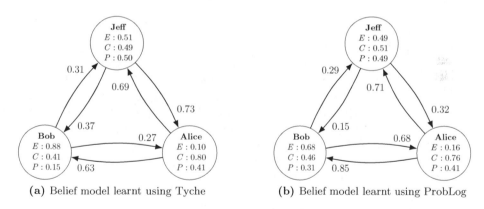

(a) Belief model learnt using Tyche   (b) Belief model learnt using ProbLog

Fig. 7. The belief models learnt from observations of received messages using the Tyche implementation **(a)**, and the ProbLog implementation **(b)**. (Color figure online)

The belief model learnt using Tyche in Fig. 7a is similar to the expected probabilities from the ground-truth model. This demonstrates that the learning capabilities of Tyche are able to accurately estimate the ground-truth belief model from observations. Conversely, the belief model learnt using ProbLog contained very close values for Jeff, but had large errors for Bob and Alice. This demonstrates that the ProbLog lfi learning was not able to accurately estimate the ground-truth belief model in this case. However, when the learning of the roles between Bob, Alice and Jeff is removed (results not included here), the ProbLog implementation produced accurate results. This suggests that the inclusion of relationships breaks the ProbLog lfi learning.

The results in Fig. 7 represent the mean of all 10 trial results, and therefore may be more accurate than the trial results themselves. However, the standard deviations of the values in the belief model learnt using Tyche had a mean of only 0.015 for the concepts, and 0.057 for the roles. This suggests that all trials of learning using the Tyche implementation achieved similar results. Additionally, each trial of learning using the Tyche implementation took an average of only 9.35 s to complete. Conversely, each trial of learning using the ProbLog implementation took an average of 1770 s, or half an hour (189x slower).

This result demonstrates Tyche's ability to efficiently learn from indirect observations about the world. We hope that Tyche may be used in similar ways to assist in the learning of unclear probabilistic properties within complex datasets.

9   Related Work

The graph representation of knowledge used in Tyche is similar to representations that can be created using the tool Protégé [13]. Protégé is a proprietary tool that supports the creation of ontologies, and the acquisition of data to build knowledge bases using them [13]. However, protégé only supports the inclusion of facts. It does not support any representations of chance. As such, it is not applicable to the problems where Tyche is intended to be used.

Tyche is tangentially related to work on other probabilistic logics such as ProbLog, Blog, and the use of fuzzy sets [4,12]. ProbLog extends the Prolog programming language with the ability to specify the probability that each independent logical clause holds in a randomly sampled program [4]. ProbLog's assumption of the independence of the probability of clauses resembles Tyche's assumption of the independence of the probabilities of concepts. However, the terms within clauses in ProbLog are not considered independent, whereas the terms in Tyche's logical sentences are considered independent. This leads to significant differences in the semantics of ProbLog and Tyche. ProbLog uses a solver to determine the probability that a query holds in a randomly sampled program, whereas Tyche recursively evaluates the probability of truth of a query, as if its terms were randomly sampled.

Blog provides a formal language to specify probability models, in a similar way to ProbLog [4,12]. However, under Blog, these probability models support worlds with unknown and unbounded numbers of objects, and identity uncertainty. This is a significant distinguishing factor from Tyche, which requires that all the individuals in the belief model, their concepts, and their roles, are defined. Additionally, Blog also requires the use of logical solvers to query its probability models, and does not consider terms in its sentences to be independent.

Fuzzy logic is used to represent degrees of truth, rather than a probability of truth as in Tyche [3]. For example, we do not usually consider objects as exclusively either "warm" or not. Instead, an object might be "slightly warm", "moderately warm", or "very warm". Fuzzy logic provides a mechanism to represent this, by representing knowledge as degrees of inclusion into a set. Therefore, fuzzy logic serves a different purpose to Tyche. However, despite this, both Tyche's logic and fuzzy logic share similar functionality in their use of t-norms [3]. T-norms represent the belief that applying an uncertain hypothesis twice will lead to a different uncertainty than applying it only once. The t-norm operator of fuzzy logic shares a lot of functional similarities with Tyche's operators, which provides evidence for the usefulness of repeated tests of a variable potentially yielding different results [3,7].

10   Conclusion

This paper has introduced Tyche, an open-source Python library to support the creation of belief models, and to facilitate the use of aleatoric description logic to reason about them. We introduced Tyche's API to succinctly construct ontological knowledge bases of individuals, the probabilistic beliefs about them, and the probabilistic relationships between them (termed belief models). The creation of logical sentences, and their use to query belief models, was discussed. We also introduced Tyche's novel observation propagation system to learn the probabilities within belief models based upon logical observations. Tyche's application for knowledge extraction was demonstrated through its use to predict the author of anonymised messages, and through its use to learn the writing tendencies of authors, and the probability distribution of who the authors receive messages from, without knowledge of who wrote any messages.

The source code of Tyche is available at https://github.com/TycheLibrary/Tyche.

References

1. Bechhofer, S., et al.: OWL web ontology language reference. Recommendation, World Wide Web Consortium (W3C), 10 February 2004. See http://www.w3.org/TR/owl-ref/
2. Brown, T.B., et al.: Language models are few-shot learners. CoRR abs/2005.14165 (2020). https://arxiv.org/abs/2005.14165
3. Cintula, P., Fermüller, C.G., Noguera, C.: Fuzzy logic. In: Zalta, E.N. (ed.) The Stanford Encyclopedia of Philosophy. Metaphysics Research Lab, Stanford University, Winter 2021 edn. (2021)
4. De Raedt, L., Kimmig, A., Toivonen, H.: ProbLog: a probabilistic prolog and its application in link discovery. In: Proceedings of the 20th International Joint Conference on Artificial Intelligence, IJCAI 2007, pp. 2468–2473. Morgan Kaufmann Publishers Inc., San Francisco (2007)
5. Demey, L., Kooi, B., Sack, J.: Logic and probability. In: Zalta, E.N. (ed.) The Stanford Encyclopedia of Philosophy. Metaphysics Research Lab, Stanford University, Summer 2019 edn. (2019)
6. French, T., Gozzard, A., Reynolds, M.: A modal aleatoric calculus for probabilistic reasoning. In: Khan, M.A., Manuel, A. (eds.) ICLA 2019. LNCS, vol. 11600, pp. 52–63. Springer, Heidelberg (2019). https://doi.org/10.1007/978-3-662-58771-3_6

7. French, T., Smoker, T.: An aleatoric description logic for probabilistic reasoning. In: CEUR Workshop Proceedings, vol. 2954 (2021)
8. Hofweber, T.: Logic and ontology. In: Zalta, E.N. (ed.) The Stanford Encyclopedia of Philosophy. Metaphysics Research Lab, Stanford University, Spring 2021 edn. (2021)
9. Ben: Applying Bayes's theorem when evidence is uncertain. https://stats.stackexchange.com/users/173082/ben. Cross validated, https://stats.stackexchange.com/q/345407 (version: 2018-05-10)
10. Jumper, J., et al.: Highly accurate protein structure prediction with AlphaFold. Nature **596**(7873), 583–589 (2021). https://doi.org/10.1038/s41586-021-03819-2
11. Krizhevsky, A., Sutskever, I., Hinton, G.E.: ImageNet classification with deep convolutional neural networks. In: Pereira, F., Burges, C., Bottou, L., Weinberger, K. (eds.) Advances in Neural Information Processing Systems, vol. 25. Curran Associates, Inc. (2012). https://proceedings.neurips.cc/paper/2012/file/c399862d3b9d6b76c8436e924a68c45b-Paper.pdf
12. Milch, B., Marthi, B., Russell, S.J., Sontag, D.A., Ong, D.L., Kolobov, A.: Blog: probabilistic models with unknown objects. In: Kaelbling, L.P., Saffiotti, A. (eds.) IJCAI, pp. 1352–1359. Professional Book Center (2005). http://dblp.uni-trier.de/db/conf/ijcai/ijcai2005.html#MilchMRSOK05
13. Musen, M.A.: The protégé project: a look back and a look forward. AI Matters **1**(4), 4–12 (2015). https://doi.org/10.1145/2757001.2757003
14. Pérez, F., Granger, B.E., Hunter, J.D.: Python: an ecosystem for scientific computing. Comput. Sci. Eng. **13**(2), 13–21 (2011). https://doi.org/10.1109/MCSE.2010.119
15. Shi, S., Chen, H., Ma, W., Mao, J., Zhang, M., Zhang, Y.: Neural logic reasoning. CoRR abs/2008.09514 (2020). https://arxiv.org/abs/2008.09514
16. Vrandečić, D., Krötzsch, M.: Wikidata: a free collaborative knowledgebase. Commun. ACM **57**(10), 78–85 (2014). https://doi.org/10.1145/2629489

Belief Revision with Dishonest Reports

Aaron Hunter[(⊠)]

British Columbia Institute of Technology, Burnaby, Canada
aaron_hunter@bcit.ca

Abstract. Belief revision operators are used to model the way that an agent's beliefs change when they acquire new information. However, if the new information comes from another agent, then we also need to be concerned with the notion of honesty. In this paper, we present a model in which agents have a memory of all reports they have received, as well as the set of agents that are assumed to be honest. We propose that all reports from dishonest agents should be ignored, while inconsistencies between reports from honest agents should be resolved through iterated revision. We demonstrate how an agent can learn about the honesty of others through direct observations, and we show how this affects the complete trajectory of beliefs. Finally, we consider the case where an agent only has partial information about the honesty of others. We demonstrate that the set of possible belief trajectories is constrained by a tree of possible revisions, which allows us to briefly explore properties that are invariant with respect to the honesty of others.

1 Introduction

Belief revision is the process of incorporating new information into a pre-existing belief state. The most influential theories of belief revision are focused on notions of minimal change, where the emphasis is on ensuring that the agent in question believes new information that is provided while abandoning as little as possible of their initial beliefs. Of course, in practice, new information often comes as a report from another agent. In this kind of situation, we can not simply take the accuracy of the new information as a given; we need to consider the notion of *trust*.

In the literature, there has been work on the notion of *knowledge-based trust* in a modal logical setting [7, 10], and also in the context of belief revision [3]. This kind of work focuses on the perceived expertise of the reporting agent, so new information is only believed when the source of the information is understood to be an authority over the relevant domain. In this paper, we are not concerned with expertise, we are concerned with trust related to *honesty*. Hence, we would like to address belief revision in a manner that takes into account the fact that the reporting agent may be deceitful.

Our goal is to present a flexible framework that allows agents to perform a kind of conditional reasoning that allows all reports to be incorporated, but also allows reports to be abandoned at any point if the reporting agent turns out to be deceitful.

1.1 Motivating Example

Alice is receiving information about the weather from two sources: Bob and Trent. We assume that these reports are coming through a messaging system, so Alice is not in the

H. Aziz et al. (Eds.): AI 2022, LNAI 13728, pp. 397–410, 2022.
https://doi.org/10.1007/978-3-031-22695-3_28

same room as the reporting agents. Initially, Alice does not have any beliefs about the weather. Suppose that she receives the following messages:

1. *It is rainy.* (from Trent)
2. *Bob is at my house.* (from Trent)
3. *It is sunny.* (from Bob)
4. *I am at the beach.* (from Bob)

It seems like someone is not telling the truth. All of the statements being reported could be easily verified by Bob and Trent; the most likely explanation for the inconsistency is therefore a non-truthful report.

The question is: what should Alice believe? If Alice simply performs ordinary belief revision, then she will reason as follows. As the statements are received, Alice will initially believe (1) and (2). When (3) and (4) are received, she will need to revise her beliefs again; this will mean that Alice no longer believes it is rainy or that Bob is at Trent's house.

But then suppose that Alice somehow learns that Bob is not honest. This could happen through some special information channel that is not included in this model. Or it could happen because Alice looks out the window and sees it is raining with her own eyes. When Alice discovers that Bob is dishonest, then everything Bob has said will be viewed with some caution. The most drastic solution is to reject everything that Bob says. But it is not sufficient to only apply this caution to future statements; it should also be applied retroactively. This means that Trent's earlier reports need to be reinstated as well, since they were only lost due to dishonest claims. In this paper, we propose a formal approach that allows this kind of reasoning to take place by keeping track of perceived honesty applied to all past reports.

2 Preliminaries

We are interested in formalizing our model in the context of propositional belief revision. As such, we assume a finite propositional vocabularly $\mathbf{F}$, and we define sentences over $\mathbf{F}$ using the usual propositional connectives $\neg, \wedge, \vee$. A *state* is a propositional interpretation of $\mathbf{F}$. A state gives a complete description of the world, by specifying exactly which atomic formulas are true or false. A *belief state* is a set of states; intuitively, it is the set of states that an agent considers to be possible. A *belief set* is a logically closed set of formulas. Since we have assumed a finite vocabulary, every consistent belief set defines a unique belief state.

The most influential approach to belief revision is the so-called AGM approach [1]. In this setting, the beliefs of an agent are represented as a deductively closed set of propositional formulas. A belief revision operator $*$ takes an initial belief set K and a formula for revision ϕ, and it returns a new belief set $K * \phi$. Informally, the new belief set incorporates ϕ while giving up as little as possible from the initial belief state. In the AGM model, the revision operator is constrained by a set of rationality postulates. It is well known that every revision operator that satisfies these postulates can be defined in terms of a minimization operation over a plausibility ranking on possible states of the world [6].

It is well known that AGM revision suffers from a major flaw, in that it is not able to capture with iterated revision. The problem is that AGM requires an ordering on possible states, but the belief set returned after a single revision does not include such an ordering. If we are interested in iterated belief revision, then the dominant approach has been the DP approach due to Darwiche and Pearl [4]. We introduce the basics of DP revision presently.

In the DP approach, the beliefs of an agent are given by an *epistemic state* $\mathbf{E}$. An epistemic state is intended to capture two things. First of all, with each epistemic state $\mathbf{E}$, there is an associated belief set $Bel(\mathbf{E})$. An epistemic state also defines a total pre-order $\preceq_{\mathbf{E}}$ over states, with the property that $Bel(\mathbf{E})$ is the set of formulas that are true in the minimal elements of $\preceq_{\mathbf{E}}$. We remark that an epistemic state is not *defined* to simply be an ordering over states, but each epistemic state does specify an ordering. The interested reader can see [8] for a discussion of categorical representations of epistemic states.

An interacted revision operator is applied to epistemic states. So $\mathbf{E} * \phi$ returns a new *epistemic state* $\mathbf{E}'$. An iterated revision operator is a DP operator if it satisfies the DP postulates. This is a set of postulates that gives required properties on both the belief set $Bel(\mathbf{E}')$ and the associated ordering following revision. For an interesting discussion of the postulates as well as some important DP operators, we refer the reader to [2].

3   Reports and Belief Change

3.1   Epistemic Histories

As noted in the previous section, we assume an finite propositional vocabulary $\mathbf{F}$. We also assume a finite set of agents $\mathbf{A}$. We use $\mathbf{E}$ to range over epistemic states over $\mathbf{F}$. So, an epistemic state $\mathbf{E}$ has an associated belief state $Bel(\mathbf{E})$ and an associated total pre-order on states $\preceq_{\mathbf{E}}$. We assume a fixed underlying DP revision operator $*$ on epistemic states.

We are interested in a situation where each $A \in \mathbf{A}$ can provide a formula ϕ as new information. In this case, we do not necessarily want to simply revise by ϕ, as it might be the case that A is dishonest. In order to address this problem, we define the following notion of a *report*.

Definition 1. *A* report *is a pair* (ϕ, A) *where* ϕ *is a propositional formula over* $\mathbf{F}$ *and* $A \in \mathbf{A}$. *A report history* $\overline{R}$ *is a finite sequence of reports.*

Hence, a report history captures all of the information that has been reported in sequence from a set of agents. Our goal is ultimately to give a precise meaning to the belief change that occurs following a sequence of reports: $(\phi_1, A_1), \ldots, (\phi_n, A_n)$.

We also introduce the notion of an *honesty assignment* over a set of agents.

Definition 2. *An* honesty assignment *is a set* $\alpha \subseteq \mathbf{A}$.

Informally, an honesty assignment represents the set of agents that are presumed to be honest. In other words, these are agents that only provide reports that they believe to be accurate. If an agent is not honest, then they may intentionally provide false reports. As such, reports should only be incorporated when they come from honest agents.

We extend the notion of an epistemic state to specifically include information about the history of past reports and observations.

Definition 3. *An* epistemic history *is a triple* $\langle \mathbf{E}, \overline{R}, \alpha \rangle$ *where* $\mathbf{E}$ *is an epistemic state,* $\overline{R}$ *is a report history, and* α *is an honesty assignment.*

An epistemic history represents the initial belief state $\mathbf{E}$ of some agent, along with all of the information that has been reported to them by other agents. We would like to define the belief state associated with an epistemic history, as we do for epistemic states.

We adopt the following natural shorthand notation for iterated revision by a sequence of formulas $\overline{\phi} = \phi_1, \ldots, \phi_n$:

$$\mathbf{E} * \overline{\phi} := \mathbf{E} * \phi_1 * \cdots * \phi_n.$$

The following definition introduces the notion of restriction to a set of honest agents.

Definition 4. *Let* $\overline{R} = (\phi_1, A_1), \ldots, (\phi_n, A_n)$ *be a report history and let* α *be an honesty assignment. We define the* restriction $\overline{R} \upharpoonright \alpha$ *to be the subsequence of* $\overline{R}$ *where* (ϕ_i, A_i) *is in the subsequence if and only if* $A_i \in \alpha$.

For example, if $\overline{R} = (\phi_1, A_1), (\phi_2, A_2), (\phi_3, A_3)$ and $\alpha = \{A_1, A_3\}$, then

$$\overline{R} \upharpoonright \alpha = (\phi_1, A_1), (\phi_3, A_3).$$

We would now like to extend Bel to epistemic histories. In other words, we would like to associate a belief set with each epistemic history. The follow definition specifies how this is done, and also specifies how we associate a corresponding ordering on states with an epistemic history.

Definition 5. *Let* $\langle \mathbf{E}, \overline{R}, \alpha \rangle$ *be an epistemic history. Then:*

$$Bel(\langle \mathbf{E}, \overline{R}, \alpha \rangle) = Bel(\mathbf{E} * (\overline{R} \upharpoonright \alpha))$$

$$\preceq_{\langle \mathbf{E}, \overline{R}, \alpha \rangle} = \preceq_{\mathbf{E} * (\overline{R} \upharpoonright \alpha)}.$$

Hence, $Bel(\langle \mathbf{E}, \overline{R}, \alpha \rangle)$ represents the belief state that an agent should have if they receive the report sequence $\overline{R}$, when α represents the set of honest agents. Similarly, $\preceq_{\langle \mathbf{E}, \overline{R}, \alpha \rangle}$ is the total pre-order over states they should have in the same situation. Note that, if $\alpha = \mathbf{A}$, then the following equalities hold for all $\mathbf{E}$ and all $\overline{R}$:

$$Bel(\langle \mathbf{E}, \overline{R}, \alpha \rangle) = Bel(\mathbf{E} * \overline{R})$$

$$\preceq_{\langle \mathbf{E}, \overline{R}, \alpha \rangle} = \preceq_{\mathbf{E} * \overline{R}}.$$

This is not surprising; if we know that all agents are honest, the belief change that occurs following a sequence of reports can be obtained by just performing normal revision.

3.2 Belief Change Due to Reports

In this section, we define some operators on epistemic histories. First, we introduce two simple operators to change the honesty assignment.

Definition 6. *For any epistemic history $\langle \mathbf{E}, \overline{R}, \alpha \rangle$ and any $\beta \subseteq \mathbf{A}$, define:*

$$\langle \mathbf{E}, \overline{R}, \alpha \rangle + \beta = \langle \mathbf{E}, \overline{R}, \alpha \cup \beta \rangle$$
$$\langle \mathbf{E}, \overline{R}, \alpha \rangle - \beta = \langle \mathbf{E}, \overline{R}, \alpha - \beta \rangle.$$

Informally, the $+$ operator is the change that occurs when an agent learns that the agents in β are all honest; similarly, the $-$ operator is the change that occurs when an agent learns that the agents in β are dishonest.

We are now in a position to define 'revision' operators for reports. In the following definition, we use $\cdot$ as a concatenation operator on report histories; so $\overline{R} \cdot r$ is just the report history obtained by adding r to the end of $\overline{R}$.

Definition 7. *Let $\langle \mathbf{E}, \overline{R}, \alpha \rangle$ be an epistemic history, let ϕ be a formula, and let $A \in \mathbf{A}$. Then:*

$$\langle \mathbf{E}, \overline{R}, \alpha \rangle *_A \phi = \langle \mathbf{E}, \overline{R} \cdot (\phi, A), \alpha \rangle.$$

The $*_A$ operator captures what happens when agent A reports the information ϕ; namely, the report history is extended.

We use the symbol $*$ for this operation because it actually represents the natural extension of revision, when we look at the associated belief states. This claim is captured in the following proposition.

Proposition 1. *For any epistemic history $\langle \mathbf{E}, \overline{R}, \alpha \rangle$ and any agent A:*

$$Bel(\langle \mathbf{E}, \overline{R}, \alpha \rangle *_A \phi) = Bel(\mathbf{E} * (\overline{R} \cdot (\phi, A)) \upharpoonright \alpha))$$

Hence, for any action history $\overline{R}$, the belief state resulting from $\langle \mathbf{E}, \overline{R}, \alpha \rangle$ followed by the report (ϕ, A) can equivalently be obtained by concatenating (ϕ, A) to $\overline{R}$ and then calculating the belief state.

In fact, the result of this revision by a report breaks into two cases, depending on whether or not A is honest.

Proposition 2. *Let α be an honesty assignment,*

- *If $A \in \alpha$, $Bel(\langle \mathbf{E}, \overline{R}, \alpha \rangle *_A \phi) = Bel(\mathbf{E} * (\overline{R}(\upharpoonright \alpha) * \phi)))$*
- *If $A \notin \alpha$, $Bel(\langle \mathbf{E}, \overline{R}, \alpha \rangle *_A \phi) = Bel(\mathbf{E} * \overline{R} \upharpoonright \alpha))$*

Essentially, reports from honest agents lead to belief revision and reports from dishonest agents are discarded.

3.3 Basic Results

The operators $*_A$, $+$ and $-$ are independent in the sense that they alter different parts of the epistemic history. This allows us to re-order the operations, as indicated in the following result.

Proposition 3. *Let* $\langle \mathbf{E}, \overline{R}, \alpha \rangle$ *be an epistemic history, let* ϕ *be a formula, let* $\beta \subseteq \mathbf{A}$. *Then:*

$$(\langle \mathbf{E}, \overline{R}, \alpha \rangle + \beta) *_A \phi = (\langle \mathbf{E}, \overline{R}, \alpha \rangle *_A \phi) + \beta$$
$$(\langle \mathbf{E}, \overline{R}, \alpha \rangle - \beta) *_A \phi = (\langle \mathbf{E}, \overline{R}, \alpha \rangle *_A \phi) - \beta$$

Hence, it does not matter if we learn an agent is dishonest before or after they provide information. In either case, the resulting epistemic history is the same. Note that none of the operators $\{*_A, +, -\}$ is commutative; we can not change the order of revisions nor can we change the order of honesty expansions/reductions. However, we can apply the proposition repeatedly to get a standard form.

Proposition 4. *If* $\langle \mathbf{E}, \overline{R}, \alpha \rangle$ *is an epistemic history,* $A_i \in A$, $\beta_i \subseteq A$, $\delta_i \subseteq A$, *then*

$$\langle \mathbf{E}, \overline{R}, \alpha \rangle *_{A_1} \phi_1 + \beta_1 - \delta_1 \cdots *_{A_n} \phi_n + \beta_1 - \delta_n$$
$$= \langle \mathbf{E}, \overline{Q}, \gamma \rangle$$

where

– γ *is obtained by starting with* α *and then iteratively adding and removing the elements of each* β_i *and* δ_i.
– $\overline{Q} = \overline{R} \cdot (\phi_1, A_1) \cdots (\phi_n, A_n)$.

As a result, every sequence of reports and honesty updates can be used to obtain a new epistemic history through mechanical operations. The new beliefs can be determined by simply applying the *Bel* operator.

3.4 Motivating Example Revisited

We return to our motivating example, involving weather reports. The set of agents in this case is $\mathbf{A} = \{B, T\}$, representing Bob and Trent respectively. The vocabulary is $\mathbf{F} = \{Sun, House\}$. The first variable is true if it is sunny, and false if it is rainy. The second variable is true if Bob is at the house, and false if he is not. Initially, all states are considered equally likely; so $Bel() = 2^{\mathbf{F}}$ where $\mathbf{E}$ represents the initial beliefs of Alice.

In this example, Alice initially believes Bob and Trent are both honest. So the initial epistemic history is $\langle \mathbf{E}, \tau, \mathbf{A} \rangle$, where τ is the empty sequence. Recall that Alice is first told $\neg Sun$ and $House$ by Trent, and then she is told Sun and $\neg House$ by Bob. In our framework, we need to calculate the following expression:

$$\langle \mathbf{E}, \tau, \mathbf{A} \rangle *_T \neg Sun *_T House *_B Sun *_B \neg House.$$

This can be written as $\langle \mathbf{E}, \overline{R}, \mathbf{A} \rangle$ where

$$\overline{R} = (\neg Sun, T), (House, T), (Sun, B), (\neg House, B).$$

Note that:

$$Bel(\langle \mathbf{E}, \overline{R}, \mathbf{A} \rangle) \models \neg House.$$

This is expected, since all agents are assumed to be honest - this just follows from regular revision.

In the original example, we added an extra step where Alice learns that Bob is dishonest. This could be captured by applying the subtraction operator.

$$\langle \mathbf{E}, \overline{R}, \mathbf{A} \rangle - \{B\} = \langle \mathbf{E}, \overline{R}, \{T\} \rangle$$

Now consider $Bel(\langle \mathbf{E}, \overline{R}, \{T\} \rangle)$. By Definition 5, this is equal to

$$Bel(\mathbf{E} * \overline{R} \upharpoonright \{T\}) = Bel(\mathbf{E} * \neg Rain * House) \models House.$$

So, in this case, Alice believes that Bob is at Trent's house. This is the desired result.

In practice, we actually learn when an agent is dishonest by comparing their reports to our own observations. Hence, a reduction to the set of honest agents should be triggered by a direct observation. We address this extension in the next section.

4   Reports from the Self

4.1   Pointed Agent Sets

In order to incorporate observations, we need to be able to designate a special agent that represents the *self*. We again assume an underlying propositional vocabulary $\mathbf{F}$. However, in this section we also assume a *pointed* agent set $\mathbf{A}$. This is a set of agents that includes a distinguished symbol O that represents the self.

A report is still a pair (ϕ, A), though the reporting agent may be the special agent O. A report of the form (ϕ, O) represents information that the agent has obtained directly. We will refer to such reports as *observations*.

Definition 8. *For a pointed agent set* $\mathbf{A}$*, an* honesty assignment *is a set* $\alpha \subseteq \mathbf{A}$ *such that* $O \in \alpha$.

The definition of an epistemic history remains unchanged, it still consists of an epistemic state, a report history and an honesty assignment; the only difference is that the report history can include observations and the honesty assignment must include O.

4.2   Observations and Pointed Agent Sets

Observations are special reports. They are special not only because the self is always considered honest, but also because observations can give us information about the honesty of other agents. Specifically, if some agent reports ψ and then we observe $\neg\psi$, then we would like to conclude that agent can not be trusted to provide accurate information.

Definition 9. *A report* (ϕ, A) *is* inconsistent *with a formula* ψ *just in case* $\phi \wedge \psi$ *is inconsistent.*

We write $(\phi, A) \perp \psi$ to indicate that (ϕ, A) is inconsistent with ψ.

Definition 10. *Given a report history* $\overline{R} = (\phi_1, A_1) \cdot \cdots \cdot (\phi_n, A_n)$ *and a formula* ψ, *let*

$$inc(\overline{R}, \psi) = \{A_i \mid \phi_i \perp \psi\} - \{O\}.$$

So inc picks out the set of agents in the report history that have reported information that is inconsistent with ψ. We remark that an agent A will be in this set if they make a single report that is inconsistent with ψ. However, we remove O from the set, because we need to handle inconsistent observations through the normal revision process.

We now extend our definition of $*_A$ to include observations.

Definition 11. *Let* $\langle \mathbf{E}, \overline{R}, \alpha \rangle$ *be an epistemic history over a pointed agent set* $\mathbf{A}$, *let* ϕ *be a formula, and let* $A \in \mathbf{A}$.

1. *If* $A \neq O$, $\langle \mathbf{E}, \overline{R}, \alpha \rangle *_A \phi = \langle \mathbf{E}, \overline{R} \cdot (\phi, A), \alpha \rangle$.
2. *If* $A = O$, $\langle \mathbf{E}, \overline{R}, \alpha \rangle *_A \phi = \langle \mathbf{E}, \overline{R} \cdot (\phi, A), \alpha - inc(\overline{R}, \psi) \rangle$.

For reports from other agents, this definition remains unchanged. In the case of observations, it indicates that we remove all agents from α that have provided information that conflicts with the new observation.

4.3 Motivating Example Revisited, Again

We return to the weather report example. If we have pointed action histories, then we formulate the motivating example using the pointed agent set $\mathbf{A} = \{B, T, O\}$, where O represents Alice (the self). The vocabulary remains unchanged. The initial epistemic history $\langle \mathbf{E}, \tau, \mathbf{A} \rangle$ is also unchanged.

In this version of the example, the four initial reports are the same. Alice is still told $\neg Sun$ and $House$ by Trent, followed by Sun and $\neg House$ by Bob. We now add one more report, $(\neg Sun, O)$. This represents an observation by Alice that it is raining outside. We need to calculate the result of the following expression:

$$\langle \mathbf{E}, \tau, \mathbf{A} \rangle *_T \neg Sun *_T House *_B Sun *_B \neg House *_O \neg Sun.$$

This can be written as $\langle \mathbf{E}, \overline{R}, \mathbf{A} \rangle$ where

$$\overline{R} = (\neg Sun, T), (House, T), (Sun, B), (\neg House, B), (\neg Sun, O).$$

Note that (Sun, B) is *inconsistent* with the report of $\neg Sun$ from O. Therefore, all observations from B will be removed when calculating the new belief state. In other words, if we let

$$\overline{R}' = (\neg Sun, T), (House, T), (\neg Sun, O)$$

then the new belief state will be

$$Bel(\langle \mathbf{E}, \overline{R}', \mathbf{A} \rangle).$$

This is equal to

$$Bel(\mathbf{E} * \overline{R} \upharpoonright \{T, O\}) = Bel(\mathbf{E} * \neg Sun * House * \neg Sun).$$

Clearly *House* is entailed by this new belief state; so Alice believes that Bob is at Trent's house, contrary to Bob's report. This result is due to the fact that Bob has made a report that conflicts with Alice's direct observations, and therefore Bob is no longer considered to be honest.

This approach to the motivating example is essentially the same as the previous treatment, except that we no longer require a direct change to the honesty assignment. Instead, the result follows from the definition of an observation.

5   Incomplete Information

5.1   Partial Honesty Assignments

Thus far, we have only considered cases where we know exactly which agents are honest. In this section, we move to the case where we only have partial information about honesty. For simplicity, we restrict attention to the case where the agent O does not appear in the set of agents.

Definition 12. *A partial honesty assignment is a set* $\Gamma \subseteq 2^{\mathbf{A}}$.

A partial honesty assignment includes all honesty assignments that are considered possible. We say that an agent A is *honest* with respect to Γ just in case $A \in \alpha$ for all $\alpha \in \Gamma$. Similarly, we say that A is dishonest just in case A is not in any element of Γ.

We define partial epistemic histories by replacing the honesty assignment with a partial honesty assignment.

Definition 13. *Let* $\mathbf{E}$ *be an epistemic history, let* $\overline{R}$ *be a report history, and let* Γ *be a partial honesty assignment. A partial epistemic history is a triple* $\langle \mathbf{E}, \overline{R}, \Gamma \rangle$.

The revision operators defined previously can be applied to partial epistemic histories in a natural manner:

$$\langle \mathbf{E}, \overline{R}, \Gamma \rangle *_A \phi = \langle \mathbf{E}, \overline{R} \cdot (A, \phi), \Gamma \rangle.$$

Similarly, $+$ can also be defined, by applying the union to each element of Γ. In other words, if we define $\Gamma + \beta = \{\gamma \cup \beta \mid \gamma \in \Gamma\}$ then we have:

$$\langle \mathbf{E}, \overline{R}, \Gamma \rangle + \beta = \langle \mathbf{E}, \overline{R}, \Gamma + \beta \rangle.$$

The same thing can be done for the $-$ operator.

We would like to associate beliefs with partial epistemic histories. But we can not actually associate a specific belief set with a particular epistemic action history. Instead, we need to associate a collection of different belief sets with each partial epistemic history.

Definition 14. *Let* $\langle \mathbf{E}, \overline{R}, \Gamma \rangle$ *be a partial honesty assignment. Then define:*

$$BS(\langle \mathbf{E}, \overline{R}, \Gamma \rangle) = \{Bel(\mathbf{E}, \overline{R}, \alpha) \mid \alpha \in \Gamma\}.$$

Hence, BS picks out the set of possible belief sets depending on which agents are actually honest. This set allows us to use credulous reasoning to identify formulas that are believed in all possible honesty assignments.

Definition 15. *Let $\langle \mathbf{E}, \overline{R}, \Gamma \rangle$ be a partial honesty assignment. We say that ϕ is believed in $\langle \mathbf{E}, \overline{R}, \Gamma \rangle$ just in case $K \models \phi$ for all $K \in BS(\langle \mathbf{E}, \overline{R}, \Gamma \rangle$.*

The set of possible belief sets can also be defined through a semantic construction, which we demonstrate in the next section.

5.2 Revision Trees

Let $\overline{\phi} = \phi_1, \ldots \phi_n$ be a sequence of formulas over the vocabularly $\mathbf{F}$.

We can define a simple tree structure that lets us visualize all possible revisions of $\mathbf{E}$ by subsequences of $\overline{\phi}$. We define the *revision tree* $T(\mathbf{E}, \overline{\phi})$ iteratively by levels $V_0, \ldots, V_n$, where V_i is the set of vertices at height i.

1. V_0 consists of a single vertex labelled with the epistemic state $\mathbf{E}$.
2. For $i > 1$, at level V_i we have two children for each vertex $v \in V_{i-1}$.
   (a) The first child is labelled $\top$.
   (b) The second child is labelled ϕ_i.

Every maximal length path in $T(\mathbf{E}, \overline{\phi})$ defines a sequence of revisions by some subsequence of $\overline{\phi}$; we simply start with $\mathbf{E}$ and then read the labels off the nodes at heights 1 up to n. Formulas that are not being included in the sequence have been labeled with $\top$. Note that we are treating $\top$ as a tautology here, to ensure revision by $\top$ does not result in any change of belief. Note also that the formulas in $\overline{\phi}$ can not be equal to $\top$, as they are defined over the vocabularly F. Of course the formulas in $\overline{\phi}$ can be tautologies, but they will still be represented as formulas over $\mathbf{F}$ without using the special symbol $\top$.

Now suppose that we have a report history $\overline{R} = (r_1, \phi_1), \ldots, (r_n, \phi_n)$. The paths in $T(\mathbf{E}, \overline{\phi})$ still correspond to revision sequences, but not every path is valid with respect to $\overline{R}$. Informally, a vertex labelled with $\top$ represents a revision that is being 'ignored' because the reporting agent is dishonest. When we consider sequences of revisions corresponding to reports, we need to ensure that we ignore *all* reports from a dishonest agent.

Definition 16. *Let $\mathbf{E}$ be an epistemic state, let $\overline{R} = (r_1, \phi_1), \ldots, (r_n, \phi_n)$ be a report history of length n. We say that a sequence $\psi_1, \ldots, \psi_n$ is valid with respect to $\overline{R}$ just in case:*

1. *For each i, ψ_i is either ϕ_i or $\top$.*
2. *If $r_i = r_j$, and $\psi_i = \top$, then $\psi_j = \top$.*

We let $val(\overline{R})$ denote set of all valid sequences with respect to $\overline{R}$.

Hence a valid sequence with respect to $\overline{R}$ is a subsequence with the following property. For each report agent r, either all formulas reported by r are in the sequence or all formulas reported by r have been replaced by $\top$.

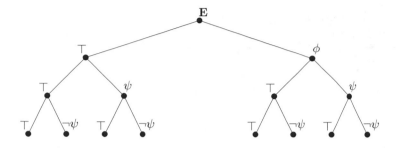

Fig. 1. A revision tree

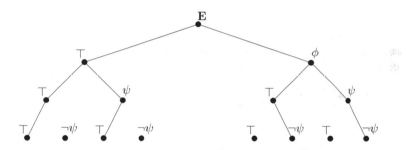

Fig. 2. Valid paths with respect to a report history

Example 1. Let $\overline{\phi} = \phi, \psi, \neg\psi$. In Fig. 1, we show the complete revision tree for $T(\mathbf{E}, \overline{\phi})$. Now suppose that $\overline{R} = (A, \phi), (B, \psi), (A, \neg\psi)$. In Fig. 2, we show the set of valid paths of the tree over $T(\mathbf{E}, \overline{\phi})$ with respect to $\overline{R}$.

An honesty assignment defines a particular modification of a report history.

Definition 17. *For any honesty assignment α and any report history $\overline{R} = (r_1, \phi_1), \ldots, (r_n, \phi_n)$, define $\overline{R}(\alpha) = (r_1, \psi_1), \ldots, (r_n, \psi_n)$ such that*

1. If $r_i \notin \alpha$, then $\psi_i = \top$.
2. $\psi_i = \phi_i$, otherwise.

The following result just ensures this is actually a valid sequence

Proposition 5. *For any honesty assignment α and any report history $\overline{R}$, the report history $\overline{R}(\alpha)$ is a valid sequence with respect to $\overline{R}$.*

The converse is also true.

Proposition 6. *Suppose that we have a path $\psi_1, \ldots, \psi_n$ in $T(\mathbf{E}, \overline{\phi})$ that is valid with respect to $\overline{R} = (r_1, \phi_1), \ldots, (r_n, \phi_n)$. Then $\alpha = \{r_i \mid \phi_i \neq \top\}$ is an honesty assignment.*

The proof of this result really just amounts to observing that α is a well defined set, which follows from the definition of a valid path.

So, given a report history $\overline{R}$, we have a one to one correspondence between honesty assignments and valid paths in the revision tree. This gives us an alternative way to look at the beliefs associated with a partial epistemic history.

Proposition 7. *Let* $\langle \mathbf{E}, \overline{R}, \Gamma \rangle$ *be a partial epistemic history. Then* $K \in BS(\langle \mathbf{E}, \overline{R}, \Gamma \rangle)$ *if and only if* $K = Bel(\mathbf{E} * \overline{\psi})$ *for some valid sequence* $\overline{\psi}$ *with respect to* $\overline{R}$.

Hence, while we can not associate a unique belief set with a partial epistemic history, the set of possible belief sets has a useful structure.

5.3   Honesty Invariants

Partial honesty assignments are useful for applications where we receive information from other agents, and we learn over time which agents are honest and which agents are not. One natural application for this work is in tracking how our beliefs are contingent upon the honesty of particular agents.

Definition 18. *Let* $\langle \mathbf{E}, \overline{R}, \Gamma \rangle$ *be a partial epistemic history. We say that* ϕ *is* honesty invariant *with respect to* $A \in \mathbf{A}$ *just in case* ϕ *is believed in* $\langle \mathbf{E}, \overline{R}, \Gamma \rangle + \{A\}$ *and* ϕ *is believed in* $\langle \mathbf{E}, \overline{R}, \Gamma \rangle - \{A\}$

When a formula ϕ is honesty invariant with respect to A, then it will be believed whether or not A is honest.

We can state a simple property of honesty invariance.

Proposition 8. *Let* $\langle \mathbf{E}, \overline{R}, \Gamma \rangle$ *be a partial epistemic history. If* $Bel(\mathbf{E}, \overline{\psi}) \models \phi$ *for every valid sequence* $\overline{\psi}$, *then* ϕ *is honesty invariant with respect to every* $A \in \mathbf{A}$.

This result states that a formula is honesty invariant with respect to all agents if it is true following every possible sequence of revisions. We give another basic result.

Proposition 9. *Let* $\langle \mathbf{E}, \overline{R}, \Gamma \rangle$ *be a partial epistemic history. Let* j *be the highest index of a report from* A *in* $\overline{R}$, *and let* k *be the highest index of a report* (B, ψ) *where* $\psi \models \phi$. *If* $k > j$ *and* $B \in \alpha$ *for all* $\alpha \in \Gamma$, *then* ϕ *is honesty invariant with respect to* A.

This result is only slightly more difficult. It says that ϕ is honesty invariant with respect to A if the last report from A is followed later by a report from an honest agent that guarantees ϕ will be true.

Using a brute force approach, in order to check if ϕ is believed in a partial epistemic history, we need to look at $|\Gamma|$ revision sequences. If there is a single report from A, then we need to look at roughly $2 \cdot |\Gamma|$ sequences to check if ϕ is honesty invariant. If there are n reports, this grows to $2^n \cdot |\Gamma|$ sequences. Hence, checking honesty invariance becomes computationally very expensive if we use an exhaustive approach; we need to use heuristics to exploit the symmetries over the set of valid sequences to do this more efficiently. We leave a complete treatment of computational issues related to honesty invariance for future work.

6   Discussion

6.1   Related Work

As mentioned in the introduction, there has been related work on the way trust impacts the dynamics of belief [3,7,10]. However, each of these frameworks is focused on trust in terms of perceived knowledge. In this context, we actually have to look at the content of a report to determine if it should be believed. Hence, a particular agent might be trusted when they report about the weather but they might not be trusted when they report about microbiology. There is no reason to dismiss an agent as dishonest, but we do need to consider their expertise over particular domains. This is a very different form of trust, and the models developed for knowledge-based trust do not apply directly to the issue of honesty addressed in this paper.

In a sense, our work is more closely related to formal models of reasoning that focus on some form of regression to maintain a consistent belief trajectory. This is the case, for example, in the case of belief evolution operators [5] or the epistemic extension of the Situation Calculus [9]. In both of these frameworks, we essentially maintain a consistent sequence of belief states by regressing change to an initial state of the world. We are following the same basic approach, by maintaining a history of all possible reports; this allows us to ensure a consistent trajectory of beliefs when we get new information about honesty.

6.2   Future Work

There are several issues to be addressed in future work. First, the theoretical foundations of our model need to be clarified. We have shown that the revision tree includes all possible belief trajectories, and that paths through the revision tree represent valid trajectories with respect to a report history. This observation provides the first step towards a representation result for honesty-sensitive revision operators. In future work, we will provide a set of rationality postulates for the $*_A$ operators. We will then attempt to prove that these postulates are fully characterized by some suitable set of paths in the revision tree.

In addition to a representation result, there are important theoretical directions to be explored in this framework. As noted previously, we would like to explore algorithms for honesty invariance. We would also like to consider the integration of honesty-based trust with knowledge-based trust.

We would also like to work on practical applications of our framework. The main example that we will pursue is the use of honesty assignments to reason about recovery following security breaches. Suppose that we have a knowledge base that has been accessed by an individual using fraudulent credentials. It is not sufficient to simply block that individual from future access; we also need to roll back any changes that they have made. This can be modelled by viewing knowledge base updates as reports associated with particular agents.

7 Conclusion

We have addressed the problem of honesty-based trust in the context of formal belief revision operators. New information comes in the form of reports from agents, and we maintain a list of all reports that have been obtained. In addition to a list of reports, we have also introduced the notion of an honesty assignment that keeps track of which agents are currently considered to be trustworthy. We have demonstrated that this allows us to modify our belief state appropriately when we obtain new reports, and also when we get new information about the honesty of other agents. By moving to pointed agent sets, we have also shown how reports from other agents interact with observations. While it is simpler conceptually when we have a single set of honest agents, we have also considered the case where our knowledge about honesty is uncertain. We have demonstrated that all reasonable belief trajectories can be defined in a simple tree structure, which is further constrained by the sequence of reporting agents. The end result is a simple model that allows us to move from a naive model of belief revision where all agents are assumed to be honest to a more realistic model where the perceived honesty of agents influences the dynamics of belief.

References

1. Alchourrón, C.E., Gärdenfors, P., Makinson, D.: On the logic of theory change: partial meet functions for contraction and revision. J. Symb. Log. **50**(2), 510–530 (1985)
2. Booth, R., Meyer, T.: Admissible and restrained revision. J. Artif. Intell. Res. **26**, 127–151 (2006)
3. Booth, R., Hunter, A.: Trust as a precursor to belief revision. J. Artif. Intell. Res. **61**, 699–722 (2018)
4. Darwiche, A., Pearl, J.: On the logic of iterated belief revision. Artif. Intell. **89**(1–2), 1–29 (1997)
5. Hunter, A., Delgrande, J.: Iterated belief change due to actions and observations. J. Artif. Intell. Res. **40**, 269–304 (2011)
6. Katsuno, H., Mendelzon, A.: Propositional knowledge base revision and minimal change. Artif. Intell. **52**(2), 263–294 (1992)
7. Liu, F., Lorini, E.: Reasoning about belief, evidence and trust in a multi-agent setting. In: An, B., Bazzan, A., Leite, J., Villata, S., van der Torre, L. (eds.) PRIMA 2017. LNCS (LNAI), vol. 10621, pp. 71–89. Springer, Cham (2017). https://doi.org/10.1007/978-3-319-69131-2_5
8. Schwind, N., Konieczny, S., Perez, R.P.: Darwiche and Pearl's epistemic states are not total preorders. In: Proceedings of the International Conference on Principles of Knowledge Representation and Reasoning (KR 2022) (2022)
9. Shapiro, S., Pagnucco, M., Lesperance, Y., Levesque, H.: Iterated belief change in the situation calculus. Artif. Intell. **175**(1), 165–192 (2011)
10. Singleton, J., Booth, R.: Who?s the expert? On multi-source belief change. In: Proceedings of the International Conference on Principles of Knowledge Representation and Reasoning (KR 2022) (2022)

Machine Learning

Active Learning for kNN Using Instance Impact

Sayed Waleed Qayyumi$^{(\boxtimes)}$ iD, Laurence A. F. Park iD, and Oliver Obst iD

Centre for Research in Mathematics and Data Science School of Computer, Data and Mathematical Sciences, Western Sydney University, Locked Bag 1797, Penrith, NSW 2751, Australia
{s.qayyumi,l.park,o.obst}@westernsydney.edu.au
https://www.westernsydney.edu.au/crmds/

Abstract. Labelling unlabeled data is a time-consuming and expensive process. Labelling initiatives should select samples that are likely to enhance the classification accuracy of the classifier. Several methods can be employed to accomplish this goal. One of these techniques is to select samples with the highest level of uncertainty in their predicted labels. Experts then label these samples. Another option is to choose samples at random. This paper proposes three methods for identifying unlabeled samples to improve predictive accuracy when they are labelled. Our study explores how to select samples when we have very few labelled samples available from manifold distributed data sets. In order to assess performance, we have compared our approaches with uncertainty sampling and random sampling. We demonstrate that our methods outperform uncertainty sampling and random sampling by using public and real-world data sets.

Keywords: Active learning · Uncertainty sampling · Unlabelled sampling · Random sampling · Incremental learning · Few shot learning · Entropy · Uncertain labels

1 Introduction

To classify complex tasks, supervised machine learning models can be used to learn complex relationships between queries and responses. For instance, machine learning models can help detect tumours at an early stage. Upon finding a tumour by these models, a specialist can examine it further. Training data must contain queries and responses created by or evaluated by specialists to train models for specialized purposes. Therefore, such data can be challenging to obtain. Many queries are available (e.g. image scans, feature vectors, videos), yet, it is hard to receive accurate responses to each of these queries. In the case of specialized data, we must hire a specialist to examine each query and provide a response. A specialist must take time to do this, which is costly for both the data modeller and the specialist. If labelling is cost-prohibitive, a smaller sample of queries is forwarded to a specialist. The selection of the samples is either accomplished randomly or using uncertainty sampling.

Table 1. Sampling/Labelling scenarios. With a large sample that is difficult to label, we resort to labelling a random sample, but is that best approach?

		Sampling	
		1: Simple	2: Difficult
Labelling	A: Simple	Label all	Label all
	B: Difficult	Label Random	Label all

This paper discusses how to sample queries for manual labelling to improve the accuracy of the machine learning models. In our research, we generally focus on small sample sizes (e.g., the few shot learning scenario and manifold distributed data). The article will proceed as follows: Sect. 2 discusses the current state of the art in the selection of the next best-unlabelled sample. Section 3 examines our approaches to the next best sample selection. Section 4 presents the results of experiments conducted on different public and real-life data sets in a few-shot learning and semi-supervised learning scenario. This section also compares our sampling techniques with active learning's uncertainty sampling and random sampling. Section 4.3 contains a list of our observations. Section 5 concludes this paper and discusses our future work.

2   Background and Related Work

In attempting to classify manifold distributed data with very few label samples, we investigated the problem of finding the best-unlabelled sample for labelling. In classifier training, there are four scenarios regarding data availability. Table 1 lists all these scenarios. This article discusses scenario 1B (easy to sample and difficult to label). In this scenario, we have to label more samples to achieve higher accuracy in classification. Labelling is a complex and costly endeavour, so choosing the right unlabelled sample is crucial. Imagine, for example, one million CT scans with only ten labels. To improve the accuracy of your classifier, you need to label another ten items. Choosing a sample that increases the accuracy of the classifier is crucial in such a scenario. This is the focus of our sample selection methods.

Active learning is the process of selecting an optimal unlabelled sample from a pool of unlabelled data. Unlabeled data is classified with a classifier, and then the observations with the most uncertain labels are identified. This process is known as uncertainty sampling. There are many methods that are available to estimate the uncertainty of a labelled sample. The active learning process consists of querying an information source, for example, an Oracle, to assign a new label to a data point. This algorithm attempts to choose the best possible sample to be labelled [16,18]. The term optimal experimental design can also refer to active learning in statistics. In situations, unlabeled data is readily available, but its labelling is costly. When such a scenario occurs, a learning algorithm can aid in identifying samples for labeling. This process is known as active learning. Choosing examples that the learner finds meaningful is generally more effective, which results in fewer examples needed than is necessary for supervised learning. Recent advances in active learning include multi-label active learning [24] and hybrid active learning [11]. These research areas combine machine learning concepts with incremental learning policies. There are three different scenarios or settings in which learners typically query instances' labels.

- The learner generates instances based on the underlying distribution in the membership query synthesis.
- In stream-based sampling, the assumption is that unlabeled samples are free to obtain. Thus, each unlabelled sample is selected one at a time. Upon reading an unlabelled instance, the learner can decide whether to query or reject. Acceptance or rejection of the instance is driven by its informativeness. A query strategy determines how informative the sample is.
- Pool-based sampling is based on the assumption that there is a large pool of unlabeled data. An informativeness measure can be applied to all samples in the pool to identify the best candidates for labeling. The proposed sampling methods described in this paper can also be referred to as pool-based sampling techniques.

The learner can utilize a variety of measures to identify the most appropriate sample. An example of one of these measures is uncertainty. The learner labels all unlabeled data using the available labelled data. Upon determining the uncertainty of each predicted label, the sample with the most uncertain label is selected and sent to Oracle for labeling. The following are three commonly used approaches to querying instances based on uncertainty sampling.

- Least Confidence: LC strategies let learners select the instance for which the learner is least confident in its most likely label.

$$U(x) = 1 - P(\hat{x}|x) \tag{1}$$

- Margin Sampling: A fundamental problem with the LC strategy is that it only considers the most probable label and disregards the other label probabilities. For this reason, the margin sampling strategy selects the instance with the minimal difference between the first and second most probable labels.

$$M(x) = P(\hat{x}_1|x) - P(\hat{x}_2|x) \tag{2}$$

- Highest Entropy: All the potential label probabilities can be computed using entropy. All instances are analyzed by calculating the entropy value of each instance and querying the instance with the highest value.

$$H(x) = -\sum_k p_k \log(p_k) \tag{3}$$

It is important to note that uncertainty sampling is dependent on predicted labels. In addition, calculating uncertainty is not straightforward for all classification methods. It is not easy, for example, to calculate the uncertainty in neural network setup [23].

Sampling plays a significant role in classifier training. In order to improve prediction accuracy, it is necessary to train a classifier with sufficient training data. Sample collection can help provide the necessary data. You can find a detailed description of most of the sampling techniques in Altmann et al. [1] and Etikan et al. [5]. In random sampling [15], each sample has an equal chance of being selected. A stratified random sampling method [14] involves dividing the population into subgroups called strata and selecting samples at random from each stratum. A systematic sample selection method, [12] is based on choosing a fixed interval and starting point. After establishing a starting point, subsequently, samples can be collected at regular intervals. Clustered sampling [6] allows drawing samples at random from some of the clusters. Clustered sampling draws samples from random groups, whereas stratified sampling selects samples from each stratum or group, allowing us to exclude entire groups from the study. The convenience sampling method [17] involves selecting a sample solely on the basis of its convenience for sampling purposes. Quota sampling [13] selects samples based on specific characteristics. There is also snowball sampling [8], which selects a sample based on the judgment of the experts who need it, and then uses it to select subsequent samples. Sampling methods are bound to be biased. A number of methods have been proposed to address bias in sampling [10,21]. All samples need not come from the same distribution. They may even come from a distribution similar to the one we study. If the main distribution is unavailable, importance sampling [22] is applied. In this scenario, we sample from another distribution by adjusting the weights of the distribution so that it represents the desired distribution. We can use information gain to select the samples. The information gain is the amount of entropy removed from the data set by splitting it. Therefore, a split with a higher information gain [2] is preferred.

Data samples are collected before classification models are built and trained. It is possible to construct classification models if enough data is available. When we do not have enough data, we can continually improve our classification models by retraining additional labelled data. During retraining, newly acquired labelled data is incorporated into the learning process. The method of learning is called incremental learning [7]. It is possible to apply several traditional classification methods to incremental learning [20]. In incremental learning, the goal is to acquire new knowledge based on new data without forgetting the existing knowledge derived from older data. The next best action recommendation is a popular marketing technique designed to retain customers. In order to determine what the best next step for a given customer is, it is necessary to compare their profile to a similar customer model [9]. Reinforcement learning determines the next best task based on this approach [4]. In a similar fashion to incremental learning, the next best task has been an active area of research [3].

Aside from uncertainty sampling, our work is also comparable to Transductive Semi-supervised Deep Learning (TSSDL) and Personalized next-best-action recommendation [3,19]. In particular, we discuss the topics of sampling, entropy, incremental learning, and recommendation of the next best task. We will discuss the relevance of these topics after providing a brief overview of these topics. We do not estimate labels for the unlabeled samples, but rather rank all unlabeled samples according to their potential influence on classification accuracy.

3   Measuring the Utility of an Instance for Training

As a rule of thumb, the performance of a classification model is contingent on how well the training data represent the population to be classified. Therefore, it is imperative to select a sufficient number of instances from the population of interest for manual labelling and inclusion in the training set. To accomplish this, we must be able to select the most appropriate sample of the population, and then manually label each instance within that sample. We should take as large a sample as possible if the labelling of each instance is a straightforward process. Furthermore, we should attempt to label all the observations in our training set. When both manual labelling and sampling are time-consuming and costly, fewer samples can be collected, and all instances will have to be labelled. There is a question regarding how to proceed when we have access to a large pool of unlabelled data but cannot label each instance. Therefore, we must determine which subset of that sample should be manually labelled and added to the training set. Our goal should be to select observations that will produce the highest increase in classification accuracy when used for training. However, the question remains as to how to choose the instances.

This section examines three candidate functions for evaluating the utility of including an unlabeled instance in a training set. Each instance is assigned a score based on its potential to influence classification accuracy. The article focuses on data with a relatively small training set (only a few cases were manually labelled), and which are manifold distributed. Thus, we use k Nearest Neighbors (kNN) as a classifier. Please also note that we have assumed that sampling from the population will be relatively straightforward, whereas labelling will be more complex. Thus, we can also safely presume that we have a large pool of candidate instances from which to choose and that it is possible to assess the utility of a selected instance; we refer to this pool of available data as the "test set". We have described our proposed methods in the following three subsections.

3.1   Neighbourhood Impact

In order for a new instance to have the potential to increase accuracy, it has to play a role in the classification of newly created instances. The training instance is only relevant if it is the nearest neighbour of the test instance in the kNN classification. Therefore, one measure of the utility of a candidate training instance is the number of data points it is closest to.

Given a set of manually labelled instances $\mathcal{X}$ and a set of unlabelled instances $\mathcal{U}$, let $N_k(u; \mathcal{X}) \subset \mathcal{X}$ be the set of k nearest neighbours of u chosen from the manually labelled set of instances, where $u \in \mathcal{U}$ and $|\mathcal{X}| > k$. We define the *Neighbourhood Impact I* of labelled instance x as

$$I(x) = \sum_{u \in \mathcal{U}} \mathbf{1}_{N_k(u;\mathcal{X})}(x). \tag{4}$$

for $x \in \mathcal{X}$, where $\mathbf{1}_A(x)$ is the indicator function ($\mathbf{1}_A(x) = 1$ if $x \in A$ or 0 otherwise).

To measure the neighbourhood impact of an unlabelled instance u, we must remove the instance from the set $\mathcal{U}$ to obtain $\mathcal{U} \setminus u$, and append it to the set of labelled instances $\{\mathcal{X}, u\}$. The neighbourhood impact for an unlabelled instance is

$$I(u) = \sum_{v \in \mathcal{U} \setminus u} \mathbf{1}_{N_k(v;\{\mathcal{X},u\})}(u). \tag{5}$$

Including an unlabelled instance in the training set will not cause the trained model's classification accuracy to improve if $I(u) = 0$. Including an unlabelled instance with a high $I(u)$ will affect the model's classification accuracy when included in the training set. The hypothesis is that if an unlabelled instance has a high $I(u)$ value, then manually labelling it and adding it to the training set will improve its accuracy.

3.2   Maximum Entropy

A neighbourhood impact refers to the potential of an instance based on its proximity to a neighbourhood. Furthermore, it is possible to examine whether the point may be able to alter the unlabelled class prediction. By calculating the entropy of the label distribution one can determine how robust the prediction is when there is a set of training labels. As a result, the notion of high entropy implies that one change in an instance label might alter a prediction, whereas the notion of low entropy requires many changes in order to change a prediction.

The class prediction for test instance u is the mode class of the set of k nearest neighbours from the labelled set $\mathcal{X}$. We define $L_k(u; \mathcal{X})$ as the set of class labels associated to the training instances $N_k(u; \mathcal{X})$. Using this, the predicted class label for instance u is $\mathrm{mode}(L_k(u; \mathcal{X}))$ and the entropy of the neighbourhood distribution is $\mathrm{Ent}(L_k(u; \mathcal{X}))$.

This potential for an unlabelled instance u to influence the class prediction is expressed by *Maximum Entropy*. Essentially, this can be defined as the maximum class distribution entropy if the example was included in the training set with a class label. The maximum entropy $H(u)$ of an unlabelled instance u can be defined as

$$H(u) = \max_{l_u \in \mathcal{L}} \sum_{v \in \mathcal{U} \setminus u} \mathrm{Ent}(L_k(v; \{\mathcal{X}, u\})). \tag{6}$$

where u is the candidate unlabelled instance, $\mathcal{U} \setminus u$ is the unlabelled set with the candidate instance removed, l_u is the label of the candidate instance, $\mathcal{L}$ is the set of all possible class labels and $\text{Ent}(X)$ is the entropy of the categorical distribution X.

3.3 Delta in Prediction

It would be ideal if we could identify which of the unlabelled instances would be suitable for labelling and inclusion in the training set. The ideal training example is the one that offers the highest accuracy. Due to the lack of labelling, we cannot examine the increase in accuracy for each candidate instance.

Instead of measuring the increase in accuracy, we can instead measure the potential increase in accuracy. The classification accuracy for kNN using training set $\mathcal{X}$ and testing set $\mathcal{U}$ is

$$\frac{1}{|\mathcal{U}|} \sum_{u \in \mathcal{U}} \mathbf{1}\left(\text{mode}\left(L_k(u; \mathcal{X})\right) = l_u\right). \tag{7}$$

where $\mathbf{1}(A)$ is an indicator function (providing 1 if A is true and 0 is A is false) and l_u is the class label of instance u.

We define the *Delta in Prediction* of labelled instance x as

$$\Delta(u) = \max_{l_u \in \mathcal{L}} \frac{1}{|\mathcal{U} \setminus u|} \sum_{v \in \mathcal{U} \setminus u} \mathbf{1}\left(\text{mode}\left(L_k(v; \{\mathcal{X}, u\})\right) = l_v\right). \tag{8}$$

where u is the candidate unlabelled instance, $\mathcal{U} \setminus u$ is the unlabelled set with the candidate instance removed, l_u is the label of the candidate instance, and $\mathcal{L}$ is the set of all possible class labels. Thus, the $\Delta(u)$ represents the maximum classification accuracy that may be obtained by including u in the training set, concerning all class labels.

In this paper, we examine the relationship between each of these functions' scores and the accuracy of classification when choosing the associated instance.

4   Experimental Setup

We have only a small training set and wish to add to it. But manual labelling is challenging, so we should choose carefully when selecting which unlabeled instances are to be labelled. This study aims to answer the question: "Does the use of instance selection functions to refine sample selection result in better accuracy than random and uncertainty based selection?". We empirically investigate this question using the data from the UCI repository.

In each run of the experiment, we follow the steps below. A random sample of instances from a given data set is chosen as the training set containing manually assigned labels. The remainder of the instances are left unlabeled. Every unlabelled observation is assigned a selection score, and the sample with the highest

Table 2. Data used for evaluating instance selection functions.

Dataset	No of classes	Characteristics	Instances	Attributes	Features
Banknotes	2	Multivariate	13,72	Real	5
Satlog	6	Multivariate	6,435	Integer	36
Segmentation	7	Multivariate	2,310	Real	19
Heart disease	5	Multivariate	303	Real	14
Diabetes	2	Multivariate, Time-series	768	Real	9
Pendigits	10	Multivariate	10,992	Integer	16

Table 3. Comparison of average classification accuracy of random and uncertainty sampling with all our methods - average of 100 iterations

Dataset	Rand samp	Uncert samp	I	Δ	H	$I\Delta$	IH	ΔH	$I\Delta H$
Banknotes	0.59	0.60	0.61‡	0.56	0.58	0.56	0.61‡	0.60†	0.61‡
Satlog	0.45	0.45	0.46‡	0.46‡	0.46†	0.46†	0.46‡	0.46‡	0.46‡
Segmentation	0.35	0.36	0.36†	0.35*	0.36†	0.35	0.36‡	0.35*	0.36‡
Heart disease	0.46	0.45	0.44	0.48‡	0.45	0.48‡	0.44	0.45	0.48‡
Diabetes	0.60	0.61	0.63‡	0.63‡	0.64‡	0.63‡	0.63‡	0.63‡	0.63‡
Pendigits	0.38	0.37	0.39†	0.38	0.38	0.37	0.39†	0.38	0.39†

Signif. codes: $\star$: $p < 0.05$, $\dagger$: $p < 0.01$, $\ddagger$: $p < 0.001$.

score is added to the labelled training set. kNN accuracy is determined before and after the new point has been added to the training set.

The experiment variables are: the candidate instance selection functions {Random selection, Neighbourhood Impact, Maximum Entropy, Delta in Prediction}, the data (shown in Table 2), the initial training set size {4, 8, 16, 32, 64, 128}, and the number of instances chosen. Initial analysis shown in Fig. 1 showed that high accuracy instances are those that provide more central scores, so we selected the instance that provided the score closest to the mean score from all observations, to include in our training set. We also expanded the candidate instance selection function set to include the sum of each combination of the three candidate function scores. The selection methods are shown in the results as: Random sampling(adding a randomly chosen instance), uncertainty sampling, Δ (Delta in prediction), H (Maximum entropy), and I (Neighbourhood impact). Whenever two or more methods have been combined, the scores for the respective methods have been added.

4.1   Choosing One Instance

In the first experiment, we examine the results of selecting one instance from the unlabeled set to be manually labelled using a randomly chosen training set size of four. The experiment is paired, i.e. each method employs the same random

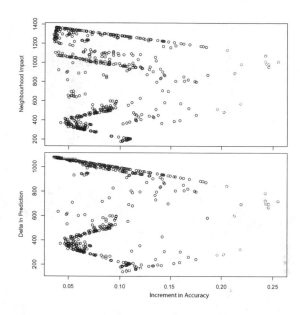

Fig. 1. Relationship of higher accuracy, neighbourhood impact and delta in prediction - Banknotes dataset. The figure shows that high accuracy is related to mean neighbourhood impact and delta in prediction

training sets. Figure 2 provides the accuracy of each method based on 100 runs, where the results are sorted by uncertainty sampling. Each point on the figure represents the average prediction accuracy after one hundred iterations. The graph has exactly 100 points, so each line represents ten thousand executions. The proposed selection technique performs much better than the benchmarks, namely uncertainty sampling and random selection. In light of our experimental findings, and the above demonstrations, we find that our proposed techniques perform very well in a few-shot learning environment. The p values of all techniques are compared in Table 3. This table also presents the average accuracy of each technique for different data sets in comparison with random sampling and uncertainty sampling.

4.2   Choosing n Instances

As the number of labelled samples increases, i.e. as we move from a few shot learning scenario to a semi-supervised learning scenario, Fig. 3 illustrates the average accuracy for Random, Oracle, uncertainty sampling and our methods. For all methods, accuracy is based on sample sizes of $4, 8, 16, 32, 64, 128$. The figure illustrates that our approaches are more accurate than random selection when the labelled sample size is small while at par with random selection as the labelled sample size increases. Our study found that as the number of labelled samples per class exceeds 32, the accuracy of selecting new unlabelled samples remains the same for all methods, including random sampling. Depending on the

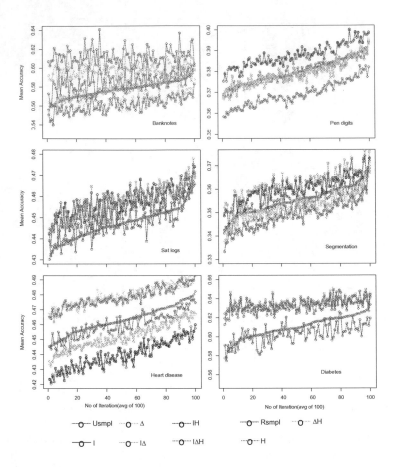

Fig. 2. Mean accuracy of classification – next best 1 sample selection – each point represent 100 executions

data set, this saturation point may vary, but it is typically close to 30 samples per class.

These techniques have been tested in a number of situations, including the next-best 1, n unlabelled samples and n labelled samples. Our results demonstrate that these techniques are on par with random selection in the next-best 1 and n unlabelled sample selection setting. Please refer to Fig. 4 that shows the results of a banknotes data set using the next-best 3, 5, 7, and 9 unlabelled samples. Next-best 3 unlabelled sample setting is one where three 3 unlabelled samples are selected to compare their accuracy.

4.3   Semi-supervised Learning Scenario

In a setting with many labelled samples, we observed similar results. This test aimed to assess performance in a semi-supervised setting. In Fig. 3, we compare

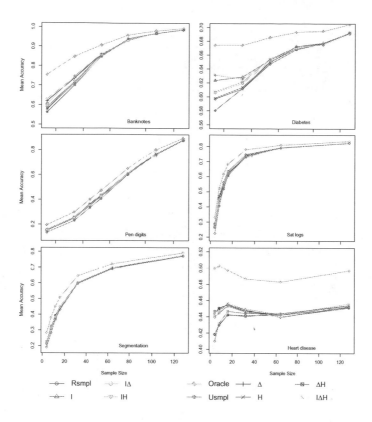

Fig. 3. Comparing the average accuracy (average of 100) for Oracle, Random, and our methods as the number of labelled samples increases

uncertainty sampling, random selection, and all of the techniques we propose using all of the data sets. A comparison is made between the average accuracy as we move from a small labelled data set (few-shot learning) to a bigger labelled data set (semi-supervised learning). The number of available samples doubles with each stage. This analysis shows that our approach performs better with 4,8,16, and 32 observations. Although our method outperforms when only a few labelled samples are available, it is still competitive when many labelled samples are available. Table 3, we present the results of ten thousand computations and compare uncertainty sampling and random with the proposed methods.

Based on our experiments conducted in few shot and semi-supervised settings, the following results were observed.

1. A systematic selection of an unlabeled sample is preferable when small labeled samples are available for the training of the classifier.
2. The methodologies we propose for systematic selection can also be applied in a semi-supervised setting where a large number of labels are available.

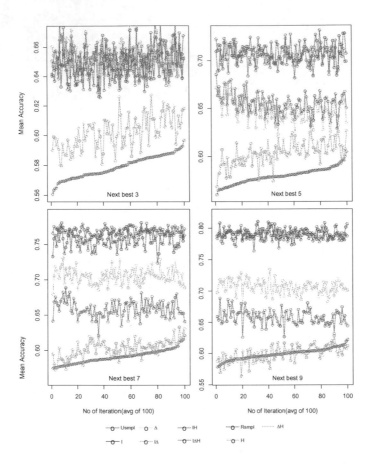

Fig. 4. Comparison of various methods for next best 3, 5, 7 and 9 sample selection –
Banknotes data set

The cost of systematic selection in this case is higher than that of random
selection.
3. There is a saturation point in terms of the number of samples that have
   been labeled. There are no differences between uncertainty sampling, random
   sampling and systematic selection beyond this point.

5   Conclusions

We present three novel approaches to selecting a good next sample in few-
shot and semi-supervised learning situations. We evaluate our proposed methods
using random sampling and uncertainty sampling as benchmarks. Performance
is evaluated by comparing the accuracy of classification before and after including the selected samples in the training set. Our evaluation of real-life, publicly

available data sets shows that our proposed sampling methods are preferable to uncertain sampling and random sampling when there are only a few labelled samples available. Furthermore, our method performs as well as the benchmarks when there are a lot of labelled samples.

References

1. Altmann, J.: Observational study of behavior: sampling methods. Behaviour **49**(3–4), 227–266 (1974)
2. Bestmann, S., et al.: Influence of uncertainty and surprise on human corticospinal excitability during preparation for action. Curr. Biol. **18**(10), 775–780 (2008)
3. Cao, L., Zhu, C.: Personalized next-best action recommendation with multi-party interaction learning for automated decision-making. arXiv preprint arXiv:2108.08846 (2021)
4. Dunn, E., Frahm, J.M.: Next best view planning for active model improvement. In: BMVC, pp. 1–11 (2009)
5. Etikan, I., Bala, K.: Sampling and sampling methods. Biomet. Biostatist. Int. J. **5**(6), 00149 (2017)
6. Fraboni, Y., Vidal, R., Kameni, L., Lorenzi, M.: Clustered sampling: low-variance and improved representativity for clients selection in federated learning. arXiv preprint arXiv:2105.05883 (2021)
7. Giraud-Carrier, C.: A note on the utility of incremental learning. AI Commun. **13**(4), 215–223 (2000)
8. Goodman, L.A.: Snowball sampling. Ann. Math. Statist. **32**, 148–170 (1961)
9. Jenkinson, A.: What happened to strategic segmentation? J. Direct Data Digit. Mark. Pract. **11**(2), 124–139 (2009)
10. Kramer-Schadt, S., et al.: The importance of correcting for sampling bias in maxent species distribution models. Divers. Distrib. **19**(11), 1366–1379 (2013)
11. Lughofer, E.: Hybrid active learning for reducing the annotation effort of operators in classification systems. Pattern Recogn. **45**(2), 884–896 (2012)
12. Madow, W.G., Madow, L.H.: On the theory of systematic sampling, I. Ann. Math. Stat. **15**(1), 1–24 (1944)
13. Moser, C.A.: Quota sampling. J. R. Statist. Soc. Ser. A (General) **115**(3), 411–423 (1952)
14. Neyman, J.: On the two different aspects of the representative method: the method of stratified sampling and the method of purposive selection. In: Kotz, S., Johnson, N.L. (eds.) Breakthroughs in Statistics, pp. 123–150. Springer Series in Statistics. Springer, New York, NY (1992). https://doi.org/10.1007/978-1-4612-4380-9_12
15. Olken, F.: Random sampling from databases. Ph.D. thesis, University of California, Berkeley (1993)
16. Rubens, N., Kaplan, D., Sugiyama, M.: Active learning in recommender systems. In: Ricci, F., Rokach, L., Shapira, B., Kantor, P.B. (eds.) Recommender Systems Handbook, pp. 735–767. Springer, Boston (2011). https://doi.org/10.1007/978-0-387-85820-3_23
17. Sedgwick, P.: Convenience sampling. BMJ. **347**, 1–2 (2013)
18. Settles, B.: Active learning literature survey (2009)
19. Shi, W., Gong, Y., Ding, C., Ma, Z., Tao, X., Zheng, N.: Transductive semi-supervised deep learning using min-max features. In: Ferrari, V., Hebert, M., Sminchisescu, C., Weiss, Y. (eds.) ECCV 2018. LNCS, vol. 11209, pp. 311–327. Springer, Cham (2018). https://doi.org/10.1007/978-3-030-01228-1_19

20. Syed, N.A., Liu, H., Sung, K.K.: Incremental learning with support vector machines (1999)
21. Syfert, M.M., Smith, M.J., Coomes, D.A.: The effects of sampling bias and model complexity on the predictive performance of maxent species distribution models. PLoS ONE 8(2), e55158 (2013)
22. Tokdar, S.T., Kass, R.E.: Importance sampling: a review. Wiley Interdiscipl. Rev. Comput. Statist. 2(1), 54–60 (2010)
23. Van Amersfoort, J., Smith, L., Teh, Y.W., Gal, Y.: Uncertainty estimation using a single deep deterministic neural network. In: International Conference on Machine Learning, pp. 9690–9700. PMLR (2020)
24. Yang, B., Sun, J.T., Wang, T., Chen, Z.: Effective multi-label active learning for text classification. In: Proceedings of the 15th ACM SIGKDD International Conference on Knowledge Discovery and Data Mining, pp. 917–926 (2009)

Multiclass Malware Classification Using Either Static Opcodes or Dynamic API Calls

Rajchada Chanajitt[1]([✉]), Bernhard Pfahringer[1]👁, Heitor Murilo Gomes[2]👁, and Vithya Yogarajan[3]👁

[1] Department of Computer Science, University of Waikato, Hamilton, New Zealand
`rajchada.ch@gmail.com`
[2] School of Engineering and Computer Science, Victoria University of Wellington, Wellington, New Zealand
[3] School of Computer Science, University of Auckland, Auckland, New Zealand

Abstract. Today's malware variants are growing at an unprecedented rate. To avoid detection by existing antivirus engines, attackers have been increasing the complexity of packers, layers of obfuscation, and encryption to obstruct the process of reverse engineering. This paper presents an automated method using static analysis for extracting opcode sequences of a length of up to 5000 and employing these sequences for classifying potential malware into eight classes, namely ransomware, trojan, backdoor, rootkit, virus, miner, benign, and other. Our empirical analysis compares four different classifiers: MLP, LSTM, GRU, and Transformer. The experimental results demonstrate that the GRU approach achieves the highest F1-score of up to 87%. In addition, we analyze dynamic API call sequences. We use a public malware dataset that comprises more than 7000 sample sequences of 342 API calls each for apps from eight different malware families. A GRU network achieves the best result for this dataset, producing an F1-score of 78%.

Keywords: Opcode · API calls · MLP · GRU · Transformer

1 Introduction

In malware detection, static analysis is usually the first choice selected by researchers due to being able to examine an executable without the need of actually executing it in an isolated virtual environment. Malware authors can develop applications to avoid detection engines by applying obfuscation techniques to safeguard the malware code and its data structures from being dissected. Acquiring static data from mnemonic instructions (opcodes) has become prevalent to prevent damage from execution. It can provide a holistic view of the application statically on what operation to perform even though manipulation of the address parameters and changes in the execution flow can be obstacles. An opcode is a part of a machine instruction that determines the function to be executed by a machine. Each instruction operates on operands that can be stored in registers, or memory, or constants stored in the instruction itself. Obtaining opcode sequences can be conducted by both static and dynamic analysis. Nevertheless, it takes less time to extract them via static analysis. For dynamic analysis, the extraction of API calls is the most popular approach for observing runtime behaviour of malware. Recent research [1] shows that machine learning approaches for sequence classification, such as multi-layer perceptrons (MLPs) and recurrent neural networks, provide satisfactory results.

This paper focuses on "`BaseOfCode`", a relative offset of code in code sections (.text) loaded into the memory. We disassemble the application, carve out the opcodes from the address of the .text section until the end of the file to statically analyze the behavior. We also present an approach to automate and extract opcodes from the binary contents. The applications used in this experiment are implemented by: UPX (Ultimate Packer for eXececutables)[1], .NET assemblies by Microsoft .Net CLI and Mono, and Zlib[2] compression. Given that all data are consecutive sequences of a length of at most 5000, the four different types of neural networks investigated here are an appropriate choice.

The main contributions of the paper are as follows:

(i) We show that using the open-source tools listed above to extract consecutive opcode sequences from binaries can provide valid empirical data.

(ii) We calculate the frequency of opcodes for each malware category and implement SHAP feature selection to obtain a good representation.

(iii) We benchmark four different neural network-based approaches to static opcodes and dynamic API calls. The experiments highlight that a GRU yields a highly accurate classifier for static and dynamic sequences.

2   Related Work

2.1   Opcodes

Azadech [7] proposed a static signature-based malware detection method based on N-gram opcode with different degrees and file signatures by using VXheaven 203 malware binaries and 216 Windows system files as benign binaries. There were three phases: extracting opcode and binary sequences from benign and malicious files, generating N-grams, and classifying files into benign and malicious groups. The results demonstrated that combining 1, 2, and 3-grams represents a feature set with an accuracy of 78%. With the proposed Top-K approach to select the topmost similar k files, the highest accuracy belongs to Top-10, at 86.63%. In the combination of opcodes and binary sequences, the K is chosen to be 3, resulting in an accuracy of 86.39%.

Regarding the static analysis, ransomware families [18] fingerprint the environment to evade the dynamic analysis explored. They collected 1787 ransomware samples from eight families: cryptolocker, cryptowall, cryrar, locky,

---

[1] https://upx.github.io/.

[2] https://docs.python.org/3/library/zlib.html.

petya, reveton, teslacrypt, and wannacry by VirusTotal, and 100 trusted software samples and obtained opcodes by using the IDA Pro disassembler. The opcode sequences were transformed to N-gram sequences and calculated by TF-IDF in descending order to select feature N-grams. Then, TF values of the feature N-grams were fed into five machine learning methods: DecisionTree, Random Forest, K-Nearest Neighbor, Naive Bayes, and Gradient Boosting Decision Tree. For each family, they performed extensive experiments with N-grams of lengths 2, 3, and 4, and with different feature dimensions ranging from 29 to 228. Overall, the Random Forest in 3-gram outperformed the other algorithms. For multi-class, the best accuracy was 91.43% when using 123 features, and the highest F1 measure of nearly 99% was achieved on wannacry. Similar to binary classification, accuracy was up to 99.3% using 180 features.

As was aforementioned above, we use open-source tools according to different compression methods to extract long sequences, compared to commercial tools in [18]. Furthermore, we use other learning models (e.g., MLP, and GRU) due to the appropriateness of input data for text classification based on consecutive long sequences to provide an alternative static input data for malware classification.

2.2 API Calls

In the work of [5], 2600 samples were run in a virtual machine on Windows XP, and API call sequences were extracted. A whole set of 534 APIs has been hooked and mapped to 26 categories. Every sequence of Windows API calls was mapped to a categorized sequence of A to Z letters. They have developed a repository of 2000 fuzzy hash signatures, 400 for each category. For each class of malware (Worm, Backdoor, Trojan-Downloader, Trojan-Dropper, and Trojan-Spy), 520 samples were selected. With n-gram analysis of the categorized sequences, class-specific patterns for all five classes of malware were retrieved. The ssdeep algorithm calculated the fuzzy hash matching score between different categorized sequences to generate a fuzzy hash-based signature from 0 to 100. A high fuzzy hash value indicates that the two binaries belong to the same class. This approach achieved an accuracy of approximately 96%.

Additionally, n-grams databases of API sequences were created using similarity score methods to discover similar characteristics among malware families. A Control Flow Graph (CFG) [13] of 15000 malware binaries for five different classes and 4000 benign PE files were generated after disassembly done by BeaEngine. This tool represents an execution flow in the form of basic blocks (nodes) and edges. A basic block is a set of instructions without a single branch or control transfer instruction. Dice coefficient, Cosine coefficients, and Tversky Index were used to calculate the similarity between the 2, 3, and 4 g databases generated from that analysis. A detection rate of 94.78% and a false positive rate of 33.51% were achieved using 3-grams and the dice coefficient.

Moreover, Yu Wang [17] introduced binary malware classification based on deep recurrent reinforcement learning (DRL) by emulating the generation of a sequence of API calls to choose one action out of two: Continue or Halt. Microsoft's production antimalware engine evaluated 75000 files to generate behavioral events after discarding files whose event sequences were shared between two classes or contained less than 50 events. This antimalware engine maps multiple low-level API calls into a single high-level event to deal with polymorphism. The engine records 114 different event types, including file IO, registry APIs, networking APIs, thread or process creation and control, inter-process communication, timing, and debugging APIs. The model's state contains three parts: the position (i.e., index) of the current event in the file, the current event ID, and the histogram of all the previous events. There are two criteria for designing a reward for each state: (1) shorter emulation sequences are assigned a higher reward, and longer sequences are given a smaller reward. (2) The closer an event prediction is to the true file label, the larger the reward will be given at that state. The model halts emulation of an unknown file and improves malware classification at 91.3% by the number of consecutive actions where the action is Halt before the DRL model stops the file's execution. Furthermore, this new model improves the true positive rate by 61.5%, at a false positive rate of 1%, compared to the best baseline classifier.

To handle adversarial learning-based attacks, Yu Wang proposed a new Actor-Critic (AC) deep reinforcement learning method [16] instead of the older DQN model [17]. Compared to the DQN approach, the new model performs better for all K and N values. For halting execution, the new model halts the execution of unknown files by up to 2.5% earlier than the DQN model and 93.6% earlier than the heuristics. For the classification task, the proposed AC model increases the true positive rate by 9.9% from 69.5% to 76.4%, at a false positive rate of 1% compared to the DQN model.

Our research makes use of sequential input data and performs multi-class malware classification. Machine learning approaches MLP, LSTM, GRU, and Transformer are used on the static opcodes to produce predictive results. The same approaches are also applied to the benchmark dynamic API call dataset [3].

3   Dataset and Feature Pre-processing

3.1   Opcodes

Samples for malware and benign applications were collected from publicly available sources. The dataset contains 3256 malicious files made up by VirusShare[3] and URLhaus[4], and 744 FileHorse[5] benign samples. The labels were obtained from submitting the files to VirusTotal[6] and using at least five or more anti-virus engines to assign a class label. The application will be assigned to the "other" class if it is not in the seven categories.

We use TextVectorization from Tensorflow for preprocessing the opcode sequences, where the input sequence length is set to a maximum of 5000 and the

---

[3] https://virusshare.com.

[4] https://urlhaus.abuse.ch.

[5] https://fileHorse.com.

[6] http://www.virustotal.com.

output sequence length to 500. We also run the experiment by setting up higher lengths and total maximum tokens; however, a maximum output sequence length of 500 can already provide good results. The 'adapt' method is also used on the initial dataset to create an index of the resulting vocabulary.

The following steps are also automated:

– Check the application implemented: UPX, Mono/.Net, Zlib.
– Disassemble the executables to obtain disassembled files.
– Search and remove flag/segment registers, interrupt instruction, illegal/(bad), and dead code.
– Extract the first 5000 opcodes.

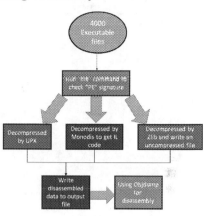

All samples are inspected to verify whether it has a "PE" signature and what type of compression algorithms is implemented. Three packer types are investigated: UPX, Zlip, and Mono/.Net. For UPX and Zlib, they are parsed to UPX unpacker and Zlib, respectively. For the

Fig. 1. Our Opcodes Extraction.

.NET Common Intermediate Language (CIL) code and metadata, "`monodis`"[7] is used to disassemble. If it successfully decompresses a file, it will output a respective logfile. Otherwise, it will use the Capstone Python library [2][8] and then "`objdump`" [8][9] with -d option to disassemble only sections expected to contain code.

Figure 1 shows the system architecture of such an automated process. The first ten opcodes of a single example binary of each category are listed here:

- **Backdoor:** [pop, mov, sub, rcr, sub, stosb, pop, mov, sub, test]
- **Miner:** [lea, lea, sub, xor, cmp, mov, mov, mov, mov, je]
- **Ransom:** [movsx, xor, dec, je, dec, jne, mov, mov, jmp, mov]
- **Rootkit:** [loopne, add, add, add, dec, add, add, add, add, push]
- **Trojan:** [mov, add, add, add, dec, add, add, add, add, push]
- **Virus:** [nop, dec, add, add, add, dec, add, add, add, add]
- **Benign:** [push, mov, sub, cmp, je, cmp, mov, jne, or, mov]
- **Other:** [nop, push, add, add, add, dec, add, add, add, add]

The statistics are presented in terms of the minimum and the maximum number of total samples, complete in 5000 lengths grouped by each type. To select the top 20 MLP significant features, SHapley Additive exPlanations (SHAP) [9] was used. The DeepSHAP function from the open-source python package was applied. To mitigate bias in a model, SHAP is one of model interpretability

---

[7] https://www.mono-project.com/docs/tools+libraries/tools/monodis/.
[8] https://www.capstone-engine.org/lang_python.html.
[9] https://man7.org/linux/man-pages/man1/objdump.1.html.

Table 1. Number of opcodes and extracted features for each malware category.

Type	Minimum/# of samples with 5,000 tokens	Total	SHAP Selected Features
Ransom-ware	1/90	194	[**n1**, n95, n3, n44, n115, n134, n50, n23, n5, n120, n10, n4, n152, n105, n40, **n55**, n2, n12, n45, n14]
Trojan	1/426	2570	[n68, n50, n5, n23, n120, **n1**, n152, **n55**, n44, n399, n115, n31, n25, n78, n40, n59, n45, n14, n66, n4]
Miner	4/63	135	[n50, n12, n3, n44, n45, **n1**, **n55**, n4, n2, n56, n115, n14, n5, n309, n8, n13, n11, n17, n33, n15]
Virus	1/26	126	[n291, n230, n170, n131, n3, n25, n2, n4, n408, n141, n209, n29, n115, n32, n411, n93, n30, **n55**, **n1**, n181]
Rootkit	31/7	16	[n44, n45, n50, n58, **n55**, n68, **n1**, n4, n23, n53, n3, n2, n18, n8, n15, n115, n63, n10, n12, n56]
Backdoor	10/64	166	[**n1**, **n55**, n50, n5, n14, n23, n120, n33, n27, n7, n152, n18, n49, n53, n40, n36, n56, n25, n8, n6]
Benign	1/374	741	[n68, n50, n393, **n55**, n33, n44, n40, n14, **n1**, n70, n224, n2, n6, n52, n90, n27, n49, n56, n17, n86]
Other	1/17	52	[**n55**, n50, n393, n33, **n1**, n68, n40, n14, n23, n2, n152, n52, n49, n6, n3, n18, n56, n4, n27, n120]

that checks all the possible combinations of features for a prediction to calculate the SHAP values. First, the learning model was trained on the full set of features, and the importance of each feature was obtained by comparing model predictions with and without the feature, which can be used to select significant features. Each feature's SHAP importance was computed individually by taking the average Shapley values and then sorted in descending order according to their importance before 20 attributes that have the most significant impact were chosen. The details are depicted in Table 1.

The features are represented as the column sequences $n1, n2, ..., n5000$ for all samples where each column contains one opcode. For example, the first selected feature for ransomware is "n1" the opcodes can be *mov, push, sub, movsx, jmp, std, daa, dec*, etc. Compared to trojan, "n68" is selected for the first feature, the opcodes can be *add, or, sub, adc, cmp, insb, xor, and*, etc.

Figure 2 presents a summary plot of the top Shapley values. The features are ordered according to their importance, where "n1" is the feature that has the most impact on the model output, and "n55" is a shared feature that appeared across all categories. Conversely, the features "n7", "n13", "n31", "n36", "n58", "n59", "n63", "n70", "n66", "n78", "n86", "n90", "n95", "n105", "n309", "n399" are used seldom, especially the most number of rarely feature use being in virus class. The top 3 opcodes for "n1" and "n55" features are listed with the total number of occurrences for each class in Fig. 4. Overall, "push" is evident in all four families and "add" is the second most common instruction.

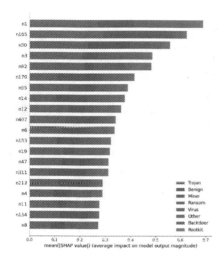

Fig. 2. SHAP summary plot of the top 20 features of opcodes for MLP.

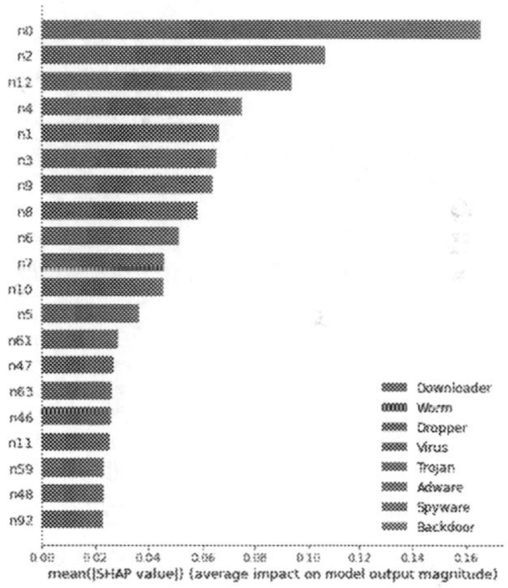

Fig. 3. SHAP summary plot of the top 20 features of APIs for MLP.

Fig. 4. Top 3 opcodes for ransomware, trojan, miner, and benign class for selected n1 and n55 features.

3.2 API Calls

Besides our opcodes, the most crucial part of the behavioral records are API call sequences. The 7107 malicious samples of [3], recording the Windows 7

API calls from application execution in an isolated environment (the Cuckoo sandbox) is used here. That paper conducted behavior-based malware analysis by collecting data from various GitHub pages with the git command-line utility. `The data belongs to multiple classes, which were obtained from the VirusTotal by searching the MD5 signatures of each malware. In particular, eight classes are considered. Five of them consist of 1001 samples, including Worm, Virus, Trojan, Backdoor, and Downloader, while the other three classes are Spyware (832), Adware (379), and Dropper (891). The objective was to build a benchmark dataset for Windows API calls of metamorphic malware, which can change the code signatures and recognize the environment to store their harmful behavior by anti-analysis techniques in the environments implemented for malware analysis [10]. Although this type of malware is hard to detect and classify with such capabilities, the meaningless opcodes with their own dissembler/assembler parts are added to observe the patterns of how metamorphic malware changes their behavior. For example, making unnecessary API calls during the behavior change can detect malware because the pattern is the same.

The Windows API is an interface for developing applications on the Windows operating system. A Windows application can make API calls to request operating system services. The data is represented with a 3-channel image obtained from grayscale image conversion and then using CNN and LSTM as the classification algorithms to detect malware families. From a sequence of 1000 API calls, the unique 342 API calls can be extracted by providing an index from 0 to 341. We also perform the SHAP selection method on these API calls, and the top 20 features are represented in Fig. 3.

4   Model

This section introduces the four machine learning algorithms employed in this work for malware detection. The methodology adopted for automating the opcode extraction and malicious behavior identification process is also described.

All models are implemented using TensorFlow, where hyper-parameters were tuned by utilizing Keras Tuner [11]. The trained classifier is tested using 10% of the data in the dataset. A 10-fold cross-validation approach is used to prevent over-fitting. Model parameters commonly affecting the predictive result are represented in terms of a list in Tables 2, 3 and 4. Their respective details are presented in the following subsections.

4.1   Multi-layer Perceptron (MLP)

The MLP [12] network follows a standard setup using nonlinear activations. It employs BatchNormalization and regularizers to avoid overfitting. An input layer sepcifying the shape of the sequences is fed into two dense layers with dropout [14] and a batchnorm layer in between consecutive dense layers. The output layer consists of eight neurons densely connected to the last batch normalization layer. All of the layers in the network use swish as their activation function. The model calculates loss based on the categorical cross-entropy and computes accuracy as the evaluation. The model is further fine-tuned by adjusting the hyperparameters to achieve the optimum results as provided in Table 2.

4.2   Recurrent Neural Networks

Recurrent Neural Networks (RNNs) are a group of neural networks designed to handle sequential data, such as text, where the data contains complex temporal dependencies and hidden information. Gated Recurrent Units (GRU) [4] and Long Short Term Memory networks (LSTM) [6] are modified versions of RNNs. For this research, we use bidirectional GRU and bidirectional LSTM, where the network considers sequences from right to left and the reverse order. We use dropout to prevent overfitting, and a softmax layer is used to output the malware family label.

Table 2. Hyperparameters tuning for MLP for opcodes and API calls

Name	Initial value	Tuning ranges	Opcode best value	APIs best value
Batch_size	15	ranges(10,50,5)	50	20
Epochs	200	ranges(100,600,50)	150	550
Optimizer	RMSprop	['Adam','RMSprop', 'Adadelta','Adagrad']	Adagrad	Adam
Learning_ rate	0.0002	[0.0001,0.001,0.01, 0.0002,0.002,0.02]	0.002	0.001
Neurons	[512,256]	range(256,512,32)	[928,896]	[480,512]
Hidden Layers	5	range(2,10)	9	3
Hidden Neurons	[256,256,256, 256,256]	range(128,512,32)	[384,352,128,416, 352,320,256,384, 224]	[160,320,480]
Bias_ regularizer	[0.002, 0.001, 0.02]	[0.0001, 0.001, 0.01, 0.0002, 0.002, 0.02]	[0.002,0.02, 0.0002]	[0.02,0.01, 0.002]
Kernel_ regularizer	[0.01,0.01, 0.02]	[0.0001, 0.001, 0.01, 0.0002, 0.002, 0.02]	[0.001,0.001, 0.001]	[0.01,0.001, 0.0001]
Kernel_ constraint	[2.0,1.5,1.5]	range(1,3,0.5)	[2.0,2.5,2.0]	[1.5,2.5,1.0]
Activity_ regularizer	[0.0001, 0.001, 0.001]	[0.0001, 0.001, 0.01, 0.0002, 0.002, 0.02]	[0.0001, 0.002, 0.0001]	[0.0002,0.01, 0.001]

4.3   Transformer

Transformers [15] are based on an attention-based encoder-decoder architecture focusing on different tokens while generating words to model opcode sequences.

The first embedding layer converted words into vectors, followed by the positional encoding layer to add the position information for each word. Then add those vectors to their corresponding input embeddings. Next, they are encoded to attention representations in the encoder layers, consisting of two sub-layers: multi-headed attention and two fully-connected layers with a ReLU activation in between. Multi-headed attention computes the attention weights for the input simultaneously. Later, a hidden state is passed to the decoding stage with an attention layer operating differently from the encoder to prevent seeing future tokens. Lastly, GlobalAveragePooling1D is added, which averages over sequence dimension and returns a fixed-length output vector before feeding into the last softmax layer to get the word probabilities. The details about optimization from the RandomSearch is provided in Table 4.

Table 3. GRU and LSTM hyperparameters tuning for API calls and opcodes.

Name	Initial value	Tuning ranges	API calls (Best)		Opcodes (Best)	
			GRU	LSTM	GRU	LSTM
batch size	15	range(10,50,5)	25	20	30	50
epochs	200	range(100,650,50)	350	500	550	200
optimizer	RMSprop	['Adam','RMSprop', 'Adadelta', 'Adagrad']	Adam	Adadelta	Adam	RMSprop
learning rate	0.0001	[0.0001, 0.001, 0.01, 0.0002, 0.002, 0.02]	0.0002	0.02	0.0001	0.0001
embedding	128	ranges(256,1024,64)	448	640	768	704
embedding regularizer	0.0001	[0.0001, 0.001, 0.01, 0.0002, 0.002, 0.02]	0.0001	0.0001	0.0002	0.0001
gru/lstm units	[512, 256]	ranges(128,512,32)	[416, 288]	[288, 384]	[224, 512]	[288, 448]
recurrent regularizer	[0.0001, 0.001]	[0.0001, 0.001, 0.01, 0.0002, 0.002, 0.02]	[0.002, 0.01]	[0.0001, 0.002]	[0.02, 0.0001]	[0.002, 0.0002]
kernel regularizer	[0.001, 0.001]	[0.0001, 0.001, 0.01, 0.0002, 0.002, 0.02]	[0.0002, 0.002]	[0.0001, 0.02]	[0.001, 0.0002]	[0.01, 0.02]
bias regularizer	[0.01, 0.02]	[0.0001, 0.001, 0.01, 0.0002, 0.002, 0.02]	[0.02, 0.01]	[0.002, 0.001]	[0.01, 0.01]	[0.0002, 0.02]
kernel constraint	[2.0, 1.5]	range(1,3,0.5)	[1.5, 2.0]	[2.0, 3.0]	[1.5, 1.5]	[1.5, 1.0]
Activity regularizer	[0.0001, 0.0002]	[0.0001, 0.001, 0.01, 0.0002, 0.002, 0.02]	[0.002, 0.02]	[0.001, 0.0001]	[0.01, 0.0002]	[0.01, 0.01]

5   Experimental Results

This section will illustrate the results of our experiment on static opcode instructions and dynamic APIs for malware detection using four learning approaches in terms of classification metrics. We repeat 30 times the 10-fold cross-validation using different seed values each time for tuning and average them to report the classification result on opcode sequences. Table 5 lists results for running all algorithms with their default parameter settings on static opcodes. Overall, a Transformer performs best among all four classifiers by achieving an accuracy of 63% for opcodes. The Transformer also provides the best predictive result for dynamic API calls [3] with default model parameters as shown in Table 6.

Table 7 shows the improvements achieved by adequately tuning the essential hyperparameters. The overall performance increases for all classification models compared to the default performances from Table 5. With the same learning approach as the default parameters, the GRU exhibits the best results by reaching the maximum accuracy at 87%. In Table 8, hyperparameter tuning is conducted on dynamic dataset [3] on the same architecture as the static opcodes. In general, the performance of all model approaches is obviously increased compared to Table 6. The GRU can provide the best predictive result among other classifiers by achieving the F1-score at 78%. In summary, the Transformer seems to be a good default architecture, but alternative architectures can outperform it with proper (and expensive) hyperparameter tuning. In our case, this is an MLP for static and a GRU network for dynamic sequences.,

Table 4. Hyperparameters tuning for transformer for opcodes and API calls

Name	Initial value	Tuning ranges	Opcode best value	APIs best value
Batch_size	10	range(10,30,5)	15	15
Epochs	100	range(100,600,50)	250	250
Optimizer	Adam	['Adam','RMSprop', 'Adadelta','Adagrad']	Adam	Adam
Learning_rate	0.0001	[0.0001,0.001,0.01, 0.0002,0.002,0.02]	0.0002	0.0002
Trans_dim	512	ranges(512,2048,32)	832	640
Num_heads	8	ranges(6,14,2)	12	12
Feed_forward_dim	512	ranges(1024,3172,32)	1536	1408
Enc_layer	5	ranges(2,10)	6	6
Dec_layer	5	ranges(1,10)	4	4
Embedding regularizer	[0.001, 0.001]	[0.0001, 0.001, 0.01, 0.0002, 0.002, 0.02]	[0.0002, 0.0002]	[0.0002, 0.0002]
Kernel regularizer	[0.001, 0.001, 0.001]	[0.0001, 0.001, 0.01, 0.0002, 0.002, 0.02]	[0.01, 0.002, 0.0002]	[0.01, 0.002, 0.0002]
Bias regularizer	[0.01, 0.02, 0.001]	[0.0001, 0.001, 0.01, 0.0002, 0.002, 0.02]	[0.001, 0.001, 0.0002]	[0.001, 0.001, 0.0002]
Kernel constraint	[2.0, 1.5, 1.0]	range(1,3,0.5)	[1.5, 2.5, 2.5]	[1.5, 2.5, 2.5]
Activity regularizer	[0.001, 0.001, 0.001]	[0.0001, 0.001, 0.01, 0.0002, 0.002, 0.02]	[0.0001, 0.01, 0.01]	[0.0001, 0.01, 0.01]

An alternative to classification accuracy is to use Precision, Recall, and F-measure metrics for imbalanced classification. Thus, the per-class performance for the respective best approach for both static opcodes and dynamic API calls is reported in Table 9. Remarkably, precision and recall on opcodes is more than 90% for trojan. Nevertheless, Rootkit has the most significant standard deviation compared to other categories. For API calls, Downloader has high precision and recall and a high standard deviation on Dropper and Spyware. In addition to the evaluation metrics, we also present the total elapsed time on the training default configuration and hyperparameter tuning with ten maximum trials for each estimator. We used stratified 10-Folds cross-validation using the GPU. Although RandomSearch considers not all possible combinations, Transformer takes the most time for optimization as shown in Table 10.

Finally, Fig. 6 and Fig. 5 depict the confusion matrix and ROC curves associated with the best model obtained from hyperparameter search space. For Fig. 6, a comparison between the predicted and expected values is calculated. The figures are less than 2% except for Trojan and Other class regarding the false positive rate. Likewise, ROC curves on opcodes as in Fig. 5 show that all but one category can yield AUC values of more than 90%.

Table 5. Model evaluation using default hyperparameter settings for opcodes

Algorithm	Avg. Test Acc	Avg. F1 Test	Avg. MAE	Avg.AUC
LSTM	0.605 ± 0.02	0.598 ± 0.05	0.111 ± 0.01	0.878 ± 0.02
GRU	0.597 ± 0.02	0.590 ± 0.03	0.112 ± 0.01	0.880 ± 0.01
Transformer	$\mathbf{0.727 \pm 0.02}$	$\mathbf{0.725 \pm 0.02}$	$\mathbf{0.083 \pm 0.01}$	$\mathbf{0.914 \pm 0.01}$
MLP	0.655 ± 0.02	0.654 ± 0.02	0.101 ± 0.01	0.859 ± 0.01

Table 6. Model evaluation using default hyperparameter settings for API calls

Algorithm	Avg. Test Acc	Avg. F1 Test	Avg. MAE	Avg.AUC
LSTM	0.449 ± 0.02	0.461 ± 0.02	0.146 ± 0.01	0.802 ± 0.01
GRU	0.426 ± 0.02	0.444 ± 0.03	0.148 ± 0.01	0.775 ± 0.01
Transformer	$\mathbf{0.461 \pm 0.02}$	$\mathbf{0.466 \pm 0.02}$	$\mathbf{0.143 \pm 0.01}$	$\mathbf{0.813 \pm 0.01}$
MLP	0.278 ± 0.02	0.227 ± 0.04	0.191 ± 0.01	0.661 ± 0.01

Table 7. Model evaluation with average 10 fold cross-validation on hyperparameter tuning for opcodes

Algorithm	Avg. Test Acc	Avg. F1 Test	Avg. MAE	Avg.AUC
LSTM	0.858 ± 0.04	0.856 ± 0.04	0.054 ± 0.01	0.977 ± 0.02
GRU	$\mathbf{0.891 \pm 0.05}$	$\mathbf{0.890 \pm 0.05}$	$\mathbf{0.036 \pm 0.01}$	$\mathbf{0.984 \pm 0.02}$
Transformer	0.818 ± 0.03	0.807 ± 0.03	0.077 ± 0.01	0.971 ± 0.02
MLP	0.826 ± 0.03	0.824 ± 0.03	0.072 ± 0.01	0.967 ± 0.02

Table 8. Model evaluation with average 10 fold cross-validation on hyperparameter tuning on API calls

Algorithm	Avg. Test Acc	Avg. F1 Test	Avg. MAE	Avg.AUC
LSTM	0.635 ± 0.06	0.642 ± 0.07	0.117 ± 0.01	0.915 ± 0.04
GRU	$\mathbf{0.736 \pm 0.09}$	$\mathbf{0.783 \pm 0.11}$	$\mathbf{0.075 \pm 0.02}$	$\mathbf{0.945 \pm 0.05}$
Transformer	0.626 ± 0.09	0.691 ± 0.10	0.102 ± 0.02	0.905 ± 0.05
MLP	0.350 ± 0.03	0.248 ± 0.04	0.185 ± 0.01	0.745 ± 0.03

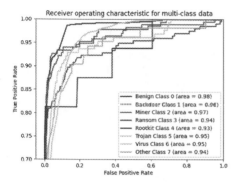

Fig. 5. Average ROC Curve for cross-validation, GRU, Opcodes

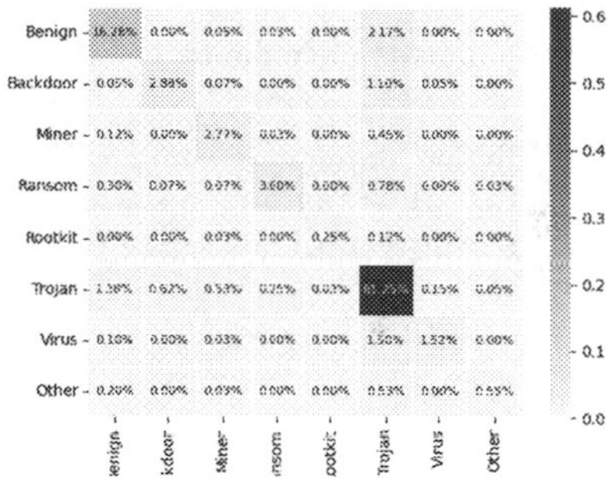

Fig. 6. Average Confusion matrix for cross-validation, GRU, Opcodes

Table 9. Precision, recall, and F1 score per class, for 10-fold cross-validation of the GRU on opcodes and API calls.

	Static opcodes				Dynamic API calls		
	Precision	Recall	F1 score		Precision	Recall	F1 score
Benign	0.90 ± 0.04	0.87 ± 0.07	0.88 ± 0.04	Adware	0.99 ± 0.02	0.67 ± 0.07	0.80 ± 0.06
Backdoor	0.81 ± 0.05	0.69 ± 0.16	0.74 ± 0.10	Backdoor	0.92 ± 0.12	0.68 ± 0.10	0.78 ± 0.10
Miner	0.80 ± 0.17	0.81 ± 0.14	0.80 ± 0.14	Downloader	0.94 ± 0.08	0.82 ± 0.10	0.88 ± 0.09
Ransom	0.92 ± 0.25	0.74 ± 0.23	0.82 ± 0.23	Dropper	0.94 ± 0.11	0.71 ± 0.14	0.81 ± 0.13
Rootkit	0.90 ± 0.46	0.56 ± 0.33	0.69 ± 0.36	Spyware	0.88 ± 0.16	0.64 ± 0.11	0.74 ± 0.12
Trojan	0.90 ± 0.04	0.95 ± 0.03	0.93 ± 0.03	Trojan	0.90 ± 0.13	0.66 ± 0.10	0.76 ± 0.11
Virus	0.88 ± 0.27	0.48 ± 0.18	0.62 ± 0.20	Worm	0.96 ± 0.06	0.64 ± 0.09	0.77 ± 0.08
Other	0.88 ± 0.4	0.42 ± 0.12	0.57 ± 0.18	Virus	0.92 ± 0.14	0.61 ± 0.10	0.73 ± 0.11

Table 10. Elapsed time

Algorithm	Opcode		API	
	Training	Tuning	Training	Tuning
LSTM	3h:20m	4h:19m	17h:48m	20h:16m
GRU	3h:04m	3h:50m	17h:25m	19h:48m
Transformer	35h:41m	1 week	27h:37m	44h:41m
MLP	1h:08m	2h:36m	2h:25m	4h:18m

6   Conclusions and Future Work

This paper uses open-source software to extract static opcode sequences and four classifiers to predict malware categories. The very same classifiers were also applied to the dynamic API call sequences. Regardless of the execution flow, the first 5000 opcode sequences extracted from the .text section seem to be sufficient for good malware prediction results. Despite the imbalanced nature of our datasets, a multiclass approach has worked well for both. A potential limitation of our approach is that opcodes may not be extracted properly by disassembling tools due to unsupported and less popular programming languages. The disassembling tools also struggle with particularly advanced techniques employed by metamorphic and polymorphic malware. As a result, the obtained opcodes may be insufficient and not reveal significant enough patterns for successful malware classification. We will try to address this issue in future work. We will also continue to collect and analyze many more malware samples, so that the results reported here can be verified on much larger datasets.

References

1. Amajd, M., Kaimuldenov, Z., Voronkov, I.: Text classification with deep neural networks. In: International Conference on Actual Problems of System and Software Engineering, pp. 364-370 2017
2. Capstone: Capstone the ultimate disassembler. https://www.capstone-engine.org/lang_python.html
3. Catak, F.O., Yazı, A.F., Elezaj, O., Ahmed, J.: Deep learning based sequential model for malware analysis using windows exe API calls. PeerJ Comp. Sci. **6**, 81 (2020)
4. Cho, K., van Merrienboer, B., Gulcehre, C., Bougares, F., Schwenk, H., Bengio, Y.: Learning phrase representations using RNN encoder-decoder for statistical machine translation. In: Conference on Empirical Methods in Natural Language Processing (2014)
5. Gupta, S., Sharma, H., Kaur, S.: Malware characterization using windows API call sequences. In: SPACE (2016)
6. Hochreiter, S., Schmidhuber, J.: Long short-term memory. Neural Comput. **9**(8), 1735–1780 (1997)

7. Jalilian, A., Narimani, Z., Ansari, E.: Static signature-based malware detection using opcode and binary information. In: Bohlouli, M., Sadeghi Bigham, B., Narimani, Z., Vasighi, M., Ansari, E. (eds.) CiDaS 2019. LNDECT, vol. 45, pp. 24–35. Springer, Cham (2020). https://doi.org/10.1007/978-3-030-37309-2_3
8. Kerrisk, M.: objdump - Linux manual page. https://man7.org/linux/man-pages/man1/objdump.1.html
9. Lundberg, S.M., Lee, S.I.: A unified approach to interpreting model predictions. In: Advances in Neural Information Processing Systems, vol. 30 (2017)
10. Maniath, S., Ashok, A., Poornachandran, P., Sujadevi, V., AU, P.S., Jan, S.: Deep learning LSTM based ransomware detection. In. 2017 Recent Developments in Control, Automation Power Engineering (RDCAPE), pp. 442–446 IEEE (2017)
11. O'Malley, T., Bursztein, E., Long, J., Chollet, F., Jin, H., Invernizzi, L., et al.: Keras Tuner (2019). https://github.com/keras-team/keras-tuner
12. Ramchoun, H., Ghanou, Y., Ettaouil, M., Janati Idrissi, M.A.: Multilayer perceptron: architecture optimization and training $4(1)$, 26–30 (2016)
13. Singh, A., Arora, R., Pareek, H.: Malware analysis using multiple API sequence mining control flow graph. arXiv preprint arxiv.org/abs/1707.02691 (2017)
14. Srivastava, N., Hinton, G., Krizhevsky, A., Sutskever, I., Salakhutdinov, R.: Dropout: a simple way to prevent neural networks from overfitting. JMLR $15(56)$, 1929–1958 (2014)
15. Vaswani, A., et al.: Attention is all you need. In: NIPS, vol. 30 (2017)
16. Wang, Y., Stokes, J., Marinescu, M.: Actor critic deep reinforcement learning for neural malware control. In: AAAI, vol. 34, pp. 1005–1012 (2020)
17. Wang, Y., Stokes, J.W., Marinescu, M.: Neural malware control with deep reinforcement learning. In: IEEE Military Communications Conference (2019)
18. Zhang, H., Xiao, X., Mercaldo, F., Ni, S., Martinelli, F., Sangaiah, A.K.: Classification of ransomware families with machine learning based onN-gram of opcodes. Future Gener. Comput. Syst. 90, 211–221 (2019)

A Novel Approach to Time Series Complexity via Reservoir Computing

Braden Thorne[1,2]([✉])(iD), Thomas Jüngling[1](iD), Michael Small[1,2](iD),
Débora Corrêa[1,2](iD), and Ayham Zaitouny[1,3](iD)

[1] University of Western Australia, Crawley, WA, Australia
[2] ARC Centre for Transforming Maintenance Through Data Science,
Crawley, WA, Australia
braden.thorne@research.uwa.edu.au
[3] University of Doha for Science and Technology, Duhail North Doha, Qatar

Abstract. When working with time series, it is often beneficial to have
an idea as to how complex the signal is. Periodic, chaotic and random
signals (from least to most complex) may each be approached in different
ways, and knowing when a signal can be identified as belonging to one
of these categories can reveal a lot about the underlying system. In the
field of time series analysis, permutation entropy has emerged as one of
the premier measures of time series complexity due to its ability to be
calculated from data alone. We propose an alternative method for calculating complexity based on the machine learning paradigm of reservoir
computing, and how the outputs of these neural networks capture similar
information regarding signal complexity. We observe similar behaviour
in our proposed measure to both the Lyapunov exponent and permutation entropy for well known dynamical systems. Additionally, we assess
the dependence of our measure on key hyperparameters of the model,
drawing conclusions about the invariance of the measure and possible
implications on informing network structure.

Keywords: Reservoir computing · Recurrent neural networks · Time
series analysis · Information entropy

1 Introduction

Time Series Complexity. When working with time series, understanding
the nature of the signal we are working with can have profound impacts on
the modelling choices we make. For example, there is little gain in applying
deep learning to a trivial periodic signal, but similarly one can waste many
hours attempting to extract patterns from a random signal that lacks relevant
information. Knowing when a signal is periodic, random or chaotic is the task
of time-series complexity (henceforth simply complexity).

There are a number of well-defined approaches to determining complexity for
known systems. Most notable among these are measures of fractal dimension,
entropy and Lyapunov exponents, all of which are very capable at establishing

H. Aziz et al. (Eds.): AI 2022, LNAI 13728, pp. 442–455, 2022.
https://doi.org/10.1007/978-3-031-22695-3_31

complexity [8,21]. However, these assume knowledge of the equations underpinning the system, or require a reasonable simulation of the system. Determining complexity of a signal from data alone is a much more demanding task, however there are a number of proposed methods that are capable of generating good approximations [9,29].

Permutation Entropy. The problem of calculating complexity from data was the focus of the work in 2002 by Bandt and Pompe [2], which introduced an entropy approximation called permutation entropy (PE). The idea of PE is to first construct a delay embedding of the signal with dimension m (called the embedding dimension) by taking m lagged states of the signal with a separation of τ time steps (called the embedding lag). From this embedding each point in time is transformed into a symbol of length m by looking at the ranking of the dimensions. For example, with delay embedding $m = 2$ and embedding lag $\tau = 1$ a series $u(t)$ can have two possible symbols; $\psi(1) = (0,1)$ if $u_t < u_{t+1}$ or $\psi(2) = (1,0)$ if $u_t > u_{t+1}$. As such, a series such as

$$u = (1, 8, -4, 5, 2)$$

would have have two $\psi(1)$ and two $\psi(2)$ symbols.

The frequency of the symbols is used to determine a probability $p(\psi)$, and the PE is then calculated as

$$H_{PE}(m) = -\sum_{i=1}^{m} p(\psi(i)) \log_2(p(\psi(i))) . \tag{1}$$

There are a number of methods for determining the two hyperparameters m and τ [6,22], however, to avoid excessive discussion we choose to set $\tau = 1$ and $m = 6$ or 12 for our trials, chosen to be sensible with respect to the systems being considered. What is important to note though is that the determination of these parameters can be nontrivial, and that the higher the values for m and τ the less symbols can be generated, leading to a problem of scale for large, complex systems. As such, the development of additional methods to estimate entropy in cases of limited data or large dimensional systems remains a field of active interest.

Permutation entropy has seen significant interest since its introduction, predominantly in the field of time series analysis where its connections to ordinal partition networks has made it a popular measure for complexity in the field of network science [17,23]. Due to this it has also seen use as a feature in application tasks such as concept drift [4,7]. These practical use cases of permutation entropy likewise encourage further research into the field of complexity estimation, particularly in the presence of noise or other artifacts of real-world processes.

We propose here the use of a reservoir computing (RC) based complexity measure as an alternative to PE due to its historically strong performance in time series prediction tasks, both with synthetic and experimental data. We detail our proposed method in Sect. 2, then explore the method in a number of qualitative and quantitative trials in Sect. 3.

2   Methods

Reservoir Computing. RC is a machine learning paradigm that emerged in the early 2000s [10, 16] as an efficient tool for time series related tasks. The defining feature of a reservoir computer is its fixed, recurrent network (the reservoir) that facilitates the need for training only at the readout step. These reservoirs need not be complicated, and in many cases can be generated randomly with only a small number of hyperparameters needing to be defined a priori. What facilitates the use of such a simple structure is the way in which the reservoir computer embeds time series into a higher dimensional space. This embedding has two key properties. Firstly, each activation state X is a function of the input U at that time and the previous states of the reservoir, resulting in what is known as the echo state property (owing to the way each state echos the prior states). Secondly, the weights given to previous terms decrease the further in the past they are, resulting in what is known as the fading memory property as the reservoir only remembers the M most recent states. These two properties result in the echo function

$$\begin{aligned} X_t &= f_\infty \left(U_t, U_{t-1}, ... \right) \\ &\approx f \left(U_t, U_{t-1}, ..., U_{t-M+1} \right) \end{aligned} \tag{2}$$

which is analog to a delay embedding with $m = M$ and thus allows reconstruction of the dynamics of the underlying system [27,28].

Because of – or despite – this simple structure, RC has exhibited impressive performance for supervised time series prediction [13,14,20] and classification [1,11] tasks. Moreover, these principles have allowed for efficient implementation of RC in hardware where physical substrates can act as a reservoir [18,24]. There is proportionally less work that has looked at RC in an unsupervised sense [12,26], however there has been more work recently looking at RC structure as a means for time series analysis without concern for the classical training setup [3,5,25], as is the context of this paper.

Implementation. We consider the discrete-time echo state network [10] implementation of a reservoir computer, with the network states $X \in \mathbb{R}^{N \times T}$ evolving such that at time t the state when driven by some input $U \in \mathbb{R}^T$ is given by

$$X_t = \tanh \left(W_{adj} X_{t-1} + W_{in} U_t + W_{bias} \right) . \tag{3}$$

We consider here the case of U being a scalar signal only, however, this implementation could be easily adapted to consider vector valued signals as well. The elements of $W_{in} \in \mathbb{R}^N$ and $W_{bias} \in \mathbb{R}^N$ are both drawn randomly from a standard Gaussian distribution, and left fixed throughout the networks lifespan. The network adjacency matrix $W_{adj} \in \mathbb{R}^{N \times N}$ is initially generated randomly with a specified average degree per node d and elements drawn from a standard Gaussian distribution. This is then scaled such that the maximal absolute eigenvalue of the adjacency matrix (called the spectral radius ρ) is of unit value to ensure consistency of the reservoir dynamics [15], and then kept fixed. The hyperbolic tangent function is chosen here as the activation function, however other sigmoid functions may be chosen with similar functionality.

The fundamental idea of RC is to generate a large ensemble of diverse non-linear transformations of the input signal U to facilitate information processing applications. The reservoir effectively acts as an embedding machine, recreating underlying structures of the system from inputs alone. As one increases the size of the reservoir N, a greater ensemble of variations is obtained and the quality of the reconstruction improves towards some limit. Since the hard work of presenting the information is already handled by the construction of the reservoir, the only training that is required is from the reservoir states to some desired output Z which is of the same size as U. The variable Z typically represents some future state of the signal $Z_t = U_{t+\Delta t}$ for forecasting tasks, a state from a different signal for prediction tasks or some label for classification tasks. The RC then approximates this target Z by an output $Y \sim Z$, following a least-squares regression that is usually implemented with Tikhonov regularisation.

$$Y = RX \tag{4}$$

$$R = ZX^\top \left(XX^\top + \beta I \right)^{-1} . \tag{5}$$

For the trials throughout this paper (unless otherwise specified), we choose to keep hyperparameters fixed with $N = 300$, $d = 5$, $\rho = 1$ and $\beta = 0.001$. These choices are somewhat arbitrary, but ensure we are operating in a regime consistent with common RC methodology.

Reservoir Readout Complexity. The concept of complexity has been considered with RC before. Pathak et al. 2017 [19] utilised a fully trained RC system as a proxy for the original system in calculating Lyapunov exponents for a number of complex systems. Carrol 2018 [5] also looked at using RC as a means of calculating Lyapunov exponents, but more prevalent to this research they also proposed a variation on PE that used the reservoir activation states in place of a delay embedding for determining symbols. They were able to show that this entropy measure could be used for informing the hyperparameter choices of the reservoir, noting a minimisation in signal classification errors when the reservoir entropy was maximised. However, there is an immediate problem here in that with even a modestly sized reservoir their are potentially a very large number of symbols when using the reservoir activation states in place of a delay embedding, far more than can be reasonably worked with. In practice the actual number of symbols should be much lower due to correlation between nodes, however problems can still arise for finitely sampled signals, in the presence of noise or in particular RC setups where larger variations between node activations occur.

We approached the idea of complexity from a more ground-up perspective, leveraging the following assumptions;

1. The more complex a signal/system, the harder the task of forecasting future values.

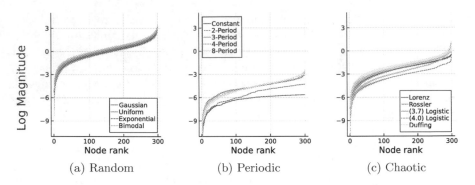

Fig. 1. Mean readout vectors R sorted by node magnitude for various random (a), periodic (b) and chaotic (c) signals. Ribbons show the 5th and 95th percentiles.

2. In RC, hard tasks require information from a larger number of nodes to simple tasks.

Thus, by considering the proportion of nodes used for time-series forecasting tasks (we opt to consider the simple one-step forecast $Y_t \sim Z_t = U_{t+1}$ here), we should be able to get a proxy for the complexity of the underlying signal. What we require then is a way of quantifying the proportion of node information used. This is analogous to looking at the magnitude of the weights assigned during the training process in presence of regularisation.

These magnitude curves for the one-step forecast across many systems are presented in Fig. 1, where the node indices have been sorted from least to most contribution. What we observe supported our initial assumptions, with clear differences in weight magnitudes being observed between periodic (low complexity), chaotic (moderate complexity) and random (high complexity) signals. Based on these results, it did not seem necessary to define a highly sophisticated measure to read information from the curves, and that instead we could simply look at a trapezoidal approximation for the area under the curves;

$$
\begin{aligned}
H_{RC} &= \log \left(\sum_{i=1}^{N-1} \frac{R_i + R_{i+1}}{2} \right) \\
&\approx \log \left(\sum_{i=1}^{N} |R_i| \right),
\end{aligned}
\tag{6}
$$

where R is sorted by magnitude as in Fig. 1(a-c) and the latter approximation holds for large N as typically used in RC. We will henceforth refer to this measure as RC complexity (RCC).

Of the hyperparameters at play here, there are three that we expect to have the largest impact on our complexity measure. Firstly, the size of the reservoir N is going to have a significant impact on the ensemble of node activations that can be chosen from during the training step. We hypothesise that as N gets large, additional nodes will add negligible additional information, thus being given weight close to 0 and not significantly impacting the complexity measure. Next, the regularisation parameter is crucial here as it ensures node selection is carried out as we expect; giving more weight to the nodes that are most informative, and punishing nodes that offer little information thus allowing us to get a good picture for what proportion of the nodes are utilised. Finally, as with any task involving linear regression, the length of the underlying signal will impact the quality of the prediction, and as such should have an impact on our complexity measure. All of these hyperparameters will be assessed in Sect. 3.

3   Results

System Analysis. To get an understanding of our proposed complexity measure, we opted to look at two well known dynamical systems; the logistic map and Rössler system.

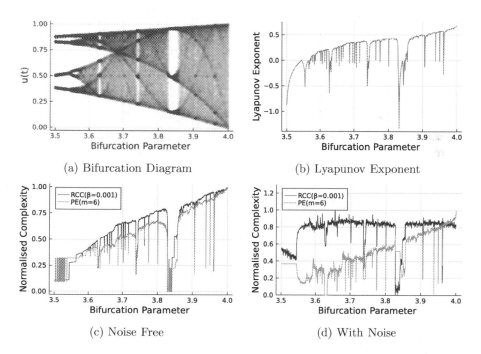

(a) Bifurcation Diagram   (b) Lyapunov Exponent

(c) Noise Free   (d) With Noise

Fig. 2. Bifurcation diagram (a), Lyapunov exponents (b) and corresponding complexity measures (c, d) for the logistic map. The PE and RCC are shown, both in the case without measurement noise (c) and with 0.4% standard Gaussian measurement noise (d).

The logistic map is a one-dimensional discrete-time map governed by the equation $u_{t+1} = ru_t(1 - u_t)$ where $r \in [3.5, 4]$ is a bifurcation parameter. Different values of r present different dynamics, from stable fixed points to orbits of various periods and even including chaos. The bifurcation diagram and corresponding Lyapunov exponents are shown in Figs. 2(a-b), respectively. The corresponding RCC and PE are presented in Fig. 2c, where we can immediately observe similarities between the spiking behaviour in the two complexity measures and the transitions in the bifurcation diagram and Lyapunov exponents.

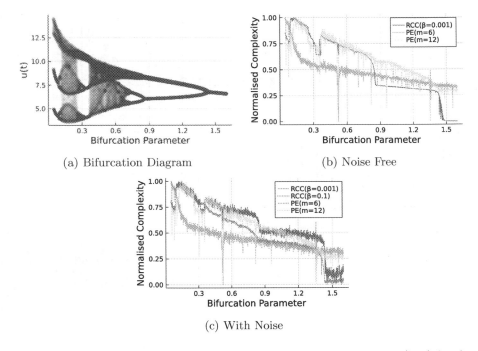

(a) Bifurcation Diagram   (b) Noise Free

(c) With Noise

Fig. 3. Bifurcation diagram (a) and corresponding complexity measures (b, c) for the Rössler system. The PE and RCC are shown, both in the case without measurement noise (b) and with 4% Gaussian measurement noise (c).

The Rössler system, unlike the logistic map, is multi-dimensional and continuous in time, however presents similar bifurcation behaviour. The equations governing the system are

$$\dot{u}_1 = -u_2 - u_3$$
$$\dot{u}_2 = u_1 + 0.2u_2 \tag{7}$$
$$\dot{u}_3 = r - u_3(u_1 - 5.7)$$

where $u = u_1$ is our signal of interest, with $r \in [0.05, 1.60]$ our bifurcation parameter. Due to the continuous nature, capturing transitions in dynamics is a harder task, and we expect factors such as sampling rate (chosen here to be $\Delta t = 0.5$) to become more critical, especially in the case of PE.

Observing the results in Fig. 3, we immediately see greater differences between the RCC and PE measures. Focusing on the noise free case, we see that the RCC responds to significantly less transitions that the PE. While concerning on face value, looking at the bifurcation diagram (particularly for the $r \in [0.9, 1.2]$ region), the Rössler system is engaged in a periodic regime with little variation with the parameters. As such, we would expect to see very little variation in our complexity measure, supporting the use of the RCC over the PE in this case. The reason for this spiking behaviour in the PE is the fixed sampling rate. As we vary our bifurcation parameter and the orbit frequency of the system changes, we encounter situations where the sampling rate leads to loss of information, and thus lower complexity hence the spiking behaviour observed.

We additionally note that with the chosen embedding dimension $m = 6$, the PE appears relatively insensitive to abrupt transitions, and generally exhibits quite high variability. Recalling the continuous dynamics and thus larger time scale of the Rössler system this prompted us to also consider the PE with an embedding dimension $m = 12$, which is generally more responsive to dynamic conditions than the PE with $m = 6$. However, it remains sensitive to the sampling rate and is still generally more variable than the RCC.

The introduction of noise seemed to impact both the PE and RCC measures in similar ways; a general increase in the variability of the measure, but maintaining similar sensitivity to transitions as in the noise-free case. The variability of the RCC appears to be impacted more by the introduction of noise than either of the PE measurements. This prompted us to consider a different choice for our regularisation parameter β, as increasing this parameter should prevent overfitting on finite size effects introduced by the noise. By increasing β we observed a significant reduction in the variability, but also note that the measure had decreased sensitivity to some of the major transitions, particularly the transitions between periodic and chaotic regimes around $r \approx 0.3$.

Hyperparameter Analysis. In this section we assess the dependence of the RCC measure on the key hyperparameters discussed in Sect. 2. To do this, we continue to look at the logistic map and Rössler systems discussed at the start of Sect. 3, choosing bifurcation parameter values of $r \in \{3.5, 3.7, 4.0\}$ and $r \in \{0.75, 0.45, 0.2\}$ for the two systems respectively, listed by increasing complexity. Additionally, we consider a signal of random noise as a comparison for both systems. In each case trials were repeated 300 times and the mean was plotted, with error bars representing the 5th and 95th percentiles. Note the use of log scales for the x-axis throughout the section.

We begin by looking at the behaviour with respect to the reservoir size N (see Fig. 4). First, it is important to note here that by the nature of this task it is not possible to keep the reservoir structure fixed, as generating reservoirs of different size requires generation of a new adjacency matrix in each case. As such we can observe the high variance resulting from different reservoir structures, which

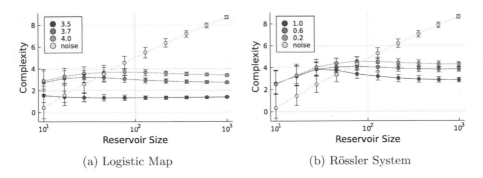

(a) Logistic Map   (b) Rössler System

Fig. 4. Variation in the RCC measure of a signal with respect to the reservoir size N for the logistic map (a) and Rössler system (b).

notably is higher than supposedly comparable results we will see in the numerical analysis for the regularisation parameter and signal length (see Figs. 5 & 6). The reservoir is kept fixed for all other investigations in this paper, which together with our choice of $N = 300$ makes the variance in the latter investigations negligible.

None the less, the observed behaviour is consistent with the general RC expectations and our assumptions. For small reservoir sizes, it is more difficult to get a good representation of the underlying signal and so there is significantly more variation in individual complexities, and greater overlap between complexities of different signals. We begin to see successful separation at around the $N = 100$ mark, and observe that the variation in complexities is almost completely eliminated by the $N = 1000$ node mark. In the context of time-series analysis, this suggests that the RCC is invariant in the limit $N \to \infty$, however the analytics required to prove such a claim falls well outside the consideration of this paper. We additionally note that the apparent invariance is even more fascinating given the variation in reservoir structure, and may suggest that for large enough reservoirs the structure is a relatively unimportant consideration outside of ensuring consistency.

Turning our attention to the regularisation parameter β (see Fig. 5), we choose to focus specifically on the Rössler system as the introduction of noise with this system proved more impactful, and as such should be more informative with regards to the role regularisation may play. There are three important behaviours we observe here related to the limit cases as $\beta \to 0$ and as $\beta \to \infty$. Considering first the $\beta \to \infty$ case, we observe a gradual convergence of the complexity for all signals with and without noise. We expect this, as increased regularisation should push the values of our readout vector towards uniform value, essentially removing any potential information that could be gained from them. We would expect this convergence to likewise reduce the variation in complexity values, which we also observe.

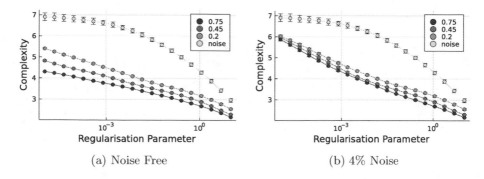

(a) Noise Free   (b) 4% Noise

Fig. 5. Variation in the RCC measure of a signal from the Rössler system with respect to the regularisation parameter β in the case of no additive noise (a) and 4% additive noise (b).

The behaviour as $\beta \to 0$ is more interesting. For the noise free case, results suggest that separation increases in the limit, and should be maximised when $\beta = 0$. Indeed, calculation of the complexities at $\beta = 0$ gives complexities of $4.93, 6.46, 6.85$ and 6.93 for the four signals in increasing expected complexity, respectively. While there is a reduction in separability, the hierarchy of complexity is maintained, and in particular the separation between the low complexity $r = 0.75$ signal and the other signals remains quite high. The explanation for this behaviour likely comes down to the mechanism of Tikhonov regularisation behind Eq. 5, which would require additional analysis which fell outside the scope for this investigation. We note however that there is no significant change to the variation as $\beta \to 0$, and that improvements in general are negligible below $\beta \approx 0.005$. The case with noise here is more telling, and we see that as regularisation falls we see a convergence of all measures to some high complexity. This aligns more with expectations, as we expect that as the regularisation gets sufficiently low, we begin to train on the finite size effects presented by the noise and as a result observe the corresponding increased complexity we have seen with other purely random signals.

Looking at the variations with respect to signal lengths (see Fig. 6), there are a number of notable features and comparisons that immediately stand out. Firstly, the RCC appears to reach an optimum at around the $T = 300$ for the random signal case. This can be explained in terms of the signal length with respect to the size of the reservoir N, the latter of which is fixed at $N = 300$. The regime $T \ll N$ is characterised by over-fitting as there are more variables than data points. This can result in seemingly optimal solutions using only a fraction of the node values that are, however, mostly finite-size effects. For $T \gg N$, the issues with trying to predict future values of random noise become apparent and the best approximation capable by the reservoir becomes mapping to the mean of the signal, which due to normalisation will always be 0. Similarly to the high regularisation case, this leads to a convergence of values in the readout vector and a loss of information. This relationship between reservoir size and signal length

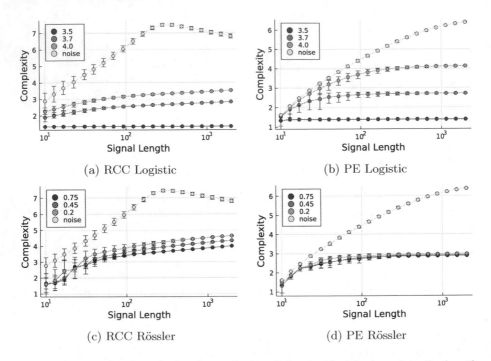

Fig. 6. Variation in the calculated complexity of the signal with respect to the length of the signal for the logistic map (a,b) and Rössler system (c,d). Both the RCC (a,c) and PE (b,d) complexities were considered.

is presented in Fig. 7, which also highlights the decreased variation as reservoir size increases. These results lead us to conclude that reservoir size and signal length should always be considered interrelated, and in the future the choice of one should not be made without consideration of the other. The behaviour in the informative signal cases is somewhat more predictable; decreased variation as signal length increases, with a gradual trend towards some steady state.

Looking again at Fig. 6 and focussing on comparisons between the RCC and PE measures without the presence of noise, there are two main conclusions. Firstly, in the limit as $T \to \infty$ the PE seems to offer comparable separation for the informative signals, but better performance for the noise signal where we do not observe the same deterioration for $T \gg N$ and instead see the eventual formation of a steady state. For low signal lengths, however, the opposite appears to be true. While the RCC appears capable of separating the four cases with as few as around 15 data points, we do not see full separation in the PE case until around 50 data points. In terms of variation of complexities, variation is almost entirely eliminated for informative signals by around 60 data points, relative to the almost 400 data points required for PE. This result is important when considering applications, as it would suggest the RC approach is generally better for determining signal complexity when data is limited. Turning attention to the the cases in the presence of noise, the PE measure appears incapable of

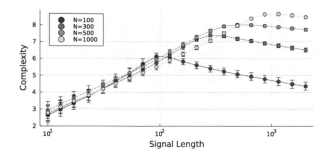

Fig. 7. RCC measure with respect to signal length T of random noise for various fixed reservoir sizes N. Note the characteristic peaks where $T \approx N$.

separating the non-random signal complexities whereas the RCC does with as few as 50 data points. This heavily suggests that RCC can offer improvements over current PE measures in this context.

4   Conclusion

We have introduced a novel method for calculating time series complexity using RC. We showed that it performs similarly to both system-driven measures such as the Lyapunov exponent, as well as the data driven measures such as PE. There are a number of implications here for future work. Firstly, the methods introduced here were not optimised; we did not look at approaches for choosing appropriate reservoir sizes or regularisation, and our choice for reading out complexity was done based on empirical observations. Further analysis may provide a more robust method that offers additional meaning than simply a proxy for complexity.

Secondly, a possible application of our method may be found in reservoir design. This research has shown a clear connection between the complexity characteristics of a signal and the structure of the reservoir computer when making predictions on that signal. Future work may look to unpack this further, and look for ways of inferring structure of the reservoir computer, such as an optimal range of connectivities d or spectral radii ρ, by first calculating an efficient complexity measures such as the PE.

Acknowledgements. BT is partially supported by the Australian Government Research Training Program and a University Postgraduate Award at The University of Western Australia. BT, MS, and DC are partially supported by the Australian Research Council through the Centre for Transforming Maintenance Through Data Science (grant number IC180100030) funded by the Australian Government.

References

1. Appeltant, L., et al.: Information processing using a single dynamical node as complex system. Nature Comm. **2**, 468 (2011). https://doi.org/10.1038/ncomms1476
2. Bandt, C., Pompe, B.: Permutation entropy: a natural complexity measure for time series. Phys. Rev. Lett. **88**, 174102 (2002). https://doi.org/10.1103/PhysRevLett. 88.174102
3. Bianchi, F.M., Scardapane, S., Lokse, S., Jenssen, R.: Reservoir computing approaches for representation and classification of multivariate time series. IEEE Trans. Neural Netw. Learn. Syst. **32**(5), 2169–2179 (2021). https://doi.org/10. 1109/TNNLS.2020.3001377
4. Cao, Y., Tung, W.w., Gao, J.B., Protopopescu, V.A., Hively, L.M.: Detecting dynamical changes in time series using the permutation entropy. Phys. Rev. E. **70**, 046217 (2004). https://doi.org/10.1103/PhysRevE.70.046217
5. Carroll, T.L.: Using reservoir computers to distinguish chaotic signals. Phys. Rev. E. **98**, 052209 (2018). https://doi.org/10.1103/PhysRevE.98.052209
6. Cellucci, C.J., Albano, A.M., Rapp, P.E.: Comparative study of embedding methods. Phys. Rev. E **67**, 066210 (2003). https://doi.org/10.1103/PhysRevE.67. 066210
7. Costa, F.G.d., Duarte, F.S., Vallim, R.M., Mello, R.F.d.: Multidimensional surrogate stability to detect data stream concept drift. Expert Syst. Appl. **87**(C), 15–29 (2017). https://doi.org/10.1016/j.eswa.2017.06.005
8. Eckmann, J.P., Ruelle, D.: Ergodic theory of chaos and strange attractors. Rev. Mod. Phys. **57**, 617–656 (1985). https://doi.org/10.1103/RevModPhys.57.617
9. Hirata, Y.: Recurrence plots for characterizing random dynamical systems. Commun. Nonlinear Sci. Numer. Simul. **94**, 105552 (2021). https://doi.org/10.1016/j. cnsns.2020.105552
10. Jaeger, H.: The echo state approach to analysing and training recurrent neural networks-with an erratum note. Ger. Natl. Res. Center for Inf. Technol. GMD Tech. Rep. **148** (2001)
11. Larger, L., Baylón-Fuentes, A., Martinenghi, R., Udaltsov, V.S., Chembo, Y.K., Jacquot, M.: High-speed photonic reservoir computing using a time-delay-based architecture: Million words per second classification. Phys. Rev. X **7**, 011015 (2017). https://doi.org/10.1103/PhysRevX.7.011015
12. Lee, G.C., Loo, C.K., Liew, W.S., Wermter, S.: Self-organizing kernel-based convolutional echo state network for human actions recognition. In: Proceedings of the European Symposium on Artificial Neural Networks, Computational Intelligence and Machine Learning (ESANN'20), pp. 591–596 (2020). https://www2. informatik.uni-hamburg.de/wtm/publications/2020/LLLW20/ES2020-99-7.pdf
13. Lu, Z., Hunt, B.R., Ott, E.: Attractor reconstruction by machine learning. Chaos **28**(6), 061104 (2018). https://doi.org/10.1063/1.5039508
14. Lu, Z., Pathak, J., Hunt, B., Girvan, M., Brockett, R., Ott, E.: Reservoir observers: model-free inference of unmeasured variables in chaotic systems. Chaos **27**(4), 041102 (2017). https://doi.org/10.1063/1.4979665
15. Lymburn, T., Khor, A., Stemler, T., Corréa, D.C., Small, M., Jüngling, T.: Consistency in echo-state networks. Chaos **29**(2), 023118 (2019). https://doi.org/10. 1063/1.5079686
16. Maass, W., Natschläger, T., Markram, H.: Real-time computing without stable states: a new framework for neural computation based on perturbations. Neural Comput. **14**(11), 2531–2560 (2002). https://doi.org/10.1162/089976602760407955

17. McCullough, M., Small, M., Iu, H.H.C., Stemler, T.: Multiscale ordinal network analysis of human cardiac dynamics. Philos. Trans. Royal Soc. A: Math. Phys. Eng. Sci. **375**(2096), 20160292 (2017). https://doi.org/10.1098/rsta.2016.0292

18. Yamane, T., et al.: Dimensionality reduction by reservoir computing and its application to IoT edge computing. In: Cheng, L., Leung, A.C.S., Ozawa, S. (eds.) ICONIP 2018. LNCS, vol. 11301, pp. 635–643. Springer, Cham (2018). https://doi.org/10.1007/978-3-030-04167-0_58

19. Pathak, J., Lu, Z., Hunt, B., Girvan, M., Ott, E.: Using machine learning to replicate chaotic attractors and calculate lyapunov exponents from data. Chaos **27**(12), 121102 (2017)

20. Pathak, J., Hunt, B., Girvan, M., Lu, Z., Ott, E.: Model-free prediction of large spatiotemporally chaotic systems from data: a reservoir computing approach. Phys. Rev. Lett. **120**, 024102 (2018). https://doi.org/10.1103/PhysRevLett.120.024102

21. Pesin, Y.B.: Dimension theory in dynamical systems contemporary views and applications. Chicago lectures in mathematics series, University of Chicago Press, Chicago (1997)

22. Qing-Fang, M., Yu-Hua, P., Pei-Jun, X.: A new method of determining the optimal embedding dimension based on nonlinear prediction. Chin. Phys. **16**(5), 1252–1257 (2007). https://doi.org/10.1088/1009-1963/16/5/014

23. Shahriari, Z., Small, M.: Permutation entropy of state transition networks to detect synchronization. Int. J. Bifurcat. Chaos **30**(10), 2050154 (2020). https://doi.org/10.1142/S0218127420501540

24. Tanaka, G., et al.: Recent advances in physical reservoir computing: a review. Neural Netw. **115**, 100–123 (2019). https://doi.org/10.1016/j.neunet.2019.03.005

25. Thorne, B., Jüngling, T., Small, M., Corrêa, D., Zaitouny, A.: Reservoir time series analysis: using the response of complex dynamical systems as a universal indicator of change. Chaos: Interdisc. J. Nonlinear Sci. **32**(3), 033109 (2022). https://doi.org/10.1063/5.0082122

26. Vertechi, P., Brendel, W., Machens, C.K.: Unsupervised learning of an efficient short-term memory network. In: Ghahramani, Z., Welling, M., Cortes, C., Lawrence, N., Weinberger, K.Q. (eds.) Advances in Neural Information Processing Systems. vol. 27. Curran Associates, Inc. (2014). https://proceedings.neurips.cc/paper/2014/file/333222170ab9edca4785c39f55221fe7-Paper.pdf

27. Verzelli, P., Alippi, C., Livi, L.: Learn to synchronize, synchronize to learn. Chaos: Interdisc. J. Nonlinear Sci. **31**(8), 083119 (2021). https://doi.org/10.1063/5.0056425

28. Verzelli, P., Alippi, C., Livi, L., Tiňo, P.: Input-to-state representation in linear reservoirs dynamics. IEEE Trans. Neural Netw. Learn. Syst. **PP**1–12 (2021). https://doi.org/10.1109/TNNLS.2021.3059389

29. Xu, X., Zhang, J., Small, M.: Superfamily phenomena and motifs of networks induced from time series. Proc. Natl. Acad. Sci. **105**(50), 19601–19605 (2008). https://doi.org/10.1073/pnas.0806082105

Boosted Self–evolving Neural Networks for Pattern Recognition

Brendon J. Woodford[✉][iD]

Department of Information Science, University of Otago, Dunedin, New Zealand
brendon.woodford@otago.ac.nz

Abstract. It has been well documented that both boosting and bagging algorithms improve ensemble performance. However, these types of algorithms have only infrequently been applied to ensembles of constructivist learners which are based on neural networks. Although there have been previous attempts at developing similar ensemble learning algorithms for constructivist learners, our proposed approach also addresses the issue of ensuring more diversity of the learners in the ensemble and offers a different approach for handling imbalanced data sets. More specifically, this paper investigates how a modified version of the AdaBoost algorithm can be applied to generate an ensemble of simple incremental learning neural network-based constructivist learners known as the Self-Evolving Connectionist System (SECoS). We develop this boosting algorithm to leverage the accurate learning of the SECoS and to promote diversity in these SECoS learners in order to create an optimal model for classification tasks. Moreover, we adopt a similar minority class sampling method inspired by RUSBoost which addresses the class imbalance problem when learning from data. Our proposed AdaBoostedSECoS (ABSECoS) learning framework is compared with other ensemble-based methods using four benchmark data sets, three of which have class imbalance. The results of these experiments suggest ABSECoS performs comparably well against similar ensemble methods using boosting techniques.

Keywords: Ensemble learning · Adaptive systems · Neural networks

One of the main problems in generating an accurate classifier that can both learn from training data and generalize well is dependent on reducing both bias and variance in its learning. Compounding this problem is the issue of class imbalance which has the effect of reducing the accuracy of the learning system.

To address this problem past solutions have applied variants on generating boosted ensembles of classifiers primarily based on algorithms such as the popular AdaBoost approach [10]. This method has been demonstrated to work well with simple weak classifiers, for example naïve Bayesian classifiers [24] and decision trees [9]. But one type of classifier which AdaBoost has not been frequently applied is the family of Evolving Connectionist Systems (ECoS) methods [13,14,16].

The motivation for investigating how ensemble methods can be applied to ECoS systems in particular is that three of the requirements for ECoS system

H. Aziz et al. (Eds.): AI 2022, LNAI 13728, pp. 456–469, 2022.
https://doi.org/10.1007/978-3-031-22695-3_32

design are 1) fast learning from a large amount of data; 2) real-time, incremental adaptation to new data; and 3) continuous improvement throughout the lifetime of the system [15]. Ensemble methods for learning systems promote diversity in its learners and ECoS methods can enhance this diversity by its constructivist algorithms which modify the structure of the network as training data is presented; normally only requiring one pass to accurately learn data. These properties of ECoS learners result in models which are hard to over-train as they learn quickly, and they are far more resistant to catastrophic forgetting than most other models as new training data are represented by adding new neurons rather than accommodating the additional data in existing neurons [27].

One specific implementation of ECoS, the aforementioned Simple-Evolving Connectionist System (SECoS) [27] uses a modified supervised learning algorithm based on [13]. To generate an AdaBoost based SECoS ensemble of learners therefore requires 1) that the original architecture of the SECoS learner be modified to become a pseudo-weak learner, 2) the AdaBoost algorithm be modified to incorporate this learner, 3) a suitable function to establish consensus amongst the ensemble of learners for each new data instance presented to it, and 4) how to deal with class imbalance if present in the data set. Satisfying all four requirements and validating this approach is the focus of this paper.

The proposed approach is compared with the performance of a single SECoS learner, Random Forest, multi-class variant of the AdaBoost algorithm, and the state-of-the-art XGBoost algorithm. Not only do the results of the experiments suggest that the SECoS-based ensemble performs comparatively well, but also provides some insight into how ensemble-based learning can be improved upon to tackle more complex problem domains especially those that require robust Machine Learning (ML) systems which can be applied for accurate pattern recognition in the horticultural domain [5].

Therefore, our original contributions of this paper are:

- We build on earlier attempts to generate ECoS-based ensembles by incorporating a modified version of the AdaBoost algorithm, addressing the problem of forcing this SECoS-based ensemble to learn "difficult" examples in the data set, controlling the growth of the SECoS, and propose a method for overcoming potential class imbalance in the data set.
- To validate our proposed approach and to provide proof of concept we compare our proposed ABSECoS algorithm with three popular ensemble-methods using four well-known ML benchmark data sets.
- Our findings indicate that our ABSECoS approach comparably well against similar ensemble methods some of which also adopt boosting techniques.

The rest of the paper is organized as follows. In Sect. 1, we briefly review relevant prior work. Section 2 details our proposed framework. Experiments applying our framework are described in Sect. 3. The results and discussion of the experiments are presented in Sect. 4. Finally, in Sect. 5 we revisit our key objectives in the context of the results we obtained leading to areas of improvement in the form of future work.

1   Background

Ensemble learning involves the combination of multiple learners to solve a particular machine learning task. The intuition behind this approach is weighing and combining several individual opinions is better than choosing the opinion of one individual [22]. In terms of a machine learning context avoiding overfitting, decreasing the risk of the ensemble obtaining a local minimum, and better coverage thus a better fit to the space the data set represents are just three reasons why ensembles can improve predictive performance [22]. Moreover, ensemble methods can overcome specific difficult challenges to machine learning algorithms such as better dealing with class imbalance. Strategies such as oversampling the minority class creating "synthetic" examples instead of sampling with replacement and under-sampling the majority class as in the Synthetic Minority Over-sampling Technique (SMOTE) [3] or randomly under-sampling the minority class like Random Under-sampling Boosting (RUSBoost) [23] can achieve better ensemble performance.

To this end the first documented attempt at proposing an ECoS-based ensemble was [29]. Here an on-line clustering algorithm [25] was used to determine the number of Evolving Fuzzy Neural Network (EFuNN)s [16] which were trained on data that these clusters were comprised. Although this architecture had potential and the results reported were favorable, there was still issues with the learning algorithm dealing with class imbalance, optimizing the parameter settings, and maintaining diversity of the individual learners in the ensemble. A similar approach was proposed by [19] who applied an on-line clustering method to partition the input space into N clusters but instead used a co-evolutionary approach to both train and optimize an ensemble of EFuNNs based on the same number of clusters. However, this framework required prior definition of parameters for the fitness function which, if poorly selected, may adversely affect ensemble performance. In [18] this issue was addressed by employing a co-evolutionary multi-objective genetic algorithm. But the high computational cost of this method would not align with one of the requirements of an ECoS-based model which is the real-time, incremental adaptation to new data [15].

Nevertheless, we still consider that an alternative architecture for an ECoS-based ensemble would be of use. This would involve reducing the complexity of the ensemble architecture since each learner in the ensemble was an Evolving Fuzzy Neural Network (EFuNN) [16]. Moreover, we need to address the issue of reducing the number of rules generated by each EFuNN whilst preserving or improving on the overall performance accuracy of the ensemble.

Therefore, we turn our attention towards the use of a related ECoS-based architecture known as the Simple-Evolving Connectionist System (SECoS) which is a less complex architecture than the EFuNN and adopts a modified supervised learning algorithm based on [13]. The main problem now is to establish how the weights for each data instance are modified and how each SECoS ensemble makes use of this information. In addition, as each SECoS learner can add more rule nodes to its hidden layer when training, then determining the maximum number of rules to add based on the classification error of the previous learner.

2   Framework Implementation

2.1   Creating an AdaBoost-Based Ensemble of SECoS Learners

Over twenty years ago [10] proposed AdaBoost as an effective means of generating robust and accurate ensembles of learners. Since then there have been many improvements to and variants of this framework. For recent treatment of AdaBoost and ensemble learning in general refer to [6,22]. The key principle of the AdaBoost algorithm is that data instances that were not correctly classified by the previous weak learner have their corresponding weights increased thus forcing the next weak learner to change its hypothesis i.e. focus on learning the "difficult" examples which have high corresponding weights.

Therefore, a solution to having an ECoS-based ensemble which employs AdaBoost is to apply a modified version of the boosting algorithm and a less complex ECoS learner to play the role of the weak learner.

2.2   The Weak Learner Used in ABSECoS

One implementation of ECoS, the SECoS, is a three layer feed-forward neural network where in the evolving (middle) layer rule nodes (neurons) are incrementally added only if a new data instance is dissimilar from prototypes represented by the existing rule nodes in this layer and one of more rule nodes are pruned (removed) if these satisfy a predefined criterion. Given the complexity of the SECoS learning algorithm one might consider the SECoS learner a strong learner when compared with a weak learner such as a decision stump. In [9], however, the authors do state that the term "weak learner" can refer to a strong learning algorithm.

To detail its learning, at the start of SECoS training the evolving layer contains no rule nodes. Data examples are incrementally added, regulated by a parameter, $sThr$, which determines how different the incoming data example is required to be before being added as a new rule node, or if not, accommodated by an existing rule node. In addition, another parameter, $errThr$, specifies how much output error the overall SECoS model can tolerate. A rule node will also be added if the new data example is similar to an existing rule node but the overall output error of the SECoS currently exceeds $errThr$. The result of adding another rule node will ensure the SECoS maintains an error below $errThr$. This means that each rule node in this evolving layer acts as a cluster centre where data examples that are similar to each other are accommodated by the same rule node and attempt to generate a model with low error. Finally, a learning rate, Lr, operates the same way as the modified backpropagation learning algorithm upon which ECoS is based [14]. We have selected the specific implementation of SECoS on the basis on the initial work by [17] adding subsequent improvements inspired by the survey of the ECoS framework conducted by [27]. For more details on the specifics of the SECoS learning algorithm, again refer to [27].

ALGORITHM 1: The ABSECoS training algorithm

Input:
Set of examples, $S\langle(x_1, y_1), \ldots, (x_m, y_m)\rangle$ with labels $y_i \in Y \leftarrow \{1, \ldots, k\}$
Number of learners, T
Weak learner, $SECoS$

Output:
A trained ABSECoS ensemble, $SECoSEnsemble$

1 Initialize $R_{max} \leftarrow k$;
2 Let $B \leftarrow \{(i, j)\} : i \in \{1, \ldots, m\}, y \neq y_i$;
3 Initialize $D_1(i, y) \leftarrow 1/|B|$ for $(i, y) \in B$;
4 Initialize $instancesIndicies \leftarrow$ number of instances for each class;
5 Initialize $sizeMinClass \leftarrow$ minimum of $instancesIndicies$;
6 **for** $t \leftarrow 1$ **to** T **do**
7 $\quad$ Initialize $ruleNodesToAdd \leftarrow 0$;
8 $\quad$ **if** $t > 1$ **then**
9 $\quad\quad$ Initialize $sampClassRows \leftarrow \{\}$;
10 $\quad\quad$ $ruleNodesToAdd \leftarrow 0$;
11 $\quad\quad$ **for** $classIndex \leftarrow 1$ **to** k **do**
12 $\quad\quad\quad$ $misClassRowsForClass \leftarrow$ the indices of $misClassified$ for class
$\quad\quad\quad\quad classIndex$;
13 $\quad\quad\quad$ Set $instInd$ to the indices of all instances of class $classIndex$;
14 $\quad\quad\quad$ $partError(classIndex) \leftarrow$
$\quad\quad\quad\quad \texttt{length}(misClassRowsForClass)/\texttt{length}(instInd)$;
15 $\quad\quad\quad$ $ruleNodesToAdd \leftarrow ruleNodesToAdd + \text{round}(partError(classIndex))$;
16 $\quad\quad\quad$ **if not** $\texttt{isempty}(instInd)$ **and**
$\quad\quad\quad\quad \texttt{length}(misClassRowsForClass) \leq sizeMinClass$ **then**
17 $\quad\quad\quad\quad$ Randomly sample $sizeMinClass$ instances from $instInd$ using
$\quad\quad\quad\quad\quad$ weights from $D_t(i, y)$ into $selectedRows$;
18 $\quad\quad\quad$ **else**
19 $\quad\quad\quad\quad$ $selectedRows \leftarrow misClassRowsForClass$;
20 $\quad\quad\quad$ **end**
21 $\quad\quad\quad$ Add $selectedRows$ to $sampClassRows$;
22 $\quad\quad$ **end**
23 $\quad\quad$ Set $theRows$ as a random permutation of $sampClassRows$;
24 $\quad$ **else**
25 $\quad\quad$ Set $theRows$ as a random permutation of S;
26 $\quad$ **end**
27 $\quad$ $R_{max} \leftarrow R_{max} + ruleNodesToAdd$;
28 $\quad$ Call $SECoS$ on $theRows$ generating a maximum of R_{max} rule nodes;
29 $\quad$ Get back a hypothesis $h_t : X \times Y \rightarrow [0, 1]$;
30 $\quad$ Calculate the pseudo-loss of h_t according to [9]:
$$\epsilon_t \leftarrow \frac{1}{2} \sum_{(i,y) \in B} D_t(i, y)(1 - h_t(x_i, y_i) + h_t(x_i, y));$$
31 $\quad$ Set $\beta_t \leftarrow \epsilon_t/(1 - \epsilon_t)$;
32 $\quad$ Update $D_t : D_{t+1}(i, y) \leftarrow \frac{D_t(i,y)}{Z_t} \cdot \beta_t^{(1/2)(1+h_t(x_i,y_i)-h_t(x_i,y))}$ where Z_t is the
$\quad\quad$ normalization constant (chosen so that D_{t+1} will be a distribution);
33 $\quad$ Test h_t on S producing h_{acc} and $misClassified$;
34 **end**

35 Output the hypothesis: $SECoSEnsemble(x) \leftarrow \arg\max_{y \in Y} \sum_{t=1}^{T} (\log \frac{1}{\beta_t}) h_t(x, y)$;

2.3   The Design of the ABSECoS Learning Algorithm

Enabling SECoS learning to be combined with the boosting algorithm requires that each SECoS learner in the ensemble focus on learning the "difficult" examples in the data set i.e. those data instance not correctly classified by the previous SECoS learner. This requirement would be satisfied by randomly sampling data instances without replacement which have high weights and these data instances subsequently used for training the next SECoS in the ensemble. To overcome the issue with class imbalance means that either an over-sampling or under-sampling strategy be adopted. There is sufficient evidence to support over-sampling the majority class results in better ensemble performance in the presence of class imbalance [23]. In addition, a method for regulating the growth of each SECoS learner also needs to be taken into consideration.

To this end we present the ABSECoS algorithm which applies the AdaBoost algorithm to an ensemble of SECoS learners incorporating a data sampling scheme which creates training data with balanced classes. Furthermore, the size of each SECoS learner in terms of how many rules nodes it can generate under SECoS learning is controlled for by the training error of the previous SECoS learner in the ensemble. This is an important step in the ABSECoS algorithm since the accuracy of a SECoS learner is highly dependent on the number of rule nodes it can generate.

The ABSECoS algorithm is presented in Algorithm 1. Initially, in Line 1 we establish that the maximum number of rule nodes that the first SECoS learner in the ensemble can generate, R_{max}, is equal to the number of classes, k, in the data set. Lines 4–5 determines which of the classes is the minority class and how many data instances are in this class. This approach is similar to the RUSBoost algorithm [23] to minimize the issue of class imbalance. However, unlike RUSBoost, ABSECoS does not require the user to specify the percentage of total instances to be represented by the minority class and can handle more than two-class problems without resorting to a One-vs-All strategy, for example.

Lines 6–34 is the main loop of the algorithm where in the first iteration of the loop the randomly ordered full data set is presented to the first SECoS learner in the ensemble. We randomly order the data instances to deliberately enforce diversity in each SECoS learner. In the second iteration and beyond we randomly select $sizeMinClass$ examples without replacement weighted by $D_t(i, y)$ from the pool of all instances for a specific class. This ensures data instances that were not correctly classified by previous SECoS learners in the ensemble have a higher probability of being selected for the data instances that are used to train the next SECoS learner as in the AdaBoost algorithm. Line 16 is present as a check to see if none of the instances for a class were correctly classified by the previous SECoS learner. If this is the case, then all the data instances associated with this class are selected for learning by the next SECoS learner in the ensemble.

One problem, however, in this approach is how to determine the number of rule nodes that could be added to the subsequent SECoS learner in the ensemble if the current SECoS learner has misclassified one or more of the training data examples for a particular class. To address this challenge, we define Eq. (1)

$$R_{max} \leftarrow R_{max} + ruleNodesToAdd, \tag{1}$$

where the value of $ruleNodesToAdd$ is calculated using Line 14 of Algorithm 1.

Therefore, in situations where the current SECoS learner has misclassified the training data, the maximum number of rule nodes the next SECoS learner can generate is incremented by an amount relative to level of per-class misclassification by the current SECoS learner. This increment can range up to a value equivalent to the number of classes in the data set. This way each subsequent SECoS learner progressively has an increased maximum number of rule nodes, R_{max}, with which to generate a classifier that will attempt to correctly classify all training data instances. However, the current SECoS learner might not generate a set of rule nodes equivalent to R_{max} so the value of R_{max} could decrease for the next SECoS learner.

Depending on the number of weak learners in the ensemble, there may not be a single weak learner that would correctly classify all training data instances but in our experiments we have found that ensembles with at least 50 SECoS learners consistently achieve high accuracy.

Similarly, we employ the weighted voting scheme to determine the winning class for unseen/test data instances according to the AdaBoost.M2 Algorithm originally proposed by [9]. Here the assumption is that information about how well an individual weak learner performed on the training data set can be used to weight the votes when the entire ensemble is used to classify a new/unseen data instance. Specifically, weak learners that performed poorly on correctly classifying training data instances (low β value) would be given less weight to its votes than weak learners which performed well (high β value). To effect this, the value of the β parameter assigned to each weak learner when training the ensemble is used to weight the vote of each weak learner when presented with a new data instance (testing example) using Eq. (2) [9].

$$h_{fin}(x) = \arg\max_{y \in Y} \sum_{t=1}^{T} \left(\log \frac{1}{\beta_t} \right) h_t(x, y), \tag{2}$$

where h_{fin} is a weighted vote (i.e. weighted linear threshold) of the weak hypotheses. To elaborate, for a given instance x, h_{fin} outputs the label y that maximizes the sum of the weights of the weak hypotheses predicting that label. The weight of hypothesis h_t is defined to be $\log(1/\beta_t)$ so that greater weight is given to hypotheses with lower error.

This means that votes as to the winning class by weak learners with a high β value are favoured over votes as to the winning class by weak learners with a low β value. Applying this weighted voting scheme produces votes by the ensemble which results in a clear winning class thus reducing the problem of deciding between two or more classes that have the same number of votes assigned to it by the ensemble. In our ABSECoS algorithm each SECoS learner is also considered to be a hypothesis, h, and treated in the same way as described in [9].

3   Methodology

Four experiments were run using this proposed learning algorithm. All experiments used four benchmark data sets; the Iris data set (*Iris*) [7], Wine data set (*Wine*) [8], the Wisconsin Breast Cancer data set (*Breast Cancer*) [28], and the Ecoli data set [12]. The breakdown of these data sets is detailed in Table 1. We selected the data sets as these were a mixture of data sets with varying number of input features, classes, and number of data instances. Regarding class imbalance in these data set, only the Iris data set did not contain any class imbalance the other three data sets had varying degree of class imbalance.

The ensemble algorithms employed were a Random Forest (*RF*) [2], AdaBoost Classifier (*AdaBoost*) [9], and the state-of-the-art XGBoost framework (*XGBoost*) [4]. For a baseline, a single SECoS classifier (*SECoS*) was used to compare with our proposed ABSECoS framework (*ABSECoS*).

In all experiments we used 50 weak learners per ensemble and performed 10-fold cross validation to remove any bias in which data examples were selected for training and testing the ensembles. In addition, stratified sampling was used to select the similar percentage of samples in each class. To ensure reproducible results the same seed was used for the random number generated used for stratified sampling. All data sets were normalized to have a mean of zero and a standard deviation of one.

Hardware to run the experiments was on a desktop using a quad-core Intel i5-6500 processor with 16GB memory running a 64 bit version of Windows 10. Both the ABSECoS and SECoS algorithms were implemented using Matlab 2021a. Python's `scikit-learn` [21] and `xgboost` packages were used for the other ensemble learning algorithms. The Python `hyperopt` package [1] was used to fine-tune the hyper-parameters for the *RF*, *AdaBoost*, and *XGBoost* ensemble learning algorithms. Here hyper-parameter optimization occurred over 80 iterations. Within each iteration ensemble generation underwent 10-fold cross validation. The average test accuracy of the result of cross validation was used as the basis to adjust the learning parameters of the ensemble methods with the objective of increasing the average test accuracy of the ensemble for the next iteration. The learning parameters for the *SECoS* learner and *ABSECoS* were set empirically also using 10-fold cross validation. For each type of ensemble method, the best performing ensemble out of the 80 generated was then selected as the model to be reported on. Performance metrics used were the established Precision, Recall, and F1-Score values [26].

After obtaining the results of each experiment, we confirmed the performance of the ensemble methods by conducting a Friedman test [11] at the 95% interval to establish whether there was a statistically significant difference between the mean performance metrics obtained by the algorithms. The *null* hypothesis was that the performance of the algorithms were equivalent i.e. had the same average rank. If the *null* hypothesis was rejected a Nemenyi post-hoc test [20] was run to determine of the corresponding average ranks differ by at least the Critical Distance (CD).

Table 1. Breakdown of the three data sets used in the experiments

Data set	Num. Instances	Num. Inputs	Num. Classes	Imbalance ratio
Iris	150	4	3	50:50:50
Wine	178	13	3	59:71:48
Breast Cancer	569	30	2	357:212
Ecoli	336	7	8	143:76:2:2:35:20:5:26

4 Results and Discussion

As a baseline we used the Iris data set which has only four explanatory variables and three classes. In addition, it exhibits no class imbalance as there are exactly the same number of data instances per class. Furthermore, the size of the data set is the smallest out of all the data sets used in the experiments.

First, Table 2 presents the results when the algorithms were used to generate ensembles to classify the Iris data set. The test performance of *ABSECoS* performed better than the other ensemble learning algorithms in terms of Precision, Recall and F1-Score. *RF*, *AdaBoost*, and *XGBoost* ensembles all achieved similar performance. Additionally, the Recall and F1-Score Standard Deviations (SD) were smaller for the *ABSECoS* ensemble. Only the Precision SD for the *RF* model was comparatively smaller. The average size of a SECoS learner in the *ABSECoS* ensemble was 7 ± 1 rule nodes.

The results of the Friedman test were $Q = 10.667$ and $p = 0.0306$. Since the p value was less than 0.05 we can reject the *null* hypothesis. The only ensemble method that exceeded the CD of 3.522 in the Nemenyi post-hoc test was the *ABSECoS* algorithm (CD=5.0).

Table 2. Test results of 10-fold cross validation on the Iris data set

Classifier	Precision	Recall	F1-Score
RF	0.9644 ± 0.0418	0.9600 ± 0.0442	0.9597 ± 0.0444
AdaBoost	0.9622 ± 0.0524	0.9600 ± 0.0533	0.9599 ± 0.0534
XGBoost	0.9622 ± 0.0524	0.9600 ± 0.0533	0.9599 ± 0.0534
SECoS	0.9474 ± 0.0595	0.9267 ± 0.0814	0.9222 ± 0.0807
ABSECoS	$\mathbf{0.9756 \pm 0.0447}$	$\mathbf{0.9867 \pm 0.0233}$	$\mathbf{0.9732 \pm 0.0467}$

Compared with the Iris data set the Wine data set is larger in both size (178 instances versus 150 instances) and number of explanatory variables (13 versus 4). Furthermore, there is class imbalance present in this data set albeit quite small. Hence, Table 3 presents the test results when the algorithms were used to generate ensembles to classify the Wine data set. The performance of *ABSECoS* was better than the other ensemble learning algorithms in terms of Recall and

F1-Score. The SD for both these metrics was also lower than the SD for the other models. However, average Precision was better for the *XGBoost* results in terms of both the average Precision value and SD. The average size of a SECoS learner in the *ABSECoS* ensemble was 6 ± 1 rule nodes.

This time the results of the Friedman test were Q = 11.467 and p = 0.0218. Again, the p value was less than 0.05 so we can reject the *null* hypothesis and through the Nemenyi post-hoc test only the ranking of the *ABSECoS* ensemble (CD=4.67) and the *XGBoost* ensemble (CD=4.33) exceeded the CD of 3.522.

Table 3. Test results of 10-fold cross validation on the Wine data set

Classifier	*Precision*	*Recall*	*F1-Score*
RF	0.9903 ± 0.0194	0.9886 ± 0.0229	0.9886 ± 0.0228
AdaBoost	0.9804 ± 0.0240	0.9775 ± 0.0276	0.9771 ± 0.0280
XGBoost	**0.9950 ± 0.0151**	0.9941 ± 0.0176	0.9941 ± 0.0176
SECoS	0.9449 ± 0.0389	0.9305 ± 0.0668	0.9166 ± 0.0697
ABSECoS	0.9944 ± 0.0176	**0.9974 ± 0.0081**	**0.9944 ± 0.0177**

Even though the Wisconsin Breast Cancer data set has only two classes it exhibits high class imbalance ratio so is a good test for an ensemble generation algorithm as not only does it need to generate a high-performing model, but it also has to deal with the issue of class imbalance.

In this case, Table 4 presents the test results when the ensemble algorithms were used to generate models to classify this data set. This time the *AdaBoost* model performance was better across the three Precision, Recall, and F1-Score metrics. The next best ensemble was the *XGBoost* model and the *ABSECoS* ensemble was third-best model. The *ABSECoS* model did obtain smaller SD values for Precision, Recall, and F1-Score metrics than the *XGBoost* model. For this ensemble, the average size of a SECoS learner in the *ABSECoS* ensemble was 5 ± 2 rule nodes.

Table 4. Test results of 10-fold cross validation on the Wisconsin Breast Cancer data set

Classifier	*Precision*	*Recall*	*F1-Score*
RF	0.9490 ± 0.0254	0.9454 ± 0.0268	0.9455 ± 0.0269
AdaBoost	**0.9787 ± 0.0176**	**0.9771 ± 0.0194**	**0.9770 ± 0.0195**
XGBoost	0.9772 ± 0.0192	0.9753 ± 0.0212	0.9752 ± 0.0214
SECoS	0.8950 ± 0.0402	0.8892 ± 0.0342	0.8891 ± 0.0356
ABSECoS	0.9688 ± 0.0176	0.9685 ± 0.0146	0.9681 ± 0.0154

For this experiment, the results of the Friedman test were Q = 11.467 and p = 0.0173. As the p value was less than 0.05 we can reject the *null* hypothesis.

Additionally, only the ranking of the *ABSECoS* ensemble (CD=5.00) and the *XGBoost* ensemble (CD=4.00) exceeded the CD of 3.522 in the Nemenyi post-hoc test.

The last set of results reveal some interesting aspects of the ensemble algorithms. This outcome is largely due to the composition of the Ecoli data set since it has seven explanatory variables and eight classes with an extremely high class imbalance ratio as presented in Table 1.

Table 5. Test results of 10-fold cross validation on the Ecoli data set

Classifier	Precision	Recall	F1-Score
RF	**0.8743 ± 0.0410**	0.8455 ± 0.0541	0.8413 ± 0.0560
AdaBoost	0.8071 ± 0.0508	0.7860 ± 0.0558	0.7785 ± 0.0490
XGBoost	0.8584 ± 0.0514	0.8664 ± 0.0532	0.8480 ± 0.0599
SECoS	0.7445 ± 0.0204	0.7265 ± 0.0825	0.6834 ± 0.0209
ABSECoS	0.8591 ± 0.1068	**0.9675 ± 0.0145**	**0.8648 ± 0.0606**

Table 5 presents the results when the algorithms were used to generate ensembles to classify the Ecoli data set. The *ABSECoS* performance was better than the other ensemble learning algorithms in terms of Recall and F1-Score. However, with respect to the *ABSECoS* the Precision SD was much larger than the equivalent Precision SD values for the other models. The *XGBoost* ensemble had the second-best performance metrics and the *AdaBoost* ensemble had the worst ensemble performance. Of note is that average size of a SECoS learner in the *ABSECoS* ensemble was 26 ± 5 rule nodes. Compared with the average number of rules generated by the *ABSECoS* models on the other three data sets this result indicates that a much larger set of rule nodes must be generated to better fit each SECoS learner in the ensemble to this data set.

The results of the Friedman test were $Q = 10.4$ and $p = 0.0342$. As the p value was less than 0.05 we can reject the *null* hypothesis. This time the *ABSECoS* ensemble (CD=4.67), *XGBoost* ensemble (CD=3.67) and *RF* ensemble (CD=3.67) was greater than the CD of 3.522 in the Nemenyi post-hoc test.

4.1   Statistical Comparison of Overall Ensemble Results

Precision, Recall, and F1-Score metrics from the evaluation of *ABSECoS* suggested that it performed well when compared with *XGBoost* and other ensemble-based methods on the same data set. As an additional test we also conducted a Friedman test at the 95% interval to establish whether there was a statistically significant difference between the mean F1-Scores obtained by the ensembles across all the data sets. The *null* hypothesis is that there is no difference between these mean F1-Score values.

The results of the Friedman test reported values of Q = 11.80 and p = 0.0189. Since the p value was less than 0.05 we can reject the *null* hypothesis that the mean F1-Scores for the ensembles were equivalent.

In order to determine the difference in performance of the ensemble methods we also conducted a Nemenyi post-hoc test. The results of running this test are depicted in Fig. 1. Here the ranking of both *ABSECoS* and *XGBoost* exceeded the CD of 3.050. This result would suggest that both *ABSECoS* and *XGBoost* are the two better performing ensemble methods over all the experiments.

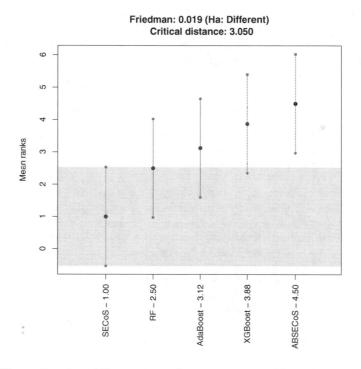

Fig. 1. Results of Nemenyi post-hoc test on ensemble performance

5 Conclusion

In this paper we have proposed the ABSECoS framework which was inspired by the seminal AdaBoost algorithm to enable ECoS-based learners to leverage a boosting framework to improve learning from a data set especially when the data set exhibits class imbalance. Although the AdaBoost framework is over twenty years old we have shown that a variant of it can be adopted for a simple ECoS-based learner in combination with a minority class-based data sampling scheme to overcome the issue with class imbalance in a data set. Experiments involving four benchmark data sets suggest *ABSECoS* performs comparably well even with the state-of-the-art *XGBoost* framework.

However, there are a number of weaknesses with the framework that are required to be addressed as future work. First, the SECoS learning algorithm needs to be adapted to make more use of the weights associated with the data instances. Second, the values of the learning parameters were sensitive to the choice of the number of learners in the *ABSECoS* ensemble. Therefore, we will investigate how altering the number of learners in an *ABSECoS* ensemble affects its performance. In turn this decision will influence what values are selected for the SECoS learning parameters to maintain acceptable ensemble performance. Fourth, the number of learners in each ensemble was set to 50. What effect an increase in the number of learners of an *ABSECoS* on overall ensemble performance has yet to be investigated. Finally, we will generalize the ABSECoS framework to be applied to regression problems and compare this extended framework with similar state-of-the-art ensemble frameworks.

References

1. Bergstra, J., Komer, B., Eliasmith, C., Yamins, D., Cox, D.D.: Hyperopt: a Python library for model selection and hyperparameter optimization. Comput. Sci. Discov. **8**(1), 014008 (2015)
2. Breiman, L.: Random Forests. Mach. Learn. **45**(1), 5–32 (2001)
3. Chawla, N.V., Bowyer, K.W., Hall, L.O., Kegelmeyer, W.P.: Smote: synthetic minority over-sampling technique. J. Artif. Intell. Res. **16**(1), 321–357 (2002)
4. Chen, T., Guestrin, C.: XGBoost: A scalable tree boosting system. In: Proc. 22nd ACM SIGKDD International Conference on Knowledge Discovery and Data Mining. pp. 785–794. KDD'16, ACM, New York, NY, USA (2016)
5. Dhiman, B., Kumar, Y., Kumar, M.: Fruit quality evaluation using machine learning techniques: review, motivation and future perspectives. Multimedia Tools and Applications , **81**, 16255–16277 (2022)
6. Dong, X., Yu, Z., Cao, W., Shi, Y., Ma, Q.: A survey on ensemble learning. Front. Comput. Sci. **14**(2), 241–258 (2019). https://doi.org/10.1007/s11704-019-8208-z
7. Fisher, R.A.: The use of multiple measurements in taxonomic problems. Annals of Eugenics 7(II), **7**, 179–188 (1936)
8. Forina, M., Lanteri, S., Armanino, C., Casolino, C., Casale, M., Oliveri, P.: PARVUS - An Extendible Package for Data Exploration, Classification and Correlation. Institute of Pharmaceutical and Food Analysis and Technologies, Tech. rep., ip. Chimica e Tecnologie Farmaceutiche ed Alimentari, Universita' di Genova (2008)
9. Freund, Y., Schapire, R.E.: Experiments with a new boosting algorithm. In: Proceedings of the Thirteenth International Conference In Machine Learning, pp. 18–156. IEEE Press (1996)
10. Freund, Y., Schapire, R.E.: A decision-theoretic generalization of on-line learning and an application to boosting. J. Comput. Syst. Sci. **55**(1), 119–139 (1997)
11. Friedman, M.: The use of ranks to avoid the assumption of normality implicit in the analysis of variance. J. Am. Stat. Assoc. **32**(200), 675–701 (1937)
12. Horton, P., Nakai, K.: A Probablistic Classification System for Predicting the Cellular Localization Sites of Proteins. In: 1996 International Conference on Intelligent Systems in Microbiology. vol. 4, pp. 109–115 (1996)

13. Kasabov, N.: ECOS: A Framework For Evolving Connectionist Systems and the ECO Learning Paradigm. In: Proceedings of the 1998 Conference on Neural Information Processing and Intelligent Information Systems, (ICONIP'1998), pp. 1232–1235. Ohmsha Ltd: Tokyo, Japan (1998)
14. Kasabov, N.: Evolving Connectionist and Fuzzy-Connectionist Systems for On-line Adaptive Decision Making and Control. In: Roy, R., Furuhashi, T., Chawdhry, P.K. (eds) Advances in Soft Computing. Springer, London (1999). https://doi.org/10.1007/978-1-4471-0819-1_3
15. Kasabov, N.: The ECOS framework and the eco learning method for evolving connectionist systems. J. Adv. Comput. Intell. **2**(6), 195–202 (1998)
16. Kasabov, N.: Evolving Fuzzy Neural Networks for Supervised/Unsupervised On-Line, Knowledge-Based Learning. In: IEEE Transactions on Systems, Man and Cybernetics, Part B: Cybernetics, vol. 31, no. 6, pp. 902–918 (2001)
17. Kasabov, N., Woodford, B.: Rule insertion and rule extraction from evolving fuzzy neural networks: algorithms and applications for building adaptive, intelligent expert systems. In: Proceedings of the 1999 IEEE Fuzzy Systems Conference. vol. 3, pp. 1406–1411. The IEEE, Kyunghee Printing Co (1999)
18. Minku, F.L., Ludermir, T.B.: EFuNN Ensembles Construction Using CONE with Multi-objective GA. In: 2006 Ninth Brazilian Symposium on Neural Networks (SBRN'06), pp. 48–53 (2006)
19. Minku, F.L., Ludermir, T.B.: EFuNNs Ensembles Construction Using a Clustering Method and a Coevolutionary Genetic Algorithm. In: 2006 IEEE International Conference on Evolutionary Computation, pp. 1399–1406 (2006)
20. Nemenyi, P.: Distribution-free Multiple Comparisons. Ph.D. thesis, Princeton University (1963)
21. Pedregosa, F., et al.: Scikit-learn: machine learning in Python. J. Mach. Learn. Res. **12**, 2825–2830 (2011)
22. Sagi, O., Rokach, L.: Ensemble learning: a survey. WIREs Data Mining Knowl. Dis. **8**(4), 241–258 (2018)
23. Seiffert, C., Khoshgoftaar, T.M., Van Hulse, J., Napolitano, A.: RUSBoost: A Hybrid Approach to Alleviating Class Imbalance. In: IEEE Transactions on Systems, Man, and Cybernetics - Part A: Systems and Humans, vol. 40, no. 1, pp. 185–197 (2010)
24. Shi, H., Lv, X.: The Naïve Bayesian Classifier Learning Algorithm Based on Adaboost and Parameter Expectations. In: 2010 Third International Joint Conference on Computational Science and Optimization. vol. 2, pp. 377–381 (2010)
25. Song, Q., Kasabov, N.: DENFIS: dynamic evolving neural-fuzzy inference system and its application for time-series prediction. IEEE Trans. Fuzzy Syst. **10**(2), 144–154 (2001)
26. Tharwat, A.: Classification assessment methods. Appl. Comput. Inf. **17**(1), 168–192 (2021)
27. Watts, M.: A Decade of Kasabov's Evolving Connectionist Systems: A Review. IEEE Trans. Syst. Man Cybern - Part C: Appl. Rev. **39**(6), 684–693 (2009)
28. Wolberg, W., Mangasarian, O.: Multisurface method of pattern separation for medical diagnosis applied to breast cytology. Proc. Nat. Acad. Sci. **87**, 9193–9196 (1990)
29. Woodford, B.J., Kasabov, N.K.: Ensembles of EFuNNs: an architecture for a multi module classifier. In: The proceedings of FUZZ-IEEE'2001. In: The 10th IEEE International Conference on Fuzzy Systems. vol. III, pp. 1573–1576. IEEE (2001)

Machine Learning Inspired Fault Detection of Dynamical Networks

Eugene Tan[1]([envelope]) [iD], Débora C. Corrêa[2,3] [iD], Thomas Stemler[1] [iD],
and Michael Small[1,3] [iD]

[1] Complex Systems Group, Department of Mathematics and Statistics,
The University of Western Australia, Crawley, WA 6009, Australia
eugene.tan@uwa.edu.au
[2] Department of Computer Science and Software Engineering,
The University of Western Australia, Crawley, WA 6009, Australia
[3] ARC Industrial Transformation Training Centre (Transforming Maintenance
through Data Science), The University of Western Australia,
Crawley, WA 6009, Australia

Abstract. Dynamical networks are a framework commonly used to model large networks of interacting time-varying components such as power grids and epidemic disease networks. The connectivity structure of dynamical networks play a key role in enabling many interesting behaviours such as synchronisation and chimeras. However, dynamical networks can also be vulnerable to network attack, where the connectivity structure is externally altered. This can cause sudden failure and loss of stability in the network. The ability to detect these network attacks is useful in troubleshooting and preventing system failure. Recently, a back-propagation regression method inspired by RNN training algorithms was proposed to infer both local node dynamics and connectivity structure from measured node signals. This paper explores the application of back-propagation regression for fault detection in dynamical networks. We construct separate models for local dynamics and coupling structure to perform short-term freerun predictions. Due to the separation of models, abnormal increases in prediction error can be attributed to changes in the network structure. Automatic detection is achieved by comparing prediction error statistics across two windows that span a period before and after a network attack. This method is tested on a simulated dynamical network of chaotic Lorenz oscillators undergoing gradual edge corruption via three different processes: edge swapping, moving and deletion. We demonstrate that the correlation between increased prediction error and the occurrence of edge corruption can be used to reliably detect both the onset and approximate location of the attack within the network.

Keywords: Complex systems · Dynamical networks · Machine learning · Recurrent neural networks · Fault detection

H. Aziz et al. (Eds.): AI 2022, LNAI 13728, pp. 470–483, 2022.
https://doi.org/10.1007/978-3-031-22695-3_33

1 Introduction

Many real-world systems can be described as a collection of identical units that communicate with their neighbours in a connected network over time. Some examples include neuron networks [9], power grids [30], epidemic spread [8], and cardiac arrhythmia [19]. Interactions with neighbours can influence individual behaviour, resulting in a variety of interesting local and global dynamics such as chimera states and synchronisation [1, 2, 38]. Systems with such a structure can be broadly described under the framework of dynamical networks.

An extension of dynamical networks is temporal dynamical networks where the connectivity structure also varies over time [15]. This is unlike its basic counterpart where the structure and individual unit behaviour of dynamical networks are static with the time varying output of signals being the main component of interest. Temporal dynamical networks have been used to describe a wide range of systems such as disease spread [11], power transmission [20] and functional brain networks [26]. The temporal variation in a dynamical network's connectivity structure can lead to significant changes in its behaviour, and in the worse case, total system failure [27, 32]. An example that exhibits such behaviour is that of edge deletion in vulnerable networks. Real world examples of network topology influence on stability include breakdown of power grids [34] and communication networks [33]. In cases where there network corruption is gradual, the early detection and location of structural changes can be valuable in guiding the diagnosis and repair of the network.

Recently, a backpropagation regression method inspired by RNN training algorithms was proposed to disentangle node signal observations and separately recover local node dynamics and network connectivity structure [25]. This method has only since been applied to the the context of constructing forecast models of dynamical networks.

This paper extends the backpropagation regression approach and investigates how the constructed models with separated local and coupling effects recovered via backpropagation regression may be used to detect structural changes in temporal dynamical networks. Specifically, we propose a method of using backpropagation regression and prediction error for the purposes of concept drift and edge corruption detection in dynamical networks. We use backpropagation regression to construct predictive models for a simulated network of chaotic Lorenz oscillators. The structure of these networks are then procedurally corrupted via three processes, edge deletion, edge moving and edge swapping. Statistical analysis of the prediction error is shown to be useful in both detecting changes in the connectivity structure and identifying the affected nodes.

2 Background

2.1 Dynamical Networks

Dynamical networks can be defined as graph $G = (f, g, C)$ with three main components: local dynamics f, coupling dynamics g and connectivity structure

C. Similar to a recurrent neural network, the node states $\mathbf{x}_i(t)$ of G vary over time according to a combination of locally defined dynamics and input from connected neighbour states. This is given by Eq. 1,

$$\dot{\mathbf{x}}_i(t) = f(\mathbf{x}_i(t)) + \sum_{i \neq j} C_{i,j} g(\mathbf{x}_i(t), \mathbf{x}_j(t)). \tag{1}$$

Analysis and prediction of dynamical networks is difficult due to the complex interplay between local dynamics and coupling effects. Applying the dynamical networks framework to real world systems requires prior information pertaining to some if not all of the components (f, g, C). In this paper, we focus on temporal dynamical networks where connectivity structure $C_{i,j}$ varies over time.

The inference of local dynamics from observed time series has been well studied. Modern approaches often utilise machine learning methods such as reservoir computers [10], recurrent neural networks [36] and radial basis networks with minimum description length [22] to replicate observed system dynamics. Many of these methods rely on the dynamical guarantees of Taken's embedding theorem [24].

The inference of connectivity structure of dynamical networks is a somewhat more difficult task and remains an open problem. Methods for inferring node connections generally rely on a direct approach [21], perturbation methods applied to a proxy system [3,4,23], or statistical or causality based arguments [12,17,29]. A more comprehensive review is provided in [25].

Recently, methods have been proposed to simultaneously identify both local dynamics and connectivity structure purely from node signals [7]. Eroglu et al. attempts to solve this problem through the construction of an 'effective network'. This network acts as an approximate proxy for the real system and can be used to infer network links and dynamical transitions. However, effective networks rely on the assumption that interactions in the network are weak and sparse, with the presence of few high degree nodes.

An alternative method by [25] explored the approach of treating dynamical networks as a recurrent neural network and likened the task of inferring of local dynamics and connectivity structure to the backpropagation and training of recurrent neural networks with an unknown activation function. This paper will focus on the application of this method in its analyses.

In real world systems, it is also possible for a network's connectivity structure to vary over time. Within the framework of dynamical networks, this results in the connectivity structure being expressed as a function of time $C(t)$. The evolution function of node states can then be rewritten as follows by replacing $C_{i,j}$ with $C_{i,j}(t)$,

$$\dot{\mathbf{x}}_i(t) = f(\mathbf{x}_i(t)) + \sum_{i \neq j} C_{i,j}(t) g(\mathbf{x}_i(t), \mathbf{x}_j(t)). \tag{2}$$

This variation of connectivity structure can also affect the overall behaviour of the network. For example, periodic edge deletion can result in disconnected nodes, which can cause cascading failure [32]. For cases where corruption occurs gradually, it has been found that network attacks can cause hysteresis and sudden changes in dynamics [14]. Therefore, the ability to detect and locate structural changes in the network would be useful in the early prevention of network failure.

Solutions of detecting faults within real-world networks vary depending on the network type. Descriptor systems are used as a framework for cyber-physical networks [18]. Pasqualetti et al. discusses the detectability and identifiability conditions for various attack types. For transmission lines, fault detection methods can be categorised into three main groups: (i) impedance and frequent methods, (ii) detection of mobile waves generated network fault points and (iii) artificial neural network (ANN) based methods [6]. Extensive work by [16] reviews the various methods of detecting faults in of wireless sensor networks.

2.2 Backpropagation Regression

A recurrent neural network is often structured with a well defined activation function σ for its forward evolution with linear diffusive coupling between neighbouring nodes. Similarly, a dynamical network can also be viewed within the framework of a recurrent neural network where the the nonlinear activation function and node couplings are given by f and g corresponding to the local dynamics and node coupling effects respectively. Training a recurrent neural network is then analogous to the simultaneous inference of both f and g. The backpropagation regression method presented by [25] approaches the problem of identifying local dynamics and network connectivity structure in 3 parts using the recurrent neural network training framework. These are initialisation, backpropagation and decoupling (see Fig. 1).

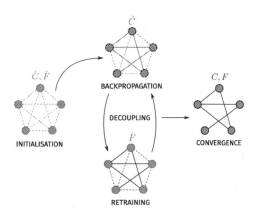

Fig. 1. Schematic of the backpropagation regression approach with three phases. Initialisation calculates an approximate local model. Backpropagation uses the local model to regress the network weights. Regressed weights are used to decouple the input signal and retrain the local model. Figure taken from [25]

During initialisation, the full observed node states are used to estimate the local vector field. A mean field approach is used to approximate the local vector field as an average across a collection of K nearby neighbours in state space,

$$\hat{\mathbf{x}}(t) = \frac{1}{K} \sum_{i=1}^{K} \frac{\dot{\mathbf{x}}^{(i)}(t + \delta t) - \dot{\mathbf{x}}^{(i)}(t)}{\delta t}. \tag{3}$$

The estimated local dynamics $\hat{f}$ is defined as a mapping $\hat{f} : \mathbf{x}(t) \to \hat{\mathbf{x}}(t)$. This mapping can be approximated by a trained simple feedforward network. The estimated coupling adjacency matrix $\hat{C}$ is also initialised in this stage with randomly selected values.

In the backpropagation stage, the estimated local model is used to regress coupling weights and improve the estimate of $\hat{C}$. Drawing inspiration from the backpropagation through time (BPTT) algorithm used to train recurrent neural networks, the backpropagation stage treats the dynamical network as an RNN, with weight training corresponding to regression of coupling weights. A forward pass consists of freerun predictions using the estimates $\hat{f}$ and $\hat{C}$ with initial values randomly selected from the observed trajectories.

Similar to BPTT, the error between the predicted values $\hat{x}_i$ and true observed trajectory x_i is calculated,

$$\mathcal{L} = \sum_{t=t_0}^{t=t_n} \sum_{i=1}^{\infty} E_i(t) = \sum_{t=t_0}^{t=t_n} \sum_{i=1}^{\infty} (\hat{x}_i(t) - x_i(t))^2. \tag{4}$$

The loss gradient is then calculated with respect to coupling weights $\hat{C}$, see Eq. 5,

$$\frac{d\mathcal{L}}{d\hat{C}} = \sum_{t=t_0}^{t=t_n} \sum_{i=1}^{\infty} \frac{\partial E_i(t)}{\partial \hat{C}} = \sum_{t=t_0}^{t=t_n} \sum_{i=1}^{\infty} 2(\hat{x}_i(t) - x_i(t)) \frac{\partial \hat{x}_i(t)}{\partial \hat{C}}. \tag{5}$$

Unlike the training of a recurrent neural network where the activation function for the forward evolution of node states is well-defined, the forward evolution functions $\hat{f}$ and g are not known. As a result, the partial derivatives required to estimate the loss gradient must be calculated recursively [25]. It is assumed that coupling only occurs in one component. The calculated loss gradients are then propagated backwards to improve the estimate of $\hat{C}$ (see Fig. 2)

The final stage of decoupling aims to improve the estimate of the local model dynamics $\hat{f}$. Because the initial local model only uses a mean field approach to counteract the effects of coupling, there will be an inherent error in the estimated vector field, particularly for the components in the coupled dimension. To improve the model $\hat{f}$ of the local dynamics, the previously regressed coupling weights $\hat{C}$ are used to decouple the observed signal, given by Eq. 6,

$$\hat{f}(x_i(t_n)) \approx \frac{x_i(t_{n+1}) - x_i(t_n) - \delta t \sum_{i \neq j} \hat{c}_{ij} g(x_i(t_n), x_j(t_n))}{\delta t}. \tag{6}$$

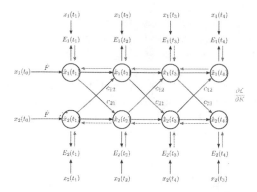

Fig. 2. Information flow during the forward and backward pass for a 2 node dynamical network. Terms with hats correspond to model predicted values.

Doing so partially filters the effect of coupling from the time series and yields a slightly improved model of the local dynamics $\hat{f}$, which is in turn used to improve the regressed estimates of $\hat{C}$. The two stages of backpropagation and regression are then repeatedly alternated until convergence is achieved and the estimated components $(\hat{f}, \hat{C}) \rightarrow (f, C)$.

Given that the main object of interest is presented as a network, there are multiple perceived similarities between backpropagation regression on dynamical networks and graph convolutional networks (GCN) methods [37]. GCNs aim to encode the graph structure of an input into the convolutional neural network framework commonly used for image processing. However, we argue that backpropagation regression tackles a different problem. In most GCN applications, information of the network structure is already known a priori and is used to calculate convolutions between neighbour signals. This knowledge is not necessarily available when inferring dynamical networks. We note however, that GCNs may be extended to tackling dynamical networks by combining with other network inference methods aimed at recovering the network connectivity structure [5, 28]. The adjacency network recovered via the latter methods may be used to construct a feedforward GCN for the purpose of n-step prediction of dynamical networks. However, this does not achieve the recovery of local node dynamics without coupling effects, which is a feature of backpropagation regression.

2.3   Concept Drift and Edge Prediction

A dynamical network with N identical nodes with d dimensional states can also be viewed as a unified dynamical system defined in $N \times d$ dimensions. Hence, learning the local and coupling dynamics, as in backpropagation regression, is equivalent to learning the combined $N \times d$-dimensional system. Therefore, changes to the connectivity structure of this network will cause a shift in the collective dynamics of the system. Concept drift is inevitable when there is temporal variation in the network's connectivity structure. Depending on the severity of the variation, the previously constructed model will become increasingly unrealiable for accurate predictions and show increases in prediction error.

This behavior though undesirable, can be used to detect network changes. Firstly, the increase in prediction error can inform and trigger a retraining of the previously constructed model. This is useful for maintaining model accuracy for cases where temporal variations in network structure are expected and not detrimental to its function. For operation critical systems where structural variations in the network are undesirable, the increased model prediction error can be used for the early detection of system failure [35]. The separation of dynamics into local and coupling effects ensures that prediction errors caused by concept drift can be entirely attributed to changes in connectivity structure.

Consequently, because coupling effects must propagate over time, changes in the network structure will first result in localised errors around nodes with altered connections. This can be used to locate failure regions in the observed dynamical network and guide the actions needed to repair the system.

3   Method

In this paper, we present a method for applying backpropagation regression to the task of concept drift and edge corruption in dynamical networks. The proposed fault detection method was tested on simulated dynamical networks of chaotic Lorenz oscillators [13]. The main aim of the analysis is twofold. Firstly, to detect the occurrence of concept drift due to structural variations in a dynamical network. Secondly, to locate these structural variations within the network. An overview of the method is provided in Fig. 3.

For testing, a dynamical network of $N = 16$ chaotic oscillators was simulated numerically with 4th order Runge-Kutta integration and timestep $dt = 0.002$ and subsampled to an effective timestep of $dt = 0.02$. The network structure was generated as an undirected random graph with edge probability $p = \log N/N = 0.173$. The dynamical networks were simulated for 25000 time steps during which the local dynamics and coupling matrix were kept constant. An additional 2000 step washout period was included to accommodate for any transient dynamics.

Backpropagation regression was used to construct a model of the local dynamics and connectivity structure. A feedforward network with 1 hidden layer and 128 nodes was used to learn the local node dynamics. Backpropagation regression was run using 10-step freerun predictions.

To simulate temporal variation of the dynamical network, simulation of the system was extended for an additional 12500 time steps with n edge corruption operations being applied at regular intervals every 2500 steps (see Fig. 4). Three types of corruption operations were tested: edge deletion, edge moving and edge swapping. In decreasing order, each edge corruption operation preserves a different amount of the network topology. Edge swapping involves the swapping of two vertices between any pair of randomly selected edges. This process preserves both the edge density and degree distribution of the network. Edge moving relocates n edges to randomly selected locations and preserves only edge density.

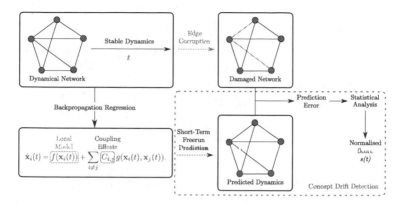

Fig. 3. Schematic overview of the proposed method of using backpropagation regression to detect concept drift and edge corruption. Backpropagation regression is used to identify the local dynamics and models and coupling weights separately. These are then used for short-term freerun predictions. Prediction errors are monitored and a normalised score $s(t)$ is calculated to detect the onset of edge corruption. Corrupted edges shown in red. (Color figure online)

Finally, edge deletion does not preserve any network measure and consists of setting n randomly selected edges to zero.

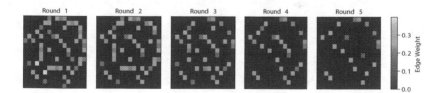

Fig. 4. Weighted connectivity matrix of the system with gradually applied edge deletion. Each round (after round 1) removes 5 randomly selected edges.

The prediction accuracy of the model was tracked by concurrently running short-term 10 step freerun predictions using initial conditions taken from previous measured states. The prediction error ϵ_i of each node was calculated using the value of the final predicted step compared to the real system trajectory,

$$\epsilon_i(t) = ||\hat{\mathbf{x}}_i(t) - \mathbf{x}_i(t)||. \tag{7}$$

The error profile for each node $\epsilon_i(t)$ was smoothed by calculating a 200 step moving average. This was done to reduce the effects of natural fluctuations in RMSE due to the local dynamics on the chaotic attractor. The collection of node error profiles $\epsilon_i(t)$ can also be summarised into a system RMSE $E(t)$ by taking the average across all nodes as given in Eq. 8,

$$E(t) = \frac{1}{N} \sum_{i=1}^{N} \epsilon_i(t). \tag{8}$$

The degree of corruption $\theta(t)$ in the connectivity structure of the dynamical network can be quantified by calculating the matrix norm of the coupling matrix,

$$\theta(t) = \frac{||C'(t) - C(t)||}{||C(t)||}, \tag{9}$$

where $C'(t)$ corresponds to the corrupted connectivity matrix due to operations on edges (i.e. deletion, moving, swapping). A similar formulation for individual nodes is calculated using the norm of the difference in the weighted degree. This quantity is not normalised to account for cases where a disjoint node becomes connected due to edge corruption,

$$\theta_i(t) = ||C'_i(t) - C_i(t)||. \tag{10}$$

To detect edge corruption from RMSE at a given time t, statistics are compared between two segments of time series correponding to recent history $(\epsilon_i(t - w_t), \epsilon_i(t))$ and a past reference $(\epsilon_i(t - w_t - \tau - w_c), \epsilon_i(t - w_t - \tau))$, where w_c and w_t correspond to the control and test window on the non-smoothed $\epsilon_i(t)$ respectively. These two windows are separated by a time lag τ during which a transition may occur. Basic edge detection is done using a normalised score $s(t)$ calculated at each time step that compares the RMSE time averages across each pair of time series segments as given in Eq. 11,

$$s(t) = \frac{\mu_t - \mu_c}{\sigma_c}, \tag{11}$$

where μ_t, μ_c and σ_c are the mean and standard deviation of the test and control segments of the time series. The presence of edge corruption will increase the prediction RMSE, which in turn results in a temporary increase in the calculated score $s(t)$. Outliers in $s(t)$ then correspond to the occurrence of potential edge corruption. Medians and maximum absolute deviation (MAD) was used for outlier detection in our analysis due to its simplicity and robustness to outliers.

4　Results

The system-wide RMSE was calculated for the simulated Lorenz network with 5 rounds of edge corruption, each round altering 5 edges. A cutoff of $k = 4.5$ for scores after being normalised by MAD was used to determine the present of edge corruption. It was found that the RMSE error correlated with the onset of overall structural changes for earlier rounds of edge corruption (see Fig. 5).

The second task of identifying the location of deleted edges can be achieved by tracking the prediction error for individual nodes. Similar to the system-wide RMSE, node prediction error was also found to correlate closely with the onset of individual node corruption (see Fig. 6). This result is arguably more

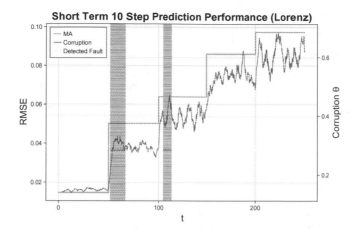

Fig. 5. Calculated RMSE over the duration of regular edge deletion averaged across all nodes of the whole network. Figure shows detected edge corruption based on irregularities (blue), and each stage of corruption $\theta(t)$ (grey). MA is the moving average of model prediction errors E(t). (Color figure online)

informative than the overall network RMSE as it provides more information onto the localised area in the network where edge corruption has occurred. Successive rounds of edge corruption is noticeably less obvious from the calculated RMSE. This is likely due to the accumulation of errors between each subsequent round, which eventually saturate across the whole range of the measured signal Fig. 7.

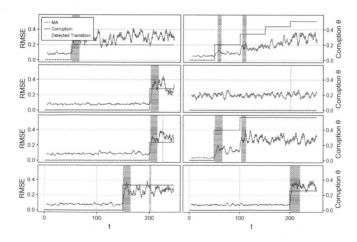

Fig. 6. Calculated individual node prediction error over the duration of regular edge deletion for the first 8 nodes of the network. The detected edge corruption based on irregularities in the prediction error are given by blue. Each stage of corruption $\theta(t)$ is given by the grey line. Detection of of corruption beyond the first instance is more inaccurate. MA is the moving average of model prediction errors $\epsilon_i(t)$. (Color figure online)

The success rate in detecting the onset of the first round of corruption in each node was also calculated for each edge corruption operation with varying levels. Sixteen iterations were calculated for each combination of edge corruption operation (i.e. swapping, moving, deletion) and number of corrupted nodes (2 to 20) and was averaged to get the final success rate. The proposed method was found to perform consistently for all operations with slightly lower performance for cases with more minor network perturbations ($n = 2$) (see Fig. 8). This can be attributed to the effects of these network changes not having a large enough impact on the dynamics of the network.

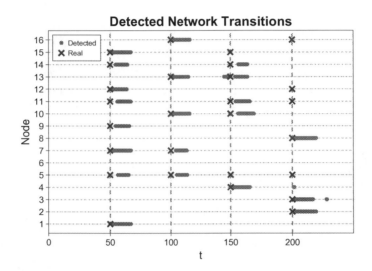

Fig. 7. Summarised plot of detected vs actual edge deletion operations from individual node RMSE for all nodes. Blue cross correspond to real instances of edge corruption. Detected edge corruptions are in red. (Color figure online)

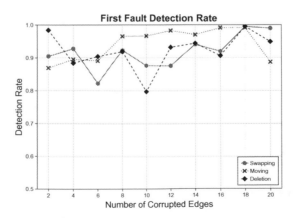

Fig. 8. Detection rates of the first applied fault in each node averaged over 16 dynamical networks for 3 different edge corruption operations.

5   Discussion and Conclusions

Dynamical networks are a useful framework for modelling networks of dynamical units that communicate with each other. Temporal dynamical networks extend this framework by accounting for temporal variations in connectivity structure. An interesting problem is the task of detecting and locating the occurrence of network attacks. This is of interest when modelling operation-critical systems such as power-grids where loss of communication between nodes can lead to sudden failure in the network [14, 27, 32–34].

This paper presents an application of backpropagation regression on dynamical networks to the problem of network fault detection. Several dynamical networks of chaotic Lorenz oscillators with randomly generated connectivity were simulated. A model of the local dynamics and connectivity structure was constructed from node signals using the method of backpropagation regression [25]. To simulate network attack, the connectivity structure underwent several rounds of edge corruption during the simulation of the dynamical network.

The detection of edge corruption was conducted by analysing the short-term prediction RMSE $\epsilon(t)$ of the originally constructed model in parallel with the simulated network. By choosing a model framework that separates local and coupling dynamics, any changes in the RMSE can be attributed to corruption in the network connectivity structure. It was found that models constructed using backpropagation regression was able to detect the early occurrence of network attack, with decreasing performance for subsequent rounds of edge corruption.

A similar approach focusing on the individual node prediction error $\epsilon_i(t)$ was also found to be sensitive in detecting edge corruption. By comparing against $\epsilon_i(t)$ profiles across multiple nodes, it is possible to identify and locate removed edges. However, similar to the system-wide RMSE, the individual node prediction error $\epsilon_i(t)$ is not as sensitive to successive rounds of attack. Open areas of research still remain, particularly in automating the detection of edge deletion from the prediction error profiles $\epsilon(t)$ and $\epsilon_i(t)$. This problem is ultimately one of automating detection of regime changes and is an active area of research in the field of signal processing [31].

Acknowledgements. E.T. is supported by a Robert and Maude Gledden Postgraduate Research Scholarship and Australian Government Research Training Program Scholarship at The University of Western Australia. M.S. and D.C.C. acknowledge the support of the Australian Research Council through the Centre for Transforming Maintenance through Data Science (grant number IC180100030), funded by the Australian Government.

References

1. Abrams, D.M., Strogatz, S.H.: Chimera states for coupled oscillators. Phys. Rev. Lett. **93**(17), 174102 (2004)
2. Andreev, A., Frolov, N., Pisarchik, A., Hramov, A.: Chimera state in complex networks of bistable Hodgkin-Huxley neurons. Phys. Rev. E **100**(2), 022224 (2019)

3. Banerjee, A., Hart, J.D., Roy, R., Ott, E.: Machine learning link inference of noisy delay-coupled networks with optoelectronic experimental tests. Phys. Rev. X **11**(3), 031014 (2021)
4. Banerjee, A., Pathak, J., Roy, R., Restrepo, J.G., Ott, E.: Using machine learning to assess short term causal dependence and infer network links. Chaos Interdisc. J. Nonlinear Sci. **29**(12), 121104 (2019)
5. Casadiego, J., Nitzan, M., Hallerberg, S., Timme, M.: Model-free inference of direct network interactions from nonlinear collective dynamics. Nat. Commun. **8**(1), 1–10 (2017)
6. Dashtdar, M., Dashti, R., Shaker, H.R.: Distribution network fault section identification and fault location using artificial neural network. In: 2018 5th International Conference on Electrical and Electronic Engineering (ICEEE), pp. 273–278. IEEE (2018)
7. Eroglu, D., Tanzi, M., van Strien, S., Pereira, T.: Revealing dynamics, communities, and criticality from data. Phys. Rev. X **10**(2), 021047 (2020)
8. Hota, A.R., Sneh, T., Gupta, K.: Impacts of game-theoretic activation on epidemic spread over dynamical networks. SIAM J. Control. Optim. **60**(2), S92–S118 (2021)
9. Izhikevich, E.M.: Dynamical systems in neuroscience. MIT press (2007)
10. Jaeger, H.: The "echo state" approach to analysing and training recurrent neural networks-with an erratum note. Bonn, Germany: German National Research Center for Information Technology Gesellschaft für Mathematik und Datenverarbeitung mbH (GMD) Technical Report 148, 13 (2001)
11. Kohar, V., Sinha, S.: Emergence of epidemics in rapidly varying networks. Chaos Solitons Fractals **54**, 127–134 (2013)
12. Kornilov, M., Sysoev, I., Astakhova, D., Kulminsky, D., Bezruchko, B., Ponomarenko, V.: Reconstruction of the coupling architecture in the ensembles of radioengineering oscillators by their signals using the methods of granger causality and partial directed coherence. Radiophys. Quantum Electron. **63**(7), 542–556 (2020)
13. Lorenz, E.N.: Deterministic nonperiodic flow. J. Atmos. Sci. **20**(2), 130–141 (1963)
14. Majdandzic, A., et al.: Multiple tipping points and optimal repairing in interacting networks. Nat. Commun. **7**(1), 1–10 (2016)
15. Masuda, N., Lambiotte, R.: A guide to temporal networks. World Scientific (2016)
16. Muhammed, T., Shaikh, R.A.: An analysis of fault detection strategies in wireless sensor networks. J. Netw. Comput. Appl. **78**, 267–287 (2017)
17. Napoletani, D., Sauer, T.D.: Reconstructing the topology of sparsely connected dynamical networks. Phys. Rev. E **77**(2), 026103 (2008)
18. Pasqualetti, F., Dörfler, F., Bullo, F.: Attack detection and identification in cyber-physical systems. IEEE Trans. Autom. Contr. **58**(11), 2715–2729 (2013)
19. Rosenblum, M., Frühwirth, M., Moser, M., Pikovsky, A.: Dynamical disentanglement in an analysis of oscillatory systems: an application to respiratory sinus arrhythmia. Phil. Trans. R. Soc. A **377**(2160), 20190045 (2019)
20. Sachtjen, M., Carreras, B., Lynch, V.: Disturbances in a power transmission system. Phys. Rev. E **61**(5), 4877 (2000)
21. Shandilya, S.G., Timme, M.: Inferring network topology from complex dynamics. New J. Phys. **13**(1), 013004 (2011)
22. Small, M., Tse, C.K.: Minimum description length neural networks for time series prediction. Phys. Rev. E **66**(6), 066701 (2002)
23. Stepaniants, G., Brunton, B.W., Kutz, J.N.: Inferring causal networks of dynamical systems through transient dynamics and perturbation. Phys. Rev. E **102**(4), 042309 (2020)

24. Takens, F.: Detecting strange attractors in turbulence. In: Rand, D., Young, L.-S. (eds.) Dynamical Systems and Turbulence, Warwick 1980. LNM, vol. 898, pp. 366–381. Springer, Heidelberg (1981). https://doi.org/10.1007/BFb0091924

25. Tan, E., Corrêa, D., Stemler, T., Small, M.: Backpropagation on dynamical networks. arXiv preprint arXiv:2207.03093 (2022)

26. Valencia, M., Martinerie, J., Dupont, S., Chavez, M.: Dynamic small-world behavior in functional brain networks unveiled by an event-related networks approach. Phys. Rev. E **77**(5), 050905 (2008)

27. Wang, J., Rong, L., Zhang, L., Zhang, Z.: Attack vulnerability of scale-free networks due to cascading failures. Physica A **387**(26), 6671–6678 (2008)

28. Wang, W.X., Lai, Y.C., Grebogi, C.: Data based identification and prediction of nonlinear and complex dynamical systems. Phys. Rep. **644**, 1–76 (2016)

29. Weistuch, C., Agozzino, L., Mujica-Parodi, L.R., Dill, K.A.: Inferring a network from dynamical signals at its nodes. PLoS Comput. Biol. **16**(11), e1008435 (2020)

30. Wu, B., Zhou, D., Fu, F., Luo, Q., Wang, L., Traulsen, A.: Evolution of cooperation on stochastic dynamical networks. PLoS ONE **5**(6), e11187 (2010)

31. Wu, H.S.: A survey of research on anomaly detection for time series. In: 2016 13th International Computer Conference on Wavelet Active Media Technology and Information Processing (ICCWAMTIP), pp. 426–431. IEEE (2016)

32. Xia, Y., Fan, J., Hill, D.: Cascading failure in Watts-Strogatz small-world networks. Physica A **389**(6), 1281–1285 (2010)

33. Xia, Y., Hill, D.J.: Attack vulnerability of complex communication networks. IEEE Trans. Circuits Syst. II Express Briefs **55**(1), 65–69 (2008)

34. Yang, L.X., Jiang, J.: Impacts of link addition and removal on synchronization of an elementary power network. Physica A **479**, 99–107 (2017)

35. Zenisek, J., Holzinger, F., Affenzeller, M.: Machine learning based concept drift detection for predictive maintenance. Comput. Ind. Eng. **137**, 106031 (2019)

36. Zhang, J.S., Xiao, X.C.: Predicting chaotic time series using recurrent neural network. Chin. Phys. Lett. **17**(2), 88 (2000)

37. Zhang, S., Tong, H., Xu, J., Maciejewski, R.: Graph convolutional networks: a comprehensive review. Comput. Soc. Netw. **6**(1), 1–23 (2019). https://doi.org/10.1186/s40649-019-0069-y

38. Zhu, Y., Zheng, Z., Yang, J.: Chimera states on complex networks. Phys. Rev. E **89**(2), 022914 (2014)

Medical AI

Multiclass Classification for GvHD Prognosis Prior to Allogeneic Stem Cell Transplantation

Md. Asif Bin Khaled$^{(\boxtimes)}$ [iD], Md. Junayed Hossain[iD], Saifur Rahman[iD], and Jannatul Ferdaus[iD]

Department of Computer Science and Engineering, Independent University, Bangladesh, Dhaka, Bangladesh
mdasifbinkhaled@gmail.com

Abstract. Hematopoietic Stem Cell Transplantation (HSCT) is an effective treatment for a variety of blood diseases, including hematologic and lymphoid malignancies, along with numerous other conditions. One of the most prevalent adverse effects of Allogeneic Stem Cell Transplantation (ASCT) in patients is Graft versus Host Disease (GvHD). Inflammation brought on by the donor's stem cells attacking the patient's body can lead to GvHD. Moreover, Acute Graft versus Host Disease (aGvHD) and Chronic Graft versus Host Disease (cGvHD) are the two possible manifestations. The patient has a chance of developing this disease even if the donor and the recipient are a perfect match. Therefore, early diagnosis of the forms of GvHD before a patient receives ASCT treatment is essential. However, it is still necessary to identify the type of GvHD even if the patient has already undergone a transplant to advise any clinical decision. In this research, the types of the GvHD are precisely predicted using a variety of multi-class classification models. The techniques utilized in this study include Random Forest, Decision Tree, K-Nearest Neighbor, Gradient Boosting, XG Boosting, LG Boosting and a feed forward Artificial Neural Network named Multilayer Perceptron. This study revealed that the Random Forest algorithm demonstrated state-of-the-art performance in multi-class classification, with an accuracy of 98.62% along with 96.38% F1-Score and an area under the ROC curve (AUC) score of 98.02%. In terms of accuracy and reduced feature dependence for predicting the multi-class target feature, this study offers a useful prognosis tool for medical experts.

Keywords: Bone marrow disease · Prognosis · Acute GvHD · Chronic GvHD · Feature selection · Classification

1 Introduction

Bone marrow is soft, adipose, spongy tissue in the body that is located in the skeletal structures. It is responsible for producing the red blood cells, white blood cells, and platelets of the human body, and it also contains hematopoietic stem

H. Aziz et al. (Eds.): AI 2022, LNAI 13728, pp. 487–500, 2022.
https://doi.org/10.1007/978-3-031-22695-3_34

cells. Bone marrow transplantation is necessary when a cancer patient's bone marrow has been damaged by radiation therapy or intense chemotherapy. When the hematopoietic stem cells are taken from another person (donor) for bone marrow transplantation, it is called an Allogeneic Stem Cell Transplantation (ASCT) [11]. ASCT remains one of the most effective treatments for acute leukemia with a high risk of relapse or in an advanced stage. It is a surgical process that replaces unhealthy bone marrow cells with healthy ones. It is used to treat cancers such as leukemia, myeloma, and lymphoma [6,9]. However, following ASCT, Graft versus Host Disease (GvHD) may occur, which is further classified into acute GvHD and chronic GvHD. Acute GvHD might surface within a day, a week, a month, or within 100 d after transplantation. It can damage several organs, such as the liver, skin, eyes, mucosa, and intestines of a patient. The patient who is afflicted may experience a variety of side effects as a result, including jaundice, hepatomegaly, rash, gastrointestinal issues, bleeding, and diarrhea. Even two or three years after the transplantation, the negative effects may start to manifest. On the other hand, chronic GvHD affects a patient's liver, gastrointestinal tract, and lungs [17]. It's interesting to note that two years after transplantation, chronic GvHD affects those who had no relapse, and thirty months or longer after the stem cell transplantation, most of these patients pass away [24].

Nowadays, Machine Learning (ML) algorithms are widely used to predict different kinds of diseases in the healthcare system. This study develops several ML and Artificial Neural Network (ANN) models that can categorize a multi-class feature by estimating the possibility that a patient will have acute GvHD, chronic GvHD, both of these, or none of these diseases. These models only rely on 9 out of the 37 features from the original dataset to make this admiringly accurate prediction. [23]. Hence, nominal data dependency and multi-class classification make these models more effective compared to other procedures.

The remainder of this study is organized starting with section two, which summarizes the relevant work. The problem statement is presented in section three, and a detailed description of the experiment results is provided in section four. Finally, the whole study is summarized and makes further suggestions for improvement.

2   Related Work

Leukemia, a blood cancer type that affects white blood cells and damages bone marrow, has become one of the most severe illnesses in recent years. Romel Bhattacharjee et al. [4] proposed a robust approach for detecting Acute Lymphoblastic Leukemia (ALL) in blood smears by segmenting pictures and comparing several classifiers of ALL in blood smears within age groups between 3 to 7 years. Another important finding is the ability to predict relapse in children acute lymphoblastic leukemia (ALL), as this disease is more vulnerable to relapse than any other [19]. Because of this, optimum therapy and follow-up planning for childhood ALL prediction is critical. Liyan et al. [19] designed a machine learning model to predict the probability of acute lymphoblastic leukemia (ALL) relapse.

Graft versus Host Disease (GvHD) is the most common reason for death in patients following their HSCT. Yasuyuki et al. [1] presented a model to predict acute GvHD after an allogeneic transplant, which is one of the types of GvHD. Serum Fibrinogen was used by Neslihan et al. [20] as a Predictive Marker for chronic GvHD in patients after allogeneic HSCT. On the other hand, Catherine et al. [14] studied the risk of acute GvHD and mortality rate after 100 d post-transplantation. Also, Ying et al. developed and verified a machine learning strategy for predicting stem cell donor availability [15].

3   Problem Statement

The chances of developing GvHD following an ASCT is difficult to predict. Failure to anticipate this disease in its early stages can lead to the development of many other diseases. One condition that is a frequent symptom of chronic GvHD is dry eye disease (DED), which has been identified as a serious side effect of ASCT [18]. In most circumstances, manually foretelling this condition is quite challenging and erroneous. Though donor gene-expression profiling [2] and a new biomarker panel [5] also can predict this GvHD but both of these methods are very time-consuming and imprecise. However, ML methods may be the best option for predicting GvHD using donor and recipient medical data, and most importantly, predicting it before the transplantation.

4   Proposed Methodology

The handling of missing values in the dataset is the initial step of the methodology. After dealing with the missing data, the two features known as acute GvHD and chronic GvHD are combined to produce a multi-class target feature. Following that, feature engineering and the feature selection approach are used to select the best features. The dataset needs to be balanced because it is unbalanced and in order to avoid overfitting problems. Finally, the machine learning models are applied for the prediction. The steps of the proposed approach are abstracted in the Fig. 1.

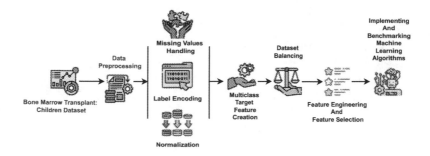

Fig. 1. Methodology employed in this study.

The following sections provide a brief description of each step of the proposed methodology.

4.1   Dataset Description

In this study, The Bone Marrow Transplant: Children dataset from the Machine Learning Repository at the University of California, Irvine [23] has been analyzed. This dataset contains information on pediatric patients who have a variety of hematologic abnormalities, including malignant and nonmalignant cases. The transplantation of hematopoietic stem cells from an unrelated allogeneic donor was performed without any modifications on the patients [23]. It comprises 187 instances, each with 37 attributes, where most of the attributes are categorical, the rest being numerical and boolean. Unfortunately, there are some missing values in this dataset. Data loss or missing data can happen for a number of reasons, such as inaccurate data entry, system failures, lost files, and even patients who choose not to participate in surveys for data collection. To resolve this issue, the Miss Forest [22] algorithm is used to impute missing values in numerical data and categorical missing values are filled using the Random Forest Classifier.

A snapshot of the statistical attributes of the dataset is abstracted in Fig. 2. There are some outliers that have been identified like ANC recovery, PLT recovery, and CD34 shown in Fig. 2. It is clear that some values lie abnormally far apart from other values. But these features with outliers have no effect on the performance of these models because the features have been dropped in the feature engineering process.

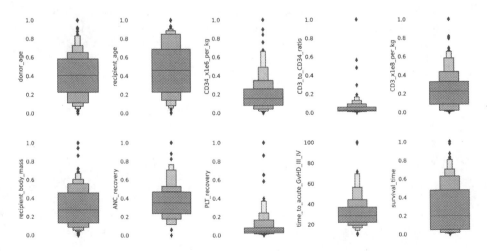

Fig. 2. Numerical features.

4.2   Normalization

Data normalization is a preprocessing technique in which the data is scaled to ensure that each feature contributes equally. Numerous research has demonstrated the significance of data normalization in improving the consistency of the information leading to the increasing effectiveness of machine learning algorithms [21]. So after the imputation procedure, some of the numerical features from the dataset are normalized to ensure that each feature contributes equally.

4.3   Target Feature Creation

The two features termed "extensive chronic GvHD" and "acute GvHD III IV" from the dataset have merged so that the ML models can perform a multi-class classification. A new multi-class attribute called "GvHD Diseases" is created using the mentioned features with binary values. Algorithm: 1 defines the process of the transformation.

Here, it is seen that the algorithm will return "Both" if both chronic and acute GvHD are positive and this indicates that the patient in concern has both types of GvHD. In accordance with the corresponding condition, all four classes are formed in the target feature in this manner. Finally, these categorical values have been converted into numerical values using a label encoder, renaming the target column as "GvHD Diseases" and appending it to the original dataset.

Algorithm 1. Target selection using features: "extensive chronic GvHD' as cGvHD and "acute GvHD III IV' as aGvHD

1: **if** $cGvHD == Yes$ && $aGvHD == Yes$ **then**
2:     return $Both$
3: **else if** $cGvHD == No$ && $aGvHD == No$ **then**
4:     return $None$
5: **else if** $cGvHD == Yes$ && $aGvHD == No$ **then**
6:     return $cGvHD$
7: **else if** $cGvHD == No$ && $aGvHD == Yes$ **then**
8:     return $aGvHD$

4.4   Balancing the Dataset

This study is aiming for a multi-class feature. There are four categories here and they are :

- Both (Acute and Chronic GvHD)
- None (No Disease or Contamination)
- Acute GvHD (aGvHD)
- Chronic GvHD (cGvHD)

Here in Fig. 3, it is seen that the category "Both" makes up 127 instances or 67.91% of the overall dataset. The remainder consists of 32 cases of "acute GvHD", 20 cases of "chronic GvHD" and 8 cases of "None" with respective percentages of 17.11%, 10.69% and 6.29%. The dataset appears to be slightly unbalanced, so Synthetic Minority Over-sampling Technique (SMOTE) technique and the "Random Over Sampler" approach have been applied to generate instances for the minor category to balance this dataset. After balancing the dataset, this is how the target feature appears in Fig. 3. Here, it is clear that all classes are now equally distributed, with 127 instances for each of these four classes.

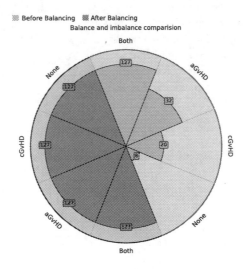

Fig. 3. Balancing the imbalance dataset.

4.5   Feature Engineering and Feature Selection

The feature engineering process is used following the normalization process. The features with a threshold value of 80% are removed from the dataset before creating the correlation matrix. Here, 80% is used as a cutoff value, meaning that if two features are 80% correlated or more, eliminating one of them won't have much of an impact on predicting the target feature. A feature selection approach is then applied among the selected features following the correlation matrix. This feature selection method makes use of lasso regression by taking α as 0.05.

Lasso regression is similar to linear regression, except it employs the "shrinkage" strategy, which reduces the coefficients of determination to zero. Lasso regression penalizes and eliminates less significant aspects of the dataset by setting their coefficients to zero. As a result, it allows to choose features and create models quickly. After chosen all the features the correlation matrix looks like Fig. 4.

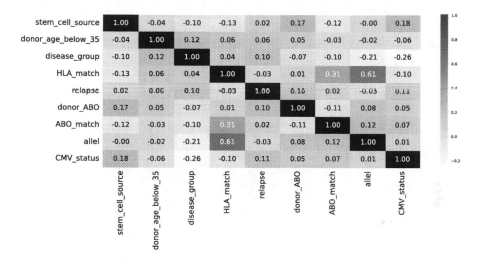

Fig. 4. Correlation matrix of the selected features.

4.6 Visualization of the Selected Features

The visualization of the dataset's finally selected features that are chosen in Fig. 5. It is clear from this that patients who have peripheral blood which is define as stem cell source = 1, are the ones who experiences chronic GvHD the most. Also if the donor is younger than 35, it is almost a guarantee that the patient won't be affected by any of these diseases. Interestingly there is fewer relapse case for the patients who don't have any types of GvHD and both types of GvHD. Contrarily, there is a big possibility that the patients will experience both forms of GvHD if the ABOs of the donor and recipient are matched. Another significant finding is that there is a strong probability that the patient won't experience any symptoms of GvHD if only one allel is matched between donor and recipient.

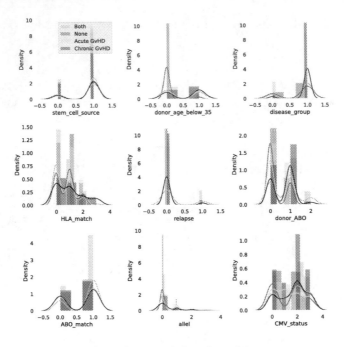

Fig. 5. Distribution of the selected features.

4.7 Applied Machine Learning and Artificial Neural Network Models

For this experiment, some supervised machine learning models have been utilized such as Random Forest [16], Decision Tree [7], K-Nearest Neighbor [25], Gradient Boosting [3], XG Boosting, LG Boosting [13], and Multilayer Perceptron [10] to predict the target feature after the preprocessing and feature engineering steps. For training, 70% of the dataset is used, and the remaining 30% is used for testing purposes. In this model, "random state" is defined as 42 for all of these methods, and the accuracy is validated using the K-Fold Cross Validation (K = 10) technique.

5   Results and Discussion

5.1   Evaluation of the Confusion Matrix

This section provides a thorough picture of the efficiency of a classification model as well as the kinds of errors it is producing.

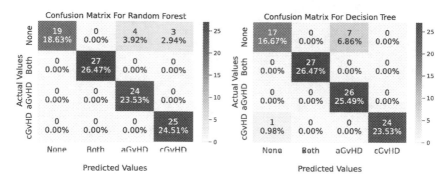

Fig. 6. Confusion matrix of random forest and decision tree algorithms.

It can be inferred from these confusion matrices in Fig. 6 that the Random Forest algorithm accurately classifies the majority of the samples where only 3.92% of instances were misclassified. On the other hand, The K-Nearest Neighbor algorithm does a terrible job of classifying the instances. A whopping 15.6% of samples were incorrectly classified by the K-Nearest Neighbor algorithm, 5.88% of which were categorized as "aGvHD" when they actually required to be categorized as "Both". Most interestingly, 3.92% of the samples are categorized as "None", indicating that these instances should not be affected by any of these diseases but these are actually in the "Both" category. As a result, the medical team's diagnosis must be muddled. In classifying the instances, other algorithms are also doing quite well. Decision Tree, Gradient Boosting, XG Boosting, and LightGBM Boosting have miss classified sample percentages of 7.84%, 7.84%, 4.9% and 7.84% respectively. On the other hand, MLP has miss classified around 17.23% of the samples. So it can be conclude that Random Forest outperforms other techniques.

5.2 Validation and Learning Curve

A valuable diagnostic tool that demonstrates the sensitivity and accuracy of a machine learning model to changes in specific model parameters is called a validation curve. Usually, a validation curve develops between the score of the model and a particular model parameter. On the other hand, a learning curve is a graph that displays how well a model learns across the size of the training sample.

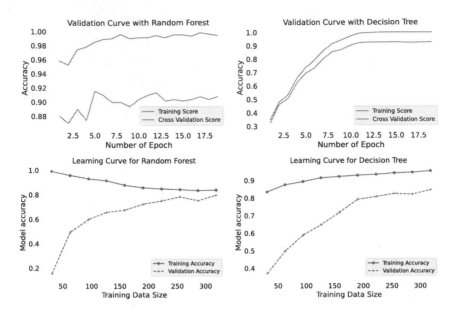

Fig. 7. Validation and learning curves of different machine learning algorithms.

Among all of the validation curves performed in this experiment, here are presented two strategies that perform well in this dataset for target prediction. It is seen that Random Forest maintains its performance as the number of epochs grows. As the number of epochs rises, the validation accuracy increases from 84% to 96%. It is clear from the learning curve that the cross-validation and training curves converge as the size of the training data increases. The cross-validation accuracy increases as more training data are added. Therefore, it is advantageous in this situation to add more training data. Since the training and validation curves meet at the expectation point, this indicates low bias and low variance. Therefore, the conclusion is that this model performs better as training data is increased. The performance of the decision tree is remarkable because its validation accuracy is nearly close to the testing accuracy. Both the training and validation curves converge as the size of the training data grows. This means that with a vast amount of training data, this model performs effectively with minimal bias and variance which indicates that there is no overfitting problem in this model.

5.3   ROC-AUC Curves

A measurement tool for binary classification issues is the Receiver Operator Characteristic (ROC) curve. Plotting the TPR (True Positive Rate) versus the FPR (False Positive Rate) at various threshold values effectively distinguishes one class from another.

Figure: 8 displays all of the ROC curves for the various machine learning techniques utilized in this study. Because the ROC curve only reflects binary classification so that it is ploted four distinct curves for a single algorithm on a single graph to display the ROC curve.

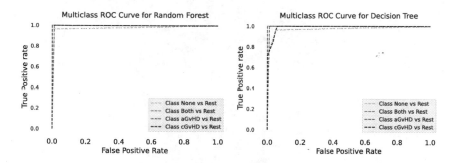

Fig. 8. ROC curves of random forest and decision tree algorithms.

The Random Forest algorithm outperforms others since it properly identifies the majority of the samples and has an AUC score of around 98.02%. It most accurately classifies "Both" and "aGvHD" and is slightly misclassified for classifying the "None" and "cGvHD" class. With the exception of identifying the category "aGvHD" and "None" decision tree performs well. Its AUC score is approximately 97.58%. However, KNN performs best when classifying only "Both" class as compared to "None", "cGvHD" and "aGvHD" classes. It has a classification accuracy rates of roughly 93.1% as it classifies the best only "Both" class. The performance of the Gradient Boosting and LG Boosting algorithms are quite similar as they classify "aGvHD" and "Both" class perfectly and their AUC score is 96.01% and 96.06% respectively. XG Boosting is the best performer for classifying the "Both" class but it is slightly misclassified for "None" and "aGvHD" classes. Its AUC score is almost 97.24%. With the 87.23% AUC score, MLP only classifies the "Both" class more accurately.

5.4   Performance Analysis

The target feature was predicted using six machine learning techniques and an ANN model named MLP, and the Random Forest classifier get the highest accuracy. Other algorithms are also perform well except the KNN and MLP. Since both precision and recall are equally important for this research so it must be needed to look up the F1-Score and again Random Forest outperforms all other algorithms in F1-Scoring. In Fig. 9 each algorithm uses in this study, accuracy, precision, recall, and F1-Score are displayed.

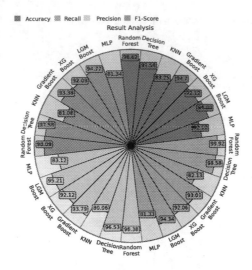

Fig. 9. Accuracy along with precision, recall and F1-score.

It can be inferred from this table that MLP performs the worst for classifying the classes, and among the machine learning algorithms KNN performs worst, while Random Forest does the best for classifying all the classes. Here traditional machine learning methods can more reliably identify unknown data than the MLP model since most of the feature contain categorical values to predict the target feature.

In below Table 1 : is comparison between this study with the related existence work. From this table it is clear that this study is new as a multi-class classification and there is no comparison analysis in multi-class classification.

Table 1. Comparative analysis with the existing works.

Study	Approach		BestResult	
	Research Area	Class	Best Algorithm	Accuracy (%)
Yasuyuki et al. [1]	acute GvHD	Binary	ADTree	76.4
Cooper et al. [8]	GvHD	Binary	Gradient Boosting	83.2
Ying et al. [15].	Predict Donor Availability	Binary	Boosted Decision Tree	82.6
Iwasaki et al. [12]	Relapse and GvHD	Binary	Random Forest	66.8
This Study	Both Types of GvHD	Multiclass	Random Forest	98.62

6   Conclusion and Future Works

Allogeneic Stem Cell Transplantation (ASCT) can save lives in cases of leukemia, myeloma, and lymphoma, however, it is also one of the causes of GvHD. So if the medical team can evaluate the possibility of adverse effects before starting the transplantation process by analyzing donor and recipient data, the procedure will have a better success rate, with a reduced odds of mortality. This study made use of methods like imputation, normalization, data balance, and feature selection. Since this study is primarily concerned with a multi-class classification, a now multi-class target feature is produced. ML techniques are often used in disease prediction to lessen human participation. It is rapidly being used in today's world in any case where prediction is needed. Subsequently, several models are benchmarked to find the best model to help the medical personnel estimate the risk of transplantation before ASCT and take the necessary precautions to lower that risk.

In the future, this work could benefit from using a larger dataset than the one used for analysis. In addition, if the amount of missing data can be decreased, the results can be even more accurate, which might result in better models.

References

1. Arai, Y., et al.: Using a machine learning algorithm to predict acute graft-versus-host disease following allogeneic transplantation. Blood Adv. **3**(22), 3626–3634 (2019)
2. Baron, C., et al.: Prediction of graft-versus-host disease in humans by donor gene-expression profiling. PLoS Med. **4**(1), e23 (2007)
3. Bentéjac, C., Csörgő, A., Martínez-Muñoz, G.: A comparative analysis of gradient boosting algorithms. Artif. Intell. Rev. **54**(3), 1937–1967 (2021)
4. Bhattacharjee, R., Saini, L.M.: Robust technique for the detection of acute lymphoblastic leukemia. In: 2015 IEEE Power, Communication and Information Technology Conference (PCITC), pp. 657–662 (2015). https://doi.org/10.1109/PCITC.2015.7438079
5. Budde, H., et al.: Prediction of graft-versus-host disease: a biomarker panel based on lymphocytes and cytokines. Ann. Hematol. **96**(7), 1127–1133 (2017). https://doi.org/10.1007/s00277-017-2999-5
6. Butturini, A., Gale, R.: Allogeneic bone marrow transplantation for leukemia. Curr. Opin. Hematol. **1**(6), 402–405 (1994)
7. Charbuty, B., Abdulazeez, A.: Classification based on decision tree algorithm for machine learning. J. Appl. Sci. Technol. Trends **2**(01), 20–28 (2021)
8. Cooper, J.P., et al.: Acute graft-versus-host disease after orthotopic liver transplantation: predicting this rare complication using machine learning. Liver Transpl. **28**(3), 407–421 (2022)
9. Gahrton, G., et al.: Allogeneic bone marrow transplantation in multiple myeloma. N. Engl. J. Med. **325**(18), 1267–1273 (1991)
10. Gardner, M.W., Dorling, S.: Artificial neural networks (the multilayer perceptron)-a review of applications in the atmospheric sciences. Atmos. Environ. **32**(14–15), 2627–2636 (1998)

11. Gupta, V., Braun, T.M., Chowdhury, M., Tewari, M., Choi, S.W.: A systematic review of machine learning techniques in hematopoietic stem cell transplantation (hsct). Sensors **20**(21), 6100 (2020)

12. Iwasaki, M., et al.: Establishment of a predictive model for gvhd-free, relapse-free survival after allogeneic hsct using ensemble learning. Blood Adv. **6**(8), 2618–2627 (2022)

13. Ke, G., et al.: Lightgbm: a highly efficient gradient boosting decision tree. Adv. Neural. Inf. Process. Syst. **30**, 3146–3154 (2017)

14. Lee, C., et al.: Prediction of absolute risk of acute graft-versus-host disease following hematopoietic cell transplantation. PLoS ONE **13**(1), e0190610 (2018)

15. Li, Y., et al.: Predicting the availability of hematopoietic stem cell donors using machine learning. Biol. Blood Marrow Transplant. **26**(8), 1406–1413 (2020)

16. Liu, Y., Wang, Y., Zhang, J.: New machine learning algorithm: random forest. In: Liu, B., Ma, M., Chang, J. (eds.) ICICA 2012. LNCS, vol. 7473, pp. 246–252. Springer, Heidelberg (2012). https://doi.org/10.1007/978-3-642-34062-8_32

17. McDonald, G.B.: Graft-versus-host disease of the intestine and liver. Immunol. Allergy Clin. North Am. **8**(3), 543–557 (1988)

18. Ogawa, Y., et al.: International chronic ocular graft-vs-host-disease (gvhd) consensus group: proposed diagnostic criteria for chronic gvhd (part i). Sci. Rep. **3**(1), 1–6 (2013)

19. Pan, L., et al.: Machine learning applications for prediction of relapse in childhood acute lymphoblastic leukemia. Sci. Rep. **7**(1), 1–9 (2017)

20. SANLI, N.M., Keklik, M., Ali, U.: Pretransplant serum fibrinogen level may be a predictive marker on chronic graft-versus-host disease (cgvhd) in patients having undergone allogeneic hematopoietic stem cell transplantation (allo-hsct). Int. J. Hematol. Oncol. **32**(1), 008–015 (2022)

21. Singh, D., Singh, B.: Investigating the impact of data normalization on classification performance. Appl. Soft Comput. **97**, 105524 (2020)

22. Stekhoven, D.J., Bühlmann, P.: Missforest-non-parametric missing value imputation for mixed-type data. Bioinformatics **28**(1), 112–118 (2012)

23. uci: uci machine learning repository: bone marrow transplant: children data set (2020). https://archive.ics.uci.edu/ml/datasets/Bone+marrow+transplant%3A+children

24. Vargas-Díez, E., García-Díez, A., Marín, A., Fernández-Herrera, J.: Life-threatening graft-vs-host disease. Clin. Dermatol. **23**(3), 285–300 (2005)

25. Wang, L.: Research and implementation of machine learning classifier based on knn. In: IOP Conference Series: Materials Science and Engineering, vol. 677, p. 052038. IOP Publishing (2019)

What Leads to Arrhythmia: Active Causal Representation Learning of ECG Classification

Shaofei Shen[1], Weitong Chen[2], and Miao Xu[1(✉)]

[1] The University of Queensland, Brisbane, Australia
{shaofei.shen,miao}@uq.edu.au
[2] University of Adelaide, Adelaide, Australia
t.chen@adelaide.edu.au

Abstract. The electrocardiogram (ECG) classification has attracted great attention as a crucial tool to detect arrhythmia which can be an early sign of heart disease. However, the key challenge of the current ECG classification methods is the lack of annotated data when applied to new patients. On the one hand, enormous ECG data are produced and they require a high labelling cost for supervised classification. On the other hand, the morphological and temporal features of ECG in individual patient can vary significantly. Therefore, the heartbeat classification models cannot be trained on adequate data and usually faced a huge performance degradation when tested on new patients without enough annotated data. Although the current works have worked on reducing labelling costs through active learning, these methods do not focus on patient differences and cannot guarantee performance when patient differences increase. Other works that aim to solve the patient differences only focus on the correlations but not the causal relations behind the data. In this paper, we firstly analyse the patient differences in ECG heartbeat in a causal view and propose **A**ctive **C**ausal **R**epresentation learning of **E**CG heartbeat **C**lassification (ACREC) to learn the stable features that have a direct causal effect on the outcome variable. The experiment results show our method can outperform other methods when handling patient differences. After active learning, our model can select the most informative data to annotate and achieve reliable performances. Moreover, we also conduct the ablation study to validate the effect of each part in our model.

Keywords: Causal learning · Deep learning · Active learning · Artificial intelligence in medicine

1 Introduction

Over the years, heart attack has become one of the major causes of sudden deaths. However, 20% of attacks are not aware by human beings even if they have made damage to the human body [21]. Although some heart attacks are silent, they can still be detected by diagnosing the arrhythmia in the electrocardiogram (ECG) signals. The ECG has become a regular physical examination item in the hospital and the ECG signal contains a series of heartbeats data. To diagnose arrhythmia in ECG, doctors with expert knowledge are required to label the heartbeat one by one and it leads to a high labelling cost. Therefore, the application of deep learning to improve the efficiency and accuracy of arrhythmia detection has attracted great attention.

To recognize arrhythmia, the ECG heartbeat data can be divided into 5 types according to the recommendation of the Association for the Advancement of Medical Instrumentation (AAMI): N (normal beats), S (supraventricular beats), V (ventricular beats), F (fusion beats), and Q (unclassifiable beats) [2]. Among these types, the N-type stands for normal heartbeats, the Q type means the unclassifiable beats that are from unknown sources and the other three types represent abnormal heartbeats. Different types of heartbeats usually have large morphological and temporal differences which can work as the classification criteria for the deep learning models to classify abnormal heartbeats [13,26]. However, the current ECG heartbeat classification faces two tough challenges that stop its application in real hospitals. On the one hand, enormous heartbeat data are generated in hospitals every day and most of them cannot be annotated. Therefore, we only have a limited number of labelled data to train the models and the traditional supervised classification methods, which require adequate labelled data, cannot perform well in real hospital scenarios. On the other hand, the morphological and temporal features varies among different patients [16,22,26]. When testing the model on the ECG data from new patients, the enormous differences among different individuals will lead to a huge drop in the prediction performance.

The previous works have tried different approaches to handle the two aforementioned challenges. One effective way to solve the high labelling cost is active learning which aims to retain high performances and reduce the size of labelled data in the meantime. [7,18] build Denoised AutoEncoder model for active learning while [23] constructs an Recurrent Neural Network (RNN). However, these works do not consider solving the patient differences from the perspective of the model itself and are highly dependent on the newly labelled data. Therefore, the performances of these active learning methods get declined when they are tested on different datasets where the patient differences increase or the size of annotated data is limited. Moreover, some inter-patient models are proposed to solve the patient differences and improve the model performances on the heartbeat data from new individuals. [8] uses an ensemble model of Support Vector Machine (SVM), and [11] and [22] choose the same CNN structure, while [22] add an additional domain adaptation module. Although the previous researchers have made great progress to solve the patient differences. These works only focus on learning the correlations but not the casual relations behind the heartbeat data via deep learning models. The correlations are easily affected by the unrelated variables and thus the model performances usually get degraded when the model is tested on new patients.

Figure 1 illustrates two different views of heartbeat classification. In Fig. 1(a), the traditional deep learning models mine the correlations between all the features X and the outcome variable Y. However, in the ECG classification, apart from the causal feature set Z that has a direct causal effect on the outcome Y, the noisy feature set C also exist in the whole feature set as shown in Fig. 1(b). The noisy features set C includes the features that lead to the patient differences. These noisy features in the different patients can affect the true causal features Z but they do not have a direct causal effect on the heartbeat types (i.e. Y). For example, the heart rate is one of the noisy features and different heart rate may lead to different morphological and temporal features, which also contains the ones that lead to different types of heartbeat. However, no direct causal relationship exists between the heart rate and heartbeat types and only the correlation exists through the mediate feature Z. The correlations between noisy heart rate and heartbeat types are misleading and result in the performance drop on the data from different patients [10].

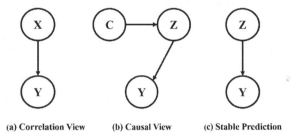

(a) Correlation View   (b) Causal View   (c) Stable Prediction

Fig. 1. Correlation view and causal view in ECG heartbeat classification. X and Y denote the features and outcome while C and Z belong to X and represent the noisy features and causal features of outcome Y individually. The feature set Z stands for the true causal features across different individuals for the predictions and the noisy feature set C represents the causes of the patient differences in heartbeat data

To overcome the aforementioned drawbacks, we can select the causal feature set Z that has a direct causal effect on the outcomes and train the models based on these causal features as shown in Fig. 1(c). Inspired by the stable learning [10], we propose the causal representation learning model and the **ACREC** method to extract the causal features across different patients and reduce the reliance on the large size of labelled data. Specifically, we minimize the causal effect of the features on all other features during the model training to learn the causal representation that has a direct causal effect on the heartbeat types. Then to improve the model performance further, we introduce active learning to achieve higher performance within a small labelled dataset. To prove the effectiveness of our methods, we conduct experiments on two different databases: MITDB and SVDB which contain heartbeats from different patients and frequencies. The code and experiment details are available in *GitHub*[1].

In summary, our contributions are in three folds:

- We invest the ECG classification problem via a causal graph to exploit the causal relations behind ECG heartbeat data.
- We propose an ECG heartbeat classification framework based on causal learning and active learning which can be more applicable than existing methods

---

[1] https://github.com/DATA-Transpose/ACREC.

in real hospital scenarios. To the best of our knowledge, this is the first work that introduces causal learning in the ECG classification field.

- We conduct experiments and compare our proposed method with other methods. The results demonstrate that our classification framework is more competitive than others. We conduct the ablation study to explore the importance of each part of our framework as well.

2   Related Work

2.1   ECG Classification

Previous works on ECG Classification have developed in three different directions: the intra-patient classifications which establish the training and test sets using the data from all individuals [24,25]; the inter-patient classifications which separate patients into different groups to set up the training and test sets [22,26]; and the active learning methods which aim to reduce the size of the labelled training set in [3,7,18,23].

For the intra-patient classification, [24] proposes a 1-D Convolutional Neural Network (CNN) and trains the model in an intra-patient way and [25] also designs a 2-D CNN model and trains it using the randomly selected data from all patients. In addition, [17] introduce the transformer for the long length signal detection. The inter-patient classification also attracts the interest of researchers. [22] applies the unsupervised domain adaptation to update the models on unlabeled data from new patients. [26] proposes a semi-supervised method through an unsupervised judgment to select N beats from an unlabelled data pool. In addition, another type of work combines active learning and inter-patient classification to reduce the reliance on the large size of the labelled data. For example, [3] ensembles the unsupervised feature extractor and different classifiers to select the most informative signals to annotate. In addition, [18] uses the Denoised Autoencoder and proposes a novel active learning workflow to select the signals to annotate and another work utilises the same active learning workflow and extracts eigenvectors of the heartbeat signals as the input features for the [7]. Moreover, [23] pre-train a graph-based CNN using a small set of the whole dataset and then update the model through an active learning approach.

In general, the supervised models can reach a rather high accuracy because the training set accounts for the majority of the whole dataset and contains adequate information about all patients. However, these assumptions cannot be held in real hospitals because of the patient differences and high labelling costs. In comparison, the inter-patient and active learning approaches are more suitable in practice. However, the inter-patient methods cannot retain high performance as the intra-patient methods and the current active learning methods only focus on selecting informative data but do not solve the patient differences.

2.2   Causal Learning

The causality describes the relationship between the cause variable and its effect on another variable. Previous works on causal learning include causal relation learning which mines the causal relationship and causal effect learning which focuses on estimating the effect of the cause variable [6].

Causal relation learning requires the model to learn the causal structure from the data. For example, [5] proposes a generative network to learn the causal structure of latent variables for domain adaptation and [15] constructs a Graph AutoEncoder to learn the large causal graph. In addition, causal relation learning also has been applied to specific areas, like the multivariate time series data [12] and the medical image data [20]. Another type of work: causal effect learning is usually based on the assumption of the specific causal structures and estimates the causal effect of noisy variables to reduce the bias and error of the models. For instance, [10] estimates the causal effect of the features on other features to reduce the correlations between features and outcomes. In addition, the causal effect learning has been widely used for debiasing [1].

3   Methodology

In this section, we introduce our novel framework ACREC based on active causal representation learning for ECG heartbeat classification. We divide the framework into two parts: the causal representation learning model and the active learning procedure.

3.1   Causal Representation Learning

Traditional ECG classification models contain two parts: a feature extractor that extracts the feature into latent representations, and a classifier that conducts the classification tasks [19, 22, 25, 26]. Previous works only focus on extracting the useful morphological and temporal features and do not pay attention to the causal relationships between the feature representations and outcome variable (i.e. Z and Y in Fig. 1).

As we have mentioned in Sect. 1, One effective way of stable representation learning is to remove the impact of the variables that have a causal effect on other variables (i.e. removing $C \rightarrow Z$ in Fig. 1). In causal inference, the inverse probability weighting (IPW) is an effective way to estimate the causal effect of one variable on other variables. Thus we can find the weights $\mathbf{w}$ of each sample in the control and treated groups and minimize the causal effect on one specific response variable to remove the causal effect of the treatment variable on the response variable [9]:

$$\mathbf{w} = \arg\min_{\mathbf{w}} \|E(\mathbf{X}|T = 1; \mathbf{w}) - E(\mathbf{X}|T = 0; \mathbf{w})\|_2^2$$

$$= \arg\min_{\mathbf{w}} \left\| \frac{\sum_{k;T_k=1} w_k \cdot X_k}{\sum_{k;T_k=1} w_k} - \frac{\sum_{k;T_k=0} w_k \cdot X_k}{\sum_{k;T_k=0} w_k} \right\|_2^2 \tag{1}$$

where T is the treatment variable, $T = 0$ and $T = 1$ represent the control and treated group, and $\mathbf{X}$ is the response variable.

In the ECG classification, we aim to remove all the causal effects inside the feature set and only preserve the direct causal effect from features to labels. Therefore, in our proposed model, we need to go through all the variables and optimize the sample weights that can reduce the average treatment effect on other variables instead of working on one fixed treatment as (1). Considering that the feature representations are in high dimensions, to learn the stable feature representations with causal effect on the predictions, we follow the global balanced weighting method in [10]. Specifically, given a feature representation $Z \in \mathbb{R}^{n \times m}$ where n denotes the number of samples and m stands for the dimension of feature representations, we first transform it into a binary format $S \in \{0,1\}^{n \times m}$ via a sign function. Then, the stable variables can be selected through a sample balance weight matrix $W \in \mathbb{R}^{n \times 1}$ by minimizing the balance loss (2) because only stable variables that have direct causal effects on Y are preserved in the sample reweighting procedure [10]. Then the causal effect of the noisy variables that are correlated to Y will be removed.

$$L_b = \sum_{k=1}^{m} \left\| \frac{S^T_{.,-k} \cdot (W \odot (S_{.,k}))}{W^T \cdot S_{.,k}} - \frac{S^T_{.,-k} \cdot (W \odot (I - S_{.,k}))}{W^T \cdot (I - S_{.,k})} \right\|_2^2 \tag{2}$$

where $S_{.,k} \in \mathbb{R}^{n \times 1}$ means the kth variable in S, and $S_{.,-k}$ means the remaining variables after excluding the kth variable in S (replacing all values of the kth variable by 0 [10]).

To regularise the weight matrix, and avoid all the weights equal to 0, the additional restrictions are required in the balance loss:

$$L_{cau} = L_b + \|W\|_2^2 + (\sum_k W_k - 1)^2 \tag{3}$$

3.2   Model Design

For the whole causal representation learning model, we propose a structure as shown in Fig. 2. This model contains a general feature extractor φ and a classifier ϕ, combined with the aforementioned causal representation learning block. The feature extractor φ contains two modules: one convolution module consisting of three convolution layers to extract morphological features and another LSTM module consisting of two LSTM layers for the temporal features. All of the CNN and LSTM layers are followed by the batch normalization layer and relu activation function. The output of the two blocks is added and then processed into low dimensions as the feature representation $\varphi(X)$. Then the feature representation is reweighted by the weight matrix W and input into the classifier ϕ consisting of two dense layers. The final output is $\phi(W \cdot \varphi(X))$, which is corresponding to four classes of heartbeats. Another part of this model is a decoder, which aims to assist in the feature extraction and reconstruct the input data from the feature representation.

As for the optimization objective, the loss function of the whole model consists of four parts: the classification loss L_{clf}, the reconstruction loss L_{con}, the regularization loss L_{reg}, and the balance weighting loss L_{cau}. For the classification loss, we employ the weighted cross-entropy loss in (4) which contains class weights V and sample weights W. The class weights aim to alleviate the class imbalance of different heartbeat types and sample weights W are used to select the stable features.

$$L_{clf} = \sum_{k=1}^{N} W_k \sum_{i=1}^{N} (V_i y_i log(\phi(W \cdot \varphi(X)))) \tag{4}$$

For the reconstruction loss L_{con}, we choose the mean quadratic loss of the original input and the reconstructed samples:

$$L_{con} = \sum_{k=1}^{m} \left\| X_K - X'_K \right\|_2^2 = \sum_{k=1}^{m} \| X_K - \omega(\varphi(X_K)) \|_2^2 \tag{5}$$

In each iteration of the training stage, we update the model parameter θ and the balance weights W in two separate steps. Firstly, we fixed all the model parameters θ and update the balance weights W by backpropagation to minimize L_{cau}. Then we fixed the balance weights W and update θ to minimize (6) where α_1, α_2, and α_3 are hyperparameters.

$$L_\theta = \alpha_1 L_{clf} + \alpha_2 L_{con} + \alpha_3 L_{reg} \tag{6}$$

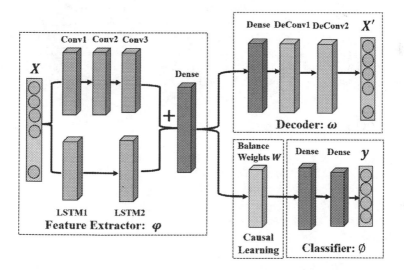

Fig. 2. The model structure of causal representation learning.

3.3   Active Learning

After pre-training our model on the specific patient set, the next part of our framework is the active learning on other patient sets, which aims to continue the causal representation learning and improve the model performance on the specific patients with the lowest label cost. Specifically, we follow the workflow in Fig. 3 to conduct the active learning procedure.

When calculating the test sample scores, we choose two metrics: the probability difference or breaking tie (BT), and the entropy as [7,18,23] do. The BT means the absolute value of the two largest predicted probabilities p_i and p_j among all classes and the entropy means the dispersion degree of the predicted probability distribution. To select the most informative sample to annotate, we choose the test samples with the lowest BT or highest entropy, which means the largest probability does not take the dominant position for the prediction and the sample is easy to be classified into other categories.

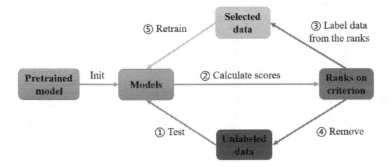

Fig. 3. The model structure of causal representation learning. Firstly, we use the pretrained model to initialise the model for active learning. Then we conduct the procedures of active learning in the sequential order: 1. test the unlabelled data using the initialised model; 2. calculate the scores by the given metric; 3. rank the scores and select the data with the highest ranks; 4. label the selected data and remove them from the unlabelled data pool; 5. retrain the model using the newly labelled data.

4   Experiments

In the experiments, we propose three research questions to verify the effectiveness of our proposed methods: **RQ1:** How is the causal representation learning working to solve the patient differences? Does it outperform other methods? **RQ2:** How is the ACREC working compared with other active learning methods on heartbeat classification? **RQ3:** How does each part of the ACREC affect the performance?

4.1   Databases

In this paper, we choose two ECG heartbeat databases for the experiments: MIT-BIH Arrhythmia Database (MITDB) [4,14], and the MIT-BIH Supraventricular Arrhythmia Database (SVDB) [4]. MITDB consists of 100420 heartbeats from 44 individuals with a frequency 360 Hz after four patients records in MITDB (#102, #104, #107 and #217) are removed to reduce the Q type of heartbeat data, which are not related to our arrhythmia detection task. Another database:

SVDB is a supplement database in the MIT-BIH Arrhythmia Database containing 78 records. This database contains 183332 heartbeats with a frequency 128 Hz. Table 1 shows the class distributions of the two databases. The detailed information is shown in Table 1.

We use the two databases because the signals recorded using different leads and frequencies can increase the patient differences and the distributions of V and S classes have enormous differences in different databases. Therefore, the experiments across databases are more similar to the real hospital scenarios to validate the effectiveness of our proposed methods. In addition, both databases are available in *Physionet*[2].

Table 1. Class distributions in three ECG datasets

Database	N	V	S	F	Patients
MITDB	89605	6998	3015	802	44
SVDB	161901	9222	12177	32	78

4.2 Preprocessing

In the preprocessing stage, we firstly use the Dual-Tree Complex Wavelet Transform (DTCWT) to denoise the data to eliminate the influence of such noises and then apply a median filter with a time window of 200 milliseconds to remove the P wave and QRS complex, then apply another median filter with a time window of 600 milliseconds to remove the T wave. Moreover, we segment the ECG signal into heartbeats using a dynamic length which means the segmentation length varies among different records and sampling periods [22,26]. Specifically, the length of each heartbeat is calculated by the average length of ±10 R-R intervals beside the aim heartbeat. We choose 90% of the length before the R peak and another 60% of the length after the R peak as to segment one single heartbeat. The reason for using dynamic length segmentation is that the heart rate is not an invariant value, which can vary in different people, and even different physiological states of different people. Therefore, using the dynamic length can lead to a more accurate segmentation of the heartbeats. Finally, we normalize the segmented data into a fixed interval by min-max normalization.

4.3 Experiments Settings

To answer the three research questions, we set up three phases of experiments.

Phase 1 Patient Differences: The first phase of experiments aims to prove the effectiveness to handle patient differences. We divide the patients in MITDB database into two sets: DS1 and DS2, which is common separation in [8,11,19, 22,23,26]. The DS1 and DS2 contains the records from different 22 patients. We regard DS1 as the training set and train our model on this dataset. Then the model is tested on DS2 and SVDB separately to prove the performance of causal representation learning to handle patient differences.

---

[2] https://physionet.org/.

Phase 2 Active Learning: To improve the model performances to a high level, we conduct the active learning on the two test sets: DS2 and SVDB and we also compare our results with other SOTA active learning algorithms [7,18,23].

Phase 3 Ablation Study: we conduct the ablation study by removing the balance weights parts and decoder parts from the proposed models. And then we compare the model performances in handling patient differences and the performance improvement during active learning.

As for the model setting and hyper-parameters, we choose to use the Adam optimisers with Nesterov momentum and set the pretraining epoch as 80 and the learning rates as 5×10^{-4}, 5×10^{-4}, and 1×10^{-4} to update the feature extractor, classifier, and decoder separately. Then we use the learning rate of 1×10^{-3} to update the balance weights W. When updating the model parameters θ, we set $\alpha_1 = 0.95$, $\alpha_2 = 0.05$, and $\alpha_3 = 1$ in (6). For the active learning steps, we set the training epoch as 50 and choose the learning rate of 1×10^{-3} to fine-tune the model parameters θ.

5   Results and Analysis

5.1   Patient Differences

Table 2. Performance of ACREC and comparisons with other inter-patients algorithms on DS2. OA, Se, Pp, Sp and F1 mean the percentage (%) of the overall accuracy, sensitivity, positive predictive value, specificity, and F1 score respectively.

	Method	OA	SVEB				VEB			
			Se	Pp	Sp	F1	Se	Pp	Sp	F1
DS2	Huang et al.[8]	93.8	**91.1**	42.2	–	57.5	93.9	**90.9**	–	92.0
	Sellami et al. [19]	95.3	82.0	30.4	92.8	43.9	92.0	72.1	97.5	80.8
	Li et al. [11]	91.4	89.0	35.4	–	50.2	95.2	90.1	–	92.4
	Zhai et al. [26]	95.2	**91.1**	26.7	90.4	41.3	87.2	80.0	98.5	83.4
	Wang et al. [22]	94.6	71.9	**82.9**	–	**77.1**	**92.0**	80.1	–	**85.6**
	Proposed	**95.4**	80.8	62.0	**99.4**	70.2	81.6	82.5	**98.7**	82.1
SVDB	Rahhal et al. [18]	–	8.8	14.3	–	10.9	65.2	9.3	–	16.3
	Wang et al. [23]	–	**25.4**	36.7	–	30.0	85.7	48.3	–	**61.8**
	Wang et al. [22]	–	15.6	46.9	–	23.4	**88.4**	35.6	–	50.8
	Proposed	**90.3**	22.9	**62.1**	**98.9**	**33.4**	67.5	**54.7**	**96.6**	60.4

The results of the first experiment are shown in Table 2. To compare with other methods, we choose the experiment results that use only DS1 as the training data to compare. In the DS2 test set, our proposed causal representation learning model can reach the overall accuracy of 95.4% and the F1 scores for SVEB (i.e. S class) and VEB (i.e. V class) of 70.2% and 82.1% respectively. In particular, the overall accuracy of our model is higher than all of the SOTA models without introducing the test data. In addition, our model has a significantly higher F1 score of S heartbeat than [8,11,19,26]. Compared with [22] which involved more than 20 convolutional layers, our model can reach a higher overall accuracy while the F1 scores of SVEB and VEB are slighter lower than the ones in their methods. However, in the SVDB test set which introduces more patient differences, our proposed method shows its strengths compared with other methods. Our overall accuracy can reach 90.3% and the F1 score of SVEB and VEB can reach 33.4% and 60.4%. In particular, our proposed method can achieve higher performances on nearly all evaluation metrics compared with [22]. The experiment results on two test sets can prove that causal representation learning can outperform other methods by selecting the stable variables in the feature representations when handling patient differences.

5.2 Active Learning

Table 3. Comparisons of ACREC and other algorithms after active learning

	Method	Metric	Labelled	SVEB			VEB		
				Se	Pp	F1	Se	Pp	F1
DS2	1DCNN [24]	BT	300	92.7	99.2	95.8	99.1	98.5	98.8
	LSTM	BT	300	89.3	99.7	94.2	98.4	98.9	98.6
	ASPP [22]	BT	300	92.0	98.1	95.0	95.4	99.7	97.5
	ACREC	Entropy	0	80.8	62.0	70.2	81.6	82.5	82.1
			100	87.1	96.9	91.7	99.3	98.7	99.0
			200	89.5	99.9	94.4	99.6	**100.0**	99.8
			300	85.0	**100.0**	91.9	99.6	**100.0**	99.8
	ACREC	BT	0	80.8	62.0	70.2	81.6	82.5	82.1
			100	88.1	98.2	92.9	99.0	99.3	99.1
			200	91.0	99.9	95.2	99.5	99.7	99.6
			300	**95.5**	99.9	**97.6**	**99.5**	100.0	**99.7**
SVDB	Rahhal et al. [18]	Entropy	300	86.4	94.3	90.2	96.0	95.7	95.8
	Rahhal et al. [18]	BT	300	78.3	92.4	84.8	90.8	86.9	88.8
	Wang et al. [23]	Fusion	350	85.1	99.6	91.8	95.2	99.6	97.4
	ACREC	Entropy	0	22.9	62.1	33.4	67.5	54.7	60.4
			100	77.3	43.2	55.4	92.8	94.8	93.8
			200	91.0	92.6	91.8	97.4	98.5	98.0
			300	**93.5**	**98.0**	**95.7**	98.7	99.0	**98.9**
	ACREC	BT	0	22.9	62.1	33.4	67.5	54.7	60.4
			100	85.5	84.2	85.0	90.9	94.3	92.6
			200	93.0	92.6	92.8	96.5	98.2	97.3
			300	92.6	96.4	94.5	97.9	**99.0**	98.4

Then to improve the model performance furtherly, we implement the active learning experiment on the three aforementioned sets. The active learning results are shown in Tables 3. In the experiments on DS2, we implement some models from the previous works for comparison with our proposed ACREC. The results show that the ACREC with BT selection score can reach the highest performance if we conduct 30 iterations of active learning and select 300 data per patient on average to annotate. The F1 scores of SVEB and VEB can reach 97.6% and 99.7%. In addition, the sensitivities of these two abnormal heartbeat types are higher than 95% and 99%. It means ACREC can detect nearly all the potential arrhythmia with less omission. And the positive predictive values can achieve nearly 100%, which shows that all the detected abnormal heartbeats are reliable by ACREC.

For the experiments on SVDB, we choose some SOTA results for comparison. Our ACREC with entropy scores can get 95.7% and 98.9% F1 scores for SVEB and VEB classifications. It can still outperform other methods when selecting the same number of data to annotate. Compared with [18] which produces one specific model for each patient in active learning, our ACREC only generates one model and this model can be tested on all patients. Moreover, we can also find only 200 labelled data per patient on average, the ACREC can get similar results that require more labelled data in other algorithms.

5.3   Ablation Study

Table 4. Performance comparison of the whole proposed model and the models without Balance Weights (W), Decoder (ω), or both ($W+\omega$).

	Method	OA	SVEB			VEB		
			Se	Pp	F1	Se	Pp	F1
DS2	Without $W + \omega$	95.1	57.1	83.1	68.1	76.2	95.0	**84.5**
	Without ω	94.6	45.9	78.0	57.8	64.2	**98.2**	77.6
	Without W	93.9	56.3	**87.4**	68.5	**81.8**	67.6	74.0
	Proposed	**95.4**	**80.8**	62.0	**70.2**	81.6	82.5	82.1
SVDB	Without $W + \omega$	77.4	25.6	50.5	**34.0**	73.5	26.9	39.4
	Without ω	85.8	**29.7**	37.5	33.2	63.7	35.4	45.5
	Without W	79.1	22.8	26.5	24.5	**77.5**	25.3	38.1
	Proposed	**90.3**	22.9	**62.1**	33.4	67.5	**54.7**	**60.4**

Compared with the traditional heartbeat classification models, our model has two additional parts: a decoder to assist the feature extraction and a balance weight module to realize the causal representation learning. To verify the effectiveness of the two parts in our proposed model, we design ablation studies based on experiments on patient differences. As shown in Table 4, we remove the decoder ω and the balance weights W separately and then remove both of the two parts.

From the experiment results, we can see the model can get the best performances on DS2 and SVDB when we preserve both ω and W. Then the model without both ω and W can also reach competitive results in the experiments on DS2. However, the performances of this model drop significantly in the SVDB experiments like other models that work on correlations in Table 2. This is because in SVDB more patient differences are introduced, for example, the frequency and record lead and then the correlations between these features and the outcomes can mislead the classification models. In addition, when we only remove ω or W, the performances on DS2 and SVDB both decrease. The decoder part and the reconstruction loss in (5) can assist the learning of latent representations while the balance weights can show their functionality to select the stable causal variables in the representations. Therefore, both the decoder ω and the balance weight W are essential for our proposed model.

6 Conclusion

In conclusion, we are the first work applying causal view analysis and causal learning in the ECG heartbeat classification field. We analyse one of the key challenges in the heartbeat classification: patient differences via a causal view and introduce the causal inference methods to select the stable features and reduce the correlations inside the features. Then we design a novel model structure for the causal representation learning, which can alleviate the impact of patient differences on the prediction models and we apply the active learning methods to reduce the size of labelled training data. Then, we conduct comprehensive experiments to prove the effectiveness of our models to handle patient differences and achieve high performances with a limited size of labelled data using active learning. Moreover, compared with other SOTA inter-patient and active learning algorithms, our model can outperform most of the current methods, especially in the SVDB test dataset. We also implement the ablation study to verify the positive influence of the decoder and balance weights parts on the classification under the situation of patient differences and active learning. From the application of causal learning to the ECG heartbeat classification, we can find the high potential and bright future of causal deep learning.

References

1. Chen, J., Dong, H., Wang, X., Feng, F., Wang, M., He, X.: Bias and debias in recommender system: a survey and future directions. CoRR abs/2010.03240 (2020). arXiv:2010.03240
2. AAMI ECAR: Recommended practice for testing and reporting performance results of ventricular arrhythmia detection algorithms, vol. 69. Association for the Advancement of Medical Instrumentation (1987)
3. Sayantan, G., Kien, P.T., Kadambari, K.V.: Classification of ECG beats using deep belief network and active learning. Med. Biol. Eng. Comput. **56**(10), 1887–1898 (2018). https://doi.org/10.1007/s11517-018-1815-2
4. Goldberger, A.L., et al.: Physiobank, physiotoolkit, and physionet: components of a new research resource for complex physiologic signals. Circulation **101**(23), e215–e220 (2000)

5. Gong, M., Zhang, K., Huang, B., Glymour, C., Tao, D., Batmanghelich, K.: Causal generative domain adaptation networks. CoRR abs/1804.04333 (2018)
6. Guo, R., Cheng, L., Li, J., Hahn, P.R., Liu, H.: A survey of learning causality with data: problems and methods. ACM Comput. Surv. **53**(4), 75:1-75:37 (2020). https://doi.org/10.1145/3397269
7. Hanbay, K.: Deep neural network based approach for ECG classification using hybrid differential features and active learning. IET Sig. Process. **13**(2), 165–175 (2019). https://doi.org/10.1049/iet-spr.2018.5103
8. Huang, H., Liu, J., Zhu, Q., Wang, R., Hu, G.: A new hierarchical method for inter-patient heartbeat classification using random projections and RR intervals. Biomed. Eng. Online **13**(1), 1–26 (2014)
9. Imbens, G.W., Rubin, D.B.: Causal Inference in Statistics, Social, and Biomedical Sciences. Cambridge University Press (2015)
10. Kuang, K., Cui, P., Athey, S., Xiong, R., Li, B.: Stable prediction across unknown environments. In: Guo, Y., Farooq, F. (eds.) Proceedings of the 24th ACM SIGKDD International Conference on Knowledge Discovery & Data Mining, KDD 2018, London, UK, 19–23 August 2018, pp. 1617–1626. ACM (2018). https://doi.org/10.1145/3219819.3220082
11. Li, F., Xu, Y., Chen, Z., Liu, Z.: Automated heartbeat classification using 3-d inputs based on convolutional neural network with multi-fields of view. IEEE Access **7**, 76295–76304 (2019). https://doi.org/10.1109/ACCESS.2019.2921991
12. Malinsky, D., Spirtes, P.: Causal structure learning from multivariate time series in settings with unmeasured confounding. In: Le, T.D., Zhang, K., Kiciman, E., Hyvärinen, A., Liu, L. (eds.) Proceedings of 2018 ACM SIGKDD Workshop on Causal Discovery, CD@KDD 2018, London, UK, 20 August 2018, vol. 92, pp. 23–47. Proceedings of Machine Learning Research. PMLR (2018). http://proceedings.mlr.press/v92/malinsky18a.html
13. Mondéjar-Guerra, V.M., Novo, J., Rouco, J., Penedo, M.G., Ortega, M.: Heartbeat classification fusing temporal and morphological information of ECGs via ensemble of classifiers. Biomed. Sig. Process. Control **47**, 41–48 (2019). https://doi.org/10.1016/j.bspc.2018.08.007
14. Moody, G.B., Mark, R.G.: The impact of the MIT-BIH arrhythmia database. IEEE Eng. Med. Biol. Mag. **20**(3), 45–50 (2001)
15. Ng, I., Zhu, S., Chen, Z., Fang, Z.: A graph autoencoder approach to causal structure learning. CoRR abs/1911.07420 (2019)
16. Niu, L., Chen, C., Liu, H., Zhou, S., Shu, M.: A deep-learning approach to ECG classification based on adversarial domain adaptation. Healthcare **8**, 437 (2020)
17. Qiu, Y., Chen, W., Yue, L., Xu, M., Zhu, B.: STCT: spatial-temporal conv-transformer network for cardiac arrhythmias recognition. In: Advanced Data Mining and Applications, ADMA 2022. LNCS, vol. 13087. Springer, Cham (2022). https://doi.org/10.1007/978-3-030-95405-5_7
18. Rahhal, M.M.A., Bazi, Y., Alhichri, H.S., Alajlan, N., Melgani, F., Yager, R.R.: Deep learning approach for active classification of electrocardiogram signals. Inf. Sci. **345**, 340–354 (2016). https://doi.org/10.1016/j.ins.2016.01.082
19. Sellami, A., Hwang, H.: A robust deep convolutional neural network with batch-weighted loss for heartbeat classification. Exp. Syst. Appl. **122**, 75–84 (2019). https://doi.org/10.1016/j.eswa.2018.12.037
20. Tomov, M.S., Dorfman, H.M., Gershman, S.J.: Neural computations underlying causal structure learning. J. Neurosci. **38**(32), 7143–7157 (2018)
21. Tsao, C.W., et al.: Heart disease and stroke statistics-2022 update: a report from the American Heart Association. Circulation **145**(8), e153–e639 (2022)

22. Wang, G., Chen, M., Ding, Z., Li, J., Yang, H., Zhang, P.: Inter-patient ECG arrhythmia heartbeat classification based on unsupervised domain adaptation. Neurocomputing **454**, 339–349 (2021). https://doi.org/10.1016/j.neucom.2021.04.104

23. Wang, G., et al.: A global and updatable ECG beat classification system based on recurrent neural networks and active learning. Inf. Sci. **501**, 523–542 (2019). https://doi.org/10.1016/j.ins.2018.06.062

24. Xu, X., Liu, H.: ECG heartbeat classification using convolutional neural networks. IEEE Access **8**, 8614–8619 (2020). https://doi.org/10.1109/ACCESS.2020.2964749

25. Zhai, X., Tin, C.: Automated ECG classification using dual heartbeat coupling based on convolutional neural network. IEEE Access **6**, 27465–27472 (2018). https://doi.org/10.1109/ACCESS.2018.2833841

26. Zhai, X., Zhou, Z., Tin, C.: Semi-supervised learning for ECG classification without patient-specific labeled data. Exp. Syst. Appl. **158**, 113411 (2020). https://doi.org/10.1016/j.eswa.2020.113411

Automated Fish Classification Using Unprocessed Fatty Acid Chromatographic Data: A Machine Learning Approach

Jesse Wood[1](✉)[ID], Bach Hoai Nguyen[1](✉)[ID], Bing Xue[1][ID], Mengjie Zhang[1][ID], and Daniel Killeen[2][ID]

[1] Victoria University of Wellington, Te Herenga Waka, Wellington, New Zealand
{jesse.wood,hoai.bach.nguyen,bing.xue,mengjie.zhang}@ecs.vuw.ac.nz
[2] New Zealand Institute for Plant and Food Research Limited, Nelson, New Zealand
daniel.killeen@plantandfood.co.nz

Abstract. Fish is approximately 40% edible fillet. The remaining 60% can be processed into low-value fertilizer or high-value pharmaceutical-grade omega-3 concentrates. High-value manufacturing options depend on the composition of the biomass, which varies with fish species, fish tissue and seasonally throughout the year. Fatty acid composition, measured by Gas Chromatography, is an important measure of marine biomass quality. This technique is accurate and precise, but processing and interpreting the results is time-consuming and requires domain-specific expertise. The paper investigates different classification and feature selection algorithms for their ability to automate the processing of Gas Chromatography data. Experiments found that SVM could classify compositionally diverse marine biomass based on raw chromatographic fatty acid data. The SVM model is interpretable through visualization which can highlight important features for classification. Experiments demonstrated that applying feature selection significantly reduced dimensionality and improved classification performance on high-dimensional low sample-size datasets. According to the reduction rate, feature selection could accelerate the classification system up to four times.

Keywords: AI applications · Classification · Feature selection · High-dimensional data · Particle swarm optimization · Multidisciplinary · Gas chromatography · Fatty acid

1 Introduction

Fish oil is rich in omega-3 polyunsaturated fatty acids, nutritionally important fats that are found at increasingly low concentrations in Western diets [21]. This has contributed to a high consumer demand for omega-3 supplements, produced from a wide range of marine biomass [17]. The suitability of a given fish species (or fish tissue) for the production of high-value omega-3 supplements depends on fatty acid composition, which is determined by an analytical chemistry technique called Gas Chromatography [6,19]. However, fatty acid data

H. Aziz et al. (Eds.): AI 2022, LNAI 13728, pp. 516–529, 2022.
https://doi.org/10.1007/978-3-031-22695-3_36

must be carefully processed and interpreted by domain experts (i.e. chemists), which is very expensive and time-consuming. Previous works using CNNs, [3,14], showed high classification accuracy on Gas Chromatography data. However, these black-box models do not produce interpretable models, making it difficult to verify/troubleshoot these models for fish processing in a factory setting.

The goal of this work is to automate the processing and interpretation of Gas Chromatography data using machine learning algorithms, to substantially increase fatty acid analysis throughput. However, it is not a trivial task to format Gas Chromatography data for existing classification algorithms. Furthermore, each Gas Chromatography data consists of almost 5000 values (features/variables), far more numerous than the number of fish samples (153). This large number of features relative to samples (the curse of dimensionality) results in a sparsely populated data space, which can result in overfitting i.e. where the built model works well on the training set but poorly on the test (unseen) set. Redundant (providing the same information as other features) or irrelevant features (providing misleading information for the classification task) are also common in this type of dataset [15], which can reduce classification performance and cause long training times. Therefore, the paper also assessed the utility of feature selection to preprocess and remove these irrelevant/redundant features.

The goals of this work are to investigate the viability of classifying different marine biomass, automate processing of raw Gas Chromatography data, improve analytical throughput and reduce labour costs, and reduce the dimensionality of Gas Chromatography data required to perform fish oil production and analysis. The contributions of this work are broken into three main steps:

- Data preprocessing: This step converts Gas Chromatography data into tabular format data appropriate as input into a machine learning algorithm. The paper finds an effective method to detect and fill the missing packets/features which improves the classification performance over using the raw data.
- Analysing classification algorithms: The second step performs experiments with five types of classification algorithms, including instance-based classifiers, probabilistic classifiers, tree-based classifiers, ensemble classifiers, and kernel-based classifiers, to classify fish samples [4,7–9,13]. Experiments find that kernel-based classifiers, particularly linear SVM, achieve high classification accuracy on the fish data. The paper visualises the learnt model and identifies that not all the data, represented as *features*, are useful, which leads to the final step.
- Feature selection: The last step applied feature selection methods to reduce the amount of collected data i.e., the number of features. The experimental results illustrate that the number of features could be reduced by almost 75% while improving the classification performance.

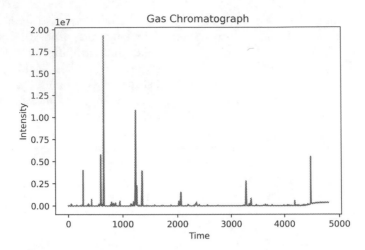

Fig. 1. Gas Chromatogram of Fatty Acid Methyl Esters from Snapper Skin.

2   Gas Chromatography

Gas Chromatography is an analytical chemistry method commonly used to investigate the fatty acid compositions of biological samples e.g. marine oils [6,19]. It works by increasing the temperature of a very narrow 'capillary' column, which separates each fatty acid from the complex mixture based on their individual chemical characteristics e.g. molecular size, volatility, and polarity. An example of Gas Chromatography for fatty acid analysis is shown in Fig. 1. The x-axis represents the time required to separate the individual fatty acids (or a packet), and the y-axis represents peak intensity (or the packet intensity), which is proportional to the concentration of each fatty acid. Chemists integrate the area under each peak to measure how much of each fatty acid is present, and use this information to understand the best use of the oil. This process can be slow, labour-intensive and expensive.

The goal of this work is to apply machine learning, particularly classification algorithms to automatically classify the fish data, a real-world problem in New Zealand. However, the current Gas Chromatography data is not readily applied to machine learning algorithms due to missing packets which are not caught by the system detector. The missing packets cause the misalignment between two samples, i.e., intensities at the same time of the two chromatographs may have different meanings. Therefore, it is necessary to detect such missing packets to align the data before applying machine learning algorithms.

Table 1. Inconsistent timestamps

	Timestamp		
	Sample 1	Sample 2	Sample 3
Packet 1	51	50	50
Packet 2	52	51	51
Packet 3	53.05	53.1	53

3 Data Preprocessing and Formation of Classification Problems

The Y-data output from the Gas Chromatography analysis consists of many packets with variable intensities. In theory, they could be used as features to classify the different fish samples, but there were a large number of inconsistencies between packets in the different fish samples. An example, focusing on these inconsistencies for three different fish samples, is shown in Table 1. Although all three samples have three packets, their timestamps are different. For example, the timestamp of the first packet of Sample 1 is 51, while the timestamp of the first packet of Sample 2 is 50. In other words, the first packet of Sample 1 does not correspond to the first packet of Sample 2, and thus it does not make sense to directly apply a classification algorithm to the raw data. Initial experiments tried KNN (K = 3), and the classification performance was only 67%, which is quite low.

Further investigation revealed that the main reason was due to the missing packets, caused by the absence of signal at the Gas Chromatography detector. For example, for Sample 1, the packet at the timestamp 50 is missed, and thus the first packet of Sample 1 is at 51. These missing packets are unavoidable for this dataset, therefore a method is needed to handle missing data. Preprocessing aligns the packets from all the samples. Firstly, all unique timestamps are collected by analysing all the possible samples in the training set. For the example given in Table 1, the set of unique timestamps is {50, 51, 52, 53, 53.05, 53.1}. Thus, there should be six packets in total, while Table 1 shows only three packets for each sample. Based on the timestamp set, the packets at {50, 53, 53.1} are missing for Sample 1. Once the missing packets are identified, these missing intensities need to be filled.

This work tried three different standard methods for missing values: filling 0, filling the average value, and filling the median value. The results show that filling 0 gives the most promising results with 83.57% on KNN (K = 3). The possible reason is that the missing packets have low intensities, which the detector might not be able to detect. Thus, the 0 value is quite close to the intensities of the missing packets. Therefore, the filling 0 method was chosen. The authors are aware that there are more complex methods for imputing missing values, [22,24], but they are not the focus of the paper and will be left for future work [23].

The processing gives 4800 packets for each sample, which meant each sample had *4800* features. The number of fish samples was 153. There is a class imbalance for the fish species dataset, where Blue cod is the majority class e.g., 68 samples are Blue Cod of the total 153 samples. There are two classification tasks associated with the data:

- To predict the fish species for each fish sample. There are four fish species: *Snapper, Gurnard, Tarakihi,* and *Blue cod.*
- To predict from which body part the fish sample is extracted. There are six body parts: *Frame, Gonad, Head, Liver, Skin,* and *Guts.*

4 Classification Performance

The following section illustrates the classification performance on the fish species and body parts.

4.1 Experiment Settings

Firstly, since the number of samples is small, the experiment uses 10-fold cross-validation to conduct the experiments. For 10-fold cross-validation, the method divides the data into 10 folds such that the proportions of the classes in each fold are representative of the proportions in the whole dataset. Each fold plays the testing role, while the remaining 9 folds are combined to form a training set. A classification algorithm is then trained on the training set, and the obtained classifier is evaluated on the test set. Finally, 10 testing accuracies are obtained, and their mean value and standard deviation are given as the final classification performance. The experiment measures the balanced accuracy, so as not to bias results towards the majority class (i.e. Blue cod for fish species).

These experiments compare five well-known classifications: K Nearest Neighbours (KNN), Naive Bayes (NB), Random Forest (RF), Decision Trees (DT), and Linear Support Vector Machines (SVM) [4,7–9,13]. The parameters are the default settings in *scikit-learn* [18].

4.2 Results and Discussion

Table 2 shows the results for KNN, RF, DT, NB, and SVM. Results are given for fish species (top), and fish part (bottom) datasets. The mean and standard deviation of balanced accuracy is given using the fish species and part datasets. For each dataset, the best accuracy is emphasized in bold.

As can be seen from the table, RF, DT and SVM achieve 100% training accuracies. However, on the test set, DT and RF do not achieve good classification performance. The main reason is that there is a small number of training samples. The trees built by DT and RF can perfectly fit the training data by creating large trees that remember all the possible training samples. Such trees do not generalise well on the test set, which is the overfitting problem in machine learning. KNN does not achieve good performance since it is a distance-based classification algorithm which suffers the most from the large number of features.

Table 2. Classification accuracies

Dataset	Method	AvgTrain ± Std	AveTest ± Std
Fish Species	KNN	83.57 ± 1.80	74.88 ± 12.54
	RF	100.0 ± 0.00	85.65 ± 10.76
	DT	100.0 ± 0.00	76.98 ± 13.12
	NB	79.54 ± 1.60	75.27 ± 4.35
	SVM	**100.0 ± 0.00**	**98.33 ± 5.00**
Body Parts	KNN	68.95 ± 3.49	43.61 ± 13.48
	RF	100.00 ± 0.00	72.60 ± 16.15
	DT	100.00 ± 0.00	60.14 ± 14.57
	NB	65.54 ± 2.69	48.61 ± 12.19
	SVM	**100.00 ± 0.00**	**79.86 ± 8.52**

Similar to KNN, NB does not achieve good performance since it assumes conditional independence between features that may not be true in the fish datasets. The SVM classifier outperforms the other classifiers on the test set, with 98.33% and 79.86% for fish species and body parts, respectively. The main reason is that SVM can handle a large number of features, so SVM is suitable to classify the fish data.

Another essential point is that the classification accuracy on the fish species is always higher than the classification accuracy on the body parts. The results suggest that classifying body parts is a more challenging problem. A possible reason is that the tissue samples from different species may have very different chemical components. Meanwhile, the tissue samples from different body parts (but on the same fish species) may have similar chemical components. Future work will investigate more sophisticated mechanisms to improve the classification performance on classifying body parts.

4.3   Interpret SVM Models

Achieving a high classification performance is great. However, in real-world applications, it is essential to analyse why the models work well. This subsection analyzes the Linear SVM model built to classify the fish species. The main idea of SVM is to build hyperplanes that separate different fish species. For SVM with linear kernels, the hyperplane is represented by a weight vector in which each weight is associated with a feature. The larger the weight, the more important the corresponding feature. After an SVM classification algorithm is trained on the training set, an SVM classifier containing a learned weight vector is obtained. This section analyses the learned weight vector to examine the contribution of each packet/feature.

Figures 2a and 2b show the coefficients of hyperplanes to separate Snapper and Blue cod from other species, respectively. The horizontal axis is the feature index and the vertical axis is the coefficient value. The negative weights are in red and the positive weights are in blue. Gas Chromatography data is

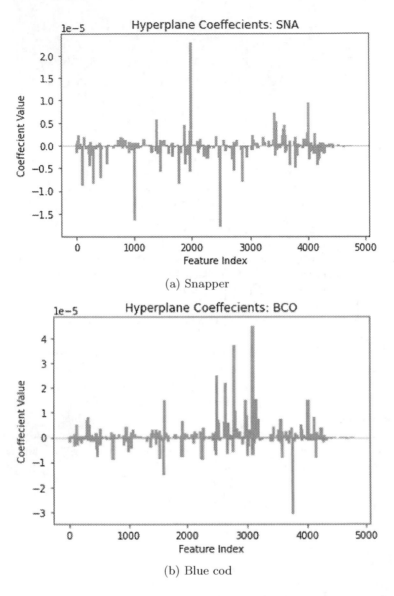

(a) Snapper

(b) Blue cod

Fig. 2. SVM hyperplane coefficients

non-negative, so only negative weights push toward the negative class, therefore positive weights are expected values, and the negative values are not. Note that when considering the feature importance, the absolute values of the weights should be considered, i.e., the longer the bar, the more important the corresponding features. Both figures demonstrate that most features have relatively small weights, which suggests not all the 4800 packets/features are needed to classify the fish data.

5   Feature Selection

5.1   Motivations

As can be seen in the SVM models, it is not necessary to use all the 4800 packets/features to perform fish classification. Therefore, the number of packets can be reduced while maintaining (or even improving) the classification performance. In an automated classification system, it would be great to significantly reduce the number of packets. Since then the system will not need to wait for a large number of packets to arrive at the end of the detector, hence significantly improving the system efficiency and throughput. The remaining question is which packets or features should be used. This question motivates us to conduct a further investigation using feature selection to select the most important packets/features.

5.2   Feature Selection Methods

In a classification problem, the classification performance relies heavily on feature quality. However, in a large set of features as in the fish data, there are usually redundant or irrelevant features that blur useful information provided by the relevant features. Feature selection aims to select an informative subset of relevant features, which is expected to significantly reduce the number of features while maintaining (or even improving) the classification performance. In a feature selection system, subset evaluation is an essential component that evaluates the quality of a feature subset. Based on the subset evaluation, the system can continuously improve the subset quality until a stopping criterion is met. The final feature subset is the output as the final solution.

This section compares four common feature selection methods:

- χ^2 (chi-square) [12] is a statistical measure that computes the independence of two variables X and Y. The formula of χ^2 is

$$\chi^2 = \sum_{k=1}^{N} \frac{(X_k - Y_k)^2}{Y_k} \tag{1}$$

where k is the index of the sample and N is the number of samples. In feature selection, χ^2 can be used to measure the independence between a feature and a class label. Since there is usually a high dependency between a relevant feature and a class label, the low χ^2 value indicates that the features are more relevant. Thus, the features can be ranked in ascending order and the top-ranked features can be selected.

- **Minimum Redundancy and Maximum Relevance (mRMR)** [5] uses mutual information to perform feature selection. Mutual information between two variables X and Y, i.e., $I(X;Y)$ calculates the dependency between two or more variables. mRMR aims to select a feature subset such that the redundancy of the selected features is minimised and the relevance between the selected features and the class label is maximised. Given a set of selected features A, the score of a feature X_i, i.e., S_i is calculated by the following formula:

$$S_i = I(Y; X_i) - \frac{1}{|S|} \sum_{X_j \in A} I(X_i; X_j) \qquad (2)$$

mRMR has many iterations where at each step mRMR will add the best feature based on Eq. (2). mRMR stops when a predefined number of features are selected.

- **ReliefF** [20] is a feature selection algorithm based on distance measures. In ReliefF, a good feature should be able to separate instances from different classes well while the instances from the same class should not be far from each other. The algorithm ranks all features based on the idea of nearest neighbours. For a feature, if the distance between two nearest instances from *different* classes (a miss) is large, the feature score is increased since the feature can separate different classes well. On the other hand, if the distance between the two nearest instances from the *same* class is large (a hit), the feature score is decreased. In ReliefF, the higher the score, the more relevant the feature. Therefore, all features are ranked in descending order, and the top-ranked features are selected.

- **Particle Swarm Optimisation (PSO)** [10,16] **for Wrapper Feature Selection** utilises the classification performance as the fitness function to achieve feature selection. The main idea is to have a swarm of particles that can explore the feature subset space in parallel. Each particle represents a feature subset. The quality of each particle is the classification performance of the corresponding feature subset. Since it is necessary to train a classification algorithm during the evaluation process, the classification algorithm is "wrapped" inside the PSO algorithm (that is why the algorithm is called Wrapper PSO). In this work, a linear SVM is used as the wrapped classification algorithm since it achieves good classification performance. Each particle records the best feature subset that it discovered so far (called personal best or *pbest*) and the best feature subset that is discovered by the whole swarm so far (called global best or *gbest*). The particle then updates its position by moving towards the two best positions. It is expected that the new subset at the new position will have better quality (i.e., higher classification performance) than the previous position. An advantage of PSO is that the particle movement is stochastic. Thus, the swarm can globally explore the feature subset search space, which is an essential point when dealing with a large and complex search space like feature selection. Therefore, PSO has gained much attention from the feature selection community recently [15].

Although there are other advanced and complicated feature selection algorithms [1,2,11,25], this work starts with the above four simple but well-known techniques. If the results are promising, future work will investigate extensions of these and/or other feature selection algorithms.

5.3   Experiment Settings

Following the same setting in the classification part, this experiment uses 10-fold cross-validation to generate the training and test sets. For each method, the balanced classification accuracy is measured with a linear SVM classification algorithm [18]. For χ^2, mRMR, and ReliefF, a hyperparameter for the number of selected features must be given. Therefore, the experiments measure the performance of the three algorithms on a wide range of the number of features: {50, 100, 150, ..., 4800} with increment 50. For PSO, the swarm size is set to 30 and the maximum number of iterations to 100. An advantage of PSO is that it does not need to specify a hyperparameter for the number of selected features. Since PSO is a stochastic algorithm, it is run 30 independent times on each classification task to make a reliable comparison.

5.4   Feature Selection Performance on Fish Species Classification

Figure 3 shows the results for χ^2 (chi2), ReliefF, mRMR and PSO on the fish species. The vertical axis is the classification accuracy and the horizontal axis is the number of selected features. As can be seen from the figures, the three algorithms χ^2, mRMR, and ReliefF perform poorly when the number of selected features is small. The main reason is that when the number of selected features is small, many relevant features are not selected, and thus essential classification information is missed. Among the three algorithms, χ^2 usually achieves the lowest classification performance since χ^2 does not reduce the feature redundancy and does not consider the interactions between features. ReliefF and mRMR achieve comparative performance. mRMR achieves its highest training and testing accuracies when the number of selected features is around 1500, which can be seen in Table 3.

As can be seen from the figure, most feature subsets evolved by PSO have from 1100 to 1500 features. The results indicate that PSO can automatically determine a good number of selected features, which cannot be achieved by the other three algorithms. As can be seen in Table 3, the highest classification performance of PSO is 99.17% which is about 1% higher than using all features. Meanwhile, PSO can remove 75% of the features, which means the classification system can be four times faster given the number of required packets/features is reduced by four times.

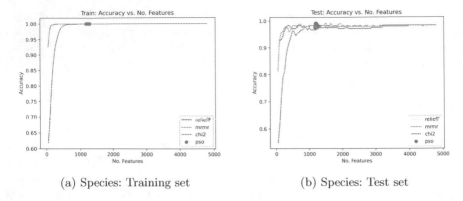

(a) Species: Training set   (b) Species: Test set

Fig. 3. Classification Accuracy of Fish Species on Different Numbers of Selected Features.

Table 3. Best accuracy on Fish Species.

Method	Number of features	Training accuracy	Testing accuracy
ReliefF	359	100.0	98.33
mRMR	**1500**	**100.0**	**99.17**
χ^2	3250	100.0	98.33
PSO	**1192**	**100.0**	**99.17**
Full	4800	100.0	98.33

5.5   Feature Selection Performance on Body Parts Classification

Figure 4 shows the results for χ^2, ReliefF, mRMR and PSO on the fish part dataset. As can be seen in Fig. 4a, χ^2, mRMR, and Relief-F witness a sharp improvement when the number of selected features is in the range [0, 500], which indicates that the 500 top-ranked features are essential to select. After that, the three approaches have a gradual incline, which peaks at 100% where all the features are selected. On the other hand, PSO selected feature subsets with sizes ranging in [1200, 1300]. Given the same classification performance, PSO usually selects a smaller number of features than the other three feature selection algorithms. The main reason is that PSO considers the interaction in the whole set of features, meanwhile, the other algorithms only consider the pair-wise interactions between feature pairs.

Table 4 illustrates the best accuracy for classifying fish body parts. As can be seen from the table, the best classification performance at 86.94% is achieved with 1500 features selected by mRMR. Thus, feature selection can also improve 7% accuracy over using all features. Meanwhile, the number of features is reduced by 2.5 times, which means the system can be 2.5 times faster. It should be noted that the testing performance of PSO is not as good as mRMR despite its superior training performance. The results indicate the potential overfitting of PSO on classifying body parts, which can be investigated more in the future.

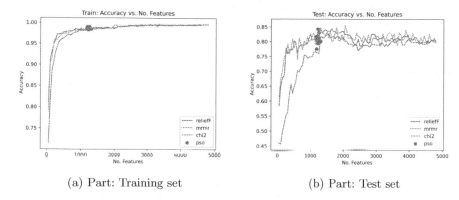

(a) Part: Training set   (b) Part: Test set

Fig. 4. Classification accuracy of Fish Body Parts on Different Numbers of Selected Features.

Table 4. Best accuracy on fish body parts

Method	Number of features	Training accuracy	Testing accuracy
ReliefF	1650	100.0	84.44
mRMR	**1500**	**100.0**	**86.94**
χ^2	1550	100.0	82.50
PSO	1223	100.0	84.31
Full	4800	100.0	79.86

5.6 Summary

In general, feature selection can significantly reduce the number of required packets/features and improve classification performance. For classifying the fish species, 75% of packets can be removed. For classifying the body parts, 60% of packets can be removed. The significant reduction means that the overall classification system can be up to 4 times faster. It should be noted that classifying the body part is more challenging than classifying the fish species. That is why classifying the body parts requires more features. Last but not least, PSO can automatically determine a good number of selected features. In general, PSO achieves good classification performance, except for some signs of overfitting which can be investigated in future.

6 Conclusions and Future Work

This paper has proposed an interpretable and effective classification process for fish oil analysis. Based on the results, it can be concluded that machine learning is a promising direction to improve the effectiveness and efficiency of the overall fish product system. In terms of accuracy, the proposed model can achieve high classification performance on classifying both fish species and body parts. However, fish species are easier to predict than body parts since there is more intra-class variation within fish species than there is a similarity between the same part from different fish. Among the considered classification algorithms, linear SVM achieves the best classification performance since it is suited to high-dimensional problems. Analysis of the SVM model demonstrates that not all packets are needed, and thus feature selection has been conducted to significantly reduce the number of packets and improve the classification performance.

It is worth noting that the classification and feature selection methods presented in this paper could be extended to further improve performance. This is particularly useful for the lower-accuracy fish part dataset. A potential direction is to improve the classification performance by constructing more informative high-level features, also known as feature construction. In addition, a more sophisticated imputation method can be developed to fill the missing packets in the fish data.

References

1. Alsahaf, A., Petkov, N., Shenoy, V., Azzopardi, G.: A framework for feature selection through boosting. Exp. Syst. Appl. **187**, 115895 (2022)
2. Alweshah, M., Alkhalaileh, S., Al-Betar, M.A., Bakar, A.A.: Coronavirus herd immunity optimizer with greedy crossover for feature selection in medical diagnosis. Knowl. Based Syst. **235**, 107629 (2022)
3. Bi, K., Zhang, D., Qiu, T., Huang, Y.: GC-MS fingerprints profiling using machine learning models for food flavor prediction. Processes **8**(1), 23 (2020)
4. Cortes, C., Vapnik, V.: Support-vector networks. Mach. Learn. **20**(3), 273–297 (1995)
5. Ding, C., Peng, H.: Minimum redundancy feature selection from microarray gene expression data. J. Bioinform. Comput. Biol. **3**(02), 185–205 (2005)
6. Eder, K.: Gas chromatographic analysis of fatty acid methyl esters. J. Chromatogr. B Biomed. Sci. Appl. **671**(1–2), 113–131 (1995)
7. Fix, E., Hodges, J.L.: Discriminatory analysis. Nonparametric discrimination: consistency properties. Int. Stat. Rev./Revue Internationale de Statistique **57**(3), 238–247 (1989)
8. Hand, D.J., Yu, K.: Idiot's bayes-not so stupid after all? Int. Stat. Rev. **69**(3), 385–398 (2001)
9. Ho, T.K.: Random decision forests. In: Proceedings of 3rd International Conference on Document Analysis and Recognition, vol. 1, pp. 278–282. IEEE (1995)
10. Kennedy, J., Eberhart, R.C.: Particle swarm optimization. In: Proceedings of the International Conference on Neural Networks, ICNN 1995, vol. 4, pp. 1942–1948. IEEE (1995)
11. Li, J., et al.: Feature selection: a data perspective. ACM Comput. Surv. (CSUR) **50**(6), 1–45 (2017)
12. Liu, H., Setiono, R.: Chi2: feature selection and discretization of numeric attributes. In: Proceedings of 7th IEEE International Conference on Tools with Artificial Intelligence, pp. 388–391. IEEE (1995)
13. Loh, W.Y.: Classification and regression trees. Wiley Interdisc. Rev. Data Min. Knowl. Discov. **1**(1), 14–23 (2011)
14. Matyushin, D.D., Buryak, A.K.: Gas chromatographic retention index prediction using multimodal machine learning. IEEE Access **8**, 223140–223155 (2020)

15. Nguyen, B.H., Xue, B., Zhang, M.: A survey on swarm intelligence approaches to feature selection in data mining. Swarm Evol. Comput. **54**, 100663 (2020)
16. Nguyen, H.B., Xue, B., Andreae, P., Zhang, M.: Particle swarm optimisation with genetic operators for feature selection. In: 2017 IEEE Congress on Evolutionary Computation (CEC), pp. 286–293 (2017). https://doi.org/10.1109/CEC.2017.7969325
17. Panse, M.L., Phalke, S.D.: World market of omega-3 fatty acids. Omega-3 Fatty Acids, pp. 79–88 (2016)
18. Pedregosa, F., et al.: Scikit-learn: machine learning in Python. J. Mach. Learn. Res. **12**, 2825–2830 (2011)
19. Restek: High-resolution GC analyses of fatty acid methyl esters (FAMEs)
20. Robnik-Šikonja, M., Kononenko, I.: Theoretical and empirical analysis of ReliefF and RReliefF. Mach. Learn. **53**(1), 23–69 (2003)
21. Simopoulos, A.P.: Evolutionary aspects of diet: the omega-6/omega-3 ratio and the brain. Mol. Neurobiol. **44**(2), 203–215 (2011)
22. Tomasi, G., Van Den Berg, F., Andersson, C.: Correlation optimized warping and dynamic time warping as preprocessing methods for chromatographic data. J. Chemom. A J. Chemometr. Soc. **18**(5), 231–241 (2004)
23. Tran, C.T., Zhang, M., Andreae, P.: Multiple imputation for missing data using genetic programming. In: The Annual Conference on Genetic and Evolutionary Computation, pp. 583–590 (2015)
24. Zhang, D., Huang, X., Regnier, F.E., Zhang, M.: Two-dimensional correlation optimized warping algorithm for aligning GC×GC-MS data. Anal. Chem. **80**(8), 2664–2671 (2008)
25. Zhang, Y., Gong, D.w., Gao, X.z., Tian, T., Sun, X.y.: Binary differential evolution with self-learning for multi-objective feature selection. Inf. Sci. **507**, 67–85 (2020)

Automated Radiology Report Generation Using a Transformer-Template System: Improved Clinical Accuracy and an Assessment of Clinical Safety

Brandon Abela[1]([✉]), Jumana Abu-Khalaf[1], Chi-Wei Robin Yang[2], Martin Masek[1], and Ashu Gupta[2]

[1] Edith Cowan University, Joondalup, WA 6027, Australia
{b.abela,j.abukhalaf,m.masek}@ecu.edu.au
[2] Fiona Stanley Hospital, Murdoch, WA 6150, Australia
ashu.gupta@health.wa.gov.au

Abstract. Radiologists are required to write a descriptive report for each examination they perform which is a time-consuming process. Deep-learning researchers are developing models to automate this process. Currently, the most researched architecture for this task is the encoder-decoder (E-D). An issue with this approach is that these models are optimised to produce output that is more coherent and grammatically correct rather than clinically correct. The current study considers this and instead builds upon a more recent approach that generates reports using a multi-label classification model attached to a Template-based Report Generation (TRG) subsystem. In the current study two TRG models that utilise either a Transformer or CNN classifier are produced and directly compared to the most clinically accurate E-D in the literature at the time of writing. The models were trained using the MIMIC-CXR dataset, a public set of 473,057 chest X-rays and 206,563 corresponding reports. Precision, recall and F1 scores were obtained by applying a rule-based labeller to the MIMIC-CXR reports, applying those labels to the corresponding images, and then using the labeller on the generated reports. The TRG models outperformed the E-D model for clinical accuracy with the largest difference being the recall rate (**T-TRG**: Precision 0.38, Recall 0.58, F1 0.45; **CNN-TRG**: Precision 0.34, Recall 0.69, F1 0.42; **E-D**: Precision 0.38, Recall 0.14, F1 0.19). Examination of the quantitative metrics for each specific abnormality combined with the qualitative assessment concludes that significant progress still needs to be made before clinical integration is safe.

Keywords: Medical text · Medical imaging · Deep learning · Templates · Encoder-decoder · CNN · Transformer

1 Introduction

Since the 1990s, the number of medical imaging examinations has increased globally at a rate significantly higher than the rate of practicing radiologists [1, 2]. This decreasing

© The Author(s), under exclusive license to Springer Nature Switzerland AG 2022
H. Aziz et al. (Eds.): AI 2022, LNAI 13728, pp. 530–543, 2022.
https://doi.org/10.1007/978-3-031-22695-3_37

radiologist to examination ratio has resulted in an excessive workload per radiologist [1], which is a major risk factor responsible for clinician burnout [3], with this burnout contributing to an increase in diagnostic errors [3–5] and loss of experienced staff [6]. A component of a radiologist's workload is the translation of medical images into diagnostics reports that describe their observations, particularly whether a region is normal, abnormal, or potentially abnormal [7]. An example report taken from the MIMIC-CXR dataset [8], a dataset consisting of chest X-rays paired with their corresponding report, has been provided in Fig. 1. Writing radiology reports is a time-consuming process that requires the expertise of a professional radiologist and therefore cannot be delegated to other clinicians [7]. This presents machine learning researchers with the opportunity to alleviate radiologist's workload through the development of automated Medical Report Generation (MRG) systems.

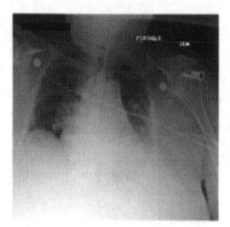

The lungs are hypoinflated with crowding of vasculature. There is progression of severe vascular enlargement with peribronchial cuffing as well as bilateral perihilar opacities with interval increase in small left pleural effusion. No right pleural effusion. No pneumothorax. Moderate cardiomegaly is stable. A right PICC tip is seen at least up to the low SVC.

Fig. 1. A chest X-ray from the MIMIC-CXR dataset [8] and the corresponding report.

1.1 Related Research

Current automated MRG research has been built atop a foundation of image captioning research and follows the same trend of improving the Encoder-Decoder (E-D) architecture [9–19]. A common theme throughout most of these studies is that the almost exclusive method of evaluating model performance is via Natural Language Processing (NLP) metrics, a group of measures that assess how well the generated output matches the target/s, with the most common metric being the BLEU [20] metric. This method primarily works by calculating the number of matching n-grams between the original and generated reports. An n-gram is a contiguous sequence of n items and when calculating a BLEU score, 1-gram (B1), 2-gram (B2), 3-gram (B3), and 4-gram (B4) metrics are commonly calculated, with a score of 1 representing a perfect match and a score of 0 representing a perfect mismatch. However, it has been shown that NLP metrics are not a good representation of the clinical content within a generated report, but rather a measure of report coherence and grammatical correctness [21, 22]. As such, a report could receive very high NLP scores, but be clinically incorrect. An example provided by Pino et al. [21] can be seen in Table 1 comparing the ROUGE-L score, a measure of the longest common subsequence, the BLEU score averaged across all n-gram values, and the clinical accuracy metrics including the precision, recall, and F1 score.

Table 1. An example of how NLP metrics are not a good representation of the quality of a generated clinical report. Italic represents correct text; underlined represents incorrect text; B = BLEU 1–4 averaged; RL = ROUGE-L; P = Precision; R = Recall.

Report	B	RL	F1	P	R
1 – **Ground Truth:** Heart size is mildly enlarged. Small right pneumothorax is seen	-	-	-	-	-
2 – Heart size is normal. No pneumothorax is seen	**0.493**	**0.715**	0	0	0
3 – *The cardiac silhouette is enlarged.* No pneumothorax	0.146	0.464	0.5	0.5	0.5
4 – *Mild cardiomegaly. Pneumothorax on right lung*	0.075	0.289	1	1	1

Boag et al. [22] were the first to assess the appropriateness of NLP metrics for healthcare-based AI research and profiled a range of models for automatic MRG, with their most notable result being the performance comparison between the conditional n-gram language model (CLM) and the CNN-LSTM encoder-decoder model. It was found that while the CNN-LSTM model outperformed the CLM on B1 (0.305 vs 0.206), B2 (0.201 vs 0.107), B3 (0.137 vs 0.057), B4 (0.092 vs 0.031) and other NLP metrics, both models performed equivalently on macro average F1 score (CNN-LSTM = 0.186 vs CLM = 0.185). This study was the first to specifically highlight the inconsistency between NLP metrics and clinical correctness.

Since publicly available datasets used to train ED models consist of unlabelled report/X-ray pairs, Boag et al. [22] obtained their precision, recall, and F1 scores by labelling the original and generated reports with the CheXpert labeller [23] and calculating their metrics with the resulting labels. The CheXpert labeller is a comprehensive rule-based labeller that provides chest X-ray reports with 14 labels (13 abnormality labels; 1 no findings label) that can subsequently be attributed to the corresponding X-ray. To assess the performance of the labeller, the authors performed three tests: mention detection, the labellers' ability to detect the mention of an abnormality, negation detection, the labellers' ability to detect when the mention of an abnormality negates its presence, and uncertainty detection, the labellers' ability to detect uncertainty within the mention of an abnormality. The labeller achieved a macro-average F1 score of 0.948 for mention detection, 0.899 for negation detection, and 0.770 for uncertainty detection.

Of the encoder-decoder studies previously cited, only 4 [14, 17–19] attempted to measure the performance of their model's clinical correctness. Alfarghaly et al. [19] had a radiologist qualitatively assess a subset (201 normal; 299 abnormal) of the generated reports and classify them as either accurate, missing detail, or false. For the normal reports, 99% were accurate, 0% were missing detail, and 1% were false. For the abnormal reports, 36.5% were accurate, 47.1% were missing detail, and 16.4% were false. From these results, it is concluded that it was harder for the model to detect abnormalities.

The other 3 studies [14, 17, 18] utilised the same approach as Boag et al. [22]. Zhang et al. [14], however, used a self-developed labeller, which performs the same task as the CheXpert labeller, and achieved 0.483 recall, 0.490 precision, and 0.478 F1. Chen et al. [17] and Lovelace et al. [18] both used the CheXpert labeller [23] with the MIMIC-CXR dataset and the performance of their models have been contrasted in Table 2. A comparison of these studies reveals information that is consistent with the findings of Boag et al. [22]; NLP metrics do not necessarily relate to clinical accuracy, with both achieving poor precision and recall.

Table 2. A comparison of the models produced by Chen et al. [17] and Lovelace et al. [18] showing that NLP metrics do not necessarily relate to clinical accuracy.

Models	B1	B2	B3	B4	RL	F1	P	R
Chen et al. (2020)	0.353	0.218	0.145	0.103	0.277	**0.276**	*0.333*	**0.273**
Lovelace et al. (2020)	**0.415**	**0.272**	**0.193**	**0.146**	**0.318**	0.228	*0.333*	0.217

While the movement away from NLP metrics is just emerging, there is still an inherent problem in the primary architecture used to generate reports; they are optimised against the structure of the original report and not the clinical content. Lovelace et al. [18] attempted to integrate clinical content learning into the current ED architecture, but they still achieved poor results, no better than Chen et al. [17] who did not optimise for clinical content.

As an alternative to the encoder-decoder architecture, Pino et al. [21] proposed a Template-based Report Generation (TRG) model that detects abnormalities using a CNN classifier and generates reports using fixed sentence templates for each abnormality. They used the same approach as Chen et al. [17] and Lovelace et al. [18] to measure clinical accuracy and achieved an F1 score of 0.428, precision of 0.381, and recall of 0.531. These results were also obtained using only the frontal X-rays and excluded the 'no findings' label provided by the CheXpert labeller which means their results cannot be directly contrasted with the study by Chen et al. [17] or Lovelace et al. [18]. The benefit of the CNN-TRG approach is that it allows the model to be directly optimised using clinical content since optimisation of a classifier is based on how well it categorises abnormalities. The limitation of this approach is that it is constrained by the labels provided and the complexity of the TRG subsystem.

The work in this paper improves upon the CNN-TRG approach by replacing the CNN with a Transformer and by implementing a more flexible TRG subsystem that retrieves templates based on probability outputs. The performance of the T-TRG and CNN-TRG models are directly compared against the ED model produced by Chen et al. [17], currently one of the most clinically accurate ED models. A qualitative assessment of the generated reports is then performed looking into the safety of integrating these deep learning models into a clinical setting. The contributions of the current study are as follows.

1. A model for generating clinical reports from chest X-rays that outperforms that current start of the art.
2. A clinical safety assessment of the current state of the art models.

2   Method

2.1   Dataset

The current study utilised the MIMIC-CXR [8] dataset, the largest public radiology dataset at the time of writing, which consists of 473,057 frontal/lateral chest X-ray images and 206,563 corresponding reports from 63,478 patients. Both the frontal and lateral X-ray images were used. The X-ray images without reports were excluded using the method employed by Chen et al. [17]. The make-up of the dataset after accounting for exclusion criteria are presented in Table 3. The training, validation, and testing splits presented in Table 3 are the official MIMIC-CXR splits after applying the exclusion criteria. The official MIMIC-CXR splits have been used since this setup is consistent with MRG literature reviewed in Sect. 1.1.

Table 3. The number of unique instances in the training, validation, and testing subsets of the MIMIC-CXR dataset after filtering out images without corresponding reports.

Dataset	Train	Val	Test
Images	279,790	2,130	3858
Reports	139,368	1,163	2,346
Patients	59,799	459	289
Avg. report length	51.24	49.63	60.76

The MIMIC-CXR dataset is an unlabelled dataset and therefore, for precision, recall, and F1 scores to be calculated, the data had to be labelled. These labels were obtained by using the CheXpert labeller [23] on the MIMIC-CXR reports. The CheXpert labeller is a comprehensive rule-based labeller that provides 13 abnormality labels to a report and provides a 14th 'no findings' label. In the study by Chen et al. [17] they chose to use the 'no findings' label and in the study by Pino et al. [21] they chose not to use it. In the current study, the 'no findings' label is not used for two reasons. First, the absence of the other 13 findings indicates 'no findings', and secondly, the addition of the 'no findings' label will confound the proposed metrics. Precision, recall, and F1 are measures that relate to the positive class and therefore representing 'no findings' with a positive label would result in the negative class leaking into the positive class. The labels were obtained for each of the reports in the dataset and were given to each of the corresponding X-ray images. These labels were used as the target values for the dataset. The resulting statistics for each abnormality are provided in Table 4.

The CheXpert dataset [23] is also used to initially train the TRG models on the same 13 abnormality labels before they are fine-tuned on the MIMIC-CXR dataset. The CheXpert dataset consists of 224,316 labelled frontal/lateral chest X-rays from 65,240 patients.

Table 4. The distribution of abnormalities within the MIMIC-CXR dataset after filtering out images without corresponding reports.

Abnormality	Train	Val	Test
No findings	48685 (18.0%)	539 (25.3%)	180 (4.7%)
Enlarged cardiomediastinum	51491 (19.0%)	192 (9.0%)	640 (16.6%)
Cardiomegaly	74209 (27.4%)	515 (24.2%)	1621 (42.0%)
Lung lesion	10892 (4.0%)	110 (5.2%)	258 (6.7%)
Lung opacity	83794 (30.9%)	593 (27.8%)	1786 (46.3%)
Edema	34205 (12.6%)	156 (7.3%)	842 (21.8%)
Consolidation	21756 (8.0%)	60 (28.2%)	429 (11.1%)
Pneumonia	31891 (11.8%)	92 (4.3%)	651 (16.9%)
Atelectasis	62111 (22.9%)	413 (19.4%)	1114 (28.9%)
Pneumothorax	26306 (9.7%)	68 (3.2%)	395 (10.2%)
Pleural effusion	64943 (24.0%)	399 (18.7%)	1421 (36.8%)
Pleural other	6034 (2.2%)	45 (2.1%)	156 (4.0%)
Fracture	11882 (4.4%)	57 (2.7%)	256 (6.6%)
Support devices	70792 (26.1%)	531 (25.0%)	1642 (42.6%)

2.2   Experiments

The CNN-TRG and T-TRG models detect abnormalities in the X-ray using the predictions from a multilabel-classifier to retrieve template sentences from a subsystem. These sentences are concatenated and used as the final report. The model structure is provided in Fig. 2.

DenseNet-121 [24] is used as the classifier for the CNN-TRG model and the original Vision Transformer (ViT) architecture [25] is used for the T-TRG model. The ViT model partitions an input image into 16x16 patches before vectorizing the patches and providing each vector with a positional encoding. The patches are then fed into a stack of multi-head attention and dense layers that utilise skip connection and normalisation. An output component of this stack is then fed into a softmax classifier which outputs a vector with the shape of the number of classes, which in the case of the current study is 13. This is a brief overview of the ViT classifier; a more in-depth description is provided in the original paper [25].

Both classifiers are first initialised with the pre-trained weights from ImageNet [26] before initially being trained on the CheXpert dataset [24] to classify the 13 abnormalities detailed in Table 4. Next the classifiers are fine-tuned on the target dataset (MIMIC-CXR). The models were trained using binary cross entropy loss for 40 epochs and early stopping was used to optimise the PR-AUC metric. PR-AUC is a single value representing the precision and recall performance of a model at all prediction thresholds.

The current study improves on the original TRG subsystem developed by Pino et al. [21] by applying class probabilities to the template retrieval process. The developed

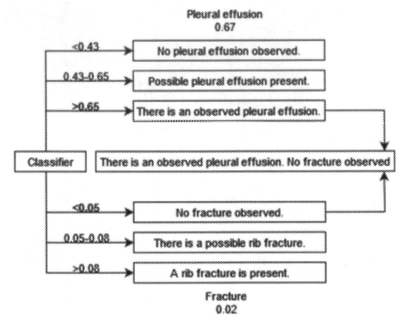

Fig. 2. An example of the T-TRG and CNN-TRG template subsystem that retrieves template sentences based on the prediction and threshold values and produces a report by concatenating them into a report.

TRG subsystem is unique to this study. The TRG component of the model consists of 3 sentences developed by a radiologist for each abnormality, with each abnormality's prediction probability retrieving a certain sentence class: either present, possibly present, or not present. The threshold for each sentence is based on the optimal PR-AUC threshold for each abnormality. A possibly present sentence is retrieved if the prediction value is up to 50% higher than the detection threshold. For example, for the standard detection threshold of 0.5, a prediction value of 0.5–0.75 would retrieve a 'possibly present' sentence, whereas a prediction value of 0.75–1.0 would retrieve a 'present' sentence. Negative predictions are also combined into individual sentences to produce more concise reports. For example, if no lung abnormalities are detected, these sentences are grouped into a single sentence, 'The lungs are clear'.

In order to evaluate our approach, a traditional E-D model was used as a baseline. The E-D developed by Chen et al. [17] was selected due to it being the most clinically accurate E-D model currently reported in the literature. The model consists of a transformer encoder, a transformer decoder, and a novel relational memory component that stores information about previously generated reports. A comprehensive description of the model is detailed in the original paper [17].

2.3 Quantitative Assessment

Both NLP and clinical accuracy metrics are used in the quantitative assessment of the models. NLP metrics have been included to highlight the inconsistency between NLP metrics and clinical accuracy and expand on current literature findings. For NLP metrics BLEU 1–4 is used, with B1 being calculated using 1-gram, B2 2-gram, B3 3-gram, and B4 4-gram. The BLEU score is calculated by taking the geometric mean of the test corpus' modified precision scores and multiplying that value by the exponential brevity penalty (BP) factor.

The modified precision score is calculated by counting the number of times each n-gram appears in the generated report, clipping those numbers to the rate of appearance in the original report, dividing the generated appearances by the original report appearances, and then adding together the results for each n-gram.

$$Pn = \frac{\sum\limits_{C \in \{Candidates\}} \sum\limits_{n\text{-}gram \in C} Count_{clip}(n\text{-}gram)}{\sum\limits_{C' \in \{Candidates\}} \sum\limits_{n\text{-}gram' \in C'} Count_{clip}(n\text{-}gram')} \tag{1}$$

The BP factor is calculated by finding the reference report with the closest length to that of the generated report, dividing that reference length with the generated length, and subtracting 1. The brevity penalty is only applied if the length of the generated report exceeds the length of the longest reference report.

In summary, we first calculate the geometric average of the modified n-gram precisions, Pn, using n-grams up to the length N and positive weights w_n summing to one. Next, with c as the length of the candidate translation and r as the effective reference length, we compute BP, then compute BLEU. The result is a value between 0 and 1, with 0 representing a perfect mismatch and 1 representing a perfect match.

$$BP = \begin{cases} 1 & if \ c > r \\ e\left(1 - \frac{r}{c}\right) & if \ c \le r \end{cases}$$
$$\text{BLEU} = \text{BP} \cdot \exp\left(\sum_{(n-1)}^{N} w_n \log p_n\right) \tag{2}$$

For the clinical accuracy metrics, the generated reports are labelled with the CheXpert labeller and the precision, recall and F1 score will be calculated by comparing the labels provided for the X-ray images to those provided for the generated reports. Precision is calculated by dividing the true positive rate by all positive predictions, recall is calculated by dividing the true positive rate by the total positive cases, and F1 is calculated by multiplying precision and recall, dividing the result by the precision plus recall, and multiplying the result by 2. Each of the TRG models are trained 5 times and the results are averaged across the 5 runs for each model.

2.4 Qualitative Assessment

For the qualitative assessment, 100 reports generated by each model were given to two radiologists to assess their clinical safety. A stratified randomisation algorithm was applied that randomly selected 100 reports that matched the 14 label distributions of the test set in Table 4 within a 25% range. The radiologists assessed the reports using 2 criteria:

1. Will the report cause immediate harm to the patient as a result of missing an abnormality or reporting an abnormality that could lead to unnecessary intervention?

2. Does the report miss any incidental findings that may not cause immediate harm, but may result in harm to the patient in the future?

An example of a report causing immediate harm is missing a pneumothorax as this has the potential for cardiorespiratory decompensation and possibly death. An example of a report causing possible long-term harm is missing a lung nodule, which on occasion may be cancerous. The proportion of generated reports that will cause immediate harm, long term harm, or both is then calculated for each model.

3 Results

The overall performance of each model is demonstrated in Table 5. The T-TRG model outperformed the CNN-TRG and E-D models on F1 score, with the T-TRG model achieving an F1 score of 0.45, the CNN-TRG model achieving a score of 0.42, and the E-D model a score of 0.19. For recall, the CNN-TRG model achieved the highest score with a value of 0.69, followed by the T-TRG model with a value of 0.58, and the E-D model only achieving 0.14 recall. Both the E-D and T-TRG model achieved the highest precision with a value of 0.38, with the CNN-TRG model close behind with a value of 0.34.

Table 5. The macro average metrics for each model contrasted against each other.

Model	B1	B2	B3	B4	P	R	F1
Transformer-TRG (Ours)	0.107	0.014	0.001	0.001	*0.38*	0.58	**0.45**
CNN-TRG (Ours)	0.192	0.045	0.008	0.001	0.34	**0.69**	0.42
Encoder-Decoder (Baseline)	**0.353**	**0.218**	**0.145**	**0.103**	*0.38*	0.14	0.19

The results for each model at the abnormality level displayed in Table 6 show findings that are consistent with the overall performance. The T-TRG model results in the highest F1 performance across all abnormalities, the CNN-TRG model results in the highest recall performance across most abnormalities, and top precision performance is shared by the T-TRG and E-D models.

Table 6. The clinical accuracy metrics per disease for each model. The highest performing model per metric is highlighted in bold – italics represent shared top performance. The % column indicate what percentage of the training data contains the corresponding abnormality.

Dataset	Transformer-TRG			CNN-TRG			Encoder-Decoder			
	P	R	F1	P	R	F1	P	R	F1	%
Enlarged cardiomediastinum	**0.22**	0.66	**0.33**	0.20	**0.81**	0.32	0.16	0.05	0.07	19.0

(*continued*)

Table 6. (*continued*)

Dataset	Transformer-TRG			CNN-TRG			Encoder-Decoder			
	P	R	F1	P	R	F1	P	R	F1	%
Cardiomegaly	**0.58**	0.83	**0.69**	0.54	**0.91**	0.67	0.50	0.42	0.46	27.4
Lung lesion	0.21	0.20	**0.21**	0.09	**0.68**	0.16	**0.42**	0.02	0.04	4.0
Lung opacity	0.52	*0.86*	**0.65**	0.51	*0.86*	0.64	**0.61**	0.23	0.34	30.9
Edema	**0.43**	0.60	**0.50**	0.39	**0.63**	0.49	0.41	0.11	0.17	12.6
Consolidation	**0.25**	0.38	**0.30**	0.19	**0.60**	0.29	0.19	0.09	0.12	8.0
Pneumonia	**0.27**	0.43	**0.33**	0.22	**0.57**	0.32	0.18	0.01	0.02	11.8
Atelectasis	**0.47**	0.66	**0.55**	0.40	**0.83**	0.54	0.44	0.18	0.26	22.9
Pneumothorax	0.17	0.40	**0.24**	0.14	**0.58**	0.22	**0.33**	0.01	0.01	9.7
Pleural effusion	0.64	**0.76**	**0.70**	0.70	0.68	0.69	**0.81**	0.18	0.29	24.0
Pleural other	**0.18**	**0.49**	**0.26**	0.10	0.38	0.16	0.00	0.00	0.00	2.2
Fracture	**0.17**	0.45	**0.25**	0.10	**0.69**	0.18	0.00	0.00	0.00	4.4
Support devices	0.86	*0.78*	**0.82**	0.81	*0.78*	0.80	**0.87**	0.56	0.68	26.1

Table 7. The results from the qualitative assessment of clinical safety. PLH = Possible Long-term Harm; IH = Immediate Harm; Both include reports that cause PLH and IH. Those in the 'Both' column are not counted in the PLH or IM column.

Model	Safe	PLH	IH	Both
Transformer-TRG (Ours)	62%	1%	34%	*3%*
CNN-TRG (Ours)	**68%**	*0%*	29%	*3%*
Encoder-Decoder (Baseline)	51%	*0%*	37%	12%

The results of the qualitative assessment are presented in Table 7 and show that the CNN-TRG model was the safest model with 68% of the reports assessed as safe, followed by the T-TRG model with a score of 62%, and the ED model with a score of 51%. Neither the CNN-TRG or ED model would cause any Possible Long-term Harm (PLH) and the T-TRG would cause PLH only 1% of the time. The CNN-TRG model would cause only Immediate Harm (IM) to the patient 29% of time, followed by the T-TRG model with a score of 34%, and the ED model with a score of 37%. Both the CNN-TRG and T-TRG models would cause both IM and PLH 3% of the time, and the ED model would cause both 12% of the time. An example report generated by each model has been provided in Fig. 3.

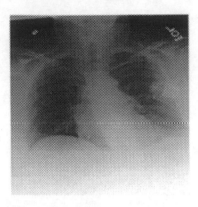

Target: There is a left pacemaker with appropriately positioned right atrial and right ventricular leads. The heart is moderately enlarged, increased in size compared to ___. There is pulmonary venous congestion with cephalization and predominantly perihilar heterogeneous opacities, perihilar heterogeneous opacities, consistent with mild interstitial pulmonary edema. No pleural effusions or pneumothorax. Possible slight loss of height of a upper mid thoracic vertebral body would be unchanged compared to ___.

T-TRG: The lungs are clear. The cardiomediastinal silhouette is possibly enlarged. The cardiac silhouette size is enlarged. No fracture. There is a noted support device.

CNN-TRG: The cardiomediastinal silhouette is possibly enlarged. The cardiac silhouette size is enlarged. No pulmonary nodules. Free of focal airspace disease. No pulmonary edema. No consolidation. No pneumonia. There is observed present atelectasis. There is a possible pneumothorax in the lungs. No pleural effusion. No pleural thickening observed. There is a possible rib fracture. There is a noted support device.

E-D: AP portable upright view of the chest. Port-a-cath resides over the right chest wall with catheter tip extending to the low svc. Lung volumes are low. limiting assessment. Allowing for this there in no focal consolidation effusion or pneumothorax. The cardiomediastinal silhouette is normal. Imaged osseous structures are intact.

Fig. 3. This figure provides an example of reports generated by each model along with the target report and corresponding X-ray image.

4   Discussion

The current study aimed to produce an automated MRG model with state-of-the-art clinical accuracy metrics and provide an assessment of whether it would be safe to integrate the model into a clinical setting. Both the T-TRG and CNN-TRG models outperformed the baseline E-D model (T-TRG: Precision 0.38, Recall 0.58, F1 0.45; CNN-TRG: Precision 0.34, Recall 0.69, F1 0.42; E-D: Precision 0.38, Recall 0.14, F1 0.19) with the T-TRG model showing the highest F1 performance. The qualitative assessment revealed the CNN-TRG model to be the safest model, followed by the T-TRG model, with the ED model being the least safe (T-TRG: Safe 62%, PLH 1%, IH 34%, Both 3%; CNN-TRG: Safe 68%, PLH 0%, IH 29%, Both 3%; ED: Safe 51%, PLH 0%, IH 27%, Both 12%).

The finding that the T-TRG model performed slightly better than the CNN-TRG model on precision but worse on recall may be due to the transformers inherent adversarial robustness (the model's ability to resist being fooled) resulting in more conservative but higher quality predictions. The work by Shao et al. [27] found that convolutions may negatively affect a model's adversarial robustness which does not affect the transformer architecture since it does not use convolutions. It was also found by Bhojanapalli et al. [28] that if pre-trained on a large enough dataset, the transformer performs significantly better on a range of out-of-distribution tests, for example resisting common image corruptions or texture-shape cue conflicting stimuli.

The study by Pino et al. [21] is the only other study that has currently attempted MRG using a CNN-TRG model and achieved an F1 of 0.428, precision of 0.381, and recall of 0.531 on the MIMIC-CXR dataset. In their study design they only used the frontal X-ray images which simplifies the task and makes it difficult to compare their results to a range of ED studies currently in the literature which utilise the frontal and lateral images. They also employed a binary TRG subsystem that only produced one of two templates: present or not present.

The results we obtained using the E-D model developed by Chen et al. [17] produced NLP metrics that aligned with those originally reported. However, the clinical accuracy metrics differed greatly due to the fact that in the original paper, they used the 14th 'no findings' label in their calculations which inflated their results due to the leaking issue discussed in Sect. 2.1. In the current study, we removed this label which accounts for the drop in performance.

The qualitative assessment revealed that the TRG models were safer than the ED model and the CNN-TRG model was safer than the T-TRG model which is consistent with the quantitative findings. The CNN-TRG model had higher recall and lower precision, which means that the model would be more liberal with its predictions resulting in fewer positive incidences being missed and the report ultimately being safer. The trade of for this safety is an increase in false positives.

5 Limitations

The current models are limited by the fact that they have only been assessed using chest X-rays. Medical imaging takes many forms and is used on many different body parts. The complexity of chest X-rays may lean in favour of the TRG approach, and the E-D approach may perform better on different parts of the body or with different imaging modalities. The current study is also limited by the fact that clinical accuracy was only assessed using the CheXpert labels which consist of 13 abnormalities.

6 Future Work

The TRG approach is limited by the rigidity of the generated reports. The current study made the reports more flexible by implementing a probability factor based on the abnormality thresholds, but there is still more that could be done. Localisation of the findings is an important component of a report, and this could be achieved by classifying heat map outputs for each class and using the predictions to retrieve a localisation template. Severity of the disease is also an important component of reports and could be achieved by replacing the binary classes with multiple classes of severity. Future TRG models could be improved using an ensemble of models based on the highest performing classifiers in the literature. And finally, future E-D models could attempt to improve clinical accuracy by integrating multi-label information to the bottleneck between the encoder and decoder.

References

1. Winder, M., Owczarek, A.J., Chudek, J., Pilch-Kowalczyk, J., Baron, J.: Are we overdoing it? Changes in diagnostic imaging workload during the years 2010–2020 including the impact of the SARS-CoV-2 pandemic. Healthcare **9**(11), 1557 (2021)
2. Ciarrapico, A., et al.: Diagnostic imaging and spending review: extreme problems call for extreme measures. Radiol. Med. (Torino) **122**(4), 288–293 (2017)
3. Harry, E., et al.: Physician task load and the risk of burnout among US physicians in a national survey. Joint Comm. J. Qual. Patient Saf. **47**(2), 76–85 (2021)
4. Bruls, R., Kwee, R.: Workload for radiologists during on-call hours: dramatic increase in the past 15 years. Insights Imaging **11**(121) (2020)
5. Owoc, J., Manczak, M., Tombarkiewicz, M., Olszewski, R.: Burnout, well-being, and self-reported medical errors among physicians. Pol. Arch. Med. Wewn. **131**, 626–632 (2021)
6. Singh, N., et al.: Occupational burnout among radiographers, sonographers and radiologists in Australia and New Zealand: findings from and national survey. J. Med. Imaging Radiat. Oncol. **61**(3), 304–310 (2016)
7. Good Practice for Radiological Reports: Guidelines from the European society of radiology (ESR). Insights Imaging **2**(2), 93–96 (2011)
8. Johnson, A., et al.: MIMIC-CXR, a de-identified publicly available database of chest radiographs with free-text reports. Sci. Data **6**, 317 (2019)
9. Shin, H., Roberts, K., L, Le., Demner-Fushman, D., Yao, J., Summers, R.: Learning to read chest X-rays: recurrent neural cascade model for automated image annotation. In: 2016 IEEE Conference on Computer Vision and Pattern Recognition (CVPR), pp. 2497–2506. IEEE Computer Society (2016)
10. Zhang, Z., Xie, Y., Xing, F., McGough, M., Yang, L.: MDNet: A semantically and visually interpretable medical image diagnosis network. In: 2017 IEEE Conference on Computer Vision and Pattern Recognition (CVPR), 3549–3557. IEEE Computer Society (2017)
11. Wang, X., Peng, Y., Lu, L., Lu, Z., Summers, R.: TieNet: Text-image embedding network for common thorax disease classification and reporting in chest X-rays. In: 2018 IEEE Conference on Computer Vision and Pattern Recognition (CVPR), pp. 9049–9058. IEEE Computer Society (2018)
12. Jing, B., Xie, P., Xing, E.: On the automatic generation of medical imaging reports. In: Proceedings of the 56th Annual Meeting of the Association for Computational Linguistics (ACL), vol. 1: Long papers, pp. 2577–2586 (2018)
13. Yuan, J., Liao, H., Luo, R., Luo, J.: Automatic radiology report generation based on multi-view image fusion and medical concept enrichment. In: Shen, D., et al. (eds.) MICCAI 2019. LNCS, vol. 11769, pp. 721–729. Springer, Cham (2019). https://doi.org/10.1007/978-3-030-32226-7_80
14. Zhang, Y., Wang, X., Xu, Z., Yu, Q., Yuille, A., Xu, D.: When radiology report generation meets knowledge graph. In: The Thirty-Second Innovative Applications of Artificial Intelligence Conference (IAAI), pp 12910–12917 (2020)
15. Syeda-Mahmood, T., et al.: Chest X-ray report generation through fine-grained label learning. In: Martel, A.L., et al. (eds.) MICCAI 2020. LNCS, vol. 12262, pp. 561–571. Springer, Cham (2020). https://doi.org/10.1007/978-3-030-59713-9_54

16. Xiong, Y., Du, B., Yan, P.: Reinforced transformer for medical image captioning. In: Suk, HI., Liu, M., Yan, P., Lian, C. (eds) Machine Learning in Medical Imaging. MLMI 2019. Lecture Notes in Computer Science, vol 11861. Springer, Cham (2019). https://doi.org/10.1007/978-3-030-32692-0_77

17. Chen, Z., Song, Y., Chang, T., Wan, X.: Generating radiology reports via memory-driven transformer. In: Webber, B., Cohn, T., He, Y., Liu, Y. (eds.) Proceedings of the 2020 Conference on Empirical Methods in Natural Language Processing, pp. 1439–1449. Association for Computer Linguistics (2020)

18. Lovelace, J., Mortazavi, B.: Learning to generate clinically coherent chest X-ray reports. In: Cohn, T., He, Y., Liu, Y.: Findings of the Association for Computational Linguistics, EMNLP, pp. 1235–1243 (2020)

19. Alfarghaly, O., Khaled, R., Elkorany, A., Helal, M., Fahmy, A.: Automated radiology report generation using conditional transformers. Inform. Med. Unlocked **24**, 100557 (2021)

20. Papineni, K., Roukos, S., Ward, T., Zhu, W.: BLEU: a method for automatic evaluation of machine translation. In: Proceedings of the 40th Annual Meeting of the Association for Computational Linguistics, pp. 311–318. Association for Computational Linguistics, Philadelphia (2002)

21. Pino, P., Parra, D., Besa, C., Lagos, C.: Clinically correct report generation from chest X-rays using templates. In: Lian, C., Cao, X., Rekik, I., Xu, X., Yan, P. (eds.) MLMI 2021. LNCS, vol. 12966, pp. 654–663. Springer, Cham (2021). https://doi.org/10.1007/978-3-030-87589-3_67

22. Boag, W., Hsu, T., McDermott, M., Berner, G., Alesentzer, E., Szolovits, P.: Baselines for chest X-ray report generation. In: Proceedings of the Machine Learning Research, pp. 126–140 (2020)

23. Irvin, J., et al.: CheXpert: A Large Chest Radiograph Dataset with Uncertainty Labels and Expert Comparison. Proc. AAAI Conf. Artif. Intell. **33**, 590–597 (2019)

24. Huang, G., Liu, Z., Van Der Maaten, L., Weinberger, K.: Densely connected convolutional networks. In: 2017 IEEE Conference on Computer Vision and Pattern Recognition (CVPR) (2017)

25. Dosovitskiy, A., et al.: An image is worth 16x16 words: transformers for image recognition at scale. In: International Conference on Learning Representations (2020)

26. Deng, J., Dong, W., Socher, R., Li, L., Kai Li, Li Fei-Fei: ImageNet: a large-scale hierarchical image database. In: 2009 IEEE Conference on Computer Vision and Pattern Recognition. (2009)

27. Shao, R., Shi, Z., Yi, J., Chen, P., Hsieh, C.: On the adversarial robustness of vision transformers. In: 2021 International Conference on Computer Vision. arXiv (2021)

28. Bhojanapalli, S., Chakrabarti, A., Glasner, D., Li, D., Unterthiner, T., Veit, A.: Understanding robustness of transformers for image classification. In: 2021 International Conference on Computer Vision. arXiv (2021)

3D Face Reconstruction with Mobile Phone Cameras for Rare Disease Diagnosis

Yiwei Liu[1(✉)], Ling Li[1], Senjian An[1], Petra Helmholz[2], Richard Palmer[2], and Gareth Baynam[2,3,4,5]

[1] School of Electrical Engineering, Computing and Mathematical Sciences, Curtin University, Kent Street, Bentley, WA, Australia
{Yiwei.Liu,Ling.Li,Senjian.An}@curtin.edu.au

[2] School of Earth and Planetary Sciences, Curtin University, Kent Street, Bentley, WA, Australia
{Petra.Helmholz,Richard.Palmer}@curtin.edu.au

[3] Rare Care Centre, Perth Children's Hospital, Perth, WA, Australia
Gareth.Baynam@health.wa.gov.au

[4] Western Australian Register of Developmental Anomalies and Genetic Services of WA, King Edward Memorial Hospital, Subiaco, WA, Australia

[5] Faculty of Health and Medical Sciences, Division of Paediatrics and Telethon Kids Institute, University of Western Australia, Perth, WA, Australia
http://www.curtin.edu.au/

Abstract. Computer vision technology is advancing rare disease diagnosis to address unmet needs of the more than 300 million individuals affected globally; one in three rare diseases have a known facial phenotype. 3D face model reconstruction is a key driver of these advances. However, the utility of 3D reconstruction from images obtained from mobile phone cameras has been questionable due to relatively low quality 2D data and need external calibration methods (e.g. visual markers) to extract accurate measurements. Herein a novel implementation pipeline, leveraging deep learning technologies, that can successfully reconstruct 3D face models from multiple 2D images taken by mobile phone cameras for clinician usage is described. Specifically, Multi-view Stereo (MVS) has been introduced to this application for providing a cost-effective pipeline of 3D face dense reconstruction. As a state-of-the-art MVS method, deep-learning based MVS has shown its strong generalization capability of using the low quality 2D face images to reconstruct 3D face models without camera calibration. The results demonstrate conceptual proof of a analytic pipeline to satisfy the clinician's needs.

Keywords: 3D face · Multi-view stereo · 3D reconstruction · Rare disease diagnosis

Supported by The Centre for Research Excellence in Neurocognitive Disorders, Neuroscience Research Australia.

1   Introduction

Rare Disease (RD) has been defined in Australia as one that affects fewer than 2000 individuals. Although this number is defined differently in many countries, the patients suffering from some form of RDs is estimated to constitute as much as 10% of the population globally. For this reason, RDs have presented a public health challenge [23]. Specifically in Australia, 2 million people live with a rare disease, and 30–50% of these individuals have experienced their first symptoms during childhood [21]. Early diagnosis is critical for supporting best medical care for people living with rare diseases (PLWRD) and their families. However, the timely and accurate RD diagnosis remains a global medical challenge. Thirty percent of RD patients are referred to an average of six specialists, and some patients wait for up to 30 or more years for a definitive diagnosis, if it is achieved at all diagnosed. On the journey towards definitive diagnosis, only half of all initial diagnoses are correct.

Many RDs can be diagnosed directly from patients' phenotype - how they appear, how they behave, what their physical and cognitive abilities are, compared to their peers [21]. Facial phenotype is recognised as an effective diagnostic parameter. However, their recognition requires expert experience, is subjective and subject to cognitive bias. Fortunately, salient aspects of facial phenotype can provide clinical diagnostic assistance [1]. Cliniface is an integrated, modular and interoperable platform of computational tools for visualising and analysing 3D facial information [20]. Cliniface is compatible with multiple 3D imaging hardware systems, for example the Vectra H1, which can perform 3D face scans with sub-millimetre accuracy. With Cliniface, clinicians can make use of anatomical facial landmarks to extract measurements of potential clinical significance, e.g., distances between certain pairs of landmarks. These facial landmarks can be detected semi-automatically using Cliniface, in which the facial landmark positions can be manually adjusted by the clinicians without revisiting the patients. The accurate measurements via Cliniface are used to assist diagnosis through referencing to databases of facial phenotype with a wide variety of demographic information (gender, date of birth, ethnicity etc.) that can be personalised to the individual being assessed. Figure 1 shows an example in Cliniface user interface.

Cliniface improves RD diagnosis compared to subjective assessments alone [20], however one of its limitations is reliance on bespoke and moderately costly medical grade 3D imaging hardware. Unfortunately, data directly acquired e.g. through using proprietary smartphone software has been shown to be unreliable and not fit-for-purpose when compared to medical grade 3D imaging hardware. The advanced 3D imaging systems are very costly and generally inaccessible to most clinicians, especially those in the remote areas. Therefore, there is a strong demand to develop novel technologies to build highly detailed and accurate 3D facial models from images taken by low-cost everyday camera systems, such as camera on smartphones, especially for patients in the remote areas and low and middle income (LMIC) settings.

Multi-view stereo (MVS) has been extensively researched [12] and achieved many 3D reconstruction results with outstanding performance, especially for static objects and large scale scenes [11]. Compared to the conventional 3D recon-

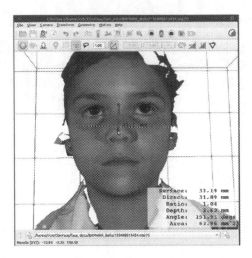

Fig. 1. An example in Cliniface user interface [20].

struction via binocular stereo images, MVS has greater flexibility for variable camera settings, and higher accuracy of 3D reconstruction results. Therefore, we applied MVS on the images taken by mobile-phone cameras to reconstruct detailed and accurate 3D facial models. For end-user convenience, we designed to deploy one smartphone camera to take a series of face photos from different angles under similar lighting conditions. The set of multi-view images is processed to be fed into an MVS algorithm to deliver 3D face reconstruction. Given that conventional MVS is challenged when processing images with significant texture-less regions and these are frequent when face images are taken from close distances by smartphones under various lighting conditions. A novel application of a deep learning method is introduced [26]. We demonstrate that this deep learning method can achieve good performance of 3D face reconstructions from a set of 2D images.

There are three major contributions of this research:

1. Develop a novel vision-based pipeline to reconstruct 3D face models from 2D face images taken by mobile phone cameras. There is a high demand on the accuracy of the reconstructed 3D face models since they are employed for medical applications.
2. Introduce a deep learning algorithm to apply the multi-view stereo for 3D face reconstruction with 2D face images taken at different times for the almost-rigid human face target. Deep-learning based dense 3D reconstruction method has shown robust performances in general 3D reconstruction tasks for rigid objects. The algorithm is deployed to show the robust 3D reconstruction capability of deep-learning based MVS dealing with almost-rigid objects.
3. Provide a vision-based computational tool to assist clinical assessments of rare diseases, such as rare neurocognitive disorders, that is incorporated into the Cliniface platform. Cliniface is free and open source to support remote and

LMIC applications, and is deployable on premise to overcome data sharing barriers, collectively promoting equitable access and scale.

2   Related Work

3D face reconstruction has been extensively investigated over the last decades, and has been widely implemented in various computer-based scientific applications [17,22] to address limitations of 2D facial analysis [18]. Compared to the 3D facial data, 2D images do not capture the complete facial geometry information, because they collapse in one dimension (losing the depth information). This may introduce large physical measurement errors. In contrary, 3D face models can provide all the face geometry details, which is invariant to pose and illumination. To successfully reconstruct 3D face models from 2D images, some prior knowledge of the targeting faces might be needed to resolve the ambiguities caused by the absence of the 3rd dimension.

3D face reconstruction is a specific 3D reconstruction task. With the conventional 3D reconstruction methods, the prior knowledge on the camera information such as the intrinsic matrix and the extrinsic matrix needs to be provided or estimated. With this prior information, the accurate 3D position can be calculated by the classical triangulation methods. To obtain the accurate prior knowledge of 2D imaging systems, camera calibrations will be applied at first stage. Based on this process, the general 3D reconstruction systems can be classified into two overarching categories, namely 3D reconstruction system with camera calibrations, and 3D reconstruction system without camera calibrations.

2.1   3D Face Reconstruction System with Camera Calibrations

Camera calibration has been used to estimate the prior knowledge of 2D face imaging systems used for 3D face reconstruction. The more accurately the camera calibration is done, the more accurate the obtained 3D face model will be. Beeler et al. 2010 [2] has applied camera calibrations in their 3D face reconstruction method with highly accurate results.

With strict camera calibrations, traditional MVS can recover the 3D objects or scenes from a set of multi-view 2D images with high accuracy. Based on the representation of reconstructed 3D models, traditional MVS can be categorized into four groups of methods: voxel-based methods, surface-evolution based methods, patch-based methods and depth-map based methods. From the perspective of 3D reconstruction implementation, depth-map based methods are more concise and flexible, compared to the other three group of methods. Patchmatch Stereo methods [8], a typical depth-map based methods, have shown robust performance with calibrated multi-view images. With Patchmatch Stereo, Galliani et al. [8] uses a red-black chessboard pattern to propagate the passing-message in parallel. Schönberger et al. [24] jointly estimates pixel-wise view selection, depth map and surface normal in his COLMAP pipeline. ACMM [29] accommodates the checkerboard sampling, multi-hypothesis joint view selection and multi-scale geometric consistency guidance with camera calibrations as well.

2.2   3D Face Reconstruction System Without Camera Calibration

There are three different kinds of strategies in 3D face reconstruction without camera calibration, namely statistical model fitting, photometric stereo, and deep learning [18].

Statistical Model Fitting Methods, as the classical approach among the three strategies, encodes the prior geometric information in a generic 3D face model, which are generated from prior knowledge of a set of 3D faces. The 3D face model will be adjusted to match the input images. Because this strategy includes geometric variations of the input faces, it can be useful for the 2D-to-3D face reconstruction with face occlusion and large pose variations. 3D face models reconstructed from the statistical model fitting methods usually consist of a mean face along with modes of variation of its geometry and appearance.

By adjusting the model parameters, a 3D facial model from a photograph can be estimated. By projecting a 3D face into the image plane with illumination, the projected images should match the given 2D face images. The most widely used statistical models of 3D faces are the 3D Morphable Models (3DMM), which were firstly proposed by Blanz and Vetter [3].

There are two recurrently noted methodological limitations of 3DMM. Firstly, as a PCA (Principal Component Analysis) based method, 3DMM algorithm estimates the principal basis vectors that model the input data globally, hence subtle information, such as wrinkles, may be discarded. Thus, reconstructing facial details by fitting a 3DMM becomes very difficult. It would pose some limitations on the local accuracy of the generated 3D face models as well. Secondly, the real general face, according to 3DMM, is not perfectly combined with the shape variations in a linear way. As a result, 3DMM cannot handle nonlinear facial variations very well.

Photometric Methods are mainly based on estimating the lighting parameters and surface normal from a set of 2D images using a Lambertian reflectance model. This approach was originally proposed by Woodham [27], who reconstructed 3D face models via estimating the surface normal from several 2D images under different lighting conditions. With the same philosophy, Ghosh et al. [9] captures high solution diffuse and specular photometric information using a multi-view face capture system, and reconstructs detailed facial geometry.

Photometric 3D face reconstruction can use images captured by simple photo-taking devices to reconstruct 3D models. However, additional prior knowledge of the 3D face model, such as template shapes [33], must be included because the RGB images are unconstrained, and the light source is unknown. For this reason, a very large number of 2D face images are needed to implement the photometric methods in real-world applications, and the 3D face reconstruction results are noisy due to lack of a geometric prior to constrain these solutions. The computational cost of processing these images is too high to be implemented on mobile devices.

Deep Learning Methods for 3D-from-2D face reconstruction needs to establish the mapping between 2D face images and 3D face models. Deep learning methods are based on the available 3D face data, learning the mapping which encodes the prior knowledge in the weights of the trained neural networks.

By applying an end-to-end deep learning method, GCNet [15] generated a stereo estimation from 3D cost volume regularization, and produces the final disparity map via a soft-argmin operation. With a coarse-to-fine approach, PSMNet [4] introduces spatial pyramid pooling (SPP) to the Cross Perspective Projection layer of the network for regularization. DeepPruner [7] builds a lightweight cost volume, which is regularized by a 3D Neural Network, via a differentiable Patchmatch module discarding most disparities.

2.3 Learning-Based MVS

As one specific subarea of deep learning methods for 3D face reconstruction, learning-based MVS has attracted much attention recently because of its high accuracy and ease of application. As it has been motioned above, voxel-based methods have clear drawbacks of a volumetric representation, which has restricted its usage to small-scale reconstructions [26]. While, many learning-based MVS are based on plane-sweep stereo [6], which uses depth maps to reconstruct 3D scenes. They regularize the cost volumes built with the warped features of multi-view images, through the 3D Convolutional Neural Networks (CNNs) and then regress the depth. Because of the heavy computational cost of 3D CNNs, these applications of deep-learning based MVS commonly apply the down-sampling operation to the cost volume. R-MVSNet [32] has tried to reduce memory by regularizing 2D cost maps. The current search efforts on learning-based MVS focus on improving efficiency and capability of estimating high-resolution depth maps. Accordingly, Cas-MVSNet [10] introduces cascade cost volumes based on a feature pyramid and estimates the depth map in a coarse-to-fine manner. Similarly, UCS-Net [5] forms cascade adaptive thin volumes by using variance-based uncertainty estimates for an adaptive construction. CVP-MVSNet [30] applies the cascading method to generate an image pyramid and constructs a cost volume pyramid.

3 Methods

2D face images of participants in this research are taken under the indoor conditions. This setting is designed to mimic real-word conditions where the final user (i.e. clinicians) are non-experts in photography and external calibration (e.g. with checkerboards or markers) is not compatible with standard clinical flow. Accordingly, Structure-from-Monition (SfM) technology and well developed key point detection algorithms are deployed to ascertain accurate information of the position and pose of the cameras and support the 3D face reconstruction. We refine PatchmatchNet [26], to specifically support 3D reconstruction using 2D face images taken by mobile phone cameras to reconstruct geometrically accurate 3D face models. Below, we describe the details of our application pipeline, covering 2D face image taking, sparse 3D face reconstruction, and dense 3D face reconstruction with deep learning technologies.

3.1   2D Face Image Acquisition

To support clinical utility, mobile phone cameras are used. Leveraged by the current camera technology, mobile phone cameras have been widely used in many applications with reasonable image qualities and high image resolutions. For example, iPhone 13 pro are using 12-megapixel cameras and Samsung S22 are using 50-megapixel cameras. An important advantage of these consumer cameras is the provision of clear facial photos at different viewpoints with auto focus. For our experiments, we applied the camera of iPhone 12 to collect multi-view 2D face photos. Lighting conditions of photo collection were set up as the general indoor environment. The light coming through windows and highlighting resources were minimised. To mimic the real application environment for the future clinical use, overall lighting conditions were far less controlled than typical 3D reconstruction settings.

Similar to previous SfM approaches, no camera calibration was performed. The background of these face photos is better to have rich texture for the sub tasks of sparse 3D face reconstruction, such as key point detection and key point mapping. As the proposed pipeline is applied with the 3D scene reconstruction method, the face must occupy the major area of the multi-view images to catch most 2D face geometry details.

3.2   Sparse 3D Face Reconstruction

To support non-calibrated 3D face reconstruction, SfM is introduced to provide a sparse 3D face reconstruction model, which has also been included in the implementation pipeline of PatchmatchNet [26]. COLMAP [24] is used to handle the process of the sparse 3D reconstruction. The initial multi-view 2D face images are fed into the COLMAP open-source platform to obtain camera relative poses and locations, referencing to the world coordinates of the scene to be reconstructed. SIFT [19] features are applied to each 2D image pairs for key point detection and matching during this process. The estimated camera intrinsic and extrinsic matrices and the matching points of multi-view images are imported into PatchmatchNet for dense 3D face reconstruction.

3.3   Dense 3D Face Reconstruction

PatchmatchNet [26] is used to complete the final dense 3D face reconstruction with the state-of-the-art deep learning technology. PatchmatchNet is designed for 3D scenes reconstruction with high-resolution multi-view stereo, and it can process high resolution imagery efficiently. For this reason, PatchmatchNet is more suited to run on resource-limited devices than competitors that employ 3D cost volume regularization. For the first time, we apply an iterative multi-scale Patchmatch in an end-to-end trainable architecture, see Fig. 2 for details. The PatchmatchNet model was trained on the DTU [13] dataset, and its 3D reconstruction results show a very competitive performance and good generalization on both the Tanks & Temples [16] dataset and the ETH3D [25] dataset.

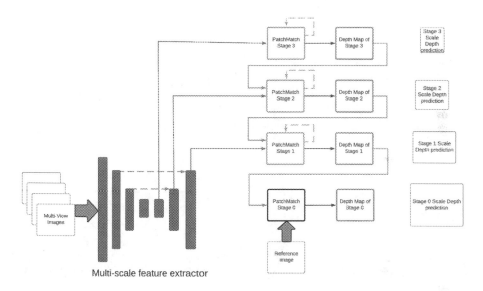

Fig. 2. Structure of PatchmatchNet [26]: multi-scale feature extractor, learning-based Patchmatch and refinement. Patchmatch process has been introduced with the coarse-to-fine manner to predict the depth map at multiple scales.

In contrast to most of cost volume regulation approach for the 3D reconstruction, PatchmatchNet proposes a strategy of adaptive sampling points for the spatial cost aggregation and extends the conventional Patchmatch ideas into deep learning methods. The original pre-trained model has shown a very robust performance on ETH3D [25] dataset under complex lighting conditions. Compared to the most popular deep-learning based MVS method - MVSNet [31], PatmatchNet has a fast performance on DTU dataset [13] with the run-time processing cost about 0.25 s per image, and less GPU memory consumption at about 2G, for images with resolution of 1152 × 864. This will allow fast reconstruction of dense 3D face models on mobile phones. According to Xiao et al. [28], model fine-tuning on the pre-trained model of MVSNet has marginal improvements in the experiments on the large extra high qualitative 3D face datasets. To simplify our 3D face reconstruction pipeline and balance the cost and efficiency of the approach, the pre-trained model of PatchmatchNet has been used directly for dense 3D face reconstruction to generate 3D face models.

4   Results and Discussion

In this section, we present experimental results on 3D face model reconstruction from images taken by mobile phones. The reconstructed 3D face models have been imported to the Cliniface platform to verify whether the reconstructed models are good enough to detect the facial landmarks correctly with the facial landmark detection function of Cliniface, which is a core part of medical diagnosing assistance. To check the robustness of the proposed method, 5 people in our research group have participated in data collection. Each person sits in their office, and multi-view face images are taken by the camera with the same iPhone 12. For all the data collected, 3D face models have been reconstructed successfully. Here, we illustrate the experimental results with the images and reconstructed 3D face model of one person.

The input face images are shown in Fig. 3. These images are taken from different views, and the image brightness varies with different view locations with different shading areas.

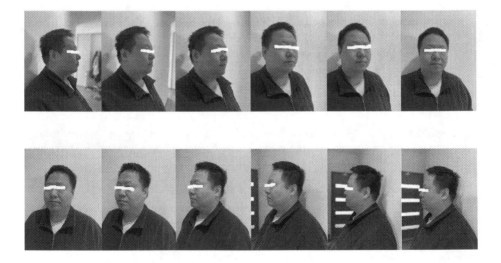

Fig. 3. Input 2D face images

With the COLMAP [24] platform, the sparse 3D face model is obtained and the result is shown in Fig. 4. It can be seen that the sparse 3D face model provides coarse face geometry information in terms of sparse point clouds. The related camera poses and positions are displayed in red in the figure.

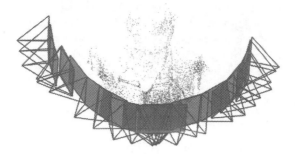

Fig. 4. Sparse 3D face reconstruction result.

Based on the sparse 3D reconstruction model, the camera intrinsic and extrinsic parameters are introduced to the PatchmatchNet framework (Fig. 2), processing the input 2D face images again to generate the final dense 3D face model. The point clouds of the reconstructed 3D face model is presented in Fig. 5. It can be seen that this 3D face model can recover the 3D face geometry more clearly. To use the 3D model in the CliniFace platform, post-processing on the point clouds model is applied with Poisson surface reconstruction [14]. The result shown in Fig. 6 has been imported to Cliniface to verify the usefulness of this 3D face model by applying the facial landmark detection on it.

Fig. 5. Dense 3D face reconstruction result.

Fig. 6. 3D face reconstruction result after post-processing with Poisson surface reconstruction.

The 3D face models of all the other participants were reconstructed successfully with the same pipeline. These 3D face models generated from iPhone input data, can then be analysed through Cliniface to provide diagnostic support. Figure 7 shows a reconstructed 3D face model and its visualisation in Cliniface platform. The facial landmarks have been detected by the embedding function of Cliniface to support conceptual proof of principle for using readily accessible camera technology for Cliniface functions, such as RD diagnostic support. The facial landmark detection result on the 3D face model generated by Vectra H1 3D imaging system is shown in Fig. 8. Qualitatively compared to this result, the facial landmarks can be detected properly on cliniface with our reconstructed 3D face model to assist the medical diagnosis.

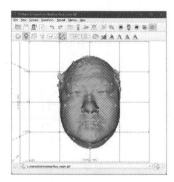

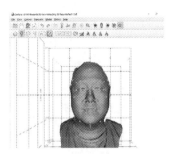

Fig. 7. 3D face model generated by our pipeline for facial landmark detection on Cliniface.

Fig. 8. 3D face model generated by Vectra H1 for facial landmark detection on Cliniface.

5   Conclusion

We have presented a novel computer vision pipeline to implement the multi-view stereo method to reconstruct 3D face models from 2D face images taken by mobile phone cameras. The PatchmatchNet, one of the modern deep-learning based 3D reconstruction methods, has been refined for deployment to address the challenges of 3D face reconstruction. This further extends the utility of Cliniface with a low cost and scalable solution, including rare disease diagnosis for remote regions and low-and-middle-income setting. Equity and inclusive access is further supported by the diverse range of normative reference data currently accessible through Cliniface, which is increasing with expanding international and community partnerships.

6   Research Ethic

This research project involves highly sensitive private information for the future clinic usage. Strict research ethic procedure has been followed since the beginning of the project. Only research team members have access to confidential data of the potential patients. Face images will not be collected without proper consent. The confidential data of individual participants in the future research will be transferred and stored according to the ethic and governance approval.

References

1. Baynam, G., et al.: 3-dimensional facial analysis-facing precision public health. Front. Pub. Health **5**, 1–6 (2017). https://doi.org/10.3389/fpubh.2017.00031
2. Beeler, T., Bickel, B., Beardsley, P., Sumner, B., Gross, M.: High-quality single-shot capture of facial geometry. In: ACM SIGGRAPH 2010 Papers, pp. 1–9 (2010)
3. Blanz, V., Vetter, T.: A morphable model for the synthesis of 3D faces. In: Proceedings of the 26th Annual Conference on Computer Graphics and Interactive Techniques, pp. 187–194 (1999)
4. Chang, J.R., Chen, Y.S.: Pyramid stereo matching network. In: Proceedings of the IEEE Conference on Computer Vision and Pattern Recognition, pp. 5410–5418 (2018)
5. Cheng, S., et al.: Deep stereo using adaptive thin volume representation with uncertainty awareness. In: Proceedings of the IEEE/CVF Conference on Computer Vision and Pattern Recognition, pp. 2524–2534 (2020)
6. Collins, R.T.: A space-sweep approach to true multi-image matching. In: Proceedings CVPR IEEE Computer Society Conference on Computer Vision and Pattern Recognition, pp. 358–363. IEEE (1996)
7. Duggal, S., Wang, S., Ma, W.C., Hu, R., Urtasun, R.: DeepPruner: learning efficient stereo matching via differentiable PatchMatch. In: Proceedings of the IEEE/CVF International Conference on Computer Vision, pp. 4384–4393 (2019)
8. Galliani, S., Lasinger, K., Schindler, K.: Massively parallel multiview stereopsis by surface normal diffusion. In: Proceedings of the IEEE International Conference on Computer Vision, pp. 873–881 (2015)
9. Ghosh, A., Fyffe, G., Tunwattanapong, B., Busch, J., Yu, X., Debevec, P.: Multi-view face capture using polarized spherical gradient illumination. In: Proceedings of the 2011 SIGGRAPH Asia Conference, pp. 1–10 (2011)
10. Gu, X., Fan, Z., Zhu, S., Dai, Z., Tan, F., Tan, P.: Cascade cost volume for high-resolution multi-view stereo and stereo matching. In: Proceedings of the IEEE/CVF Conference on Computer Vision and Pattern Recognition, pp. 2495–2504 (2020)
11. Haendel, M., et al.: How many rare diseases are there? Nat. Rev. Drug Discov. **19**(2), 77–78 (2020). https://doi.org/10.1038/d41573-019-00180-y
12. Hartley, R., Zisserman, A.: Multiple View Geometry in Computer Vision, 2nd edn. Cambridge, New York (2003)
13. Jensen, R., Dahl, A., Vogiatzis, G., Tola, E., Aanæs, H.: Large scale multi-view stereopsis evaluation. In: Proceedings of the IEEE Conference on Computer Vision and Pattern Recognition, pp. 406–413 (2014)
14. Kazhdan, M., Bolitho, M., Hoppe, H.: Poisson surface reconstruction. In: Proceedings of the 4th Eurographics Symposium on Geometry Processing, vol. 7 (2006)
15. Kendall, A., et al.: End-to-end learning of geometry and context for deep stereo regression. In: Proceedings of the IEEE International Conference on Computer Vision, pp. 66–75 (2017)
16. Knapitsch, A., Park, J., Zhou, Q.Y., Koltun, V.: Tanks and temples: benchmarking large-scale scene reconstruction. ACM Trans. Graph. (ToG) **36**(4), 1–13 (2017)
17. Kung, S., et al.: Monitoring of therapy for mucopolysaccharidosis type i using dysmorphometric facial phenotypic signatures. In: Zschocke, J., Baumgartner, M., Morava, E., Patterson, M., Rahman, S., Peters, V. (eds.) JIMD Reports, Volume 22. JR, vol. 22, pp. 99–106. Springer, Heidelberg (2015). https://doi.org/10.1007/8904_2015_417

18. Morales, A., Piella, G., Sukno, F.M.: Survey on 3D face reconstruction from uncalibrated images. Comput. Sci. Rev. **40**, 100400 (2021). https://doi.org/10.1016/j.cosrev.2021.100400
19. Ng, P.C., Henikoff, S.: SIFT: predicting amino acid changes that affect protein function. Nucleic Acids Res. **31**(13), 3812–3814 (2003)
20. Palmer, R.L., Helmholz, P., Baynam, G.: Cliniface: phenotypic visualisation and analysis using non-rigid registration of 3D facial images. Int. Arch. Photogram. Remote Sens. Spat. Inf. Sci. - ISPRS Arch. **43**(B2), 301–308 (2020). https://doi.org/10.5194/isprs-archives-XLIII-B2-2020-301-2020
21. Poulton, C., Thomas, Y.: Scanning a new landscape, 22–24 March 2020
22. Rai, M.C.E.L., Werghi, N., Al Muhairi, H., Alsafar, H.: Using facial images for the diagnosis of genetic syndromes: a survey. In: 2015 International Conference on Communications, Signal Processing, and their Applications, ICCSPA 2015, pp. 1–6. IEEE (2015)
23. Rubinstein, Y.R., et al.: The case for open science: rare diseases. JAMIA Open **3**(3), 472–486 (2020). https://doi.org/10.1093/jamiaopen/ooaa030
24. Schonberger, J.L., Frahm, J.M.: Structure-from-motion revisited. In: Proceedings of the IEEE Conference on Computer Vision and Pattern Recognition, pp. 4104–4113 (2016)
25. Schops, T., et al.: A multi-view stereo benchmark with high-resolution images and multi-camera videos. In: Proceedings of the IEEE Conference on Computer Vision and Pattern Recognition, pp. 3260–3269 (2017)
26. Wang, F., Galliani, S., Vogel, C., Speciale, P., Pollefeys, M.: PatchmatchNet: learned multi-view patchmatch stereo. In: Proceedings of the IEEE/CVF Conference on Computer Vision and Pattern Recognition, pp. 14194–14203 (2021)
27. Woodham, R.J.: Photometric method for determining surface orientation from multiple images. Opt. Eng. **19**(1), 191139 (1980)
28. Xiao, Y., Zhu, H., Yang, H., Diao, Z., Lu, X., Cao, X.: Detailed facial geometry recovery from multi-view images by learning an implicit function. arXiv preprint arXiv:2201.01016 (2022)
29. Xu, Q., Tao, W.: Multi-scale geometric consistency guided multi-view stereo. In: Proceedings of the IEEE/CVF Conference on Computer Vision and Pattern Recognition, pp. 5483–5492 (2019)
30. Yang, J., Mao, W., Alvarez, J.M., Liu, M.: Cost volume pyramid based depth inference for multi-view stereo. In: Proceedings of the IEEE/CVF Conference on Computer Vision and Pattern Recognition, pp. 4877–4886 (2020)
31. Yao, Y., Luo, Z., Li, S., Fang, T., Quan, L.: MVSNet: depth inference for unstructured multi-view stereo. In: Ferrari, V., Hebert, M., Sminchisescu, C., Weiss, Y. (eds.) ECCV 2018. LNCS, vol. 11212, pp. 785–801. Springer, Cham (2018). https://doi.org/10.1007/978-3-030-01237-3_47
32. Yao, Y., Luo, Z., Li, S., Shen, T., Fang, T., Quan, L.: Recurrent MVSNet for high-resolution multi-view stereo depth inference. In: Proceedings of the IEEE/CVF Conference on Computer Vision and Pattern Recognition, pp. 5525–5534 (2019)
33. Zeng, D., Zhao, Q., Long, S., Li, J.: Examplar coherent 3D face reconstruction from forensic Mugshot database. Image Vis. Comput. **58**, 193–203 (2017)

Non-linear Continuous Action Spaces for Reinforcement Learning in Type 1 Diabetes

Chirath Hettiarachchi[1](✉)(iD), Nicolo Malagutti[1](iD), Christopher J. Nolan[1](iD), Hanna Suominen[1,2](iD), and Elena Daskalaki[1](iD)

[1] Australian National University, Canberra, Australia
{chirath.hettiarachchi,nicolo.malagutti,christopher.nolan,
hanna.suominen,eleni.daskalaki}@anu.edu.au
[2] University of Turku, Turku, Finland

Abstract. Artificial Pancreas Systems (APS) aim to improve glucose regulation and relieve people with Type 1 Diabetes (T1D) from the cognitive burden of ongoing disease management. They combine continuous glucose monitoring and control algorithms for automatic insulin administration to maintain glucose homeostasis. The estimation of an appropriate control action—or—insulin infusion rate is a complex optimisation problem for which Reinforcement Learning (RL) algorithms are currently being explored due to their performance capabilities in complex, uncertain environments. However, insulin requirements vary markedly according to sleep patterns, meal and exercise events. Hence, a large dynamic range of insulin infusion rates is required necessitating a large continuous action space which is challenging for RL algorithms. In this study, we introduced the use of non-linear continuous action spaces as a method to tackle the problem of efficiently exploring the large dynamic range of insulin towards learning effective control policies. Three non-linear action space formulations inspired by clinical patterns of insulin delivery were explored and analysed based on their impact to performance and efficiency in learning. We implemented a state-of-the-art RL algorithm and evaluated the performance of the proposed action spaces *in-silico* using an open-source T1D simulator based on the UVA/Padova 2008 model. The proposed exponential action space achieved a 24% performance improvement over the linear action space commonly used in practice, while portraying fast and steady learning. The proposed action space formulation has the potential to enhance the performance of RL algorithms for APS.

Keywords: Reinforcement learning · Glucose regulation · Continuous action space

1 Introduction

Reinforcement Learning (RL) is a class of machine learning algorithms where an intelligent agent learns to act in an underlying environment to maximise a cumulative reward [25]. The reward is formulated to reflect a desired objective. RL algorithms have been successfully applied in games, where they have demonstrated superhuman performance capability/potential [20]. However, the application of RL to real-world problems is challenging due to complexities and constraints such as critical safety requirements, lack of knowledge for the formulation of reward functions, delays in sensors or actuators, partial observability, and high-dimensional continuous state or action spaces [6].

The problem of glucose regulation in Type 1 Diabetes (T1D) features all of the above challenges in RL. In healthy individuals, insulin secretion is performed by the islet β-cells of the pancreas. During periods of fasting (e.g., during sleep), a low basal rate of insulin secretion is required, whereas after meals surges in insulin secretion superimposed on basal secretion are necessary to maintain normal blood glucose concentrations [19]. In people with T1D, the autoimmune destruction of the β-cells of the pancreas results in complete insulin deficiency [5]. As a result, external insulin administration is vital to maintain glucose homeostasis [17]. Efforts to estimate the right rates of insulin infusion in T1D are challenged by delays in continuous glucose monitoring (CGM) sensing and insulin action; high inter- and intra-population variability; and critical safety constraints that cannot be compromised. Current advancements in T1D management methods include insulin administration by a continuous subcutaneous insulin infusion (CSII) pump alongside CGM in open-loop or hybrid closed-loop systems. In both cases, insulin delivery is divided into two distinct infusion patterns: low-range, almost-continuous basal pattern of delivery, and a pattern of intermittent high-range (or bolus) delivery of insulin used mainly to counter the glucose elevation due to meals [17]. In an open-loop setting, both basal and bolus insulin rates are calculated based on patient-specific characteristics (e.g., total daily insulin requirement, carbohydrate ratio) combined with the estimated amount of carbohydrate (CHO) content of a meal [17], as well as on insulin pharmacokinetic and pharmacodynamic properties [23]. In hybrid closed-loop schemes, basal insulin infusion rates are automatically estimated by a control algorithm according to CGM inputs, while insulin bolus dosing is manually calculated and administered by the user prior to meals. For decades, research interest has been on developing a fully automated Artificial Pancreas System (APS) (Fig. 1A) [4]. An APS consists of a CGM, a CSII pump, and a control algorithm to calculate automatically the insulin infusion rate for all circumstances in an effort to improve the total time in the normoglycemic range and relieve the people living with T1D from the heavy cognitive burden included in the manual calculation of meal CHO and insulin bolus doses [2].

Current APS research is investigating the use of RL algorithms due to their capability to perform well in uncertain and complex dynamic environments with disturbances [1]. However, one of the main challenges faced by RL algorithms in the APS context is the large and continuous insulin action space, which differs from the discrete actions present in game environments. According to the basal-bolus scheme, the low-range basal insulin actions account for the vast majority of the total insulin actions, while the large bolus actions are intermittent.

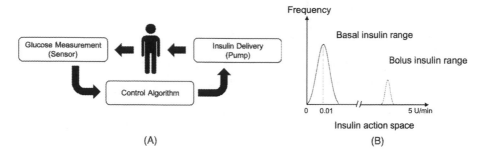

Fig. 1. (A) Artificial Pancreas System, (B) Frequency distribution of insulin action based on a clinical perspective (not to scale).

Typical clinical patterns of insulin delivery can hence be interpreted as bi-modal in the frequency (Fig. 1B). This challenges the RL algorithm by requiring efficient exploring of the entire continuous insulin action space to learn suitable control strategies for different situations which necessitates varying amounts of insulin out of a large dynamic range. The application of RL algorithms to continuous action spaces is not straightforward compared to low-dimensional discrete action spaces, while high-dimensional actions further increase the learning difficulty [13]. The complexity of continuous action spaces can be sub-optimally solved by discretizing the action space. However, this may not be suitable for high-precision control problems such as glucose regulation, as it could eliminate required information regarding the structure of the action space. Current available CSII pumps are discretized with fine resolution (e.g., Medtronic Minimed Pump basal increment of 0.025U/hr [18]), hence allowing the assumption of a continuous insulin action space. An alternative approach could be the introduction of two separate actions for the RL algorithm to focus on the clinical conventions of basal and bolus insulin separately. However, this increases the degrees of freedom in the algorithm and, due to the very sparse use of large insulin doses, could add complexity to learning.

In this study, we introduced the use of non-linear continuous action spaces as a method to overcome the challenges associated with efficiently exploring the dynamic range of insulin to learn effective glucose regulation strategies. Three non-linear translation functions were designed to map the RL action to the insulin infusion rate, inspired by the basal-bolus pattern of clinical insulin treatment practice. We implemented a state-of-the-art RL algorithm used in continuous control (e.g., 3D humanoid motion problems, physics simulations [22]) to evaluate the learning performance and efficiency of the proposed non-linear continuous action spaces. We evaluated our approach *in-silico* using an open-source T1D simulator based on the FDA approved UVA/Padova 2008 model [9]. We demonstrated that a linear action space is not suitable for the problem of glucose regulation in T1D and show that the proposed non-linear continuous action spaces improve the performance while portraying fast and steady learning.

2   Related Work

The use of RL for continuous control has gained much attention in the recent past due to applications such as locomotion, self-driving, and dexterous manipulation tasks [13]. In the problem of glucose regulation in T1D, the majority of the RL-based approaches focus on hybrid systems, which only control basal insulin levels, while bolus insulin infusion is carried out manually by the user [7,14,31,32]. These studies use a small set of handcrafted discrete actions to represent basal insulin [31,32]. A handful of studies have sought to control both basal and bolus insulin without any user input [7,8,12]. In particular, [7,12] used a discrete action distribution to control insulin. However, the action space discretization could lead to a loss of information, while a continuous action space is expected to enable more flexible RL agents that can learn more robust control strategies. [8] is the only research which focused on a continuous insulin action space. They divided the action space to two equal regions, where one represented no insulin administration and the other mapping the action linearly to the insulin pump. This strategy encouraged sparse insulin dosing and was evaluated on reasonably low CHO meals which required insulin rates <0.5 U/min. However, in real-life scenarios the presence of meals with large CHO content results in a large insulin action space ([0, 5] U/min).

According to the RL-related literature, applications with a large continuous actions space can present challenges related to its efficient exploration [11] and the existence of redundant or irrelevant actions [30]. [11] used a novel actor-critic algorithm based on a Sequential Monte Carlo approach to improve the exploration, while [30] proposed an approach which combined the RL algorithm with an action elimination network to eliminate sub-optimal actions.

In the present study, we design and develop a fully automated APS based on a RL algorithm. We introduce a challenging meal protocol including meals of large CHO content, which translates to the need of a large insulin action space, reflective of a real-life scenario. To address this challenge, we propose the use of non-linear representations of the RL algorithm action spaces. This results in a non-uniform resolution across the action space which guides the RL algorithm to explore insulin delivery patterns observed in clinical treatment. To the best of our knowledge this is the first attempt to explore action space representations to tackle complexities in large continuous action spaces associated with glucose regulation in T1D.

3   Method

3.1   Problem Formulation

The glucose control problem can be formulated as a Partially Observable Markov Decision Process (POMDP), where perfect state information is unavailable and limited to noisy sensor measurements. This POMDP can be defined as a 6-tuple $(S^\star, S, O, A, P, R)$, where $s^\star \in S^\star$ denotes the true states, $s \in S$ the noisy states observed by an observation function O, $a \in A$ the actions, $P : (s^\star, a) \to s'$ the transition function where s' denotes the next state, and $R : (s, a) \rightarrow r \in \mathbb{R}$ the reward function. The reward function is designed based on glucose risk indices proposed in [10], where dangerous glucose levels are penalised and normal glucose levels encouraged. We define an observation function $O : s_t^\star \rightarrow g_{t-k:t}, i_{t-k:t}$ which maps the true state $s_t^\star$ at current time t to glucose sensor observation g_t and administered insulin i_t augmented by their past k historical values. Hence, the observed state space is formulated as $s_t = (g_{t-k:t}, i_{t-k:t})$ where past k samples encompass the information related to glucose dynamics and the effect of insulin.

3.2   Action Space

We take a policy gradient approach for designing the RL algorithm since it is more suited for continuous action spaces and for learning stochastic policies [25]. In this formulation, the RL algorithm is required to predict a distribution over the actions ($\pi(a|s)$) for a given state (s). We use a normal distribution ($\mathcal{N}(\mu, \sigma)$) where the RL algorithm learns both μ & σ parameters. The final predicted action is bounded to the range $[-1, 1]$ which is then mapped to the insulin infusion rate of the insulin pump ($I_{pump} \in [0, 5]$ U/min) based on a translation function T. As discussed earlier, the common practice in RL is to map the predicted action linearly to the underlying actuator [3,26,27] as shown in Eq. 1, where I_{max} corresponds to the maximum insulin.

$$I_{pump} = I_{max} \cdot \frac{(a + 1)}{2}, a \in [-1, 1]. \tag{1}$$

3.3   Proposed Translation Functions

Glucose regulation requires frequent use of very small insulin doses for basal insulin compared to the less frequent larger doses, resulting in a skewed concentration in the action space, as opposed to the uniform resolution provided by the linear mapping (Eq. 1). In order to capture this property, we explore three non-linear translation functions; (1) quadratic, (2) proportional-quadratic, and (3) exponential to formulate non-linear action spaces in order to provide better resolution to the important target insulin ranges (Fig. 2).

Quadratic. The translation function T is a quadratic function of the RL action a. This formulation integrates the two distinct actions (basal-bolus) used in typical insulin treatment to a single continuous action space avoiding the complexity of using multiple actions. The action space is divided into two segments, where actions in $[-1, 0]$ are translated to a basal dose range with a maximum basal insulin of $\delta_1(0.05)$ and actions in $(0,1]$ considered as the bolus range with a maximum bolus insulin of I_{max}. This results in a duplication of the basal range $[0, 0.05]$ in the bolus range $[0, 5]$, which could be considered negligible due to the low resolution of the bolus range.

$$I_{pump} = \begin{cases} \delta_1 \cdot a^2 & -1 \leq a \leq 0 \\ I_{max} \cdot a^2 & 0 < a \leq 1 \end{cases}. \tag{2}$$

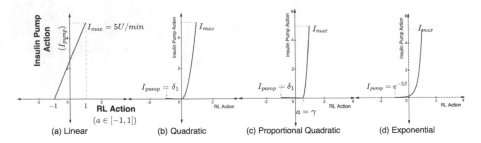

Fig. 2. Translation functions used to map RL action to the insulin infusion rate: (a) linear, (b) quadratic, (c) proportional-quadratic, and (d) exponential.

Proportional-Quadratic. This function is a modification of the Quadratic function where the parameter $\gamma(0.5)$ is introduced to adjust the resolution of the two dose ranges and δ_1 is set to 0.5.

$$
I_{pump} = \begin{cases} \frac{\delta_1}{(\gamma+1)^2} \cdot (a - \gamma)^2 & -1 \leq a \leq \gamma \\ \frac{I_{max}}{(\gamma-1)^2} \cdot (a - \gamma)^2 & \gamma < a \leq 1 \end{cases}. \tag{3}
$$

Exponential. The translation function T is an exponential function of the RL action $a \in [-1, 1]$ with a tuneable parameter $\beta(4.0)$ which ensures $I_{pump} \in (0, 5]$. This increases the resolution of the basal dose range while ensuring the action space is continuous without any duplication of actions. This formulation provides more flexibility for the RL algorithm to use the fully continuous structure of the action space and avoids any instabilities in learning, which might be caused by action duplication.

$$
I_{pump} = I_{max} \cdot e^{\beta(a-1)}, a \in [-1, 1]. \tag{4}
$$

3.4   Algorithm

The RL algorithm was designed based on PPO [22], which is one of the state-of-the-art on-policy RL methods used in continuous control problems. We formulate the glucose control problem as a continuing (not episodic) task, where the goal of the RL algorithm is to maximise the average reward r [16,25] while following a control policy π defined as,

$$
r(\pi) \doteq \lim_{h \to \infty} \frac{1}{h} \sum_{t=1}^{h} \mathbb{E}[R_t | S_0, A_{0:t-1} \sim \pi]. \tag{5}
$$

The PPO algorithm consists of a policy network (π_θ) and a value network (V_ϕ) which we have implemented using recurrent and dense neural network layers. The main objective of the policy network is to learn a suitable policy while the value network learns the n-step expected return being in a given state (s_t).

The PPO algorithm imposes constraints on policy updates to avoid excessive changes between the old policy ($\pi_{\theta_{old}}$) and the new policy (π_θ) by clipping the probability ratios of the new and old policies at $1 - \epsilon$ or $1 + \epsilon$ as shown in the policy objective below:

$$L^{policy}(\theta) = \hat{\mathbb{E}}_t \left[min(\frac{\pi_\theta(a_t|s_t)}{\pi_{\theta_{old}}(a_t|s_t)} \hat{A}_t, \right.$$

$$\left. clip(\frac{\pi_\theta(a_t|s_t)}{\pi_{\theta_{old}}(a_t|s_t)}, 1 - \epsilon, 1 + \epsilon)\hat{A}_t) + \beta_s H(\pi(\cdot|s_t)) \right], \quad (6)$$

where $\hat{A}_t$ is the advantage function [21] estimate at timestep t. The entropy term $H(\pi(\cdot|s_t))$ facilitates exploration, where β_s is a hyperparameter. We used n-step returns to compute the advantage function and value function targets (V_t^{target}). The value network is optimised using the objective:

$$L^{value}(\phi) = \hat{\mathbb{E}}_t \left[\frac{1}{2}(V_\phi(s_t) - \hat{V}_t^{target})^2 \right]. \quad (7)$$

3.5   Simulation Protocol

The UVA/Padova T1D simulator was used for the conduction of our study [9]. This is the only FDA-approved T1D simulator and can be used as a replacement of animal studies prior to clinical evaluation in humans. The simulator comprises a cohort of 30 *in-silico* subjects of three age categories (adults, adolescents and children) as well as models of different CGM and CSII pumps available in the market. In order to allow for reproducability of our results by the community, we used an open-source Python implementation of this simulator [29]. We conducted the evaluation using the adolescent cohort (10 subjects) due to their highly complex individual dynamics and glucose variability which create a very challenging glucose control environment. The Guardian RT glucose sensor and the Insulet pump with a sampling time of 5 min was used for the experiments. A challenging meal scenario was defined for the training and testing of the RL algorithm. For the training phase, the meal scenario consisted of three random meals (breakfast, lunch, and dinner) which were randomised based on the amount of CHO, time, and probability of occurrence (Table 1). The testing scenario spanned 24 h starting at 00:00 hrs and was fixed with three meals: 40 g of CHO for breakfast at 8:00 h, 80 g of CHO for lunch at 13:00 h, and 60 g of CHO for dinner at 20:00 h. Simulations which recorded glucose levels that exceeded the detectable range (39–600 mg/dL) of the glucose sensor were terminated and considered as a *catastrophic failure*.

3.6   Implementation Details and Data Analysis

The simulations were carried out on a workstation machine with $2 \times$ NVIDIA 3090 GPUs. Each action space representation was evaluated for three random

Table 1. Training meal protocol.

Meal type	Time (hours)	Probability	Carbohydrates (g)
Breakfast	7.00–9.00	0.95	30–60
Lunch	12.00–14.00	0.95	70–100
Dinner	19.00–21.00	0.95	50–110

seeds per subject, where all other hyperparameters were kept fixed. The RL algorithms were trained for $500,000$ interactions ($1,736$ human days, 1 interaction = insulin action taken every $5\,\text{min}$), which was identified as sufficient to reach convergence for the above proposed meal protocol. Upon the conclusion of training, $1,500$ testing simulations were also conducted for each subject. The best performing action space representation was compared against the benchmarking linear action space by conducting statistical significance tests for each individual subject. A Shapiro-Wilk Test [24] was performed to check the normality and a Mann-Whitney U Test [15] was conducted to evaluate significance using a confidence level of 0.05.

3.7   Evaluation Metrics

The evaluation was twofold, and included analysis of the final performance after training using the results of the testing simulations and analysis of the learning efficiency derived from the training phase. For the final performance assessment, we used as metrics the total reward achieved by the RL algorithm as a percentage of the maximum achievable reward (PR) and the Time In Range (TIR) calculated as the average percentage of time that the glucose levels were maintained in the normoglycemic range (70–180 mg/dL) during a simulation. In addition we calculated the Failure Rate (FR) as the percentage of simulations which resulted in catastrophic failures over the total testing simulations. The clinical objective is to increase TIR and reduce the FR.

The aim of the RL algorithm during training is to iteratively improve a control policy reasonably fast (relative to the application's time scale), while avoiding excessive changes (smooth learning) which could lead to sudden unanticipated behaviour and even result in catastrophic failures. To evaluate the learning efficiency, we first defined reward thresholds of 25%, 50%, 70%, and 80% of the maximum achievable reward. The average number of interactions required to reach the threshold as a percentage of total interactions (PI) was used to compare the learning efficiency between the candidate action spaces. Furthermore, we qualitatively assessed the learning smoothness through visual inspection of the shape and fluctuations of the reward curves during training.

4   Results

The exponential action space improved the performance in terms of PR by 24%, while reducing the FR by 42% compared to the benchmark linear action space, on average across the subjects. The improvement was statistically significant ($p < .001$) for all the subjects. The performance of the candidate action spaces for all subjects, based on PR and FR metrics is summarised in Table 2. An inter-subject variability in performance improvements was identified, with substantial performance improvements for some subjects (Adolescent8). All the proposed non-linear action spaces were able to outperform the linear action space. The proposed non-linear functions all performed similarly in terms of TIR and FR (Fig. 3). Figure 3 also highlighted the variability in glucose control present among the subjects.

The exponential action space was the most efficient in reaching the 80% reward threshold in 34.90% PI, for all 10 subjects. Table 3 summarises the PI required for identified reward thresholds and the number of subjects reaching the target threshold. The linear action space was unable to reach the 80% reward threshold and also took more PI to cross lower reward thresholds. The quadratic and proportional-quadratic functions also performed better compared to the linear action space. The structure of the reward graphs (Fig. 4) for the exponential and quadratic action spaces gave evidence of steady learning and better convergence for all subjects. The reward graphs of adolescent 3 and 5 clearly indicated unsteady learning for the linear action space where sudden large reward fluctuations are observed during training. The linear action space failed to achieve convergence in some subjects (Adolescent 0, 8) and convergence to sub-optimal reward levels was observed in some subjects (Adolescent 1, 6, 9).

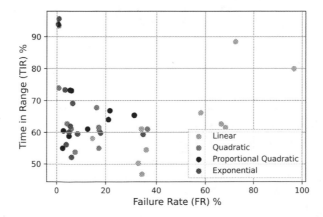

Fig. 3. The percentage time in normoglycemic range (TIR) and Failure Rate (FR) for candidate action spaces (Each dot represents an adolescent subject and each color a candidate action space.) (Color figure online)

Table 2. Adolescent Cohort Summary Results—Total Reward Achieved As A Percentage of Maximum Achievable Reward (PR) & Failure Rate (FR) for the candidate functions: Linear(**L**), Quadratic(**Q**), Proportional-Quadratic(**PQ**), Exponential(**E**).

Adolescent ID	Reward (PR)				Failure rate (FR)			
	L	Q	PQ	E	L	Q	PQ	E
0	52.55%	97.08%	97.33%	97.52%	72.53%	0.87%	0.60%	0.87%
1	66.70%	79.23%	81.75%	79.36%	35.87%	7.53%	2.47%	6.07%
2	76.09%	80.82%	77.92%	87.71%	32.67%	17.33%	21.73%	5.60%
3	58.48%	88.27%	75.20%	86.95%	34.13%	4.27%	31.07%	5.07%
4	68.21%	85.74%	76.81%	84.57%	58.13%	5.93%	21.13%	8.47%
5	70.20%	81.93%	88.96%	87.89%	33.93%	16.20%	6.00%	6.53%
6	54.96%	71.19%	82.78%	76.20%	68.27%	36.33%	2.80%	17.87%
7	75.94%	74.63%	81.40%	80.75%	14.53%	17.20%	5.07%	3.93%
8	25.17%	93.43%	90.72%	91.97%	96.67%	0.87%	5.53%	3.47%
9	59.65%	78.29%	80.90%	74.05%	66.53%	17.13%	12.47%	34.60%
Average (mean ± std)	60.79% ± 14.98%	83.06% ± 8.12%	83.38% ± 6.93%	84.70% ± 7.24%	51.33% ± 24.97%	12.37% ± 10.82%	10.89% ± 10.33%	9.25% ± 9.99%

Table 3. Efficiency analysis—Average Number of Interactions Required to reach the reward threshold as a percentage of total interactions (PI) and the number of adolescents achieving identified reward thresholds.

Translation function	Reward threshold			
	25%	50%	70%	80%
Linear	64.55% (10)	71.91% (9)	80.28% (5)	None
Quadratic	15.65% (10)	19.91% (10)	26.71% (10)	41.37% (8)
Proportional quadratic	38.75% (10)	45.79% (10)	52.51% (10)	63.44% (9)
Exponential	16.96% (10)	21.22% (10)	25.97% (10)	34.90% (10)

5   Discussion

The application of RL algorithms to problems with large continuous action spaces is challenging and currently being tackled through the design of efficient exploration algorithms [11] and irrelevant/redundant action elimination [30]. The common practice in continuous control RL tasks present in OpenAI Gym [3], DeepMind Control Suite [26], and MuJoCo physics environments [27] is to use a linear action space. Inspired by clinical treatment methods for T1D, in this study we introduced non-linear continuous action space representations to tackle the challenge of the large, continuous, and non-uniform insulin action space. To the best of our knowledge this is the first study to explore non-linear action space formulations to compensate the challenges present in continuous action spaces associated to glucose regulation in T1D.

The proposed exponential action space outperformed the linear action space, with statistically significant ($p < .001$) improvements in PR and FR metrics for

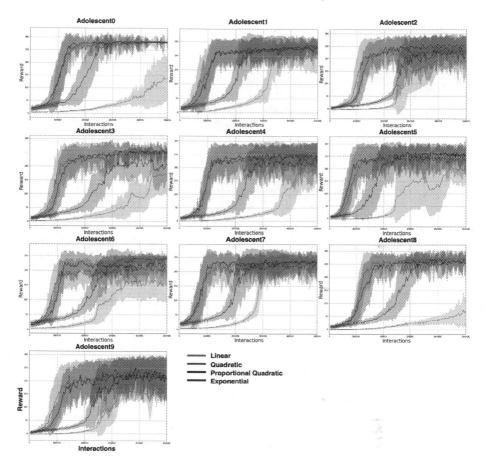

Fig. 4. Comparison of candidate action spaces during training. The mean and standard deviation of the total testing reward (3 random seeds × 20 testing scenarios) achieved for each candidate action space is presented against 500, 000 learning interactions.

all subjects. It also exhibited steady convergence properties and was the most efficient in reaching the 80% reward threshold out of all the candidate action spaces. Unsteady learning (Adolescent 3, 5) and convergence to sub-optimal reward levels (Adolescent 1, 6, 9) was observed for subjects under a linear action space. The linear action space showed very poor performance for some subjects (Adolescent 0, 8) as it was unable to converge within the target number of training interactions. However, it is expected that the linear action space based RL algorithm might converge if the number of training interactions are increased. Meanwhile, all the proposed non-linear action spaces were able to converge within the target training interactions. The subjects who achieved convergence under the linear action space only converged to a sub-optimal level which indicates that increasing training interactions might not be beneficial for these subjects. The learning efficiency achieved in our proposed approach indicates efficient exploration by the

RL algorithm. This also reduces the computational time requirements for training, which is very valuable in the design and development phase of RL algorithms for glucose control, as often multiple iterations of designs are explored and tested. Increasing the complexity in the RL algorithmic architecture or the simulation protocol is expected to result in increased compute times until convergence is achieved. Hence, faster learning can become not only desirable but also vital for the experimental design of future RL algorithms for glucose regulation.

The successful real-world application of a RL-based APS would require online continual learning to adapt the control strategy based on biological variability (e.g., ageing, hormonal disturbances) of the user. Hence, the steady learning observed in the exponential and quadratic action spaces is vital to ensure safety by avoiding sudden excessive changes. The proposed approach illustrated favourable characteristics in this regard, while further future research is required. Our approach can also be applied to other medical applications with similar action space properties. The application of propofol dosing in general anaesthesia is such an application where a non-uniform action distribution is observed and RL currently being explored [28].

We selected the PPO algorithm due to its suitability towards the glucose regulation application which requires continuous control and steady learning. Hence, the performance improvement through non-linear action spaces were only analysed based on PPO. It is expected that the identified benefits would also be applicable to other similar on-policy RL algorithms. The results draw a promising line of research to explore the contribution of non-linear action spaces on other on-policy and off-policy RL algorithms. The use of a linear action space does not impose any bias in the learning process as all actions are equally probable. In contrast, our approach adds prior knowledge about the insulin action distribution to facilitate better learning which imposes the bias of the current clinical practice in T1D insulin treatment. This introduced bias was demonstrated as necessary for the glucose regulation task to achieve effective and efficient control. However, the effect of the bias in the proposed approach could be detrimental in problem domains with limited expert knowledge, and future research could focus on methods to identify the most suitable translation functions to design the target non-linear action spaces. The inter-subject variability in performance observed in the analysis highlights that the design of personalised action spaces using clinically recognised parameters of individuals may be beneficial, thus a potential area of future research. The designed RL-based system can be further improved to increase the TIR and reduce the FR. In future work, we aim to explore reward function formulations and algorithmic improvements to enhance the performance, while focussing on aspects such as safety, explainability, and transferability to real-life, which are vital for a robust APS.

6   Conclusions

This study proposed the use of non-linear action space representations for a RL-based APS with the aim to address the challenge of the large, continuous, and non-uniform insulin action space and enhance the learning efficiency and performance of glucose control strategies. Our results demonstrated superior performance of the non-linear action spaces compared to the standard linear one with faster and smoother convergence and higher final reward. This research is expected to contribute to the development of RL-based fully-autonomous APS.

Acknowledgments. This research was funded in part by the Australian National University and the Our Health in Our Hands initiative.

Code Availability. A repository of code used in this study, and further supplementary material, is available at https://github.com/chirathyh/G2P2C.

References

1. Bothe, M.K., Dickens, L., et al.: The use of reinforcement learning algorithms to meet the challenges of an artificial pancreas. Expert Rev. Med. Devices **10**(5), 661–673 (2013)
2. Brew-Sam, N., Chhabra, M., et al.: Experiences of young people and their caregivers of using technology to manage type 1 diabetes mellitus: systematic literature review and narrative synthesis. JMIR Diabetes **6**(1), e20973 (2021)
3. Brockman, G., et al.: OpenAI gym. arXiv Eprint arXiv:1606.01540 (2016)
4. Cobelli, C., Renard, E., Kovatchev, B.: Artificial pancreas: past, present, future. Diabetes **60**(11), 2672–2682 (2011)
5. DiMeglio, L.A., Evans-Molina, C., Oram, R.A.: Type 1 diabetes. Lancet **391**(10138), 2449–2462 (2018)
6. Dulac-Arnold, G., Mankowitz, D., Hester, T.: Challenges of real-world reinforcement learning. arXiv preprint arXiv:1904.12901 (2019)
7. Fox, I., Wiens, J.: Reinforcement learning for blood glucose control: challenges and opportunities. In: Reinforcement Learning for Real Life (RL4RealLife) Workshop in the 36th International Conference on Machine Learning (2019)
8. Fox, I., et al.: Deep reinforcement learning for closed-loop blood glucose control. In: Machine Learning for Healthcare Conference, pp. 508–536. PMLR (2020)
9. Kovatchev, B.P., Breton, M., et al.: In silico preclinical trials: a proof of concept in closed-loop control of type 1 diabetes. J. Diabetes Sci. Technol. **3**(1), 44–55 (2009)
10. Kovatchev, B.P., Clarke, W.L., et al.: Quantifying temporal glucose variability in diabetes via continuous glucose monitoring: mathematical methods and clinical application. Diabetes Technol. Ther. **7**(6), 849–862 (2005)
11. Lazaric, A., Restelli, M., Bonarini, A.: Reinforcement learning in continuous action spaces through sequential monte carlo methods. In: Advances in Neural Information Processing Systems, vol. 20 (2007)
12. Lee, S., Kim, J., et al.: Toward a fully automated artificial pancreas system using a bioinspired reinforcement learning design: in silico validation. IEEE J. Biomed. Health Inform. **25**(2), 536–546 (2020)
13. Lillicrap, T.P., Hunt, J.J., Pritzel, A., et al.: Continuous control with deep reinforcement learning. arXiv preprint arXiv:1509.02971 (2015)
14. Lim, M.H., Lee, W.H., et al.: A blood glucose control framework based on reinforcement learning with safety and interpretability: in silico validation. IEEE Access **9**, 105756–105775 (2021)

15. Mann, H.B., Whitney, D.R.: On a test of whether one of two random variables is stochastically larger than the other. Ann. Math. Stat. **18**, 50–60 (1947)
16. Naik, A., Shariff, R., et al.: Discounted reinforcement learning is not an optimization problem. arXiv preprint arXiv:1910.02140 (2019)
17. Nathan, D., Genuth, S., et al.: The effect of intensive treatment of diabetes on the development and progression of long-term complications in insulin-dependent diabetes mellitus. N. Engl. J. Med. **329**(14), 977–986 (1993)
18. Online: Insulin pump comparison. http://www.betterlivingnow.com/forms/Insulin-Pump-Comparison.pdf. Accessed 24 Mar 2022
19. Rorsman, P., Eliasson, L., Renstrom, E., Gromada, J., Barg, S., Gopel, S.: The cell physiology of biphasic insulin secretion. Physiology **15**(2), 72–77 (2000)
20. Schrittwieser, J., Antonoglou, I., et al.: Mastering Atari, go, chess and shogi by planning with a learned model. Nature **588**(7839), 604–609 (2020)
21. Schulman, J., Moritz, P., Levine, S., et al.: High-dimensional continuous control using generalized advantage estimation. arXiv preprint arXiv:1506.02438 (2015)
22. Schulman, J., Wolski, F., et al.: Proximal policy optimization algorithms. arXiv preprint arXiv:1707.06347 (2017)
23. Shah, R.B., Patel, M., et al.: Insulin delivery methods: past, present and future. Int. J. Pharm. Investig. **6**(1), 1–9 (2016)
24. Shapiro, S.S., Wilk, M.B.: An analysis of variance test for normality (complete samples). Biometrika **52**(3/4), 591–611 (1965)
25. Sutton, R.S., Barto, A.G.: Reinforcement Learning: An Introduction. MIT Press, Cambridge (2018)
26. Tassa, Y., Doron, Y., Muldal, A., Erez, T., et al.: DeepMind control suite. arXiv preprint arXiv:1801.00690 (2018)
27. Todorov, E., Erez, T., Tassa, Y.: MuJoCo: a physics engine for model-based control. In: 2012 IEEE/RSJ International Conference on Intelligent Robots and Systems, pp. 5026–5033. IEEE (2012)
28. Vajapey, A.: Predicting optimal sedation control with reinforcement learning. Ph.D. thesis, Massachusetts Institute of Technology (2019)
29. Xie, J.: Simglucose v0. 2.1 (2018). https://github.com/jxx123/simglucose. Accessed 13 Jan 2022
30. Zahavy, T., et al.: Learn what not to learn: action elimination with deep reinforcement learning. In: Advances in Neural Information Processing Systems, vol. 31 (2018)
31. Zhu, T., Li, K., Georgiou, P.: A dual-hormone closed-loop delivery system for type 1 diabetes using deep reinforcement learning. arXiv preprint arXiv:1910.04059 (2019)
32. Zhu, T., Li, K., Herrero, P., Georgiou, P.: Basal glucose control in type 1 diabetes using deep reinforcement learning: an in silico validation. IEEE J. Biomed. Health Inform. **25**(4), 1223–1232 (2020)

Cognitive Impairment Prediction by Normal Cognitive Brain MRI Scans Using Deep Learning

Justin Bardwell[1], Ghulam Mubashar Hassan[1(✉)] (iD), Farzaneh Salami[1,2] (iD), and Naveed Akhtar[1] (iD)

[1] Department of Computer Science and Software Engineering,
The University of Western Australia, Perth, Australia
{ghulam.hassan,naveed.akhtar}@uwa.edu.au
[2] Department of Industrial Engineering, Alborz Campus, University of Tehran,
Tehran, Iran
farzane.salami@ut.ac.ir

Abstract. Alzheimer's disease is a neurodegenerative disease without a cure and is one of the leading causes of death across the world. The early detection of cognitive impairment could prove crucial for reducing the occurrence of Alzheimer's disease in the future. Significant research into detecting the disease from MRI images has already been performed and has produced encouraging results. However, there has been very limited work on predicting conversion from normal cognition to cognitive impairment. This study is aimed at producing a deep learning model to predict whether a subject will remain cognitively normal or progress to a state of cognitive impairment in the future. We found that the use of a patch-based approach combined with pre-trained ResNet-50 model using 3D MRI scans provide better results as compared to equivalent whole brain voxel-based approach and other state-of-the-art CNN models. Our proposed model achieved an accuracy of 90% and an area under the receiver operating characteristic curve of 0.99, which are better than the existing state-of-the-art results.

Keywords: Alzheimer's · CNN · Deep learning · MRI · Cognition prediction

1 Introduction

Alzheimer's disease (AD) is a progressive neurodegenerative disease which destroys memory and other important cognitive functions. It is the most common form of dementia affecting up to 70% of those diagnosed [30]. In the United States alone, it is estimated that there are over 6 million people living with AD [2]. Dementia, including AD, has been the second leading cause of death in Australia since 2013 [3]. Currently, there is no cure for AD.

At the present time, there is no single test that can be used to identify AD and the diagnosis can only be confirmed by examination of the subject's brain tissue after death. Clinical diagnosis of AD usually involves a combination of

H. Aziz et al. (Eds.): AI 2022, LNAI 13728, pp. 571–584, 2022.
https://doi.org/10.1007/978-3-031-22695-3_40

psychiatric assessment, cerebrospinal fluid testing, APOE genotyping, positron emission tomography (PET) and magnetic resonance imaging (MRI). However, the expert use of these tests only leads to a 77% diagnostic accuracy for clinical diagnosis of AD [24].

Mild cognitive impairment (MCI) is a condition in which people experience significant memory loss but do not yet meet the criteria for a clinical diagnosis of AD [23]. Davis et al. [5] estimated that, at age 65, 8% of people will progress from normal cognition to MCI and 22% of patients with MCI will progress to a clinical diagnosis of AD annually. For a cohort of 100 cognitively normal (CN) patients at age 65, it was found that a 20% reduction in the progression rate from normal cognition to MCI would avoid 5.7 cases of MCI and 5.6 cases of AD in the future.

Non-pharmacological interventions such as diet, exercise and cognitive exercise have been shown to have an influence on reducing the incidence of development of MCI and dementia [32]. As such, early intervention should lead to increased life expectancy and less time spent in severe AD health status.

Lee et al. [19] found that radiological examinations' retrospective error rate is approximately 30%. There are a range of factors that can cause errors to be made but it was found that errors were mainly due to fatigue and radiologists' inherent biases while performing diagnoses. These diagnostic errors are estimated to account for up to 80,000 annual deaths in the US. As such, the use of automated systems for diagnosis is a growing field of interest. Due to the recent advancement in machine learning, many methods are being researched and utilised to improve the medical practices.

MRI is a medical imaging technique that is commonly used for diagnosing brain injuries and diseases. It is one of the most useful and effective tools a physician can use to assist in the diagnosis of AD [8].

Dukart et al. [7] provided some of the first research into using machine learning techniques to detect AD using MRI data. The study used a 50% split of AD patients and CN subjects taken from a 56 subject sample of the Alzheimer's Disease Neuroimaging Initiative (ADNI) dataset. The volume of several regions of interest of the brain were extracted from the MRI data. A support vector machine (SVM) was fit to this volumetric data and was able to distinguish between AD and CN patients with an accuracy rate of 80.4%. Gray et al. [10] explored the ability of a random forest (RF) to classify AD using MRI volumes. The results from this study suggested that the RF was able to predict AD with a higher degree of accuracy in comparison to previous studies which used SVMs. A subset of the ADNI data was used and included 37 AD patients, 75 MCI patients and 35 CN persons. The study reported that the random forest model was able to detect AD patients with 82.5% accuracy. The study also reported accuracy of 67.3% to classify between MCI patients and CN.

Recently, deep learning has been used to solve many complex medical problems such as detecting Multiple Sclerosis, Alzheimer's disease and various types of cancer [21,25]. The interest in deep learning for medical imaging has largely been due to the ability of convolutional neural networks (CNNs) to learn useful representations of complex images. The first study to explore the use of deep learning for AD classification was performed by Suk and Shen in 2013 [28]. Unlike previous methods which used simple features extracted from MRI such as brain tissue volumes, the deep learning method implemented here was proposed to be able to extract more complicated patterns from the data. A stacked auto-encoder (SAE) with three hidden layers was utilised and the output layer of the auto-encoder is used to represent the class label of the input data. This implementation of the SAE acted as a classifier and was able to detect between healthy controls and AD patients with 85.7% accuracy. The SAE could also distinguish MCI patients from healthy controls with 70.6% accuracy. Gunawardena et al. [11] compared the use of SVMs and CNNs in the detection of AD from MRI data. The data consisted of 1615 images from the ADNI dataset. The proposed CNN model was constructed using two convolution layers, a pooling layer and a fully connected layer. The SVM used in this study predicted AD with an accuracy of 84.4% while the CNN model had an accuracy of 96.0%. This difference in performances confirms that deep learning methods can greatly improve the ability to automatically detect AD from MRI.

Farooq et al. [9] investigated the AD detection performance of transfer learning by using pre-trained CNN models with ImageNet weights. 355 MRI volumes from the ADNI dataset were used, with each scan being broken down into many 2D slices. Binary classification between CN patients and patients with AD or MCI was performed effectively by these models. ResNet-18 and GoogLeNet were the top two performing models with both achieving accuracy rates of over 99%. The performance of the selected models exceeded most other state-of-the-art technologies present in the literature at the time but the data did not include entire ADNI dataset.

Although three-dimensional neural networks are more computationally intensive than their 2D counterparts, the ability of 3D CNNs to extract discriminative features from MRI data is found to be superior [31].

Similarly, the use of ensembles of various deep learning networks has been tested and has shown promising results. Dua et al. [6] implemented an ensemble technique to combine CNN, recurrent neural network and a long short-term memory model to improve on each model's individual performance. An ensemble of five different 3D DenseNets provided state-of-the-art results for the three-way classification of CN, MCI and AD [29]. On its own, the best performing 3D DenseNet in this study provided a three-way classification accuracy of 94.77% while an ensemble method improved this to an accuracy of 98.83%.

In addition to classification, early diagnosis of AD and MCI due to AD could help slow or halt patients' cognitive decline [18]. Methods which allow us to detect early signs of cognitive impairment or predict further cognitive decline are more valuable compared to classification approaches which only seek to differentiate between a CN subject and a subject suffering from heavily progressed AD [8]. Shen et al. [26] used a 2D CNN to extract features from the MRI scans of 165 MCI patients recorded in ADNI. After feature extraction, a support vector machine was used to classify these features and predict whether the patients would be diagnosed with AD within 12 months. The classification accuracy using an RBF kernel was 92.3%.

Albert et al. [1] investigated the ability to predict individuals' progression from normal cognition to MCI. Their study used baseline data from 224 CN subjects, of which 75% had a first degree relative with dementia. At the end of the 5 year study, 46 of these subjects had progressed to having MCI. The study found that age, right hippocampus volume, right entorhinal cortex thickness, APOE genotype and cerebrospinal fluid testing results were significant for the purposes of prediction of a patient's progression from normal cognition to MCI. This study's use of MRI scans was limited to measuring the specific parts of brain which include brain's right hippocampus volume and entorhinal cortex thickness. For the prediction of progression from CN to MCI, the single-modal use of the MRI domain technique in this study provided an AUC of 0.740. When combined with other demographic and genetic variables the AUC improved to 0.849.

In this study, we propose deep learning based approach to predict the cognitive decline of a patient from CN to MCI or clinical diagnosis of AD. To date, there has been limited research into predicting individuals' progression from normal cognition to cognitive impairment as mentioned above, and there are substantial opportunities to improve this practice.

2   Materials and Methods

All computational experiments have been performed in a Google Colab environment which provides access to GPUs. Implementation of the experiments was performed using Python 3.7 and TensorFlow 2.6.

2.1   Datasets

In 2010, Alzheimer's Disease Neuroimaging Initiative (ADNI) dataset was released. It provided a compilation of MRI and PET scans of 819 elderly subjects [14]. Since the release, the amount of data has been steadily increasing. Ebrahimighahnavieh et al. [8] found that 90% of the studies into Alzheimer's detection and classification uses ADNI dataset.

The Open Access Series of Imaging Studies (OASIS) is a recently released neuroimaging dataset which focuses on subjects' cognitive decline. OASIS-3 was released in 2018 and is an openly available dataset containing MRI and PET imaging for 1,098 subjects [17]. 850 of the participants entered the study as CN while there were 248 participants who entered with some form of cognitive impairment. Throughout the study, 245 of the patients who were initially CN had converted to a state of cognitive impairment. Over the course of the study, there were a total of 2,168 MRI scans produced. As the OASIS-3 images were obtained over a period of more than 10 years and on a range of different scanners, there were several different file types storing the data. However, a single standard format has been provided and all data files have been converted to NifTI format files.

The OASIS-3 dataset also includes related clinical data as well as post-processed outputs and regional segmentations of the brain. Clinicians assessed the participants and provided a dementia diagnosis which included categories of "cognitively normal", "AD dementia" and "vascular dementia".

The Clinical Dementia Rating (CDR) is a commonly used measure in longitudinal studies of AD [22]. Patients' CDR is based on their impairment in memory, orientation, judgment and problem solving, community affairs, home and hobbies, and personal care. A CDR of 0 corresponds to a cognitively normal patient, 0.5 represents very mild dementia, 1 represents mild dementia, 2 represents moderate dementia and 3 represents severe dementia.

The groups of patients in OASIS-3 have been defined as "Stable Controls", "Converters" and "Dementia at aging". Stable controls are individuals who begun with a CDR score of 0 and remained on the same score. Converters are those subjects who started with CDR of 0 and progressed to a CDR greater than 0. Subjects defined as dementia at aging were those who initially had a CDR greater than 0.

In both the ADNI and OASIS datasets, subjects completed clinical assessment protocols in line with the National Alzheimer Coordinating Center Uniform Data Set. The main difference between the OASIS-3 and ADNI datasets is that the majority of the OASIS-3 participants were initially categorised as being cognitively normal and their potential decline was followed through longitudinal progression. ADNI, however, primarily enrolled patients who already had some form of dementia or MCI. Thus, we found OASIS-3 to be optimal dataset for our study as it has the potential to greatly improve early detection methods of MCI and AD.

2.2   Image Data

For each subject in OASIS-3, we have access to T1-weighted MRI images. We used the post-processed MRI images from OASIS-3 that come in the form of FreeSurfer files. These files contain a 3D image of the brain with the skull stripped and have a shape of $256 \times 256 \times 256$ voxels.

After reading in the FreeSurfer files, we performed further processing steps in order to have a useful format for feeding the images into our deep learning model. These steps included cropping the image so that the majority of blank space around the brain is removed and then resizing this cropped image to $128 \times 128 \times 128$ voxels. After the cropping and resizing, each image was normalised so that each voxel has an intensity value between 0 and 1.

2.3   Subject Selection

There were two classes of subjects that we were interested in for our experiments. These were the subjects who remained CN for the foreseeable future and the

Table 1. Split of number of subjects in each class of the OASIS-3 dataset after selection process.

Class	Training	Validation	Test
Healthy controls	30	10	10
Converters	30	10	10

subjects who converted from CN to cognitive impairment. For our subjects that remained CN throughout the study, we selected those who have received a clinical assessment with a CDR of 0 at least 3000 days after their initial scan.

For the class of subjects who converted to cognitive impairment, we selected subjects who initially had a CDR of 0 but at a later assessment received a CDR of 0.5 or higher. As the scans of each subject were not usually taken on the date of their clinical assessment, we ensured that the scans we selected were taken at a point in time when the subject is still likely to be CN. To do this, we selected scans of subjects that have occurred within 365 days of receiving a CDR score of 0 in an assessment and receive a CDR rating of 0.5 or above within the next 1,000 days after the scan. We also ensured that the date of the scan was closer to the subject's CDR 0 assessment in comparison to their CDR $\geq$ 0.5 assessment.

A 50/50 split of subjects in each class was obtained as suggested in earlier studies [7,9]. 60% of our data was used in the training set, 20% for the validation set and 20% for the test set [16].

At the end of selection process, we had 204 subjects remaining that satisfied our selection criteria. However, only 50 of these were subjects who would go on to convert to cognitive impairment. In order to create a balanced dataset, we reduced the number of subjects that remained CN in our dataset from 154 to 50. After performing our train, validation and test split, we were left with the subjects as detailed in Table 1.

2.4 Evaluation Metrics

For the purposes of calculating evaluation metrics in this study, subjects that convert from normal cognition to cognitive impairment are referred to as the positive class. Subjects remaining CN throughout the study are referred to as the negative class.

The evaluation metrics used were accuracy, sensitivity, specificity, area under the receiver operating characteristic curve (AUC) as suggested in the literature [1]. These metrics are calculated using the following equations.

$$Accuracy = \frac{TP + TN}{TP + TN + FP + FN}$$

$$Sensitivity = \frac{TP}{TP + FN}$$

$$Specificity = \frac{TN}{FP + TN}$$

where TP, TN, FP and FN represent true positives, true negatives, false positives and false negatives respectively.

AUC is calculated as the area under the *Sensitivity* (True Positive Rate) curve plotted against $1 - Specificity$ (False Positive Rate).

2.5 Model Architecture

There are two popular methods for dealing with 3D images in literature: voxel-based approach [13] and patch-based approach [20]. We propose a deep learning network that uses both the approaches to examine their effectiveness. The overall architectures for the two approaches are very similar with a difference that voxel-based approach uses voxel intensity values from the whole image while a patch-based approach breaks down the whole image into several small three-dimensional cubes.

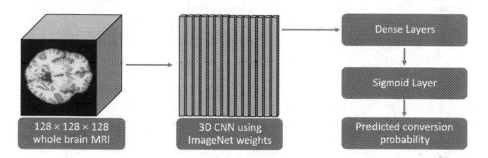

Fig. 1. Overview of the architecture of our voxel-based approach

Our voxel-based architecture as presented in Fig. 1 takes an input for each subject of a $128 \times 128 \times 128$ whole brain MRI. This input image is fed into a pre-trained 3D CNN using ImageNet weights. The 3D ImageNet weights we used were proposed by Solovyev et al. [27] for their winning submission to a machine learning competition regarding identification of stalled brain capillaries. As the pre-trained networks are built to classify between 1,000 classes in the ImageNet problem, it produces 1,000 output probabilities. As we are only interested in a binary classification output, we need to transform this output to a single probability.

To achieve binary classification, we added three dense layers consisting of 1,000, 500 and 200 units respectively as well as a sigmoid layer for the output layer. The use of these layers takes the 1,000 output probabilities from the pre-trained CNN and produces a single output probability which corresponds to the predicted probability of a subject converting from normal cognition to cognitive impairment.

Figure 2 presents our method of implementing a patch-based approach which is similar to the voxel-based approach. The first difference between the two

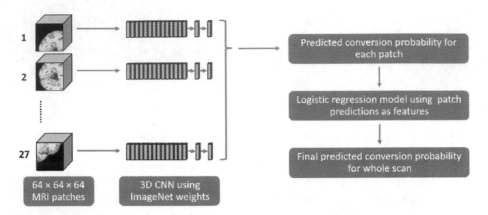

Fig. 2. Overview of the architecture of our patch-based approach

approaches is the input to the model. Instead of using the $128 \times 128 \times 128$ whole brain MRI, we split this image into $3 \times 3 \times 3$ patches. Each patch of voxels has a 50% overlap with its neighbouring patch [4]. This patch-based method results in the generation of 27 uniform-sized patches for each MRI.

Similar to voxel-based approach, after feeding the inputs into a pre-trained 3D CNN, the same dense layers and sigmoid layer are applied in order to produce a single output probability for each patch. A stacked model has been applied by using the outputs from each patch based model as features to feed in to a logistic regression model for us to make our final prediction.

The dense layers and sigmoid layer following the pre-trained CNN were fine-tuned in a brief training process. The parameters for the training of the CNNs in the voxel-based and patch-based models are shown in Table 2. The only difference between the parameters used for the models is the batch size. The reduced batch size for the voxel-based method is necessary due to the memory limits of the GPUs we used for the experiments.

Table 2. CNN parameters for our voxel-based and patch-based approaches

Parameter	Voxel-based	Patch-based
Input shape	64, 64, 64	128, 128, 128
Pre-trained weights	ImageNet	ImageNet
Batch size	20	5
Loss	Binary cross-entropy	Binary cross-entropy
Learning rate	0.001	0.001
Maximum epochs	1,000	1,000
Early stopping	50 epochs	50 epochs
Optimizer	Adam	Adam

2.6   Experiment 1 - Voxel-Based vs. Patch-Based Approaches

In our first experiment, we compared the performance of a voxel-based model to a patch-based model. These models were built according to the specifications detailed in Sect. 2.5. As mentioned earlier, we used a 3D ResNet-50 model initialised using ImageNet weights for both models.

2.7   Experiment 2 - CNN Architectures

We compared the performance of two of the historically best performing CNN architectures for medical imaging problems, ResNet-50 and DenseNet-121. In addition to this, ResNet-18 was also tested as it has previously shown similar performance to ResNet-50 for the binary classification of AD while having significantly less parameters [9].

Squeeze-and-excitation networks were introduced by Hu et al. [12]. The addition of squeeze-and-excitation blocks to previously developed and well established architectures such as ResNet provided improvements in performance with only a slight increase in computational intensity. In their testing of the model on ImageNet, SE-ResNet-50 was found to outperform the original ResNet-50 architecture improving the top-5 error from 7.48% to 6.62%. SE-ResNeXt-50 showed further improvements with a top-5 error of only 5.49% on ImageNet. In the field of medical imaging, Jiang et al. [15] found that the use of an SE-ResNet module provided improvements to the state-of-the-art results for the task of breast cancer classification. We wanted to explore whether the use of squeeze-and-excitation networks could provide any improvement over our ResNet-50 model implemented in Experiment 1 of this study. As such, we tested the performance of SE-ResNet-50 and SE-ResNeXt-50 in a patch-based approach.

An ensemble of the five models selected for comparison was tested to determine whether a combination of models could provide additional improvements.

3   Results and Discussion

As mentioned in the earlier section, we conducted two sets of experiments. The aim of the first experiment was to compare and evaluate the performance of voxel-based and patch-based approaches. This helped to select a better approach for the next experiment. The aim of the second experiment was to obtain the best model for our prediction problem.

3.1   Experiment 1 - Voxel-Based vs. Patch-Based Approaches

After fine-tuning the voxel-based and patch-based networks' weights on the training and validation set, the two models produced the results which are presented in Table 3.

In Experiment 1, the voxel-based approach predicted that all the subjects in the test set would remain CN. However, the patch-based approach achieved an accuracy of 90% with only 2 subjects wrongly classified. These 2 subjects were predicted by the model to remain CN while they actually converted to a state of cognitive impairment.

From the accuracy results seen in Table 3, it is evident that the patch-based approach greatly outperforms the voxel-based approach. The result shows that the patch-based model will be able to provide more accurate predictions than the voxel-based model for our next experiment.

Table 3. Model performance on test set for Experiment 1

Model	Accuracy	AUC	Sensitivity	Specificity
Voxel-based approach	0.50	0.50	0.00	1.00
Patch-based approach	0.90	0.99	0.80	1.00

3.2   Experiment 2 - CNN Architectures

After finding the patch-based approach to be optimal for our prediction problem, we tested a range of patch-based 3D CNNs as presented in Table 4.

From the comparison of the five individual CNN architectures presented in Table 4, we found that ResNet-50 and SE-ResNet50 produced the most accurate predictions, with both models correctly predicting the class of the test subject 90% of the time. These two models resulted in the same predictions for each subject in the test set, suggesting that there was no improvement made by using the squeeze-and-excitation version of the network.

Interestingly, ResNet-50 greatly outperformed ResNet-18 in all metrics. Previously, Farooq et al. [9] had shown that for the binary classification of AD and CN subjects, ResNet-18 outperformed ResNet-50. For our task of predicting a subject's conversion to cognitive impairment, the use of ResNet-50 produced an accuracy of 90% and an AUC of 0.99 in comparison to ResNet-18's accuracy of 65% and AUC of 0.80. It is important to note that Farooq et al.'s study used 2D CNNs while we have used 3D CNNs. While it is obvious that the deeper model of ResNet produced more accurate results in our experiments, our results may also suggest that the use of a deeper model is beneficial for 3D image classification tasks.

In Experiment 2, we also tested the performance of DenseNet-121. In other studies using 3D medical imaging data, DenseNet models have performed strongly and exceeded the results produced by other CNN architectures [27,29]. Based on our testing, we found that DenseNet-121 outperformed ResNet-18 but did not outperform ResNet-50 in regards to model accuracy and AUC. DenseNet-121 did, however, produce the highest sensitivity of all the models tested with 100% of converting subjects predicted correctly in the test set.

Lastly, we also observed that an ensemble method which uses all of the CNN architectures tested did not provide the best accuracy for our predictions. Its performance in terms of accuracy was better than ResNet-18 but lower than DenseNet-121 and ResNet-50 models.

Table 4. Model performance on test set for experiment 2

Model	Accuracy	AUC	Sensitivity	Specificity
ResNet-50	0.90	0.99	0.80	1.00
SE-ResNet-50	0.90	0.99	0.80	1.00
SE-ResNeXt-50	0.75	0.85	0.90	0.60
ResNet-18	0.65	0.80	0.70	0.60
DenseNet-121	0.85	0.97	1.00	0.70
Ensemble model	0.80	0.97	0.90	0.70

Overall from Experiment 1 and Experiment 2, we found that the models producing the highest accuracy and AUC were the patch-based implementations of ResNet-50 and SE-ResNet50. Due to the fact that ResNet-50 has slightly fewer parameters and thus is less computationally intensive, we recommend this is as our most promising model for future applications.

From the literature, we found that the only published article that has attempted to predict whether a CN subject will convert to cognitive impairment in the near future, was performed by Albert et al. [1]. Table 5 presents the model performance of Albert et al. in comparison to our model's performance.

On the test set of our data, our approach has produced an AUC of 0.99 while only making use of MRI data. When using MRI data only, Albert et al.'s [1] approach produced an AUC of 0.74 for the prediction of conversion to cognitive impairment in a five year time frame. We acknowledge that we can not directly compare the results of our study to the results of Albert et al. [1] as the study used a private dataset that was gathered by the Geriatric Psychiatry Branch of

Table 5. Comparison of our model's performance to that of Albert et al. [1]

	Model	Time to outcome	AUC	Sensitivity	Specificity
Albert et al. [1]	Cox regression (MRI data only)	5 years	0.740	0.641	0.710
		7 years	0.722	0.662	0.659
		9 years	0.705	0.616	0.678
	Cox Regression (MRI + demographic + genetic variables)	5 years	0.850	0.804	0.740
		7 years	0.843	0.815	0.724
		9 years	0.831	0.764	0.759
Our approach	**Patch-based ResNet50**	**1,000 days**	**0.990**	**0.800**	**1.000**

the National Institute of Mental Health and as such, we did not have access to this data. However, we believe that our patch-based 3D ResNet-50 model has the potential to improve the previously established methods of predicting a subject's probability of conversion to cognitive impairment.

4   Conclusion

AD is a leading cause of death across the world. It has been found previously that a reduction in the progression rate from normal cognition to MCI would reduce the number of cases of AD in the future. Due to the potential for medical interventions to reduce the incidence of development of MCI, the ability to accurately predict whether a person will convert from normal cognition to a state of cognitive impairment could be important in the future.

In this paper, we developed an approach to detect whether a person will progress from normal cognition to a state of cognitive impairment in the future. While there is still work to be done in order to validate the results of this study on a wider cohort of subjects, the application of our approach to the OASIS-3 dataset appears to produce impressive results. In our testing, we found that a 3D patch-based approach has outperformed an equivalent voxel-based approach. We tested various CNN architectures and found that a pre-trained ResNet-50 using ImageNet weights produced the greatest performance with a prediction accuracy of 90%.

In future, this study can be extended to predict the time to conversion of CN to MCI instead of predicting a probability of the conversion. The provision of more data and additional of clinical data in the input is expected improve the results.

Acknowledgements. Data was provided by OASIS-3: Principal Investigators: T. Benzinger, D. Marcus, J. Morris; NIH P50 AG00561, P30 NS09857781, P01 AG026276, P01 AG003991, R01 AG043434, UL1 TR000448, R01 EB009352. AV-45 doses were provided by Avid Radiopharmaceuticals, a wholly owned subsidiary of Eli Lilly.

References

1. Albert, M., et al.: Predicting progression from normal cognition to mild cognitive impairment for individuals at 5 years. Brain **141**(3), 877–887 (2018)
2. Alzheimer's Association: 2021 Alzheimer's disease facts and figures (2021)
3. Australian Bureau of Statistics: Causes of death, Australia, 2019 (2020)
4. Cheng, D., Liu, M.: Classification of Alzheimer's disease by cascaded convolutional neural networks using PET images. In: Wang, Q., Shi, Y., Suk, H.-I., Suzuki, K. (eds.) MLMI 2017. LNCS, vol. 10541, pp. 106–113. Springer, Cham (2017). https://doi.org/10.1007/978-3-319-67389-9_13
5. Davis, M., et al.: Estimating Alzheimer's disease progression rates from normal cognition through mild cognitive impairment and stages of dementia. Curr. Alzheimer Res. **15**(8), 777–788 (2018)

6. Dua, M., Makhija, D., Manasa, P., Mishra, P.: A CNN-RNN-LSTM based amalgamation for Alzheimer's disease detection. J. Med. Biol. Eng. **40**(5), 688–706 (2020). https://doi.org/10.1007/s40846-020-00556-1

7. Dukart, J., et al.: Meta-analysis based SVM classification enables accurate detection of Alzheimer's disease across different clinical centers using FDG-PET and MRI. Psychiatry Res. Neuroimaging **212**(3), 230–236 (2013)

8. Ebrahimighahnavieh, M.A., Luo, S., Chiong, R.: Deep learning to detect Alzheimer's disease from neuroimaging: a systematic literature review. Comput. Methods Programs Biomed. **187**, 105242 (2020)

9. Farooq, A., Anwar, S., Awais, M., Alnowami, M.: Artificial Intelligence based smart diagnosis of Alzheimer's disease and mild cognitive impairment. In: 2017 International Smart Cities Conference (ISC2), pp. 1–4. IEEE (2017)

10. Gray, K.R., Aljabar, P., Heckemann, R.A., Hammers, A., Rueckert, D., Initiative, A.D.N., et al.: Random forest-based similarity measures for multi-modal classification of Alzheimer's disease. Neuroimage **65**, 167–175 (2013)

11. Gunawardena, K., Rajapakse, R., Kodikara, N.: Applying convolutional neural networks for pre-detection of Alzheimer's disease from structural MRI data. In: 2017 24th International Conference on Mechatronics and Machine Vision in Practice (M2VIP), pp. 1–7. IEEE (2017)

12. Hu, J., Shen, L., Sun, G.: Squeeze-and-excitation networks. In: Proceedings of the IEEE Conference on Computer Vision and Pattern Recognition, pp. 7132–7141 (2018)

13. Islam, J., Zhang, Y.: Brain MRI analysis for Alzheimer's disease diagnosis using an ensemble system of deep convolutional neural networks. Brain Inf. **5**(2), 1–14 (2018)

14. Jack Jr, C.R., et al.: The Alzheimer's disease neuroimaging initiative (ADNI) MRI: methods. J. Magn. Reson. Imaging Official J. Int. Soc. Magn. Reson. Med. **27**(4), 685–691 (2008)

15. Jiang, Y., Chen, L., Zhang, H., Xiao, X.: Breast cancer histopathological image classification using convolutional neural networks with small SE-ResNet module. PLoS ONE **14**(3), e0214587 (2019)

16. Kruthika, K., Maheshappa, H., Initiative, A.D.N., et al.: CBIR system using capsule networks and 3D CNN for Alzheimer's disease diagnosis. Inf. Med. Unlocked **14**, 59–68 (2019)

17. LaMontagne, P.J., et al.: OASIS-3: longitudinal neuroimaging, clinical, and cognitive dataset for normal aging and Alzheimer disease. medRxiv (2019)

18. Langa, K.M., Levine, D.A.: The diagnosis and management of mild cognitive impairment: a clinical review. JAMA **312**(23), 2551–2561 (2014)

19. Lee, C.S., Nagy, P.G., Weaver, S.J., Newman-Toker, D.E.: Cognitive and system factors contributing to diagnostic errors in radiology. Am. J. Roentgenol. **201**(3), 611–617 (2013)

20. Liu, M., Zhang, J., Nie, D., Yap, P.T., Shen, D.: Anatomical landmark based deep feature representation for MR images in brain disease diagnosis. IEEE J. Biomed. Health Inform. **22**(5), 1476–1485 (2018)

21. Lundervold, A.S., Lundervold, A.: An overview of deep learning in medical imaging focusing on MRI. Z. Med. Phys. **29**(2), 102–127 (2019). https://doi.org/10.1016/j.zemedi.2018.11.002

22. Morris, J.: Current vision and scoring rules the clinical dementia rating (CDR). Neurology **43**, 2412–2414 (1993)

23. Petersen, R.C., et al.: Current concepts in mild cognitive impairment. Arch. Neurol. **58**(12), 1985–1992 (2001)

24. Sabbagh, M.N., Lue, L.F., Fayard, D., Shi, J.: Increasing precision of clinical diagnosis of Alzheimer's disease using a combined algorithm incorporating clinical and novel biomarker data. Neurol. ther. 6(1), 83–95 (2017)
25. Salami, F., Bozorgi-Amiri, A., Hassan, G.M., Tavakkoli-Moghaddam, R., Datta, A.: Designing a clinical decision support system for Alzheimer's diagnosis on oasis-3 data set. Biomed. Signal Process. Control **74**, 103527 (2022)
26. Shen, T., Jiang, J., Li, Y., Wu, P., Zuo, C., Yan, Z.: Decision supporting model for one-year conversion probability from MCI to AD using CNN and SVM. In: 2018 40th Annual International Conference of the IEEE Engineering in Medicine and Biology Society (EMBC), pp. 738–741. IEEE (2018)
27. Solovyev, R., Kalinin, A.A., Gabruseva, T.: 3D convolutional neural networks for stalled brain capillary detection. arXiv preprint arXiv:2104.01687 (2021)
28. Suk, H.-I., Shen, D.: Deep learning-based feature representation for AD/MCI classification. In: Mori, K., Sakuma, I., Sato, Y., Barillot, C., Navab, N. (eds.) MICCAI 2013. LNCS, vol. 8150, pp. 583–590. Springer, Heidelberg (2013). https://doi.org/10.1007/978-3-642-40763-5_72
29. Wang, H., et al.: Ensemble of 3D densely connected convolutional network for diagnosis of mild cognitive impairment and Alzheimer's disease. Neurocomputing **333**, 145–156 (2019)
30. World Health Organization: Global action plan on the public health response to dementia 2017–2025 (2017)
31. Yagis, E., Citi, L., Diciotti, S., Marzi, C., Atnafu, S.W., De Herrera, A.G.S.: 3D convolutional neural networks for diagnosis of Alzheimer's disease via structural MRI. In: 2020 IEEE 33rd International Symposium on Computer-Based Medical Systems (CBMS), pp. 65–70. IEEE (2020)
32. Yao, S., et al.: Do nonpharmacological interventions prevent cognitive decline? A systematic review and meta-analysis. Transl. Psychiatry **10**(1), 1–11 (2020)

A Text-Independent Forced Alignment Method for Automatic Phoneme Segmentation

Bryce Wohlan[1], Duc-Son Pham[1(✉)], Kit Yan Chan[1], and Roslyn Ward[2]

[1] School of Electrical Engineering, Computing and Mathematical Sciences,
Curtin University, Perth, WA, Australia
dspham@ieee.org
[2] School of Allied Health, Curtin University, Perth, WA, Australia

Abstract. Phoneme segmentation is important for many healthcare applications, such as the diagnosis and monitoring of children with speech sound disorders (SSDs). This is usually addressed by performing forced alignment (FA), which essentially annotates an audio file to provide information on what has been uttered and where. While many FA tools exist, very few can work automatically without the assistance of a transcription. This work aims at providing a novel text-independent FA tool by using two models, namely wav2vec 2.0 and an unsupervised segmentor known as UnsupSeg. To provide labels to the segments, the class regions that are obtained by nearest-neighbour classification with wav2vec 2.0 labels pre-CTC collapse as the reference points. Maximal overlap between the class regions and the segments determines class label. Additional post-processing steps, such as over-fitting cleaning and application of voice activity detection, are also performed to further improve the segmentation performance. All the models used to create the tool are self-supervised, and thus can leverage great amounts of unlabelled data to reduce the need for labelled data. When evaluated on the TIMIT dataset, our implementation achieved a harmonic mean score of 76.88%, competitive against other alternatives.

Keywords: Forced alignment · Phoneme segmentation · Transformer · Self-supervised learning · Connectionist temporal classification · Voice activity detector · Speech sound disorder · Speech processing · Deep learning

1 Introduction

The term Speech Sound Disorder (SSD) is used as an umbrella term to describe a common communication disorder in young children. The prevalence is reported to range from approximately 1% to 4% in children aged 4–5 years of age, higher in younger children [15]; and accounts for up to 70% of a speech-language pathologist's (S-LPs) caseload [4]. A SSD is defined as "... any difficulty or combination of difficulties with perception, motor production or phonological representation

H. Aziz et al. (Eds.): AI 2022, LNAI 13728, pp. 585–598, 2022.
https://doi.org/10.1007/978-3-031-22695-3_41

of speech sounds ..." [1], resulting in difficulty saying sounds in words correctly, thus affecting speech intelligibility [14]. Access to timely intervention is required to mitigate the well documented cascading consequences of a SSD, that contribute to long term educational and employment difficulties [3]. The assessment and management of SSDs is usually performed by a qualified S-LP, with phonetic transcription of the child's speech integral to the process of identifying the nature of the SSD [14]. Phonetic transcription entails the use of phonetic symbols to represent speech sounds and requires knowledge of the International Phonetic Alphabet [17].

Although clinical practice guidelines recommend phonetic transcription of a speech sample form the "first step" towards diagnosis of a SSD, in a recent survey of Australian and British S-LPS, it was reported only 39.5% and 45.4%, respectively were using broad transcription. Challenges to the use of phonetic transcription included lack of proficiency, time challenges and service delivery issues [17]. The authors concluded there is a need for further resources to support S-LPs in their use of phonetic transcription.

To date, recommendations to support S-LPs with phonetic transcription have focused on the development of competencies through resources, training, and tutorials [16]. Another possible solution includes computer assisted tools, such as automatic speech analysis [13]. To date, computer assisted tools have focused predominantly on anomaly detection and designed to be used in automatic speech recognition (ASR) contexts. These tools, however, do not provide analysis at the phoneme level, as is required for differential diagnosis, by an S-LP. Further, the models used in ASR are based on older machine learning techniques or statistical models. It has been observed that these tools fare poorly on non-complaint utterances, having phone phoneme label accuracies as low as 46.42% [12]. This can be attributed to two factors;

- The tools are dependant on being able to first correctly predict the spoken word, then break the predicted word into graphemes (phonetic symbols) which are assigned to each individual segment of audio. Non-complaint speech will certainly have some issues with predicting the correctly spoken word, and therefore be unable to predict the correct phonemes.
- The models are built to recognise contextual clues and are not frame independent - for a given phoneme, it may have some idea what should come next, raising the probability that it predicts those phonemes[1].

Fortunately, there have been many new developments in the field of deep learning regarding sequence modelling. Sequence models are designed to operate on sequential data, such as text or audio. One may be able to apply, adjust or re-purpose some of these new sequence modelling techniques to become flexible enough to label the utterances of a SSD patient. A few techniques, such as bi-directional long short-term memories (BLSTMs) [21] paired with connectionist temporal classification (CTCs) [6] are noted for having a sense of conditional independence, meaning all outputs are independent of each-other, given the input. This means greater flexibility at the cost of contextual awareness - a bad thing in

---

[1] This refers to conditional dependence on the previous states in a sequence.

other applications but an opportunity here. New techniques, such as `wav2vec 2.0` [2] are showing outstanding performance using self-supervision, which reduces the amount of labelled data required to create a strong performing model.

Ultimately, the goal is to make a tool which can take audio speech data of a non-compliant (SSD) speaker, detect sections which belong to phonemes and correctly label the phonemes, then output the data in a form which is manipulatable by SLPs. This task is called *text-independent forced alignment*, as it performs forced alignment without the aid of text information. Fortunately, advances in machine learning and sequence modelling techniques have resulted in very powerful models which are showing excellent results in tasks such as automatic speech recognition (ASR). This work uses `wav2Vec 2.0 (w2v2)`, which is based on the transformer sequence model [2]. `w2v2` is purpose-built for ASR tasks. It is considered as the state-of-the-art obtain the orthographic and phonetic transcriptions of an audio file without any external support. This is used to obtain the phonetic information from the audio file. Recent work has been done to attain the segments of phonemes without looking the phoneme labels [9,10,22]. This can be used to acquire accurate timings, improving temporal accuracy. Whilst we also use `w2v2` for the speech representation part, we present an alternative approach that is equally competitive whilst being more intuitive and extendable.

The paper is organised as follows. Section 2 details how to build the model and how to obtain metrics which meet the specifications. The results and validation section, Sect. 3, performs a series of experiments to gauge how each component of the tool interacts with each other and how the resulting tool configuration performs. Finally, Sect. 4 contains concluding remarks. The implementation of the proposed method presented in this paper will be made publicly available at https://github.com/dsphamgithub/fatool.

2 Methodology

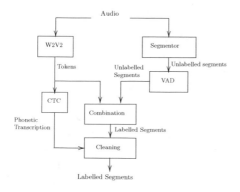

Fig. 1. Data flow

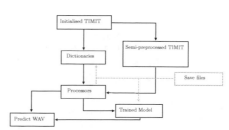

Fig. 2. Dependencies within the Wav2vec 2.0.

2.1   Proposed Forced Aligner

The data flow of our proposed forced aligner is show in Fig. 1. Our approach relies on two important components: an unsupervised segmenter (UnsupSeg) [9] and the well-known `wav2vec 2.0` (`w2v2`) model [2] for speech representation. Our novelty lies in the specific adaptation of `w2v2` from automatic speech recognition (ASR) to phoneme segmentation and a scheme to combine the tokens generated by `w2v2` and unlabeled segments from the unsupervised segmenter. We develop a novel algorithm that combines the information to generate accurate boundaries and assigns the correct phonetic transcription using the output of `w2v2`'s CTC. Some extra elements, such as voice activity detector and cleaning, are also required to improve the overall performance. We now describe the motivation for the proposed system and the details of individual components.

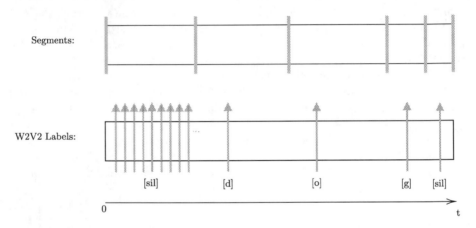

Fig. 3. The output of `wav2vec 2.0` and the segmentation function visualised. Green arrows represent "impulses" of labels pre CTC-collapse. Grey arrows represent PAD tokens that are present throughout the whole sequence, but are only shown at the beginning of the diagram for visual clarity. (Color figure online)

We start with the outputs of `w2v2` and UnsupSeg, which are shown in Fig. 3. Here, the UnsupSeg provides segments, while `w2v2` provides labels and *weak* positional information. To overcome this issue, one could opt to ignore the segments and split using the `w2v2` labels. However, this would most likely violate the tight 20 ms tolerance that speech language pathologists need for a human-level alignment. Another option is to segment the audio file on the segments provided by UnsupSeg, and then use `w2v2` to annotate each snippet to form a number of annotations. However, this is not ideal, as contextual information regarding the whole utterance would be lost. For example, knowing vowel sounds usually follows consonant sounds - annotating each segment independently would lose the contextual knowledge.

Another temptation is to use the w2v2 labels to annotate the segments by labelling segements using the impulses at that time. That is to say, if there is a segment from t_1 and t_2 and a label L exists between t_1 and t_2 then that segment's label is L. This would be an effective strategy, since every segment may not have a w2v2 label within its boundaries, or the label might have several conflicting w2v2 labels within its boundaries. This issue is solved by the proposed technology presented in this paper. Instead of directly using the labels to vote, we can use a decision boundary to annotate labels. Supposed we used a nearest neighbour approach, where we annotate a segment with 0 duration (i.e. an impulse). The region where the nearest neighbour approach will deterministically assign label π_i is noted. This region is described as

$$R_i : x \rightarrow \pi_i \; \forall \; R_i \; \epsilon x \qquad (1)$$

where R_i is region i belonging to class π_i, and x is the term to be classified. The decision boundary is the limits of this space where two regions meet. It is considered ambiguous as to what class will be assigned. The implementation in this project assigns both classes to the boundary. This has no bearing on performance due to the computation method of class labels which is described ahead. We use these class regions to calculate the overlap with each segment. The class label with the greatest degree of overlap with the segment is determined to be the segment's label. For example, if a segment x spanned a number of regions belonging to class labels $\pi =$ [a, b, c] in proportions of 40%, 30% and 30% respectively, then we would decide x's label as **a**. The region boundaries are calculated by using the midpoint of two successive class label impulses. The fomula for region boundaries is detailed in Eq. 2.

$$Boundary(R_1 : R_2) = (t_1 + t_2)/2 \; \forall t_1 < t_2 \wedge \nexists \; t_1 \leq t_x \leq t_2, \; \forall x, \qquad (2)$$

where R_i is region i belonging to class π_i, t_i is the time of label impulse i with $t_i \leq t_{i+1}$, and t_x is the time of any segment that is not t_1 or t_2. However, w2v2 might assign labels to be closer to the start or end of the true segment. This is compensated by introducing a bias factor to calculate the class region boundaries. The bias factor pushes the boundary closer (for $\beta \rightarrow 1$) or further (for $\beta \rightarrow 0$) from the uppermost segment.

$$BiasedBoundary(R_1 : R_2) = (1 - \beta)t_1 + \beta t_2 \; \forall \; t_1 < t_2 \wedge \nexists \; t_1 \leq t_x \leq t_2 \qquad (3)$$

where β is a bias factor. Edge cases such as the last and start segment will extend towards the end of the speech sample. It is also beneficial to "clean up" successive segments which have the same label by amalgamating the two segments. The start time of the earlier segment and the end time of the latter segment becomes the boundaries for the new segment, when the class label is preserved.

Algorithm 1 forms the most critical part of the proposed system. A visual illustration is given in Fig. 4.

With the boundary finding algorithm above, Algorithm 2 (also see Fig. 1) describes the working of our forced aligner. As can be seen, there are few more

---

Algorithm 1. Boundary Calculation

1: **procedure** DECISIONBOUNDARYCALC(timedTokenList, seconds, bias)
2:     *timedTokenList is a list of tuples of labels and their timings*
3:     *seconds is the duration of the speech sample in seconds*
4:     *bias is the bias factor which is positive and smaller than 1.*
5:     DCB ← new List
6:     **for** ii in range of length timedTokenList **do**
7:         **if** ii equals length of timedTokenList - 1 **then**
8:             upper ← seconds
9:             lower ← timedTokenlist[ii - 1][time])*(1-bias) + timedToken-List[ii][time])*(bias)
10:        **else if** ii equals 1 **then**
11:            upper ← timedTokenlist[ii +1 1][time])*(bias) + timedTokenList[ii][time])*(1-bias)
12:            lower ← 0
13:        **else**
14:            upper ← timedTokenlist[ii +1 1][time])*(bias) + timedTokenList[ii][time])*(1-bias)
15:            lower ← timedTokenlist[ii - 1][time])*(1-bias) + timedToken-List[ii][time])*(bias)
16:        **end if**
17:        Append tuple (timedTokenList[ii][label], lower, upper)
18:     **end for**
19:     **return** DCB
20: **end procedure**

---

additional functions that are also needed for it to work (details are omitted due to lack of space):

- The voice activity detector (VAD) function is implemented to remove any unnecessary segments. The segments returned from the Unsupervised Segmentor are stored in a list called segVect.
- The TokensToTimedTokens function turns a 1D array of class labels into a list of tuples containing the class label and the respective time in the sample at which that class label corresponds to. Afterwards, tokens used for CTC are removed, and includes the pad, unknown and delimiting tokens.
- The maxDCBInitDict is used to initialise a blank dictionary. It uses the string-to-unicode dictionary as its template, but the values are replaced by zeroes. It is effectively a string to 0 dictionary.
- The MaxContribution function takes the segment vector, decision boundaries and the MaxDCBInitDict to analyse the contribution, or overlap of each class within each class boundaries. It labels the entire segment with the dominant class label from the overlapping class regions. It returns a label list, which is a list with the class labels in order, so the first label in the list represents the region between segments i and $i + 1$. This is converted to a list of 3 length tuples, which include the phone label, start segment time, and end segment time.

– Finally, the cleaning function `cleanSegs` returns a list with fewer overfitting labels. Cleaning is a post processing step which aims to increase the accuracy of the predictions by removing overfitted labels. It works by doing the following steps: 1) Get the word spoken **with** CTC collapse; 2) Calculate transitions based on every two letters, i.e. cat = (c-a, a-t); and 3) Scan through the labelled segments. If two labels are the same but aren't a permissible transition, amalgamate them.

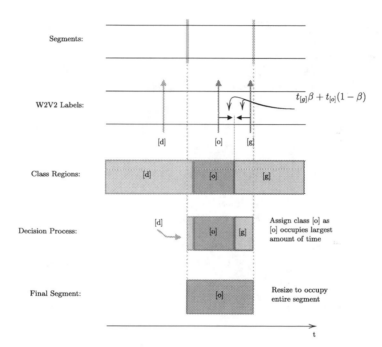

Fig. 4. Example of the process of determining the label of an unlabelled segment.

The list of tuples is converted to a list of dictionaries for easier use when evaluating. This is effectively a list of segments with their labels. This completes the forced alignment process.

2.2   Pre-processing

In order for the forced aligner to work, important pre-processing steps are needed. In this paper, `wav2vec 2.0` (`w2v2`) [2] is implemented using the Hugging-Face implementation (see Fig. 2). Whilst there is existing documentation on using `w2v2` in the context of automatic speech recognition (ASR), it is substantial work to extend it to phoneme prediction. The method for fine-tuning `w2v2` for phoneme recognition involves taking the pre-trained model and further

Algorithm 2. Forced Aligner

1: **procedure** LABELLED SEGMENTER(wavPath)
2:     signal, samplingFreq ← *soundfile*.read(wavPath)
3:     seconds ← length of signal / SamplingFreq
4:     wp ← Wav2Vec2PredictiorObject
5:     tokens ← wp.predictWavNoCollapse(wavPath)
6:     segPredictor ← UnsupervisedsegmenterPredictorObject
7:     segVect ← segPredictor.predict(wavPath, CheckpointPath)
8:     segVect ← VADFilterSegments(wavPath, SegVect)
9:     segVect ← toList(segVect)
10:    timedTokens ← tokensToTimedTokens(signal, samplingFreq, tokens)
11:    filteredTimedTokens ← new List
12:    **for** timedToken in timedTokens **do**
13:        **if** timedToken[label] is not "[pad]" or "[unk]" or "—" **then**
14:            Append timedToken to filteredTimedTokens
15:        **end if**
16:    **end for**
17:    decisionBoundaries ← decisionBoundaryCalc(filteredTimedTokens, seconds, bias)
18:    strToUnicodeDict ← Read in from wav2vec2 object save
19:    MaxDCBinitdict ← dictionary fromkeys(strToUnicodeDict, 0)
20:    Insert 0 at index 0 to segVect
21:    Append seconds value to the end of segVect
22:    labelList ← MaxContribution(segVect, maxDCBInitDict, DCB)
23:    segList ← new List
24:    **for** ii in range of length of labelList **do**
25:        Append tuple (LabelList[ii], segVect[ii], segVect[ii+1]) to segList
26:    **end for**
27:    segList ← cleanSegs(segList)
28:    Convert list of tuples to list of dictionaries
29:    **return** segList
30: **end procedure**

---

training it on TIMIT [5]. We only retain relevant phonetic details for the study. As required by Hugging Face, we also needed to convert the *encoding* of phone labels in ARPABET form to something more atomic

$$ARPABET \xleftarrow{\ PhonetoUnicodeDict\ } Unicode \xleftarrow{\ UnicodetoNumericDict\ } NumericID$$

A dictionary which maps a phoneme to a token ID (numeric) must be created for the tokenizer. To achieve this, a list of unique phone labels is created by scanning the entire TIMIT dataset, and adding every new phone seen to a list. A numeric ID exists for each phone label and is created by enumerating each value in the aforementioned phone label list. This numeric value is mapped to each phone label in the list to form a dictionary.

Due to the limitations of the current version of w2v2 , the phonemes in the dictionary cannot be represented by multi-character strings - `"aa""`:32 would not work, but `"a"`:32 would; labels are not appropriate to encode the sounds i.e. bait - 'b'-'**ay**'-t', 'ay' would cause issues. This is further complicated by Python 3.8's handling of strings as lists of characters - the tokenizer class cannot differentiate between a list of strings and a list of characters. Therefore, the phone labels had to be transformed to a single character representation by adding an intermediary unicode step in the dictionary encoding. While IPA is a tempting choice due to its universal use for representing phonemes, it is problematic it uses character accents - which behind-the-scenes are actually separate characters. Unfortunately, this violates the restriction to single character representations. Therefore, the phones were instead successfully encoded to unicode emojis (U+1F600 unward) due to the enormous selection available which enables easy one to one mapping between phones and emojis.

A w2v2 tokenizer is created from the Unicode to numeric dictionary, and a w2v2 feature extractor is declared with: feature size = 1, sampling rate = 16 kHz, padding value = 0, and normalise = False. The processor is a combination of a tokenizer and a feature extractor. The processor used the dictionary from within the w2v2 to process the dataset. This embedded tokenizer within the processor is the cause of the intermediary step of converting the dataset to unicode - w2v2 's trainer and data collator requires a processor to be used. Hence, one can't manually code their own tokenizer that works with strings.

A map is applied to the dataset using the processor, converting the *phonetic detail* phone labels to its numeric equivalent and assigning that to a feature called *target phones*. The target phones acts as the supervision of the trainer as it tries to minimise the edit distance between its predictions and the target phones. We used the data collator suggested in [19].

Fine tuning is required to use w2v2 for a specific "down-stream" task as the pre-trained model `wav2vec2-large-xlsr-53` that we used was performed by Facebook in a self-supervised mannner on generic datasets covering 53 different languages. Fine-tuning in a supervised manner helps w2v2 adapt for our use case, in order to achieve accurate predictions about the phonemes in an audio file.

Phoneme Error Rate (PER) is used to evaluate the performance of the forced aligner. It is a metric derived from the Levenshtein distance [7], which is the smallest number of substitutions, removals or insertions of characters to make one string equal to another. The PER is calculated as the Levenshtein distance all divided by the number of characters/phones. This metric is similar to the word error rate (WER), but operates on a character level, rather than the word level. It only depends on the order of the sequence, and not the exact timing. While this metric is used primarily for evaluating w2v2 and other sequence modelling tools, it is also somewhat relevant regarding the overarching forced alignment tool which implements w2v2 , as the PER of w2v2 will act as a soft ceiling for the accuracy metric of the FA tool. Note that the PER can be greater than 1, as the number of operations can exceed the number of phones. The loss function for training the aligner is based on PER (Table 1).

Table 1. An example of each atomic operation

Operation	String	Target	Usage	No. of operations
Substitution	bat	cat	b → c	1
Removal	cat**t**	cat	del $\mathbf{t}^{(4)}$	1
Insertion	at	cat	insert **c**	1

3   Experiments

A well performing forced alignment tool depends on: 1) the ability to predict the correct *labels* and the ability to *position* the predictions accurately. The former is measured with precision, recall and F_1 score, and the latter is measured with offset timing errors, Δt_{end} and Δt_{stop}. The proportion of matched predictions correct $P_{GroundTruthCorrect}$ is a way of assessing how *precise* our classifier is. The proportion of ground truths correctly classified $P_{MatchedPredictionsCorrect}$ is a way of assessing our algorithm's ability to *recall* the correct answer. The harmonic mean between these two metrics is called the *Harmonic Mean* score:

$$(P_{GroundTruthCorrect}^{-1} + P_{MatchedPredictionsCorrect}^{-1})^{-1}$$

In the speech language research, a *midpoint method* is also often used to evaluate forced alignment methods [11]. From each utterance, several metrics are obtained, such as start offset time, end offset time, %-match and accuracy. Each segment in the ground truth is compared with each segment in the prediction list. If temporal mid-point of the ground truth is both greater than a predicted segment's start time, and smaller than the predicted segments end time, then it is stated that the prediction has "matched" the manual alignment. Of the matched segments, the absolute difference in time of the segment boundaries, $\Delta_i = |t_{i,predict} - t_{i,truth}|$, is noted for both the end times and the start times separately.

We use the database TIMIT [5] to validate the proposed model for all experiments. TIMIT contains 4620 instances on the training portion, and 1680 instances on the test portion. We first evaluate the two core components of the forced aligner.

- **wav2vec 2.0**: Fine-tuning the proposed model took 151.2 min on a system with a single RTX 3080. The fine-tuned w2v2 model used in this paper attained an phoneme error rate (PER) of 10.6% when trained on TIMIT's training set and evaluated on TIMIT's test set. It achieved a minimum PER of 13.2% when applied validation set.
- **Unsupervised Segmentation (UnsupSeg)** was trained on TIMIT+ using the publicly available repository [8] and produced an R-val of 0.83.

We next validate the overall forced alignment performance of the proposed model. An illustration is shown in Fig. 5. First, we study how the forced aligner performs when the bias is varied. We examine Harmonic Mean and the

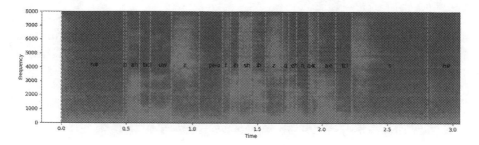

Fig. 5. The segments obtained by using the tool with VAD, hard cleaning, equal/no bias. The utterance says "Bubbles, fishes and cats".

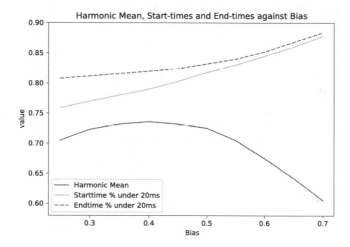

Fig. 6. Performance with varied bias

proportion of Start-times and End-times being less than 20 ms compared to groundtruth. The result is shown in Fig. 6. From Fig. 6, a bias of 0.5 (otherwise unbiased) yields good performance. Its visible that the harmonic mean is inverse parabolic with a maximum of 0.736 at a bias of 0.4. Timing performance seems inversely correlated with bias, so higher bias yields greater timing performance. Selecting a bias would involve weighing up the trade-offs - accept a small hit in timing performance for optimal harmonic mean accuracy, or concede a small loss in accuracy for a increase in timing performance. A bias of 0.45 is a reasonable compromise between the aforementioned trade-offs. Keeping the timing metrics above 0.8 is preferable, and is achieved for all bias values greater than 0.4, but 0.45 bias leaves a good amount of margin for little expense.

Next, we examine the forced alignment performance with the bias being 0.45. We also use the following hard cleaning scheme: if w2v2 specified that a duplicate transition (i.e. "ah" → "ah") was allowed to occur at the end of the sequence, but the cleaning segment found one at the start of the sequence; it will amalgamate it anyway.

The accuracy and timing errors are given in Tables 2 and 3 respectively. It is observed that when VAD is not used, the label prediction accuracy improves slightly, whilst both Segment Start and End Times are similar. Overall, we achieve a maximum Harmonic Mean score of 76.88%, about 82% of the segments have Start Time within 20 ms of groundtruth and 88% of the segments have End Time within 20 ms of groundtruth.

Table 2. Forced alignment results

Metric	With VAD	Without VAD
Accuracy of predictions	71.9%	82.35%
Proportion of manual labels correctly classified	71.4%	72.02%
Harmonic mean	71.6%	76.88%

Table 3. Proportions of the segment boundary errors in milliseconds

	<20 ms	<40 ms	<60 ms
Segment start time (w/VAD)	81.73%	94.21%	97.22%
Segment end time (w/VAD)	87.53%	96.72%	98.70%
Segment start time (w/o VAD)	81.50%	94.07%	97.16%
Segment end time (w/o VAD)	88.33%	96.87%	98.71%

Table 4. A comparison with other text-independent aligners

Model/Tool	P	R	F_1	Source
FAVE	0.57	0.59	0.58	[20]
Gentle	0.49	0.46	0.48	[18]
W2V2-CTC-20 ms	0.31	0.30	0.31	[22]
W2V2-FS-20 ms	0.40	0.42	0.41	[22]
W2V2-FC-20 ms-Libris	0.57	0.59	0.58	[22]
Ours	0.62	0.54	0.58	This paper

The comparison with other text independent models presented in Table 4 shows how this tool performs relative to similar specified models when evaluated on the TIMIT dataset. We note that previous work did not publish the segment start and end times information. In addition, we also needed to use more traditional metrics (precision, recall, and F_1 used in previous work). A tolerance of 20 ms is used, as a segment within 20 ms of the ground truth is considered indistinguishable from the manual alignment. We observe that the precision is excellent compared to other models, but the recall is slightly behind. It is possible to fine-tune the tool to achieve better recall. However, the tool is overall competitive to other recent tools and models.

4   Conclusions

We have presented a new tool capable of text-independent alignment, based on an unsupervised segmenter and `wav2vec 2.0`. This paper provides a novel way of using 1-NN class regions to annotate unlabelled segments provided by the Unsupervised segmenter; it also uses the class labels provided by `w2v2` 's output before being sent to CTC, which has weak temporal accuracy and is unfit to label segments in that form. When evaluated, the tool shows reasonable power. When evaluated on TIMIT using the onset transitions (with a tolerance of 20 ms), the tool achieved a precision of 0.62, which means 62% of the predictions made by the tool were indistinguishably accurate. When evaluated using the midpoint method, the tool captures $\approx$72% of the ground truth midpoints, and 82.35% of the predictions were hits. This tool could potentially be used in clinical settings as an assistive technology for the diagnosis of speech-sound disorders.

Acknowledgement. This work has been supported by the Western Australian Future Health Research and Innovation Fund, which is an initiative of the WA State Government". This work is being conducted to inform a larger research program, being led by a team of researchers at Curtin university. The research program is focused on the development of an application that will provide objective kinematic and acoustic measurements, to support speech language pathologists in the diagnosis of speech sound disorders. The authors would also like to thank Pawsey supercomputing centre for their support.

References

1. Speech sound disorders-articulation and phonology (2022). https://www.asha.org/practice-portal/clinical-topics/articulation-and-phonology/
2. Baevski, A., Zhou, Y., Mohamed, A., Auli, M.: wav2vec 2.0: a framework for self-supervised learning of speech representations. In: Proceedings of Advances in Neural Information Processing Systems, vol. 33, pp. 12449–12460 (2020)
3. Daniel, G.R., McLeod, S.: Children with speech sound disorders at school: challenges for children, parents and teachers. Aust. J. Teach. Educ. **42**(2), 81–101 (2017)
4. Furlong, L., Serry, T., Erickson, S., Morris, M.E.: Processes and challenges in clinical decision-making for children with speech-sound disorders. Int. J. Lang. Commun. Disord. **53**(6), 1124–1138 (2018)
5. Garofolo, J.S., et al.: TIMIT Acoustic-Phonetic Continuous Speech Corpus (1993). https://catalog.ldc.upenn.edu/LDC93S1
6. Graves, A., Schmidhuber, J.: Framewise phoneme classification with bidirectional LSTM and other neural network architectures. Neural Netw. **18**(5–6), 602–610 (2005)
7. Konstantinidis, S.: Computing the edit distance of a regular language. Inf. Comput. **205**(9), 1307–1316 (2007)
8. Kreuk, F.: Self-Supervised Contrastive Learning for Unsupervised Phoneme Segmentation (INTERSPEECH 2020) (2021)
9. Kreuk, F., Keshet, J., Adi, Y.: Self-Supervised Contrastive Learning for Unsupervised Phoneme Segmentation. arXiv:2007.13465 [cs, eess, stat] (2020)

10. Kreuk, F., Sheena, Y., Keshet, J., Adi, Y.: Phoneme boundary detection using learnable segmental features. In: Proceedings of ICASSP, Barcelona, Spain, pp. 8089–8093. IEEE (2020)
11. Mahr, T., Berisha, V., Kawabata, K., Liss, J., Hustad, K.: Performance of forced-alignment algorithms on children's speech. Technical report, PsyArXiv (2020)
12. McKechnie, J., Ahmed, B., Gutierrez-Osuna, R., Monroe, P., McCabe, P., Ballard, K.J.: Automated speech analysis tools for children's speech production: a systematic literature review. Int. J. Speech Lang. Pathol. **20**(6), 583–598 (2018)
13. McKechnie, J.G.: Exploring the use of technology for assessment and intensive treatment of childhood apraxia of speech. Ph.D. thesis (2019)
14. Mcleod, S., Baker, E.: Speech-language pathologists' practices regarding assessment, analysis, target selection, intervention, and service delivery for children with speech sound disorders. Clin. Linguist. Phon. **28**(7–8), 508–531 (2014)
15. McLeod, S., et al.: Profile of Australian preschool children with speech sound disorders at risk for literacy difficulties. Aust. J. Learn. Diffic. **22**(1), 15–33 (2017)
16. McLeod, S., et al.: Tutorial: speech assessment for multilingual children who do not speak the same language (s) as the speech-language pathologist. Am. J. Speech Lang. Pathol. **26**(3), 691–708 (2017)
17. Nelson, T.L., Mok, Z., Eecen, K.T.: Use of transcription when assessing children's speech: Australian speech-language pathologists' practices, challenges, and facilitators. Folia Phoniatr. Logop. **72**(2), 131–142 (2020)
18. Oschshorn, R., Hawkins, M.: Gentle (2017)
19. von Platen, P.: Fine-Tune Wav2Vec2 for English ASR in Hugging Face with huggingface Transformers (2021). https://huggingface.co/blog/fine-tune-wav2vec2-english
20. Rosenfelder, I., et al.: Fave (forced alignment and vowel extraction) suite version 1.1.3 (2014). https://doi.org/10.5281/zenodo.9846
21. Schuster, M., Paliwal, K.: Bidirectional recurrent neural networks. IEEE Trans. Signal Process. **45**(11), 2673–2681 (1997)
22. Zhu, J., Zhang, C., Jurgens, D.: Phone-to-audio alignment without text: a semi-supervised approach. In: Proceedings of ICASSP, pp. 8167–8171. IEEE (2022)

Multi-componential Emotion Recognition in VR Using Physiological Signals

Rukshani Somarathna[(✉)], Aaron Quigley, and Gelareh Mohammadi

School of Computer Science and Engineering, University of New South Wales, Kensington, Australia
r.somarathna@student.unsw.edu.au,
{a.quigley,g.mohammadi}@unsw.edu.au

Abstract. Emotion recognition affords new approaches ranging from context-awareness to the efficiency of system interaction with the ability to perceive and express emotions. While most studies are dominated by *discrete* and *dimensional* theoretical models of emotion, neuroscience analysis aligns with the multi-component interpretation of emotional phenomena. One such componential theory is the Component Process Model (CPM), with five synchronized components: *appraisal, motivation, physiology, expression* and *feeling*. However, limited attention has been paid to the systematic investigation of emotions assuming a full CPM. Therefore, we induced various emotions in this preliminary analysis using 27 interactive Virtual Reality (VR) games. We measured the manifestation of 28 participants across CPM components, 20 *discrete* emotion terms, heart activity, skin conductance, and facial electromyography. Our work aims to analyze the relationship between *discrete* theory-based emotions and the theoretically defined components with physiological measures. Further, we analyze the correlation between subjective *expression* terms with objective facial expressions. Our Machine Learning (ML) analysis reveals a significant relationship between emotions and full componential features with physiological signals. Further, our study presents the role of each CPM component in emotion differentiation.

Keywords: Emotion · Component process model · Physiological responses · Computational modeling

1 Introduction

Human emotions are multifaceted phenomena that are fundamental in individual development. Emotions are cultural and psychobiological adaptation mechanisms that play a significant role in flexible and dynamic communication with internal and external contingencies [34]. Therefore, emotion recognition is increasingly important in numerous domains and affords new approaches ranging from context-awareness to the efficiency of system interaction with the ability to perceive and express emotions [29]. Recent digital developments such as virtual agents and intelligent machines have heightened the need for emotional intelligence with emotion understanding, context awareness, and better interaction.

© The Author(s), under exclusive license to Springer Nature Switzerland AG 2022
H. Aziz et al. (Eds.): AI 2022, LNAI 13728, pp. 599–613, 2022.
https://doi.org/10.1007/978-3-031-22695-3_42

Existing research considers one or a combination of emotion models as *Discrete*, *Dimensional*, and *Appraisal* [13]. However, considerable literature has grown around *discrete* models that define emotions by distinct elements such as happiness, fear, or *dimensional* models, theorizing emotions by valence and arousal dimensions [40]. As these two models primarily consider the *feeling* component, it leads to major theoretical neglect without an explanation for the temporal evolution of emotions with complex processes. Although *appraisal* models have inferred emotion evolution based on temporal processes, research in this domain has been mostly confined to limited data-driven studies [27]. This may be partially explained by the complex definitions of *appraisal* theories that require extensively defined experimental settings. Nevertheless, as emotions are assumed to be complex mechanisms with many processes [27,33], which converge with the *appraisal* model, it is necessary to consider a process-based model to determine emotion formation and analyze the underlying components.

Recent investigations have led to a renewed interest in theories based on the *appraisal* model. One such model is the Component Process Model (CPM), with five interrelated components: *appraisal, motivation, physiology, expression* and *feeling* [33]. Recent works have confirmed the efficacy of using CPM in emotion understanding but lack involving active participation [24,26], evaluating *physiology* and *expression* components with objective measures (physiological and facial signals) [25] and understanding a wider range of emotion variations. Therefore, exploring the potential of CPM through a data-driven approach with active participation and objective measures may lead to the resolution of controversies.

In this work, we use a data-driven approach with multimodal analysis to explore the relationship between the full CPM and emotions using an immersive medium. We induced a wider range of emotions using interactive Virtual Reality (VR) games and collected physiological and facial signals to enrich our understanding of CPM components with objective measures. This study was designed to understand the relationship between full CPM with objective measures and the *discrete* model of emotions using Machine Learning (ML) algorithms. Further, we analyze the correlation of facial EMG expressions with the *expression* component. Our findings can inform the importance and role of each component and modality in encoding emotion features. Our insights reveal which elements can be used to improve context awareness and system interactions in domains such as adaptive interfaces, game designing, healthcare, e-learning, entertainment, and other disciplines.

The structure of the paper is as follows: Sect. 2 begins by laying out the theoretical dimensions of the research, Sect. 3 describes our methodology, Sect. 4 presents the results, followed by Sect. 5 with discussion, and, finally, Sect. 6 concludes the analysis.

2   Background and Related Works

Given the importance of affect recognition, numerous studies have been based on *discrete* and *dimensional* theories of emotions. Despite evidence from the neuroscience studies of emotion, *appraisal* theory-based models have largely been neglected, possibly due to the complexity of their architectures. Recently, an increase in articulation based on CPM, a variant of the *appraisal* model, can be seen.

CPM comprises five components: *appraisal, motivation, physiology, expression,* and *feeling* [33] to define the emotional phenomena. According to the CPM, the *appraisal* component is an initiator for event assessment at several cognitive levels defined by four objectives 1) Relevance: "Does this event relevant?", 2) Implications: "What are the consequences of this event?", 3) Coping potential: "What is the possibility that I can overcome these consequences?" and 4) Normative significance: "Is this event important with respect to social norms and values?". Based on that, the *appraisal* component activates interdependent processes on other components. Then the *motivation* component initiates action tendencies (e.g., fight or flight). The outcomes of both these components are attended by the *expression* and *physiology* components and define changes in expressive motor behavior and body, respectively. The *feeling* component is an awareness of these integrated changes and represents an emotional experience that we know by categorical and verbal labels. To evaluate CPM, the GRID instrument was developed as a questionnaire with shortened versions as Core-GRID and MiniGRID [35].

Most research on CPM has been carried out considering a single or combination of components [6,10,30,32]. However, due to the component synchronization and interconnection [33,34], assuming a full CPM is more rational to understand the underlying mechanism of emotions, the role of each component in *discrete* emotion differentiation and the correlation between components. For instance, using film clips, Mohammadi and Vuilleumier [26] have elicited a range of *discrete* emotions to study the efficacy of CPM in defining emotion attributes. The authors showed that CPM could be used to distinguish *discrete* emotions, and components contribute differently to each emotion. Further, using hierarchical clustering, they clearly differentiated between the positive and negative emotions in the CPM space.

CPM components such as *physiology* and *expression* can be objectively measured by physiological signals, facial signals, and body gestures. Moreover, the inclusion of multimodal objective measures is more reliable than self-reports due to the representation of conscious and unconscious responses [5]. Menétrey [23] attempted to explore the CPM by collecting self-reports and multimodal signals such as heart activity, skin conductance and respiration. The authors used film clips as the stimuli and conducted the experiments inside a functional Magnetic Resonance Imaging (fMRI) machine. They showed the possibility of ML techniques in identifying the patterns of CoreGRID terms and physiological signals in characterizing emotions. Similarly, using the XGBoost, the authors show the possibility of developing a personalized model in multi-componential space [42]. Despite the research validity of CPM, these works were limited in providing a better ecologically valid experimental setup due to mobility constraints inside fMRI. Furthermore, using passive emotional stimulation methods such as movies reduces active participation, so the user is more like an observer [23,26,40].

Recent CPM-based studies using VR as an active mechanism to induce emotion are considered at an exploratory level [25,39,41]. A recent investigation by Meuleman and Rudrauf [25] used seven VR games to trigger *discrete* emotions. They showed the possibility of VR games in investigating the CPM features, multi-componentiality of emotional response, and importance of the *appraisal* component in explaining emotion variations. However, the authors focused more on exploring the characteristics of CPM using self-reports rather than objective measures. Furthermore, their study was more focused on the *appraisal* component; hence, additional research on other components is required.

3   Methodology

3.1   Proposed System

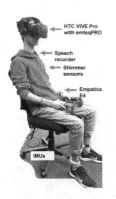

Fig. 1. Experimental setup

Our objective is to develop a data-driven framework to inform emotion recognition systems and serve as a testbed to understand the underlying mechanism of emotional formation more fully, assuming a full CPM. Accordingly, our research is designed to generate a wide range of emotions to span the componential space via VR games, eliciting ecologically valid content with the active contribution of participants. As shown in Fig. 1, we used a HTC VIVE Pro headset with controllers for game visualization and play. To measure facial expressions by EMG muscle activity, we used an emteqPRO embedded into the VR headset [11]. We used an Empatica E4 wristband to record Heart Rate (HR), Electrodermal Activity (EDA), Blood Volume Pulse (BVP), Inter Beat Intervals (IBI), skin temperature, and acceleration. In addition, we used Shimmer sensors to collect respiration and Electrocardiography (ECG) and Inertial Measurement Unit (IMU) [8] to measure involuntary body gestures. Before each game, for calibration purposes, participants were instructed to remain in a neutral position, maximally smiling, frowning, and raising their eyebrows.

In the current study, we have only analyzed the self-reports, emteqPRO EMG and E4 signals, as the primary aim is to explore the relationship between *discrete* theory-based emotions and the theoretically defined components with physiological measures. Further analysis of other objective measures will be conducted in later stages.

3.2   Material and Assessment

We used 27 interactive VR games from the Steam platform, which were used in the literature [1,25,37,40]. We annotated each game using the 20 emotion terms from Geneva Emotion Wheel (GEW) [38] based on literature [1,25,37], researcher experience and reviews from game review sites. Nevertheless, participants' emotional experience is based on their cognitive evaluations, so we did not assume that participants have the same experience for each game, like our pre-labeling [43]. We evaluated participants' emotional experiences using 20-item GEW and 51-item CoreGRID questionnaires [26,38,41].

3.3 Procedure

After ethics approval from the University of New South Wales Human Research Ethics Committee (HC200809), we conducted a preliminary experiment with 28 participants (10 females, 18 males, mean age 24.1, SD = 5.1 years). First, we conducted a VR training session for each participant and then organized three data collection sessions on separate dates. They all had a normal or corrected vision and were given a $30 gift voucher per session. In each session, we explained the procedure, instructed them to immerse themselves in the game, express their feelings, and report their experiences with each game. We developed an application to present games and questionnaires, time sync each event, and save data. Based on the game pre-labeling, we randomized the sequence of games across participants and sessions; however, we made sure that games with similar emotional content did not recur in a single session, so we got a balanced distribution of different emotions in each session. Further, we randomized the order of each questionnaire item. We evaluated participants' demography, personality, and mood using surveys. Before each game, we calibrated the devices by collecting several expressions. Each participant played a three-minute VR game wearing the HTC VIVE Pro with emteqPRO, E4, Shimmer, IMU sensors, and speech recorder. Then GEW and GRID questionaries [35] were presented to rate their emotional experience on a 5-points Likert scale (*1-Not at all, 5-Strongly*). This iterative process continued until each participant completed the 27 games. We collected 756 observations (27 games × 28 participants) from all participants but considered 749 observations due to technical problems such as slow loading and game updates while collecting the rest. Moreover, we used the Emteq SuperVision[1] application to get collected EMG signals' expression, arousal, and valence insights.

This analysis presents our results from self-reports, E4 biosignals (BVP, EDA, skin temperature, HR) and filtered EMG signals from the emteqPRO. To preserve the signal quality, EMG signals are sampled at 1000 Hz [22,28]. Therefore, we resampled all the signals to 1000 Hz. We up-sampled EDA (4 Hz), skin temperature (4 Hz) to 1000 Hz and applied a Savitzky-Golay filter [31] with a *window_length* of 31 and *polyorder* of 2 [4,44]. Then we up-sampled BVP (64 Hz) and HR (1 Hz) to 1000 Hz and applied a median filter [4]. We used filtered EMG signals collected at 2000 Hz from emteqPRO and downsampled to 1000 Hz [22]. Finally, for each biosignal time-series data, we extracted the mean, median, max, min, and standard deviation [4] and used them to train our ML models.

---

[1] https://support.emteqlabs.com/monitoring-tools/supervision.

4   Analysis and Results

4.1   Exploratory Analysis of Facial EMG

To understand the possibility and power of facial EMG activations in representing emotions, we conducted a correlation analysis with the CoreGRID's *expression* items. For that, we uploaded the calibration data (neutral, smile, frown
and eyebrow raise) and three minutes game data to the SuperVision application
and retrieved intensities for each of the neutral, smile, frown and eyebrow raise
expressions. As shown in Fig. 2, we analyzed the Spearman correlation between
the average intensity of each EMG expression insight with the eight self-reporting
items of the *expression* component.

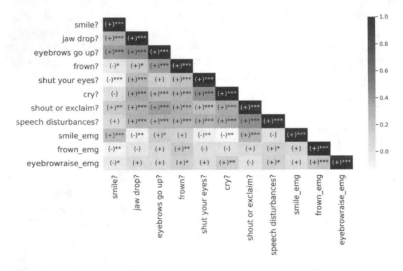

Fig. 2. Correlation matrix of average intensity of each EMG expressions (smile_emg,
frown_emg, eyebrowraise_emg) (black font) with the CoreGRID *expression* component
(blue font). (+) shows a positive correlation, and (−) shows a negative correlation.
Asterisks indicate the significance of results at p-value: * $p < 0.1$, ** $p < 0.01$, ***
$p < 0.001$. (Color figure online)

According to Fig. 2, a positive correlation of "smile_emg" is highly significant with CoreGRID's "smile?", "shout or exclaim?" ($p < 0.001$), and moderately significant with "eyebrows go up?" ($p < 0.1$). The "smile_emg" shows
a significant negative correlation with "jaw drop?", "shut your eyes?" and
"cry?" ($p < 0.01$). The "frown_emg" shows a significant positive correlation with
"frown?" ($p < 0.01$), "speech disturbances?" ($p < 0.1$) and a significant negative
correlation with "smile?" ($p < 0.01$). The "eyebrowraise_emg" shows the highest positive correlation with "frown_emg" ($p < 0.001$) and moderate significance
with "cry?" ($p < 0.01$), "frown?" ($p < 0.1$), "speech disturbances?" ($p < 0.01$),
but not with "eyebrows go up?". The observed difference may be due to the
applications' inaccuracy in capturing these expressions. The correlation between
"eyebrowraise_emg" and "smile?" is negatively significant ($p < 0.01$).

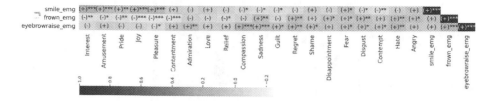

Fig. 3. Correlation matrix of average intensity of each EMG expressions (smile_emg, frown_emg, eyebrowraise_emg) (black font) with the GEW emotions (blue font). (+) shows a positive correlation, and (−) shows a negative correlation. Asterisks indicate the significance of results at p-value: * $p < 0.1$, ** $p < 0.01$, *** $p < 0.001$. (Color figure online)

Similarly, we performed a Spearman correlation with GEW items and average expression intensities to find the encoding pattern of facial expressions with emotions. Figure 3 shows the correlation between GEW items and three expressions. Accordingly, "smile_emg" shows a highly significant positive correlation with interest, amusement, joy, and pleasure ($p < 0.001$), moderate significant correlation with pride ($p < 0.01$) and fear ($p < 0.1$). A significant negative correlation with "smile_emg" can be seen with contempt ($p < 0.01$), disgust, guilt, sadness, and compassion ($p < 0.1$). With "frown_emg", a positive correlation can be seen with sadness, regret, disgust, hate ($p < 0.01$) and disappointment, fear, and anger ($p < 0.1$). All the positive emotions show a negative correlation with "frown_emg", and some are significant at several p-levels as joy, pleasure, contentment ($p < 0.001$), interest, pride ($p < 0.01$), and amusement, relief ($p < 0.1$). The "eyebrowraise_emg" shows a high positive significant correlation with compassion, sadness, "frown_emg" ($p < 0.001$). Further, it demonstrates a significant positive correlation with admiration, regret, and hate ($p < 0.01$). At a lower significant level ($p < 0.1$), the correlation of "eyebrowraise_emg" with guilt, shame, fear, disgust, and contempt is positively significant. However, "eyebrowraise_emg" shows a significant negative correlation with pleasure ($p < 0.1$). Overall, these results suggest that EMG features recorded through the emteqPRO device in VR can be used to identify facial expressions and emotional characteristics where face recording is not feasible.

4.2 Interpretation of Emotional Experience Through the CoreGRID and Physiological Changes

Next, to analyze the possibility of using ML techniques to find the features of emotions assuming a full CPM, we trained several classifiers such as Random Forest (RF), Support Vector Machines and XGBoost. For that, we z-normalized the 51 CoreGRID items and statistical features (mean, median, max, min, and standard deviation) of resampled BVP, EDA, HR, skin temperature, and filtered EMG signals and used them as features. We treated each emotion as the target. We converted our problem to a binary classification by dividing emotion ratings into two classes (high and low), using the median of each emotion as the split point. We used random sampling to reduce the imbalance (refer to 'Class Dis'

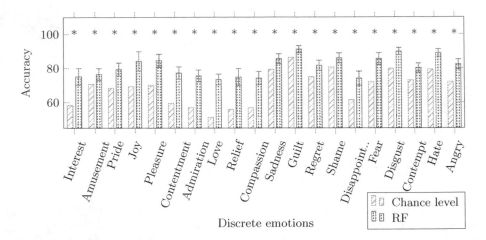

Fig. 4. Accuracy of the Random Forest (RF) binary classifiers for the differentiation of each emotion. Accuracy is compared with the chance level (majority class portion). Error bars show the standard deviation. Asterisks indicate the significance of results at p-value: *$p < 0.001$. The chance level is used as the baseline.

column in Table 1) in the dataset. We trained and tested the model using a stratified 10-fold cross-validation and evaluated average performance after ten iterations. Although we trained several classification methods, we report results from RF because it performed better than other methods.

Figure 4 shows the results of training the RF classifier for each emotion. We used the chance level (majority class prediction) as the baseline and one-sample t-test to compute the significance. Accordingly, the accuracy of all emotions is significantly ($p < 0.001$) higher than the chance level. Although, the performance is significantly higher than the chance level for all emotions, the margin is smaller for emotions that are more skewed towards one class. This is due to having less representative for that class which hinders the model to find the discriminating pattern. Overall, these results show the efficacy of objective measures and CoreGRID items in differentiating emotion features.

4.3   Role of Each CPM Component and Physiological Changes in Differentiating Discrete Emotions

To explore the role of each CPM component with objective measures, we trained ML models like the above section. Each model was trained to predict *discrete* emotions' high and low values and evaluated the average performance after stratified 10-fold cross-validation. Table 1 presents the results of our ML models using the individual component descriptors.

Accordingly, the *feeling* component is significant for interest, joy, pleasure, contentment, admiration, love, relief, compassion, fear ($p < 0.001$) and anger ($p < 0.01$). However, the *feeling* component is not significant with most negative emotions. For the *physiological* component, first, we trained models only with CoreGRID *physiology* questions and then with both questions and statistical features of BVP, EDA, HR, and skin temperature values. We found that models using self-reports and signal data performed better (2.3%–11.4% improvement) than only self-report for all emotions except compassion and admiration. So, we report only the multimodal analysis. The *physiology* component performed better in contentment, love, relief, compassion, guilt, fear, contempt ($p < 0.001$) followed by interest, admiration, sadness, regret, disappointment, and disgust ($p < 0.01$). Like *physiology*, we used filtered EMG statistical data and self-reports to find the role of the *expression* component in CPM. We report those data because accuracy using both signals and self-reports was better (1.5%–8.8% improvement) than unimodal except for love. These accuracy increments using subjective and objective measures implicitly show the importance of using objective measures rather than only relying on subjective evaluations for such components which people may not be conscious about. The *expression* component analysis shows significant performance for all emotions except amusement, pride, and sadness, and has a lower significance for joy, and pleasure. The *motivation* component performs significantly for interest, admiration, love, hate ($p < 0.001$) and compassion, disappointment, and disgust ($p < 0.01$); and surprisingly did not appear as significant for anger and fear which is in contrast with previous findings [26]. This could be due to not having enough samples with high levels of fear and anger. Our results show that the *appraisal* component is significant for all the emotions except amusement, sadness, and shame. Given that the performances are still higher than the chance level for those emotions, we suggest this could be the effect of not having samples with extreme values of amusement, sadness, and shame.

5 Discussion

Our first correlation analysis with facial EMG expressions (smile, frown, and eyebrow raises) and CoreGRID *expression* items shows the patterning of facial activations to the events trigged using VR games. Accordingly, EMG smile expression is highly correlated with "smile?" and "shout or exclaim?" items. Although the correlation of EMG smile with self-reported "smile?" is expected [3,9,12], the correlation to "shout or exclaim?" may be due to the zygomaticus muscle (mouth area) activations such as involuntary speech and orofacial movements [36]. As the SuperVision application currently only provides four facial expressions (neutral, smile, frown, eyebrow raise), our correlation analysis cannot detect the difference between speech and smiles as both are related to the same muscle areas. In contrast, EMG smile negatively correlates with "jaw drop?", "shut your eyes?" and "cry?". The analysis of EMG frown positively correlates with "frown?" and "speech disturbances?" which may be due to the involvement of corrugator supercilli muscles (near the medial end of eyebrows) that are contributing to frowning [14,15]. Moreover, EMG frown activation is negatively correlated with self-reported "smile?" possibly due to the opposite activations. The EMG activations of eyebrow raises showed positive correlations to EMG frown expression and "cry?", "frown?" and "speech disturbances?" CoreGRID

Table 1. Accuracy of Random Forest binary classifiers for the differentiation of each emotion using individual components. Asterisks indicate the significance of results at p-value: *p < 0.01, **p < 0.001. The chance level is used as the baseline. (The physiology component involves CoreGRID questions and BVP, EDA, HR, and skin temperature values. The expression component contains CoreGRID questions and filtered EMG values. FEEL: Feeling, PHYI: Physiology, EXPRE: Expression, MOTI: Motivation, APPR: Appraisal, BASE: Chance level, Class Dis: Class distribution (#high, #low))

	FEEL	PHYI	EXPRE	MOTI	APPR	CPM	BASE	Class Dis
Interest	67.15**	65.15*	69.57**	70.48**	68.49**	75.03**	58.21	(436, 313)
Amusement	71.42	70.77	72.35	68.89	72.10	76.36**	70.63	(529, 220)
Pride	72.22	71.57	71.16	65.28	75.30**	79.31**	68.36	(512, 237)
Joy	79.29**	72.77	77.70*	70.36	81.44**	84.11**	69.29	(519, 230)
Pleasure	82.24**	71.83	77.83*	69.56	81.58**	84.38**	69.83	(523, 226)
Contentment	70.49**	66.09**	68.75**	60.90	69.55**	77.16**	59.41	(445, 304)
Admiration	73.17**	64.21*	68.89**	64.35**	70.50**	75.43**	56.88	(426, 323)
Love	67.57**	70.76**	64.21**	64.35**	69.96**	73.30**	51.40	(385, 364)
Relief	66.09**	69.03**	66.90**	57.81	64.23*	74.77**	55.94	(419, 330)
Compassion	63.95**	66.74**	69.43**	62.48*	69.56**	74.11**	56.74	(324, 425)
Sadness	81.99	83.71*	83.05	78.90	83.05	85.45**	79.31	(155, 594)
Guilt	87.85	89.72**	90.79**	86.65	89.06*	91.06**	86.25	(103, 646)
Regret	73.44	78.77*	80.37**	73.17	79.71*	81.58**	74.90	(188, 561)
Shame	82.78	82.51	84.78**	79.70	83.17	85.85**	80.64	(145, 604)
Disappoint	66.49	65.68*	69.96**	66.21*	68.75**	73.96**	61.42	(289, 460)
Fear	78.91**	82.38**	78.77**	75.97	78.37**	85.45**	71.96	(210, 539)
Disgust	81.30	83.72*	87.19**	85.45*	86.38**	89.72**	79.71	(152, 597)
Contempt	76.10	77.84**	78.11**	73.03	77.44**	80.10**	72.90	(203, 546)
Hate	81.84	82.90	85.30**	83.31**	87.57**	88.91**	79.31	(155, 594)
Angry	76.36*	77.31	79.18**	75.43	77.97*	82.24**	71.96	(210, 539)

items. This correlation may be due to stimulating frontalis muscles (brow lowerer) in frowning situations [7]. So, the SuperVision application cannot find the differences in these expressions. Although EMG eyebrow raise shows a positive correlation to "eyebrows go up?", the correlation is not significant, probably due to some oversights in the SuperVision application.

Our following analysis involves exploring the correlation between GEW items and EMG expressions. Agreeing with the literature [3,15], EMG's smile correlates positively with positive emotions except admiration, relief, and compassion. This negative association can be explained by the less arousal and control involved with admiration, relief, and compassion [38], so minimal zygomaticus activations may be expected. Also, it shows a negative correlation with negative emotions except for shame, fear, and anger. This positive correlation of some negative emotions (shame, fear, and anger) may be possible due to the facial action units such as lip tightener, lip stretcher, and jaw drop [7]. As expected, all the positive emotions are negatively correlated with EMG frown expression. This is due to the less activation of corrugator supercilli muscles in positive emotional phenomena and high activation in negative ones [12]. Supported by literature [18], EMG frown positively correlates with all negative emotions except guilt. Guilt is a complex emotion [19], defined as social or self-conscious emotions based on human evolution and development [17]; it might be challenging to induce such emotions. Further, in the current dataset, guilt emotion ratings were skewed towards the lower, so the data may not be enough for modeling. Nevertheless, the correlation of EMG frown with negative emotions agrees with Cacioppo, et al. [3], where authors show higher corrugator muscle activation in unpleasant situations than in pleasant. The observed correlation of EMG eyebrow raises is positive for most GEW items except for interest, pride, joy, and pleasure. This can be explained by the high arousal involving these positive emotions (interest, pride, joy, and pleasure), where frontalis EMG shows lower activation in such events [3]. However, the high positive correlation between compassion and sadness may be due to the frontalis muscle activation due to the lowering of the eyebrows and inner brow raiser [7,18]. Overall, both correlation analyzes with EMG facial expressions and the CoreGRID *expression* component shows that EMG encodes facial activations. Further, it validates the principles of the CPM *expression* component.

Our subsequent analysis focused on modeling the emotional features using subjective and objective measures. All our classifiers performed significantly better than the chance level, showing the efficacy of ML models in differing high and low emotions segments. However, we noticed that emotions ratings are skewed to the lower side for guilt and shame. It is possible due to the complexity of these emotions [19], which are difficult to evoke using VR games. Further, as VR games are more oriented towards entertainment, finding such game content is more challenging.

Finally, we analyzed the role of individual CPM components in emotion differentiation. The less significance of the *feeling* component with most of the negative emotions may be due to the skewness of these emotion ratings to the lower end as VR appears to be more pleasant even in challenging and scary content [25]. On another note, this could be because the *feeling* component or selected CoreGRID items for this component do not capture the emotional experience well. The *physiology* component performed better for most of the negative emotions and some of the positive emotions. This may be due to the variations in collected heart rate and skin conductance, which are associated with negative emotions [20]. So, our physiological signals are suited to capturing those emotions. The *expression* component's higher performance to most emotions demonstrates the capacity of selected CoreGRID items and facial EMG to encode the emotional experience relatively well. Therefore, as facial expressions are the most natural way of non-verbal communication [2,21,28], most emotions can be identified by facial EMG expressions and subjective *expression* analysis. Furthermore, the lower performance of the *expression* component for amusement agrees with previous work using film clips [26]. Since the *motivational* component was related to positive and negative emotions, it could be explained by approach and avoidance [16]. Also, this performance can be partially justified by previous works [26], which showed significant performance for fear, anxiety, anger and disgust. The importance of the *appraisal* component in predicting fear [25], disgust [26], anger [42], and joy [25] agrees with the literature. In general, the model with all CPM self-reports and signals performs better in all the emotions. These results show that each CPM component could have a specific role in emotion differentiation which has not been studied in previous works. For example, expression may not be a reliable signal for detecting all emotions. Further, considering only one component will limit our understanding of emotion formation.

6   Conclusion and Future Work

This investigation aimed to explore the CPM as a model to frame emotion by interconnected components and processes. For that, we conducted a preliminary study using 27 interactive VR games and collected self-reports, physiological signals, facial signals, and body gestures. Our correlation analysis with the facial EMG expressions and emotions, facial EMG expressions and CoreGRID *expression* items showed the possibility of using facial EMG in analyzing motor expressions and *discrete* emotions. All the classifiers were able to differentiate the features of each of the categorical emotions significantly. Moreover, our analysis demonstrates the significant role of each CPM component in understanding the emotional experience. The current findings clearly support the importance of assuming a full CPM. These analyzes showed the possibility of using ML methods to model the CPM emotion theories. Overall, our results strengthen the importance of considering full CPM to enrich our understanding of emotional phenomena. Additionally, our work implies the theoretical limitations in literature which considers only one component.

In future, our research is determined to increase the stability of ML models by accompanying diverse participants and using data augmentation. More broadly, we plan to determine the contribution of other physiological signals and their features rather than statistical elements.

References

1. Bassano, C., et al.: A VR game-based system for multimodal emotion data collection. In: MIG 2019. Association for Computing Machinery, New York (2019)
2. Boot, L.: Facial expressions in EEG/EMG recordings. Thesis (2009)
3. Cacioppo, J.T., Petty, R.E., Losch, M.E., Kim, H.S.: Electromyographic activity over facial muscle regions can differentiate the valence and intensity of affective reactions. J. Pers. Soc. Psychol. **50**(2), 260 (1986)
4. Chandra, V., Priyarup, A., Sethia, D.: Comparative study of physiological signals from Empatica E4 wristband for stress classification. In: Singh, M., Tyagi, V.,

Gupta, P.K., Flusser, J., Ören, T., Sonawane, V.R. (eds.) ICACDS 2021. CCIS, vol. 1441, pp. 218–229. Springer, Cham (2021). https://doi.org/10.1007/978-3-030-88244-0_21

5. Domínguez-Jiménez, J.A., Campo-Landines, K.C., Martínez-Santos, J.C., Delahoz, E.J., Contreras-Ortiz, S.H.: A machine learning model for emotion recognition from physiological signals. Biomed. Signal Process. Control **55**, 101646 (2020)

6. Dupré, D., Tcherkassof, A., Dubois, M.: Emotions triggered by innovative products: a multi-componential approach of emotions for user experience tools. In: 2015 International Conference on Affective Computing and Intelligent Interaction (ACII), pp. 772–777 (2015)

7. Ekman, P., Friesen, W.V.: Facial action coding system. Environ. Psychol. Nonverbal Behav. (1978)

8. Elvitigala, D.S., Matthies, D.J.C., Nanayakkara, S.: Stressfoot: uncovering the potential of the foot for acute stress sensing in sitting posture. Sensors **20**(10), 2882 (2020)

9. Frank, M.G., Ekman, P., Friesen, W.V.: Behavioral markers and recognizability of the smile of enjoyment. J. Pers. Soc. Psychol. **64**(1), 83 (1993)

10. Gentsch, K., Beermann, U., Wu, L., Trznadel, S., Scherer, K.: Temporal unfolding of micro-valences in facial expression evoked by visual, auditory, and olfactory stimuli. Affect. Sci. **1**(4), 208–224 (2020)

11. Gnacek, M., et al.: EmteqPRO-fully integrated biometric sensing array for non-invasive biomedical research in virtual reality. Front. Virtual Reality **3** (2022)

12. Granato, M., Gadia, D., Maggiorini, D., Ripamonti, L.A.: An empirical study of players' emotions in VR racing games based on a dataset of physiological data. Multimedia Tools Appl. **79**(45), 33657–33686 (2020)

13. Grandjean, D., Sander, D., Scherer, K.: Conscious emotional experience emerges as a function of multilevel, appraisal-driven response synchronization. Conscious. Cogn. **17**, 484–95 (2008)

14. Gruebler, A., Berenz, V., Suzuki, K.: Emotionally assisted human-robot interaction using a wearable device for reading facial expressions. Adv. Robot. **26**(10), 1143–1159 (2012)

15. Inzelberg, L., Rand, D., Steinberg, S., David-Pur, M., Hanein, Y.: A wearable high-resolution facial electromyography for long term recordings in freely behaving humans. Sci. Rep. **8**(1), 1–9 (2018)

16. Izard, C.E.: Basic emotions, natural kinds, emotion schemas, and a new paradigm. Perspect. Psychol. Sci. **2**(3), 260–280 (2007)

17. Izard, C.E.: Emotion theory and research: highlights, unanswered questions, and emerging issues. Annu. Rev. Psychol. **60**, 1–25 (2009)

18. Kehri, V., Awale, R.: A facial EMG data analysis for emotion classification based on spectral kurtogram and CNN. Int. J. Digital Signals Smart Syst. **4**(1–3), 50–63 (2020)

19. Kory, J.M., D'Mello, S.K.: Affect elicitation for affective computing. In: The Oxford Handbook of Affective Computing, p. 371 (2014)

20. Kreibig, S.D.: Autonomic nervous system activity in emotion: a review. Biol. Psychol. **84**(3), 394–421 (2010)

21. Mavridou, I., et al.: FACETEQ interface demo for emotion expression in VR. In: 2017 IEEE Virtual Reality (VR), pp. 441–442 (2017)

22. Mavridou, I., Seiss, E., Hamedi, M., Balaguer-Ballester, E., Nduka, C.: Towards valence detection from EMG for virtual reality applications. In: 12th International Conference on Disability, Virtual Reality and Associated Technologies (ICDVRAT 2018). ICDVRAT, University of Reading, Reading, UK (2018)

23. Menétrey, M.: Assessing the Component Process Model of Emotion using multivariate pattern classification analyses. Thesis (2019)
24. Menétrey, M.Q., Mohammadi, G., Leitão, J., Vuilleumier, P.: Emotion recognition in a multi-componential framework: the role of physiology (2021). https://doi.org/10.1101/2021.04.08.438559
25. Meuleman, B., Rudrauf, D.: Induction and profiling of strong multi-componential emotions in virtual reality. IEEE Trans. Affect. Comput. **12**(1), 189–202 (2018)
26. Mohammadi, G., Vuilleumier, P.: A multi-componential approach to emotion recognition and the effect of personality. IEEE Trans. Affect. Comput. **13**(3), 1127–1139 (2020)
27. Ojha, S., Vitale, J., Williams, M.A.: Computational emotion models: a thematic review. Int. J. Soc. Robot. **13**(6), 1253–1279 (2020)
28. Perusquía-Hernández, M., Hirokawa, M., Suzuki, K.: Spontaneous and posed smile recognition based on spatial and temporal patterns of facial EMG. In: 2017 Seventh International Conference on Affective Computing and Intelligent Interaction (ACII), pp. 537–541. IEEE (2017)
29. Picard, R.W.: Affective Computing. MIT Press, Cambridge (2000)
30. van Reekum, C., Johnstone, T., Banse, R., Etter, A., Wehrle, T., Scherer, K.: Psychophysiological responses to appraisal dimensions in a computer game. Cogn. Emot. **18**(5), 663–688 (2004)
31. Savitzky, A., Golay, M.J.E.: Smoothing and differentiation of data by simplified least squares procedures. Anal. Chem. **36**(8), 1627–1639 (1964)
32. Scherer, K., Dieckmann, A., Unfried, M., Ellgring, H., Mortillaro, M.: Investigating appraisal-driven facial expression and inference in emotion communication. Emotion **21**(1), 73 (2019)
33. Scherer, K.R.: The dynamic architecture of emotion: evidence for the component process model. Cogn. Emot. **23**(7), 1307–1351 (2009)
34. Scherer, K.R.: Emotions are emergent processes: they require a dynamic computational architecture. Philos. Trans. R. Soc. Lond. Ser. B Biol. Sci. **364**(1535), 3459–3474 (2009)
35. Scherer, K.R., Fontaine, J.R.F., Soriano, C.: CoreGRID and MiniGRID: Development and Validation of Two Short Versions of the GRID Instrument. Oxford University Press, Oxford (2013)
36. Schilbach, L., Eickhoff, S.B., Mojzisch, A., Vogeley, K.: What's in a smile? Neural correlates of facial embodiment during social interaction. Soc. Neurosci. **3**(1), 37–50 (2008)
37. Shumailov, I., Gunes, H.: Computational analysis of valence and arousal in virtual reality gaming using lower arm electromyograms. In: 2017 Seventh International Conference on Affective Computing and Intelligent Interaction (ACII), pp. 164–169 (2017)
38. Shuman, V., Schlegel, K., Scherer, K.: Geneva Emotion Wheel Rating Study (2015)
39. Somarathna, R., Bednarz, T., Mohammadi, G.: An exploratory analysis of interactive VR-based framework for multi-componential analysis of emotion. In: 2022 IEEE International Conference on Pervasive Computing and Communications Workshops and other Affiliated Events (PerCom Workshops), pp. 353–358 (2022)
40. Somarathna, R., Bednarz, T., Mohammadi, G.: Virtual reality for emotion elicitation - a review. IEEE Trans. Affect. Comput. 1–21 (2022)
41. Somarathna, R., Bednarz, T., Mohammadi, G.: Multi-componential analysis of emotions using virtual reality. In: Proceedings of the 27th ACM Symposium on Virtual Reality Software and Technology, Article 85. Association for Computing Machinery (2021)

42. Somarathna, R., Vuilleumier, P., Bednarz, T., Mohammadi, G.: A machine learning model for analyzing the multivariate patterns of emotions in multi-componential framework with personalization. Available at SSRN 4075454
43. Subramanian, R., Wache, J., Abadi, M.K., Vieriu, R.L., Winkler, S., Sebe, N.: Ascertain: emotion and personality recognition using commercial sensors. IEEE Trans. Affect. Comput. 9(2), 147–160 (2018)
44. Val-Calvo, M., Álvarez-Sánchez, J.R., Ferrández-Vicente, J.M., Fernández, E.: Affective robot story-telling human-robot interaction: exploratory real-time emotion estimation analysis using facial expressions and physiological signals. IEEE Access 8, 134051–134066 (2020)

Liver Disease Classification by Pruning Data Dependency Utilizing Ensemble Learning Based Feature Selection

Md. Asif Bin Khaled(✉) , Md. Mahin Rahman , Md. Golam Quaiyum ,
and Sumiya Akter

Department of Computer Science and Engineering, Independent University,
Bangladesh, Dhaka, Bangladesh
mdasifbinkhaled@gmail.com

Abstract. Liver disease is responsible for over 2 million additional deaths globally each year. Therefore, early detection and treatment may lower the likelihood of liver disease-related death. Many researchers have been using artificial intelligence to detect liver disease. Inaccurate and disorganized data, however, make it difficult for them to choose an approach for determining the condition. Additionally, disproportionate data worsens dataset biases, reducing the validity of the research. As a result, it becomes necessary to develop techniques for dealing with this sort of challenge. This study suggests a methodology that integrates approaches for classifying liver disease by reduction of data dependency, which gives the advantage of getting more accurate predictions even with less data. Two imputation strategies were employed to tackle missing value and were contrasted with each other. Despite showing slight differences, no statistically significant distinctions between them were found. Machine Learning (ML) methods such as Random Forest, Extra Trees, Support Vector Machine, and K-Nearest Neighbor and neural network such as Multilayer Perceptron were employed to categorize liver diseases. The Extra Trees classifier outperformed other approaches in both of the imputed datasets, achieving accuracy of 98.37% and 99.18%, F1-Score of 98.37% and 99.17% while achieving 99.3% and 99.4% area under the ROC curve (AUC) respectively. This unorthodox method delivers cutting-edge accuracy with few feature dependencies. Hence, the suggested technique will make it easier for medical practitioners to identify liver diseases more quickly, resulting in a classification with lower data reliance that is less susceptible to error.

Keywords: Liver disease · Missing value · Data imbalance · Data dependency · Feature selection · Classification

1 Introduction

Liver diseases are the major causes of morbidity and mortality in many parts of the world. According to the World Health Organization (WHO) [14], it is one of

H. Aziz et al. (Eds.): AI 2022, LNAI 13728, pp. 614–627, 2022.
https://doi.org/10.1007/978-3-031-22695-3_43

the top ten principal causes of death in developing and underdeveloped countries. The liver is in charge of metabolism and protein absorption. It is affected due to a variety of conditions, such as cirrhosis, fibrosis, and hepatitis. Many essential physical functions that are crucial for survival are supported by the liver. As a result, it is imperative to have precise methods for the diagnosis and detection of liver diseases. However, the liver's functioning doesn't start to deteriorate until more than 75% of the liver tissue has been damaged. If their liver somehow malfunctions, the affected cannot live more than a few days. Therefore, the necessity for precise liver disease detection and diagnostic procedures becomes critical. The traditional methods of detection with extensive amounts of clinical tests might be expensive and unavailable to patients. The application of machine learning and data mining approaches can be effective in addressing this issue. In various inferential and decision-making applications, researchers and lab workers have used a variety of methodologies, including statistical techniques and machine learning approaches [6,19]. Regardless, there are challenges in data analysis for medical diagnostics, including dimensionality, insufficient data, missing data, choosing the right target and feature variables, and many more [21]. For researchers, many of these obstacles are more frequent than others, and they could be challenging to overcome. Missing data is a troublesome and persistent concern in medical research, resulting from various factors such as study subjects refusing to continue or laboratory worker errors. Statistical power gets reduced because of missing data, which may lead to bias in research [5]. The imputation techniques such as Simple Imputation, Multiple Imputation by Chained Equation (MICE), and MissForest are proven to be effective in resolving this problem [17,22]. These imputation techniques handles the missing spots by replacing them with some proxy value which helps to keep information at possible maximum. Furthermore, imbalanced data is another barrier. Data augmentation techniques like Synthetic Minority Oversampling Technique (SMOTE) and Adaptive Synthetic (Adasyn) sampling approaches are used to bring balance to data [2,4].

With the support of artificial intelligence and data mining techniques in healthcare, a lot of progress can be brought in the diagnosis and treatment of diseases. However, collecting medical records in large quantities is often too hard to complete. Since numerous tests are necessary, they could also be costly and time-consuming. Therefore, this study aims to reduce data dependency by removing features that are less contributing to model training through selective feature selection processes. Additionally, to study their impact, two different imputation techniques were compared. Moreover, many algorithms were employed on two differently imputed datasets to gauge models and propose a better-performing model for the detection of liver disease.

The remainder of this paper comprises four more sections where Sect. 2 reviews the related works. The following section starting with Sect. 3 outlines the methodology used for the problem and Sect. 4 analyzes the investigated result. Finally, the whole work was abstracted in the last section.

2 Related Works

Machine Learning (ML) models can be applied to predict the presence of liver disease by analyzing data collected from various medical datasets such as blood samples, tissue samples, and histologically stained slides. Based on an Indian Liver Patient dataset, Muthuselvan et al. suggested a technique to diagnose liver cancer illness using four classifiers, Naive Bayes, J48, Random Forest, and K-Star respectively. The findings revealed that Random Forest had a higher success rate [13]. Pasha and Fatima detected a way of detecting liver cancer based on the Indian Liver Patient Dataset. They applied Grading Meta-Learning, Ada Boost, Logit Boost, and Bagging algorithms. Compared to other methodologies, their findings revealed that Grading Meta-Learning performed better [15]. Vijayarani and Dhayanand found that Support Vector Machines (SVM) were able to perform better than Naive-Bayes in identifying the types of liver cancer from liver function test results of patients [23]. Additionally, Jaganathan et al. employed SVM with an optimal descriptor set in predicting drug-induced hepatotoxicity [7].

Moreover, previous studies also used different Deep Learning (DL) approaches. An artificial neural network was used by Rau et al. in the process of predicting the patients who have liver cancer with type 2 diabetes [16]. Chakraborty et al. showed that using a deep convolutional neural network helps distinguish between different types of liver cancer with the help of imaging data sets [11]. Additionally, Saillard et al. proposed a neural network model that has been able to anticipate the mortality rate of a patient after their liver tumor was removed, which used data that contains images of stained and processed tissues originating in Biopsies [20]. Although DL methods perform very well for large datasets, however, for a limited dataset, traditional machine learning algorithms often outperform DL methods. Subsequently, the previous researches didn't shed sufficient light upon the fact of data dependency reduction. Therefore, the need for this study became salient.

3 Methodology

3.1 Data Description

The University of California, Irvine (UCI) Machine Learning Repository provided the Hepatitis C Prediction (HCV) dataset utilized in this study [10]. This dataset comprised laboratory (clinical) and demographic information for 615 hepatitis C patients and blood donors. Furthermore, Aspartate Aminotransferase (AST), Alkaline Phosphatase (ALP), Albumin (ALB), Alanine Aminotransferase (ALT), Creatinine (CREA), Bilirubin (BIL), Cholesterol (CHOL), Choline Esterase (CHE), Gamma-glutamyl Transferase (GGT), and Total Protein (PROT) were among the 10 laboratory attributes included in the dataset along with age and sex (gender) as demographic features.

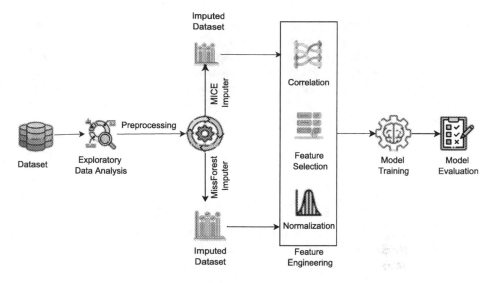

Fig. 1. Study procedures

3.2 Preprocessing of Data

Initially, it was discovered that some of the values from the dataset were missing. Subsequently, those missing values were imputed separately using two different imputation techniques.

Multiple Imputation by Chained Equation (MICE). Multiple Imputation by Chained Equations is one of the most reliable and informative ways to deal with missing specifics in datasets [17]. It was utilized to impute the missing values of the dataset in the beginning.

MissForest Imputation Technique. MissForest [22] is a random forest-based missing data imputation technique. The dataset was additionally imputed using the MissForest technique to see if there is a significant improvement in the accuracy of models over MICE imputed data.

3.3 Data Dependecy Reduction by Feature Selection

Random Forest, which is an ensemble learning method, has a built-in function to calculate the importance of features [1]. It is possible to compute it using both mean decrease impurity and mean decrease accuracy. In each set of trees constructed, there are nodes and leaves. How the data will be divided based on their similarity is a decision taken by selected features in the nodes. These features are selected based on criteria such as Gini impurity or information gain. The amount of impurity reduced by each feature can be calculated, and the average of entire trees in the forest is employed to determine feature importance. The Recursive Feature Elimination (RFE) technique internally ranks the features by using a fitted coefficient of a model or attributes found from feature importance of model and recursively keeps eliminating the weak features till a given condition is satisfied while it tries to eradicate dependencies and collinearity. Consequently, the Recursive Feature Elimination (RFE) method was used to select the features once the Random Forest algorithm had determined the feature's importance, internally using Random Forest algorithm. So the combined feature selection process is utilizing the ensemble learning method, in this case which is Random Forest.

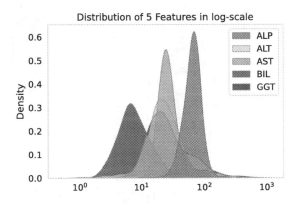

Fig. 2. Distribution of the selected features

3.4   Applied Machine Learning Methods

A few supervised classification models was used for this investigation. Following is a brief discussion of the models used.

Random Forest Classifier. Random Forest is a supervised machine learning algorithm. For the purpose of regression and classification, it can be used. Random Forest algorithm is based on the concept of ensemble learning method [1].

Extra Trees Classifier. Another variant of the ensemble learning technique that produces a classification result by adding the results of various pre-distorted decision trees assembled in a forest is the Extremely Randomized Trees Classifier, also known as the Extra Trees Classifierr [3].

K-Nearest Neighbor (KNN). The K-Nearest Neighbour (KNN) algorithm [9] can be used to address classification and regression problems. It is widely used in the industry for categorization and prediction challenges.

Support Vector Machine (SVM). Support Vector Machine (SVM) [8] creates a decision boundary known as a hyperplane that can partition n-dimensional space into different sets to put incoming data points into the correct one. SVM chooses the maximal points/vectors that help create the hyperplane.

Multilayer Perceptron (MLP). A Multilayer Perceptron (MLP) is an utterly linked class of feedforward artificial neural network (ANN). At least three layers of nodes exist in an MLP. Those three layers are known as a layer of input, a hidden layer, and a layer of output. A supervised learning method that uses gradient descent, called backpropagation, gets utilized in MLP for training. MLP can discriminate data that is separable non-linearly [8].

3.5   Evaluation Metrics

Following model training, measurement was carried out to assess the models performance using a variety of performance metrics.

Table 1. Confusion matrix

Confusion matrix	Actual class		
Predicted class	Label	0	1
	0	TP	FN
	1	FP	TN

The confusion matrix structure is presented in Table 1. In the matrix, the actual target values are contrasted with those that the machine learning model predicted. It assists in calculating Precision, F1-Score, Recall, Accuracy, and other metrics. In this context, true positive, true negative, false positive, and false negative are each defined by TP, TN, FP, and FN, respectively. The assumption that each class is equally important was employed in accuracy. Sometimes it's more crucial to correctly diagnose a patient with liver illness than it is to diagnose a non-patient with the condition. Thus, to scale accurate forecasts, Precision, and Recall values were also utilized. Furthermore, Precision and Recall were used to gauge the F1-Score. For a healthcare based decision, it is more appropriate that precision and recall is given more priority over accuracy since correctly classifying a patient is the goal. So F1-Score is considered the go to metric by many researcher and hence is given priority in this study too.

4   Results and Discussion

To comprehend the structures, and distributions, and gather information, several data visualization approaches were taken. Figure 3 shows the types of liver disease concerning gender. With 377 instances of males and 238 instances of

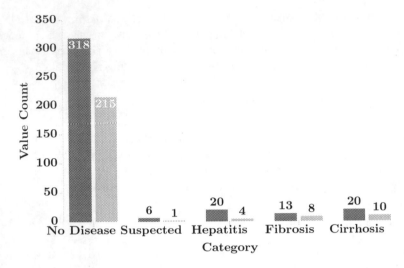

Fig. 3. Count plot for the types of liver disease according to Sex (Gender, Blue: Male, Yellow: Female). (Color figure online)

females, the category was classified as 0 for "No Liver Disease" (Blood Donor and Suspected Blood Donor) and 1 for "Liver Disease" (Hepatitis, Fibrosis, Cirrhosis). 86.67% of the data fell into the "No Liver Disease" category, while 13.33% of the data fell into the "Liver Disease" category. The result indicates that the dataset was highly imbalanced. Since the majority of the hematological features were highly right-skewed and leptokurtic, hence the data were not normally distributed.

Exploratory data analysis revealed ALB, ALP, ALT, CHOL, and PROT had a few missing observations. The dataset was imputed using MICE with linear regressor as estimator and MissForest separately which produces two datasets. In Fig. 4, all of the datasets were visualized by using the letter value plots. Since it is medical data, a letter value plot would be more insightful as it could capture more values as input features. Also, it will leave only a few too extreme values as outliers. A few extreme values were detected for the following features: ALT, AST, CREA, and GGT. However, some of the donors' records may have had elevated amounts of these features because of secondary, non-liver causes. Furthermore, it seems to be quite possible that a lab error occurred during the initial data acquisition. For handling these, 95% of the data were considered to be correct, and the rest were treated as outliers. Robust scaling was the choice for the normalization technique that would handle those values above 95%.

Feature Selection. During the process, Pearson's Correlation was calculated to measure the relationship between features. The target variable Category had a fairly positive relationship with the variables AST, BIL, and GGT in both datasets, and to a lesser extent, a negative relationship with the variables ALB,

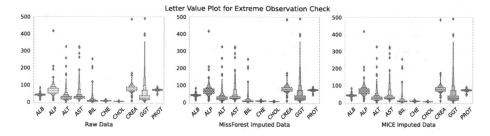

Fig. 4. Letter value plot for visualizing the state before and after imputation.

CHOL, and CHE in the dataset created using Iterative Imputer. ALP was also added to the list of negative relationships in the dataset created using MissForest. AST, BIL, and GGT all had a correlation of 0.62, 0.4, and 0.44 respectively, with AST having the strongest relationship with category features.

The variance inflation factor (VIF) was calculated to see if there is any multicollinearity, even if the correlation matrix does not show much hint of it. Any feature having a VIF higher than five was temporarily removed to observe if the multicollinearity was reduced. Five features were found under the specified VIF value, ALP, ALT, AST, BIL, and GGT. When compared for feature importance, those features were contributing the most in feature importance. To visualize the feature importance, a random forest feature importance test was conducted, shown in Fig. 5a and Fig. 5b. The feature importance was calculated using Gini importance. It was discovered that CHE and BIL were engaging in the process practically identically. From Pearson's correlation, it was confirmed that BIL had a positive correlation whereas CHE had a negative correlation with the target

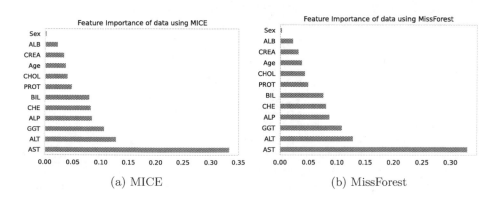

(a) MICE   (b) MissForest

Fig. 5. Random Forest feature importance plot (left) of the imputed dataset using MICE; (right) imputed dataset using MissForest.

feature. Therefore only one of the similarly important features was decided to be discarded picking only 5 features. The Recursive Feature Elimination approach was used to confirm the reduction and selection of features, where Random Forest was choosen as internal model and it was discovered that ALP, ALT, AST, BIL, and GGT were chosen by the algorithm, bearing almost 73.11 % and 73.07 % importance for MICE and MissForest imputed data, respectively. Then both the dataset was divided into individual training (80%, 492 instances) and testing (20%, 123 instances) set and normalization were applied using a robust scaling method due to the probable outliers and different measuring units presence. While analyzing classification performance on two data sets, model overfitting occurred.

An oversampling method known as SMOTE was applied separately to training datasets to address this problem and produce a balanced dataset. Figure 6 was used to visualize the impact of SMOTE between any two random features (in this case, the two most contributing feature AST and GGT) and all the features together respectively. From those plots, it was clear to see that the minor sample increased in numbers and SMOTE didn't change the distribution entirely. SMOTE was used only on the training datasets to prevent data leakage.

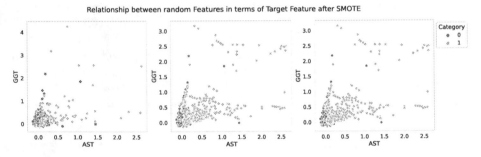

Fig. 6. Scatter plot for (left) the raw data, (mid) the imputed dataset using MICE, (right) the imputed dataset using MissForest.

Model Evaluation. Prior to the use of classification algorithms on testing datasets, the models were trained using train datasets, and before the training, the hyperparameter tuning was performed for each of the models with each of the train sets. Hyperparameter tuning was conducted using different kind of combinations of parameters on a trial-and-error basis method to obtain the best model. After acquiring information on hyperparameters of the five models, they were trained. Consequently, they were tested with the test data for different datasets, which were completely unseen by those models. Table 2 shows the obtained accuracy of each model in different datasets.

Table 2. Comparison of model accuracies (measured in percentage) across imputed datasets.

Model Data	Random Forest Classifier	Extra Trees Classifier	Multilayer Perceptron	SVM Classifier	KNN Classifier
MICE	95.12	98.37	97.56	95.12	93.49
MissForest	98.37	99.18	96.74	95.93	95.12

From Table 2, it was clear that all the models except MLP, performed better in terms of accuracy with the MissForest imputed dataset. Among the models, Extra Trees Classifier outshined other models in both datasets, with an accuracy of 98.374% and 99.187% using MICE and MissForest imputed datasets respectively. Random Forest achieved moderate accuracy in MICE imputed set but relatively high accuracy with MissForest imputed data. MLP seemed to perform well with MICE imputed data. SVM and KNN performed poorly in terms of accuracy considering the accuracy of other models. To gain confidence in the result, nested cross-validation was performed with 5-fold in the outer layer and 10-fold in the inner layer, which achieved similar (± 0.005 to ± 0.01) mean average with respect to each model's accuracy. The results from the confusion matrix showed each model's ability to correctly classify the instances. For each of the datasets, the confusion matrix of the model achieved the highest accuracy as shown in Fig. 7a and Fig. 7b.

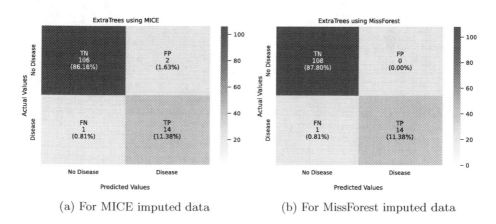

(a) For MICE imputed data   (b) For MissForest imputed data

Fig. 7. Extra trees classifier confusion matrix

From the confusion matrix, it was found that Extra Trees using MICE imputed data misclassified one instance as "No Disease" when the actual label is "Disease" and two instance as "Disease" when the actual label is "No Disease". However, Extra Trees using MissForest imputed data misclassified one instance as "No Disease" when the actual label is "Disease" but classified all the actual "No Disease" categories correctly.

In Table 3, the scores are calculated using weighted average. As the test data was imbalanced, a weighted average would be an ideal choice. Extra Tree with MissForest shows the highest F1-Score of 99.17%. Moreover, Extra Tree with MissForest achieved the highest precision, and recall too among the models with values of 99.19% and 99.18% respectively. There weren't many fluctuations in these scores for Extra Trees. On the other hand, KNN with MICE performed poorly on all the metrics and also in comparison to all the models. The AUC-ROCs score further supported these findings. The AUC-ROC curves of the classification models here in Fig. 8a and Fig. 8b, validate the applied techniques. Interesting observations were made here, where Extra Trees with MICE achieved a very high ROC-AUC score of 99.41%. Random Forest with MICE also achieved a good area under curve score, topping Extra Trees by obtaining 99.44%. In another observation, Extra Trees with MissForest achieved the highest AUC score of 99.38%.

Table 3. Summary of evaluation metric scores.

Metric	Imputed Dataset	Extra Trees Classifier (%)	Random Forest Classifier (%)	Multilayer Perceptron (%)	Support Vector Machine (%)	K - Nearest Neighbor (%)
Precision	MICE	98.374	94.931	97.655	95.946	95.074
	MissForest	99.194	98.404	96.748	96.454	95.946
Recall	MICE	98.374	95.122	97.561	95.122	93.496
	MissForest	99.187	98.374	96.748	95.935	95.122
F1-Score	MICE	98.374	94.97	97.595	95.363	93.945
	MissForest	99.175	98.323	96.748	96.091	95.363
AUC-ROC	MICE	99.414	99.444	96.975	97.901	94.599
	MissForest	99.383	99.29	97.716	98.21	95.802

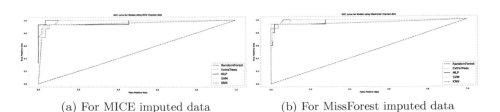

(a) For MICE imputed data   (b) For MissForest imputed data

Fig. 8. AUC-ROC curve

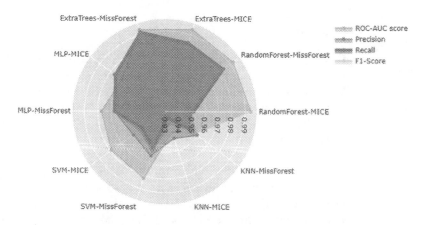

Fig. 9. Radar Plot to visualize the comparison different evaluation metrics of models created

From all these evaluations, with comparison visualization in Fig. 9 it was found that Extra Trees performed overall better in terms of F1-Score, Precision, Recall, and Accuracy, achieving higher values than previously established models from researchers. Therefore, data dependency reduction was successful with Extra Trees Classifier being the proposed model with an accuracy of 99.18%. Additionally, it was observed that for a small and skewed dataset like this, non-parametric imputation techniques such as MissForest perform slightly better on average than MICE with a linear regressor as its estimator. But to find if it was statistically significant, a two-sample t-test was performed for two groups of data. The null hypothesis was that the mean of the two groups is equal, hence these two imputation impacts are equal. With alpha $\alpha = 0.05$, the p-value was $0.3587 >$ alpha. Therefore, the null hypothesis could not be rejected, hence it was concluded that there was not sufficient evidence to claim that the impact of one imputation algorithm is significantly different from the other.

Table 4. Comparison with other works

Author name	Feature selection	Best model	Accuracy (%)
Mostafa et al. [12]	Yes, PCA	Random Forest	98.14
Safdari et al. [18]	No	Random Forest	97.29
Proposed approach	Yes RF Feature Importance & RFE	Extra Trees	99.18

Comparison with Other Related Works. There are a few work with the same dataset for binary liver disease classification which are compared with proposed study in Table 4. From Table 4, it is clear that proposed study achieved higher accuracy than other works.

5   Conclusion

Experienced healthcare professionals carefully analyze collected hematological elements and distinguish between normality and abnormality among the results. But the process is very sophisticated. The application of Machine Learning (ML) techniques can help in sophisticated processes like the diagnosis of liver disease. A similar approach to this study can be taken and used to assist to some extent in the work of healthcare professionals. Also, this approach can classify liver disease patients with a higher probability while using much fewer data. With feature importance calculation, only five out of ten hematological features were used to achieve better accuracy than previously stated works. These five features together contain most of the information needed to classify liver disease. It was observed that a non-parametric imputation method like MissForest is of great help while conducting research with a nonnormal kind of dataset. It was observed that, in terms of evaluation metrics, the best-performing model was the Extra Trees Classifier. Nonetheless, this study had some hindrances. The dataset was really small to generalize. Dataset was heavily imbalanced. Furthermore, without using SMOTE, the dataset mostly referred to the majority class and overfitting occurred. Without imputing the missing values, the alternative could be dropping the rows containing missing values which would result in decreasing the minority class more and the dataset would be more imbalanced.

In the future, the proposed methodology can be tested with other advanced algorithms and more data to observe its applicability. It is recommended to advance into multiclass classification as it will open up more possibilities for investigation and help to identify the exact liver disease.

References

1. Breiman, L.: Random forest. Mach. Learn. **45**(1), 5–32 (2001)
2. Chawla, N.V., Bowyer, K.W., Hall, L.O., Kegelmeyer, W.P.: SMOTE: synthetic minority over-sampling technique. J. Artif. Intell. Res. **16**, 321–357 (2002)
3. Geurts, P., Ernst, D., Wehenkel, L.: Extremely randomized trees. Mach. learn. **63**(1), 3–42 (2006)
4. He, H., Bai, Y., Garcia, E.A., Li, S.: ADASYN: adaptive synthetic sampling approach for imbalanced learning. In: 2008 IEEE International Joint Conference on Neural Networks (IEEE World Congress on Computational Intelligence), pp. 1322–1328. IEEE (2008)
5. Hughes, R.A., Heron, J., Sterne, J.A., Tilling, K.: Accounting for missing data in statistical analyses: multiple imputation is not always the answer. Int. J. Epidemiol. **48**(4), 1294–1304 (2019)
6. Islam, M., Rab, R., et al.: Analysis of CT scan images to predict lung cancer stages using image processing techniques. In: 2019 IEEE 10th Annual Information Technology, Electronics and Mobile Communication Conference (IEMCON), pp. 0961–0967. IEEE (2019)
7. Jaganathan, K., Tayara, H., Chong, K.T.: Prediction of drug-induced liver toxicity using SVM and optimal descriptor sets. Int. J. Mol. Sci. **22**(15), 8073 (2021)

8. Joloudari, J.H., Saadatfar, H., Dehzangi, A., Shamshirband, S.: Computer-aided decision-making for predicting liver disease using PSO-based optimized SVM with feature selection. Inf. Med. unlocked **17**, 100255 (2019)
9. Keller, J.M., Gray, M.R., Givens, J.A.: A fuzzy k-nearest neighbor algorithm. IEEE Trans. Syst. Man Cybern. **SMC-15**(4), 580–585 (1985)
10. Lichtinghagen, R., Klawonn, F., Hoffmann, G.: UCI Machine Learning Repository: HCV data Data Set (2020). https://archive.ics.uci.edu/ml/datasets/HCV+data
11. Midya, A., et al.: Deep convolutional neural network for the classification of hepatocellular carcinoma and intrahepatic cholangiocarcinoma. In: Medical Imaging 2018: Computer-Aided Diagnosis, vol. 10575, pp. 501–506. SPIE (2018)
12. Mostafa, F., Hasan, E., Williamson, M., Khan, H.: Statistical machine learning approaches to liver disease prediction. Livers **1**(4), 294–312 (2021)
13. Muthuselvan, S., Rajapraksh, S., Somasundaram, K., Karthik, K.: Classification of liver patient dataset using machine learning algorithms. Int. J. Eng. Technol **7**(3.34), 323 (2018)
14. World Health Organization: The top 10 causes of death (2014). https://www.who.int/news-room/fact-sheets/detail/the-top-10-causes-of-death
15. Pasha, M., Fatima, M.: Comparative analysis of meta learning algorithms for liver disease detection. J. Softw. **12**(12), 923–933 (2017)
16. Rau, H.H., et al.: Development of a web-based liver cancer prediction model for type ii diabetes patients by using an artificial neural network. Comput. Methods Programs Biomed. **125**, 58–65 (2016)
17. Royston, P., White, I.R.: Multiple imputation by chained equations (MICE): implementation in stat. J. Stat. Softw. **45**, 1–20 (2011)
18. Safdari, R., Deghatipour, A., Gholamzadeh, M., Maghooli, K.: Applying data mining techniques to classify patients with suspected hepatitis C virus infection. Intell. Med. (2022)
19. Sahoo, A.K., Pradhan, C., Das, H.: Performance evaluation of different machine learning methods and deep-learning based convolutional neural network for health decision making. In: Rout, M., Rout, J.K., Das, H. (eds.) Nature Inspired Computing for Data Science. SCI, vol. 871, pp. 201–212. Springer, Cham (2020). https://doi.org/10.1007/978-3-030-33820-6_8
20. Saillard, C., et al.: Predicting survival after hepatocellular carcinoma resection using deep learning on histological slides. Hepatol. **72**(6), 2000–2013 (2020)
21. Smiti, A.: When machine learning meets medical world: current status and future challenges. Comput. Sci. Rev. **37**, 100280 (2020)
22. Stekhoven, D.J., Bühlmann, P.: Missforest-non-parametric missing value imputation for mixed-type data. Bioinformatics **28**(1), 112–118 (2012)
23. Vijayarani, S., Dhayanand, S.: Liver disease prediction using SVM and Naïve bayes algorithms. Int. J. Sci. Eng. Technol. Res. (IJSETR) **4**(4), 816–820 (2015)

Optimization

Optimizing the Feature Set for Machine Learning Charitable Predictions

Greg Lee[1]([✉]), Jordan Pippy[1], and Mark Hobbs[2]

[1] Acadia University, Wolfville, Canada
glee@acadiau.ca
[2] Fundmetric, Halifax, Canada

Abstract. For most charities, there is a lack of features describing their constituents to be used for machine learning predictions about charitable behaviour. But as charities learn to collect and synthesize more data, the feature sets have grown and these sets should be optimized. We investigate several methods for optimizing the feature set for charitable predictions. We first systematically remove different types of data (e.g., education) and then remove individual features, using Pearson and Spearman correlation coefficients, random forests and removal of "unimportant" features as determined by the GINI measure of decision trees. Ultimately, for one prediction we found that only 3 features were needed for machine learning algorithms to achieve an accuracy equal to the accuracy achieved with the full feature set. This finding should help charities focus on using accurate predicted lists instead of trying to determine themselves which features of a constituent matter.

Keywords: Machine learning · Charitable giving · Feature engineering

1 Introduction

Charities seek to target constituents (people in their databases) with relevant appeals in order to increase the chances of the constituents donating to the charity. Machine learning can be used to predict which donors are likely to give to a cause at a particular time. While it is possible to ask every constituent to donate to every cause or appeal, this can lead to donor fatigue and churn (attrition), as the donors are inundated with solicitations [3]. Thus, targeted lists are preferred by charities in order to ask donors for a donation at the best possible time and to help ensure more donation dollars go to the intended recipients. This is an important problem, as in the USA, charities raised $484.85 billion in 2021 [4].

In machine learning problems, having more features on which to learn and form a model generally helps improve accuracy. But charities often do not have hundreds of features describing their constituents – they may have as few as 10, and they seek to augment their data with any features they can find or derive. Feature sets bloated with irrelevant or correlated features could lead to less accurate models that take (sometimes prohibitively) longer to train and/or

H. Aziz et al. (Eds.): AI 2022, LNAI 13728, pp. 631–645, 2022.
https://doi.org/10.1007/978-3-031-22695-3_44

run. There is thus a need to optimize the feature set for charitable giving with respect to machine learned predictions, to optimize the accuracy of the learned models and the speed of the training process.

The more accurate the machine learning model for a given charitable prediction, the more likely charities using the list will send solicitations to the best constituents for each solicitation. This will increase charitable revenue, while reducing donor churn. We seek to learn which features and characteristics of feature relationships (e.g., correlations) lead to the best dataset on which to learn various models for various charities, in order to help inform charities and the machine learning community. To do this, we experiment with removing different types of data (e.g., donation), removing unimportant correlated features, and using only the most important features to train machine learning models.

The rest of the paper is organized as follows. We next review related work, followed by the problem formulation. We then describe our approach, followed by the empirical evaluation. We conclude with discussion and future work.

2   Related Research

Feature selection has been studied previously, in domains outside of charitable giving. In [7], the authors build a dataset of customer review features from five different categories (structural, lexical, syntactic, semantic, and meta-data) and analyze which have the most impact on model accuracy, using an SVM regression algorithm. We do similar work by analyzing which categories of data (demographic, donation, behavioural, or education) have the most impact on charitable machine learning models as we describe later.

Feature engineering has also been studied extensively outside of the charitable domain [16]. Model optimization has been studied through reducing the number of features in large datasets based on their importance [6]. This leads to faster run times where resulting accuracies can range from small increases to acceptable decreases. We make use of common feature selection techniques in this work, such as calculating feature correlations and importance, using decision trees' GINI importance measure [11].

Machine learning for customer behaviour prediction has been well-studied [2, 5]. Some lessons can be learned from this research, but ultimately, people give to charities for different reasons than why they make purchases [1,12,14].

Machine learning for charitable giving has been studied with respect to the donor journey [9,10]. The donor journey is the series of actions a charity and a constituent take that leads to a donation from the constituent, which involves predictions on chronological data, while we focus on point in time predictions. Combining data across charities in order to improve accuracy of point in time predictions has been explored [8]. In other charitable predicition work, the effect of combining National Survey of Student Engagement data with donation data for predicting telephone donations was explored in [13]. Gaussian Naïve Bayes classifiers, random forests, and support vector machine algorithms were used to seek new donors in [15]. These bodies of research did not seek to optimize the feature set available for charitable prediction, which is the subject of our work.

3   Problem Formulation

The problem we are attempting to solve is optimizing the feature set for building machine learning models for charitable giving. The four predictions we are optimizing are shown in Table 1 The positive and negative labels for each prediction are provided in the original data.

Table 1. Four predictions whose feature sets we seek to optimize, and their corresponding positive and negative data sets.

ID	Prediction	Positive set	Negative set
P1	Donors likely to become major donors	Major donors	Non-major donors
P2	One-time donors likely to repeat	Repeat donors	One-time donors
P3	Lapsed donors likely to return	Current donors	Lapsed donors
P4	Current donors likely to lapse	Lapsed donors	Current donors

These predictions are made on three universities, whose data is described in Table 2. This data comes from Fundmetric (www.fundmetric.com), a machine learning platform that provides anonymized data that mirrors the real world completeness of most data sets for nonprofits. Positive examples are examples of the behaviour we are trying to predict (e.g., who is going to give a second gift who has only given one gift? – P2) Negative examples are the "prospects" for the charity, being the constituents the charity thinks may start exhibiting the positive behaviour. The predictions are meant to tell the charities which of these prospects is most likely to exhibit the positive behaviour. In terms of defining the groups, a donor is generally considered lapsed after not having donated for a period of 2 years. A donor is considered a *major* donor if they have given at least one gift at or above the major gift threshold for the charity, which is typically in the range of $25,000–$50,000.

Table 2. Data for the three university charities whose feature sets we attempt to optimize. Positive and negative set sizes are separated by "/".

Charity	Major donor	Likely to lapse	Likely to return	Likely to repeat
C1	1436/107490	39071/18362	18885/54833	52517/13654
C2	367/11236	4217/2136	2253/5369	7336/845
C3	2537/80489	18842/20523	21362/38633	40333/15792

For each prediction, the same feature set is available, but some features must be removed as they are *giveaway* features. These are features that provide the machine learning algorithms with immediate answers to the question in the prediction and produce a model that does not help the charity build useful appeal lists. An example is the feature *maximum donation* in P1 (Table 1), where the algorithm could learn the simple rule, "if the *maximum* donation is greater than the major gift threshold, major donor, else, not major donor". While this type of feature will often produce a 100% accurate model, it does not provide any useful information to charities seeking *prospects*, and thus we seek to eliminate giveaway features from each prediction's data set.

The data available for each constituent can be divided into four types: demographic, donation, behavioural, and education. A subset of the features used in our experiments is shown in Table 3.

Table 3. A subset of the features used in experiments.

Type	Examples
Demographic	Solicit mail/email/phone, contact mail/email/phone, postal/zip code, age, prefix,
Donation	Standard deviation, days since first donation, largest donation, number of small gifts, giving frequency, donation method
Behaviour	Percentage of emails opened, number of volunteer activities, links clicked, videos started, days since last answered phone call
Education	Number of degrees, year of graduation, last school graduated from

3.1   Demographic Data

Demographic data primarily describes a constituent's location and contact preferences. Whether they can be contacted by email, or solicited by phone are features recorded by charities to ensure smooth communication. While age, income, and employment status are features that would fit in this category, most charities do not have access to this information for more than half their constituents.

3.2   Donation Data

Charities record donation amounts and dates, and from these two features many features can be derived. These calculated features include maximum/ minimum/ mode/ mean donations, donation lifetime, and the slope of the line of best fit of a constituent's donations in chronological order to inform the machine learning algorithms (roughly) whether this constituent's donations are increasing (positive slope) or decreasing (negative slope).

3.3   Behavioural Data

Constituents interact with charities in ways that do not involve a financial transaction and these actions can indicate constituent preferences. Whether constituents open emails, the percentage of emails they open, and whether they attend events can all be used by machine learning algorithms to help learn a model concerning their donation habits.

3.4   Educational Data

Educational institutions provide more data on their constituents than many other charities, recording what degree the constituent earned, how many degrees they earned and if this institution was where they earned their ultimate degree.

3.5   Experimental Design Being Optimized

We seek to optimize the feature set so as to maximize prediction accuracy given the following setup. The data is divided into training and testing data, and balanced by using the full smaller of the two sets and a random selection of the larger set, which is repeated 10 times. All non-numeric features are one-hot encoded and all accuracies reported are on an unseen test set. This process is used with each charity (Table 2) and each algorithm. In the next section, we detail how we vary the data set used in this process to understand how to optimize said data set.

4   Our Approach

We seek to optimize the feature set by discovering and eliminating strongly correlated features, discovering which features are important and creating models with only those features deemed "important" by the algorithm at hand.

In our Experiment 2, we calculate the correlation coefficients between features using two methods – Spearman and Pearson. The Spearman correlation coefficient is calculated as:

$$p = 1 - \frac{6 \sum d_i^2}{n(n^2 - 1)} \tag{1}$$

where p is Spearman's rank correlation coefficient, d_i is the difference between each observation, and n is the number of observations.

The Pearson correlation coefficient is calculated as:

$$r = \frac{\sum (x_i - \bar{x})(y_i - \bar{y})}{\sqrt{\sum (x_i - \bar{x})^2 \sum (y_i - \bar{y})^2}} \tag{2}$$

where r is the correlation coefficient, x_i is the value of the current x variable, $\bar{x}$ is the mean of the values of the x variable, y_i is the value of the current y variable in a sample, and $\bar{y}$ is the mean of the values of the y variable.

The Pearson method determines whether a linear relationship exists between any of the variables. It depends on all of the features being continuous, making it useful for many regression-based problems. The Spearman method accepts discrete or continuous variables and can be used to determine the monotonic relationship between them. For the Pearson method, the relationship between x and y must be described linearly. We investigate correlations calculated with both Pearson and Spearman methods in the next section.

For feature importance calculations in Experiments 3 and 4, we use the GINI measure, or Mean Decrease in Impurity (MDI), of decision trees/random forests. Features that show no decrease in impurity are deemed to be zero importance features, and we experiment with the removal of these features from the data set. We also consider the *most* important features according to this measure, building sets only from these features and noting the change in accuracy in models compared to training with the full set.

5   Empirical Evaluation

In this section, we detail the experiments run to help optimize the feature set for the four predictions given in Table 1. We first experiment with the removal of entire *types* of data (e.g., donation) and observe the effects. Next, we calculate the difference in findings for the Spearman and Pearson correlation coefficients, then use the Spearman coefficient along with the GINI measure of decision trees/random forests to eliminate correlated features and observe the difference. We then take this a step further and eliminate features deemed not "important" using the GINI measure of a baseline random forest model and observe the change in accuracy across three machine learned models. Finally, we compare the performance of these algorithms using only the most important features, compared to using the full feature set.

Various parameterizations of three machine learning algorithms were used. These are k-nearest neighbours (KNNs), artificial neural networks (ANNs) and random forest classifiers (RFCs). The best performing architectures were used in each experiment. Ten cross validation folds were used for each experiment. We show results for all predictions where the results vary by prediction, otherwise we show representative results.

5.1   Experiment 1: Exploring Removal of Data Set Types

Before exploring the effect of removing individual features, we first explored the removal of each of the types of data described in Sect. 3 – demographic, donation, behavioural, and education. Figure 1 shows the effect of the removal of each type of data on C3 on the "Donors Likely to Lapse" prediction using a KNN classifier. This effect was similar across all charities and predictions so we present Fig. 1 as a representative example. The y-axis labels indicate which types of data have been dropped from the set, so the 3rd bar from the top would be the set with no behavioural, demographic, or donation data, and thus only education data, as an example. Not surprisingly, the sets with donation data are the most important (the bars with the highest accuracy do not have donation data dropped). Given that we did not see any other significant patterns, we moved on to more granular feature removal experiments.

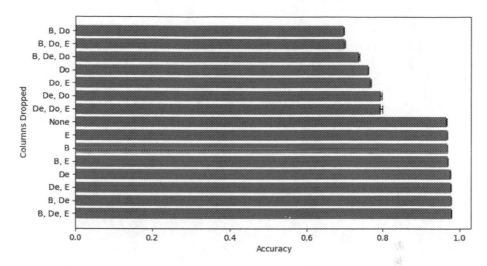

Fig. 1. The effect of the removal of various data types for the "Donors Likely to Lapse" prediction for C3 using a KNN classifier. Here "B" = behavioural, "De" = demographic, "Do" = donation, "E" = educational and "None" shows the full data set being used.

5.2   Experiment 2: Calculating Spearman and Pearson Coefficients

For each prediction and charity, we calculated the coefficient matrix using both the Spearman and Pearson methods. We provide a sample of a Spearman correlation matrix in Fig. 2 with a subset of the feature set, and the same sample using a Pearson correlation matrix in Fig. 3. The same pattern holds across all predictions and charities with the full feature set, with many features correlated, most of which are explainable. In Fig. 2, distinct years of giving is heavily correlated with donation lifetime. This make sense, as the former is a count of the number of different years the donor gave and the latter is the number of years since the donor started giving. It is quite possible that machine learning algorithms do not need both features to build an accurate model, even if they can provide quite distinct information.

Note that while both methods are able to calculate positive correlations between features, the Spearman coefficient calculations also capture negative correlations more reliably. Figure 3 is almost completely devoid of negative correlations. Thus, we used the Spearman coefficient going forward in our experiments.

Using the GINI method for determining feature importance and the correlation coefficients as determined in Fig. 2, we removed the less important feature of any correlated pair and compared the performance of KNNs, ANNs, and RFCs using the two sets on the "Donors Likely to Lapse" prediction. Results are shown in Table 4. Here, for pairs of features to be correlated, they must have a Spearman coefficient of $|0.75|$.

In most cases, removing the unimportant feature of a correlated pair did not have a large effect on accuracy. ANNs benefitted the most, with a 2.4–2.6% boost

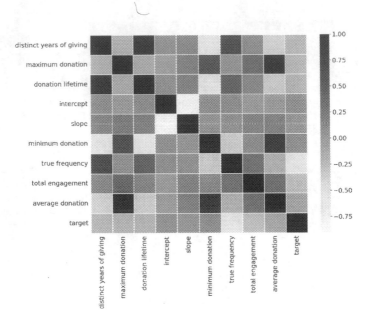

Fig. 2. A Spearman correlation matrix on a sample of the features available to the machine learning algorithms.

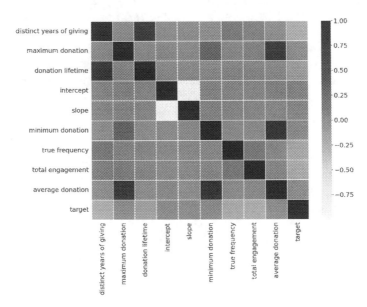

Fig. 3. A Pearson correlation matrix on a sample of the features available to the machine learning algorithms.

in accuracy with the C2 and C3 data sets. Since this data set change did not show great promise, we next tried to remove more unimportant features. This pattern holds across predictions and charities.

Table 4. Comparing accuracy using full data set to set with unimportant correlated features removed on the likely to lapse prediction.

Algorithm	Set	Accuracy	Time	Charity
RFC	Full	98.8 ± 1.9E-04	127	C1
	Trimmed	**99.0 ± 1.6E-04**	**41**	C1
KNN	Full	98.9 ± 8.0E-05	131	C1
	Trimmed	**99.1 ± 3.2E-04**	**101**	C1
ANN	Full	93.2 ± 1.2E-02	437	C1
	Trimmed	**96.0 ± 1.6E-02**	**73**	C1
RFC	Full	97.4 ± 3.3E-04	**12**	C2
	Trimmed	**98.3 ± 1.0E-03**	13	C2
KNN	Full	89.6 ± 3.7E-03	**5**	C2
	Trimmed	**91.0 ± 3.4E-03**	9	C2
ANN	Full	70.8 ± 1.5E-02	27	C2
	Trimmed	**73.4 ± 1.3E-02**	**12**	C2
RFC	Full	**99.6 ± 1.6E-04**	122	C3
	Trimmed	99.0 ± 7.0E-05	**51**	C3
KNN	Full	96.6 ± 4.9E-04	138	C3
	Trimmed	**97.4 ± 6.0E-04**	**113**	C3
ANN	Full	82.8 ± 2.4E-02	431	C3
	Trimmed	**85.2 ± 3.7E-02**	**111**	C3

5.3   Experiment 3: Removing Zero Importance Features

Using the GINI importance measure of random forests, we compared the accuracy of training a model with the full set of features to the accuracy of models trained with all of the zero importance features removed. Table 5 shows the change in accuracy for the "Donors likely to lapse" prediction. The ANN benefits for C1, but in general the removal of the unimportant features does not have a large effect on accuracy. The run times for these algorithms are all lowered, though, and when the data is scaled up, this could become a factor in determining the significance of zero importance feature removal. This pattern holds across predictions and charities.

5.4   Experiment 4: Comparing Top 3 Feature Set to Full Set

After determining values for features using the GINI measure of random forests, we ran models using only the top 3 most important features. For each prediction,

Table 5. Comparing accuracy using full data set to set with unimportant features removed on the likely to lapse prediction.

Algorithm	Set	Accuracy	Run time	Charity
RFC	Full	98.8 ± 1.9E-04	127	C1
	Trimmed	**99.1 ± 2.8E-04**	**106**	C1
KNN	Full	98.9 ± 8.0E-05	131	C1
	Trimmed	**99.0 ± 3.2E-04**	**118**	C1
ANN	Full	93.2 ± 1.2E-02	437	C1
	Trimmed	**95.5 ± 1.5E-02**	**319**	C1
RFC	Full	97.4 ± 3.3E-04	12	C2
	Trimmed	**97.8 ± 7.7E-04**	12	C2
KNN	Full	89.6 ± 3.7E-03	5	C2
	Trimmed	**92.9 ± 4.0E-03**	**4**	C2
ANN	Full	70.8 ± 1.5E-02	**27**	C2
	Trimmed	**82.0 ± 2.7E-02**	28	C2
RFC	Full	99.6 ± 1.6E-04	122	C3
	Trimmed	**99.8 ± 1.0E-04**	**107**	C3
KNN	Full	**96.6 ± 4.9E-04**	138	C3
	Trimmed	**96.6 ± 6.1E-04**	**126**	C3
ANN	Full	82.8 ± 2.4E-02	431	C3
	Trimmed	**85.9 ± 3.5E-02**	**330**	C3

we ran experiments with the full data set and compared the accuracy and run time to using the same algorithms with only the top 3 most important features. Tables 6, 7, 8 and 9 show the comparison.

Donors Likely to Lapse: The top 3 features are: frequency (number of gifts/days since first donation), days since first donation, and total donations (sum of the constituent's donation). ANNs benefit from using only the top 3 features, increasing their accuracy by 6 to 21%, while cutting their run time by one third for the larger sets. RFC and KNN accuracies increase as well, but to a point where they produce almost no prospects (no false positives). So, while using only the set of 3 features helps to produce more accurate models, it does not help a charity *practically* in terms of finding people in danger of lapsing.

Donors Likely to Become a Major Donor: For the major donor prediction, the top 3 features were not able to capture a sufficient amount of information to make accurate predictions. The features were top engagement type (the manner in which they engaged most with the charity), total donations per total engagement (how often they gave compared to how often they engaged), and age. In almost all cases there was a drop in accuracy when using only the top 3 features compared to the full set. This would indicate that the major donor prediction is likely more nuanced than other predictions where accuracy is unaffected or even increased by considering only the top 3 features.

Table 6. Comparing accuracy and run times using the top three features (Donors Likely to Lapse).

Charity	Algorithm	Set	Accuracy	Time(s)
C1	RFC	Full	98.8 ± 0.0002	112.9
		Limited	**99.9 ± 2.14E-05**	**54.7**
	KNN	Full	98.9 ± 0.0025	127.1
		Limited	**99.9 ± 6.69E-05**	**6.4**
	ANN	Full	93.2 ± 0.00002	361.1
		Limited	**99.4 ± 0.00184**	**139.4**
C2	RFC	Full	97.4 ± 0.0003	12.0
		Limited	**99.5 ± 0.00042**	**8.5**
	KNN	Full	89.6 ± 0.0037	5.0
		Limited	**99.8 ± 0.00043**	**0.9**
	ANN	Full	71 ± 0.0146	**25.7**
		Limited	**92.6 ± 0.04320**	28.6
C3	RFC	Full	99.6 ± 0.00015	114.9
		Limited	**99.9 ± 5.8E-05**	**56.8**
	KNN	Full	96.6 ± 0.00049	135.4
		Limited	**99.9 ± 0.00016**	**6.5**
	ANN	Full	82.8 ± 0.02374	370.6
		Limited	**99.8 ± 0.00075**	**218.9**

Table 7. Comparing accuracy and run times using the top three features (Donors likely to become major donors).

Charity	Algorithm	Set	Accuracy	Time(s)
C1	RFC	Full	**97.1 ± 0.00245**	19.4
		Limited	89.9 ± 0.00383	**16.2**
	KNN	Full	**87.8 ± 0.00586**	42.6
		Limited	84.9 ± 0.00595	**18.6**
	ANN	Full	**88.0 ± 0.01176**	29.2
		Limited	75.2 ± 0.04341	**15.7**
C2	RFC	Full	**96.3 ± 0.00511**	5.2
		Limited	83.7 ± 0.01320	**5.0**
	KNN	Full	**85.3 ± 0.01709**	3.5
		Limited	73.1 ± 0.03067	**2.3**
	ANN	Full	60.8 ± 0.02246	4.7
		Limited	**61.9 ± 0.04433**	**3.8**
C3	RFC	Full	**97.9 ± 0.00079**	21.8
		Limited	89.7 ± 0.00621	**19.0**
	KNN	Full	**89.5 ± 0.00324**	46.7
		Limited	87.4 ± 0.01146	**16.2**
	ANN	Full	**91.3 ± 0.00704**	41.3
		Limited	86.9 ± 0.00656	**25.9**

One Time Donors Likely to Repeat: Similar to the major donor prediction, the top 3 features set was not informative enough to compete with the full data set for any algorithm. The top 3 most important features for this prediction were days since last donation, days since last interaction, and minimum donation.

Table 8. Comparing accuracy and run times using the top three features (One Time Donors Likely to Repeat).

Charity	Algorithm	Set	Accuracy	Time(s)
C1	RFC	Full	**99.9 ± 2.9E-06**	**33.8**
		Limited	94.5 ± 0.00067	38.4
	KNN	Full	**99.5 ± 0.00044**	64.9
		Limited	92.4 ± 0.00023	**5.5**
	ANN	Full	**99.8 ± 0.00044**	157.1
		Limited	80.1 ± 0.01555	**88.0**
C2	RFC	Full	**99.9 ± 0.00010**	**4.9**
		Limited	0.94833 ± 0.00600	5.0
	KNN	Full	**94.7 ± 0.00328**	2.4
		Limited	88.2 ± 0.00559	**0.9**
	ANN	Full	**80.7 ± 0.03251**	12.7
		Limited	70.5 ± 0.05369	**11.6**
C3	RFC	Full	**99.9 ± 4.6E-06**	**46.4**
		Limited	86.7 ± 0.00197	55.9
	KNN	Full	**99.2 ± 0.00015**	93.1
		Limited	81.4 ± 0.00172	**5.6**
	ANN	Full	**99.7 ± 0.00071**	202.3
		Limited	58.2 ± 0.00801	**80.4**

Lapsed Donors Likely to Return: The lapsed donors likely to return prediction had the greatest increase in accuracy by using only the top 3 features of all the predictions. In addition to this, the accuracies are not almost 100%, thus meaning that there are still prospects on the list for charities. The top 3 features for this prediction were frequency, distinct years of giving count (in how many different years did the constituent give a gift), and donation lifetime.

6   Discussion and Future Work

In this work, we evaluated several methods for choosing features to include for training charitable giving models using machine learning. Removing less important correlated features did not have a large effect. Removing zero importance features did help accuracy improve in some cases. This is likely because the

Table 9. Comparing accuracy and run times using the top three features (Likely to Return).

Charity	Algorithm	Set	Accuracy	Time(s)
C1	RFC	Full	89.8 ± 0.00139	129.4
		Limited	**94.3 ± 0.00073**	**69.5**
	KNN	Full	91.0 ± 0.00218	235.2
		Limited	**94.1 ± 0.00131**	**11.6**
	ANN	Full	88.9 ± 0.00329	471.5
		Limited	**94.6 ± 0.01068**	**104.9**
C2	RFC	Full	86.0 ± 0.01128	12.8
		Limited	**90.3 ± 0.00463**	**9.7**
	KNN	Full	79.2 ± 0.00208	6.8
		Limited	**93.2 ± 0.00469**	**1.6**
	ANN	Full	70.0 ± 0.03086	23.0
		Limited	**88.5 ± 0.03408**	23.0
C3	RFC	Full	85.8 ± 0.00125	149.8
		Limited	**91.2 ± 0.00098**	**86.7**
	KNN	Full	85.4 ± 0.00158	211.1
		Limited	**91.2 ± 0.00091**	**9.9**
	ANN	Full	77.0 ± 0.05961	330.1
		Limited	**94.0 ± 0.00162**	**95.7**

algorithms we used are robust to irrelevant features, in particularly ANNs and RFCs. On the major donor and one time likely to repeat predictions, using only the top 3 most important features as determined by the GINI measure did not provide machine learning algorithms with sufficient information needed to build an accurate model.

Using only the top 3 most important features had a significant effect on the lapsed donors likely to return prediction - models were all more accurate when training only on these 3 features compared to training on the full data set. These top 3 features are *affinity* features – measuring how long the donor had been giving (donation lifetime), how often they gave in that lifetime (frequency), and whether their giving was regular (distinct years of giving). The importance of these features should be noted by charities seeking to create rules for choosing which lapsed donors to contact in hopes of having them return to giving to the charity. On the machine learning side, this shows an example of a problem that can be solved with a small set of features and where seemingly informative features may be irrelevant and misleading.

In the future, we would like to experiment with the (anonymous) combination of data, to see if these methods have the same effect on a more general set of charitable data. More individual charities will be used for data sets as well, including non-university data, to see whether the lessons learned in this current work hold in other charitable verticals (e.g., disease or sports charities).

References

1. Andreoni, J.: Giving with impure altruism: applications to charity and Ricardian equivalence. J. Polit. Econ. **97**(6), 447–58 (1989)
2. Apte, C., Bibelnieks, E., Natajaran, R., Pednault, E., Tipu, F., Campbell, D.: Segmentation-based modeling for advanced targeted marketing. In: Proceedings of the Seventh ACM SIGKDD International Conference on Knowledge Discovery and Data Mining, pp. 408–413 (2001)
3. Bekkers, R., Wiepking, P.: A literature review of empirical studies of philanthropy: eight mechanisms that drive charitable giving. Nonprofit Voluntary Sector Q. **40**(5), 924–973 (2011). http://journals.sagepub.com/doi/10.1177/0899764010380927
4. Benefactor: Giving USA 2022. benefactorgroup.com/givingusa2022/. June 2022
5. Burez, J., Van den Poel, D.: CRM at a pay-TV company: using analytical models to reduce customer attrition by targeted marketing for subscription services. Expert Syst. Appl. **32**, 277–288 (2005)
6. Chen, R.-C., Dewi, C., Huang, S.-W., Caraka, R.E.: Selecting critical features for data classification based on machine learning methods. J. Big Data **7**(1), 1–26 (2020). https://doi.org/10.1186/s40537-020-00327-4
7. Kim, S.M., Pantel, P., Chklovski, T., Pennacchiotti, M.: Automatically assessing review helpfulness. In: Proceedings of the 2006 Conference on EMNLP, pp. 423?430. EMNLP 06, Association for Computational Linguistics, USA (2006)
8. Lee, G., Adunoor, S., Hobbs, M.: Machine learning across charities. In: Proceedings of the 17th Modeling Decision in Artificial Intelligence Conference (2020). in press
9. Lee, G., Raghavan, A.K., Hobbs, M.: Deep learning the donor journey with convolutional and recurrent neural networks. In: Wani, M.A., Raj, B., Luo, F., Dou, D. (eds.) Deep Learning Applications, Volume 3. AISC, vol. 1395, pp. 295–320. Springer, Singapore (2022). https://doi.org/10.1007/978-981-16-3357-7_12
10. Lee, G., Raghavan, A.K.V., Hobbs, M.: Improving the donor journey with convolutional and recurrent neural networks. In: Wani, M.A., Luo, F., Li, X.A., Dou, D., Bonchi, F. (eds.) 19th IEEE International Conference on Machine Learning and Applications, ICMLA 2020, Miami, FL, USA, 14–17 December 2020, pp. 913–920. IEEE (2020). https://doi.org/10.1109/ICMLA51294.2020.00149
11. Menze, B.H., Kelm, B.M., Masuch, R., Himmelreich, U., Bachert, P., Petrich, W., Hamprecht, F.A.: A comparison of random forest and its Gini importance with standard chemometric methods for the feature selection and classification of spectral data. BMC Bioinformatics **10**(1), 213 (2009). https://doi.org/10.1186/1471-2105-10-213
12. Patras, L., Martínez-Tur, V., Gracia, E., Moliner, C.: Why do people spend money to help vulnerable people? PLoS ONE **14**(3), e0213582 (2019)
13. Rau, N.: Predictive Modeling of Alumni Donors: an engagement model for fundraising in postsecondary education. Ph.D. thesis, James Madison (2014)
14. Shockley, C.C.: The Relationship Between Student Engagement and Alumni Giving at Higher Education Institutions: A comparative case study analysis. Ph.D. thesis, Department of Education, Delaware State University (2019)

15. Ye, L.: A Machine Learning Approach to Fundraising Success in Higher Education. Master's thesis, University of Victoria (2017)
16. Yuan, R., Xue, D., Xu, Y., Xue, D., Li, J.: Machine learning combined with feature engineering to search for BaTiO3 based ceramics with large piezoelectric constant. J. Alloys Compounds **908**, 164468 (2022). https://www.sciencedirect.com/science/article/pii/S0925838822008593

Operation-based Greedy Algorithm for Discounted Knapsack Problem

Binh Thanh Dang$^{(\boxtimes)}$ [ID], Bach Hoai Nguyen [ID], and Peter Andreae [ID]

School of Engineering and Computer Science, Victoria University of Wellington,
Kelburn, Wellington 6012, New Zealand
{binh.dang,bach.nguyen,peter.andreae}@vuw.ac.nz
https://www.wgtn.ac.nz/

Abstract. The discounted knapsack problem (DKP) is an NP-hard combinatorial optimization problem that has gained much attention recently. Due to its high complexity, the usual solution combines a global search algorithm with a greedy local search algorithm to repair candidate solutions. The current greedy algorithms use a heuristic that ignores the items already in a candidate solution. This paper presents a new greedy algorithm for DKP that uses an expanded set of operators and better heuristics that are more effective at considering the selected items. Experimental results show that the proposed greedy algorithm has superior performance to three well-known greedy algorithms for DKP, both when operating independently and when combined with global search algorithms.

Keywords: Combinatorial optimization · Discounted knapsack problem · Greedy algorithm · Local search

1 Introduction

The Knapsack problem (KP) [3,13] is a combinatorial optimization problem that has many real-world applications in logistics, energy usage optimization, financial system modeling, cryptographic systems, etc [1]. The classic KP assumes that there exists a container of a given capacity and a set of predefined items, each with a value and a weight. The task is to select items that maximize the total value while their total weight does not exceed the container's capacity.

In the Discounted Knapsack Problem (DKP) [5,18] variant, the task is to select a set of items with costs and values, maximizing the total value, subject to a fixed maximum cost. There are discounts on the costs of certain groups of items if all items in a group are selected. DKP has been applied to many real-world applications, including purchasing decisions, investment, project selection, and budget control [20]. For example, in purchasing decisions, a buyer can get a cost discount if he/she buys a group of items together. The DKP is challenging because of the increased interactions between the items in a group which any heuristic rule needs to consider.

H. Aziz et al. (Eds.): AI 2022, LNAI 13728, pp. 646–660, 2022.
https://doi.org/10.1007/978-3-031-22695-3_45

In the standard data sets for DKP, the groups of items with a discount all have exactly two items. A common formulating approach of DKP is to model the groups as if they had three items: x_{i1} and x_{i2} are the real items belonging to the i-th group, and x_{i3} is a fake item representing the case where both x_{i1} and x_{i2} are selected. The value and cost of item x_{ij} are v_{ij} and c_{ij}, respectively. Note that, $v_{i3} = v_{i1} + v_{i2}$ and $c_{i3} = c_{i1} + c_{i2} - d_i$, where d_i is the discount of the i-th group. DKP can be formulated as the following optimization problem:

$$Maximize \quad \sum_{i=1}^{n}(s_{i1}v_{i1} + s_{i2}v_{i2} + s_{i3}v_{i3}) \tag{1a}$$

$$s.t. \quad s_{i1} + s_{i2} + s_{i3} \leq 1, \tag{1b}$$

$$\sum_{i=1}^{n}(s_{i1}c_{i1} + s_{i2}c_{i2} + s_{i3}c_{i3}) \leq C, \tag{1c}$$

$$s_{i1}, s_{i2}, s_{i3} \in \{0,1\}, \forall i \in \{1,2,\ldots,n\} \tag{1d}$$

where n is the number of groups, $s_{ij} = 0$ if item x_{ij} is not selected, and $s_{ij} = 1$ if x_{ij} is selected. Equation (1b) is the constraint that, for each group, at most one of the three items is selected. Equation (1c) is the constraint that the total cost of the selected items may not exceed the maximum cost C. A binary vector $S = (s_{11}, s_{12}, s_{13}, s_{21}, s_{22}, s_{23}, \ldots, s_{n1}, s_{n2}, s_{n3}) \in \{0,1\}^{3n}$ is a candidate solution of DKP. If constraints in Eqs. (1b) and (1c) are satisfied, S is a feasible solution, otherwise S is an infeasible solution.

Existing greedy algorithms for DKP [8,10–12,16] build a solution in terms of a set of items by carrying out a sequence of operations (adding/removing items) on the current solution. The operations are usually ordered by a heuristic rule. However, the existing heuristic rule does not consider the items already selected in the current solution. For example, adding a specific item has a single position in the ordered list, regardless of whether or not the other item in its group has already been in the solution. Thus, the operation order is inaccurate, which might not result in a good item set.

This paper presents an improved greedy algorithm for DKP. It considers the interactions of the items in groups and distinguishes the cases that can happen to the group. New heuristics are developed to evaluate these different cases, which allow the algorithm to make better choices at the right time. The results show that the proposed greedy algorithm gives better results when running independently and when combined with a heuristic global search.

The rest of this paper is organized as follows. Section 2 presents a literature survey of existing greedy algorithms for DKP and analyzes their limitations. Section 3 describes our proposed greedy algorithm in detail. Section 4 discusses the experimental design and analyzes the results. Finally, Sect. 5 concludes the paper.

2   Existing Greedy Algorithms for DKP

This section discusses three well-known greedy algorithms for DKP: GR-DKP [8, 16], NROA [7,11,12,19], and D-GROA [2,10,17]. GR-DKP uses an H list of $3n$ items $(x_{11}, \ldots, x_{n3})$ sorted descending by their value/cost ratios in order to choose items greedily. It repairs and improves a candidate solution in three steps. First, it checks the solution to ensure that only one item is selected for each group. If more than one item is selected, only the item with the highest value/cost is kept. Second, suppose the cost of the solution is greater than the container's capacity (C). In that case, it progressively removes the item with the lowest value/cost ratio until the total cost is less than or equal to C. The first two steps ensure that the obtained solution or item set is feasible while keeping the solution's quality as high as possible. Finally, it traverses H again in forwarding order, adding items with high value/cost as long as they do not make the solution infeasible.

NROA uses an integer representation where each solution is a vector of n integer values corresponding to n groups. Each value is in $\{0, 1, 2, 3\}$, which indicates the selected item from a group. For example, the value of 0 means no item is selected, while the value of 1 means the first item is selected. Such representation allows at most 1 item to be selected for each group; thus, the first constraint (Eq. (1b)) is always satisfied. Like GR-DKP, NROA also sorts the items based on their value/cost ratios, resulting in the H list. NROA first steps through H, adding up the costs of the items in the candidate solution. Any item that takes the total cost over C is removed from the solution. The candidate solution is then guaranteed to be feasible. NROA then traverses H again, adding an item if none of the items in its group are selected, and the new total cost does not exceed C. One problem of NROA is that it may leave a very low value/cost item in the solution as long as it has a low cost, even if it prevents the addition of a higher value/cost item.

D-GROA is very similar to NROA but avoids the problem noted above. D-GROA changes the order in which items are removed to ensure that if an item is removed from the solution to meet the maximum cost constraint, then all other items in the solution with a lower value/cost have already been removed.

The solution representation models of GR-DKP, NROA, and D-GROA share one thing: they all use fake items for the option of selecting both items. Unfortunately, this approach leads to a distorted view of the DKP problem because it does not consider the inherent relationship between item 1, item 2, and item 3. Ignoring this relationship can lead to errors in deciding what actions to take, as we explain in what follows.

Furthermore, existing greedy algorithms do not consider the current state of the groups in the candidate solution. Let us consider the situation of adding item 1 of the i-th group to the solution. If group i has no items already selected, adding item 1 will increase the solution's total value by v_{i1} and the total cost by c_{i1}. If item 2 is already in the solution, existing greedy algorithms will remove item 2, then adds item 1. As a result, the total value of the solution will change by $v_{i1} - v_{i2}$. Similarly, the total cost will change by $c_{i1} - c_{i2}$. Existing greedy

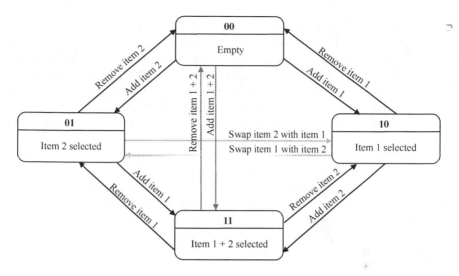

Fig. 1. State transition diagram of a group.

algorithms do not separate these two cases clearly, and both are weighed by the original value and cost of item 1. In other words, the value/cost ratios do not accurately reflect the benefits of adding and removing items. Our proposed algorithm aims to address these problems.

3   Operation-Based Greedy Algorithm for DKP

This section first presents our choice of solution representation model, then introduces new lists of operations that are sorted using new heuristic models. The section concludes with the overall proposed algorithm.

3.1   Solution Representation

Our goal is a solution representation scheme without fake items because a better representation is essential for making the best greedy decisions. Given that the DKP problem has $2n$ real items, we represent solutions as binary vectors containing $2n$ values. Each group has two corresponding bits where the first bit represents the first item and the second bit represents the second item. A bit value of 0 means the item is not selected, while the value of 1 means the item is selected. If both bits are 1, both items of the group are selected.

This representation shows that a group can be in four different states: 00—no items selected (the group is empty), 10—item 1 selected, 01—item 2 selected, and 11—both item 1 and item 2 are selected (the group is full). A problem with the traditional approach is the inadequate recognition of transitions between these states. The existing algorithms' actions include three adding operations (add item 1, add item 2, and add item 3) and three removal operations (remove item 1, remove item 2, and remove item 3). Adding an item results in the group having only that item, and removing the item results in an empty group. There is currently no model for adding an item to a group with the other item already selected to create item 3, and no model for removing a sub-item from item 3 so that the solution still contains the remaining item of the corresponding group. To address the problem, this paper distinguishes all possible states and events related to a group and organizes them into a state transition diagram in Fig. 1.

As depicted in Fig. 1, there are three types of operations: adding, removing, and swapping. An adding operation adds item(s) to an empty group or a group that already has another item. A removing operation removes item(s) from the solution, resulting in an empty group or, in the case of removing an item from a group with two items selected, a group with the remaining item. Finally, swapping operations are to swap item 1 with item 2 and vice versa.

Table 1. List of operations for a group.

No.	Name	State 1	State 2	Used for
1	Op{00}-{10}	00	10	Adding/Removing
2	Op{00}-{01}	00	01	Adding/Removing
3	Op{00}-{11}	00	11	Adding/Removing
4	Op{01}-{11}	01	11	Adding/Removing
5	Op{10}-{11}	10	11	Adding/Removing
6	Op{01}-{10}	01	10	Swapping
7	Op{10}-{01}	10	01	Swapping

3.2   List of Operations

Using the states and events in Fig. 1, we propose a list of operations as in Table 1. In Table 1, State 1 and State 2, similar to the definitions in Fig. 1, serve as the applying condition and the resulting group state. Although Fig. 1 defines five adding operations, five removing operations, and two swapping operations, removing operations are indeed the reverse of adding operations. Therefore, Table 1 includes only seven operations, of which the first five are for adding and removing. If one of these five operations is used to add item(s) to the solution, it will transform the related group from State 1 to State 2. Similarly, if that operation is used to remove item(s), it will convert that group from State 2 to State 1.

With the current greedy algorithms, the phases of adding items to and removing items from the solution occur separately, and this list fits perfectly into that approach. In the cases of item swapping operations, the greedy algorithm only needs to consider transferring the related group from State 1 to State 2.

Since each group has seven entries in the operation list as above, the complete list for n groups of a DKP instance has $7n$ entries.

3.3   New Heuristic Models

Contribution-based Value/cost Heuristic. The operations in the proposed list should be ordered by an appropriate measure. Instead of using each item's original value and cost, we propose that the contribution to the solution should determine these values and costs. For example, let us consider adding item 1 of group i into the solution. There are two possible cases:

- If no other items in group i were already selected, the value and cost of this operation is v_{i1} and c_{i1}, respectively.
- If item 2 is already selected and adding item 1 will upgrade the selection to the group item (both item 1 and item 2 are selected), the value and cost of this operation should be v_{i1} and $c_{i1} - d_i$, respectively.

Then, the value/cost ratio is calculated. By doing that, the current state of the solution is taken into account, and the value/cost ratios reflect what they contribute to the solution. Applying this heuristic to adding/removing operations in Table 1, we obtain the add/remove operation list (AROL). Table 2 describes the AROL list entries for the i-th group.

Table 2. AROL entries of the i-th group.

Name	Value	Cost	Value/Cost	State 1	State 2
Op{00}-{10}	v_{i1}	c_{i1}	v_{i1}/c_{i1}	00	10
Op{01}-{11}	v_{i1}	$c_{i1} - d_i$	$v_{i1}/(c_{i1} - d_i)$	01	11
Op{00}-{01}	v_{i2}	c_{i2}	v_{i2}/c_{i2}	00	01
Op{10}-{11}	v_{i2}	$c_{i2} - d_i$	$v_{i2}/(c_{i2} - d_i)$	10	11
Op{00}-{11}	$v_{i1} + v_{i2}$	$c_{i1} + c_{i2} - d_i$	$\frac{v_{i1}+v_{i2}}{c_{i1}+c_{i2}-d_i}$	00	11

Table 3. Comparison of value/cost descending and dual-ranking increasing sorting.

Operation	Value	Cost	Value/cost approach		Dual-ranking approach			
			Value/Cost	Order	Value rank	Cost rank	Sum rank	Order
O_1	2	4	0.5	2	3	2	5	3
O_2	10	5	2	1	1	3	4	2
O_3	4	-2	-2	3	2	1	3	1

It can be seen that the change in value and cost of the solution of the adding and removing operations are the same, except that one is adding value and cost to the solution, while the other deducts value and cost from it. This confirms what was discussed above that we do not need a list for adding and another for removing.

Dual-ranking Heuristic. The contribution-based value/cost heuristic is not used for swapping operations because it risks incorrect prioritization. While the existing approach does not provide any real value/cost values for the swapping operations, the new contribution-based value/cost system does provide them. Specifically, the swapping operation of item 1 of group i to item 2 will have a value and cost of $(v_{i2} - v_{i1})$ and $(c_{i2} - c_{i1})$, respectively. Similarly, the swap of item 2 to item 1 will have a value and cost of $(v_{i1} - v_{i2})$ and $(c_{i1} - c_{i2})$, respectively. Let us consider the following situations:

- Item 1 and item 2 have the same value and same cost. Therefore, the value/cost is undefined (0/0).
- Item 1 and item 2 have the same value, but their costs are not equal. As a result, the value/cost is always zero no matter how different the costs are, resulting in inaccurate ordering of the swapping operations.
- Assume a swap operation has a value of 2 and a cost of −2. Thus, its value/cost is −1. Since the list is sorted descending by value/cost, this operation has a low priority when it should be given a high priority because it gives a better value with a lower cost.

Table 4. SOL entries of the i-th group.

Name	Value	Cost	Sum rank	State 1	State 2
Op{01}-{10}	$v_{i1} - v_{i2}$	$c_{i1} - c_{i2}$	sum rank of Op{01}-{10}	01	10
Op{10}-{01}	$v_{i2} - v_{i1}$	$c_{i2} - c_{i1}$	sum rank of Op{10}-{01}	10	01

The above three cases show the limitations of applying the value/cost calculation to contribution-based value and cost. In some cases, the results are incorrect or even provide a misleading view of the priority of operations. Thus, we need another heuristic that satisfies two requirements:

- Using the new contribution-based value and cost.
- Replacing value/cost ratio by a different metric solving the above problems.

We propose a new dual-ranking system using a Borda [4] voting rule with two voters (value rank and cost rank). Specifically, operations' values are ranked descending, and their costs are ranked ascending. This ranking approach considers the operations' differences in values and costs. Lower costs and higher values result in higher ranks. An operation's ranking positions are then added to reflect its combined benefit.

The example in Table 3 shows how differently the dual-ranking system affects the operations' priority compared to the existing approach. Using the value/cost-based descending sorting, although having a very good value and cost, operation O_3 has the last position in the sorted list, which means that it has the lowest priority to be performed. On the other hand, if using the dual-ranking heuristic, O_3 has the first position, which better reflects its contribution to the solution.

Applying the dual-ranking heuristic for swapping operations, we have a swapping operation list (SOL). Table 4 lists two entries of SOL for the i-th group. Note that the dual-ranking system is not applied to AROL because our experiment shows that the combination of an AROL list with the contribution-based value/cost and an SOL list with dual-ranking gives better results than a single $7n$ operation list with dual-ranking.

3.4   The Proposed Operation-based Greedy Algorithm (OGA)

Algorithm 1 presents the pseudo-code of our proposed OGA algorithm for DKP. The input parameters of OGA include the cost vector, value vector, and n, which is the number of groups of the problem. The greedy algorithm works by applying operations in SOL and AROL lists in order of their priority.

Our algorithm consists of four steps. In step 1, a candidate solution X, the AROL list, and the SOL list are generated. For proper operation of the greedy algorithm, AROL needs descending value/cost sorting, and SOL needs ascending sum rank sorting.

Step 2 is for swapping operations, which are considered in the order of the SOL list. If a swap operation is applicable and makes the related group better, it will be executed. A group is better if the difference between its total value in state 2 and its total value in state 1 is more significant than the difference between its total cost in state 2 and its total cost in state 1.

Algorithm 1: Operation-based Greedy Algorithm (OGA) for DKP

Input: n: number of groups, C: maximum cost, cost vector, value vector
Output: A feasible solution X: set of items (of size $2n$) and its total value
% Step 1: Initialization
Initialize a candidate solution $X = [x_1, x_2, \ldots, x_{2n}], x_i \in \{0, 1\}, i \in \{1, 2, \ldots, 2n\}$
Build $5n$ *AROL* list in decreasing contribution-based value/cost
Build $2n$ *SOL* list in increasing sum rank
Calculate total value and total cost of X
% Step 2: Swapping phase
foreach *operation Op in SOL* **do**
    if *Op is applicable* **then**
        if *the group of Op is better after applying Op* **then**
            Transform the group in X specified by Op from State 1 to State 2
            Update total value and total cost

% Step 3: Adding phase
if *total cost $< C$* **then**
    foreach *operation Op in AROL* **do**
        if *using Op for adding is applicable* **and** *total cost + cost of Op $\leq C$*
        then
            Transform the group in X specified by Op from State 1 to State 2
            Update total value and total cost

% Step 4: Removing phase
if *total cost $> C$* **then**
    foreach *operation Op in reverse order of AROL* **do**
        if *using Op for removing is applicable* **then**
            Transform the group in X specified by Op from State 2 to State 1
            Update total value and total cost
            if *total cost $\leq C$* **then**
                break

return *X, total value of X*

After that, if the solution has a total cost lower than the maximum cost, the algorithm will activate step 3. In this step, in the order of the AROL list, applicable adding operations will be performed if they do not make the total cost exceed the maximum cost until no more of them can be done. Conversely, if the solution's total cost is greater than the maximum cost, step 4 will remove items gradually from the solution using remove operations in the reverse order of the AROL list until the total cost is less than or equal to the maximum cost.

3.5 Complexity Analysis

The complexity of an algorithm is often assessed through two aspects: the time it takes to solve a particular problem and the space, precisely the amount of memory it needs to run.

OGA algorithm avoids exponentially increasing the number of computation operations. First, OGA traverses SOL once. Then, OGA traverses AROL once, with the direction depending on whether the total cost of the candidate solution is greater than C or not. Another factor to consider is that the AROL and SOL lists need sorting. The best case (lists ordered already) takes linear time. Otherwise, $O(n * log(n))$ comparisons are needed to sort an array of n elements in the worst case. Thus, our algorithm has a linearithmic complexity of $O(n * log(n))$. This complexity is similar to those of existing greedy algorithms for DKP. On the other hand, OGA requires more memory than existing greedy algorithms since we use larger-sized lists. However, OGA has no spatial barriers with the existing data sets.

4 Results and Discussion

4.1 Experimental Design

The experiments are to answer the following questions:

1. When running independently, is the OGA algorithm better than existing greedy algorithms for DKP (including GR-DKP, NROA, and D-GROA)? OGA, GR-DKP, NROA, and D-GROA will fix and improve input candidate solutions generated randomly under two cases:
   - Case 1: A total of 5 percent of groups in a solution will have an item selected while the remaining groups are empty. This case checks how well the greedy algorithms work when they have more container space to do what they are designed for.
   - Case 2: Candidate solutions of length $2n$ binary bits are randomized without any restrictions and then converted to the format used by each greedy algorithm.

In each of the two cases, each algorithm performs 100 runs per data set instance.

2. Which type of input candidate solution works better with OGA? Should the candidate solution be empty (which forces the greedy algorithm to add items to it) or overfull (a lot of items are in it, and the greedy algorithm mostly removes items from the solution)?

The proposed OGA algorithm will be tested in the following cases:

- Case 1: The candidate solution is empty.
- Case 2: All groups have the combined item (item 1 + item 2) selected.
- Case 3: All groups have the best value/cost item selected.

The first case helps examine how well OGA adds items to an empty candidate solution. Following are the two cases where the container is overfull, i.e., the maximum cost is very far exceeded.

3. When the greedy algorithms under consideration are combined with a global search algorithm for DKP to form hybrid algorithms, which resulting hybrid algorithm gives better results?

The group theory optimization algorithm (GTOA) [10] is the global search algorithm of our choice. To our knowledge, the hybrid between GTOA and D-GROA provides the best results that a metaheuristic could achieve in solving DKP until now. The following combinations will be examined:

- GTOA as the global search and D-GROA as the greedy repair operator.
- GTOA as the global search and OGA as the greedy repair operator.

For fairness, these algorithms use the same values for shared parameters. The population size is 50, and the maximum iteration is 1000. The mutation probability, a parameter used by GTOA, has the value of 0.008, which [10] suggested. The test runs 30 times separately.

A total of 40 DKP data set instances available at [9] are used. They are classified into four categories: strongly correlated (SDKP), weakly correlated (WDKP), uncorrelated (UDKP), and inverse correlated instances (IDKP). Correlation is strong when the value and cost of an item are closely related and weak when the relationship is loose. Each data set types consists of 10 instances with the number of groups ranges from 100 to 1000. The test algorithms are programmed in Python and run on VUW's Rāpoi HPC cluster.

4.2   OGA, GR-DKP, NROA, and D-GROA Run Individually

Table 5 shows the resulting statistical results. To save space, we include only test results of instances of sizes 2 and 8 (IDKP2, IDKP8, SDKP2, SDKP8, UDKP2, UDKP8, WDKP2, and WDKP8) because they can represent instances with a small and large number of dimensions. The Instance column indicates the instance's name along with the optimum value for that instance. The next four columns show the statistical results of case 1, and the last four are for case 2. The

Table 5. Mean total values and standard deviation of returned solutions of 100 runs.

Instance	Case 1				Case 2			
	GR-DKP	NROA	D-GROA	OGA	GR-DKP	NROA	D-GROA	OGA
IDKP2	116,981.8	116,923.6	117,021.7	**117,834.4**	96,790.3	96,791	96,710.6	97,374.7
118,268	±445.6	±441.6	±403.2	±142.4	±1,718.5	±1,490.3	±1,997	±1,876.1
IDKP8	528,510.5	528,471.1	528,267.8	**532,544.5**	438,036.1	437,246	437,852.6	**465,002.9**
533,841	±899.2	±901.8	±933.5	±251.5	±4,748.2	±3,914.3	+4,098.2	+11,903.1
SDKP2	157,546.7	157,671.8	157,563.4	**160,394.6**	137,443.3	137,174.5	137,353.3	**149,711.3**
160,805	±585.2	±557	±522.5	±181.1	±2,194.4	±2,215.9	±2,340.8	±3,553
SDKP8	657,395.7	657,319.1	657,338.2	**669,587.5**	579,852.9	579,601.2	579,980.5	**632,389.3**
670,697	±1,064.7	±965.9	±1,037.2	±331.1	±3,865.4	±4,357.6	±3,961	±3,997.5
UDKP2	141,015.3	141,097.7	141,138.7	**162,969.8**	113,191.4	113,157	113,912.7	**139,422.3**
163,744	±1,026.4	±834.5	±991.3	±521.5	±4,601.5	±4,775.2	±5,393.2	±7,675.2
UDKP8	552,595.9	553,113.7	553,090.7	**647,948.8**	491,782.6	491,065.2	491,337.3	**596,095.9**
650,206	±2,279.1	±2,197.1	±2,160.8	±840.4	±5,906.2	±6,149.6	±5,787.8	±4,374.1
WDKP2	136,374.5	136,440	136,419.8	**137,874.5**	114,003.7	114,406.6	114,516	**121,210.1**
138,215	± 479.1	± 454.5	± 492.2	± 149.4	± 1971.8	± 1974.7	± 1660.3	± 4173.8
WDKP8	569,239.4	569,289.3	569,250.2	**575,940.7**	483,947.6	482,473.9	482,193.5	**529,318.1**
576,959	± 872.4	± 953.3	± 976.5	± 354.5	± 4448.9	± 4075.5	± 4382.3	± 6271.4

four sub-columns of each case represent the four greedy algorithms evaluated. The data shown for each algorithm is the mean value of 100 total values returned after 100 runs, along with the standard deviation of these 100 total values, preceded by a ± sign. This statistical table also integrates the results of the Wilcoxon rank-sum test [6] with a significant level of 0.05. The highlighted values are statistically significantly different from those of the remaining algorithms.

Table 5 shows the superiority of the proposed OGA algorithm. It has better mean total values in all cases and instances. In terms of standard deviation, OGA has very small standard deviation values in tests of case 1, which means its output solutions are primarily at the same performance level. In case 2, when the candidate solutions' total costs are close to the maximum and greedy algorithms could not do many things, such a difference in standard deviation could not be repeated. Although the paper's length limitation prevents us from showing the test results in all 40 instances, looking at them also gives interesting facts. Our greedy algorithm obtains the optimal value in three instances (WDKP1, UDKP2, UDKP3). In most cases, our algorithm has established quite a distance in performance compared to existing greedy algorithms and is close to the optimum values. Observing the results of combining OGA with a suitable global search algorithm will be interesting.

4.3   Effects of Solution Initialization Strategy

To save space, Table 6 shows the OGA's results for instances of sizes 2 and 8 only. The Instance column specifies the instance name, the OPT column stores the instance's optimal total value, and the three final columns store the total

Table 6. OGA's results on different types of starting candidate solution.

Instance	OPT	Case 1	Case 2	Case 3
IDKP2	118,268	**118,232**	**118,232**	**118,232**
IDKP8	533,841	**533,816**	**533,816**	**533,816**
SDKP2	160,805	159,327	**159,636**	159,327
SDKP8	670,697	662,633	**664,203**	662,633
UDKP2	163,744	144,552	**160,381**	144,552
UDKP8	650,206	580,800	**638,912**	579,784
WDKP2	138,215	137,457	**137,747**	137,457
WDKP8	576,959	574,358	**575,221**	574,358

value of the solution returned by OGA in each test case. Note that in case 1, the solution is initialized empty (every bit is 0). Case 2 is when the solution is initialized with both item 1 and item 2 of all groups selected. Finally, in case 3, the solution starts with the best value/cost item of each group selected.

Test results show that, if started with a solution in which all groups have both item 1 and item 2 selected, OGA often gives better results by removing items until the solution is within the maximum cost. This conclusion may suggest later algorithms for DKP a promising strategy for generating candidate solutions.

4.4 The Combination of Greedy Algorithms with Global Search

Finally, we investigate how OGA and other greedy algorithms for DKP contribute to the final result in cooperation with a global search algorithm. In such a combination, global search algorithms cover the solution space, while greedy

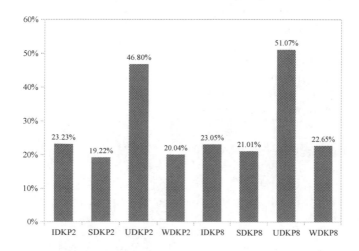

Fig. 2. Difference of mean total values of GTOA-O to GTOA-D.

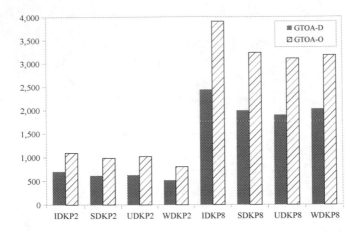

Fig. 3. Computational time (in seconds).

algorithms act as repair operators which modify and optimize candidate solutions given by the global search. In this experiment, the global search algorithm is GTOA [10], while OGA and D-GROA will be the repair operators. For simplicity, we refer to the hybrid algorithm between GTOA and D-GROA as GTOA-D and the hybrid algorithm between GTOA and OGA as GTOA-O.

Figure 2 shows the percentage differences of the mean total values between GTOA-O and GTOA-D. To obtain these values, for each hybrid algorithm and test instance, we calculate the mean value of 30 total values of 30 best solutions returned after 30 runs. After that, the difference between the mean values of GTOA-O and GTOA-D is calculated for each test instance. To save space, we only include the results of instances of sizes 2 and 8. The GTOA and OGA hybrid algorithm gives much better results than the combination of GTOA and D-GROA. The least difference is 19.22% in the case of SDKP2, while it is 51.07% in the case of UDKP8.

Figure 3 illustrates the average running time of GTOA-D and GTOA-O over 30 runs. These results reflect the natural differences between D-GROA and OGA. D-GROA is more uncomplicated and runs faster than OGA. In the worst case, GTOA-O needs 60% more time than the GTOA-D. This compromise is necessary to achieve much better results in most cases.

5   Conclusions

This paper has proposed a new operation-based greedy algorithm (OGA) for DKP. Our algorithm integrates many improvements: new heuristics based on value and cost re-perceptions, consisting of the new contribution-based value/cost coefficients and a new dual-ranking system; and most importantly, a more reasonable system of operations to replace the existing list of items. The experiments have shown that OGA is a robust greedy algorithm for DKP. No matter how bad the input candidate solutions could be, OGA consistently produces high-quality solutions. Furthermore, OGA has impressive results compared to existing greedy algorithms when functioning as a repair operator for global search algorithms. Of course, more completeness comes at the expense of higher computational time, but clearly, the cost is worth it.

For such a problem of a hardness level as DKP, greedy local search algorithms are usually not enough to find optimal solutions. Instead, hybrid metaheuristics in which global optimizers such as swarm-based algorithms [14,15] cooperate with local greedy algorithms should be an answer. For this work focuses on greedy algorithms only, more studies of how a global search algorithm should best use OGA will be necessary. Thus, we plan to develop a new global optimization algorithm for DKP in the future.

References

1. Cacchiani, V., Iori, M., Locatelli, A., Martello, S.: Knapsack problems - an overview of recent advances. Part I: single knapsack problems. Comput. Oper. Res. **143**, 105692 (2022)
2. Dang, B.T., Truong, T.K.: Binary salp swarm algorithm for discounted $\{0-1\}$ knapsack problem. PLOS ONE **17**(4), 1–28 (2022)
3. Dantzig, G.B.: Discrete-variable extremum problems. Oper. Res. **5**(2), 266–288 (1957)
4. Emerson, P.: The original Borda count and partial voting. Social Choice Welfare **40**(2), 353–358 (2013)
5. Guldan, B.: Heuristic and exact algorithms for discounted knapsack problems. University of Erlangen-Nürnberg, Germany (2007)
6. Haynes, W.: Wilcoxon Rank Sum Test, pp. 2354–2355. Springer, New York (2013). https://doi.org/10.1007/978-1-4419-9863-7_1185
7. He, Y.C., Wang, X.Z., Li, W.B., Zhang, X.L., Chen, Y.Y.: Research on genetic algorithms for the discounted $\{0-1\}$ knapsack problem. Jisuanji Xuebao/Chin. J. Comput. **39**(12), 2614–2630 (2016)
8. He, Y.C., Wang, X.Z., He, Y.L., Zhao, S.L., Li, W.B.: Exact and approximate algorithms for discounted $\{0-1\}$ knapsack problem. Inf. Sci. **369**(C), 634–647 (2016)
9. He, Y.: Four kinds of D$\{0-1\}$KP instances. ResearchGate (2019)
10. He, Y., Wang, X.: Group theory-based optimization algorithm for solving knapsack problems. Knowl.-Based Syst. **219**, 104445 (2021)
11. He, Y., Wang, X., Gao, S.: Ring theory-based evolutionary algorithm and its application to D0–1 KP. Appl. Soft Comput. **77**, 714–722 (2019)
12. Li, Y., He, Y., Liu, X., Guo, X., Li, Z.: A novel discrete whale optimization algorithm for solving knapsack problems. Appl. Intell. **50**, 3350–3366 (2020)
13. Mathews, G.B.: On the partition of numbers. Proc. Lond. Math. Soc. **s1–28**(1), 486–490 (1896)
14. Nguyen, B.H., Xue, B., Andreae, P., Zhang, M.: A new binary particle swarm optimization approach: momentum and dynamic balance between exploration and exploitation. IEEE Trans. Cybern. **51**(2), 589–603 (2021)
15. Nguyen, B.H., Xue, B., Zhang, M.: A survey on swarm intelligence approaches to feature selection in data mining. Swarm Evol. Comput. **54**, 100663 (2020)

16. Sulaiman, A., Sadiq, M., Mehmood, Y., Akram, M., Ali, G.A.: Fitness-based acceleration coefficients binary particle swarm optimization to solve the discounted knapsack problem. Symmetry **14**(6), 1208 (2022)
17. Truong, T.K.: different transfer functions for binary particle swarm optimization with a new encoding scheme for discounted $\{0-1\}$ knapsack problem. Math. Prob. Eng. **2021** (2021)
18. Wilbaut, C., Hanafi, S., Coelho, I.M., Lucena, A.: The knapsack problem and its variants: formulations and solution methods. In: The Palgrave Handbook of Operations Research, pp. 105–151. Springer, Heidelberg (2022),https://doi.org/10.1007/978-3-030-96935-6_4
19. Wu, C., Zhao, J., Feng, Y., Lee, M.: Solving discounted $\{0-1\}$ knapsack problems by a discrete hybrid teaching-learning-based optimization algorithm. Appl. Intell. **50**, 1872–1888 (2020)
20. Zhu, H., He, Y.C., Wang, X., Tsang, E.C.: Discrete differential evolutions for the discounted $\{0-1\}$ knapsack problem. Int. J. Bio-Inspired Comput. **10**(4), 219–238 (2017)

Dynamic Bus Holding Control Using Spatial-Temporal Data – A Deep Reinforcement Learning Approach

Yuguang Zhao[1](✉) , Gang Chen[1] , Hui Ma[1] , Xingquan Zuo[2] ,
and Guanqun Ai[2]

[1] Victoria University of Wellington, Wellington, New Zealand
zhaoyugu@myvuw.ac.nz, {aaron.chen,hui.ma}@ecs.vuw.ac.nz
[2] Beijing University of Posts and Telecommunications, Beijing, China
zuoxq@bupt.edu.cn

Abstract. This paper proposes a deep reinforcement learning (DRL) approach that dynamically determines the dispatching of bus services at the starting bus stop for a high-frequency bus service line. Most previous studies focus on planning bus timetables in advance based on expected future passenger demand. They often ignore real-time data and are therefore not competent at handling unexpected passenger demand fluctuations. To address this issue, we propose a Spatial-Temporal data driven Dynamic Holding (STDH) approach in this paper to dispatch bus on the fly at any decision granularity, e.g., every minute. Both spatial and temporal information regarding bus fleet and passengers are captured in a newly designed state matrix. STDH further employs a Deep Q-Network (DQN) based learning system to optimize timetabling decisions dynamically. Our DQN features the use of a newly designed self-attention network architecture to facilitate effective processing of spatial-temporal data, enabling DRL to make desirable bus dispatching decisions in accordance with real-time passenger flow. Experiments have been conducted using real-world data collected in Xiamen China. Our experiments show that STDH can effectively learn a control policy to dynamically dispatch bus services in a high-frequency urban line.

Keywords: Dynamic bus holding control · Spatial and temporal information

1 Introduction

The bus transportation system is a critical public service in many cities and an effective solution to alleviate urban congestion, protect the environment and save social resources [11]. However, the service quality is unstable in many cities. Common problems are low punctuality, long in-vehicle travel times, crowded buses or stranded passengers during peak hours [3]. Therefore, improving the quality of service of bus transit systems and operational efficiency is necessary to attract more citizens from private vehicles to public transit.

© The Author(s), under exclusive license to Springer Nature Switzerland AG 2022
H. Aziz et al. (Eds.): AI 2022, LNAI 13728, pp. 661–674, 2022.
https://doi.org/10.1007/978-3-031-22695-3_46

Traditionally, bus timetables are determined through tactical planning [11]. When bus trips deviate from planned *headways* that measure the time intervals between consecutive bus dispatches, *local bus holding control* is applied where bus drivers are instructed to move on or hold at a bus station [7,21]. While being simple, local holding control has several limitations. It ignores "global" information including bus occupancy and passenger demand. In addition, it does not consider the impacts of current holding on future bus trips.

Most existing bus holding control strategies are also static in nature [4]. They rely on expectations of passenger demand and fleet operation, which is hard to hold in practice due to the unstable nature of bus operations [19]. Moreover, it is common that minor disruptions of one trip can cause cascading effects on multiple subsequent trips [18]. To address these issues, leveraging on the fast advancement of Intelligent Transportation Systems (ITS) and Automatic Fare Collection (AFC) Systems, *dynamic bus holding control* has attracted increasing attention recently. Existing research showed that dynamic holding is essential for high-frequency bus line services (i.e., bus lines with headways of no more than 10–15 minutes during peak hours [6,10]). This is because passengers tend not to look at published timetables before arriving at bus stops [6] for high-frequency lines.

With the rising popularity of *Deep Reinforcement Learning* (DRL) [15], the technology has been applied to many challenging real-time control problems with great success. In recent years we are witnessing the trend of transforming traditional human-involved or static controls to DRL-based autonomous control in traffic signal control (TSC) [20], autonomous driving [9] and railway scheduling [14]. The dynamic nature of bus operations and passenger flow make the bus holding problem a good candidate for DRL approaches. Several DRL-based approaches for dynamic bus holding control have been developed successfully [1,2,5,18]. For example, [5] proposed a Q-Learning based technique to evenly distribute all on-trip buses. A similar approach was proposed by [2] using DQN algorithm. [18] proposed a decentralized DRL algorithm to achieve headway equalization and vehicle coordination.

All these research works assumed a fixed number of running buses and sufficient bus capacity (i.e., no stranded passengers), which may not be always practical. To address this issue, a recent study [1] explored a different approach for bus holding control at the starting stop. However, the effectiveness of this approach depends critically on the knowledge of future passenger demand and bus occupancy. Furthermore, temporal and spatial information is not explored extensively in prior DRL-based approaches. For example, in [18], the number of currently waiting passengers at each stop is included in state information. However, the number of waiting passengers in the past at adjacent downstream stops are not considered for optimal bus holding control. Furthermore, prior dynamic control approaches often ignored the variability of passenger demand due to the unrealistic assumption of sufficient bus capacity (i.e., no stranded passengers). Operational cost is also neglected, as the total number of bus trips is assumed to be either fixed [2,18] or unlimited [19].

Following [1], this paper aims to propose a new *Spatial-Temporal data-driven Dynamic Holding* (STDH) approach that can quickly and dynamically respond to riding demand changes. It makes real-time dispatching decisions on the first stop of a high-frequency bus line, taking into account both spatial and temporal information regarding bus fleet and passengers. STDH also adopts a combined objective that balances passenger demand fulfilment and bus operational efficiency, subject to constraints on the maximum number of running buses, bus capacity limitation, as well as the minimum and maximum headways. We focus on the single-stop bus holding problem because it significantly impacts the total operational cost of a high-frequency bus line. The dispatching times of all trips at the first stop effectively determine the total number of trips for a service day. This paper has the following contributions:

1. We design a new *state matrix* to capture both spatial and temporal data regarding passenger and fleet status. Historical information regarding on-going bus trips and past passenger arrivals is important in identifying the riding demand trend. Spatial information regarding bus locations are also important for bus holding control, since adjacent bus trips are highly correlated and should be considered collectively for the holding control agent.
2. We propose a new reward function to achieve a good trade-off between passenger satisfaction and bus operational efficiency. We use positive rewards to indicate the fulfilment of passenger riding demand and negative rewards to indicate the bus operational cost, passenger waiting time as well as the corner cases when passengers get stranded or give up waiting.
3. We develop a self-attention neural network architecture [17] to effectively process the spatial-temporal data embedded in the state matrix, in order to determine the degree of attentions on bus trips with strong spatial and temporal dependencies. It enables the *holding control agent* to learn quickly and reliably in our experiments.
4. We implement a comprehensive simulator based on real-world data collected in Xiamen China. Our simulator considers the realistic resource constraints without relying on the common assumption of unlimited bus trips. In addition, passenger stranding due to limited bus capacity and passengers who give up are also supported the first time by our simulator.

2   Problem Description

In this section, we describe the dynamic bus holding control problem with realistic constraints.

Single Stop Bus Holding Control. In this paper we consider a *bus line* as an one-way corridor with a fixed number (N) of bus stops $S \in (s_0, \ldots s_{N-1})$, where N is the total number of stops. To service passengers' riding demands multiple *bus trips* are dispatched every operational day. All buses are dispatched from stop s_0 and travel through all stops until the last stop s_{N-1} where all passengers must alight. Single stop *bus holding control* aims to dynamically hold and dispatch every bus trip at the first stop (s_0). Figure 1 illustrates the overall holding control process. At every minute (or any other decision granularity), the holding control agent dynamically determines bus headways by making "dispatch" or "hold" control decisions at stop s_0, with the objective to to minimize operational cost while optimizing customer riding experience (defined by the reward function in Sect. 3.1).

Bus holding control is subject to several key constraints. Specifically, all buses have a predefined capacity constraint and can only carry a certain number of passengers at maximum. In addition, the bus company can only support a specific number of ongoing bus trips at any time. There are various headway ranges used in previous studies focusing on high-frequency buses [6,10]. We adopt a common range of 3–15 minutes in our study. The following steps are performed periodically every minute.

Step 1: A snapshot of bus locations and passenger counts during the last minute is collected;

Step 2: A state matrix representing spatial-temporal information is constructed based on the newly collected and historical information;

Step 3: An action ("dispatch" or "hold") is selected by the holding control agent;

Step 4: The front bus at the first stop (s_0) operates according to the control decision in Step 3;

Step 5: The system progresses for one minute, all running buses move ahead and interactions between buses and passengers take place.

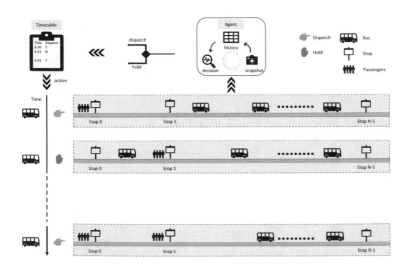

Fig. 1. A DRL-based single stop holding control framework.

As shown in Fig. 1, the holding control agent either dispatches a new bus trip or holds the front bus at the first stop based on the state matrix, which is updated every minute in accordance with the bus dispatch times, fleet operational data, and passenger flow data (see Subsect. 3.1). In Sect. 3, we develop the STDH approach that formalizes this learning problem as a single-agent reinforcement learning problem.

3   The STDH Approach for Dynamic Bus Holding Control

3.1   Problem Setup

We formally model the single-stop timetabling problem as a reinforcement learning problem in this paper. Specifically, the holding control agent has the goal to learn an optimal control policy (π^*) that takes agent observation (o_t) at any time t as its input and outputs a bus holding decision ($a_t = \pi^*(o_t)$), such that the expected total rewards obtainable within an operational day can be maximized:

$$\pi^* = \underset{\pi}{\mathrm{argmax}}\mathbb{E}[\sum_{t=T_b}^{T_e} R(o_t, \pi(o_t))],\tag{1}$$

where T_b and T_e refer to the earliest and latest service time of the bus line respectively. $R(o_t, \pi(o_t))$ gives the immediate reward for the holding control agent based on its current decision at time t. o_t is the agent's observation described below.

Observation Space Design. In a reinforcement learning problem, there are two types of state information: *environment states* and *agent observations*. Environment states are not directly accessible to the agent. An example is passenger destination, which is not known to the agent until a passenger alights from a bus. Agent observation refers to state information exposed to the agent at each step. In STDH, we design the observation as a $M \times (3 * N + 1)$ state matrix in Eq. (2), where M stands for the maximum number of running buses at any time. The state matrix contains information of the last $M - 1$ trips plus the next trip which is yet to be dispatched. Particularly, each row of the state matrix captures the information of one trip across all bus stops (from s_0 to s_{N-1}).

$$Obs(t) = (Trip_{(M-1)}, ...Trip_1, Trip_0)^T\tag{2}$$

$Trip_0$ in Eq. (2) denotes the trip subject to holding control at the first stop at time t. $Trip_i$ is the i-th trip. Each trip has several stops s_{ij} at which passengers can board and alight. We model a trip $Trip_i$ with a set of attributes, i.e., $Trip_i = (HW(i), [LD(i, 0), BD(i, 0), AL(i, 0)], \ldots, [LD(i, N-1), BD(i, N-1), AL(i, N-1)])$. Each attribute of $Trip_i$ is explained below:

- $HW(i)$: Dispatch headway of trip i. It measures the difference between the dispatching time of this trip and its preceding trip at the first stop s_0. For $Trip_0$ under holding control, we set it to the period between the dispatch time of $Trip_1$ and the current time;

- $LD(i,j)$, $j \in \{0,\dots,N-1\}$: Bus load information of $Trip_i$. It represents the number of passengers in the bus of $Trip_i$ between stop s_j and stop s_{j+1};
- $BD(i,j)$, $j \in \{0,\dots,N-1\}$: Boarding numbers at stop s_j for $Trip_i$;
- $AL(i,j)$, $j \in \{0,\dots,N-1\}$: Alighting numbers at stop s_j for $Trip_i$.

Bus loads (LD), Boarding (BD) and Alighting (AL) of $Trip_i$ in the state matrix are all set to (-1) whenever the corresponding bus stops have not been visited by trip $Trip_i$. For the example state matrix in Table 1, for $Trip_0$, dispatch headway $HW(0) = 3$, meaning that the bus waiting at stop 0 has been held for 3 min after the previous trip. Previous trips are either finished (i.e. $Trip_{M-1}$ and $Trip_{M-2}$) or still running. For example, $Trip_1$ has passed stop s_0 and is cruising towards stop s_1 since the passenger counts at stop s_1 $LD(1)$, $BD(1)$ and $AL(1)$ are -1.

Table 1. A Sample Observation Space Representation.

Trip	HW	LD(0)	BD(0)	AL(0)	LD(1)	BD(1)	AL(1)	...	AL(N − 1)
$Trip_{M-1}$	8	4	4	0	7	3	0	...	7
$Trip_{M-2}$	4	7	7	0	8	1	0	...	9
$Trip_{...}$	4	3	3	0	4	2	1	...	17
$Trip_1$	7	10	10	0	−1	−1	−1	...	−1
$Trip_0$	3	−1	−1	−1	−1	−1	−1	...	−1

Spatial and temporal information regarding both the latest and upcoming trips are captured in the state matrix. Particularly, each row gives the spatial location of every bus trip, which can be easily determined from the state matrix by checking the index of the first -1 value along the row. Meanwhile, each column provides temporal/historical information regarding the bus load, boarding and alighting passengers with respect to every bus stop. They jointly enable the holding control agent to determine the relationship among all ongoing bus trips as well as the future riding demand trend. Our state matrix can be constructed by using AFC and history data, making it practical for real-world use.

Action Space. The action space of our algorithm is very small, including two alternative actions, i.e., "dispatch" and "hold". Accordingly, the headway is determined after every "dispatch" action. For example, a 5 min headway is determined by a "dispatch" action following four consecutive "hold" actions.

Reward Function Design. We design the reward function as a weighted sum of the bus operational cost and passenger satisfaction, as presented below:

$$R(o,a) = w_b \times N_{boarding} + w_a \times N_{alighting}$$
$$+ w_d \times N_{dispatching} + w_w \times T_{waiting} + w_s \times N_{stranding} + w_l \times N_{leaving} \quad (3)$$

where $w_b, w_a > 0$ are coefficient for positive reward components. $w_d, w_w, w_s, w_l <$ 0 are coefficients for negative reward components. All these coefficients are configured in accordance with their practical significance (See Subsect. 4.1 for more details in our experiments). Each reward component in Eq. (3) is explained below:

1. $N_{boarding}$: The number of total boarding passengers during time period $(t, t+1)$.
2. $N_{alighting}$: The number of total alighting passengers during time period $(t, t+1)$.
3. $N_{dispatching}$: Dispatch cost, 1 if a new bus trip is dispatched and 0 otherwise.
4. $T_{waiting}$: Total waiting time across all passengers waiting at all stops during time period $(t, t+1)$.
5. $N_{stranding}$: The total number of stranded passengers due to limited bus capacity during time period $(t, t+1)$.
6. $N_{leaving}$: The total number of passengers who gave up after waiting for long time during period $(t, t+1)$.

In the above, positive reward components reflect successful fulfillment of passenger riding demands, which will be maximized. Negative reward components refer to the operational cost and other losses to be minimized.

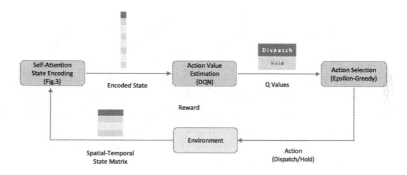

Fig. 2. Training methodology.

3.2   Holding Control Agent

We develop a holding control agent based on the double DQN algorithm, where Q-values are estimated using "self-attention" embedding of states matrices. The overall behavioral of our holding control agent is depicted in Fig. 2 with three key steps:

- **State encoding:** the agent collects and encodes state information using a self-attention based neural network;
- **Action value estimation:** the Q values are estimated using the double DQN algorithm proposed in [8];
- **Action selection:** the agents makes "dispatch" or "hold" control decisions, guided by the learned Q values. Steps 1 and 3 are discussed in more details below.

Self-Attention State Encoding. We propose a *self-attention based* neural network architecture (Fig. 3) for state encoding. The "attention" mechanism was initially introduced in an encoder-decoder architecture to process sequential data with both spatial and temporal dependencies [17]. Instead of only looking at the last "hidden" state as classical RNNs, self-attention neural networks can achieve state-of-the-art performance by looking at elements of the input sequence at all positions and learning to "attend" to important ones.

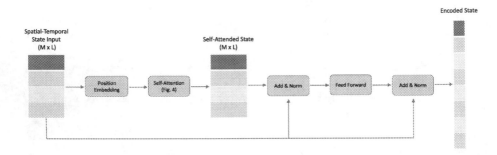

Fig. 3. Self-attention state encoding.

In this paper we define *Attention* as a function that maps a *query vector* (q) and each *key vector* (k) of a set of key-value pairs to weighting numbers that quantify the "compatibility" between the query and the key. The output of the attention function is a weighted sum of the original *value vector* (v) based on the calculated weighting numbers. *Self Attention* is the attention function where q, k and v are identical. Furthermore, the commonly used dot-product attention function is realized through matrix multiplication, where sequences of q, k, and v are packed into matrices Q, K and V respectively: $Attention(Q, K, V) = softmax(QK^T)V$.

We depict the self-attention network architecture adopted in STDH in Fig. 4. Specifically, Query matrix (Q), Key Matrix (K) and Value matrix (V) in our attention network are identical copies of the state matrix as we defined in Eq. (2) with a size of the $M \times L$, where $L = 3 * N + 1$. Each row of the three matrices (q, k, v) represents the feature vector of one trip. To calculate the "compatibility" between any pair of trip vectors, we project both q and k vectors into different feature spaces (L') before applying dot-product attention below:

$$Attention(Q, K, V) = softmax((QW_q)(KW_k)^T)V \tag{4}$$

where $Q, K, V \in \mathbb{R}^{M \times L}$, and $W_q, W_k \in \mathbb{R}^{L \times L'}$.

Our self-attention function (Eq. 4) uses two trainable projection matrices (W_q and W_v) to project the original query and key features from an L-dimensional space into a smaller L'-dimensional space (we used $L' = 6$ in our experiments, where the original query and key dimensions are $L = 109$). $softmax[(QW_q) \cdot (KW_k)^T] \in \mathbb{R}^{M \times M}$ is a matrix of the attention weights between all pairs of

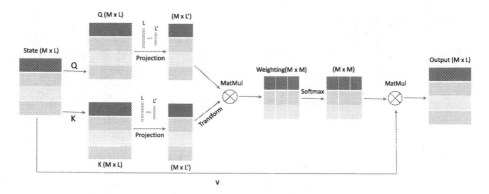

Fig. 4. The self-attention network of STDH.

trips. The output of the self-attention network is of the dimension $M \times L$, where each row contains attention weights of all M trips with respect to one trip in the state matrix. Therefore, the left part of the self-attention network in Fig. 4 focuses on processing spatial information in the state matrix. The right part of this network focuses on processing temporal information.

Positional Embedding. For positional embedding, we adopt one of the widely used "learnable pattern" [16] shown in Eq. (5) that adds an L-dimensional vector to each row of the state matrix.

$$embedded(Obs(t)) = Obs(t) + (\mathcal{E}(1), \ldots, \mathcal{E}(M))^T \tag{5}$$

$\mathcal{E}(m)$ is a trainable embedding function that maps every integer $m \in \{1, \ldots, M\}$ to an L-dimensional vector.

Action Selection. Building on the high-level features extracted by the self-attention network from the input state matrices, Q-values are further estimated using a feed-forward neural network trained by the double DQN algorithm [8]. Using the estimated Q-values, the behavioral policy used during Q-learning is determined according to the ϵ-greedy strategy shown below.

$$\pi(s) \leftarrow \begin{cases} \text{argmax}_{a \in A} Q(s, a) \text{ with probability of } 1 - \epsilon \\ random(A) \text{ with probability of } \epsilon \end{cases} \tag{6}$$

Additional constraints are also considered by the holding control agent during action selection. Specifically, the agent must select its action to meet the constraints on the minimum and maximum headway. Meanwhile, we prevent the holding control agent from selecting the "dispatch" action whenever no bus is available to dispatch at any time t.

4   Experiment

We use data collected from a specific bus line (BRT Line 2 in Xiamen city, China) as experimental data. Xiamen's Bus Rapid Transit (BRT) is a bus rapid transit system in Xiamen, China. It was established in 2008 and is one of the earliest elevated BRT networks in China that allow buses to operate at high speed with small headways [12]. The implementation code and experiment data are available here.

Table 2. Learning performance of three DQN models.

DQN architecture	Average reward	Max reward	Training time
Self-attention DQN	**1158**	**1891.6**	4 h 25 min
Fully connected DQN [1]	584.4	1690.8	**1 h 5 min**
CNN-DQN [8]	−1629	1201.2	2 h 29 min

4.1   Experiment Data

There are two cruising directions for BRT Line 2 (upward and downward). We take the direction upward as the experimented bus line. The environment is simulated based on real passenger swiping records and bus travel time information collected on a random day in June 2018. Some pre-processing steps are performed to simulate passenger arrival times based on corresponding boarding times since it is required for passenger waiting time calculation and is not directly recorded in AFC. A uniform distribution of arrivals during two consecutive buses at each stop is assumed as a valid estimation because passengers are more likely to arrive randomly for high-frequency bus lines without planning ahead [6]. An overview of the daily operational data is summarized as following:

1. There are 36 stops in total for the BRT Line 2 upward direction.
2. The first bus was dispatched at 6:15 and the last at 22:30. A bus trip takes between 51 and 64 min, with an average of 61 min.
3. The total number of passengers is 4346 for the day. The average number of hourly passengers is 256, ranging from 59 to 634. Peak hours are between 7:00–10:00 and 16:00–19:00, with more than 300 passengers per hour.

In the simulated environment for the holding control agent, we set the maximum number M of running buses at any time to 13 and the capacity C of buses to 30 according to the real world operational data of BRT Line 2. Following [1], the minimum headway $H_{min} = 3$ min and the maximum headway $H_{max} = 15$ min. We also set the maximum waiting time for all passengers $W = 15$ min.

The various coefficients of the reward function in Eq. (3) are configured as follows: boarding coefficient $w_b = 1$, alighting coefficient $w_a = 1$, dispatching coefficient $w_d = -30$, waiting time coefficient $w_w = -0.2$, stranding coefficient $w_s = -2$, and leaving passenger coefficient $w_l = -5$. In our experiments, the coefficient of waiting time is set to −0.2, and the boarding coefficient is set to

1.0, such that the cost of waiting time is on par with the boarding reward when a passenger waited for 5 min. Any longer waiting time is considered to hurt passengers' satisfaction [6, 10]). When one passenger is finally alighted at their destination, a point of 1.0 is rewarded to the agent for the transportation of this passenger. Since the operational cost of one bus trip is set to −30, a trip serving more than 30 passengers (the capacity limit of any bus) is considered beneficial with overall positive rewards.

Table 3. Comparison of four competing approaches on the original passenger demand.

Policy	Reward	Dispatches	Boarded	Avg waiting	Stranded	Left
STDH	1891.6	124	4345	3.52	1	1
SH	1703.8	141	4343	3.11	17	3
NH	−23594.4	124	4083	6.07	150	263
EH	510.6	105	4287	4.99	170	59

Table 4. Comparison of four competing approaches on the passenger demand with shifted peak hours.

Policy	Reward	Dispatches	Boarded	Avg waiting	Stranded	Left
STDH	1567.6	118	4341	4.01	33	5
SH	1193.2	141	4329	3.49	62	17
NH	−23869	124	4068	6.31	146	278
EH	1046.6	105	4322	4.82	79	24

4.2 Experiment Results

In this subsection we report the hyper-parameter settings of the holding control agent and then present and discuss the performance results.

Hyper-Parameter Settings. In our experiment, every episode of training data for the holding control agent contains a fixed number of records (1050 in our experiments based on recorded bus services between 6:15 AM and 23:45 PM during an operational day). The capacity limit of the replay buffer is set to 10,000 records and the batch size is set to 64. Thus at each training iteration, 64 records are selected randomly from the replay buffer to train the holding control agent. Each experiment involves 300k training iterations. ϵ in Eq. (6) is initialized as 1.0 and decreased linearly at each iteration until it reaches its minimum value of 0.01. Adam optimizer with a learning rate of 0.001 is configured as the gradient descent optimizer.

Main Results. We evaluated three DQN network architectures in our experiments. Besides the self-attention DQN model proposed in Subsect. 3.2, we used convolutional DQN (CNN-DQN) [13] and fully connected DQN [8] as competing approaches.

Table 2 summarizes the respective learning performance. The total rewards reported in the table are obtained by testing the final trained DQN models on the original passenger data. The self-attention DQN model clearly outperformed the other two DQN models in terms of both the average and max total rewards. Any statistically significant results verified by a pairwise T-test with a significance level of 0.02 are bolded in the table.

The total training time on a Linux desktop computer with Intel i7 processor and 16 GB of memory is about one hour for the fully connected DQN and two hours for CNN-DQN. Our self-attention DQN takes longer time to train (approx. 4 h). Despite of the difference in training time, the training time required by all models are considered acceptable since the trained models can be used for dynamic bus holding control on many operational days.

Further Analysis. To demonstrate how the STDH approach can dispatch buses responding to real-time passenger demand fluctuations, We compare our STDH approach against three commonly practiced bus holding approaches for comparison:

– The proposed STDH control approach (*STDH*): A policy trained using the proposed STDH approach.
– Naive equal headway holding control (*EH*): A "hard" holding control approach determined by a pre-fixed headway [5], such that dispatching time is evenly distributed for a day's service. We set this to 10 min in our experiments.
– Naive schedule-based holding control (*SH*): Buses departure intervals are set differently for peak hours and non-peak hours respectively based on expectations [18]. We use 8 min headway during peak hours and 15 min headway during non-peak hours in our experiments.
– No holding control (*NH*): The original dispatching times at the first station are used directly to mimic the original timetable generated by human operators.

Tables 3, 4, and 5 compare these approaches in terms of the total number of dispatched bus trips, the total number of boarded passengers, the total amount of passenger waiting time, the total number of stranded passengers, and the total number of passengers who gave up with respect to the three passenger demand scenarios.

Table 5. Comparison of four competing approaches on the passenger demand with temporal passenger surge.

Policy	Reward	Dispatches	Boarded	Avg waiting	Stranded	Left
STDH	1383.4	125	4799	3.94	221	38
SH	1157	141	4754	3.55	163	83
NH	−24183.4	124	4518	6.35	351	319
EH	−39.4	105	4698	5.27	316	139

To further understand the effectiveness of our STDH approach, we have examined its performance with respect to three passenger demand scenarios. These scenarios are depicted in Fig. 5 for a single operational day where the number of passenger arrivals during every time interval of 15 mins is presented together as a histogram. Meanwhile, the solid lines in the figure represent the total capacity of bus services dispatched by STDH and manually designed timetables (i.e., SH approach) respectively for the same time period. Clearly, STDH can generate demand-responsive bus flows compared with SH. Whenever the passenger demand changes, the bus flows can change quickly to match the demand, as evidenced in these figures.

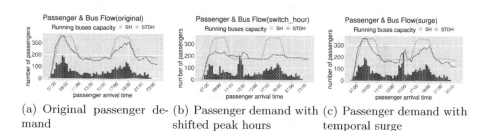

(a) Original passenger demand

(b) Passenger demand with shifted peak hours

(c) Passenger demand with temporal surge

Fig. 5. The passenger demand and the corresponding running bus capacity achieved by STDH and SH.

In comparison to other approaches, STDH achieved the best overall balance between passenger satisfaction and bus operational cost. Especially, with unexpected changes to the passenger demand (i.e., shift peak hours and temporal passenger surge), STDH shows much better performance in total rewards. For example, in Table 4 where peak hour passenger demand is switched with non-peak hour, on average 17 passengers gave up due to prolonged waiting time caused by using SH, compared with only 5 upon using STDH. Furthermore, SH is less efficient with an average of 141 dispatched bus trips, compared with only 118 upon using STDH.

5   Conclusions

This paper developed a DRL-based STDH approach for demand-responsive bus holding control. We proposed a new state matrix and a self-attention neural network to effectively process dynamic information regarding passenger riding trend and bus service status.

STDH also adopts a newly designed reward function in order to achieve a desirable balance between passenger satisfaction and bus operational costs. Extensive experiments using real-world data showed that STDH can effectively solve the dynamic bus holding control problem, in comparison to existing DRL-based approaches and manually designed bus holding control techniques.

References

1. Ai, G., Zuo, X., Wu, B., et al.: Deep reinforcement learning based dynamic optimization of bus timetable. arXiv preprint arXiv:2107.07066 (2021)
2. Alesiani, F., Gkiotsalitis, K.: Reinforcement learning-based bus holding for high-frequency services. In: 21st International Conference on Intelligent Transportation Systems. ITSC, pp. 3162–3168 (2018)
3. Asgharzadeh, M.A., Shafahi, Y.: Real-time bus-holding control strategy to reduce passenger waiting time. J. Transp. Res. Board **2647**(1), 9–16 (2017)
4. Bartholdi, J.J., Eisenstein, D.D.: A self-coördinating bus route to resist bus bunching. Transp. Res. Part B Methodol. **46**(4), 481–491 (2012)
5. Chen, W., Zhou, K., Chen, C.: Real-time bus holding control on a transit corridor based on multi-agent reinforcement learning. In: 2016 IEEE 19th International Conference on Intelligent Transportation Systems (ITSC), pp. 100–106 (2016)
6. Esfeh, M.A., Wirasinghe, S.C., Saidi, S., Kattan, L.: Waiting time and headway modelling for urban transit systems - a critical review and proposed approach. Transp. Rev. **41**(2), 141–163 (2021)
7. Fu, L., Yang, X.: Design and implementation of bus-holding control strategies with real-time information. Transp. Res. Rec. **1791**(1), 6–12 (2002)
8. Hasselt, H.V., Guez, A., Silver, D.: Deep reinforcement learning with double q-learning. In: Proceedings of the Thirtieth AAAI Conference on Artificial Intelligence, AAAI 2016, pp. 2094–2100. AAAI Press (2016)
9. Haydari, A., Yilmaz, Y.: Deep reinforcement learning for intelligent transportation systems: a survey. IEEE Trans. Intell. Transp. Syst. **PP**, 1–22 (2020)
10. Hickman, M.D.: An analytic stochastic model for the transit vehicle holding problem. Transp. Sci. **35**(3), 215–237 (2001)
11. Ibarra-Rojas, O., Delgado, F., Giesen, R., Muñoz, J.: Planning, operation, and control of bus transport systems: a literature review. Transp. Res. Part B Methodol. **77**, 38–75 (2015)
12. Li, Z., Hensher, D.A.: Performance contributors of bus rapid transit systems: an ordered choice approach. Econ. Anal. Policy **67**, 154–161 (2020)
13. Mnih, V., et al.: Playing atari with deep reinforcement learning (2013)
14. Ning, L., Li, Y., Zhou, M., Song, H., Dong, H.: A deep reinforcement learning approach to high-speed train timetable rescheduling under disturbances. In: 2019 IEEE Intelligent Transportation Systems Conference (ITSC), pp. 3469–3474 (2019)
15. Silver, D., et al.: Mastering the game of go without human knowledge. Nature **550**(7676), 354–359 (2017)
16. Tunstall, L., von Werra, L., Wolf, T.: Natural Language Processing with Transformers, Revised Edition. O'Reilly Media. California (2022)
17. Vaswani, A., et al.: Attention is all you need. In: Advances in Neural Information Processing Systems (2017)
18. Wang, J., Sun, L.: Dynamic holding control to avoid bus bunching: a multi-agent deep reinforcement learning framework. Transp. Res. Part C Emerg. Technol. **116**, 102661 (2020)
19. Xuan, Y., Argote, J., Daganzo, C.F.: Dynamic bus holding strategies for schedule reliability: optimal linear control and performance analysis. Transp. Res. Part B Methodol. **45**(10), 1831–1845 (2011)
20. Yau, K.L.A., Qadir, J., Khoo, H.L., Ling, M.H., Komisarczuk, P.: A survey on reinforcement learning models and algorithms for traffic signal control. ACM Comput. Surv. **50**(3), 1–38 (2017)
21. Zhao, J., Dessouky, M.M., Bukkapatnam, S.T.S.: Optimal slack time for schedule-based transit operations. Transp. Sci. **40**, 529–539 (2006)

Latent Pattern Identification Using Orthogonal-Constraint Coupled Nonnegative Matrix Factorization

Anandkumar Balasubramaniam[1] (ID), Thirunavukarasu Balasubramaniam[2] (ID), Anand Paul[1(✉)] (ID), and Richi Nayak[2] (ID)

[1] School of Computer Science and Engineering, Kyungpook National University, Daegu, South Korea
paul.editor@gmail.com
[2] School of Computer Science and Centre for Data Science, Queensland University of Technology, Brisbane, QLD, Australia
{thirunavukarasu.balas,r.nayak}@qut.edu.au

Abstract. The recent advancements and developments in Intelligent Transportation Systems (ITS) lead to the generation of abundant spatio-temporal traffic data. Identifying or understanding the latent patterns present in these spatio-temporal traffic data is very much essential and also challenging due to the fact that there is a chance of obtaining duplicate or similar patterns during the process of common pattern identification. This paper proposes an Orthogonal-Constraint Coupled Nonnegative Matrix Factorization (OC-CNMF) method and studies how to effectively identify the common as well as distinctive patterns that are hidden in the spatio-temporal traffic-related datasets. The distinctiveness of the patterns is achieved by the imposition of the orthogonality constraint in CNMF during the process of factorization. The imposition of the orthogonal constraint helps to ignore similar/duplicate patterns among the identification of the common patterns. We have shown that imposing orthogonality constraint in CNMF improves the convergence performance of the model and is able to identify common as well as distinctive patterns. Also, the performance of the OC-CNMF model is evaluated by comparing it with various performance evaluation measures.

Keywords: Coupled nonnegative matrix factorization · Orthogonal constraint · Spatiotemporal · Vehicular traffic pattern mining

1 Introduction

With the recent advancements in ITS [1–3], the general road traffic data is not just restricted to the count of vehicles on road but other contextual information such as location and time [4, 5] becomes available. Incorporating such spatio-temporal context is challenging and learning traffic behavior patterns becomes a complex and challenging task. Therefore, in this research, the main focus is on learning peoples' traffic-related mobility patterns that will help government and related organizations in traffic planning, road construction, etc. During the COVID-19 pandemic, the traffic patterns have

© The Author(s), under exclusive license to Springer Nature Switzerland AG 2022
H. Aziz et al. (Eds.): AI 2022, LNAI 13728, pp. 675–689, 2022.
https://doi.org/10.1007/978-3-031-22695-3_47

drastically changed, and understanding the behavior changes becomes challenging [6, 7]. Previously, by using NMF, Balasubramaniam [8] proposes an NMF-based architecture to understand the variations in spatio-temporal traffic patterns during COVID-19, whereas NMF [9–11] is one of the matrix factorization-based dimensionality reduction techniques. A case study has been conducted observing the variations in peoples' spatio-temporal mobility patterns in Great Britain. In addition to this, the author has shown that the independent patterns are present in 2019 and 2020 datasets i.e. before pandemic and during pandemic traffic datasets, respectively.

To achieve the independent patterns, NMF must run individually for the two datasets as proposed in [8]. The interesting question here is: How can we find common as well as distinctive patterns? One way is: to compare the similarities between the two patterns achieved by running two NMF models. This may sound good, but this leads to time complexity. Another way is: finding the solution to identify the common patterns without running two NMF models. The latter way leads to the motivation for proposing OC-CNMF in which CNMF identifies only the common patterns that are present in the provided datasets, whereas CNMF [12–14] is the modified version of NMF. In addition to this, most of the papers dealing with COVID-19 scenarios are focusing mainly on what has changed due to COVID-19 but less focusing on identifying the common patterns or behaviors before and during COVID-19. However, the works focused on comparing the common behaviors are computationally expensive. This problem has also been addressed in this work by the implementation of proposing the novel approach of the OC-CNMF model on vehicular traffic data. The main contributions of this research are (i) Designing a Coupled Nonnegative Matrix Factorization architecture to learn common patterns from two different, but related datasets. (ii) Formulate a CNMF with orthogonal constraint to learn the common, but distinct patterns. (iii) Avoid multiple NMF models through the proposed OC-CNMF and improve the convergence performance. (iv) Evaluate the convergence performance of the proposed OC-CNMF by comparing it against NMF and CNMF models. (v) Identify the common but distinctive patterns on different datasets. (vi) Evaluating the proposed OC-CNMF model performance by conducting various performance measures. The outputs of the analyzed common, as well as distinctive traffic pattern behaviors, will be useful in the fields of traffic management and in various stages of unprecedented events concerning road traffic.

2   Related Works

Coupled matrix representation is a way to fuse and represent two data together along one mode. To learn the common features together from the coupled matrix representation, coupled NMF can be applied. There are several research works performed by utilizing CNMF technique. Some of those related works are as follows. The authors in [16] dealt with the problem of acquiring spatio-temporal resolution images, especially spatio-temporal fusion problems. They proposed the concept of fusing the hyperspectral and multispectral images in every time series by using CNMF. However, in their work, CNMF is used as only the combining tool to get hyperspectral images. In another work [17], the authors proposed spatiotemporal constraint nonnegative matrix factorization model to deal with the problem to identify intra-urban mobility patterns from taxi trips and learn inherent spatiotemporal characteristics of human intra-urban movements. The authors in [18] presented the nonlinear loose coupled nonnegative matrix factorization approach mainly to describe the features that are common in the images with different resolutions. They addressed the problem of the potential relationship and mapping process between the high-resolution images and low-resolution images. Another research article [19] is focused on utilizing nonnegative coupled matrix tensor factorization, which is implemented on smart device-generated spatio-temporal data. Their work mainly focuses on analyzing additional contexts with the data to generate useful patterns. However, this concept of utilizing coupled nonnegative matrix factorization is used only on the tensors. In the article [20], the authors proposed the idea of solving the multi-view representation of the data by using coupled matrix factorization. Alongside this, the authors come up with the approach of variable selection-based greedy coordinate descent algorithm to improvise the computational efficiency. However, there is no orthogonality constraint involved in the quoted articles. To the best of knowledge, CNMF models with the implication of orthogonality constraint do not exist that can deal with spatio-temporal traffic-related datasets.

3 Proposed Orthogonal Constraint Coupled NMF (OC-CNMF)

3.1 OC-CNMF & Its Architecture

Inspired by the capability of Coupled NMF in learning shared features, in this research we model a coupled NMF architecture to learn common features across two different matrices. However, one drawback in off-the-shelf usage of Coupled NMF is that the features learned are not necessarily distinct. This shows that there is a high chance that similar patterns are identified. To overcome this, we innovatively introduce Orthogonality to the features learned as the features are distinct from each other. The concept of orthogonality is inspired by one of the existing works on NMF [21], which clearly shows the differentiation between NMF and orthogonal NMF. This is not for Coupled NMF, however, the concept is inspired and implemented for Coupled NMF which leads to the proposal of OC-CNMF for identifying latent, common as well as distinct patterns. Figure 1 shows the proposed OC-CNMF architecture in which the overall OC-CNMF process is performed on the datasets to identify common, as well as distinct patterns. As depicted in Fig. 1, the datasets are imported to the experimental setup and initial data pre-processing is carried out. In a follow-up to the data pre-processing, matrix files with respect to the datasets are generated based on the non-shared modes and shared modes among the datasets. The coupling process of imposing orthogonality constraint is performed on the OC-CNMF block. After performing the coupling process by OC-CNMF, three lower-dimensional factor matrices are generated among which two factor matrices (H_1 & H_2) are generated for the two non-shared modes and one common factor matrix (W) is generated for the shared mode. From these generated factor matrices, common and distinct temporal patterns are identified using the shared mode factor matrices which are common to the spatial behavior of the respective datasets.

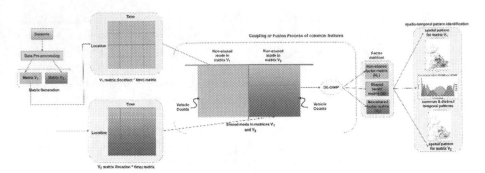

Fig. 1. The proposed OC-CNMF architecture

3.2 OC-CNMF Mathematical Formulation

The objective function of the Coupled NMF is given based on the NMF objective functions. Assume, matrix V_1 is factorized into two factor matrices, namely, W_1 and H_1; and matrix V_2 is factorized into two factor matrices, namely, W_2 and H_2.

$$\min f_1(W_1, H_1) = \left\| V_1 - W_1 H_1^T \right\|, \text{ s.t. } W_1 \geq 0 \text{ and } H_1 \geq 0 \tag{1}$$

$$\min f_2(W_2, H_2) = \left\| V_2 - W_2 H_2^T \right\|, \text{ s.t. } W_2 \geq 0 \text{ and } H_2 \geq 0 \tag{2}$$

In coupled NMF, instead of learning W_1 and W_2 independently, it is learned commonly. Therefore, only one factor matrix (W) will be learned for the shared mode. The objective function of Coupled NMF is formulated as,

$$\min f(W, H_1, H_2) = \min f_1(W_1, H_1) + \min f_2(W_2, H_2) \tag{3}$$

Substituting Eq. 1 and 2 in Eq. 3, we get

$$\min f(W, H_1, H_2) = \left\| V_1 - W_1 H_1^T \right\| + \left\| V_2 - W_2 H_2^T \right\|,$$
$$\text{s.t. } W \geq 0, \ H_1 \geq 0, \text{ and } H_2 \geq 0. \tag{4}$$

The coupled NMF objective function as given in Eq. 4 can be viewed as a concatenation of two NMFs with only exception of learning one common factor matrix. Therefore, the update rule for W can be derived from the standard multiplicative update rule that is defined as,

$$W \leftarrow W - \left(\delta_{f_1} \left(\frac{\partial f_1}{\partial W} \right) + \delta_{f_2} \left(\frac{\partial f_2}{\partial W} \right) \right) \tag{5}$$

where δ_{f_1} and δ_{f_2} are learning rates, $\frac{\partial f_1}{\partial W}$ is the gradient of function f_1 (Eq. 1) with respect to W, and $\frac{\partial f_2}{\partial W}$ is the gradient of function f_2 (Eq. 2) with respect to W. The gradients $\frac{\partial f_1}{\partial W}$ and $\frac{\partial f_2}{\partial W}$ can be calculated by first-order derivatives solved as follows,

$$\frac{\partial f_1}{\partial W} = WH_1 H_1^T - V_1 H_1^T \tag{6}$$

$$\frac{\partial f_2}{\partial W} = WH_2H_2{}^T - V_2H_2{}^T \tag{7}$$

While the learning rate δ can be defined with any scalar value, the multiplicative update rule defines a dynamic learning rate that changes in every iteration. The learning rate is defined as,

$$\delta_{f_1} = \frac{W}{WH_1H_1{}^T} \tag{8}$$

$$\delta_{f_2} = \frac{W}{WH_2H_2{}^T} \tag{9}$$

By substituting Eq. 6 and 8 in $\delta_{f_1}\left(\frac{\partial f_1}{\partial W}\right)$ of Eq. 5, we can measure the change in W with respect to matrix V_1. This is given as

$$\delta_{f_1}\left(\frac{\partial f_1}{\partial W}\right) = W - \frac{W}{WH_1H_1{}^T}\left(WH_1H_1{}^T - V_1H_1{}^T\right) \tag{10}$$

By simplifying Eq. 10, the update rule for updating the factor matrix W with respect to V_1 is given as,

$$W = \left(\frac{W \times (V_1 \times H_1{}^T)}{W \times \left(H_1 \times H_1{}^T\right)}\right) \tag{11}$$

where W and H_1 represent the factor matrices with respect to V_1. To avoid, divide by zero error, the denominator in the above Eq. 11 is set to a small scalar value, ϵ, if the denominator is zero. Elementwise, it is given by,

$$W_{ij} \times \left(H_{1j} \times H_{1j}^T\right) = W_{ij} \times \left(H_{1j} \times H_{1j}^T\right)\forall i,j \text{ if } W_{ij} \times \left(H_{1j} \times H_{1j}^T\right) \neq 0 \tag{12a}$$

$$W_{ij} \times \left(H_{1j} \times H_{1j}^T\right) = \epsilon\forall i,j \text{ if } W_{ij} \times \left(H_{1j} \times H_{1j}^T\right) = 0 \tag{12b}$$

Since the above Eq. Updates W with respect to matrix V_1 only, a further update must be made to account for changes in W with respect to matrix V_2. By substituting Eq. 7 and 9 in $\delta_{f_2}\left(\frac{\partial f_2}{\partial W}\right)$ of Eq. 5, we can measure the change in W with respect to matrix V_2. This is given as

$$\delta_{f_2}\left(\frac{\partial f_2}{\partial W}\right) = W - \frac{W}{WH_2H_2{}^T}\left(WH_2H_2{}^T - V_2H_2{}^T\right) \tag{13}$$

This measure of change in the factor matrix W with respect to V_2 is added to Eq. 11 to make the final update rule for W. By simplifying Eq. 13, the update rule for updating the factor matrix W with respect to V_2 becomes,

$$W = W + \left(\frac{W \times (V_2 \times H_2{}^T)}{W \times \left(H_2 \times H_2{}^T\right)}\right) \tag{14}$$

where W and H_2 represent the factor matrices with respect to V_2. Similar to Eq. 12, to avoid, divide by zero error, the denominator in the above Eq. 14 is set to a small scalar value, ϵ, if the denominator is zero. Elementwise, it is given by,

$$W_{ij} \times \left(H_{2j} \times H_{2j}^T \right) = W_{ij} \times \left(H_{2j} \times H_{2j}^T \right) \forall i, j \text{ if } W_{ij} \times \left(H_{2j} \times H_{2j}^T \right) \neq 0 \quad (15a)$$

$$W_{ij} \times \left(H_{2j} \times H_{2j}^T \right) = \epsilon \forall i, j \text{ if } W_{ij} \times \left(H_{2j} \times H_{2j}^T \right) = 0 \quad (15b)$$

Equation 14 can also be given as,

$$W = \left(\frac{W_1 \times (V_1 \times H_1{}^T)}{W_1 \times (H_1 \times H_1{}^T)} \right) + \left(\frac{W_2 \times (V_2 \times H_2{}^T)}{W_2 \times (H_2 \times H_2{}^T)} \right) \quad (16)$$

3.3 Orthogonal Constraint Objective Function

Different from the update rule of Coupled NMF, in this paper, the update rule is modified to reflect the orthogonality. We will now see the mathematical formulation of the updating W for the proposed OC-CNMF. Firstly, the objective function of OC-CNMF is formulated as,

$$\min f(W, H_1, H_2) = \left\| V_1 - WH_1^T \right\| + \left\| V_2 - WH_2^T \right\|,$$
$$\text{s.t. } W \geq 0, \ H_1 \geq 0, \ \text{and } H_2 \geq 0; \ WW^T = I, \quad (17)$$

where $WW^T = I$ is the imposed orthogonal constraint, and I stands for the identity matrix. The orthogonality constraint imposed on the factor matrix W will ensure that the rows and columns of the matrix are distinct (orthonormal vector). Since the orthogonal constraint is imposed only on one factor matrix (W) in Eq. 17, the optimization problem of W must be specially treated and it is given as:

$$\max tr(W^T (V_1 H_1)(V_1 H_1)^T W) \quad (18)$$

For the constrained optimization, the objective function of OC-CNMF can be reformulated using the Lagrangian function as,

$$L(W) = \left\| V_1 - WH_1^T \right\| + \left\| V_2 - WH_2^T \right\| + Tr\left[\lambda \left(W^T W - I \right) \right],$$
$$\text{s.t. } W \geq 0, \ H_1 \geq 0, \ \text{and } H_2 \geq 0; \ WW^T = I, \quad (19)$$

Equation 19 can be rewritten as,

$$L(W) = Tr\left[V_1^T V_1 - 2WH_1^T V_1 H_1 + H_1^T H_1 W^T W \right]$$
$$+ Tr\left[V_2^T V_2 - 2WH_2^T V_2 H_2 + H_2^T H_2 W^T W \right] + Tr[\lambda \left(W^T W - I \right)] \quad (20)$$

Lagrangian value is represented as λ and the gradient of the Lagrangian function with respect to the factor matrix W can be calculated using first-order partial derivative as,

$$\frac{\partial L}{\partial W} = -V_1 H_1 + W {H_1}^T H_1 + W\lambda + (-V_2 H_2 + W {H_2}^T H_2 + W\lambda) \qquad (21)$$

The Lagrangian function as given in Eq. 20 can be solved using the auxiliary function proposed in [22] to derive the update rules as follows:

Update Rule for W with Orthogonal Constraint

The update rule for W with respect to V_1 alone is given as (W_1),

$$W \leftarrow \left(\frac{W \times \sqrt{(V_1 \times {H_1}^T)}}{\sqrt{((W \times {W_1}^T) \times (V_1 \times {H_1}^T))}} \right) \qquad (22)$$

Since in Coupled NMF framework the update rule must also reflect the changes W with respect to V_2 also, the update becomes,

$$W \leftarrow W + \left(\frac{W \times \sqrt{(V_2 \times {H_2}^T)}}{\sqrt{((W \times W^T) \times (V_2 \times {H_2}^T))}} \right) \qquad (23)$$

The complete update rule for W in OC-CNMF using the gradient calculated in Eq. 21 can be rewritten as,

$$W = \left(\frac{W \times \sqrt{(V_1 \times {H_1}^T)}}{\sqrt{((W \times W^T) \times (V_1 \times {H_1}^T))}} \right) + \left(\frac{W \times \sqrt{(V_2 \times {H_2}^T)}}{\sqrt{((W \times W^T) \times (V_2 \times {H_2}^T))}} \right) \qquad (24)$$

The importance of OC-CNMF can be seen from the Eq. 24. Instead of learning two factor matrices individually, a single factor matrix is learned.

Update Rules for $H1$ and $H2$:

Since the optimisation process of any factorization algorithm is a non-convex optimisation, alternatively updating the least squares is the key to learning factor matrices. In other words, once the factor matrix W is updated as per Eq. 24, the other factor matrices can learn in a traditional update rule framework similar to Eq. 12 and 14 but with the respective gradients calculated. The update rule for H_1 and H_2 are the same as the traditional NMF. However, the interesting fact is that the factor matrices H_1 and H_2 are learned from the same W factor matrix. The gradients of H_1 and H_2 can be calculated as,

$$\frac{\partial f}{\partial H_1} = {H_1}^T W^T W - {V_1}^T H_1 \qquad (25)$$

$$\frac{\partial f}{\partial H_2} = {H_2}^T W^T W - {V_2}^T H_2 \qquad (26)$$

With the calculated gradients, the update rules for H_1 and H_2 can be derived using the multiplicative update rule with a dynamic learning rate. The update rule for H_1 is given as,

$$H_1 = \frac{H_1 \times (W^T \times V_1)}{(H_1 \times V_1{}^T) \times (W \times H_1)} \tag{27}$$

The update rule for H_2 is given as,

$$H_2 = \frac{H_2 \times (W^T \times V_2)}{(H_2 \times V_2{}^T) \times (W \times H_2)} \tag{28}$$

Once the factor matrices are updated using Eq. 24, 27 and 28, it is repeated until convergence or stopped. This will ensure the learning is converging to the best minimum solution.

4   Experiment Section

4.1   Dataset

To evaluate the performance of the proposed OC-CNMF model, various other datasets have been analyzed. As a total, OC-CNMF performance on the five datasets is compared with NMF and CNMF models using various evaluation measures such as Root Means Square Error, Silhouette Score, Pattern Distinctiveness, Calinski-Harabasz, Davies-Bouldin, and Karzanowski and Lai. The five datasets used in the experiments consist of vehicular traffic volume count records based on location and time in different regions and in different time periods. Dataset (D1) [15] consists of the vehicle count records in Great Britain based on 12656 locations, 12 h of time period and 6537 locations, 12 h of time period during 2019 and 2020. Dataset (D2) [24] is the vehicle count records in New York City based on 203 locations, 24 h of time and 156 locations, 24 h of time during 2012 and 2013. Dataset (D3) [24] records the vehicle counts based on 273 locations, 24 h of time and 83 locations, 24 h of time during the January and February months of 2012. Dataset (D4) [15] is the vehicle count records based on 12178 locations, 12 h of time and 12656 locations, 12 h of time during 2018 and 2019. Similarly, dataset (D5) [15] is the record of the vehicle counts based on 12178 locations, 12 h of time and 6537 locations, 12 h of time during the years 2018 and 2020 in Great Britain. Each of these location and time features from datasets D1 to D5 are grouped as two matrices considering the time feature as the common shared mode among the two matrices to perform the coupling process of these matrices from datasets D1 to D5.

4.2   OC-CNMF Theoretical Convergence Analysis

As per the convergence analysis made in an orthogonal constraint NMF research [23], the objective function will monotonically decrease, and it will attain the best minimal solution when a constraint $WW^T = I \geq 0$. This theory holds for OC-CNMF as well, as the orthogonal constraint imposed in (Eq. 17) will provide $WW^T = I \geq 0$ and satisfies the condition for monotonically decreasing the cost function or the objective function. However, the optimization will not guarantee to reach the best solution if no orthogonality constraint is imposed. Therefore, the traditional CNMF will not reach to better solution as fast as the OC-CNMF can.

4.3 OC-CNMF Computational Complexity Analysis

Suppose there are two matrices $V_1 \in \mathbb{R}^{N \times M}$ and $V_2 \in \mathbb{R}^{N \times P}$. Suppose V_1 is factorized into two factor matrices $W_1 \in \mathbb{R}^{N \times K}$ and $H_1 \subset \mathbb{R}^{K \times M}$; V_2 is factorized into two factor matrices $W_2 \in \mathbb{R}^{N \times K}$ and $H_2 \in \mathbb{R}^{K \times P}$; The time complexity of updating the factor matrices using NMF on V_1 is $O(NK + MK)$, which can further be simplified as, $O(K(N + M))$. Similarly, the time complexity of updating the factor matrices using NMF on V_2 is $O(NK + PK)$, which can further be simplified as, $O(K(N + P))$. Therefore, the time complexity of updating the factor matrices when running two NMF becomes $O(K(N + M)) + O(K(N + P)) = O(K(2N + M + P))$. Now, to prove the efficiency achieved using Coupled NMF, we will prove that the time complexity of updating factor matrices is less than that of NMF. Unlike NMF, in coupled NMF and OC-CNMF there are three factor matrices to update. Suppose $W_1 \in \mathbb{R}^{N \times K}, H_1 \in \mathbb{R}^{K \times M}$, and $H_2 \in \mathbb{R}^{K \times M}$ are the factor matrices. The time complexity of updating the factor matrix W is $O(NK)$, H_1 is $O(MK)$, and H_2 is $O(PK)$. The total time complexity becomes, $O(NK) + O(MK) + O(PK)$. This will be simplified as, $O(K(N + M + P))$, which is less than that of $O(K(2N + M + P))$. Thus, proving the computational efficiency of Coupled NMF and OC-CNMF in comparison to NMF, which are tabulated in Table 1.

Table 1. Computational complexity analysis.

Method	Computational complexity
NMF	$O(K(2N + M + P))$
CNMF & OC-CNMF	$O(K(N + M + P))$

4.4 Baseline Method and Evaluation Criteria

The standard CNMF is used as the baseline to compare the performance of the proposed OC-CNMF. Since the main contribution of this paper is to show how OC-CNMF is converging fast to a better solution, and how it is learning common features, it is only fair if it is compared against NMF and CNMF. To evaluate the convergence performance of the proposed OC-CNMF, the root means square error (RMSE), Silhouette Score, Pattern Distinctiveness, Calinski-Harabasz, Davies-Bouldin, and Karzanowski and Lai are used. As the goal of both CNMF and OC-CNMF are set to minimize the objective functions defined using Euclidean distance, the usage of RMSE is more appropriate. In RMSE, the original matrices are compared against the approximated matrix. For matrix V_1, the approximated matrix based on the factor matrices W and H_1 is given by,

$$\widehat{V_1} = W H_1{}^T \tag{29}$$

Similarly, for matrix V_2, the approximated matrix based on the factor matrices W and H_2 is given by,

$$\widehat{V_2} = WH_2^T \tag{30}$$

Now, the original matrix and the approximated matrix can be compared to calculate the RMSE. Let us take V_1 as an example. The RMSE equation is defined as:

$$RMSE = \sqrt{\frac{\sum_{i=1}^N V_{1i} - \widehat{V_{1i}}}{N}} \tag{31}$$

Silhouette score is utilized to evaluate the quality of clusters in terms of how well similar samples cluster with each other. Cluster-specific Silhouette scores are determined for each sample. Pattern distinctiveness score is used to measure the similarity of each pattern with other patterns such that the quality of the learned patterns is evaluated. In pattern distinctiveness, the lower the score, the better the performance. Calinski-Harabasz score is also termed to be as Variance Ratio Criterion. Calinski-Harabasz evaluation is the ratio of the sum of between-cluster dispersion and within-cluster dispersion. In Calinski-Harabasz evaluation, the higher the score, the better the performance. In Davies-Bouldin score, the score is the average measure of how similar each cluster is to its most similar cluster. Similarity is the ratio of the distances within a cluster to the distances between clusters. So, a better score will come from clusters that are farther apart and less spread out. In Davies-Bouldin evaluation, the lower the score, the better the performance. These evaluation measures are used to evaluate the proposed model of OC-CNMF in comparison with NMF and CNMF.

5   Results

5.1   Identification of Rank

Before analyzing the convergence performance, sensitivity analysis is performed to determine the rank. Rank is determined by conducting evaluation measures (within-cluster dispersion and between-cluster dispersion) on the matrices. Based on the respective sensitivity analysis, the rank value is determined for running NMF, CNMF and OC-CNMF models on the respective matrices of the datasets. According to the sensitivity analysis performed on the matrices of the datasets, the rank values are determined to be 4, 7, 5, 7, 5 for the datasets D1, D2, D3, D4, D5, respectively.

5.2 Convergence Performance

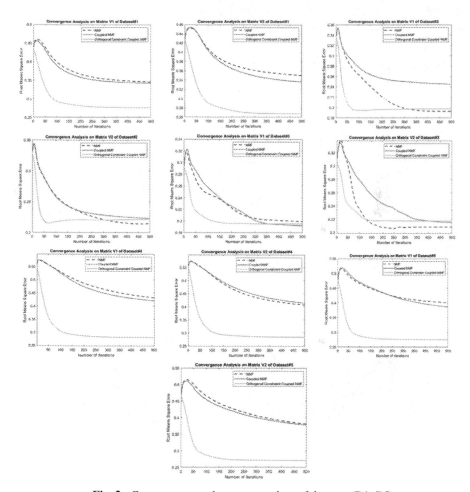

Fig. 2. Convergence analyses on matrices of datasets (D1–D5)

As shown in Fig. 2, the convergence performance of OC-CNMF outperforms NMF and CNMF by a huge margin in various datasets. If we have a closer look at images in Fig. 2, RMSE of NMF and coupled NMF starts to increase in the initial few iterations and then starts to decrease gradually. On the other hand, the RMSE of OC-CNMF starts to reduce smoothly from the first iteration. This is because of the orthogonal constraint imposed. While the OC-CNMF converged to a better solution at around 100 to 150 iterations, coupled NMF cannot reach that level even after 300 iterations in most cases. This shows that the convergence of OC-CNMF is at least 2 times faster than that of the CNMF. This is a significant result for the approximation of the input matrices. Such performance guarantees that the OC-CNMF will perform well on most applications of factorization, such as recommendation systems, link prediction, image reconstruction, etc. It is also evident that the OC-CNMF achieve similar performance in the approximation of the second input data, whereas standard NMF and CNMF cannot achieve better convergence. This proves that the OC-CNMF is suitable for simultaneously approximating both the matrices of the respective datasets.

5.3   Common Pattern Identification

Figure 3a & 3b show the common pattern identification using the CNMF and OC-CNMF models. Figure 3a shows the common patterns obtained by running the stand-alone CNMF model on the 2019 and 2020 vehicular traffic datasets (D1). According to Fig. 3a, though the CNMF model identifies the common patterns, we can see that CNMF also identifies similar (duplicate) patterns. This can be proved by considering the patterns (p2 and p3) in Fig. 3a in which the before-mentioned patterns are following a similar kind of pattern structure during the time from 12 pm to 6 pm. In addition to this, the similar pattern structure is found in the patterns (p1 and p2) of Fig. 3a during the time from 10 am to 3 pm. This shows that though CNMF helps to find common patterns, but they are not distinctive patterns. Figure 3b shows the four patterns (namely p1, p2, p3, and p4) that are learned from the proposed OC-CNMF model on the 2019 and 2020 vehicular traffic datasets (D1).

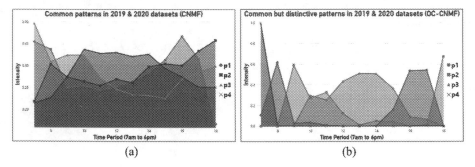

(a)   (b)

Fig. 3.   Common patterns identified using CNMF and OC-CNMF

The objective of the proposed model is to identify the common but distinctive latent (hidden) patterns that are present in the given two matrices. As shown in Fig. 3b, there are 4 common patterns present in the 2019 and 2020 matrices of D1 for the period before pandemic and during pandemic. Also, patterns are not following the similar structure such that the identified common patterns are distinctive and unique.

5.4   Performance Measure Comparison

Table 2. Performance measure comparison.

Evaluation measures		Root means square error (lower the better)	Silhouette score (higher the better)	Pattern distinctiveness (lower the better)	Calinski-Harabasz (higher the better)	Davies-Bouldin (lower the better)	Karzanowski and Lai (higher the better)
D1	NMF	0.37	0.50	0.98	6.69	0.92	0.05
	CNMF	0.35	0.40	0.97	6.05	0.97	0.05
	OCCNMF	**0.28**	**0.73**	**0.10**	**18.59**	**0.46**	**0.67**
D2	NMF	**0.19**	0.32	0.59	15.46	0.77	**0.76**
	CNMF	0.24	0.39	0.59	23.89	0.82	0.46
	OCCNMF	0.20	**0.91**	**0.06**	**50.48**	**0.30**	0.51
D3	NMF	0.20	0.47	0.86	13.05	0.68	0.34
	CNMF	**0.19**	0.54	0.76	15.87	0.63	0.42
	OCCNMF	0.20	**0.74**	**0.12**	**27.60**	**0.54**	**0.97**
D4	NMF	0.43	0.29	0.96	2.23	0.59	0.13
	CNMF	0.42	0.16	0.96	1.67	0.45	0.13
	OCCNMF	**0.28**	**0.91**	**0.09**	**15.56**	**0.41**	**0.48**
D5	NMF	0.40	0.36	0.96	4.05	0.89	0.11
	CNMF	0.39	0.58	0.97	7.26	**0.41**	0.05
	OCCNMF	**0.28**	**0.60**	**0.08**	**13.38**	0.58	**0.79**

Table 2 shows the performance comparison results of the proposed OC-CNMF model in comparison with NMF and CNMF models on five datasets (D1 to D5). The performance evaluation is carried out using various evaluation or performance measures such as Root Means Square Error, Silhouette score, Pattern Distinctiveness score, Calinski-Harabasz score, Davies Bouldin score and Karzanowski score. The entire evaluation measures are carried out in a MATLAB environment and the evaluation score results are tabulated as above for comparison. According to the table, it is very clear that the overall performance of the proposed OC-CNMF model is out-performing NMF and CNMF models based on the various evaluation score results.

6   Conclusion

This paper proposed the novel approach of OC-CNMF model that helps to find the common as well as distinctive patterns that are present in the provided datasets. The concept of orthogonality is applied in the CNMF architecture to learn the common but distinctive patterns or features. This proposed approach eliminates the usage of running individual NMF models which further leads to improving the convergence performance of the proposed OC-CNMF model. One key observation that we like to highlight is that the repeated patterns are not identified as individual patterns in the OC-CNMF model. This is because of the orthogonal constraint imposed on the dataset during the factorization process. The orthogonal constraint aims to learn the common patterns exist in both the matrices of the dataset, and at the same time its constraint itself learns distinctive patterns. Therefore, similar patterns will be ignored, and the OC-CNMF model will provide us with the unique and common patterns that are hidden in the datasets.

Acknowledgement. This research was supported by the National Research Foundation of Korea (Grant No. 2020R1A2C1012196).

References

1. Mohandu, A., Kubendiran, M.: Survey on big data techniques in intelligent transportation system (ITS). Materials Today Proceedings **47**, 8–17 (2021)
2. Lamssaggad, A., Benamar, N., Hafid, A.S., Msahli, M.: A survey on the current security landscape of intelligent transportation systems. IEEE Access **9**, 9180–9208 (2021)
3. Liu, C., Ke, L.: Cloud assisted Internet of things intelligent transportation system and the traffic control system in the smart city. J. Control and Decision, 1–14 (2022)
4. Wang, S., Cao, J., Yu, P.: Deep learning for spatio-temporal data mining: a survey. IEEE Tran. Knowl. Data Eng. (2021)
5. Xia, D., et al.: Discovering spatiotemporal characteristics of passenger travel with mobile trajectory big data. Phys. A **578**, 126056 (2021)
6. Saeed, F., Paul, A., Ahmed, M.J.: Forecasting COVID-19 cases using multiple statistical models. In: 8th International Conference on Orange Technology (ICOT), pp. 1–5. IEEE (2020)
7. He, P., Jiang, G., Lam, S.-K., Sun, Y.: Learning heterogeneous traffic patterns for travel time prediction of bus journeys. Inf. Sci. **512**, 1394–1406 (2020)
8. Balasubramaniam, A., Balasubramaniam, T., Jeyaraj, R., Paul, A., Nayak, R.: Nonnegative matrix factorization to understand spatio-temporal traffic pattern variations during COVID-19: a case study. In: Xu, Y., Wang, R., Lord, A., Boo, Y.L., Nayak, R., Zhao, Y., Williams, G. (eds.) AusDM 2021. CCIS, vol. 1504, pp. 223–234. Springer, Singapore (2021). https://doi.org/10.1007/978-981-16-8531-6_16
9. Lei, S., Zhang, B., Wang, Y., Dong, B., Li, X., Xiao, F.: Object recognition using non-negative matrix factorization with sparseness constraint and neural network. Information **10**(2), 37 (2019)
10. Pauca, V.P., Piper, J., Plemmons, R.J.: Nonnegative matrix factorization for spectral data analysis. Linear Algebra Appl. **416**(1), 29–47 (2006)
11. Cheung, V.C.K., Devarajan, K., Severini, G., Turolla, A., Bonato, P.: Decomposing time series data by a non-negative matrix factorization algorithm with temporally constrained coefficients. In: Proceedings of the Annual International Conference of the IEEE Engineering in Medicine and Biology Society, EMBS (2015)
12. Yang, F., Ma, F., Ping, Z., Xu, G.: Total variation and signature-based regularizations on coupled nonnegative matrix factorization for data fusion. IEEE Access **7**, 2695–2706 (2019)

13. Priya, K., Rajkumar, K.K: Multiplicative iterative nonlinear constrained coupled non-negative matrix factorization (MINC-CNMF) for hyperspectral and multispectral image fusion. Int. J. Adv. Comput. Sci. Appl. **12**(6) (2021)

14. Ahmad, T., Lyngdoh, R.B., Anand, S.S., Gupta, P.K., Misra, A., Raha, S.: Robust coupled non-negative matrix factorization for hyperspectral and multispectral data fusion. In: IEEE International Geoscience and Remote Sensing Symposium IGARSS (2021)

15. Vehicle counts recorded on major and minor roads. https://roadtraffic.dft.gov.uk/#6/55.254/-6.053/basemap-regions-countpoints. Accessed 04 Mar 2022

16. Chang, P.C., Lin, J.T., Lin, C.H., Tang, P.W., Liu, Y.: Optimization-based hyperspectral spatiotemporal super-resolution. IEEE Access **10**, 37477–37494 (2022)

17. Gao, Y., Liu, J., Xu, Y., Mu, L., Liu, Y.: A spatiotemporal constraint non-negative matrix factorization model to discover intra-urban mobility patterns from taxi trips. Sustainability **11**, 4214 (2019)

18. Zhao, Y., Wang, C., Pei, J., Yang, X.: Nonlinear loose coupled non-negative matrix factorization for low-resolution image recognition. Neurocomputing **443**, 183–198 (2021)

19. Balasubramaniam, T., Nayak, R., Yuen, C.: Nonnegative coupled matrix tensor factorization for smart city spatiotemporal pattern mining. In: Machine Learning Optimization and Data Science, pp. 520–532. Springer, Berlin, Germany (2019) https://doi.org/10.1007/978-3-030-13709-0_44

20. Luong, K., Balasubramaniam, T., Nayak, R.: A novel technique of using coupled matrix and greedy coordinate descent for multi-view data representation. In: Hacid, H., Cellary, W., Wang, H., Paik, H.-Y., Zhou, R. (eds.) WISE 2018. LNCS, vol. 11234, pp. 285–300. Springer, Cham (2018). https://doi.org/10.1007/978-3-030-02925-8_20

21. Jiho, Y., Choi, S.: Nonnegative matrix factorization with orthogonality constraints. J. Comput. Sci. Eng. **4**(2), 97–109 (2010)

22. Lee, D.D., Seung, H.S.: Algorithms for non-negative matrix factorization. Advances in Neural Information Processing Systems, 13 (2001)

23. Ding, C., Li, T., Peng, W., Park, H.: Orthogonal nonnegative matrix factorizations for clustering. In: Proceedings of the 12th ACM SIGKDD International Conference on Knowledge Discovery and Data Mining, pp. 126–135. ACM (2006)

24. 2011–2013 NYC Traffic Volume Counts Kaggle. https://www.kaggle.com/hanriver0618/2011-2013-nyc-traffic-volume-counts. Accessed 15 Mar 2022

Efficiency and Truthfulness in Dial-a-Ride Problems with Customers Location Preferences

Martin Aleksandrov[1(✉)] ⓘ and Philip Kilby[2] ⓘ

[1] Freie Universität Berlin, Berlin, Germany
martin.aleksandrov@fu-berlin.de
[2] Australian National University, Canberra, Australia
philip.kilby@anu.edu.au

Abstract. We look at the dial-a-ride problem through the lens of mechanism design, where the goal is to design mechanisms, toward natural objectives, in strategic settings, where customers act rationally. For this problem, we consider customer preferences for detour times and waiting times of services, as well as mechanisms for minimising the detour times, waiting times, and detour plus waiting times of customers on vehicles. We characterise mechanisms that are economically efficient and game-theoretically truthful. With detour-time preferences, we show that there are mechanisms that are both efficient and truthful in all instances. With waiting-time preferences, we show that there are mechanisms that are efficient but not truthful in some instances.

Keywords: Multi-agent systems · Social choice · Vehicle routing

1 Introduction

Let us consider the dial-a-ride problem [5]. On the one hand, when minimising customer costs in this problem, customers rank minimising their detour times and waiting times as two of the most important criteria. This is perhaps natural because doing so is often more convenient for customers and they normally pay lower costs for shorter trips [12]. On the other hand, drivers tend to pick longer trips so that they maximise their profit. This often makes routing decisions biased towards customers with longer routes and, thus, leave customers with shorter routes less satisfied with their services. In this paper, we look at dial-a-ride models where vehicle locations are known to customers and customer costs for services are not explicitly modelled. For example, taxi companies such as Uber and Lyft reveal the locations of as many of their cars as possible simply because this extends the number of travel options, which in turn makes it more attractive for customers to use their services and can additionally make it easier for cars to avoid making longer tours by, for example, not making U-turns whenever customers can cross roads. The absence of customer costs is useful either when the services are free-of-charge or fixed-priced. This is the case whenever we use such vehicles for social services.

The original version of this chapter was revised: Table 1 and the equation on page 696 have been corrected. The correction to this chapter is available at
https://doi.org/10.1007/978-3-031-22695-3_57

H. Aziz et al. (Eds.): AI 2022, LNAI 13728, pp. 690–703, 2022.
https://doi.org/10.1007/978-3-031-22695-3_48

According to the Federal Statistical Office of Germany, nearly 16 million Germans are disabled. Subsidising vehicles such as those at Uber and Lyft could promote mobile opportunities for these people at low costs. Also, in Berlin, more than 210 000 commuters use public transportation daily. Sharing Uber and Lyft vehicles could promote commuting alternatives for these people at the price of a public transportation ticket, which is a common goal for smart cities: see e.g. the Smart City Mobility Bamberg project. In such settings, customers also have *detour-* and *waiting*-time preferences that do not relate explicitly to their income, gender, or race. Under such preferences, our goal is to deliver efficiency (i.e. we cannot provide better services to all customers) and truthfulness (i.e. we motivate customers to be sincere when submitting their locations).

Customer location misreports have been considered widely in facility location problems, see e.g. [16], but have only recently been seen as an issue in routing problems [15]. In response, we let customers have some private locations. In general, customers have preferences over their private locations, which can be more complex than accounting solely for their detour times or waiting times. For example, private locations might be their current locations or nearby locations, where they prefer moving and submitting requests. That can be around the corner if they want to go there for health reasons or do some shopping before taking a lift to a party with friends. However, we consider customer location preferences for detour times or waiting times. In this context, customers can strategically report nearby locations where they can move from their private locations and, thus, decrease their detour times or waiting times.

On the plus side, customers might thus receive quicker services. To see this, suppose that minimising the overall waiting time for customers a and b requires that a vehicle picks up first a and then b. This is efficient for both a and b. However, b can strategically report some nearby location that is closer to the vehicle location and where they can move quickly in order to receive a quicker pickup. For example, if b would have to wait 10 min for the vehicle at their private location because the vehicle would have to pick up first a, then b might wish to request a ride from a nearby location which is 2 min away by foot from their private location and where the vehicle could pick them up in the next 3 min before it picks up a. Thus, moving from their private location to the reported nearby location plus waiting at that location would take $2 + (3 - 2) = 3$ min for b, which would be strictly less than 10 min.

On the minus side, manipulators might make others wait longer for services. In the example above, by reporting the nearby location and moving there, customer b makes customer a wait longer because a is picked up after b. However, unlike manipulators, some customers may generally not be able to move to better locations, e.g. senior citizens and home patients. Such customers might feel less socially included because they could be disadvantaged due to the manipulations of others. Therefore, in addition to achieving efficiency, achieving truthfulness seems to further improve the social inclusion of such customers. This is in line with the goal of the 2022 International Transport Forum to design mobility solutions for inclusive societies. Thus, we investigate the following question:

Under customer preferences for their detour times and waiting times, when and how can we achieve economic efficiency, game-theoretic truthfulness, or, whenever possible, both of them?

We tackle this question for customers that remained non-serviced by a given time point during the day. Their detour times are the travel times between their origins and destinations minus the shortest travel times between these locations, respectively. Their waiting times are the times between booking trips and visiting their origins, respectively. We consider detour-time preferences (*d-utilities*), waiting-time preferences (*w-utilities*), and detour-plus-waiting-time preferences (*(d+w)-utilities*). Satisfying detour-time preferences seems to be more relevant in Uber and Lyft settings where, when submitting requests via their devices, customers can already select vehicles with the shortest estimated arrival times (ETAs), which minimises their waiting times. On the contrary, satisfying waiting-time preferences seems thus to be more relevant in cab settings where customers do not know the ETAs of the vehicles. For each of these preference types, we consider generally contradicting but nevertheless desired objectives such as maximising service efficiency and service fairness. Service efficiency is often measured in terms of minimising the total time travelled by all vehicles [6]. We consider three other measures for it, i.e. minimising the *total* detour time (totDET), waiting time (totWAIT), and detour plus waiting time (totDETWAIT) for all vehicles. Service fairness is often measured in terms of minimising the maximum time travelled by any vehicle [10]. We consider three other measures for it as well, i.e. minimising the *maximum* detour time (maxDET), waiting time (maxWAIT), and detour plus waiting time (maxDETWAIT) of any vehicle.

In our work, we consider the classes of mechanisms that minimise these six objectives, respectively. Mechanisms can be based on exact methods such as the branch-and-cut algorithm [3] or inexact methods such as the large-neighborhood-search metaheuristic [7]. In particular, we are interested in mechanisms that are efficient for given customer preferences. We use *Pareto optimality* [13] as an efficiency criterium. Efficient mechanisms guarantee thus that we cannot provide better services to all customers. Also, we are interested in mechanisms that encourage customers to report their private locations sincerely. *Truthful* mechanisms do exactly this and, thus, elicit these locations. With waiting-time preferences, in the instance from the above example, we can notice that achieving both Pareto optimality and truthfulness is not possible. At the same time, with detour-time preferences, we will prove that it is possible to achieve both criteria in every instance. Doing so requires that the requests are serviced one after another *and* the customers do not take detours. The former condition is widely realistic with cab, Uber, and Lyft vehicles, where drivers do not start servicing new customers before finishing the services of their current customers. However, the latter condition is often violated by cab drivers since they try to take longer detours that maximise their profits, which often leads to increased customer dissatisfaction [9]. But, with Uber and Lyft drivers, customers prepay fixed costs and so drivers are not tempted to detour but quickly move to the next customers. So, in such settings, the latter condition seems to hold as well.

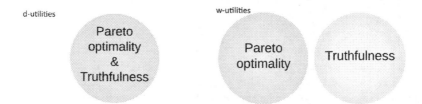

Fig. 1. General possibility results for Pareto optimality and truthfulness in our setting: (left) d-utilities (detour-time preferences); (right) w-utilities (waiting-time preferences).

Figure 1 shows our two main axiomatic results. These results tell us *when* (i.e. under which preferences) efficiency and truthfulness can be achieved but they do not tell us *how* (i.e. under which mechanisms) to do it. This is what we investigate in the rest of the paper. Firstly, we discuss literature gaps in Sect. 2. In Sect. 3, we give formal preliminaries. In Sect. 4, we define the preferences. In Sect. 5, we define the classes of mechanisms for totDET, maxDET, totWAIT, maxWAIT, totDETWAIT, and maxDETWAIT. After that, in Sects. 6 and 7, we show that mechanisms for totDET and maxDET are Pareto-optimal and truthful with detour-time preferences, as well as only mechanisms for totWAIT and totDETWAIT are only Pareto-optimal with waiting-time preferences and detour-plus-waiting-time preferences, respectively. Finally, we present a summary of results and prospects in Sect. 8.

2 Related Work

Surprisingly, customer preferences and objectives have received limited attention in the vast VRP literature [15]. Regarding customer preferences, a few works considered costs but not utilities: Bei and Zhang [4] analysed trip assignments of minimal costs in ride-sharing; Li et al. [8] looked at minimising shared taxi costs for people and parcels; Santos and Xavier [14] also studied this task but only for people. Regarding customer objectives, Nucamendi, Cardona-Valdes, and Angel-Bello Acosta [11] considered the total waiting time in the context of one vehicle and visit requests of unit demands. In contrast, we study this objective, denoted as totWAIT, with multiple vehicles and heterogeneous requests of arbitrary positive demands. In addition, we study totDET and totDETWAIT. Workload equity has recently been announced to be underexplored [10]. In response, we investigate maxDET, maxWAIT, and maxDETWAIT that relate to equity because they minimise the workload of any vehicle, which is measured by means of the total detour time, waiting time, or detour plus waiting time of the customers on the vehicle. Notably, unlike the existing works above, we transfer tools such as efficiency and truthfulness from the mechanism design field to the customer side in the vehicle routing field. From this perspective, our analysis seems to be highly novel to the best of our knowledge.

3   Model

For $t \in \mathbb{N}_{>0}$, we let $[t]$ denote $\{1, \ldots, t\}$. We also let $L \subset \mathbb{R}^2$ denote a finite set of at least two locations. We write $D = [d(l, l')]_{|L| \times |L|}$, where $d(l, l') \in \mathbb{R}_{\geq 0}^{<\infty}$ is the distance between $l, l' \in L$. The road network is $N = (L, D)$. We consider *vehicle* set $V = \{v_i | i \in [n]\}$, where each v_i begins at $b_i \in L$, ends at $e_i \in L$, has available capacity $q_i \in \mathbb{N}_{>0}$, and moves with average velocity $\nu_i \in \mathbb{R}_{>0}$. We also consider *request* set $R = \{r_j | j \in [m]\}$, where each non-serviced $r_j = (p_j, d_j, m_j)$ is submitted earlier by agent j, and is for transporting $m_j \in \mathbb{N}_{>0}$ customers between two different locations from L (i.e. $p_j \neq d_j$) or visiting m_j customers at some location from L (i.e. $p_j = d_j$). In practice, groups split into sub-groups and submit multiple requests whenever they do not fit in vehicles. Thus, we let each r_j be feasible for some v_i, i.e. $\max_{r_j \in R} m_j \leq \max_{v_i \in V} q_i$ hold.

Furthermore, for each v_i, we let $R_i \subseteq R$ denote a set of requests assigned to v_i and $\mathcal{R}_i = (s_1(i), \ldots, s_{2|R_i|}(i))$ denote a route through the service (i.e. pick-up/drop-off/visit) locations of the requests from R_i. A plan $\mathcal{P} = \{\mathcal{R}_1, \ldots, \mathcal{R}_n\}$ is a set of routes, where route $\mathcal{R}_i$ is assigned to v_i. We let $\mathcal{I} = (N, V, R)$ denote an instance. Let us recall that the requests are non-serviced by some given time point during the day. At that point, we want to plan how to service them subject to several constraints. In particular, we look at *feasible plans* for $\mathcal{I}$ that satisfy completeness constraints (i.e. each request is assigned to some vehicle), disjointness constraints (i.e. no request is assigned to more than one vehicle), ordering constraints (i.e. each origin is visited before its destination), and capacity constraints (i.e. no vehicle capacity is ever exceeded): see e.g. [2].

4   Preferences

Let us consider instance $\mathcal{I}$. Further, pick feasible plan $\mathcal{P}$ for $\mathcal{I}$, route $\mathcal{R}_i$ from $\mathcal{P}$, and request r_j from R_i. The detour distance for agent j in $\mathcal{R}_i$ is the tour distance travelled by v_i between p_i and d_j minus the shortest distance between p_j and d_j: $d_{ij} = [\sum_{s_l(i) \in (p_j, \ldots, d_j), s_l(i) \neq d_j} d(s_l(i), s_{l+1}(i))] - d(p_j, d_j)$. The latency distance for agent j in $\mathcal{R}_i$ is the distance travelled by v_i between b_i and p_j: $\delta_{ij} = d(b_i, s_1(i)) + [\sum_{s_l(i) \in (s_1(i), \ldots, p_j), s_l(i) \neq p_j} d(s_l(i), s_{l+1}(i))]$.

The plan $\mathcal{P}$ induces detour time and waiting time for agent j. Their detour time is d_{ij}/ν_i. Their waiting time is $w_{ij}/\nu_i = (\tau_j + \delta_{ij}/\nu_i)$, where $\tau_j \in \mathbb{R}_{\geq 0}$ is the tardiness time agent j waits between submitting their request and planning the problem and δ_{ij}/ν_i is the vehicle travel time between planning the problem and being picked up at their origin. We use these times to formally define the utility $u_{ij}(\mathcal{P})$ of agent j in $\mathcal{P}$ where they are serviced by v_i.

For this purpose, we also assume that agent j has some private start and finish locations $s_j, f_j \in L$. This is because they might be using our system for partial planning of their journey. To capture this, we let $T_j \in \mathbb{R}_{\geq 0}$ denote the additional travel time between s_j and p_j and between d_j and f_j. Thus, agent j can travel T_j units of time by mixed means of transport. This allows them to use our system in a strategic manner with other systems such as Google Maps.

We next define $u_{ij}(\mathcal{P})$ as a quasi-linear function with respect to (wrt) T_j. That is, for some $a_j, b_j, c_j \in \mathbb{R}_{\geq 0}$, $u_{ij}(\mathcal{P}) = -m_j \cdot [(a_j \cdot d_{ij}/\nu_i) + (b_j \cdot w_{ij}/\nu_i)]$ if $T_j \leq c_j$ and $u_{ij}(\mathcal{P}) = -m_j \cdot [(a_j \cdot d_{ij}/\nu_i) + (b_j \cdot w_{ij}/\nu_i) + T_j]$ if $T_j > c_j$. The weights a_j, b_j, and c_j regulate the importance of the detour time, the waiting time, and the additional time for agent j, respectively. We do not assume to know the functional form of a_j, b_j, or c_j.

For example, in a full system, agent j might want to detour at most $10\,\mathrm{min}$ and wait at most $5\,\mathrm{min}$. In this case, a_j and b_j could be quasi-linear functions wrt these thresholds, i.e. $a_j = 0$ if $(d_{ij}/\nu_i) \leq 10$ and $a_j = 1 - (10 \cdot \nu_i/d_{ij})$ if $(d_{ij}/\nu_i) > 10$; $b_j = 0$ if $(w_{ij}/\nu_i) \leq 5$ and $b_j = 1 - (5 \cdot \nu_i/w_{ij})$ if $(w_{ij}/\nu_i) > 5$. However, for simplicity of elicitation, we instead only assume that agents who want to use our system care for at least one of these times, i.e. $a_j + b_j > 0$.

We note that $u_{ij}(\mathcal{P}) \leq 0$ holds because of $d_{ij} \geq 0$ and $w_{ij} \geq 0$. We say that the agents have *d-utilities* (detour-time preferences) if $\forall j \in [m] : b_j = 0$ hold; *w-utilities* (waiting-time preferences) if $\forall j \in [m] : a_j = 0$ hold; *(d+w)-utilities* (detour-plus-waiting-time preferences) if $\forall j \in [m] : a_j = b_j$ hold. We also say that the agents are *sincere* if $\forall j \in [m] : p_j = s_j$ and $d_j = f_j$ (i.e. $T_j = 0$) hold; *strategic* if $\exists j \in [m] : p_j \neq s_j$ or $d_j \neq f_j$ (i.e. $T_j > 0$) hold.

5   Mechanisms

Based on the detour times of customers, we define two natural *detour-time* routing objectives in feasible plans: (totDET): $\min_{\mathcal{P}: \text{ feasible}} \sum_{v_i \in V} \sum_{r_j \in R_i} m_j \cdot (d_{ij}/\nu_i)$; (maxDET): $\min_{\mathcal{P}: \text{ feasible}} \max_{v_i \in V} \sum_{r_j \in R_i} m_j \cdot (d_{ij}/\nu_i)$. They are useful when we cannot start any new service before finishing the current service.

Based on the waiting times of customers, we define two natural *waiting-time* routing objectives in feasible plans: (totWAIT): $\min_{\mathcal{P}: \text{ feasible}} \sum_{v_i \in V} \sum_{r_j \in R_i} m_j \cdot (w_{ij}/\nu_i)$; (maxWAIT): $\min_{\mathcal{P}: \text{ feasible}} \max_{v_i \in V} \sum_{r_j \in R_i} m_j \cdot (w_{ij}/\nu_i)$. These objectives are useful when we only care about how much time agents wait.

The waiting-time objectives differ from the detour-time objectives. For example, minimising the former is shown to be generally intractable [1], whereas minimising the latter requires servicing the requests one after another and is, therefore, tractable. Also, minimising the former might not be minimising the latter. We illustrate this in Example 1.

Example 1. Let us consider points $A(0,0)$, $B(0,1)$, $C(1,0)$ and the straight-line metric. Further, let us consider one vehicle and two requests. We let these be v_1 with $b_1 = C$, $e_1 = C$, $q_1 = 2$, $\nu_1 = 1$, and $r_1 = (A, B, 1)$, $r_2 = (A, C, 1)$. Also, we let $\tau_1 = \tau_2 = 0$ hold. Minimising the detour times requires not picking up both agents at once. We next confirm this.

If the vehicle v_1 services r_1 before r_2, then following A, B, A, C minimises the detour times but generates waiting times: $d_{11} + d_{12} = 0 + 0$ and $w_{11} + w_{12} = 1 + 3 = 4$. If the vehicle services r_2 before r_1, then following A, C, A, B gives $d_{11} + d_{12} = 0 + 0$ and $w_{11} + w_{12} = 3 + 1 = 4$. The agent-wise time d-profile is $(0,0)$, w-profile is $(3,1)$, and (d+w)-profile is $(3,1)$.

Next, suppose that v_1 picks up both agents. Following A, A, B, C minimises the waiting times but generates detour times: $d_{11} + d_{12} = 0 + \sqrt{2} > 0$ and $w_{11} + w_{12} = 1 + 1 = 2 < 4$. Also, following A, A, C, B gives symmetric outcomes: $d_{11} + d_{12} = \sqrt{2} + 0 > 0$ and $w_{11} + w_{12} = 1 + 1 < 4$. The agent-wise time d-profile is $(\sqrt{2}, 0)$, w-profile is $(1, 1)$, and (d+w)-profile is $(\sqrt{2} + 1, 1)$.   □

We further consider two more objectives that allow agents to be indifferent among feasible plans where their detour plus waiting times are equal, as long as the additive sum of these times is minimised across all vehicles and per any vehicle, respectively: (totDETWAIT): $\min_{\mathcal{P}: \text{ feasible}} \sum_{v_i \in V} \sum_{r_j \in R_i} m_j \cdot ((d_{ij} + w_{ij})/\nu_i)$; (maxDETWAIT): $\min_{\mathcal{P}: \text{ feasible}} \max_{v_i \in V} \sum_{r_j \in R_i} m_j \cdot ((d_{ij} + w_{ij})/\nu_i)$.

In Example 1, minimising totDETWAIT or maxDETWAIT minimises further the waiting-time objectives but not the detour-time objectives. By comparison, there are also settings where minimising totDETWAIT or maxDETWAIT minimises further the detour-time objectives but not the waiting-time objectives. We confirm this in Example 2.

Example 2. Let us consider points $A(0,0)$, $B(0,1)$, $D(1,1)$ and the straight-line metric. Further, let us consider vehicle v_1 with $b_1 = D$, $e_1 = D$, $q_1 = 2$, $\nu_1 = 1$ and requests $r_1 = (D, B, 1)$, $r_2 = (A, B, 1)$. Also, let $\tau_1 = \tau_2 = 0$ hold. Minimising the detour plus waiting times requires that v_1 services the agent at D before the agent at A.

Indeed, if the vehicle transports first the agent from D to B and then the other agent from A to B, then the corresponding plan gives zero detour times but generates waiting times: $d_{11} + d_{12} = 0$ and $w_{11} + w_{12} = 0 + 2$. The agent-wise time d-profile is $(0, 0)$, w-profile is $(0, 2)$, and (d+w)-profile is $(0, 2)$.

By comparison, let us suppose that the vehicle picks up the agent at D, drives then to A where it picks up the other agent, and heads towards B. This plan minimises the waiting times but gives greater detour times: $d_{11} + d_{12} = \sqrt{2} + 0$ and $w_{11} + w_{12} = 0 + \sqrt{2} < 2$. The agent-wise time d-profile is $(\sqrt{2}, 0)$, w-profile is $(0, \sqrt{2})$, and (d+w)-profile is $(\sqrt{2}, \sqrt{2})$.   □

Minimising totDET, totWAIT, and totDETWAIT measure service efficiency because they minimise the overall times for all vehicles whereas minimising maxDET, maxWAIT, and maxDETWAIT measure service fairness because they aim at achieving similar time workloads across all vehicles.

For given instance $\mathcal{I}$, we write $\text{FP}(\mathcal{I})$ for the non-empty set of all feasible plans for instance $\mathcal{I}$. A *feasible* mechanism $\mathcal{M}$ maps $\mathcal{I}$ into some plan $\mathcal{M}(\mathcal{I})$ from $\text{FP}(\mathcal{I})$. In the rest of the paper, we consider feasible mechanisms for totDET, maxDET, totWAIT, maxWAIT, totDETWAIT, and maxDETWAIT.

6   Efficient Mechanisms

Let us consider sincere agents. A mechanism $\mathcal{M}$ is *Pareto-optimal (PO)* if, for each instance $\mathcal{I}$, there is no mechanism $\mathcal{M}'$ such that $\forall j \in [m] : u_{i_j j}(\mathcal{M}(\mathcal{I})) \leq u_{k_j j}(\mathcal{M}'(\mathcal{I}))$ and $\exists j \in [m] : u_{i_j j}(\mathcal{M}(\mathcal{I})) < u_{k_j j}(\mathcal{M}'(\mathcal{I}))$. With d-utilities, we next characterise the class of Pareto-optimal mechanisms.

Theorem 1. *With sincere agents and d-utilities, a feasible mechanism min-imises totDET/maxDET if and only if it is Pareto-optimal.*

Proof. As the agents are sincere, $T_j = 0$ holds for each agent $j \in [m]$. Also, as they have d-utilities, $a_j > 0$ and $b_j = 0$ hold for each agent $j \in [m]$. Let us consider a mechanism $\mathcal{M}$ that minimises totDET/maxDET in each instance $\mathcal{I}$. Hence, $d_{ij} = 0$ holds for each $i \in [n]$ and $j \in [m]$. As a result, $u_{ij}(\mathcal{M}(\mathcal{I})) = 0$ also holds. In other words, each agent receives the maximum possible utility. The mechanism is Pareto optimal. Let us consider a mechanism $\mathcal{M}$ that is Pareto-optimal in each instance $\mathcal{I}$. For the sake of contradiction, suppose that it does not minimise totDET/maxDET in a given instance $\mathcal{I}$. Hence, $d_{ij} > 0$ and $u_{ij}(\mathcal{M}(\mathcal{I})) < 0$ must hold for some $i \in [n]$ and $j \in [m]$. For each $k \in [n]$ and $l \in [m]$ such that $u_{kl}(\mathcal{M}(\mathcal{I})) < 0$ does not hold, $u_{kl}(\mathcal{M}(\mathcal{I})) = 0$ holds because $u_{kl}(\mathcal{M}(\mathcal{I})) \leq 0$ holds by definition. In contrast, a mechanism that minimises tot-DET/maxDET induces d-utility of zero for each agent. Hence, such a mechanism Pareto dominates $\mathcal{M}$ and, therefore, $\mathcal{M}$ cannot be Pareto-optimal. $\square$

In Example 1, each mechanism for totWAIT/maxWAIT/totDETWAIT/maxDETWAIT does not minimise totDET/maxDET. Supposing that the agents are sincere and have d-utilities, *no* mechanism for any of the former objectives is Pareto-optimal in this instance.

Furthermore, in Example 1, each mechanism for totDET/maxDET does not minimise totWAIT/maxWAIT/totDETWAIT/maxDETWAIT. Supposing that the agents are sincere and have w-utilities/(d+w)-utilities, *no* mechanism for any of the former objectives is Pareto-optimal in this instance.

In Example 2, each mechanism for totDETWAIT/maxDETWAIT does not minimise totWAIT/maxWAIT. Supposing that the agents are sincere and have w-utilities, *no* mechanism for the former objectives is Pareto-optimal.

In Example 3, we show that each mechanism for totWAIT/maxWAIT does not minimise totDETWAIT/maxDETWAIT. Supposing that the agents are sincere and have (d+w)-utilities, it follows that *no* mechanism for any of the former objectives is Pareto-optimal.

Example 3. Let us consider $A(0,0)$, $B(0,1)$, and $C(1,0)$: $d(A,B) = 1$, $d(A,C) = 1$, and $d(B,C) = \sqrt{2}$. Further, let us consider one vehicle and two requests. We let these be v_1 with $b_1 = C$, $e_1 = C$, $q_1 = 2$, $\nu_1 = 1$, and $r_1 = (B,A,1)$, $r_2 = (A,C,1)$. Also, we let $\tau_1 = \tau_2 = 0$ hold.

The plan where the vehicle begins at C, services A, then B, A, and ends at C minimises the waiting time: $w_{11} + w_{12} = 2 + 1 = 3$. This plan generates detour time because the agent at A travels through B and A on their way to C. Indeed, $d_{11} + d_{12} = 0 + 2 = 2$. The agent-wise time d-profile is $(0,2)$, w-profile is $(2,1)$, and (d+w)-profile is $(2,3)$.

By comparison, the plan where the vehicle begins at C, services B, then A, and ends at C generates zero detour time and waiting time of $w_{11} + w_{12} = \sqrt{2} + (1 + \sqrt{2}) \approx 3.8 \in (3,5)$. The agent-wise time d-profile is $(0,0)$, w-profile is $(\sqrt{2}, \sqrt{2} + 1)$, and (d+w)-profile is $(\sqrt{2}, \sqrt{2} + 1)$. For $a_1 > 0$ and $a_2 > 0$, the utility (d+w)-profile $(-\sqrt{2}a_1, -(1 + \sqrt{2})a_2)$ Pareto dominates $(-2a_1, -3a_2)$. $\square$

Nevertheless, with w-utilities, we can partially characterise Pareto optimality in terms of mechanisms that minimise totWAIT. This characterisation is strict. That is, with w-utilities, not every feasible mechanism that is Pareto-optimal also minimises totWAIT. To see this, we refer the reader to Example 3. In this setting, with w-utilities, the mechanism that minimises totDETWAIT is Pareto-optimal, but it does not minimise totWAIT.

Theorem 2. *With sincere agents and w-utilities, a feasible mechanism for tot-WAIT is Pareto-optimal.*

Proof. As the agents are sincere, $T_j = 0$ holds for each agent $j \in [m]$. Also, as they have w-utilities, $a_j = 0$ and $b_j > 0$ hold for each agent $j \in [m]$. Let us consider a mechanism $\mathcal{M}$ that minimises totWAIT in each instance $\mathcal{I}$. For the sake of contradiction, let us suppose that $\mathcal{M}$ is not Pareto-optimal in a given instance $\mathcal{I}$. Hence, there is another mechanism $\mathcal{M}'$ that Pareto dominates $\mathcal{M}$ in $\mathcal{I}$. That is, $\forall j \in [m] : u_{i_j j}(\mathcal{M}'(\mathcal{I})) \geq u_{k_j j}(\mathcal{M}(\mathcal{I}))$ and $\exists j \in [m] : u_{i_j j}(\mathcal{M}'(\mathcal{I})) > u_{k_j j}(\mathcal{M}(\mathcal{I}))$, where $i_j, k_j \in [n]$. For each $j \in [m]$, we note that $m_j > 0$ holds. We derive $\forall j \in [m] : -w_{i_j j}/v_{i_j} \geq -w_{k_j j}/v_{k_j}$ and $\exists j \in [m] : -w_{i_j j}/v_{i_j} > -w_{k_j j}/v_{k_j}$. It follows $\forall j \in [m] : w_{k_j j}/v_{k_j} \geq w_{i_j j}/v_{i_j}$ and $\exists j \in [m] : w_{k_j j}/v_{k_j} > w_{i_j j}/v_{i_j}$. This implies that $\mathcal{M}$ does not minimise totWAIT in $\mathcal{I}$. This is a contradiction with the choice of the mechanism. The result follows.   □

With (d+w)-utilities, a similar characterisation result holds for Pareto optimality and mechanisms that minimise totDETWAIT. This characterisation is also strict. That is, with (d+w)-utilities, not every feasible mechanism that is Pareto-optimal also minimises totDETWAIT. To see this, we refer the reader to Example 2. In this setting, with (d+w)-utilities, the mechanism that minimises totWAIT is Pareto-optimal, but it does not minimise totDETWAIT.

Theorem 3. *With sincere agents and (d+w)-utilities, a feasible mechanism for totDETWAIT is Pareto-optimal.*

Proof. As the agents are sincere, $T_j = 0$ holds for each agent $j \in [m]$. Also, as they have (d+w)-utilities, $a_j = b_j > 0$ hold for each agent $j \in [m]$. Let us consider a mechanism $\mathcal{M}$ that minimises totDETWAIT in each instance $\mathcal{I}$. From now on, the proof uses an analogous line of reasoning as the proof of Theorem 2, with the exception that one needs to consider (d+w)-utilities instead of w-utilities. Thus, we can derive a contradiction. The result follows.   □

With just one vehicle, maxWAIT/maxDETWAIT coincides with totWAIT/totDETWAIT, respectively. In such instances, it follows by Theorems 2 and 3 that a feasible mechanism for maxWAIT/maxDETWAIT is Pareto-optimal whenever agents have w-utilities/(d+w)-utilities; see also Example 1.

With strictly more vehicles, all mechanisms for maxWAIT/maxDETWAIT may *violate* Pareto optimality whenever agents have w-utilities/(d+w)-utilities. It follows that service fairness and service efficiency are generally incompatible in these domains unlike with d-utilities. We demonstrate this in Example 4.

Example 4. For $\epsilon > 0$, let us consider locations $A_1(-\epsilon, 0)$, $A_2(0, \frac{1}{2})$, $B_1(1,0)$, $B_2(1, \frac{1}{2})$, $B_3(1,1)$, and the straight-line metric. Also, let us consider 3 vehicles and 2 requests. For $i \in \{1,2\}$, we let v_i be such that $b_i = A_i$, $e_i = A_i$, $q_i = 1$, and $\nu_i = 1$ hold. We let v_3 be such that $b_3 = B_3$, $e_3 = B_3$, $q_3 = 2$, and $\nu_3 = 1$ hold. We let $r_1 = (B_1, A_1, 1)$ and $r_2 = (B_2, B_1, 1)$. Also, we let $\tau_1 = \tau_2 = 0$ hold.

The plan that sends vehicle v_2 for service at B_2 and B_1, and vehicle v_3 for service at B_1 and A_1 generates zero detour times and minimises maxWAIT (maxDETWAIT) to 1. The waiting (detour plus waiting) times of agents 1 and 2 are 1 and 1, respectively. Hence, their sincere w-utilities ((d+w)-utilities) are $-b_1$ and $-b_2$, respectively. The total waiting (detour plus waiting) time in this plan is 2.

This plan is not Pareto-optimal with w-utilities ((d+w)-utilities). Indeed, the plan that minimises totWAIT (totDETWAIT) Pareto dominates it. This plan asks v_3 to pick up the agent at B_2 and head towards B_1 to pick up the other agent, and then towards A_1. The maximum (total) vehicle waiting (detour plus waiting) time is $\frac{3}{2} \in (1,2)$.

At the same time, this plan does not minimise maxWAIT (maxDETWAIT) because v_3 travels strictly more than 1 unit of time. The waiting (detour plus waiting) times of agents 1 and 2 are 1 and 1/2. Hence, their sincere w-utilities ((d+w)-utilities) are $-b_1$ and $-b_2/2$, respectively. For $b_2 > 0$, the agent-wise utility profile $(-b_1, -b_2/2)$ Pareto dominates $(-b_1, -b_2)$.   □

To sum up, with d-utilities, we might use mechanisms for minimising totDET and maxDET because they are Pareto-optimal. By comparison, with w-utilities ((d+w)-utilities), we might use mechanisms for minimising totWAIT (totDET-WAIT) because they are Pareto-optimal whereas mechanisms for minimising maxWAIT (maxDETWAIT) may not be Pareto-optimal.

7   Truthful Mechanisms

A mechanism $\mathcal{M}$ is *truthful* (TR) if, for each instance $\mathcal{I}$, there is no other instance $\mathcal{I}'$ where $R' = R \setminus \{r_j\} \cup \{r'_j\}$ for some agent $j \in [m]$ and some request r'_j such that $u_{ij}(\mathcal{M}(\mathcal{I})) < u_{kj}(\mathcal{M}(\mathcal{I}'))$ holds for some $i, k \in [n]$, supposing that agent j have complete knowledge of $\mathcal{I}$.

For sincere agents, $T_j = 0$ holds for each $j \in [m]$. As a result, $u_{ij}(\mathcal{M}(\mathcal{I})) = -m_j \cdot [(a_j \cdot d_{ij}/\nu_i) + (b_j \cdot w_{ij}/\nu_i)]$. For strategic agents, $T_j > 0$ holds for some $j \in [m]$. As a result, $u_{ij}(\mathcal{M}(\mathcal{I})) = -m_j \cdot [(a_j \cdot d_{ij}/\nu_i) + (b_j \cdot w_{ij}/\nu_i)]$ if $T_j \leq c_j$ and, otherwise, $u_{ij}(\mathcal{M}(\mathcal{I})) = -m_j \cdot [(a_j \cdot d_{ij}/\nu_i) + (b_j \cdot w_{ij}/\nu_i) + T_j]$.

As we discussed previously, the pick-up and drop-off locations are normally private in practice and agents may attempt to lie about them in order to receive quicker services. Truthful mechanisms are robust to such manipulations. With d-utilities, we next characterise the class of truthful mechanisms.

Theorem 4. *With d-utilities, a feasible mechanism minimises totDET/ maxDET if and only if it is truthful.*

Proof. As agents have d-utilities, $a_j > 0$ and $b_j = 0$ hold for each agent $j \in [m]$. Let us consider mechanism $\mathcal{M}$ which minimises totDET/maxDET in each instance $\mathcal{I}$. We let $\mathcal{M}(\mathcal{I})$ denote one such minimal plan for $\mathcal{I}$. It follows that the detour time for each vehicle v_i is zero. Hence, $d_{ij} = 0$ and $u_{ij}(\mathcal{M}(\mathcal{I})) = 0$ hold for each agent j. This utility value is the maximum possible. As a result, agent j cannot increase it by acting strategically. Let us consider some truthful mechanism $\mathcal{M}$. Suppose that it does not minimise totDET/maxDET in instance $\mathcal{I}$. This means that $\mathcal{M}$ induces $d_{ij} > 0$ for some $i \in [n]$ and $j \in [m]$. This implies $p_j \neq d_j$ and $u_{ij}(\mathcal{M}(\mathcal{I})) < 0$. If agent j reports pick-up location p_j, misreports drop-off location $d'_j = p_j$, and assigns some value to c_j such that $T_j \leq c_j$ holds, then their utility in each feasible plan is zero. In this case, we note that $\mathcal{M}$ still sends a vehicle for agent j because $\mathcal{M}$ is feasible. Hence, $\mathcal{M}$ cannot be truthful. This is a contradiction. □

With d-utilities, each mechanism that minimises the waiting times is not truthful. Agents could increase their waiting times by misreporting their private locations. However, this could lead to a significant decrease in their detour times, despite the fact they may have to travel a bit longer to the new locations.

Theorem 5. *With d-utilities, there are instances where no feasible mechanism for totWAIT or maxWAIT is truthful.*

Proof. For $\epsilon \in (0, \frac{1}{2})$, let us consider locations $A(2\epsilon, 1)$, $B(0,0)$, $C(1,0)$ and $D(1,1)$. Mark the distances as $d(A,B) = 1 + 2\epsilon$, $d(B,C) = 1$, $d(C,D) = 1 - \epsilon$, $d(A,D) = 1 - 2\epsilon$, $d(B,D) = \sqrt{2} + \epsilon$, and $d(A,C) = \sqrt{2} + 2\epsilon$. We consider v_1 with $b_1 = D$, $e_1 = D$, $q_1 = 3$, and $\nu_1 = 1$. Further, we consider $r_1 = (C, B, 1)$, $r_2 = (B, A, 1)$, and $r_3 = (A, D, 1)$. Also, we let $\tau_1 = \tau_2 = \tau_3 = 0$ hold.

Suppose that agents are sincere. As $d(A,D) < d(C,D)$, one minimal plan sends v_1 to A, then to B and C. The waiting time in it is $6 - 2\epsilon$. For agent 3, the detour time is at least 2 and the waiting time is $1 - 2\epsilon$. This happens when v_1 goes to D after C. Otherwise, it is easy to see that their detour time can only increase. Hence, the 3's sincere utility is bounded from above by $-2 \cdot a_3 < 0$.

Suppose next that agent 3 is strategic and submits $r'_3 = (A', D, 1)$ such that (1) $A'(0, 1)$ and (2) $d(A', A) = 2\epsilon$, $d(A', D) = 1$, $d(A', B) = 1$, $d(A', C) = \sqrt{2}$. As $d(A', D) > d(C, D)$, one minimal plan now sends v to C, then to B and A'. The waiting time in it is $6 - 3\epsilon$. The only remaining service location is D. Hence, v_1 visits it.

For agent 3, the detour time is $0 < 2$ and the waiting time is $3 - \epsilon$. We note that $3 - \epsilon > 1 - 2\epsilon$ for $\epsilon \in (0, \frac{1}{2})$. Thus, by being strategic, agent 3 decreases their detour but waits longer. However, with d-utilities, they do not mind waiting. Nevertheless, they still need to travel from A to A' in order to catch the vehicle there. We will show that thus they will still increase their utility.

Indeed, as they have to travel from A to A', T_3 is proportional to the distance between these locations. Thus, we let $T_3 = \frac{2\epsilon}{\overline{\nu}}$ hold for some velocity $\overline{\nu} \in \mathbb{R}_{>0}$. This means that T_3 and their least utility $-T_3$ go to zero as ϵ goes to zero. Hence, for $\epsilon \in (0, \min\{\frac{1}{2}, \overline{\nu} \cdot a_3\})$, this utility is strictly greater than their maximum sincere utility of $-2 \cdot a_3$. □

With w-utilities, each mechanism that minimises the waiting times is also not truthful. Agents could increase their detour times by misreporting their private locations. However, this could lead to a significant decrease in their waiting times, even though they have to travel a bit longer to the new locations.

In addition, agents might wish to decrease their detour plus waiting times via location manipulations. They can do this as well. In fact, such manipulations are easy. The key idea is that getting closer to vehicle locations might decrease their waiting times and not change their detour times.

Theorem 6. *With w-utilities or (d+w)-utilities, there are instances where* no *feasible mechanism for totDET, maxDET, totWAIT, maxWAIT, totDETWAIT, or maxDETWAIT is truthful.*

Proof. Let us consider $A(0,0)$ and $B(0,1)$. We let $d(A,B) = 1$ hold. Further, pick one vehicle v_1. We let v_1 be such that $b_1 = A$, $e_1 = A$, $q_1 = 1$, and $\nu_1 = 1$. Pick also one request $r_1 = (B, A, 1)$. Suppose that $\tau_1 = 0$ holds. We consider two cases depending on whether agent 1 misreports their sincere pick-up location.

In the first case, suppose that agent 1 is sincere. Hence, $T_1 = 0$. Each feasible mechanism sends v_1 to B and then to A. This plan minimises the detour time to 0 and the waiting time to 1. The utility of agent 1 is $-b_1$. This utility is strictly negative because of $b_1 > 0$.

In the second case, suppose that agent 1 is strategic and submits $r'_1 = (B', A, 1)$ with (1) $B'(0, 1 - \epsilon)$ and (2) $d(A, B') = 1 - \epsilon$, $d(B', B) = \epsilon$, where $\epsilon \in (0, 1)$. Each feasible mechanism sends v_1 to B' and then to A. This plan minimises the detour time to 0 and the waiting time to $1 - \epsilon$. The least utility of agent 1 is now $-b_1 + b_1 \cdot \epsilon - T_1$.

However, to make it up for their decreased waiting time, agent 1 needs to travel from B to B'. The travel time is $T_1 = \frac{\epsilon}{\bar{\nu}}$ for some velocity $\bar{\nu} \in \mathbb{R}_{>0}$. As $\epsilon > 0$ and $b_1 > 0$, we conclude that their utility increases strictly for $\bar{\nu} \in (\frac{1}{b_1}, \infty)$. We can always find such $\bar{\nu}$. □

This result holds even if agents know only the vehicle locations, namely in Uber and Lyft settings. Perhaps, this justifies partly why shuttle companies such as GoCarma and GoAirlink do not reveal these locations. In these applications, customer location manipulations are therefore not possible.

Furthermore, the result applies also to systems where agents submit arbitrary combinations of preferences as long as at least one agent cares about their waiting time, i.e. at least one agent has a waiting-time preference and each other agent has either a detour-time preference or a waiting-time preference.

Lastly, agents can manipulate mechanisms for the waiting-time objectives, and deciding whether manipulations strictly increase their utilities relates to computing feasible plans. As we mentioned, doing so is shown to be intractable in general [1]. Consequently, agents might give up such manipulations.

8   Result Summary and Future Work

In this paper, we studied routing problems where vehicles service customers. For these problems, we looked at minimising the detour times and waiting times of the customers on vehicles. Additionally, we formulated the efficiency and truthfulness of mechanisms for this task. Thus, we characterised mechanisms that satisfy these properties wrt the customers' preferences for their detour times, waiting times, and detour plus waiting times. Table 1 contains the results.

Table 1. totDET/maxDET - minimising the total/maximum detour vehicle time; totWAIT/maxWAIT - minimising the total/maximum waiting vehicle time; totDET-WAIT/maxDETWAIT - minimising the total/maximum detour plus waiting vehicle time; PO - Pareto optimality; TR - truthfulness; d-utilities - detour-time preferences; w-utilities - waiting-time preferences; (d+w)-utilities - detour-plus-waiting-time preferences; ✓ - "satisfied in all instances"; × - "not satisfied in some instances".

totDET/maxDET	totWAIT/maxWAIT	tot/maxDETWAIT
PO for d-utilities, ✓ (Theorem 1)	PO for d-utilities, × (Example 1)	PO for d-utilities, × (Example 1)
PO for w-utilities, × (Example 1)	PO for w-utilities, ✓ (Theorem 2)/× (Example 4)	PO for w-utilities, × (Example 2)
PO for (d+w)-utilities, × (Example 1)	PO for (d+w)-utilities, × (Example 3)	PO for (d+w)-utilities, ✓ (Theorem 3)/× (Example 4)
TR for d-utilities, ✓ (Theorem 4)	TR for d-utilities, × (Theorem 5)	TR for d-utilities, × open
×, TR for w-utilities, (d+w)-utilities (Theorem 6)		

In the future, we will look at special cases such as the real line and trees and more general cases with time windows and service times. Also, we will study the equilibrium strategies of customers in our setting. Furthermore, we will study the loss in economic efficiency due to strategic behaviour. For this purpose, we will look at how we might apply the price of anarchy to our setting and quantify this loss.

Acknowledgements. Martin Aleksandrov was supported by the DFG Individual Research Grant on "Fairness and Efficiency in Emerging Vehicle Routing Problems" (497791398). Martin Aleksandrov thanks Prof. Christoph Benzmüller for their brief feedback on the motivation. We thank the reviewers for their valuable and constructive comments.

References

1. Afrati, F., Cosmadakis, S., Papadimitriou, C.H., Papageorgiou, G., Papakostantinou, N.: The complexity of the travelling repairman problem. RAIRO Theor. Inform. Appl. Inform. Théorique et Appl. **20**(1), 79–87 (1986). http://www.numdam.org/item/ITA_1986__20_1_79_0/

2. Aleksandrov, M.D.: Fleet fairness and fleet efficiency in capacitated pickup and delivery problems. In: Proceedings of the 32nd IEEE Intelligent Vehicles Symposium (IV21), 11–17. Nagoya, Japan, pp. 1156–1161. IEEE Xplore (2021). https://doi.org/10.1109/IV48863.2021.9576002

3. Araque G, J.R., Kudva, G., Morin, T.L., Pekny, J.F.: A branch-and-cut algorithm for vehicle routing problems. Ann. Oper. Res. **50**(1), 37–59 (1994). https://doi.org/10.1007/BF02085634

4. Bei, X., Zhang, S.: Algorithms for trip-vehicle assignment in ride-sharing. In: Proceedings of the AAAI Conference on Artificial Intelligence, vol. 32, no. 1 (2018). https://ojs.aaai.org/index.php/AAAI/article/view/11298

5. Cordeau, J.F., Laporte, G.: The dial-a-ride problem (DARP): models and algorithms. Ann. Oper. Res. **153**, 29–46 (2007). https://doi.org/10.1007/s10479-007-0170-8

6. Dantzig, G.B., Ramser, J.H.: The truck dispatching problem. Manage. Sci. **6**, 80–91 (1959). https://doi.org/10.1287/mnsc.6.1.80

7. Gschwind, T., Drexl, M.: Adaptive large neighborhood search with a constant-time feasibility test for the dial-a-ride problem. Transp. Sci. **53**, 480–491 (2019). https://doi.org/10.1287/trsc.2018.0837

8. Li, B., Krushinsky, D., Reijers, H.A., Woensel, T.V.: The share-a-ride problem: people and parcels sharing taxis. Eur. J. Oper. Res. **238**(1), 31–40 (2014). https://doi.org/10.1016/j.ejor.2014.03.003

9. Liu, M., Brynjolfsson, E., Dowlatabadi, J.: Do digital platforms reduce moral hazard? The case of uber and taxis. Manage. Sci. **67**(8), 4665–4685 (2021). https://doi.org/10.1287/mnsc.2020.3721

10. Matl, P., Hartl, R.F., Vidal, T.: Workload equity in vehicle routing problems: a survey and analysis. Transp Sci. **52**(2), 239–260 (2018). https://doi.org/10.1287/trsc.2017.0744

11. Nucamendi, S., Cardona-Valdes, Y., Angel-Bello Acosta, F.: Minimizing customers' waiting time in a vehicle routing problem with unit demands. J. Comput. Syst. Sci. Int. **54**(6), 866–881 (2015). https://doi.org/10.1134/S1064230715040024

12. Paquette, J., Bellavance, F., Cordeau, J.F., Laporte, G.: Measuring quality of service in dial-a-ride operations: the case of a Canadian city. Transportation **39**(3), 539–564 (2012). https://doi.org/10.1007/s11116-011-9375-4

13. Pareto, V.: Cours d'économie politique. Œuvres complètes publiées sous la direction de giovanni busino. tomes 1 et 2 en un volume (1897). https://doi.org/10.3917/droz.paret.1964.01, 9782600040143

14. Santos, D.O., Xavier, E.C.: Dynamic taxi and ridesharing: a framework and heuristics for the optimization problem. In: Proceedings of the Twenty-Third International Joint Conference on Artificial Intelligence IJCAI 2013, pp. 2885–2891. AAAI Press (2013). https://dl.acm.org/doi/10.5555/2540128.2540544

15. Vidal, T., Laporte, G., Matl, P.: A concise guide to existing and emerging vehicle routing problem variants. Eur. J. Oper. Res. **286**(2), 401–416 (2020). https://doi.org/10.1016/j.ejor.2019.10.010

16. Zhang, M.: Mechanism design in facility location games. In: Proceedings of the 20th International AAMAS Conference, pp. 1850–1852. AAMAS 2021, International Foundation for Autonomous Agents and Multiagent Systems, Richland, SC (2021). https://dl.acm.org/doi/10.5555/3463952.3464262

Reinforcement Learning

Human-Autonomous Teaming Framework Based on Trust Modelling

Wenhao Ma[✉], Yu-Cheng Chang, Yu-Kai Wang, and Chin-Teng Lin

Australian Artificial Intelligence Institute, University of Technology Sydney,
Ultimo, NSW 2007, Australia
Wenhao.Ma@student.uts.edu.au

Abstract. With the development of intelligent technology, autonomous
agents are no longer just simple tools; they have gradually become our
partners. This paper presents a trust-based human-autonomous teaming
(HAT) framework to realize tactical coordination between human and
autonomous agents. The proposed trust-based HAT framework consists
of human and autonomous trust models, which leverage a fusion mechanism to fuse multiple performance metrics to generate trust values in
real-time. To obtain adaptive trust models for a particular task, a reinforcement learning algorithm is used to learn the fusion weights of each
performance metric from human and autonomous agents. The adaptive
trust models enable the proposed trust-based HAT framework to coordinate actions or decisions of human and autonomous agents based on their
trust values. We used a ball-collection task to demonstrate the coordination ability of the proposed framework. Our experimental results show
that the proposed framework can improve work efficiency.

Keywords: Trust model · Human-autonomous teaming ·
Reinforcement learning

1 Introduction

The application of artificial intelligence technology is gradually reflected in
almost every aspect of our real world. At the same time, many scholars in the
field of machine learning have begun to pay attention to the problems brought
about by these applications. Many researchers believe that when autonomous
agents make decisions, proper control/decision by human agents is crucial, since

This work was supported in part by the Australian Research Council (ARC) under
discovery grant DP210101093 and DP220100803. Research was also sponsored in part
by the Australia Defence Innovation Hub under Contract No. P18-650825, US Office
of Naval Research Global under Cooperative Agreement Number ONRG - NICOP
- N62909- 19-1-2058, and AFOSR - DST Australian Autonomy Initiative agreement
ID10134. We also thank the NSW Defence Innovation Network and NSW State Government of Australia for financial support in part of this research through grant
DINPP2019 S1-03/09 and PP21-22.03.02.

autonomous agents cannot make correct decisions in all cases [12]. In this case, there are two major problems occurred when humans need to participate in an autonomous system of human-autonomy teams. 1) How to build an efficient human-machine interface. 2) How to assess human decisions to be reliable. To solve the two problems mentioned above, a trust model designed for human and autonomous agents is proposed in this paper to optimize the decision-making process among agents during the interaction.

Human-autonomous teaming framework has a broad application prospect, since it combines the advantages of both humans and machines, and is more suitable for complex industrial scenes and some dangerous and hazardous environments for human beings [8], such as rescue and disaster relief missions in extreme environments, sophisticated medical operations, and military missions, etc. Many approaches based on human-autonomous collaboration have achieved remarkable results [6,7]. In the case of human agents and autonomous agents working together, evaluating the performance and capabilities of both human and autonomous agents becomes a critical issue as it is a key factor in coordinating agents or assigning tasks. To enhance the credibility of human-autonomous teaming in these tasks, a few trust models based on statistical methods have been created.

In [13], Spencer et al. leveraged Switched Linear Quadratic Regulator (SLQR) to switch the control of the autonomous vehicle between human and autonomous based on the autonomous performance and human workload. Shahrdar et al. [11] proposed a data collection-based trust evaluation approach to study the effects of autonomous driving in different scenarios on human trust. To investigate the effects of human, autonomous agent, and environment factors on trust modeling, [3] made a qualitative analysis of the problem in the human-robot interaction domain. Wang et al. [15] introduced a mutual trust model according to the robot and human performance in the presence of faults and then designed a trigger control strategy based on this trust model. Their trust model considers fewer performance metrics, so it may not suitable for complex scenarios. [10] presented a time-series trust model for human-robot collaboration task. The trust values of human and robot are evaluated respectively according to the established performance models of robot and human. This method is similar to our trust model, but their performance models only apply to specific scenarios. The studies mentioned above are either based on statistical data or prior knowledge and therefore are not possible to adjust the trust metrics parameter configuration adaptively according to individual changes.

To eliminate the above limitations, this paper proposes a trust-based human-autonomous teaming (HAT) framework which consists of human and autonomous trust models. The trust models leverage a fusion mechanism to fuse multiple performance metrics to generate trust values in real-time. To generate adaptive trust models for specific tasks, a reinforcement learning algorithm is used to learn the fusion weights of each performance metric from human and autonomous agents without prior knowledge. The adaptive trust model enables the proposed trust-based HAT framework to coordinate actions or decisions of human and autonomous agents according to their trust values.

The rest of this paper is organized as follows. Section 2 describes the method of establishing the trust models, including the specific trust evaluation metrics contained in trust models and the trust metric fusion function we designed. Section 3, the experimental part, introduces the experimental scenario we established to verify the trust-based human-autonomous teaming (HAT) framework first, then the experimental procedure is described in detail, and followed by experimental results. Finally, the conclusion is given in Sect. 4.

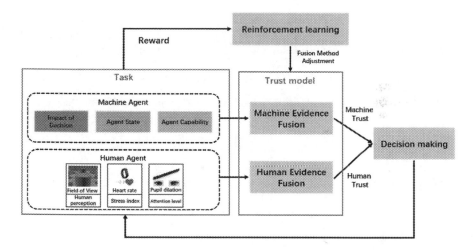

Fig. 1. Structure of the trust model based human-autonomous teaming framework

2   Method

This section presents the proposed trust model of autonomous and human agents in the HAT system respectively. Figure 1 shows the proposed HAT trust model framework. The trust model is implemented based on the agents' trust metrics including agent capability, agent state, the impact of decisions for machine agent and the human perception ability, the stress index and the attention level for the human agent, which are all acquired during the human-machine interaction. Then, we combine these trust metrics into a single trust value based on the evidence theory and fusion mechanism. Thus, in the following subsections, we first introduce the specific trust evaluation metrics contained in the human and autonomous agents trust model in detail, and then we further describe how to fuse these trust metrics through the trust metrics fusion function.

2.1   Trust Model for Human and Autonomous Agents

Trust Evaluate Metrics for Human Agents. The trust model for human agents takes three pieces of human trust evidence into consideration to evaluate human trust value in real-time. The three pieces of evidence are the attention level, stress index and human perception.

A. Attention level

Human attention level is employed as the first important metric for human state evaluation. The attention level is calculated based on the real-time pupil response of human agents, as Hoeks et al. [4] proposed that pupil response is an important feature for estimating the human attention concentration degree. To collect and monitor the human agent pupil response, Tobii Pro X2-30 screen-based eye tracker is used in our trust model study. The Hoeks-Levelt pupillary model [4] was then used to measure human agents' attention level $x(t)$:

$$y(t) = h(t) * x(t), \tag{1}$$

where $y(t)$ represents the pupillary response, $h(t)$ is defined as the impulse response, which is a system constant, $x(t)$ is the attention and $*$ is the convolution operator. The variables y, h and x are functions of time t. In this case, identical attention will be generated as the pupil changes identically every time. The relation between the pupillary response and attention is described by the impulse response constant $h(t)$. As presented by Hoeks's work, $h(t)$ is computed as:

$$h(t) = s \times (t^n) \times e^{\left(\frac{-n \times t}{t_{max}}\right)} \tag{2}$$

where n indicates the number of layers, t is the response time, t_{max} is used to represent the subject's maximum response time and s is a weight parameter for the pupillary response function scaling. In our trust model approach, we set $n = 10.1$, $t_{max} = 5000$ ms, $s = \frac{1}{10^{33}}$ respectively.

B. Stress index

The second metric is the human stress index. The reason for choosing the human stress index as a trust evaluation metric is that a person's performance on a task is affected by the level of stress she/he is exposed to. To obtain the accurate stress index of the participant, we choose heart rate variability (HRV) as the stress index evaluation signal in this paper. HRV is a common index used to assess the human's autonomic regulation [9]. It is obtained by measuring the interval of a series of consecutive cardiac cycles, named the inter-beat interval (IBI). Normally, under the regulation of the parasympathetic nervous system, the resting heart rate of a normal person is maintained at 60–70 beats per minute. When a person's stress increases, the sympathetic nervous system becomes more active, which in turn affects IBI duration and heartbeat rate. Our trust model assesses participants' stress levels by using distribution analyzes of IBI through the geometric method. We use the Empatica-E4 wristband to collect the participants' IBI data and then we use Baevsky's [1] stress index (SI) function to calculate the stress index value:

$$SI = \frac{AMo}{2 \times Mo \times MxDMn}, \tag{3}$$

where AMo is the amplitude modulation index in percent, Mo is the highest frequency of the RR interval mode selected and $MxDMn$ is the RR interval distribution range, which expresses the range of variation in the RR interval. To calculate the stress index of each participant with Baevsky's method, We divide the RR interval by 50 milliseconds, as shown in Fig. 2, AMo is calculated by dividing the number of beats contained in each interval by the total number of beats, Mo is defined as the median of the RR interval and $MxDMn$ is the size of the range from the largest RR interval to the smallest.

C. Human perception

Human perception is defined as the third metric. In the HAT scenario, human perception of the environment will be affected by factors such as viewing angle, obstacles, etc. In different visual perception states, human confidence in making decisions is also different. Thus, human perception is used to measure how confident human participants are in executing decisions. Human participants can observe two types of states in HAT scenarios, namely autonomy state and environment state. Since our HAT task in this study is target seeking, therefore, two autonomy situations and two environment situations are combined in pairs to form four different situations in the scenario, as listed in Table 1. For the autonomy state, we classify situations as agents or no agents based on whether human participants can observe the autonomous agent. In the same way, the environment state can be divided into two situations, target or no target, depending on whether humans can observe the target. The autonomy state is related to its position, orientation and distance, and the environment state is related to the distance from the target, so we define four attributes to measure the perception ability of humans, namely position ($S1$), orientation ($S2$), distance ($S3$) and view angle ($S4$). From this we can derive the human perception evaluation formula:

$$E_a = f(S1, S2, S3, S4) \qquad (4)$$

where E_a is the human perception value, and $f(\cdot)$ represents perception evaluation function. The specific calculation formula will be elaborated in the experimental part according to the experimental scenario design.

Table 1. Four situations of human perception.

Situations	Human perception
1	Agent + Target
2	No Agent + Target
3	Agent + No Target
4	No Agent + No Target

Trust Evaluate Metrics for Autonomous Agents. Just like the trust model for human autonomous agents, the trust model of autonomous agents also evaluates the trust values of each agent based on three pieces of metrics: agent state, the impact of decisions, and agent capability.

A. Agent state

The agent state is evaluated by the current state and the expected state. A good agent state is that the agent can reach the expected next state after receiving the current control command. Conversely, we define a bad agent state as when the agent cannot reach the next state as expected. The state of the agent may be affected by internal and external factors. We leverage the particle filter method to measure the agent state. Specifically, the particle filter can be used to predict the next timestep's state of the agent based on the agent's current state information, e.g., position and velocity. We calculate the agent state evaluation value by comparing the error between the predicted state and the actual state.

B. Impact of decision

The impact of a decision is the second metric of trust for autonomous agents. The agent typically makes decisions based on the environment states and its own state information, which are collected by various sensors equipped on it. The decision quality of the agent directly affects the performance of the overall system. An inappropriate decision often leads to chaos in an otherwise harmonious collaborative system. In our human-autonomous teaming system, we commonly choose distance sensor data as the basis for judging the quality of robot decisions. For example, we can evaluate the quality of a robot's decision by comparing the distance between the robot and the obstacle d_o and the corresponding threshold τ_d. We formally define the impact of decision metrics as follows:

$$E_t^p = \frac{\alpha_p + 1}{\alpha_p + \beta_p + 2} \tag{5}$$

where α_p and β_p are both updatable parameters, they are updated according to the value of d_o. If $d_o < \tau_d$, which means collisions may occur and the decisions of the autonomous agent may be anomalous. Thus, β_p is updated as $\beta_p \leftarrow \beta_p + 1$, which would cause E_t^p to decrease. On the contrary, if $d_o \geq \tau_d$, α_p is updated, $\alpha_p \leftarrow \alpha_p + 1$, which leads E_t^p to increase, indicating that the decision of the autonomous agent is reasonable.

C. Agent capability

The last metric is agent capability. Generally speaking, the capability of an agent is positively correlated with its experience, that is, an agent that has completed more tasks or explored a wider task space tends to be more capable. In our human-autonomous teaming system, we define the capability of an autonomous agent to be related to the area it has explored. As the agent continues to explore new areas, its experience is also accumulating. The more explored areas, the stronger its capability. Specifically, we divide the task space where cooperative tasks are performed into N_{env} blocks, and the number of blocks explored by the autonomous agent is $N_{explored}$. Then we can derive the agent capability equation as

$$E_t^e = \frac{N_{explored}}{N_{env}} \tag{6}$$

2.2 Trust Metric Fusion Function

Once we obtain the trust metrics of humans and autonomous agents, we can then generate the final human and autonomous agent trust values through fusion mechanism [2] based on these metrics. These trust values could help calibrate the final decision of the human and autonomous agents' cooperative task. The fusion mechanism in our trust model can be described as mapping multiple trust metrics values in the range $[0, 1]$ to the final trust value. Thus, the trust metric fusion method is defined as $F(\boldsymbol{E}) : [0, 1]^n \rightarrow [0, 1]$. We then leverage the fusion mechanism to combine the evidence with weights to generate trust values. Assume that the fusion result $F(\boldsymbol{E})$ meets the inequality $min(E_1, E_2, ..., E_n) \leq F(\boldsymbol{E}) \leq max(E_1, E_2, ..., E_n)$. We can define an aggregation function as follows:

$$F(\boldsymbol{E}) = \sum_{i=1}^{n} (E_i - E_{i-1}) * w_i \tag{7}$$

where $E = (E_1, E_2, ..., E_n) \in [0, 1]^n$ is an increasing permutation, that is, $0 \leq E_1 \leq E_2 \leq ... \leq E_n$, where $E_0 = 0$; $w = [w_1, w_2, ..., w_n]$ is the fusion weight vector, and $w_1 + w_2 + \cdots + w_n = 1$

Reinforcement Learning

In this part, we introduce the method of updating the fusion weight vector for the trust model. As mentioned above, the final output of the trust model is a single value obtained by weighted summation of each trust metric, the fusion weight vector is crucial for the model. We choose reinforcement learning methods [14] to deal with this problem since reinforcement learning does not depend on the environment model, as long as a reasonable reward function is set, it can learn by itself and get the best reward. Specifically, we use Q-learning to select the most appropriate weight vector for each agent in real-time, so that the output of the trust model can well reflect the actual state of the agent. The Q value is updated according to action i and state s at each step by the following formula:

$$Q(s, i) \leftarrow Q(s, i) + \alpha \times ((\omega(s, i) \times \nabla + \gamma \times \sum_{j=1}^{n} (\omega(s', j) \times Q(s', j)) - Q(s, i)), \tag{8}$$

where $\alpha, \gamma \in (0, 1]$, ∇ denote the learning rate, discount factor and reward in reinforcement learning algorithm respectively; the next state when taking action i at state s is represented by s' and $\omega(s, i)$ is the weight at step s and perform action i, which is updated by following method:

$$\omega'(s,i) \leftarrow \omega(s,i) + \begin{cases} (1 - \omega(s,i)) \times \delta \times (\frac{1}{1+e^{-a \times Q(s,i)+b}}), & if\, i = \arg\max_j Q(s,j) \\ (0 - \omega(s,i)) \times \delta \times (\frac{1}{1+e^{-a \times Q(s,i)+b}}), & otherwise \end{cases}$$

(9)

where $\delta \in (0,1]$ is a constant and $\sum_{i=1}^{n} \omega(s,i) = 1$. Next, we elaborate on designing reward functions in Q-learning.

Reward Function Design for Reinforcement learning

To explain the reward function, we take the trust fusion of autonomous agents as an example to illustrate the specific design in detail. First, we define two indicator functions to reflect the changes in the trust metrics and trust value at the current timestep and the previous timestep, respectively.

$$tag_1 = sgn(E_t^p - E_{t-1}^p) + sgn(E_t^e - E_{t-1}^e)$$

(10)

$$tag_2 = sgn(Tr_t - Tr_{t-1})$$

(11)

where $sgn(*)$ stands for the sign function; E_t^p and E_t^e represent the evaluated perception value and capability value of the autonomous agent at time t, respectively; Tr_t is the trust value at time t. The value of tag_1 reflects the change of the trust metrics evaluation value at time t. According to the characteristics of the sign function, the value of tag_1 is -2, 0, or 2. The value of tag_2 is -1, 0 or 1 according to the change of the trust value at time t. Based on the settings of $tag1$ and $tag2$, we define the reward function:

$$r_{fusion} = \begin{cases} reward_1, & if\, |tag_1 + tag_2| = val_1 \\ reward_2, & if\, tag_1 = val_2 \\ reward_3, & otherwise \end{cases}$$

(12)

Here, $val_1 = 3$ and $val_2 = 0$, The reward function means that, when both the trust value and the trust metrics have the same trend of change, the reward is $reward1 = 3$; when the changes of two trust metrics are different, the reward is $reward_2 = 0$ and otherwise the reward is $reward_3 = -3$. As for the human agent trust value fusion, we also implement it through a similar indicator function setting and reward function design.

3   Experiment

In this section, we first introduce the human-autonomous collaboration scenario designed for our experiment, then the experimental procedure is described in detail. Finally, we analyze the advantage of our method based on the experimental results.

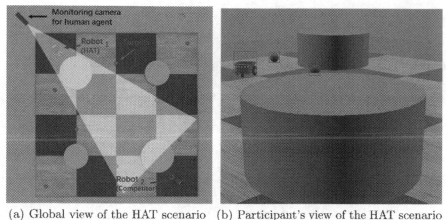

(a) Global view of the HAT scenario (b) Participant's view of the HAT scenario

Fig. 2. Scenario of the HAT task

3.1 Scenario Design

We design a human-autonomous agents collaboration scenario to evaluate our trust model in the Webots [5] robot simulation platform, as shown in Fig. 2. In this scenario, humans and autonomous agents need to cooperate to complete the task of searching and collecting balls. Figure 2(a) is the global view of the entire scenario, the green dots are the balls that need to be collected, and the gray cylinders are obstacles. Robot$_1$ represents teammates autonomous agents which collaborate with human agents as a HAT team and robot$_2$ represents an autonomous agent competitor that performs tasks independently. We set a fixed number of balls in the scenario as goals to be collected. During the experiment, the autonomous agent (robot$_1$) needs to cooperate with the human agent as a team and make decisions through the trust model in real-time to collect as many balls as possible, while the competitor (robot$_2$) performs the task under the control of the automatic controller, so our scenario is a mix of competition and cooperation.

3.2 Experimental Procedure

During the experiment, the human agent can only observe the state of the environment and the autonomous agents through the monitoring perspective as shown in Fig. 2(b). Since this perspective has limitations on the field of view, the credibility of human instruction will fluctuate depending on the environment state and the autonomy state in the field of view. The robot$_1$ can act based on its sensory information when no instructions are made by the human agent. The human agent plays the role of the supervisor to observe and monitor the environment. The autonomous agent, acting as executor and explorer, searches for targets based on instruction from human or their own decision according to the on-robot sensor data. The action of robot$_1$ is determined by trust-based decision-making in real-time. The other autonomous agent (robot$_2$) controlled by itself plays a competitor against the HAT team in this ball collection task. Both autonomous agents can find the ball through the on-robot camera, and the human agent can find the ball through the observing monitor on the side of the field. The goals can be identified through the autonomous agents' camera or human eyes. A team will gain one point after collecting one ball. The simulation ends when all balls arc collected and the side with the most balls wins.

3.3   Experimental Result

Table 2 shows the 10 trials of simulation results and one simulation result without the human agent involved. Games 1 to 10 present the results with the HAT team, while the game without human presents the results produced without human participation. All robots are controlled by themselves when no human is participating in the simulation. The HAT team took seven out of ten games; two games resulted in a draw score, and Robot$_2$ took one game. The average time to finish the task across Games 1 to 10, which involves the HAT team, is 127.5 sec. In comparison, the time consumed by the game excluding the human agent is 136.8 sec. We, therefore, can conclude that the efficiency of the task was improved by adding trust-based HAT decision-making.

Table 2. Experiment results.

Game	Without human	1	2	3	4	5	6	7	8	9	10	Average across Game 1 to 10
Time to complete the task (sec)	136.8	128.3	120.3	130.6	125.9	132.7	127.0	128.4	122.6	124.8	133.9	127.5
Winner	Draw	HAT	HAT	Draw	HAT	Draw	HAT	HAT	HAT	HAT	Robot$_2$	–

4   Conclusion

This paper developed a trust model for the HAT system, which is based on the trust metrics of humans and autonomous agents. The fusion function is used to fuse trust metrics together to generate the trust value. Moreover, we leverage reinforcement learning to adaptively adjust the weight and obtain the best weight for each trust metric. Thus, the trust value generated in this way can accurately reflect the real-time status and trustworthiness of human and autonomous agents. To verify the effectiveness of our trust model, we designed an experimental scenario for human-autonomous agents collaboration in the simulation environment. The experimental results show that our trust model can effectively improve the efficiency of the HAT tasks. Compared with purely autonomous agents, human-autonomous collaborative work can take advantage of both human and autonomous agents, and dynamically adjust decision-making according to human and autonomous agents' state fluctuations caused by changes in the environment in real-time. In future work, we will consider to introducing more trust metrics into our trust model, especially those metrics that can better reflect the real-time state of human agents. For example, employing electrical brain waves (EEG-signals) as evidence for real-time trust evaluation of human agents is a promising direction.

References

1. Baevsky, R., Bennett, B., Bungo, M., Charles, J., Goldberger, A., Nikulina, G.: Adaptive responses of the cardiovascular system to prolonged spaceflight conditions: assessment with holter monitoring. J. Cardiovasc. Diag. Proc. **14**(2), 53–57 (1997)
2. Dimuro, G.P., et al.: The state-of-art of the generalizations of the choquet integral: from aggregation and pre-aggregation to ordered directionally monotone functions. Inf. Fusion **57**, 27–43 (2020)
3. Hancock, P.A., Billings, D.R., Schaefer, K.E., Chen, J.Y., De Visser, E.J., Parasuraman, R.: A meta-analysis of factors affecting trust in human-robot interaction. Hum. Fact. **53**(5), 517–527 (2011)
4. Hoeks, B., Ellenbroek, B.A.: A neural basis for a quantitative pupillary model. J. Psychophysiol. **7**, 315–315 (1993)
5. Michel, O.: Cyberbotics ltd. webots™: professional mobile robot simulation. In: J. Adv. Rob. Syst. **1**(1), 5 (2004)
6. Mutlu, B., Terrell, A., Huang, C.M.: Coordination mechanisms in human-robot collaboration. In: Proceedings of the Workshop on Collaborative Manipulation, 8th ACM/IEEE International Conference on Human-Robot Interaction, pp. 1–6. Citeseer (2013)
7. Nicora, M.L., Ambrosetti, R., Wiens, G.J., Fassi, I.: Human-robot collaboration in smart manufacturing: Robot reactive behavior intelligence. J. Manuf. Sci. Eng. **143**(3) (2021)
8. O'Neill, T., McNeese, N., Barron, A., Schelble, B.: Human-autonomy teaming: a review and analysis of the empirical literature. Human Fact., 0018720820960865 (2020)
9. van Ravenswaaij-Arts, C.M., Kollee, L.A., Hopman, J.C., Stoelinga, G.B., van Geijn, H.P.: Heart rate variability. Ann. Internal Med. **118**(6), 436–447 (1993)
10. Sadrfaridpour, B., Saeidi, H., Burke, J., Madathil, K., Wang, Y.: Modeling and control of trust in human-robot collaborative manufacturing. In: Mittu, R., Sofge, D., Wagner, A., Lawless, W.F. (eds.) Robust Intelligence and Trust in Autonomous Systems, pp. 115–141. Springer, Boston, MA (2016). https://doi.org/10.1007/978-1-4899-7668-0_7
11. Shahrdar, S., Park, C., Nojoumian, M.: Human trust measurement using an immersive virtual reality autonomous vehicle simulator. In: Proceedings of the 2019 AAAI/ACM Conference on AI, Ethics, and Society, pp. 515–520 (2019)
12. Shively, R.J., Lachter, J., Brandt, S.L., Matessa, M., Battiste, V., Johnson, W.W.: Why human-autonomy teaming? In: Baldwin, C. (ed.) AHFE 2017. AISC, vol. 586, pp. 3–11. Springer, Cham (2018). https://doi.org/10.1007/978-3-319-60642-2_1
13. Spencer, D.A., Wang, Y.: SLQR suboptimal human-robot collaborative guidance and navigation for autonomous underwater vehicles. In: 2015 American Control Conference (ACC), pp. 2131–2136. IEEE (2015)

14. Sutton, R.S., Barto, A.G.: Reinforcement Learning: An Introduction. MIT press, Cambridge (2018)
15. Wang, Y., Shi, Z., Wang, C., Zhang, F.: Human-robot mutual trust in (semi)autonomous underwater robots. In: Koubaa, A., Khelil, A. (eds.) Cooperative Robots and Sensor Networks 2014. SCI, vol. 554, pp. 115–137. Springer, Heidelberg (2014). https://doi.org/10.1007/978-3-642-55029-4_6

Using Uncertainty as a Defense Against Adversarial Attacks for Tabular Datasets

Poornima Santhosh[1], Gilad Gressel[1(✉)], and Michael C. Darling[2]

[1] Center for Cybersecurity Systems & Networks Amrita Vishwa Vidyapeetham,
Amritapuri, India
`gilad.gressel@am.amrita.edu`
[2] Sandia National Laboratories, Albuquerque, USA
`michael.darling@sandia.gov`

Abstract. Adversarial examples are a threat to systems that use machine learning models. Considerable research has focused on adversarial exploits using homogeneous datasets (vision, sound, and text) while primarily attacking deep learning models. However, many industries such as healthcare, business analytics, finance, and cybersecurity rely upon heterogeneous (tabular) datasets. The attacks which perform well on homogeneous datasets do not extend to heterogeneous datasets due to feature constraints. Therefore, tabular datasets require different forms of attack and defense mechanisms. In this work, we propose a novel defense against adversarial examples built from tabular datasets. We use an uncertainty metric, the Minimum Prediction Deviation (MPD), to detect adversarial examples generated by a tabular evasion attack algorithm, the Feature Importance Guided Attack (FIGA). Using MPD as a defense we are able to detect 98% of the adversarial samples with a 7.8% false positive rate on average.

Keywords: Machine learning · Adversarial defenses · Adversarial examples · Tabular datasets · Uncertainty

1 Introduction

Machine learning (ML) is quickly becoming ubiquitous in the world of technology in all aspects, such as business, healthcare, and transportation. However, little attention is being given to the security of machine learning models, even though researchers have discovered a variety of ways to exploit machine learning models. Adversarial attacks on machine learning models come in many forms, such as data poisoning, model inversion, evasion attacks, and membership inference, each of which requires a different form of defense.

Generally speaking, an adversarial attack is any form of exploit executed by a malicious party to cause harm or evade a machine learning system. The most common type of attack is known as the evasion attack. In an adversarial evasion attack, the attacker crafts a small perturbation that causes the machine learning model to misclassify when added to the input. Thus the input sample 'evades'

H. Aziz et al. (Eds.): AI 2022, LNAI 13728, pp. 719–732, 2022.
https://doi.org/10.1007/978-3-031-22695-3_50

the model. The samples generated in these attacks are known as 'adversarial examples'.

Adversarial examples have been shown to elude models with high success rates, making it problematic to rely upon machine learning models in security-critical areas. In recent years, several studies examined various adversarial attacks and potential defense mechanisms, but a good defense against adversarial attacks is still an unsolved problem [3].

Most research has focused on deep neural networks (DNN) and their associated homogeneous datasets, such as images, sound, and text. While many adversarial attacks and defenses have been formulated for these homogeneous datasets and DNNs, very few studies have been undertaken on tabular heterogeneous datasets [11, 18–20].

Heterogeneous datasets are tabular datasets, and due to the feature constraints on these datasets, we need to use modified or different adversarial algorithms. For example, most adversarial attacks on images assume that all pixels can be modified (the perturbation may be performed on any or all pixels). However, in tabular datasets, some features may be immutable and impossible for an adversary to perturb. Consider medical records; an immutable feature would be the patient's blood type. It can be digitally manipulated in feature space, but a blood type cannot be modified in real life. Similarly, any perturbation must be feasible. A person's birth date cannot occur after their death date [11]. These types of constraints require additional efforts when crafting adversarial examples.

Recent studies have proposed various defense mechanisms for detecting adversarial examples. However, they are restricted to DNNs and computer vision. The defense methods focus on one of two main objectives: detecting adversarial examples [5, 12] or increasing the robustness of the defending model [7, 16]. The defenses are either designed to mitigate a particular attack or attack agnostic, where prior knowledge of the attack is not required. A popular strategy is adversarial training, creating a dataset with a mix of clean and perturbed samples (correctly labeled) to enhance a model's robustness to attack [7, 16]. Other strategies, such as gradient masking, feature denoising, and defense distillation, are all only suitable for DNNs and homogeneous datasets.

This paper uses uncertainty to detect adversarial attacks on tabular datasets. We select the Feature Importance Guided Attack (FIGA) as our evasion attack against tabular datasets [8]. We adopt the Minimum Prediction Deviation (MPD) as an uncertainty metric, which we use to detect the adversarial examples [4]. We show that a sample that has been perturbed has an increased MPD score (higher uncertainty) which allows for detection.

We validate our approach by attacking a phishing dataset that contains 348,739 samples. A random forest model trained on this dataset obtains 97.2% recall and 97.1% f1-score. When we send the adversarial examples to this model, its recall is lowered to 19.3%, which indicates a 77.9% success rate of the attack. Using MPD as a detection method, we can detect 99.4% of the adversarial examples, with a 7.8% false positive rate on average.

Further, we show that false positives generated are highly uncertain samples that the random forest classifier obtains a 73.1% f1-score in classifying in contrast to its baseline 97.1% f1-score. The model struggles to classify these highly uncertain samples. This demonstrates a side-effect of using uncertainty to detect adversarial samples; even in failure, it is discovering difficult to classify samples that warrant a deeper inspection. We believe that MPD is a promising defense mechanism to detect adversarial examples and improve the model's overall performance by flagging highly uncertain samples.

We see the following as our novel contributions:

- We apply MPD to the detection of tabular adversarial examples, establishing that adversarial samples contain a higher level of uncertainty than normal samples
- We demonstrate the effectiveness of MPD as a detection mechanism over a wide range of attack strengths.
- We show that samples with high values of MPD are difficult to classify, even if they are not adversarial.

2   Related Work

It has been found that adversarial examples easily compromise all machine learning algorithms. While a significant amount of research has been performed on the creation of adversarial attacks, a much smaller amount of research has explored defense mechanisms against adversarial attacks. Creating a practical defense against adversarial examples has been challenging for the following reasons:

- The majority of attack algorithms generate adversarial samples through a complex optimization process. As a result, developing a framework for defending against all adversarial samples is challenging because there is always a new unique attack on the horizon, which evades the defense [7,9].
- Many defenses which increase the robustness of a model to adversarial attacks degrade the model's overall performance. There is an inherent trade-off between being adversarially robust and having high-performance [21].

Numerous adversarial defenses have been presented, but shortly after, the same defenses have been quickly broken and found to be unadaptable to the new attacks. With a basic understanding of the defense architecture, an attacker may quickly develop an attack to overcome it [2].

While we are focusing on defending against adversarial attacks on heterogeneous datasets, there are very few heterogeneous attacks and no known defenses. Therefore, we will present the most popular defenses against homogeneous attacks and their shortcomings.

2.1   Adversarial Training

Adversarial training is one of the earliest defenses. The main objective is to train a model by injecting correctly labeled adversarial examples and clean samples into the training dataset to increase the model's robustness. It has been described as a conventional brute force strategy due to the large amount of data used for the defending model.

Madry et al. presented a theoretical analysis on adversarial training [10] where they trained a neural network using the local first-order information against a strong adversary, the Projected Gradient Descent (PGD). They use PGD to demonstrate the robustness of a deep learning model on the MNIST dataset. However, later studies have shown the framework's limitations [13].

Gau et al. [6] provided a theoretical justification for why adversarial training produces minimal resilience loss using Neural Tangent Kernels and online learning methods. They showed that extra model capacity was necessary for robust interpolation. However, their method only applied to networks of exponential width and runtime.

Even though adversarial training makes a model more resilient, it is still a non-adaptive strategy because it needs either realistic attack scenarios or knowledge to train the model. As a result, it does not apply to undiscovered or novel attacks.

This defense mechanism was ineffective in the case of a black-box attack scenario because the adversarial examples were generated using a locally trained substitute model. Zhang et al. [22] demonstrated that adversarial training defense was vulnerable to blind-spot attacks. The presence of blind spot attacks makes adversarial training defense difficult due to a scarcity of training sets and the curse of dimensionality. The defense was also vulnerable to more complex multi-staged attacks in which random perturbations are transferred and attacked with any classical attack technique such as FGSM, etc.

2.2   Feature Denoising

This defense strategy identifies noise-free images and uses their information to reduce noise in the pixel space while preserving pixel details. It is added during adversarial training to improve classification accuracy on adversarial images. Xie et al. [21] proposed a feature denoising defense strategy to improve the robustness of Convolutional Neural Networks (CNN) against adversarial attacks. The CNN was constructed using non-local means filters to denoise layer outputs. Adversarial-generated samples were used for end-to-end network training to minimize feature-map perturbations.

One disadvantage of this method was that it was time-consuming. It also demonstrated that the trained model was differential, making it vulnerable to white-box attacks.

2.3   Uncertainty Metrics

Uncertainty in a model, particularly in deep learning models, is measured by adding randomness to the model or removing it via the network's hidden layers. Most researchers leverage dropout to measure the uncertainty of a neural network.

Smith and Gal [15] examined several uncertainty metrics such as mutual information, predictive entropy, and the softmax variance to detect adversarial examples. Probabilistic ensemble models had improved the quality of uncertainty estimators. However, dropout alone did not provide a convincing defense against adversarial examples as it fails to capture the entire Bayesian uncertainty in the visualizing gaps of model uncertainty. They used softmax variance to assess epistemic uncertainty as a substitute for mutual information and were found to perform better than other uncertainty measures.

Sheikholeslami et al. [14] introduced a technique where the hidden layers of the deep learning models were randomly selected based on the Bayesian Uncertainty where the distance of a deep learning model's uncertainty differs from clean data to the in-distribution of the adversarial examples. In order to estimate uncertainty, they introduced a layer-wise minimum variance solver and used a mutual information-based threshold. The uncertainty of the input image was computed at the inference time using the outputs from the hidden layers. If the mutual information of the input sample exceeds the threshold, then the sample is declared adversarial.

In order to detect adversarial samples, O.F Tuna et al. [17] proposed a defense mechanism using aleatoric and scibilic uncertainty. They proved that using moment-based prediction uncertainty along with the closeness between the input sample's representation to that of the predictive class distribution in the subspace of the last hidden layer activation can be a successful defense mechanism against adversarial attacks.

From all the above work, it is to be noted that the existing and ongoing research is focused on deep learning environments using homogeneous datasets.

In this work, we propose a novel defense mechanism using the Minimum Prediction Deviation (MPD) uncertainty metric. We use this metric to measure the uncertainty of a phishing detection model's predictions on samples from a tabular dataset. One strong advantage of MPD is that it is used as a detection method. Therefore it can harden all machine learning models.

3   Background

3.1   Homogeneous vs Heterogeneous Datasets

Adversarial attacks and their defenses have different requirements for homogeneous or heterogeneous datasets. Homogeneous datasets are composed of sound, vision, and text data with semantically identical features. For instance, all features of an image are pixels whose values are continuous and bounded at the same range (0–255). Conversely, heterogeneous (tabular) datasets contain categorical, numerical, and nominal features, which are often missing values. Some domains that utilize heterogeneous datasets are healthcare assessments, business data analytics, finance, and cybersecurity [11].

Adversarial attacks are more easily performed on homogeneous datasets, where the perturbation of the feature can be applied equally to all features. In short, the attack can focus on one feature (e.g., a pixel) and be applied to the entire sample. However, it is different in the case of heterogeneous datasets, where features are not standardised. The perturbation may not necessarily be viable for all features in a tabular dataset because perturbations may create invalid features. For tabular features to remain valid, there may be some limitations, such as the age of a person cannot be negative, and certain features may be immutable in a tabular dataset. Creating an attack on tabular datasets requires modifying the perturbations to be valid for all feature columns.

3.2 Feature Importance Guided Attack (FIGA)

Feature Importance Guided Attack (FIGA) is an evasion attack designed for heterogeneous (tabular) datasets [8]. FIGA is model agnostic and gradient-free. It neither requires prior knowledge of any learning algorithm nor any gradient information. It takes three parameters: fi, the feature importance ranking algorithm, n, the total number of features to be perturbed, and ϵ, the amount of perturbation to be added to the input sample.

The algorithm is divided into two steps. The first phase ranks the most significant features of the dataset and finds the direction in which features must be perturbed. The intuition is that we would like to move the input class toward the target class. Therefore, the direction of the perturbation is determined by comparing the input class's mean feature values to the target class's mean feature values.

The second phase of the algorithm involves adding a perturbation to the important features of the input samples. We select n features, which are evenly manipulated by adding a perturbation of size ϵ in the direction of the target class. This has the effect of "moving" the input samples in the direction of the target class. Gressel et al. demonstrated that FIGA can achieve 97% success rates against four different datasets trained with five different machine learning models. For full details of the attack, please refer to [8].

3.3 Minimum Prediction Deviation (MPD)

Minimum Prediction Deviation (MPD) is a metric that measures the uncertainty of a machine learning model's prediction of a single sample [4]. This measure of uncertainty is based on the distribution of probabilistic predictions generated by an ensemble of bootstrapped estimators. MPD measures the inconsistencies within the model's predictions in a quantifiable manner. Consider the scenario where two ML models may have varying probabilistic distributions. The distributions could be bi-modal and uniform, yet both share the same mean. MPD allows us to quantify these variations in the distributions as a single metric.

MPD has two key steps:

– *Bootstrapping*: First, we create different sample subsets from the original dataset with replacement. We create and train an ensemble of n classifiers using these datasets. After constructing the ensemble of classifiers, we will use their probabilistic predictions to calculate the MPD score per sample.

– *ProbabilityDistribution*: In order to calculate our MPD score we iterate over all classifiers in the ensemble and obtain a vector of n probabilistic predictions per sample.

Algorithm 1. Bootstrapping w/ Replacement

    Input: $X, y, Classifier, n$
    Output: *ensemble*

1: *ensemble* ← []
2: **for** $i \leftarrow 0, n$ **do**
3:     $X^*, y^* \leftarrow resample(X, y, replace = True)$
4:     $clf \leftarrow clone(classifier)$ ▷ new unfitted estimators
5:     $C* \leftarrow clf.train \ (X^*, y^*)$
6:     *ensemble*.append($C*$)
7: **end for**
8: **return** *ensemble*

MPD is a modified form of the standard deviation equation. Standard deviation is the square root of the expectation E of the squared deviation of a random sample x from its mean. Equation (1) is the standard deviation.

$$\sigma(q(x)) = \sqrt{E[(q(x) - q^*(x))^2]} \tag{1}$$

The goal is to determine the uncertainty of a sample x classified from the probability distribution $q(x)$. Therefore, the equation of the standard deviation (1) is modified such that it quantifies the deviation of the distribution $q(x)$ from 0 and 1 class labels.

$$U_0(x) = \sqrt{E[(q(x) - 0)^2]} \tag{2}$$

$$U_1(x) = \sqrt{E[(q(x) - 1)^2]} \tag{3}$$

U_0 (2) and U_1 (3) quantify the deviation of the probabilistic distribution $q(x)$ from 0 and 1, which represents the two class labels.

The Minimum Prediction Deviation (MPD) is defined as the minimum of these two deviations, such that a low MPD score represents low uncertainty and a high MPD represents high uncertainty.

$$U_q(x) = \min[U_0(x), U_1(x)] \tag{4}$$

If $U_0(x) < U_1(x)$, the distribution is clustered closer to 0 and if $U_1(x) < U_0(x)$, then the distribution mass is clustered closer to 1. Whichever value is lower (U_0 or U_1) represents the uncertainty of the prediction that will be made by the classifier. Therefore, U_q (the MPD) represents the uncertainty of the prediction

---

Algorithm 2. Minimum Prediction Deviation (MPD)

Input: $x, ensemble$
Output: $U_q(x)$

1: $Prob\_predict \leftarrow []$
2: **for** clf in $ensemble$ **do**
3:     $p \leftarrow clf.predict\_proba(x)$
4:     $Prob\_predict.\text{append}(p)$
5: **end for**
6: $q(x) \leftarrow Prob\_predict$ ▷ $Prob\_predict$ is the sampled $q(x)$
7: $U_0(x) = \sqrt{E[(q(x) - 0)^2]}$
8: $U_1(x) = \sqrt{E[(q(x) - 1)^2]}$
9: $U_q(x) = \min[U_0(x), U_1(x)]$
10: **return** $U_q(x)$

---

for a single sample. Darling M. proves that U_q increases as the uncertainty increases [4].

The general steps for creating an ensemble of classifiers are shown in Algorithm 1. The input consists of a sample X, a label vector y, and n, the number of bootstrapped models to be created. The algorithm is an n-step loop in which we bootstrap a dataset, fit a model classifier, and save it for later use. We compute the MPD score by using the prediction of the ensemble of classifiers created. We show the steps required to compute the MPD score in Algorithm 2. It accepts a classifier ensemble as input and x as a data point. We iterate over all classifiers in the ensemble to obtain a probabilistic prediction on sample x to calculate the MPD.

4   Methodology

4.1   Approach

Machine learning models will be more uncertain about adversarial examples than normal samples. We want to use MPD to detect adversarial examples. To test the hypothesis, we measure the MPD of each sample present in the dataset and compare them to the MPD of the adversarial examples (i.e., samples when perturbed) created. Then we calibrate a threshold value such that if the MPD of the adversarial examples is higher than the threshold value, those samples are considered adversarial. We measure the performance of the MPD detector with f1-score, precision, recall, and the false-positive rate. We use a phishing dataset as our testbed to perform all our experiments. Finally, we build a phishing detection classifier with a random forest model. We use this classifier to evaluate the strength of the FIGA attack. We require knowledge of the FIGA attack's success rate to evaluate the MPD defense.

4.2 Data Collection

We use the dataset collected by Gressel et al. [8]. They have collected benign and phishing website data using a Selenium-based web crawler. The phishing data URLs in the datasets were taken from PhishTank during the year 2019–2020. PhishTank.com is a website where users report and validate phishing sites; it is considered a ground truth in the academic phishing community. The legitimate URLs in the dataset were collected from Tranco, which is combined from four URL ranking lists: Alexa, Majestic, Quantcast, and Umbrella. The logic is that websites that are frequently visited will be benign. 348,739 URLs were collected, and 52 features were extracted from the source code.

5 Experiments

5.1 Experiment 1: Relationship Between MPD and FIGA

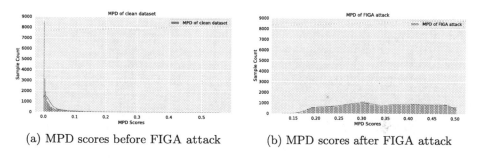

(a) MPD scores before FIGA attack   (b) MPD scores after FIGA attack

Fig. 1. MPD scores before and after a FIGA attack

In Fig. 1 we display a histogram of MPD scores on clean unperturbed data and compare it to a histogram of MPD scores on FIGA perturbed data (all data was perturbed). The distribution of MPD scores shifts dramatically to the right when under attack; that is, the scores increase. When there is no attack, the MPD distribution is primarily near 0, with over 7500 samples containing zero uncertainty. When FIGA perturbs the same data, the MPD mass increases and nearly all samples have a score of more than .18. This indicates that an MPD threshold can be used to detect adversarial attacks successfully.

5.2 Experiment 2: Evaluation of MPD Detection

We prepare a test set of 69,748 samples. 50% of the samples are benign, and 50% malicious. We use FIGA to perturb half of the phishing samples. This creates a final set that has 25% adversarial samples, 25% phishing samples, and 50% benign samples. The exact numbers are given in Table 1.

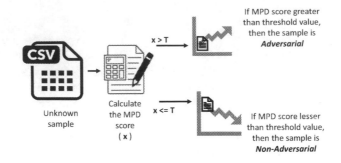

Fig. 2. Detecting if an unknown sample is adversarial or not

We want to select an MPD threshold value that can be used to classify an unknown sample as adversarial or not. If the uncertainty of a sample is greater than the threshold, it will be labeled as adversarial; if it is lower than the threshold, it will be labeled as normal. To determine the best threshold value for a given attack, we conduct experiments over a range of thresholds varying from 0.01 to 0.77 (the maximum MPD score possible) while varying the strength of the attack.

Table 1. Breakdown of Test Dataset used for Evaluation

	Benign	Phishing	Adversarial Phishing	Total
# Samples	42,053	13,848	13,847	69,748

We calculated the MPD scores for each attack for the entire test set. Next, we select an MPD threshold and divide the samples into positive (adversarial) and negative (normal) classes. We then use these predictions to compute a set of metrics to evaluate the performance of the MPD detection. The process is illustrated in Fig. 2

Note that both unperturbed phishing and benign samples are not adversarial and are considered the negative class for this experiment. We aim only to detect the positive class, the adversarial phishing samples.

We examined the recall, precision, f1-score, and false positive rate of the samples MPD selected as adversarial. The recall score tells us exactly how many adversarial samples were detected. However, maximizing recall while sacrificing the false positive rate is easy. In order to have a balance between detection and precision, we selected the threshold which maximized the f1-score, the harmonic mean of precision and recall. This would ensure a reasonable detection rate with a high level of precision (correlating to a lower level false-positive rate).

In order to understand if the ideal threshold rate would change based on the strength of the attack, we iteratively increased the strength of the attack while testing all threshold values - selecting the threshold which maximizes the

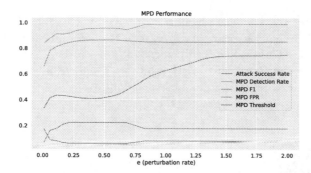

Fig. 3. The security evaluation curve for the FIGA attack shows us how MPD responds as the attack strength increases. We set $n = 10$ (10 perturbed features) and we increased the amount of perturbation e from 0.01, to 2.0. The larger e grows, the stronger the attack. The success rate is measured against a random forest phishing detection classifier

f1-score. This is a security evaluation curve, which examines the detection performance as the attack strength increases [1]. In our case, we examine both performances of detection and the ideal threshold level. This experiment allows users to determine the best threshold for their risk model.

6   Results and Analysis

The security evaluation curve is presented in Fig. 3. We used the FIGA parameters n=10 and applied a range of e from 0.01 to 2.0 as e grows, the perturbation amount increases. The attack success rate (calculated by sending attack samples to the random forest classifier) also increases.

There are several interesting things to note. The f1-score and detection rate remain stable as the attack strength and success rate increase. The MPD f1-score averages .83, and the MPD detection rate (recall) at .96. This is satisfactory as it indicates that 96% of adversarial samples are detected.

Table 2. We examine the performance of the false positives generated by the MPD detection. PDC F1: The f1-score for the phishing detection classifier.

		PDC F1	
	MPD FPR	Only MPD's FP	Full Data
Mean	7.9%	0.738	0.976

The average false positive rate is 7.8%. However, the samples that are falsely classified as adversarial are highly uncertain samples (they have a high MPD). In Table 2, we drill down into the false positives selected by MPD. When classifying *only* the false positive samples, the phishing detection classifier (PDC) yields an f1-score of .738, which is below its typical .976. This indicates that the false positive samples are, in fact, difficult to classify due to their higher uncertainty. This is an added benefit of using MPD as a detection mechanism; when it falsely labels samples, those samples should be examined regardless due to their higher uncertainty.

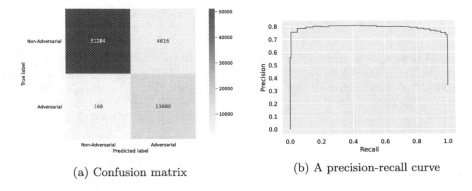

(a) Confusion matrix   (b) A precision-recall curve

Fig. 4. (a) Confusion matrix generated with the perturbation rate $e = 1.5$ and MPD threshold $= 0.178$. (b) A precision-recall curve of the perturbation rate $e = 1.5$. This curve is generated by varying the MPD decision threshold

Figure 4(a), is a confusion matrix obtained with a perturbation rate of 1.54 and an MPD threshold of 0.18. It is helpful to examine the breakdown of errors made. We see that our method has more false positives than false negatives-this is important as it is evidence of a very low false negative rate and high recall.

In Fig. 4(b), we display a precision recall curve generated by varying the MPD threshold. Looking at the curve, we see that we can achieve a reasonable AUC. We can achieve a high recall without sacrificing much precision.

7   Conclusion and Future Work

Adversarial examples elude models with high evasion rates. These vulnerabilities make it challenging to deploy machine learning models into security-critical areas. It is an ongoing research problem to find effective defensive mechanisms to mitigate the threat of adversarial examples. In this work, we developed a novel defense for tabular datasets.

This work utilized Minimum Prediction Deviation (MPD) as a defensive mechanism against adversarial examples. By using the MPD, we quantify the uncertainty per sample so that we can differentiate between adversarial examples and normal samples. We demonstrated that even the false positives from the MPD detection are worth examining since they contain a high uncertainty, and the phishing detection model performed poorly at classifying them.

We strongly believe that MPD has a good defense for detecting adversarial samples on many datasets, not only the phishing testbed we experimented with. Further, we would like to see an adoption of uncertainty analysis for all samples, regardless of a perceived threat. We believe this would increase the effectiveness of all machine learning models.

In future work, we would like to explore using MPD to detect adversarial samples in additional tabular datasets and discover if MPD can be useful for homogeneous datasets.

References

1. Biggio, B., Roli, F.: Wild patterns: ten years after the rise of adversarial machine learning. In: Proceedings of the 2018 ACM SIGSAC Conference on Computer and Communications Security (2018)
2. Carlini, N., Wagner, D.: Adversarial examples are not easily detected: bypassing ten detection methods. In: Proceedings of the 10th ACM Workshop on Artificial Intelligence and Security, pp. 3–14 (2017)
3. Chen, K., et al.: A survey on adversarial examples in deep learning. J. Big Data **2**(2), 71 (2020)
4. Darling, M.C.: Using uncertainty to interpret supervised machine learning predictions. In: (2019)
5. Deng, Z., et al.: Libre: a practical bayesian approach to adversarial detection. In: Proceedings of the IEEE/CVF Conference on Computer Vision and Pattern Recognition, pp. 972–982 (2021)
6. Gao, R., et al.: Convergence of adversarial training in overparametrized neural networks. In: Advances in Neural Information Processing Systems, vol. 32 (2019)
7. Goodfellow, I., Shlens, J., Szegedy, C.: Explaining and harnessing adversarial examples. In: International Conference on Learning Representations (2015)
8. Gressel, G., et al.: Feature importance guided attack: a model agnostic adversarial attack. arXiv preprint arXiv:2106.14815 (2021)
9. Lyu, C., Huang, K., Liang, H.-N.: A unified gradient regularization family for adversarial examples'. In: IEEE International Conference on Data Mining, vol. 2015, pp. 301–309. IEEE (2015)
10. Madry, A., et al.: Towards deep learning models resistant to adversarial attacks. In: International Conference on Learning Representations (2018)
11. Mathov, Y., et al.: Not all datasets are born equal: on heterogeneous tabular data and adversarial examples. Knowl.-Based Syst. **242**, 108377 (2022)
12. Qin, Y., et al.: Detecting and diagnosing adversarial images with class conditional capsule reconstructions. In: International Conference on Learning Representations (2020)
13. Sharma, Y., Chen, P.-Y.: Attacking the madry defense model with L_1-based adversarial examples (2018)
14. Sheikholeslami, F., Jain, S., Georgios, B., Giannakis: Minimum uncertainty based detection of adversaries in deep neural net- works. In: Information Theory and Applications Workshop (ITA), vol. 2020, pp. 1–16. IEEE (2020)
15. Smith, L., Gal, Y.: Understanding measures of uncertainty for adversarial example detection. In: Uncertainty in Artificial Intelligence (2018)
16. Szegedy, C., et al.: Intriguing properties of neural networks. CoRR abs/1312.6199 (2014)

17. Tuna, O.F., Catak, F.O., Eskil, M.T.: Closeness and uncertainty aware adversarial examples detection in adversarial machine learning. Comput. Electr. Eng. **101**, 107986 (2022)
18. Vinayakumar, R., Soman, K.P., Poornachandran, P.: Detecting malicious domain names using deep learning approaches at scale. J. Intell. Fuzzy Syst. **34**(3), 1355–1367 (2018)
19. Vinayakumar, R., Soman, K.P., Poornachandran, P.: Evaluating deep learning approaches to characterize and classify malicious URL's. J. Intell. Fuzzy Syst. **34**(3), 1333–1343 (2018)
20. Vinayakumar, R., et al.: A deep-dive on machine learning for cyber security use cases. In: Machine Learning for Computer and Cyber Security, pp. 122–158. CRC Press (2019)
21. Xie, C., et al.: Feature denoising for improving adversarial robustness. In: Proceedings of the IEEE/CVF Conference on Computer Vision and Pattern Recognition, pp. 501–509 (2019)
22. Zhang, H., et al.: The limitations of adversarial training and the Blindspot attack. In: International Conference on Learning Representations (2019)

Autonomous UAV Navigation in Wilderness Search-and-Rescue Operations Using Deep Reinforcement Learning

Muhammad Talha[✉][iD], Aya Hussein[iD], and Mohammed Hossny[iD]

School of Engineering and Information Technology, UNSW, Canberra, Australia
m.talha@student.unsw.edu.au

Abstract. Wilderness Search and Rescue (WiSAR) operations require navigating large unknown environments and locating missing victims with high precision and in a timely manner. Several studies used deep reinforcement learning (DRL) to allow for the autonomous navigation of Unmanned Aerial Vehicles (UAVs) in unknown search and rescue environments. However, these studies focused on indoor environments and used fixed altitude navigation which is a significantly less complex setting than realistic WiSAR operations. This paper uses a DRL-powered approach for WiSAR in an unknown mountain landscape environment. To manage the complexity of the problem, the proposed approach breaks up the problem into five modules: Information Map, DRL-based Navigation, DRL-based Exploration Planner (waypoint generator), Obstacle Detection, and Human Detection. Curriculum learning has been used to enable the Navigation module to learn 3D navigation. The proposed approach was evaluated both under semi-autonomous operations where waypoints are externally provided by a human and under full autonomy. The results demonstrate the ability of the system to detect all humans when waypoints are generated randomly or by a human, whereas DRL-based waypoint generation led to a lower recall of 75%.

Keywords: Autonomous navigation · Curriculum learning · Search and rescue

1 Introduction

WiSAR operations often occur after natural disasters or accidents to find missing victims. WiSAR operations are time-critical where the objective is to find all victims within the shortest amount of time. Recent studies have been exploring the use of UAVs in search and rescue (SAR) missions where access to humans and ground vehicles is limited [8,15,27]. Due to their greater field of view, ease of deployment and navigation, and low manufacturing cost, UAVs lend themselves to complex SAR missions, including WiSAR [2,30].

H. Aziz et al. (Eds.): AI 2022, LNAI 13728, pp. 733–746, 2022.
https://doi.org/10.1007/978-3-031-22695-3_51

In the last decade, most field studies have primarily focused on the use of manually controlled UAVs in SAR. However, manual control puts significant workload demands on human operators in such stressful contexts which results in errors. Thus, the interest in autonomous UAV operation has recently increased to allow for more effective SAR operations [4,5,8,14] and allow for better use of the available manpower.

In typical WiSAR operations, no accurate model of the environment exists. This raises the need for developing algorithms to enable a UAV to autonomously make its decisions based on its evolving perception of the environment without relying on continuous human interaction. Reinforcement Learning (RL) has been used to enable autonomous SAR operations [7] as it enables the UAV to learn efficient navigation strategies without requiring a model of the environment. Nevertheless, due to the complexity of SAR problems, past studies using RL considered simplified settings by focusing on indoor environments. Another key limitation of past RL studies is fixing the altitude of the UAV to reduce the action space of RL. However, such simplifications reduce the applicability of the algorithms to realistic WiSAR operations that are characterised by high environmental clutter, partial observability, and variable landscape elevation.

This paper aims to address this gap by proposing an approach to autonomous UAV operation in complex WiSAR environments. This work proposes a modular approach consisting of five modules: Information Map, Navigation module, Exploration Planner, Obstacle Detection module, and Human Detection module. The Information Map is used to mitigate partial observability of the environment by serving as a memory for storing information collected by the UAV. The Navigation module is an RL algorithm concerned with taking the UAV through the shortest obstacle-free path towards a given location. Curriculum learning is used for training the Navigation RL agent to facilitate learning whilst allowing the UAV to change its altitude in mountain environments. The Exploration Planner is another RL agent that provides high-level planning of the WiSAR operation by generating a sequence of waypoints for the UAV. The Obstacle Detection and Human Detection modules use the UAV sensor feed to detect obstacles and humans in the environment and estimate their 3D positions. These modules are described in detail in Sect. 5. The proposed approach has proved its viability both under semi-autonomous operations (where waypoints are externally provided by a human) and under full autonomy (where waypoints are generated by the Exploration Planner Module). Detailed results are provided in Sect. 6.2. The next section gives a brief review of autonomous UAV operations.

2   Related Work

Many studies have focused on the autonomous navigation of UAVs in SAR operations. Tomic et al. [29] introduced a framework for the autonomous execution of SAR operations using aerial robots. They used laser and stereo vision odometry fused with an inertial measurement unit in an extended Kalman filter to enable seamless navigation. A successful test was performed on a quadrotor platform using four cameras and a laser scanner, but the system did not have any collision avoidance capabilities. Scherer et al. [24] proposed a UAV-based architecture for SAR missions in outdoor environments. The architecture used a distributed control system for coordinating the activities between multiple UAVs. The main aim was to detect a target through its colour, shape, or texture and live stream the aerial video for remote monitoring. The project used a swarm of UAVs to act as a communication relay. But the main limitation of these non-RL solutions is that the environment model was already known at the start of the task which is typically not the case for SAR operations.

Recently, RL has been extensively used in developing autonomous algorithms for UAV navigation, path planning, and trajectory optimisation [3,13]. Kersandt et al. [16] used DRL for the self-training of drones in a fully autonomous flight where depth images were used as the observation vector. After 3 days of training, the DRL agent was deemed effective by achieving 80% success in test flights. Hodge et al. [10] described a navigation algorithm that uses onboard sensor data (i.e. GPS and compass) to guide the drone to a target location in a static environment. They used Proximal Policy Optimisation (PPO) DRL combined with incremental curriculum learning and Long Short-term Memory (LSTMs) Neural Networks to implement their algorithm. A key limitation in both of these works is that the altitude of the drone was kept constant. This is done to reduce the action space of the DRL agent to facilitate learning, however, this limits the applicability of the algorithms to fixed-elevation environments.

Pham et al. [22] used RL to train a UAV to navigate safely to a target point and locate the immobile human (if present at that location) using a combination of Proportional-Integral-Derivative(PID) and Approximated Q-learning algorithm. The authors used a discrete state space containing the relative distance of the UAV from the target and the distances to the nearest obstacle in the North, South, East, or West direction. The limitation of this study was a constant UAV altitude throughout its flight. An RL-based framework was presented in [7] that enabled a UAV to autonomously observe the environment and map a trajectory for the fastest localisation of multiple objects in a SAR mission. The approach divided the search environment into M cells and the cell center was used as a state vector for the agent. The framework was implemented in two phases. In the first phase, the UAV was controlled by a human operator through an initial scan trajectory to get the number of terrestrial objects and to train the agent online. Then, in the second phase, the UAV autonomously controlled its movements using RL to minimize the average location errors of all objects. The need for a human to teleoperate the UAV during the initial scan is problematic as it can result in significant inefficiencies.

Most of these studies have focused on SAR operations for indoor and less cluttered environments. A common limitation was that the altitude of the UAV was kept constant during training and testing. Furthermore, the victims' locations were already known at the start of the SAR operation via GPS so these algorithms will fail in scenarios where the location is unknown. In contrast to all the aforementioned developments, this work proposes a fully-autonomous UAV that is not only capable of autonomously navigating in the wild with varying altitudes but can also create a map of the environment that has positional information for the obstacles and humans in the environment. This map can help the SAR team localize the victims easily.

3   Preliminaries

RL is formulated as a MDP with state space S, action space A, reward function R, and discounting factor γ. By repeatedly interacting with the environment through trial and error, an RL agent aims to learn a policy that maximises the accumulated reward; where a policy π provides the mapping from states to actions $\pi : S \rightarrow A$. In many real-world problems, the true state of the environment can not be deterministically sensed by the agent. The problem can then be formulated as partially observable MDP (POMDP) by using agent observations instead of the true states of the environment. As such, it becomes crucial for the performance to account for this partial observability (e.g., by maintaining a belief of the current state or by equipping the agent with memory).

Tasks with huge state and action spaces have represented a significant challenge for classic RL. Deep RL (DRL) aims to address this challenge by using Neural Networks(NN) as function approximators. DRL has produced excellent results in various complex tasks including Atari games [20], robotic control [12], and nuclear fusion [6]. This paper uses a widely used DRL algorithm called proximal policy optimisation PPO [25].

4   Problem Definition

The SAR problem addressed in this paper consists of a UAV D with position $P_D = (x_D, y_D, z_D)$ and a group of N scattered humans $(h_1, h_2, ...h_N)$ with positions $P_{h_i} = (x_{h_i}, y_{h_i}, z_{h_i})$. The SAR operation is conducted in a geofenced continuous-space environment of dimensions $L \times L \times H$. The geofence has O obstacles in it. The environment has a designated starting position for the UAV given as P_D^0. The maximum linear velocity of the UAV is V_D and the field of view of its camera is θ. The UAV is assumed to be spatially aware of its position P_D^t but the positions of humans are unknown. The objective is to find all humans within a predefined amount of time T_{max}. The task is finished when all humans have been detected or when T_{max} is reached, whichever happens first. The task is considered successful if and only if all lost humans are found.

5   Methodology

A modular approach is proposed in this paper to solve the problem defined above. The proposed system consists of five different modules that are integrated to enable finding an efficient solution to the main problem. A schematic overview of the proposed system is given in Fig. 1. The UAV Navigation module is responsible for controlling the flight of the UAV and guiding it towards a given target (generated by the Exploration Planner module) while avoiding collisions. The Human Detection and Obstacle Detection modules are used to detect humans and obstacles, respectively along the way and calculate their position in the environment. This information is stored by the Information Map module and is then used by the Exploration Planner module for generating waypoints for the UAV.

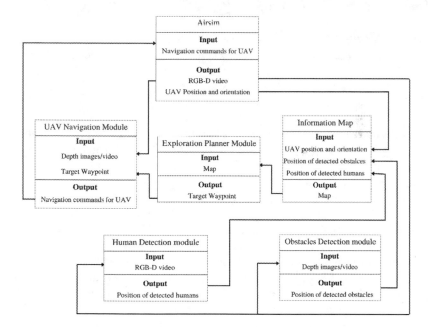

Fig. 1. Schematic overview of the system modules.

5.1   Simulator

Airsim is used in this work to simulate the UAV in a *Landscape Mountains* environment, as shown in Fig. 2a. Airsim is a photo-realistic simulator for drones and ground vehicles developed on top of Epic's Unreal Engine 4 (UE4) by Microsoft Research. This simulator is selected because it closely simulates real-world dynamics [26]. The quadcopter model provided by Airsim supports sensors such as RGB and depth cameras, IMUs, GPS, and LIDARS. Airsim has Python APIs that can be used for both controlling the UAV and obtaining its sensor readings.

5.2   Information Map

The Information Map is a 3D occupancy map (a voxel map) containing information about UAV detections and past trajectories. This map serves as an input for the Exploration Planner module. The initial size of the map is $10 \times 10 \times 3$

(a) Screenshot of the simulation environment   (b) Vision volume of the UAV

Fig. 2. UAV navigation.

cells and it increases during runtime based on the region explored by the UAV. Each voxel in the map represents $1\,m^3$ volume. A virtual geofence is defined within the scene to mark the region that the UAV should be exploring. Each cell can take one of these states: unknown, free, contains obstacle, contains human, previously visited by the UAV, or currently occupied by the UAV.

Initially, all cells are marked as unknown. The state of a cell can change depending on the UAV path (calculated using GPS through Airsim) and observations. A pyramid-shaped vision volume is modelled to keep track of the observed cells as shown in Fig. 2b. The vision range of $(\pi/2)$ is selected to generate a realistic vision volume that can be achieved from the drone's faced-down camera in Airsim. If a previously unknown cell comes within the vision of the UAV, it will be marked as either free, obstacle or human based on information from the Obstacle and Human Detection modules. The cells that lie behind any obstacle are not detected by the UAV and their states are kept unchanged.

5.3   UAV Navigation Module

The Navigation module is responsible for the UAV movement and is controlled by a DRL agent. The goal state for this module is set based on the current target waypoint generated by the Exploration Planner module. The observation space for the agent consists of a depth image of the environment and the distance and direction of the waypoint relative to the UAV. The action space consists of five discrete actions: fly forward 0.25 m, rotate left or right 10°, and move up or down 0.25 m. This module aims to navigate through the shortest obstacle-free path to the target waypoint. The episode is terminated when one of the four conditions is met: reaching waypoint, collision, going outside geofence, or T_{max} is reached. A reward of +5 is given for reaching the waypoint and a reward of −5 for collision or going outside the geofence. Besides, a shaping reward of +0.1/−0.1 is given for moving towards/away from the waypoint. A small negative reward of −0.05 is also included for rotation and moving up/down to discourage the UAV from doing these actions while staying at the same location. This was deemed necessary based on some initial experiments in which the UAV preferred to rotate and ascend/descend at the same location rather than moving forward.

The UAV takes off at a predefined starting location which remains the same in each episode. During training, the target waypoint is set according to a simple-to-complex curriculum learning to allow for efficient learning [11]. In RL domains, curriculum learning can be used during the training phase by starting the training with easy scenarios where the probability of success is high before proceeding with more complex scenarios. This enables the RL agent to receive useful learning signals early on, which expedites learning. Curriculum learning is used in the training of the Navigation module as follows. The waypoint position is generated randomly within a maximum distance of d_{max} from the UAV. Initially, d_{max} is set to 5 m. After each successful training episode, d_{max} is increased by 0.1 m. After training is complete, the waypoint is set by the Exploration Planner module and can take any location within the geofence.

The RL algorithm used for the Navigation module is PPO [25] which is an on-policy algorithm that learns only from its current batch without using a replay buffer. PPO uses an actor-critic architecture in which the actor learns the policy through policy gradient, while the critic learns the value function for each state-action pair and helps with selecting high-value actions. PPO is used in this paper due to its efficiency and simplicity. It is more stable and converges faster than deep-Q networks and vanilla policy gradient. Also, it is simpler than trust region policy optimisation (TRPO).

The neural network used for the training of this module consists of both a visual encoder and a vector encoder to handle the depth image and distance and orientation components of the observation space respectively. The visual encoder consists of a stack of two convolutional layers with (16, 32, 64) filters per layer followed by a max-pooling layer while the vector encoder only consists of two fully connected layers each of size 32 neurons. The two encoders run in parallel and their outputs are then concatenated. The concatenated output is then passed through two shared fully connected hidden layers of size (32, 32) which split the network into the actor and critic head. The critic head is a single neuron with no activation function. The solution uses mini batch gradient descent with the Adam optimizer [17] and entropy regularisation [1]. *Stable-baselines 3* has been used for the implementation of PPO with the neural network as described above. The learning rate is set to 1e−5; other hyper-parameters are set to their default values in *Stable-baselines 3*.

5.4   Exploration Planner

The Exploration Planner module deals with exploring the environment by generating a sequence of target waypoints (within the geofence) for the UAV. These waypoints are used by the pre-trained Navigation module to guide the UAV through the environment. The input for this module is the Information Map which gives information regarding the positions of the obstacles, detected humans, and previously visited areas in the environment. The objective is to create a policy for the waypoint generation which maximises the exploration of the environment within a given time limit. This module is formulated as DRL and the main inspiration for this waypoint-based exploration was the work done in [31] and [19].

The state space of the RL environment for the exploration module comprises the Information Map. The action is defined as generating a waypoint for the agent, described by the displacement (dx, dy) from the UAV's current position. After receiving the action, the pre-trained navigation policy controls the UAV until it safely reaches the waypoint and the Information Map is updated continuously during this task. To maximise the exploration, the reward function for this module is based on the detection of previously unseen cells on the map. The reward for each new waypoint is equal to the number of unknown cells detected by the UAV, inversely scaled with the number of steps taken to reach the waypoint and the size of the vision volume. A similar approach was used in [21]. An extra reward of +10 is also provided to the agent if it detects a new human during exploration. The reward function R_E for exploration is given as:

$$R_E = \frac{number\ of\ new\ cells\ detected}{vision\ volume\ \cdot\ number\ of\ steps\ taken} + \begin{cases} 10 \text{ if new human detected} \\ 0 \text{ otherwise} \end{cases}$$

(1)

PPO is used for training this module similar to the UAV Navigation module. The NN and hyper-parameters used are also almost the same. The differences are that the NN consists of a visual encoder only due to the spatial nature of the information map. In the convolutional stack, max pooling is not used to avoid losing important information. Also, a continuous action head is used due to the continuous action space.

5.5   Obstacle Detection Module

This module is responsible for detecting obstacles such as trees, rocks, and mountains using the onboard depth camera of the UAV. Using a 64X64 depth image as input, the algorithm checks the depth at a given number of grid points and estimates the corresponding global coordinates based on the pinhole camera model [28]. A 3D point with global coordinates (X_a, Y_a, Z_a) is represented in the pixel coordinate system (u, v) by the equations:

$$u_a = k_u(f\frac{X_a}{Z_a} + x_0)$$

(2)

$$v_a = k_v(f\frac{Y_a}{Z_a} + y_0)$$

(3)

where f is the focal length and k_u and k_v are the pixel densities in the u and v directions, respectively. From these equations, it is easy to reconstruct the coordinates of a 3D point relative to the UAV using a depth image. These coordinates are then transformed into absolute global coordinates using the position and orientation of the UAV and the result is stored in the Information Map.

5.6   Human Detection Module

This module uses 1024×1024 RGB images to detect humans and calculate their 3D positions. This is enabled by using YOLO [23] which is a widely used algorithm for real-time object detection. In YOLO, an image is first divided into grids of equal dimensions. Each grid cell makes B bounding boxes and provides the confidence scores for detection by using a single bounding box regression that predicts the height, width, center, and class of the objects. Intersection over unions (IOU) ensures that the predicted boxes are equal to the real boxes of the objects. This is done by eliminating the unnecessary bounding boxes that do not match the height and width of the object. YOLO is used in the current work as it outperforms other object detection algorithms [18]. When YOLO detects a human, it makes a bounding box around it within the image. The location of the human in the image is calculated as the center (x_c, y_c) of the projected box. This 2D point can be re-projected to 3D space using the same method used in the Obstacle Detection module. The 3D points are then stored in the Information Map to keep track of the detected humans.

6   Experimental Results

The training and evaluation experiments have been conducted on a Linux system (1080 Nvidia GPU) using the parameter settings listed in Table 1. The RL training results are first presented in Sect. 6.1 then the system evaluation is presented in Sect. 6.2.

Table 1. Parameter setting used for training and evaluation.

Parameter	P_D^0	Geofence	N	O	V_D	θ	T_{max}
Value	(0,0,0)	$60 \times 60 \times 15\,\mathrm{m}^3$	4	10	0.5 m/s	$\pi/2$	512 steps

6.1   Training Results

UAV Navigation Agent. Figure 3a shows the performance of the Navigation PPO agent along training time. During the early training epochs, the agent learnt to avoid negative penalties by avoiding collisions and staying inside the geofence. Then the agent learnt to reach waypoints more frequently as indicated by the positive rewards. When a target waypoint was too close to the geofence boundary, the agent kept crossing the geofence which resulted in a negative reward. However, including such waypoints in the training was necessary to ensure that the resulting agent can maintain the geofence constraints regardless of the position of the target waypoint.

Exploration Planner Agent. The training results of the Exploration PPO agent are shown in Fig. 3b. In the early training epochs, the agent frequently generated waypoints too close to obstacles which led to early episode termination due to collisions. Although there was no explicit negative reward for collision in the Exploration Planner RL agent, collisions prevented the agent from exploring more portions of the environment and detecting more humans and receiving the associated positive rewards. Hence, over time the agent learnt to avoid generating waypoints too close to obstacles.

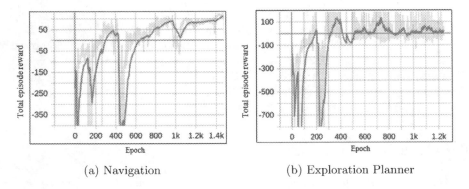

(a) Navigation   (b) Exploration Planner

Fig. 3. Training results. Epoch = 2048 time steps.

6.2   Evaluation

After the training of the RL agents is completed, evaluation has been conducted in two phases. The first phase aims to evaluate system performance in semi-autonomous operations, whereas the second phase evaluates the system under full autonomy.

Evaluation Under Semi-autonomous Operations. In today's real-life scenarios of robot operation, human supervision is necessary for critical operations. In these human-robot teaming settings, high-level planning is conducted by humans while plan execution is left for the robot [9]. To evaluate the proposed system under such settings, the waypoints are provided by the authors in a way that would allow the UAV vision volume to cover all humans. The authors conducted evaluation experiments to assess the system performance in this setting where the waypoints are externally generated (rather than generated by the Exploration Planner module), and the system otherwise operates as depicted in Fig. 1.

Five evaluation experiments have been conducted during which the positions of the detected obstacles and humans have been logged for analysis. A human is considered to be found if they are detected and their position is estimated within $1.5\,\mathrm{m}^2$ from their actual position. Similar to the literature on information retrieval, *recall* is used as a measure for reflecting the system's ability to retrieve humans' positions. However, instead of using the *precision* measure that is suitable for classification settings, we use *position estimation error* for measuring the accuracy of calculating the positions of humans found. In our setting, recall is calculated as the ratio of humans found, and position estimation error is calculated as the distance between the estimated and actual positions. A recall of 100% has been achieved in the 5 evaluation runs. The average position estimation error was $0.48\,\mathrm{m}$ with a standard deviation of $0.27\,\mathrm{m}$. Figure 4 presents a 3D visualisation of the map obtained during one of the evaluation runs. The yellow cubes represent the detected humans, green represents the flight path of the UAV and red represents the detected obstacles/floor of the environment. The figure shows more than one yellow cube at three out of four human locations. This was due to multiple location estimates generated by the UAV for three humans whereas the fourth one was detected only once.

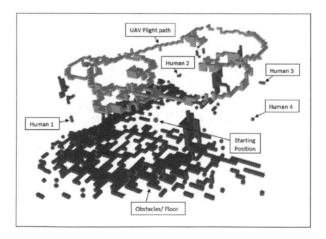

Fig. 4. Visualisation of the results of the semi-autonomous system operation.

Evaluation Under Full Autonomy. The full system has been evaluated under full autonomy: first using the pre-trained Exploration Planner RL agent for waypoint generation and then using random waypoints. Figure 5 shows the average recall over five evaluation experiments. When the Exploration Planner RL agent is used for waypoint generation, a recall of 75% is achieved within the time limit, T_{max}. Meanwhile, randomly generated waypoints resulted in a 100% recall.

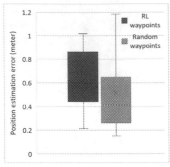 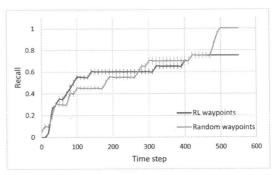

(a) Position estimation error   (b) Average recall. Error bars represent standard error

Fig. 5. Results of five evaluation runs under full system autonomy.

7 Conclusion and Future Work

The autonomous UAV navigation in WiSAR is a highly complex problem due to cluttered environments and partial observability. Previous studies using RL focused on SAR for indoor environments and used fixed altitude navigation which is a significantly less complex setting than WiSAR. This paper addresses the complexity of WiSAR by proposing a modular DRL-powered system that consists of five modules: 1. Information Map, 2. Navigation, 3. Exploration Planner, 4. Obstacle Detection, and 5. Human Detection.

The system has been evaluated both under semi-autonomy (i.e. waypoints externally provided to the UAV) and full autonomy (i.e. waypoints autonomously generated by the system). In the semi-autonomous experiments, the system achieved a 100% recall rate whilst achieving collision-free navigation. Under full autonomy, random waypoint generation resulted in a higher recall than DRL-based waypoint generation. This suggests that the Exploration Planner RL agent was not useful in the experimental environment and should be replaced by a simple random waypoint generator in similar settings. However, under more complex settings (e.g. dynamic environments), human-generated waypoints are expected to result in the highest performance.

Using the Airsim simulator for system development and evaluation has important practical implications as the AirLib library (main library in Airsim) can be compiled and deployed on a real drone [26]. By using a model of a commercially available quad-rotor for evaluation, we show that the proposed system has no unrealistic assumptions about the computational capabilities of the drone.

Future studies can extend this work in several directions. First, the current work considered static humans, however in the general case humans can be moving in the environment. Second, environmental clues (e.g. human belongings) can be utilised for guiding the search operation. Third, extending the system to allow for operating a swarm of UAVs will enable reducing the search time and covering larger WiSAR areas.

Acknowledgement. This work is partially supported by the Australian Research Council Grant DP200101211.

References

1. Achiam, J.: Proximal policy optimization (2018). https://spinningup.openai.com/en/latest/algorithms/ppo.html
2. Adams, S.M., Friedland, C.J.: A survey of unmanned aerial vehicle (uav) usage for imagery collection in disaster research and management. In: 9th International Workshop on Remote Sensing for Disaster Response, vol. 8, pp. 1–8 (2011)
3. Bayerlein, H., De Kerret, P., Gesbert, D.: Trajectory optimization for autonomous flying base station via reinforcement learning. In: 2018 IEEE 19th International Workshop on Signal Processing Advances in Wireless Communications (SPAWC), pp. 1–5. IEEE (2018)
4. Becerra, V.M.: Autonomous control of unmanned aerial vehicles (2019)
5. Carlson, D.F., Rysgaard, S.: Adapting open-source drone autopilots for real-time iceberg observations. MethodsX **5**, 1059–1072 (2018)
6. Degrave, J., et al.: Magnetic control of tokamak plasmas through deep reinforcement learning. Nature **602**(7897), 414–419 (2022)
7. Ebrahimi, D., Sharafeddine, S., Ho, P.H., Assi, C.: Autonomous UAV trajectory for localizing ground objects: a reinforcement learning approach. IEEE Trans. Mob. Comput. **20**(4), 1312–1324 (2020)
8. Eyerman, J., Crispino, G., Zamarro, A., Durscher, R.: Drone efficacy study (des): Evaluating the impact of drones for locating lost persons in search and rescue events brussels. DJI and, Belgium (2018)
9. Hocraffer, A., Nam, C.S.: A meta-analysis of human-system interfaces in unmanned aerial vehicle (UAV) swarm management. Appl. Ergon. **58**, 66–80 (2017). https://doi.org/10.1016/j.apergo.2016.05.011
10. Hodge, V.J., Hawkins, R., Alexander, R.: Deep reinforcement learning for drone navigation using sensor data. Neural Comput. Appl. **33**(6), 2015–2033 (2021)
11. Hussein, A., Petraki, E., Elsawah, S., Abbass, H.: Autonomous swarm shepherding using curriculum-based reinforcement learning. In: Proceedings of the 2022 International Conference on Autonomous Agents and MultiAgent Systems, May 2022
12. Ibarz, J., Tan, J., Finn, C., Kalakrishnan, M., Pastor, P., Levine, S.: How to train your robot with deep reinforcement learning: lessons we have learned. Int. J. Robot. Res. **40**(4–5), 698–721 (2021)
13. Imanberdiyev, N., Fu, C., Kayacan, E., Chen, I.M.: Autonomous navigation of UAV by using real-time model-based reinforcement learning. In: 2016 14th International Conference on Control, Automation, Robotics and Vision (ICARCV), pp. 1–6. IEEE (2016)
14. Kanellakis, C., Nikolakopoulos, G.: Survey on computer vision for UAVs: current developments and trends. J. Intell. Robot. Syst. **87**(1), 141–168 (2017)
15. Karaca, Y., et al.: The potential use of unmanned aircraft systems (drones) in mountain search and rescue operations. Am. J. Emerg. Med. **36**(4), 583–588 (2018)
16. Kersandt, K., Muñoz, G., Barrado, C.: Self-training by reinforcement learning for full-autonomous drones of the future. In: 2018 IEEE/AIAA 37th Digital Avionics Systems Conference (DASC), pp. 1–10. IEEE (2018)
17. Kingma, D.P., Ba, J.: Adam: a method for stochastic optimization. arXiv preprint arXiv:1412.6980 (2014)

18. Lee, Y.H., Kim, Y.: Comparison of CNN and yolo for object detection. J. Semiconductor Display Technol. **19**(1), 85–92 (2020)
19. Mnih, V., et al.: Asynchronous methods for deep reinforcement learning. In: International Conference on Machine Learning, pp. 1928–1937. PMLR (2016)
20. Mnih, V., et al.: Playing ATARI with deep reinforcement learning. arXiv preprint arXiv:1312.5602 (2013)
21. Persson, E., Heikkilä, F.: Autonomous mapping of unknown environments using a UAV (2020)
22. Pham, H.X., La, H.M., Feil-Seifer, D., Van Nguyen, L.: Reinforcement learning for autonomous UAV navigation using function approximation. In: 2018 IEEE International Symposium on Safety, Security, and Rescue Robotics (SSRR), pp. 1–6. IEEE (2018)
23. Redmon, J., Divvala, S., Girshick, R., Farhadi, A.: You only look once: unified, real-time object detection (2015). https://doi.org/10.48550/ARXIV.1506.02640
24. Scherer, J., et al.: An autonomous multi-UAV system for search and rescue. In: Proceedings of the First Workshop on Micro Aerial Vehicle Networks, Systems, and Applications for Civilian Use, pp. 33–38 (2015)
25. Schulman, J., Wolski, F., Dhariwal, P., Radford, A., Klimov, O.: Proximal policy optimization algorithms. arXiv preprint arXiv:1707.06347 (2017)
26. Shah, S., Dey, D., Lovett, C., Kapoor, A.: AirSim: high-fidelity visual and physical simulation for autonomous vehicles. In: Hutter, M., Siegwart, R. (eds.) Field and Service Robotics. SPAR, vol. 5, pp. 621–635. Springer, Cham (2018). https://doi.org/10.1007/978-3-319-67361-5_40
27. Silvagni, M., Tonoli, A., Zenerino, E., Chiaberge, M.: Multipurpose UAV for search and rescue operations in mountain avalanche events. Geomat. Nat. Haz. Risk **8**(1), 18–33 (2017)
28. Sturm, P.: Pinhole Camera Model, pp. 610–613. Springer, Boston (2014). https://doi.org/10.1007/978-0-387-31439-6_472
29. Tomic, T., et al.: Toward a fully autonomous UAV: research platform for indoor and outdoor urban search and rescue. IEEE Robot. Autom. Mag. **19**(3), 46–56 (2012)
30. Torresan, C., et al.: Forestry applications of UAVs in Europe: a review. Int. J. Remote Sens. **38**(8–10), 2427–2447 (2017)
31. Wijmans, E., et al.: DD-PPO: learning near-perfect pointgoal navigators from 2.5 billion frames. arXiv preprint arXiv:1911.00357 (2019)

Active Learning Using Difficult Instances

Bowen Chen[✉], Yun Sing Koh, and Ben Halstead

The University of Auckland, Auckland, New Zealand
bche264@aucklanduni.ac.nz, {y.koh,ben.halstead}@auckland.ac.nz

Abstract. Active learning systems achieve high accuracy with a low labeling budget by annotating high utility instances incrementally. In uncertainty sampling, labels of instances with maximal uncertainty are queried; however, redundant instances with similar features are often selected during the sampling process. We proposed a novel difficulty-based active learning framework that constructs decision boundaries by sampling instances with maximal classification difficulty. We propose three instance level difficulty measures, specifically *base classifier count*, *fluctuation score* and *individual error score*, in a boosted ensemble setting to identify difficult to classify instances. In real-life settings, obtaining labeled data is often expensive and requires domain experts; unlike other difficulty measures that assume complete label knowledge, the proposed measures need only limited labeled data. Experiments with real-world and synthetic datasets show that difficulty-based sampling requires significantly fewer labeled instances to achieve high accuracy than uncertainty sampling.

Keywords: Complexity measures · Active learning · Boosting

1 Introduction

Recently there has been a shift from model-centric to data-centric machine learning [12], as improving the data quality often yields better results than algorithm improvements. Active learning systems aim to achieve high performance by selecting the most representative and informative instances for training according to a utility score. The utility score is often quantified in terms of predictive uncertainty. However, a single probabilistic distribution to represent existing knowledge does not entail the reasons for uncertainty [14]. The sources of uncertainty can be found by studying the composition of the instance within the dataset. Current research measures the overall complexity of a dataset [7]. There are three major drawbacks of data complexity measures: (1) the majority of measures only characterize a dataset's overall complexity but cannot give indications of instance complexity; (2) they focus on specific cases of difficult-to-classify instances (*e.g.* outliers [16], border points [1]); and (3) they often require complete label knowledge.

Supplementary Information The online version contains supplementary material available at https://doi.org/10.1007/978-3-031-22695-3_52.

To mitigate these limitations, we propose a new active learning framework, Difficulty based Active Learning (DAL) framework, that samples instances based on the classification difficulty of each instance. Instead of incrementally sampling instances with maximal uncertainty, a measurement of the whole system (model and data), we sample instances with maximal difficulty based on the composition of instances within the dataset. Our approach leverages the label fluctuation of instances during the construction process of an ensemble model as an indication of label difficulty, with instances that require a weaker learner and experience more fluctuations given higher scores. The method of our proposed framework is outlined in Fig. 1.

The difficulty metrics heavily influences the instances sampled by the active learning framework. In addition to DAL, we propose a novel framework that measures different classification difficulty types and the process outlined in Fig. 2. Three difficulty measures are proposed based on the label fluctuations in an ensemble setting. The classifier count metric measures the minimum description length of an instance. The fluctuation score is a measurement of label consistency. The individual error score indicates label consistency, independent of inputs from other ensemble members. Unlike a single probabilistic distribution such as uncertainty, our difficulty measures can differentiate between the reducible and irreducible portions of classification difficulty represented by classifier count and fluctuation score. Reducible classification difficulty is when the difficulty measure can be reduced if provided with additional data; irreducible difficulty is when the difficulty measure cannot be reduced with more data. The distinction between the two types of classification difficulty allows for high model performance with less labeled training data and helps researchers interpret why particular instances are selected.

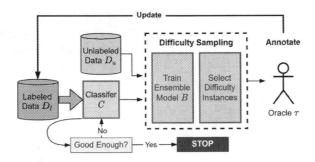

Fig. 1. Difficulty based active learning (DAL) framework.

Our main contributions are summarized as follows. Firstly, we propose an active learning framework, a difficulty-based active learning (DAL) framework, that samples difficult-to-classify instances to be annotated so that future difficult instances can be correctly classified and fewer instances are required to reach high accuracy. Our framework finds difficult to classify instances based

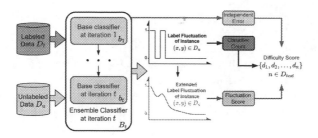

Fig. 2. Classification difficulty framework.

on their composition within the dataset in a boosted ensemble setting. Lastly, We propose three difficulty measures, namely *Base Classifier Count, Fluctuation Score, Individual Error Score*, to provide insight into why some instances are more difficult to classify.

2 Related Work

Several methods have been proposed to identify instances that are difficult to classify. The dataset complexity measures from Ho and Basu [7] have been widely used to analyze the classification difficulty problem. Ho and Basu noted that the classification difficulty mainly originates from a combination of three sources: 1) Class ambiguity which is when for a given dataset, the classes cannot be distinguished regardless of the learning algorithm used; 2) The dimensionality and sparsity of the dataset, and 3) The complexity of the decision boundary. In addition, they present several measures that focus on the complexity of the classification boundary, such as the Fisher's Discriminant Ratio (F1). This idea of class ambiguity is also shared by Hüllermeier and Waegeman [8], at which they discuss the distinction between two sources of uncertainty: aleatoric and epistemic. Similar to class ambiguity, aleatoric uncertainty is the irreducible uncertainty where even with the precise knowledge of the optimal hypothesis, the prediction is still uncertain, and epistemic uncertainty is reducible uncertainty caused by a lack of data. Other complexity measure [6,9,15] that capture similar aspects of complexity has been proposed. However, these methods all assume the availability of labeled data and are designed for supervised classification problems, limiting their applicability in areas with limited labeled data, such as active learning.

3 Active Sampling with Varying Difficulty Metrics

In this section, we first describe the difficulty-based active learning (DAL) setup. Next, we introduce our novel difficulty measures explicitly designed to measure the classification difficulty of instances in a semi-supervised setting.

3.1   Difficulty Based Active Learning (DAL)

Given a dataset that contains a small set of labeled data D_l and a large pool of unlabeled data D_u, the problem we are tackling is to sample and label instances that minimize labeling cost and maximize performance gain. We propose that the most difficult to classify instances should be queried. For each instance $D_u = \{x_1, \ldots, x_n\}$, we denote the classification difficulty of the instances as $\{d_1, \ldots, d_n\}$. The difficulty score d_n is given by the ensemble model B with training set D_l. We use the ensemble model B semi-supervised to output a difficulty score. Based on the difficulty scores, we select δ difficult to classify instances in addition to $(\beta - \delta)$ random instances from D_u to be queried by an Oracle τ, such that the difficult to classify instance can adjust the decision boundary and correctly classify future difficult to classify instances. The process is outlined in Algorithm 1.

Algorithm 1. Difficulty Based Active Learning

Input: labeled data D_l, unlabeled data D_u, classifier θ, ensemble B, batch size β,
    difficult batch size η, oracle τ
Output: classifier θ
1: **while** $D_u \neq \emptyset$ **do**
2:    $\theta \leftarrow \theta.\text{train}(D_l)$
3:    $B \leftarrow B.\text{train}(D_l)$
4:    $scores_{diff} \leftarrow \{B.\text{diff}(x)|x \in D_u\}$ ▷ Get difficulty score for unlabeled instances
5:    $x_{diff} \leftarrow$ Top η difficult instances given $scores_{diff}$
6:    $y_{diff} \leftarrow \{\tau.\text{Query}(x)|x \in x_{diff}\}$
7:    $D_l \leftarrow D_l \cup \{(x,y)|x \in x_{diff}, y \in y_{diff}\}$ ▷ Update D_l
8:    $D_u \leftarrow D_u \setminus x_{diff}$ ▷ Update D_u
9:    $x_{rand} \leftarrow (\beta - \eta)$ randomly sampled instances from D_u
10:   $y_{rand} \leftarrow \{\tau.\text{Query}(x)|x \in x_{rand}\}$
11:   $D_l \leftarrow D_l \cup \{(x,y)|x \in x_{rand}, y \in y_{rand}\}$
12:   $D_u \leftarrow D_u \setminus (x_{rand})$
13: **end while**

3.2   Instance Difficulty

All instances in a dataset have different degrees of classification difficulty. An instance's classification difficulty largely depends on the learning algorithm used for classification and the composition of the instance within the dataset. We use a two-dimensional example as demonstrated in Fig. 3. Instances x_1, x_2, x_3, and x_4 are instances of the same class. Instances x_1 and x_4 are considered outliers; however, in this scenario, instance x_1 would almost always be correctly classified, and instance x_4 would almost always be incorrectly classified. Instances near the decision boundary, such as instances x_2 and x_3, would be more difficult to classify correctly than inlier instances that are further away from the decision boundary.

Given a dataset $D = \{(x_1, y_1), \ldots, (x_n, y_n)\}$ of size n with n_c unique class labels, the classification difficulty d_i of an instance x_i is defined as the amount of resources required for a learner θ to learn the underlying concept and output the ground truth label y_i. We measure both time and memory computation resources. For our proposed method, we train one weak base leaner as the unit for measuring the classification difficulty of an instance. Specifically, we deliberately overfit to difficult to classify instances and measure the number of weak learners required for an instance to be correctly classified. For example, instance x_3 would require more weak learners to be classified correctly than instance x_2. In comparison, instance x_4 would require a more robust overall model with even more weak learners to be correctly classified. This process of over-fitting by including more weak learners is fundamentally an increase in time and space resources.

AdaBoost [3] is chosen over other boosting methods, whereas, other boosting methods [2,5] do not penalize misclassified instances but instead use a loss function. Overall Adaboost model consists of *total* weak learners $B_{total}(x) = \sum_{t=1}^{total} b_t(x)$ where b_t is the t weak learner that takes instance x as input and returns a score indicating the predicted label of the instance. The boosting process consists of *total* iterations where a weak classifier is boosted and added to the ensemble at each iteration using information gained from previous iterations. Intuitively, as the number of weak classifiers increases, the amount of performance gained from each additional weak learner decreases. Following the definition of classification difficulty, difficult instances would need more resources to be correctly labeled. For each instance, we monitor the changes in the label, correctly or incorrectly labeled denoted by 0 and 1 respectively, as new weak classifiers are added. Given a Adaboost model B_{total} consisting of $b_1, \ldots, b_{total}$ weak learners. The label fluctuation sequence of an instance x_i is $fluc_{total}(x_i)$ denoted by $fluc_{total}(x_i) = \{f_t\}_{t=1}^{total}(x_i) = \{f_1(x_i), \ldots, f_{total}(x_i)\}$ where $f_t(x_i)$ is 1 if x_i is incorrectly labeled by the ensemble model at iteration t and 0 otherwise. Here $f_t(x_i) = I(B_t(x_i))$ where $I(\cdot)$ is an indicator function:

$$I(f(x_i)) = \begin{cases} 0, & \text{if } f(x_i) == y_i \\ 1, & \text{otherwise.} \end{cases} \tag{1}$$

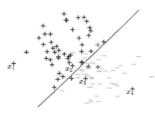

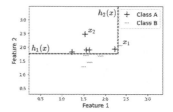

Fig. 3. 2-dimensional example dataset.

Fig. 4. Example of base classifier count score.

Intuitively more difficult to classify instances will require more weak base classifiers to be correctly classified than easy ones. In the next section, we present three difficulty measures that measure the following: the computation resources required for a correct and consistent label, the consistency of the label, and the state of the ensemble when the label is misclassified. All proposed difficulty measures are aimed to extract meta-information that provides indications of classification difficulty from the fluctuations in label; hence these measures are not limited to only labeled data.

3.3   Difficulty Measures

This section introduces three difficulty measures that extract meta-information and provide indications of classification difficulty based on the fluctuations in the label. We also distinguish between two types of classification difficulty, namely, label difficulty and correctness difficulty.

Base Classifier Count C. This measure approximates the minimum computation required to correctly label an instance, which is similar to estimating an instance's description length. The base classifier count of an instance is the number of weak classifiers necessary to build an ensemble that can predict the correct label, denoted by the following:

$$C(x) = \frac{\operatorname{argmin}_t B_t(x) == y, B_t(x) \in B_{total}(x)}{total} \tag{2}$$

where y is the true label of x, and $total$ is the total number of weak classifiers.

Given that the boosting model $B_{t_{total}}$ is a linear combination of t_{total} weak learners, each learner can be treated as a weak hypothesis $h_t \in \mathcal{H}$, where $\mathcal{H}$ is the hypothesis space consisting of all possible hypotheses h mapping instances x_i to labels y_i, and the overall boosting model is a linear combination of the set of these hypotheses $\{h_1, \ldots, h_{t_{total}}\}$. Difficult to classify instances will require a greater number of weak hypotheses to be labeled correctly, and easy to classify instances will need less. As an example, in Fig. 4, instance x_1 is considered more difficult to classify compared to instance x_2, as it requires an additional hypothesis h_2 to be correctly classified.

Fluctuation Score F. The Fluctuation Score is a measurement of label consistency. From the fluctuations in the label, we apply a sliding window to produce a fluctuation graph such that small label fluctuations contribute less to the overall fluctuations in label. Given a fluctuation sequence $fluc_{t_{total}}$, the extended fluctuation sequence is a series containing averages of different subsets of the fluctuation sequence. Each subset is defined by a sliding window that moves across the fluctuation sequence. The size of the sliding window increases at the beginning of the sequence up to a predefined window size and decreases at the endpoints such that the average is taken over only the elements that fill the window. The extended fluctuation sequence with a sliding window of size l is defined as $S_l(fluc_{total}(x)) = \{s_{1,1}, s_{1,2}, \ldots, s_{1,l}, s_{2,l}, \ldots, s_{total-1,2}, s_{total,1}\}$ where $s_{t,l}$ is

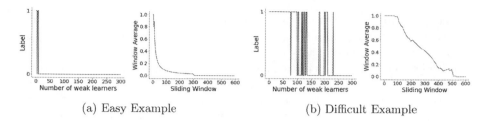

(a) Easy Example   (b) Difficult Example

Fig. 5. The label fluctuation sequence of an easy to classify instance (a) and a difficult to classify instance (b).

the window average of the tth fluctuation in the label fluctuation sequence, with a window size of l, and it is given by:

$$s_{t,l} = \frac{1}{l} \sum_{j=t}^{t+l-1} f_j, f_j \in fluc_{total}(x).$$

(3)

Figure 5 shows extended fluctuation graph, where the sliding window size is the same as the size of the fluctuation sequence ($total = 300$), and the window size increases from 1 to 300, then back down to 1, resulting in an extended fluctuation sequence that contains 600 window averages. The fluctuation score of an instance is a ratio between the area under curve and the area above curve of the fluctuation graph given by:

$$F(x) = 1 - \text{abs}\left(\frac{2 \times \sum S_l(x)}{|S_l(x)|} - 1\right).$$

(4)

The F value has a range between $[0, 1]$. A value closer to 1 indicates that the instance is correctly classified and misclassified for a similar number of iterations and the label of the instance is inconsistent. If the instance is labeled correctly classified and incorrectly classified for the same duration, the fluctuation score will be zero.

Individual Error Score IE. Instead of the cumulative prediction made by the ensemble model $B_{t_{total}}$, the prediction of the weak learner b_t is used to construct the individual fluctuation sequence of an instance denoted by $\hat{fluc}_{t_{total}}(x_i)$, and it is given by: $\hat{fluc}_{t_{total}}(x_i) = \{\hat{f}_t\}_{t=1}^{total}(x_i) = \{\hat{f}_1(x_i), \ldots, \hat{f}_{total}(x_i)\}$ where $\hat{f}_t(x_i)$ is 1 if x_i is incorrectly labeled by the weak learner at iteration t and 0 otherwise. It is denoted by $\hat{f}_t(x_i) = I(b_t(x_i))$. Individual Error Score measure the number of times an instance is incorrectly labeled by each individual weak classifier of the ensemble defined as:

$$IE(x) = \frac{\sum_{t=1}^{t_{total}} I(b_t(x))}{t_{total}}.$$

(5)

Each weak classifier of the ensemble has different sample weights $w_{i,t}$ determined by the previous weak classifier. Intuitively more difficult instances with a higher IE value will be misclassified more often compared to easier instances.

Label Difficulty vs Correctness Difficulty. Similar to epistemic uncertainty and aleatoric uncertainty [8], a distinction can be made between label difficulty and correctness difficulty. Label difficulty is the difficulty of correctly labeling an instance, and similar to epistemic uncertainty, this difficulty can be reduced if provided with enough relevant data. On the other hand, correctness difficulty is the difficulty in determining if an instance is correctly labeled or incorrectly labeled. Both an instance that is incorrectly classified for all iterations and another that is correctly classified for all iterations are considered to have low correctness difficulty as we can be confident that the label is either correct or incorrect. This is similar to aleatoric uncertainty, where the difficulty cannot be reduced with additional data. In our case, the label fluctuations-based difficulty measures F is a measurement of correctness difficulty and cannot distinguish between an instance that is incorrectly classified for all the iterations and another that is correctly classified for the entire duration. In comparison, the classifier counts C as a metric of label difficulty and can distinguish between two such instances. The individual error scores IE while being a measurement label difficulty, it is affected by nearby high fluctuation instances and indicates correctness difficulty.

3.4   Pseudo True Label Assumptions

Given that the difficulty measures are based on label fluctuations, in the absence of labeled data we can assume 1) the most probable label or 2) the most consistent label as the pseudo true label. The most probable label is the label with the highest prediction probability given by $\hat{y} = \{y|Pr(y) = max(Pr(y))\}$.

The most consistent label is the label that remains the same for the longest number of iterations during the boosting process. We monitor the fluctuation between correct (0) and incorrect labels (1). We are interested in the fluctuation between all possible labels. We denote the prediction sequence as $P_{t_{total}}(x_i) = \{p_t\}_{t=1}^{t_{total}}(x_i) = \{B_t(x_i), \ldots, B_{total}(x_i)\}$ where p_t is the prediction output of the ensemble model with t weak learners. The value range of the prediction sequence is dependent on the number of unique class labels, for example, the most consistent label of an instance with a prediction sequence of $\{0, 1, 1, 1, 2, 0, 0\}$ is labelled 1 as the label is maintained for three continuous iterations.

Complexity Analysis Discussion. Our method consists of an ensemble boosting algorithm and some difficulty measure calculations. Given a dataset D with training set D_X, the time complexity of training a decision stump base classifier b_t is $\mathcal{O}(|D_X| \times f)$ where f is the number of features. The cost of training an ensemble with t base classifiers is therefore $\mathcal{O}(t \times |D_X| \times f)$ and the test time complexity is $\mathcal{O}(t \times f)$. In addition to the ensemble testing cost, given that all difficulty measures are based on the label fluctuations, each difficulty measure can be computed in linear time in respect to t. Additionally, the training cost of the ensemble can be effectively reduced by scaling down the number of base learners included in the ensemble.

4   Experiments

We perform several experiments to answer the following questions: (RQ 1) How does difficulty based sampling perform in conjunction with different baseline sampling methods? (RQ 2) How does difficulty based sampling compare under different pseudo true label assumptions? (RQ 3) How does difficulty based sampling compare under different difficulty measures? (RQ 4) How does difficulty based sampling compare with state-of-the-art active learning baselines?

Datasets. Experiments are conducted on both controlled synthetic and real-world datasets. Given that class skew magnifies any sources of classification difficulty present in the instances [11], we used random tree and random RBF data generators [10] to build a series of imbalanced datasets such that 75% of the dataset are instances belonging to the same class, and instances from other classes make up the remainder 25% of the dataset. A series of real-world datasets from the UCI repository are also used in our experiments. Some are modified to be imbalanced with the majority class consisting of 75% of all instances, and they are denoted by a *Imb* suffix.

Evaluation Metrics. Plotting the Cohen's kappa as a function of the size of the labeled set produces a learning curve, and the ALC is the area under this curve. Cohen's Kappa Score is used instead of accuracy as accuracy provides misleading information given the high imbalance ratio for some of the datasets. We use the normalized Area Under Learning Curve to evaluate the classification results. The normalized ALC is given by $ALC = \frac{ALC-ALC_{rand}}{ALC_{max}-ALC_{rand}}$ where ALC_{rand} is the area under learning curve given random predictions, and ALC_{max} is the area under learning curve given perfect predictions (area under curve with constant 100% accuracy). In addition to ALC, we also use the percentage of training instances ($|L|$) required to achieve the highest kappa score K_{max} as another evaluation metric. In this case K_{max} is the kappa score of the test set given all instances from the query set all included in the training set. $|L|$ has a range between 0–1, with higher scores indicating more instances are required to achieve the same K_{max}.

Baseline Methods. We compare difficulty sampling with random sampling (RS), uncertainty sampling (US), and query by committee [13] with three ($QBC(3)$) committee members as the base active learning model. Query by committee uses the disagreement between a diverse set of committee members with different hypotheses for sampling [4]. For each dataset we use 50% of the dataset as the test set and the remaining dataset is then split into the initial training set and query set. As the overall budget increases, instances are drawn from the query set evenly divided over 100 iterations. For example, given that the query set contained $1,000$ instances, 10 instances are queried and added to the training set in each iteration for 100 iterations.

Parameter Settings. Experiments are also carried out to examine the effect of different parameter settings such as training size, batch size and the number of weak learners. The results are provided in supplementary materials and our code is available at https://anonymous.4open.science/r/Difficulty-Based-Active-Learning-BB79/Active%20Learning%20Curve%20NEW.py.

Table 1. Comparison of random sampling with/without difficulty sampling.

Dataset	Metric	Baseline	Probable			Consistent		
		RS	C+RS	F+RS	IE+RS	C+RS	F+RS	IE+RS
Nursery	ALC	0.94 ± 0.01	0.94 ± 0.01	0.95 ± 0.00	0.94 ± 0.01	0.94 ± 0.01	$\mathbf{0.95 \pm 0.00}$	$\mathbf{0.95 \pm 0.01}$
	$|L|$	0.92 ± 0.07	0.90 ± 0.08	0.77 ± 0.10	0.93 ± 0.09	0.85 ± 0.09	$\mathbf{0.76 \pm 0.09}$	0.85 ± 0.13
Adult	ALC	0.48 ± 0.00	0.48 ± 0.00	0.48 ± 0.00	0.48 ± 0.00	0.48 ± 0.00	0.48 ± 0.00	0.48 ± 0.00
	$|L|$	$\mathbf{0.22 \pm 0.11}$	0.24 ± 0.14	0.29 ± 0.22	0.33 ± 0.27	0.26 ± 0.25	0.29 ± 0.22	0.28 ± 0.27
Mushroom	ALC	1.00 ± 0.00	1.00 ± 0.00	$\mathbf{1.00 \pm 0.00}$	1.00 ± 0.00	1.00 ± 0.00	$\mathbf{1.00 \pm 0.00}$	1.00 ± 0.00
	$|L|$	0.32 ± 0.13	0.14 ± 0.12	$\mathbf{0.10 \pm 0.07}$	0.30 ± 0.26	0.13 ± 0.07	$\mathbf{0.10 \pm 0.07}$	0.25 ± 0.21
Mammo	ALC	0.49 ± 0.03	0.52 ± 0.03	0.51 ± 0.03	0.52 ± 0.02	0.52 ± 0.03	0.51 ± 0.03	$\mathbf{0.52 \pm 0.03}$
	$|L|$	0.50 ± 0.32	$\mathbf{0.24 \pm 0.23}$	0.32 ± 0.33	0.35 ± 0.32	0.25 ± 0.28	0.32 ± 0.33	0.28 ± 0.37
Car	ALC	0.80 ± 0.02	0.82 ± 0.02	0.81 ± 0.02	0.81 ± 0.01	$\mathbf{0.82 \pm 0.02}$	0.82 ± 0.01	0.81 ± 0.01
	$|L|$	0.71 ± 0.14	0.62 ± 0.21	0.58 ± 0.14	0.62 ± 0.16	$\mathbf{0.52 \pm 0.10}$	0.60 ± 0.12	0.56 ± 0.16
Drybean	ALC	0.86 ± 0.00	0.86 ± 0.00	0.86 ± 0.00	0.86 ± 0.00	0.86 ± 0.00	0.86 ± 0.00	$\mathbf{0.86 \pm 0.00}$
	$|L|$	0.40 ± 0.17	0.45 ± 0.19	0.39 ± 0.13	0.47 ± 0.24	0.42 ± 0.19	0.38 ± 0.20	$\mathbf{0.30 \pm 0.21}$
Thyroid	ALC	0.95 ± 0.01	0.96 ± 0.01	$\mathbf{0.96 \pm 0.00}$	0.96 ± 0.01	0.96 ± 0.01	0.96 ± 0.01	0.96 ± 0.01
	$|L|$	0.70 ± 0.25	0.20 ± 0.17	$\mathbf{0.14 \pm 0.06}$	0.21 ± 0.14	0.25 ± 0.20	0.18 ± 0.08	0.21 ± 0.12
Nursery Imb	ALC	0.90 ± 0.01	0.91 ± 0.01	0.91 ± 0.01	0.91 ± 0.00	$\mathbf{0.91 \pm 0.01}$	0.91 ± 0.01	0.91 ± 0.01
	$|L|$	0.78 ± 0.18	0.47 ± 0.11	$\mathbf{0.43 \pm 0.12}$	0.58 ± 0.29	0.52 ± 0.18	0.44 ± 0.97	0.66 ± 0.26
Drybean Imb	ALC	0.82 ± 0.01	0.83 ± 0.01	0.83 ± 0.01	0.83 ± 0.01	0.83 ± 0.01	$\mathbf{0.83 \pm 0.01}$	0.83 ± 0.01
	$|L|$	0.29 ± 0.15	0.17 ± 0.10	0.20 ± 0.12	0.24 ± 0.14	0.21 ± 0.16	$\mathbf{0.13 \pm 0.07}$	0.17 ± 0.07
RT1 Imb	ALC	0.83 ± 0.01	0.86 ± 0.01	0.87 ± 0.02	0.85 ± 0.01	0.87 ± 0.01	0.87 ± 0.01	$\mathbf{0.87 \pm 0.01}$
	$|L|$	0.93 ± 0.07	0.75 ± 0.22	$\mathbf{0.65 \pm 0.30}$	0.72 ± 0.19	0.65 ± 0.28	0.73 ± 0.22	0.65 ± 0.34
RBF1 Imb	ALC	0.70 ± 0.01	0.71 ± 0.01	0.71 ± 0.01	0.71 ± 0.01	0.71 ± 0.01	$\mathbf{0.72 \pm 0.01}$	0.71 ± 0.01
	$|L|$	0.70 ± 0.10	$\mathbf{0.46 \pm 0.08}$	0.53 ± 0.12	0.50 ± 0.14	0.55 ± 0.18	0.48 ± 0.18	0.50 ± 0.09
RT2 Imb	ALC	0.79 ± 0.01	0.84 ± 0.01	0.85 ± 0.01	0.83 ± 0.01	0.84 ± 0.01	$\mathbf{0.85 \pm 0.01}$	0.84 ± 0.01
	$|L|$	0.92 ± 0.09	$\mathbf{0.40 \pm 0.21}$	0.44 ± 0.25	0.55 ± 0.16	0.44 ± 0.21	0.43 ± 0.21	0.40 ± 0.20
RBF2 Imb	ALC	0.68 ± 0.01	0.71 ± 0.01	0.71 ± 0.01	$\mathbf{0.71 \pm 0.01}$	0.70 ± 0.01	0.71 ± 0.01	0.71 ± 0.01
	$|L|$	0.84 ± 0.14	$\mathbf{0.49 \pm 0.12}$	0.54 ± 0.20	0.51 ± 0.17	0.60 ± 0.24	0.50 ± 0.18	0.54 ± 0.16

*Our methods are denoted by C (Classifier Count), F (Fluctuation Score), IE (Individual Error Score). We also denote $C + RS$, $F + RS$, $IR + RS$ for difficulty with random sampling

(RQ 1). Tables 1 and 2 show the results of difficulty sampling combined with random and uncertainty sampling. For each iteration, 50% of the instances are first selected with RS/US, and the remainder 50% are sampled with difficulty sampling. We observe that baseline methods with additional difficult sampling can achieve much greater performance than those without difficult sampling. When compared with difficulty sampling (Table 3), the performance of both difficulty sampling combined with random and uncertainty sampling is shown to output higher performance for balanced datasets and lower performance for more difficult imbalanced datasets.

Table 2. Comparison of uncertainty sampling with/without difficulty sampling.

Dataset	Metric	Baseline US	Probable C+US	F+US	IE+US	Consistent C+US	F+US	IE+US
Nursery	ALC	0.95 ± 0.00	0.94 ± 0.01	**0.95 ± 0.00**	0.95 ± 0.00	0.94 ± 0.01	0.95 ± 0.00	0.95 ± 0.00
	$|L|$	0.92 ± 0.09	0.90 ± 0.09	**0.79 ± 0.08**	0.85 ± 0.15	0.87 ± 0.11	0.81 ± 0.10	0.90 ± 0.11
Adult	ALC	0.48 ± 0.01	**0.48 ± 0.00**	0.48 ± 0.01	0.48 ± 0.01	**0.48 ± 0.00**	0.48 ± 0.01	0.48 ± 0.00
	$|L|$	0.35 ± 0.22	**0.23 ± 0.15**	0.26 ± 0.27	0.23 ± 0.23	0.25 ± 0.21	0.28 ± 0.28	0.31 ± 0.25
Mushroom	ALC	1.00 ± 0.00	1.00 ± 0.00	1.00 ± 0.00	**1.00 ± 0.00**	1.00 ± 0.00	**1.00 ± 0.00**	**1.00 ± 0.00**
	$|L|$	0.55 ± 0.20	0.17 ± 0.11	0.21 ± 0.10	0.28 ± 0.16	**0.16 ± 0.10**	0.21 ± 0.10	0.28 ± 0.16
Mammo	ALC	0.47 ± 0.03	0.51 ± 0.03	0.52 ± 0.03	0.51 ± 0.02	0.51 ± 0.03	**0.51 ± 0.03**	0.51 ± 0.03
	$|L|$	0.57 ± 0.23	0.29 ± 0.36	0.21 ± 0.26	0.37 ± 0.35	0.36 ± 0.33	**0.20 ± 0.28**	0.34 ± 0.34
Car	ALC	0.79 ± 0.01	0.81 ± 0.01	0.82 ± 0.02	0.81 ± 0.01	0.81 ± 0.01	**0.82 ± 0.01**	0.82 ± 0.01
	$|L|$	0.76 ± 0.15	0.61 ± 0.20	0.53 ± 0.13	0.59 ± 0.12	0.56 ± 0.16	**0.52 ± 0.15**	0.54 ± 0.12
Drybean	ALC	0.86 ± 0.00	0.86 ± 0.00	0.86 ± 0.00	0.86 ± 0.00	0.86 ± 0.00	0.86 ± 0.00	**0.86 ± 0.00**
	$|L|$	0.52 ± 0.25	**0.34 ± 0.14**	0.38 ± 0.24	0.40 ± 0.24	0.39 ± 0.19	0.40 ± 0.22	0.37 ± 0.31
Thyroid	ALC	0.95 ± 0.01	**0.96 ± 0.01**	**0.96 ± 0.01**	0.96 ± 0.01	0.96 ± 0.01	0.96 ± 0.01	0.96 ± 0.01
	$|L|$	0.65 ± 0.26	0.31 ± 0.09	**0.15 ± 0.05**	0.21 ± 0.08	0.19 ± 0.07	0.18 ± 0.05	0.18 ± 0.07
Nursery Imb	ALC	0.90 ± 0.00	0.91 ± 0.01	0.91 ± 0.00	0.91 ± 0.01	**0.92 ± 0.01**	0.91 ± 0.00	0.91 ± 0.01
	$|L|$	0.81 ± 0.15	0.45 ± 0.07	**0.43 ± 0.13**	0.52 ± 0.25	0.43 ± 0.21	0.44 ± 0.12	0.65 ± 0.27
Drybean Imb	ALC	0.82 ± 0.01	0.83 ± 0.01	0.83 ± 0.01	0.83 ± 0.01	0.83 ± 0.01	**0.83 ± 0.01**	0.83 ± 0.01
	$|L|$	0.30 ± 0.13	0.20 ± 0.12	0.22 ± 0.10	0.20 ± 0.12	**0.19 ± 0.09**	0.20 ± 0.08	0.20 ± 0.11
RT1 Imb	ALC	0.83 ± 0.01	0.86 ± 0.01	0.87 ± 0.02	0.85 ± 0.01	0.87 ± 0.01	0.86 ± 0.01	**0.87 ± 0.01**
	$|L|$	0.88 ± 0.08	0.72 ± 0.27	**0.49 ± 0.32**	0.78 ± 0.19	0.66 ± 0.35	0.76 ± 0.26	0.64 ± 0.28
RBF1 Imb	ALC	0.70 ± 0.01	0.71 ± 0.01	**0.71 ± 0.01**	0.71 ± 0.01	0.71 ± 0.01	0.71 ± 0.01	0.71 ± 0.01
	$|L|$	0.69 ± 0.11	0.48 ± 0.12	**0.44 ± 0.08**	0.47 ± 0.08	0.53 ± 0.15	0.49 ± 0.11	0.47 ± 0.14
RT2 Imb	ALC	0.80 ± 0.01	0.84 ± 0.01	**0.85 ± 0.01**	0.83 ± 0.01	0.84 ± 0.01	0.85 ± 0.01	0.84 ± 0.01
	$|L|$	0.85 ± 0.11	0.38 ± 0.23	0.36 ± 0.22	0.64 ± 0.19	0.53 ± 0.23	**0.32 ± 0.11**	0.41 ± 0.21
RBF2 Imb	ALC	0.68 ± 0.01	0.70 ± 0.01	0.70 ± 0.01	**0.71 ± 0.01**	0.70 ± 0.01	0.71 ± 0.00	0.70 ± 0.01
	$|L|$	0.78 ± 0.10	0.53 ± 0.10	0.51 ± 0.18	0.52 ± 0.17	0.59 ± 0.17	**0.49 ± 0.19**	0.54 ± 0.18

*We denote $C + US$, $F + US$, $IR + US$ for difficulty with uncertainty sampling

(RQ 2). Figure 6 shows the difficulty measures classifier count score (C) and fluctuation score (F) have similar rankings with individual error (IE) having the worst ranking. Additionally, Table 3 shows none of the difficulty measures are consistently the best across different datasets. This suggests that these datasets have various sources of difficulty. For example, the best performing difficulty measure for dataset RT1 Imb is classifier count score C, with a lower fluctuation F and individual error (IE) using ALC. This indicates that difficult instances from this dataset are difficult to classify due to label difficulty and require more data to become correctly classified. For RBF1 Imb, both fluctuation and individual error scores have higher ALC and lower $|L|$ than classifier count score. In this case, the instances are difficult due to their frequent changes in the predicted label and are within the overlapping regions of the decisions.

(RQ 3). Figure 6 shows that there is no significant difference between the two types of pseudo true label assumptions. However, it is also noticeable that the same difficulty measure using the most consistent label as the pseudo true label always has a higher $|L|$ rank than using the most probable label as the pseudo true label. This suggests that assuming the most consistent label as the true label

Table 3. The ALC and $|L|$ results of difficulty based sampling, the highest score is highlighted in bold.

Dataset	Metric	Baseline			Probable			Consistent		
		RS	US	QBC(3)	C	F	IE	C	F	IE
Nursery	ALC	0.94±0.01	0.95±0.00	**0.96±0.00**	0.94±0.01	0.93±0.01	0.93±0.01	0.94±0.01	0.93±0.01	0.94±0.01
	$|L|$	0.92±0.07	0.92±0.09	**0.63±0.16**	0.80±0.12	0.72±0.05	0.90±0.09	0.81±0.15	0.73±0.04	0.92±0.05
Adult	ALC	0.48±0.00	0.48±0.01	**0.49±0.00**	0.48±0.00	0.48±0.00	0.45±0.01	0.48±0.00	0.48±0.01	0.46±0.01
	$|L|$	0.22±0.11	0.35±0.22	0.23±0.21	0.28±0.29	**0.21±0.14**	0.58±0.42	0.24±0.16	**0.21±0.14**	0.51±0.44
Mushroom	ALC	1.00±0.00	1.00±0.00	**1.00±0.00**	1.00±0.00	1.00±0.00	1.00±0.00	1.00±0.00	1.00±0.00	1.00±0.00
	$|L|$	0.32±0.13	0.55±0.26	**0.06±0.01**	0.20±0.09	0.14±0.07	0.65±0.21	0.23±0.10	0.14±0.07	0.66±0.22
Mammo	ALC	0.49±0.03	0.47±0.03	0.51±0.03	0.52±0.03	**0.52±0.03**	0.51±0.03	0.52±0.03	**0.52±0.03**	0.51±0.03
	$|L|$	0.50±0.32	0.57±0.23	0.29±0.32	0.37±0.42	**0.14±0.16**	0.32±0.28	0.21±0.28	**0.14±0.16**	0.42±0.31
Car	ALC	0.80±0.02	0.79±0.01	**0.84±0.01**	0.81±0.02	0.82±0.02	0.82±0.01	0.83±0.02	0.83±0.01	0.82±0.01
	$|L|$	0.71±0.14	0.76±0.15	**0.36±0.07**	0.53±0.09	0.45±0.04	0.54±0.06	0.54±0.17	0.48±0.09	0.53±0.15
Drybean	ALC	0.86±0.00	0.86±0.00	0.85±0.00	0.86±0.00	0.86±0.00	0.86±0.00	**0.86±0.00**	0.86±0.00	0.86±0.00
	$|L|$	0.40±0.17	0.52±0.25	0.83±0.05	0.43±0.21	0.52±0.23	0.50±0.18	**0.36±0.15**	0.51±0.31	0.39±0.26
Thyroid	ALC	0.95±0.01	0.95±0.01	0.95±0.01	0.96±0.01	**0.97±0.01**	0.96±0.01	0.96±0.01	0.96±0.01	0.96±0.01
	$|L|$	0.70±0.25	0.65±0.26	0.40±0.29	0.11±0.04	**0.10±0.02**	0.20±0.16	0.17±0.13	0.10±0.02	0.13±0.12
Nursery Imb	ALC	0.90±0.01	0.90±0.00	0.92±0.01	0.91±0.01	0.92±0.01	0.91±0.01	0.92±0.01	**0.92±0.01**	0.91±0.01
	$|L|$	0.78±0.18	0.81±0.15	0.32±0.13	0.37±0.22	0.26±0.04	0.81±0.29	0.33±0.23	**0.26±0.04**	0.71±0.35
Drybean Imb	ALC	0.82±0.01	0.82±0.01	0.83±0.01	0.83±0.01	0.83±0.01	**0.83±0.01**	0.83±0.01	0.83±0.01	0.83±0.01
	$|L|$	0.29±0.15	0.30±0.13	0.25±0.20	0.19±0.12	0.16±0.05	**0.14±0.06**	0.17±0.09	0.14±0.07	0.14±0.10
RT1 Imb	ALC	0.83±0.01	0.83±0.01	0.88±0.01	0.88±0.02	0.88±0.01	0.86±0.01	**0.88±0.01**	0.88±0.01	0.88±0.01
	$|L|$	0.93±0.07	0.88±0.08	0.73±0.12	0.58±0.30	0.59±0.29	0.70±0.21	**0.37±0.21**	0.42±0.25	0.63±0.25
RBF1 Imb	ALC	0.70±0.01	0.70±0.01	0.72±0.01	0.71±0.01	0.72±0.01	**0.72±0.01**	0.71±0.01	0.71±0.01	0.72±0.01
	$|L|$	0.70±0.10	0.69±0.11	0.49±0.26	0.54±0.13	0.46±0.16	**0.39±0.15**	0.47±0.16	0.43±0.15	0.44±0.11
RT2 Imb	ALC	0.79±0.01	0.80±0.01	0.84±0.01	**0.86±0.01**	0.85±0.01	0.84±0.01	0.85±0.01	0.85±0.01	0.85±0.01
	$|L|$	0.92±0.09	0.85±0.11	0.63±0.18	**0.25±0.18**	0.33±0.32	0.52±0.25	0.47±0.29	0.25±0.23	0.36±0.25
RBF2 Imb	ALC	0.68±0.01	0.68±0.01	0.71±0.01	0.71±0.01	0.71±0.01	**0.72±0.01**	0.71±0.01	0.71±0.01	0.71±0.01
	$|L|$	0.84±0.14	0.78±0.10	0.49±0.21	0.45±0.20	0.40±0.16	**0.38±0.12**	0.53±0.23	0.41±0.16	0.42±0.18

results in more informative instances being sampled and higher active learning performance. Figure 7 shows the learning curve plots of difficulty sampling with the most probable labels, compared to RS and US for the generated synthetic imbalanced datasets. The learner's accuracy when all training instances are used for training is given as the dotted line.

(RQ 4). Table 3 presents the performance (ALC and $|L|$) comparison between the baseline methods and difficulty-based sampling under both pseudo true label approaches. In addition, Fig. 6 shows the critical difference between the different methods at 95% confidence level. The results show that there is no significant difference between RS and US for all datasets. In contrast, difficulty sampling outperforms both RS and US regardless of the difficulty measure used for imbalance datasets (*Imb*). The effectiveness of difficulty based sampling is dependant on the difficulty of the dataset. The performance gain (higher ALC and lower $|L|$) is much greater for difficult and imbalanced datasets (*e.g.* Nursery Imb) and not as effective when the dataset is easy (*e.g.* Nursery). For simple datasets, all instances will have similar difficulty scores which results in random sampling. However, even for the easier to classify datasets with similar ALC scores (*e.g.* Mushroom), difficulty sampling requires fewer instances (lower $|L|$) to achieve high performance when compared to uncertainty sampling. In comparison to difficulty-based sampling, $QBC(3)$ has significantly higher performance for easier datasets (*e.g.* Nursery and Adult), with difficulty-based sampling methods

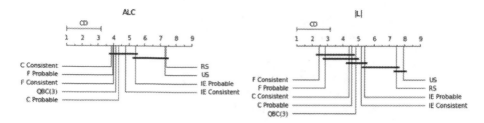

Fig. 6. Critical difference for ALC and $|L|$ results in Table 3.

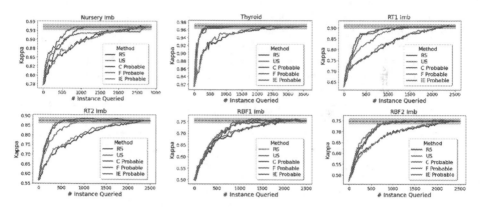

Fig. 7. Examples of difficult sampling in comparison with baseline methods.

achieving a higher ALC value and lower $|L|$ score for imbalanced datasets. Experiments are also carried out to examine the performance of QBC with five committee members and included in the supplementary section as it an extreme case that requires all committee members to be retrained at each query iteration and is considerably more computationally expensive than other methods.

5 Conclusion

We proposed a novel method, difficulty-based active learning (DAL), to find it difficult to classify instances within a dataset based on their label fluctuations through a boosting process. The boosting model is constructed through an additive process with an increasing number of weak learners per iteration. We show the effect of incorporating the proposed difficulty measures, classifier count, fluctuation score and individual error score, into the active learning sampling process, such that sampling difficult-to-classify instances can construct a more robust decision boundary to achieve high model performance with less labelled training data. The experiment results show that difficulty based sampling outperforms uncertainty sampling and query by committee with three committee members when the dataset is imbalanced and contains difficult-to-classify instances. In future work, we plan to explore the potential of using the proposed difficulty measures for performance estimation of datasets and as a filter to remove noisy instances from a dataset.

References

1. Armano, G., Tamponi, E.: Experimenting multiresolution analysis for identifying regions of different classification complexity. Pattern Anal. Appl. **19**(1), 129–137 (2016)
2. Chen, T., Guestrin, C.: XGBoost: a scalable tree boosting system. In: Proceedings of the 22nd ACM SIGKDD International Conference on Knowledge Discovery and Data Mining (2016)
3. Freund, Y., Schapire, R.E.: A decision-theoretic generalization of on-line learning and an application to boosting. J. Comput. Syst. Sci. **55**(1), 119–139 (1997)
4. Friederich, P., Häse, F., Proppe, J., Aspuru-Guzik, A.: Machine-learned potentials for next-generation matter simulations. Nat. Mater. **20**(6), 750–761 (2021)
5. Friedman, J.H.: Stochastic gradient boosting. Comput. Stat. Data Anal. **38**, 367–378 (2002)
6. Garcia, L.P., de Carvalho, A.C., Lorena, A.C.: Effect of label noise in the complexity of classification problems. Neurocomputing **160**, 108–119 (2015)
7. Ho, T.K., Basu, M.: Complexity measures of supervised classification problems. IEEE Trans. Pattern Anal. Mach. Intell. **24**(3), 289–300 (2002)
8. Hüllermeier, E., Waegeman, W.: Aleatoric and epistemic uncertainty in machine learning: an introduction to concepts and methods. Mach Learn. 1–50 (2021)
9. Lorena, A.C., Costa, I.G., Spolaôr, N., De Souto, M.C.: Analysis of complexity indices for classification problems: cancer gene expression data. Neurocomputing **75**(1), 33–42 (2012)
10. Montiel, J., Read, J., Bifet, A., Abdessalem, T.: Scikit-multiflow: a multi-output streaming framework. J. Mach. Learn. Res. **19**(72), 1–5 (2018)
11. Pungpapong, V., Kanawattanachai, P.: The impact of data-complexity and team characteristics on performance in the classification model. Int. J. Bus. Anal. (2022)
12. Sambasivan, N., Kapania, S., Highfill, H., Akrong, D., Paritosh, P., Aroyo, L.M.: "Everyone wants to do the model work, not the data work": data cascades in high-stakes AI. In: Proceedings of the 2021 CHI Conference on Human Factors in Computing Systems, pp. 1–15 (2021)
13. Seung, H.S., Opper, M., Sompolinsky, H.: Query by committee. In: Proceedings of the 5th Annual Workshop on Computational Learning Theory, pp. 287–294 (1992)
14. Sharma, M., Bilgic, M.: Evidence-based uncertainty sampling for active learning. Data Min. Knowl. Disc. **31**(1), 164–202 (2017)
15. Smith, M.R., Martinez, T., Giraud-Carrier, C.: An instance level analysis of data complexity. Mach. Learn. **95**(2), 225–256 (2014)
16. Wang, H., Bah, M.J., Hammad, M.: Progress in outlier detection techniques: a survey. IEEE Access **7**, 107964–108000 (2019)

Reinforcement Learning for Collective Motion Tuning in the Presence of Extrinsic Goals

Shadi Abpeikar$^{(\boxtimes)}$, Kathryn Kasmarik, and Matt Garratt

School of Engineering and Information Technology, University of New South Wales, Canberra, ACT, Australia
s.abpeikar@unsw.edu.au

Abstract. Recent work has shown the possibility for groups of robots to self-bootstrap collective motion behaviours in open arenas. However, for real-world swarm robotics missions, such as package delivery, the environment is likely to be cluttered with obstacles and the swarm will have a target goal. This paper proposes an architecture for self-bootstrapping collective motion behaviours in the presence of such extrinsic goals. The architecture takes a reinforcement learning approach and combines a generic 'intrinsic' reward for collective motion tuning, with a mission specific 'extrinsic' reward. Two instances of extrinsic goals including the target search and obstacle avoidance have been considered in this paper. We demonstrate that our reinforcement learner can tune the behaviour of randomly moving groups so that structured collective motion emerges while avoiding an obstacle or searching for a target. We compare our approach to behaviour bootstrapping in open arenas to show that the presence of external environmental constraints does not affect the quality of the bootstrapped behaviours, with respect to the number of collisions, as well as group, and order metrics.

Keywords: Deep reinforcement learning · Collective motion · Self-bootstrapping behaviour · Extrinsic goal

1 Introduction

Recent work has shown that it is possible for groups of robots to self-bootstrap collective motion behaviours in open arenas [1, 2]. Self-bootstrapping collective motion refers to the procedure that agents form a collective motion by themselves without any external effort. Also, collective motion behaviour refers to the behaviour of the flock of birds, herds of animals, and schools of fish, which the agents interact with each other, and form a patterned motion to perform a special task [3, 4]. In future, we envisage industries and businesses will want to construct ad hoc robot swarms from available off-the-shelf platforms, without the need for manual programming of the collective motions. This may be useful for exploration in dangerous environments, like finding chemical residues [2, 5] or performing search and rescue procedures after natural disasters [6, 7].

Existing work has addressed collective motion tuning in open arenas. However, the environment is likely to be cluttered with objects in real-world swarm robotics missions

© The Author(s), under exclusive license to Springer Nature Switzerland AG 2022
H. Aziz et al. (Eds.): AI 2022, LNAI 13728, pp. 761–774, 2022.
https://doi.org/10.1007/978-3-031-22695-3_53

[8, 9]. This raises new challenges of how to represent such external goals in a format that the robots can understand, and how this format should be integrated with intrinsic goals for collective motion.

This paper proposes an architecture for self-bootstrapping collective motion behaviours in the presence of an extrinsic goal. The architecture takes a reinforcement learning approach to bootstrapping collective motion and combines generic 'intrinsic' reward for collective motion, with mission specific 'extrinsic' reward for avoiding obstacles or target search in an environment. Multi-Objective Reinforcement Learning (MORL) is a generalized technique of reinforcement learning, that the reward signal is extended to multiple reward functions [10]. Our proposed method does not necessarily follow this concept of MORL, since not only both the intrinsic and extrinsic goal should be reached at the same time, but also, none of the goals surpasses the other.

We demonstrate that our reinforcement learner can tune the behaviour of randomly moving groups so that structured collective motion emerges, that is suited to both the robot and the externally imposed constraints of its environment. We compare our approach to behaviour bootstrapping in open arenas to show that the presence of external environmental constraints does not affect the quality of the bootstrapped behaviours in terms of the number of collisions between agents, group metric, and order metric. The contributions of the paper are as follows:

- An architecture for behaviour bootstrapping in the presence of both intrinsic reward for collective motion, and extrinsic reward describing environmental constraints.
- A reward signal that combines the if-then rule engine of the intrinsic reward signal integrated with the numerical, distance-based metrics for defining environmental constraints.
- A demonstration that we can tune collective motion behaviour while satisfying environmental constraints in a point-mass simulator, in less than 20 actions.

The reminder of this paper is organised as follows: Sect. 2 presents background and related work. Section 3 describes our approach of collective motion tuning in the presence of extrinsic goals. Section 4 discusses the experimental setup and the results. Section 5 concludes the paper and examines the directions for future work.

2   Background and Related Work

2.1   Boid Guidance Algorithm

The boid Guidance Algorithm (BGA) is an extension of Reynold's boid model [11] suitable for robots with a mass and movement constraints. Boid is short for Bird Android. Reynold's boid model is the first computer-based system inspired by flocks of birds. The simulated motion uses three rules which ensures movement with the same direction (alignment), close to each other (attraction/cohesion) and without running into each other (repulsion/separation).

In simulation for a set of N boids $B^i \in \{B^1, B^2, \ldots, B^N\}$ at timestep t, the separation force s_t^i, alignment force a_t^i, and cohesion force c_t^i can be computed as Eq. (1), (2), and (3), [12].

$$\vec{s_t^i} = \frac{\sum_k x_t^k}{\left|(N_s)_i^t\right|}, \quad s_t^i = x_t^i - \vec{s_t^i} \tag{1}$$

$$a_t^i = \frac{\sum_k V_i^k}{\left|(N_a)_i^t\right|} \tag{2}$$

$$\vec{c_t^i} = \frac{\sum_k x_t^k}{\left|(N_c)_i^t\right|}, \quad c_t^i = \vec{c_t^i} - x_t^i \tag{3}$$

By considering R_s, R_a, R_c as the radii in which the separation, alignment and cohesion forces are applicable, $(N_s)_i^t, (N_a)_i^t, (N_c)_i^t$ are the subsets of boids within the corresponding radius ranges of boid B^i at timestep t. Moreover, x_t^i and V_t^i are the position point and velocity vector of boid B^i at timestep t. Then in each timestep, the boids position will be updated using Eq. (4).

$$x_{t+1}^i = x_t^i + v_{t+1}^i \tag{4}$$

where the velocity vector will be updated using the following equation.

$$v_{t+1}^i = v_t^i + W_c c_t^i + W_a a_t^i + W_s s_t^i \tag{5}$$

2.2 Reinforcement Learning

Reinforcement learning (RL) is a trial-and-error learning approach, which learns to accumulate reward by choosing an appropriate action to change the current state to reach more rewarding states [13, 14]. The main parameters of the reinforcement learning are states (S), actions (A), and reward (R), which makes a tuple (S, A, R) in each step of each learning episode. Each episode contains a finite number of steps, limited by a termination criterion. The termination criterion occurrence is when the current step matches the objective or when the episode has passed a step length criterion [15, 16]. In each step, the environment is acted on by the agent, and changes it to a new state. A reward value shows how effective this action was, then the new state and computed reward are sent to the agent for the next step [13, 17]. The RL agent should exploit a good action which received a high reward in the past but should also explore to find better actions in future. A policy is the structure which stores a record of good actions for the current state that will maximize future reward [13].

2.3 Actor-Critic Reinforcement Learning for Collective Motion Tuning

Reinforcement learning for Collective Motion Tuning (CoMoT), [1, 2] uses a rule engine as the reward signal generator. This rule engine is extracted from a decision tree trained with human labelled data [18]. Applying this rule engine permits automatic collective motion tuning, since tuning is a simple task for humans and this rule engine imitates human perception of collective motions [19, 20]. The consequence of if-then rules is a binary decision of "collective motion" or "random" behaviour [18]. Using this rule engine, the reward signal generator can recognize collective motion from random motions for the purpose of rewarding state-action pairs which results in collective motion. Both the action space and state space are continuous working with collective behaviour parameters discussed in Sect. 2.1. The new state results in a new movement pattern of boids in the point-mass simulator. The average decision of the rule engine for at least 10 timestep of monitoring the current state, generates the reward/penalty signal. The trained reinforcement learning is able to automatically tune a collective motion from any random motion in an open arena with less than 10 actions. For more information on the reinforcement learning of CoMoT please see e.g., [1].

This paper addresses the question of whether CoMoT can learn structured behaviours in the presence of external goals. Our methodology is presented in the next section.

3   Actor-Critic Reinforcement Learning for Collective Motion Tuning in Presence of Extrinsic Goal

This section discusses the methodology to implement the reinforcement learning with intrinsic goal of collective motion tuning, and existence of extrinsic goal. As presented in Fig. 1, the collective motion tuning is called the intrinsic goal (driven by intrinsic reward) because it depends only on the physical properties of the agent and their communications. Also, the obstacle avoidance or target search goals are two instances of extrinsic goal because each refers to properties of the environment external to the agent. We anticipate that in future these goals could be set by an external entity such as a human or computer mission controller.

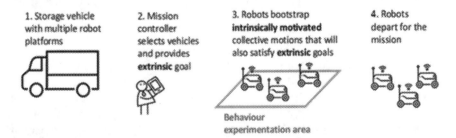

Fig. 1. Sources of intrinsic and extrinsic reward

The state space for the intrinsic goal is a continues space, consisting of the situational awareness parameters like separation, alignment, and cohesion forces, and the probability of applying these forces on boids motion in each simulation timestep. To design an architecture which could cover both the intrinsic and extrinsic goal, some new parameters are required in this state space. These additional parameters pay attention to the extrinsic goal and enhance the capability of the state space for bootstrapping both intrinsic and extrinsic goals. A distance between the boids and the goal (for example an obstacle or a target), a force to avoid of or be attract to the goal in two sample scenarios of obstacle avoidance and target search, and a radii to apply this force on boids, are some instances of these parameters.

The reward signal is the component which should compromise between these two parameter sets, and their corresponding intrinsic and extrinsic goals. Therefore, for a bootstrapping architecture, we need to choose the reward values for extrinsic goals in a way that they could compete with the intrinsic goal, but do not suppress it. To do this we defined a general reward signal as Eq. (6). The constant value $\alpha \geq 1$ enforces higher reward with the same rate of importance for both goals when both intrinsic and extrinsic goals are satisfied. Doing multiple experiments, large α values lead to overfitting, while using low α values the policy could not be trained [1]. Therefore, a moderate value of α could better compromise between the reward signals. Moreover, the punishment should be a negative reward, and in a way that the extrinsic goal does not surpass the intrinsic goal. To do this, a $\beta > 0$ value is required to balance the significance degree of punishment of extrinsic goal and intrinsic goal. In Eq. (6), R_{in} and R_{ex} are the reward signal value for intrinsic and extrinsic goals, respectively. G_i is the goal of collective motion tuning with existence of an extrinsic goal i. Satisfactory bootstrapping in this equation refers to generating an automatic collective behaviour, while reaching the extrinsic goal. More detailed information on the reward signal generator of the intrinsic goal has been provided in [1]. Expanding this reward signal to deal with extrinsic goals is one of the main contributions of this paper.

$$
R_{G_i} = \begin{cases} (R_{in} + R_{ex}) \times \alpha & \text{if bootstrapping of } G_i \text{ is satisfactory} \\ -\left(\frac{1}{R_{in}} + \frac{1}{R_{ex} \times \beta}\right) & \text{otherwise} \end{cases} \tag{6}
$$

This paper considers two instances of using this general reward signal generator, where the goal is tuning collective motion subject to extrinsic goals of: 1) obstacle avoidance, and 2) target search. The adaptation of a general reward signal for bootstrapping intrinsic reward and each of the instances of extrinsic reward are mentioned in Eq. (7) and (8). In these equations, $\overline{D} = \{ mean(D_{i_t}) | 1 \leq i \leq N_b \, \& \, 0 < t \leq 10\}$ is the average score of the if-then rule engine for number of boids (N_b) in 10 timesteps (t). It is the reward signal for intrinsic goal of collective motion tuning. The rule engine is extracted from a decision tree trained on a binary dataset of collective motion recognition [21]. It is able to recognize collective behaviour of boids by monitoring the temporal state values (Sect. 2.1) for maximum of 10 timesteps. The score for "collective motion" is 5 ($D_{i_t} = 5$), and for "not collective motion" is -1 ($D_{i_t} = -1$). For more information see e.g., [1, 2, 18]. Moreover, to compromise between the intrinsic goal and each of the instances of extrinsic goal the moderate value of $\alpha = 100$ is selected for the positive reward. Also, $\beta = 10$ for extrinsic goal of obstacle avoidance, and $\beta = N_b$ for extrinsic goal of target search, could manage the negative reward signal in a way that extrinsic goal does not surpass the intrinsic goal. In Eq. (7) δ_o is the average distance of boids from obstacles, which is controlled by avoidance force from obstacle, and $|N|_c$ is the number of collisions between boids and the obstacle within the 10 timesteps. In Eq. (8), $|N|_a$ is the average number of boids attracted to the target as the result of attraction force, within the attraction radius of R_T over 10 timesteps.

$$R_{G_1} = \begin{cases} (\overline{D} + \delta_o) \times 100 & \text{if bootstrapping of } G_1 \text{ is satisfactory} \\ -\left(\frac{1}{\overline{D}} + \frac{1}{(|N|_c \times 10)}\right) & \text{otherwise} \end{cases} \tag{7}$$

$$R_{G_2} = \begin{cases} (\overline{D} + |N|_a) \times 100 & \text{if bootstrapping of } G_2 \text{ is satisfactory} \\ -\left(\frac{1}{\overline{D}} + \frac{1}{(|N|_a \times N_b)}\right) & \text{otherwise} \end{cases} \tag{8}$$

The action space is also a continues space which could change the current state with respect to each goal. In the designed architecture, the bootstrapping policy is learned by an actor-critic deep neural network. The actor network is a deep neural network of fully connected layers of Rectified Linear Units (ReLU) layer, tangent hyperbolic layer and scaling layer. The critic network estimates Q-values and approximates the reward signal of the current state and action pair. The critic network is a fully connected deep neural network of a ReLU layer and tangent hyperbolic layer. The architecture of the actor and critic deep neural networks are based on the default MATLAB configurations, and the number of nodes equals to the number of state parameters.

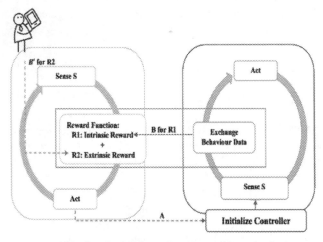

Fig. 2. Reinforcement learning framework for collective motion tuning in the presence of extrinsic goal. The learning cycle (left) can modify the parameters of the robot centric (right) and satisfy both intrinsic reward and extrinsic reward (A is the action, S is the state, B is the temporal state data and B' is the live positioning of agents)

Training consists of 2000 episodes and a maximum of 100 steps in each episode. In each episode, the initial state is a random motion, which boids move randomly with no embedded pattern and without paying attention to the extrinsic goal. Then an action changes the current state S to a new state. The new state produced by the learning cycle (left of Fig. 2) results in a new movement pattern of boids in the point-mass simulator (robot centric cycle on the right of Fig. 2). Then the learner collects temporal state values which discussed in Sect. 2.1 for all boids ($N_b = 200$), and in a given number of timesteps (10 timesteps). The collected temporal state values are used to compute the reward signal. If the current state receives both intrinsic and extrinsic reward, then the training of current episode ceases, and the new episode starts. If each of the intrinsic or extrinsic goal is not rewarded, then the procedure iterates for a maximum of 100 steps, until the condition for both intrinsic and extrinsic goals are met.

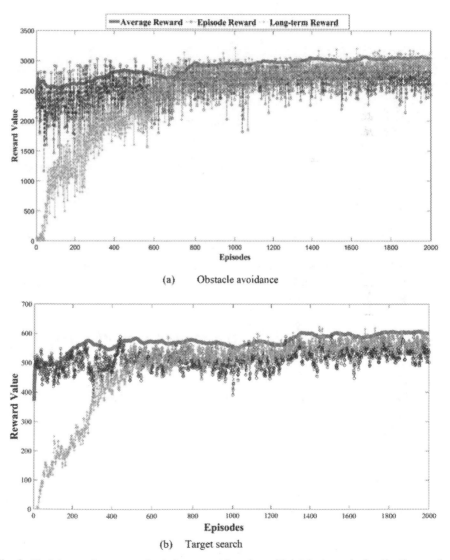

(a)   Obstacle avoidance

(b)   Target search

Fig. 3. Training performance of reinforcement learning with intrinsic goal of collective motion tuning and extrinsic goal of (a) obstacle avoidance and (b) target search (average over 10 runs)

The average over 10 runs of training the reinforcement learning with intrinsic reward of collective motion tuning and extrinsic reward of obstacle avoidance and target search is presented in Fig. 3. As shown in this figure, the average reward converges to the long-term reward, which confirms that the policy has been trained by 2000 episodes. The training metrics are defined as follows:

- Episode Reward is the cumulative reward of all the steps of each episode of the reinforcement learning.
- Average Reward is the average over current and all previous episode rewards.
- Long-term reward (Episode $Q0$) is the critic estimate of the discounted long-term reward based on the initial state of each episode.

4 Experimental Study

In this section we run two experiments to evaluate the trained reinforcement learning. The experiments are as follows:

1. The first experiment tunes collective motion with extrinsic goal for 100 random scenarios. The aim of this experiment is to investigate the performance of the trained reinforcement learner in tuning collective motion in the presence of an extrinsic goal starting from previously unseen random motions.
2. The second experiment examines the swarm metrics, including the number of collisions between boids, group, and order for tuned motions in an open arena, in an environment with an obstacle, and in an arena with target. The aim of this experiment is to evaluate the quality of collective motion in cluttered environments in the presence of an extrinsic goal.

4.1 Experiment 1: Evaluation of the Trained Reinforcement Learner for Collective Motion Tuning in Presence of Extrinsic Goal

All the 10 trained reinforcement learning methods show a very good performance of reaching intrinsic and extrinsic goal, therefore one of the 10 trained reinforcement learning methods of Sect. 3, is picked to tune boids' behaviour in the following experiments. To evaluate the trained learner, 100 random movements are first generated, which satisfy both of the following properties:

- They are not recognised as a collective motion by the if-then rule engine,
- They do not satisfy the extrinsic goal.
- For obstacle avoidance: At least one collision is detected between the boids and the obstacle.
- For target search: the target is not reached by all the boids.

Exp 1–1: Extrinsic Goal of Obstacle Avoidance
By applying the trained reinforcement learning from Fig. 3(a) the reward values for each of these 100 random configurations with existence of obstacle is a positive value, which indicates that the trained reinforcement learner tunes the random configurations,

so that the behaviour is a collective motion and there is no collision between the boids and the obstacle. The 100 initial random configurations are non-collective behaviour, and there is at least one collision with the obstacle. These tuned behaviours could be achieved so fast and in less than 6 actions (each action could be achieved in less than a second). Figure 4, presents some examples of tuned behaviours. The direction of boids movement is shown by blue arrows in this figure. As these arrows show, the boids move with a pattern, approaching the obstacle while avoiding it by maintaining a predefined distance threshold. The areas visited by boids in 10 timesteps is presented in Fig. 5. This figure shows the visited areas of the four samples of Fig. 5. The most visited grid cells (visited by more than 30 boids) during these 10 timesteps are presented with the darkest colour, while the least visited ones (visited by less than 2 boids) are presented with the lightest colour. As shown in this figure, for tuned behaviours, the grid cells around the obstacle within the radius of R_O are always represented by the lightest colour. This indicates that the boids avoid this area during their collective movement.

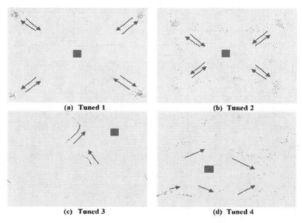

Fig. 4. Some examples of tuned motions by the reinforcement learner with intrinsic goal of collective motion and extrinsic goal of obstacle avoidance (The small black dots are boids, the red squares are obstacle, and the blue arrows show the direction of boids motion) (Color figure online)

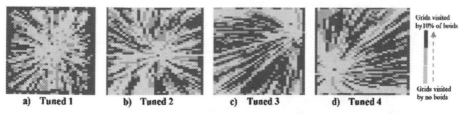

Fig. 5. Grids covered by boids when tuned by the reinforcement learner with intrinsic goal of collective motion and extrinsic goal of obstacle avoidance (The red rectangle is the obstacle) (Color figure online)

Exp 1–2: Extrinsic Goal of Target Search

Again, by applying one of the trained reinforcement learnings of Fig. 3(b), the final reward values for collective motion tuning with extrinsic goal of target search is positive for all 100 scenarios. The 100 initial random configurations are non-collective behaviour, and not all the boids converge to the target position. However, the tuned collective motions could be achieved very fast and in less than 20 actions (each action could be achieved in less than a second). Figure 6 shows some of the tuned motions of these 100 scenarios. The blue arrows indicate the direction of boids' movement. In this figure the boids provide a collective motion (a pattern is visible in their motion). Moreover, the direction of movement shows that they are moving toward the target. The second snapshot of each part of this figure shows a set of boids which reach the target and wiggle around it. Moreover, Fig. 7 shows the grid cells which are visited by boids within 10 timesteps, for each of the motions of Fig. 6. In this figure, the most visited grids are presented with the darkest colour. The grid cells which are least visited are presented with the lightest colour. As shown in these figures, the boids mostly move around the target during the last 10 timesteps of simulation, confirming that tuning has occurred.

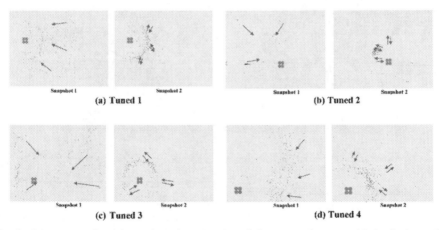

Fig. 6. Some examples of tuned motions by the reinforcement learner with intrinsic goal of collective motion and extrinsic goal of target search (The small black dots are boids, the green circles represent a target, and the blue arrows show the direction of boids motion) (Color figure online)

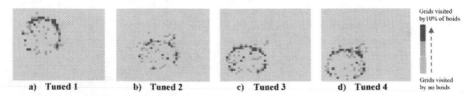

Fig. 7. Grids covered by boids when tuned by the reinforcement learner with intrinsic goal of collective motion and extrinsic goal of target search (The red rectangle is the target) (Color figure online)

4.2   Experiment 2: Effect of Extrinsic Goal on Collective Motion Metrics

In this experiment we aim to compare the quality of collective motions generated in the presence of extrinsic goals to those generated in open arena. By this experiment, we aim to investigate how collective motions could be compromised by adding an extrinsic goal, and how reliable is the reinforcement learner in taking care of both intrinsic and extrinsic goals. There exist different indicators to measure collective behaviour. Group, order, and the number of collisions between boids are some popular indicators to evaluate a collective motion [22], which are defined as follows:

- The group metric is an indicator that measures the cohesion of the collective behaviour, based on the average distance between position of boids u_i and the average over boids' positions $u_{average}$ [23], and is computed by Eq. (9).

$$Group = \frac{1}{N} \sum_{i=1}^{N} \| u_{average} - u_i \| \tag{9}$$

- Order metric is an indicator which measures the alignment of a collective behaviour. This metric uses average distance between velocity of boids V_i from the average velocity $V_{average}$, [24]. The order metric is computed by Eq. (10).

$$Order = \frac{1}{N} \sum_{i=1}^{N} \| V_{average} - V_i \| \tag{10}$$

- Number of collisions among boids is another indicator which measure the separation of a collective behaviour. This is computed based on the distance between boids. If the distance between boids is less than a threshold, then the number of collisions increases.

For this experiment, we use the same 100 random scenarios of Experiment 1, then for each of them the reinforcement learning tunes policies for collective motions: 1) in absence of extrinsic goals [1], 2) in the presence of obstacle, and 3) in the presence of a target. Then the three metrics are computed for the tuned collective motions for each of these examined reinforcement learners. To compute the metrics, each of the tuned collective motions are monitored for 100 timesteps. The average group values for tuned collective motions by each of the reinforcement learning methods is presented in Fig. 8(a), and the average order values are presented in Fig. 8(b). As is visible in these figures, all methods have similar average group values, which indicates that adding the extrinsic goal to the reinforcement learning, does not make the boids less cohesive. Also, as Fig. 8(b) shows, slightly larger order values are achieved when the extrinsic goal is added to the reinforcement learning. This increase is significant at the 95% confidence level and is likely due to the need for boids to change direction when an obstacle or a target is detected. Moreover, the average over number of collisions between boids for these 100 tuned scenarios are computed for two different distance thresholds. The first threshold considers a very small distance of 0.5 between boids. For this threshold the average value is 1 for all methods, which indicates that there is an average of 1 collision per episode among boids. Then a greater threshold is considered in Fig. 8(c). The threshold applied is 25. This value is the lowest separation radius to handcraft a collective motion, as discussed by Khan, et al., [12]. As it is presented in this figure, adding the extrinsic goal provides a tighter formation of boids, to assist them to avoid an obstacle or attract to a target. However, the intrinsic goal compromises efficiently, so that the boids avoid collision with each other within this tight formation.

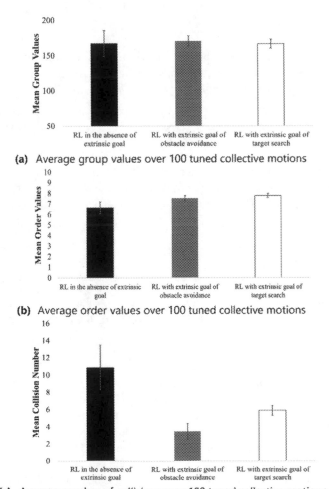

(a) Average group values over 100 tuned collective motions

(b) Average order values over 100 tuned collective motions

(c) Average number of collisions over 100 tuned collective motions

Fig. 8. Collective motion metric values over 100 scenarios and 100 timesteps of monitoring with 95% confidence interval

5   Conclusion and Future Work

This paper has presented an architecture for self-bootstrapping collective motion behaviours in the presence of extrinsic goal relevant for transport tasks, specifically obstacle avoidance and target search. We have demonstrated that our reinforcement learner can tune the behaviour of randomly moving groups so that structured collective motion emerges in the presence of externally imposed goals. We compared our approach to behaviour bootstrapping in open arenas to show that the presence of external environmental constraints does not affect the quality of the bootstrapped behaviours. Some avenues for future work are as follows:

- Learning in more complicated environment with dynamic obstacle or target: Although in the current work the position of obstacle or target in different scenarios are different, but they are static during each experiment. Therefore, we need to consider a situation when the obstacle or target could dynamically change its position during the tuning procedure.
- Learning on real robots: The work in this paper is based on simulated boids, however the implementation of the RL tuning method on real robots which have different global positioning methods, could be a more challenging task to do. In this case, mature RL or Multi-agent RL will be helpful.

References

1. Abpeikar, S., Kasmarik, K., Garratt, M., Hunjet, R., Khan, M.M., Qiu, H.: Automatic swarm behaviour tuning using actor-critic deep reinforcement learning. Swarm Evol. Comput. **72**, 101085 (2022)
2. Abpeikar, S., Kasmarik, K., Tran, V., Garratt, M., Anavatti, S., Khan, M.M.: Tuning swarm behaviour for environmental sensing tasks represented as coverage problems. In: Asadnia, M., Razmjou, A., Beheshti, A. (eds.) Artificial Intelligence and Data Science in Environmental Sensing, pp. 155–178. Elsevier (2022)
3. Navarro, I., Matía, F.: An introduction to swarm robotics. ISRN Robotics **2013**, 1–10 (2013)
4. Kolling, A., Walker, P., Chakraborty, N., Sycara, K., Lewis, M.: Human interaction with robot swarms: a survey. IEEE Trans. Hum.-Mach. Syst. **46**(1), 9–26 (2016)
5. Amjadi, A.S., Raoufi, M., Turgut, A.E., Broughton, G., Krajník, T., Arvin, F.: Cooperative pollution source localization and cleanup with a bio-inspired swarm robot aggregation (2019). arXiv:1907.09585
6. Niroui, F., Zhang, K., Kashino, Z., Nejat, G.: Deep reinforcement learning robot for search and rescue applications: exploration in unknown cluttered environments. IEEE Robot. Autom. Lett. **4**(2), 610–617 (2019)
7. Berman, I., et al.: Trustable environmental monitoring by means of sensors networks on swarming autonomous marine vessels and distributed ledger technology. Frontiers in Robotics and AI **7**, 70 (2020)
8. Rogers, K., Wiles, J., Heath, S., Hensby, K., Taufatofua, J.: Discovering patterns of touch: a case study for visualization-driven analysis in human-robot interaction. In: 2016 11th ACM/IEEE International Conference on Human-Robot Interaction (HRI), pp. 499–500 (2016)
9. Croitoru, A.: Deriving low-level steering behaviors from trajectory data. IEEE Int. Conf. Data Min. Workshops **2009**, 583–590 (2009)
10. Van Moffaert, K., Nowé, A.: Multi-objective reinforcement learning using sets of pareto dominating policies. J. Mach. Learn. Res. **15**(1), 3483–3512 (2014)

11. Reynolds, C.W.: Flocks, herds, and schools: a distributed behavioral model. Computer Graphics **21**(4), 25–34 (1987)
12. Khan, M., Kasmarik, K., Barlow, M.: Autonomous detection of collective behaviours in swarms. Swarm Evol. Comput. **57**, 100715 (2020)
13. Sutton, R.S., Barto, A.G.: Reinforcement Learning: An Introduction, 2nd edn. Adapt Comput Mach Le (2018)
14. Busoniu, L., Babuska, R., De Schutter, B.: A comprehensive survey of multiagent reinforcement learning. IEEE Trans. Syst. Man Cybern. C **38**(2), 156–172 (2008)
15. Fernández, F., Veloso, M.: Probabilistic policy reuse in a reinforcement learning agent. In: Proceedings of the Fifth International Joint Conference on Autonomous Agents and Multiagent Systems, pp. 720–727 (2006)
16. Haarnoja, T., Zhou, A., Abbeel, P., Levine, S.: Soft actor-critic: off-policy maximum entropy deep reinforcement learning with a stochastic actor. In: International Conference on Machine Learning, pp. 1861–1870. PMLR (2018)
17. Wang, R.Q.: Reinforcement learning: an introduction. In: Proceedings of 2006 International Conference on Artificial Intelligence, pp. 632–637 (2006)
18. Abpeikar, S., Kasmarik, K., Tran, P.V., Garratt, M.: Transfer learning for autonomous recognition of swarm behaviour in UGVs. In: Australian Joint Conference on Artficial Intelligence, to appear, Sydeny, Australia, vol. 13151, pp. 531–542 (2022)
19. Harvey, J., Kasmarik, K., Abbass, H.A.: Application of chaos measures to a simplified boids flocking model. Swarm Intell. **9**(1), 23–41 (2015)
20. Harvey, J., Kasmarik, K., Abbass, H.A.: Assessing human judgment of computationally generated swarming behavior. Frontiers in Robotics and AI **5**, 13 (2018)
21. Abpeikar, S., Kasmarik, K., Barlow, M., Khan, M.M.: Swarm Behaviour Dataset on UCI Data Repository. http://archive.ics.uci.edu/ml/datasets/Swarm+Behaviour
22. Hussein, A., et al.: Characterization of indicators for adaptive human-swarm teaming. Frontiers in Robotics and AI **9** (2022)
23. Ferrante, E., Turgut, A.E., Duenez-Guzman, E., Dorigo, M., Wenseleers, T.: Evolution of self-organized task specialization in robot Swarms. Plos Comput. Biol. **11**(8), e1004273 (2015)
24. Vicsek, T., Czirók, A., Ben-Jacob, E., Cohen, I., Shochet, O.: Novel type of phase transition in a system of self-driven particles. Phys. Rev. Lett. **75**(6), 1226–1229 (1995)

Enhancing the Speed of Hierarchical Learning Automata by Ordering the Actions - A Pioneering Approach

Rebekka Olsson Omslandseter[1]([⊠]), Lei Jiao[1], and B. John Oommen[2]

[1] Department of Information and Communication Technology, University of Agder,
4879 Grimstad, Norway
{rebekka.o.omslandseter,lei.jiao}@uia.no
[2] School of Computer Science, Carleton University, Ottawa K1S 5B6, Canada

Abstract. For the past six decades, the operation of Learning Automata (LA) has involved states and action probabilities. These have been central to "remembering" the quality of the actions chosen during the learning. The latest enhancements have also incorporated estimates of the actions' reward probabilities. However, a phenomenon that has never been used to-date is that of considering how these actions *themselves*, can be *ordered*. Ordering the actions in traditional LA is rather meaningless unless one resorts to invoking the theory of Random Races [1]. However, we show that such an ordering makes sense if the automata operate hierarchically, within a tree, with the actions being placed at the leaves. In this paper, we shall show that when the LA are arranged "in a tree formation", and when the learning is achieved within such a tree, the hierarchical LA has a superior performance if the actions located at the leaves of the tree are arranged suitably. While this concept can be incorporated in *any* hierarchical LA, we demonstrate its power for the most recent machine, i.e., the Hierarchical Discretized Pursuit Automaton (HDPA). These strategies can also be included in the Hierarchical Continuous Pursuit Automaton (HCPA), and to both of these which utilize traditional Maximum Likelihood (ML) or Bayesian estimates [2]. The experimental results presented here are very impressive, and so, if we consider the chronology of LA from FSSA through VSSA, the Estimator schemes, and the recent hierarchical LA, our modest claim is that the inclusion of the ADE represents the state-of-the-art which is not easily surpassed.

Keywords: Learning automata · Reinforcement learning · Hierarchical learning automata

1 Introduction

The field of Learning Automata (LA)[1] concerns non-human agents learning the optimal action from a set of actions through the principles of Reinforcement

---

[1] The term LA is used interchangeably to address the field of Learning Automata or the Learning Automata themselves, depending on the context.

H. Aziz et al. (Eds.): AI 2022, LNAI 13728, pp. 775–788, 2022.
https://doi.org/10.1007/978-3-031-22695-3_54

Learning (RL). The learning agent in LA is often referred to as the *Learning Automaton*. The system that the learning agent operates in and learns through interactions with, is often referred to as the *Environment*. The learning agent selects an action, and the Environment responds with a feedback to the automaton. The feedback can be a set of discrete responses or from a continuous range, but most commonly, the feedback is binary, consisting of either a reward or a penalty. Depending on the LA's learning policy, the LA updates its knowledge based on the feedback, and hopefully, learns to output the action that yields the highest probability of getting a reward over time. The LA learns in a *semi-supervised* manner, meaning that it does not need examples of solutions but learns via the feedback in a trial-and-error mode. The metric of quantifying the performance of LA is the average number of *iterations* it takes, over an ensemble of experiments. For the Variable Structure Stochastic Automata (VSSA) type of LA used in this paper, the algorithm converges once the LA attains a specific probability level that is arbitrarily close to unity.

1.1   Memory Considerations

Learning cannot be achieved without remembering certain quantities during the process. We shall informally refer to these as the *"memory"*. However, even if one remembers all the pertinent information in a very efficient manner, the learning will not succeed if the algorithm that utilizes the memory, is poor. To bring out the salient features of the pioneering contribution of this paper, we briefly itemize the respective components of the different families.

- **FSSA**: Fixed Structure Stochastic Automata (FSSA) are LA, where the memory is encapsulated in states which are identical to those possessed by Finite State Machines or flip flops. Examples of these are the Tsetlin, Krinsky, and Krylov LA [3]. In each case, the learning algorithm directs the LA to move across the states based on the response from the Environment, and each of the latter boast their own individual strategy. Correspondingly, they all have different convergence characteristics.
- **VSSA**: Unlike FSSA, in VSSA, the memory is contained in the action probability vector, $P(n)$. The action is chosen based on $P(n)$, which is then communicated to the Environment. $P(n+1)$ is obtained in the next step, and is based on $P(n)$, the action chosen, $\alpha(n)$, and the feedback that the Environment provides, $\beta(n)$. The updating algorithm, on the other hand, can be varied and includes, among others, the Linear Reward-Penalty (L_{R-P}) scheme, the Linear Reward-Inaction (L_{R-I}) scheme, the Linear Inaction-Penalty (L_{I-P}) scheme, and the Linear Reward-ϵPenalty ($L_{R-\epsilon P}$) [4,5].
- **Estimator Algorithms**: In Estimator LA, the memory resides in the action probability vector and running estimates of the reward probabilities. In the Pursuit algorithm, the learning algorithm now increases the probability of the currently recorded best action and not of the action that is chosen. In the Pursuit algorithm, only the probability of the best action is increased. In the Generalized Pursuit, the action probabilities of a subset of actions

are increased, while the rest of the probabilities are decreased. One has to mention that the estimates can be done in an ML or Bayesian manner [2], and the updates done in a continuous or a Discretion paradigm.

- **Hierarchical LA**: In the family of hierarchical LA, the machines are arranged in a tree structure, and the actions are at the leaves. These individual LAs can be VSSA or can be Pursuit machines themselves. More details of this are included in the next section.

The above bullets briefly encapsulates the entire prior art.

1.2   Action Ordering Considerations

The reader will observe that throughout the above discussions, the ordering of the actions has remained insignificant. This is valid because the ordering is unknown unless one resorts to a prior Random Race competition that is not relevant to our present study [1]. Of course, the ordering of the actions in an action probability vector is also meaningless.

The hypothesis of this study is that there is an advantage to ordering the actions. Clearly, such an ordering can only be enforced if there are crude estimates of the reward probabilities. If they are arranged linearly, ordering the actions can enhance the corresponding choice by resorting to a fast searching mechanism, as opposed to a linear search. We shall not elaborate on that issue here.

However, let us consider the case when the automata operate in a hierarchical manner. The actions then are placed at the leaves of the tree, and the decisions of the individual LA trickle up to the root. Our hypothesis is that rather than keeping the leaves completely unordered, information gleaned during the initial learning phase can be used to order them, and to yield a superior performance. This is precisely the hypothesis and contribution of our paper.

1.3   Contributions of This Paper

The contributions of this paper can be summarized as follows:

- Unlike the prior art, we show that there is an advantage in considering the ordering of the actions when the LA operate in a hierarchical manner.
- We demonstrate this, by considering the most recent machine in the field, the HDPA.
- We confirm the hypothesis, by reporting the results of extensive simulations in different Environments and a host of distributions.

As mentioned above, the concept presented in this paper is true, not only for the HDPA, but also for any other type of machine utilizing a hierarchical structure.

2   Related Work

The paradigm of LA originated in the 1960s with Michael Lvovitch Tsetlin and his innovation of learning agents and, ultimately, the Tsetlin Automata [6]. Later advancements followed, and the types of LA are generalized into two categories, namely FSSA and the VSSA. In VSSA, we have the Linear Reward-Penalty(L_{R-P}) scheme, the Linear Reward-Inaction(L_{R-I}) scheme, the Linear Inaction-Penalty(L_{I-P}) scheme, and the Linear Reward-ϵPenalty($L_{R-\epsilon P}$) [4,5]. In these different schemes, the probability vector is updated linearly. The updating can also be done in a non-linear manner [4,5,7]. At the same time, VSSA schemes can be continuous or discrete [8]. Continuous type VSSA updates the probability in a multiplicative manner with a factor, while the Discretion type updates the probability with a constant in each update. Due to the multiplicative updating, the continuous type can experience slower algorithm speeds than the Discretion type. Thus, when an action selection probability gets closer to unity, the change in its probability becomes less and less. The continuous and discrete updating functionality has been investigated mathematically in [9,10].

Another major discovery in the field of LA was the Estimator-based Algorithms (EAs), which significantly increased the convergence speed of VSSA [11]. The concept of EAs is the utilization of estimation. In more detail, the LA keeps reward estimates while in operation, using these estimates to pursue the currently most promising action (referred to as *Pursuit* in the Literature) [12]. Researchers combined the Pursuit concept with Discretion updating, leading to the paradigm of Discrete Estimator Algorithms (DEAs) [13], superior to earlier VSSA variants in terms of convergence speed.

Although all of the advances mentioned above increased the applicability and efficiency of LA dramatically, VSSA still had an impediment as the number of actions (possible solutions to a problem) became *large* (e.g., more than ten [14]). Therefore, the authors of [14] proposed the HCPA, bringing structure to the domain of VSSA. The HCPA was a quantum step to the field of LA, making VSSA able to handle a large number of actions. However, as the accuracy requirement to the HCPA became large, e.g., above 0.98, the HCPA suffered from its multiplicative property of updating its action selection probabilities, resulting in sluggish convergence. In [15], the HDPA was proposed, combining VSSA, Discretion updating, the Pursuit concept, and structure. The HDPA provides a solution to problems with many actions and high accuracy requirements, constituting the state-of-the-art for generic LA, being significantly faster when compared to the HCPA.

3   Incorporating Ordering into an Hierarchical LA

Although the HDPA improved the convergence speed significantly for high accuracy requirements, we experienced that the convergence speeds were substantially dependent on the action distribution on the leaf level of the tree. Linked to the concept of Random Races [1], where the LA is modeled to find an ordering of the actions in an ascending/descending order, we hypothesize that we can order the actions in a manner that is beneficial to the algorithm. More specifically, we propose that we can use an Estimation Phase in the HDPA process and also the estimated reward probabilities to reorder the actions at the leaf level to yield an improved performance. Consequently, in this paper, we propose the Action Distribution Enhancing (ADE) approach for enhancing the convergence speed of the HCPA and HDPA. The improvement in the convergence speed becomes more noticeable as the number of actions at the leaf level increases. Therefore, organizing the actions in an improved manner can significantly reduce the number of iterations before the convergence is achieved. While this was our intended hypothesis, as demonstrated through extensive simulations documented later in the paper, we show that the ADE approach is, indeed, beneficial compared to randomly initializing the actions at the leaf level of the tree. The reader should note that the problems that LA can solve are random in nature, and for a real problem, no information about the reward probabilities can be known *a priori*. Therefore, understandably, the Estimation Phase is needed to obtain an improved ordering.

3.1   Motivating Arguments

To motivate the development of our new paradigm, we consider a problem involving four actions $\mathcal{A} = \{\alpha_1, \alpha_2, \alpha_3, \alpha_4\}$ with the corresponding estimated action probability vector $\hat{D} = \{\hat{d}_1, \hat{d}_2, \hat{d}_3, \hat{d}_4\}$, taken over an initial estimation phase of 20 iterations. The reader must please observe that because the number of iterations are small, the corresponding estimates will be inaccurate. Also, before we proceed with the arguments, it is wise to see how these estimates will effect the learning process. Further, in the interest of simplicity, we proceed with the discussion by considering the case of the HDPA instead of any other arbitrary hierarchical LA.

At the leaf levels, the four actions are to be placed in one of the 4! positions. It is also obvious that the descending and ascending orders of the placements of the actions are merely mirror reflections of each other. On a deeper examination of an Hierarchical Pursuit LA, one observes that from every level, the estimates of the most suitable actions chosen at that level will be trickled up. This implies that the automata at each level will be dealing with problems of different complexities. However, the most important automaton is the one placed at the root, because that governs, or rather dictates, the operations of all the automata below it.

Consider the following figure in which the four actions are placed at the leaves in the descending order (Fig. 1a). From Fig. 1a, we see that the LA $A_{1,1}$ has to deal with actions whose reward estimates are $\frac{15}{20}$ and $\frac{11}{20}$. Similarly, when the actions are in a more random order, Fig. 1b, the corresponding reward estimates for the LA $A_{1,1}$ are $\frac{15}{20}$ and $\frac{3}{20}$. If all goes well as in a perfect world, the trickled up estimates are those of the optimal actions of their corresponding children. The root level, which is encountering the most important task, has now to deal with distinguishing between the reward estimates $\frac{15}{20}$ and $\frac{7}{20}$ in Fig. 1a, and $\frac{15}{20}$ and $\frac{11}{20}$ in Fig. 1b. This means that the root automaton, that has to solve the most

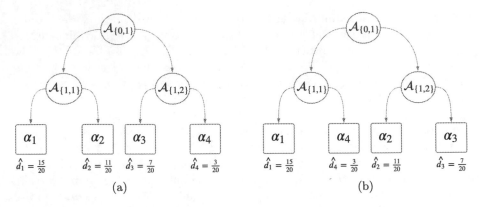

Fig. 1. Example of two hierarchical tree structures for four actions with different action distributions at the leaf level.

discriminating problem of all the automata, has to resolve actions α_1 and α_2, in the case of Fig. 1b, which is much more difficult than the problem in Fig. 1a.

When we use the euphemistic expression above "if all goes well in a perfect world", we emphasize that it is statistically not at all unrealistic. This is because by the law of large numbers or the Estimation Phase in the Bayesian case, the estimates of the reward probabilities will converge to their true values with an arbitrarily high accuracy. Thus, the asymptotic arguments (and probabilities) of the trees of Fig. 1a and Fig. 1b will still be valid[2].

4   The Action Distribution Enhancing (ADE) Approach

As mentioned earlier, such an ADE approach applies to both the HCPA and the HDPA, and indeed, to any hierarchical LA. However, because the HDPA has demonstrated superior performance to the HCPA for high accuracy requirements, and we have limited space, we only highlight the ADE approach for the HDPA, by understanding that the principles for the HCPA are analogous. The ADE approach for the HCPA is similar to the approach explained for the HDPA.

The ADE approach concerns distributing the actions at the leaf level of the hierarchical tree in an improved manner. For a practical, real-life problem, there can be little information as *a priori* information about the actions. This is why the distribution of the actions at the leaf level is an intricate problem which has not yet been considered in the Literature. The ADE approach is two-pronged. The first prong is the *Estimation Phase*, used for estimating the action reward probabilities. The second concerns distributing the actions in an improved manner, and is referred to as the *Reallocation Process*.

---

[2] The proof that the ADE approach represents a superior solution compared with unordered solutions, will be proven in the extended version of the paper [16].

The first part of the ADE approach concerns the Estimation Phase. In the HDPA, the estimated reward probability and the action selection probabilities of LAs throughout the tree structure are initialized as 0.5, according to [14] and [15]. Thus, the HDPA estimates the reward probabilities as per the Pursuit concept (maintaining an estimated reward probability vector). We propose a standalone Estimation Phase with the ADE approach prior to the HDPA starting its regular operation. This means that in this phase, we include θ iterations per action for estimating their reward probabilities. These estimates are further utilized as to initialize the corresponding values in the regular operation of the HDPA, which happens after the Reallocation Process.

The second part of the ADE approach concerns the Reallocation Process, which distributes the actions in an improved manner. For a two-action LA configured tree, such an organization can be achieved by ordering the actions according to their estimated reward probabilities in either an ascending or descending order. In this way, asymptotically, the optimal and second optimal actions will share the same LA at the level below the root. Thus, asymptotically, the algorithm will have achieved a correct estimation of the reward probabilities, and the actions will be distributed in an improved manner. Clearly, due to the stochastic nature of the problem, the Estimation Phase might not always yield a perfect interpretation of the reward estimates and the corresponding trees. This phenomenon and its consequences are explained further in the section with the experimental results (Sect. 5).

The Reallocation Process uses the estimated reward probabilities from the Estimation Phase to reallocate the actions to the tree's leaves in an ascending/descending order. Thus, after the Estimation Phase, the actions are given a new location at the leaf level. The reader should note that the estimates also need to be updated according to this new ordering. By maintaining these estimates, the information from the Estimation Phase is also retained. After the Reallocation Process, the HDPA starts its *normal* operation, by utilizing the reward estimates from the Estimation Phase[3].

5   Experimental Results

To demonstrate the effectiveness of the proposed ADE approach presented in this paper, we conducted experiments with various action distributions, numbers of actions, and Environments. For quantifying the effectiveness of the LA algorithms, we recorded the number of iterations required before convergence, as this is the most common evaluating measurement [5]. As mentioned earlier, the convergence requirement for VSSA is that one of the action selection probabilities attains a certain threshold (T). In our experiments, we tested different convergence criteria, and report the results for the most utilized convergence threshold to be 0.995. The reader should note that this metric is helpful because

---

[3] Although the algorithm have been explained in details verbally in this paper, a more detailed programmatic description of the algorithm will be presented in an extended version of the paper [16].

it will remain identical, regardless of the computing power on the machine used for the experiments, or the efficiency of the code or language used[4].

Paired together with this, is the accuracy of the algorithm. Due to the stochastic nature of the problem, one often conducts more experiments and reports the average performance. In terms of accuracy, we usually measure this as the percentage of experiments which have converged to the optimal action. Consequently, when a 100% accuracy is achieved, all the experiments conducted in an ensemble of experiments have converged to the optimal action (i.e., the action with the highest reward probability). The reader should note that the number of iterations used for the Estimation Phase is also reckoned into the overall number of iterations in our experimental results.

In LA, the tuning of the learning parameter leads to a trade-off between the accuracy and the speed. Generally, as the learning parameter becomes smaller, the number of iterations increases, and at the same time, the algorithm performs more accurately. The same dilemma applies to the algorithm proposed in this paper. As demonstrated in the experiments, placing the optimal and sub-optimal actions in opposite parts of the tree at the leaf level requires more iterations than establishing them as entities in the same part of the tree (for example, i.e., with an ascending/descending ordering). Thus, the ascending/descending orders generally require substantially less number of iterations before convergence.

With our experiments, we wanted to demonstrate the behavior of the HDPA for different action distributions at the leaf level, thereby demonstrating the improved performance with the ADE approach. In practice, we have little or no *a priori* information about the reward estimates in a real-life scenario. In our simulations, we know that with a knowledge of the exact reward probabilities, we are able to execute the programs for different action distributions with and without the ADE approach. By demonstrating the phenomena for different configurations, we intend to highlight the improved performance that can be achieved in a real-life scenario by using the ADE approach with the Estimation Phase and Reallocation Process before the actual LA learning is performed.

In our simulations, we denote the *real* reward probabilities, i.e., the probability of getting a reward for selecting a certain action, as d_j, where $j \in \{1, 2, ..., 2^K\}$. If nothing else is specified, for the ADE HDPA, we used a descending ordering in the Reallocation Process. To help understand the ordering at the leaf level, we present visualizations of the action distributions with their corresponding reward probabilities. Please note that these visualizations show the real reward probability of α_1 to α_{2^K}, i.e., d_1 to d_{2^K}, for the actions at the leaf level. Consequently, the HDPA without the ADE will run the experiments with the

---

[4] In our experiments, we have configured the convergence criterion as being achieved once *any* of the LA has attained a certain threshold of choosing one of the actions in its action probability vector. However, in [15], they defined the convergence as being achieved only when all the LA along the path to a leaf action had attained the prescribed threshold. Thus, the convergence criterion in this paper is different, i.e., it utilizes the "logical or" instead of the "logical and", making the algorithms (i.e., both the HDPA without/with the ADE) attain a faster convergence.

actions ordered as displayed by the configuration. Conversely, the ADE HDPA will reallocate the actions at the leaf level in an ascending/descending order based on the corresponding reward estimates.

5.1   Simulation 1: 8 Actions

Let us first consider the results for Simulation 1 presented in Tables 1 and 2, which involve Environments of 8 actions. The difference between the two tables is that Table 1 does not incorporate the ADE, and Table 2 does. The categoric superiority of the results in Table 2 demonstrates the power of the ADE.

Table 1. Experimental results for different action distributions without the ADE approach for eight actions with $T = 0.995$ as the convergence criterion. The results were averaged over $1,000$ experiments, with $\Delta = 9e-5$.

Config.	d_1	d_2	d_3	d_4	d_5	d_6	d_7	d_8	Avg	Std	Acc.
1	0.99	0.95	0.87	0.6	0.43	0.54	0.67	0.51	6,727.40	42.76	100%
2	0.87	0.6	0.43	0.54	0.67	0.51	0.99	0.95	6,791.75	56.81	100%
3	0.87	0.6	0.99	0.95	0.43	0.54	0.67	0.51	6,730.42	42.03	100%
4	0.87	0.6	0.43	0.99	0.95	0.54	0.67	0.51	7,082.38	215.46	100%
5	0.99	0.6	0.43	0.54	0.67	0.51	0.87	0.95	7,109.98	220.78	100%
6	0.95	0.99	0.51	0.67	0.54	0.43	0.6	0.87	6,791.15	57.52	100%
7	0.99	0.95	0.87	0.67	0.6	0.54	0.51	0.43	6,713.85	42.49	100%

Table 2. Experimental results for different action distributions with the ADE HDPA with eight actions and $T = 0.995$ as the convergence criterion. The results were averaged over $1,000$ experiments, with $\Delta = 9e-5$.

Config.	d_1	d_2	d_3	d_4	d_5	d_6	d_7	d_8	Avg	Std	Acc.
1	0.99	0.95	0.87	0.6	0.43	0.54	0.67	0.51	6,810.10	44.66	100%
2	0.87	0.6	0.43	0.54	0.67	0.51	0.99	0.95	6,824.48	51.26	100%
3	0.87	0.6	0.99	0.95	0.43	0.54	0.67	0.51	6,823.52	49.19	100%
4	0.87	0.6	0.43	0.99	0.95	0.54	0.67	0.51	6,821.79	49.51	100%
5	0.99	0.6	0.43	0.54	0.67	0.51	0.87	0.95	6,825.32	49.49	100%
6	0.95	0.99	0.51	0.67	0.54	0.43	0.6	0.87	6,813.32	47.15	100%
7	0.99	0.95	0.87	0.67	0.6	0.54	0.51	0.43	6,810.91	44.67	100%

In Simulation 1, we ran 1,000 experiments with similar reward probabilities associated with the Environment, but with different action distributions. In Tables 1 and 2 we have marked the optimal and optimal action in at the leaf level with a different background color. For example, we can observe Config. 5, where the optimal and sub-optimal actions are located in opposite parts of the tree. We had earlier explained that this configuration was the hardest for the

HDPA to handle. As opposed to this, having the actions with regard to the reward probabilities in an ascending/descending order, like the case of Config. 7, would be easier. This is confirmed from the results in Table 1, where the HDPA with the ADE has a superior performance. Config. 4 and Config. 5 required the highest number of iterations and had high standard deviations (Std). However, for Config. 7, where the actions were ordered in a descending order according to the reward probabilities, the number of iterations was the lowest. This is obvious since this is the most optimal setting. Again, the results for the ascending and descending orders are almost identical.

In Table 2 we present the results for the experiments for the ADE HDPA for Simulation 1. As we observe from the table, some configurations required more iterations than the other configurations without the ADE approach. However, the ADE HDPA was more consistent in the number of iterations, and had a more stable and smaller standard deviation. Config. 2, which is the reversed form of Config. 6, yielded a little higher standard deviation than the others. We also, tested Config. 2 for the ADE HDPA with an ascending ordering in the Reallocation Process, and these experimental results were similar to those of Config. 6. More specifically, for 1,000 experiments with the ADE HDPA and an ascending ordering in its Reallocation Process, the algorithm required 6,818.14 iterations on average and a standard deviation of 48.46 yielded 100% accuracy.

In real-life, we do not know the underlying reward probabilities. Therefore, we can only tune the number of tests, θ, that is used in the Estimation Phase. Testing each action for a larger number of iterations in the Estimation Phase will make the ADE more certain that it has estimated the reward probabilities correctly, and it will, thus, order them correctly as well. However, in most cases, because we want a fast convergence, a rough estimate might be sufficient. In Simulation 1, we used $\theta = 12$. If we had *perfect* estimation of the reward probabilities, we would have similar results to Config. 7 without the ADE.

5.2   Simulation 2: 16 Actions

In Simulation 2, we increased the number of actions to 16. Again, we set the value of θ to be 12. The different original action distributions are visualized in Fig. 2, where we focus on the optimal and sub-optimal actions' locations in the different configurations. However, the actions in between them were distributed more or less *randomly*. In subsequent simulations, we shall highlight what happens when we have more constructed forms in the original distributions.

In Tables 3 and 4, we present the results for Simulation 2. Analogous to Simulation 1, we see that the number of iterations is more consistent, and that the standard deviation is smaller for the ADE HDPA (Table 4). The HDPA without ADE still had better results for the configurations where the actions were manually ordered in an ascending/descending order, which is quite understandable. Thus, the ADE HDPA did probably not achieve the *perfect* estimation of the reward probabilities, since the number of iterations used for the estimation, where $\theta = 12$ and $R = 16$ ($\theta R = 192$), was too small.

Furthermore, the simulation demonstrated a bigger gain using the ADE approach for 16 actions when compared with the 8 actions case. As an example, for Row No. 1, the ADE HDPA used approximately 11,550 iterations before convergence, while the HDPA without the ADE used approximately 12,700 iterations, which yielded a superiority of more than 1,000 iterations. Thus, the ADE had an approximately 9.95% better performance in terms of the number of iterations. In comparison, for Simulation 1, the biggest gain of using the ADE approach was approximately 4.25%.

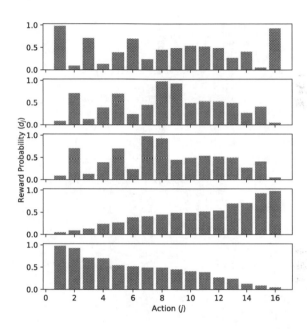

Fig. 2. The figure shows the action distribution in accordance with the reward probabilities for Simulation 2. Config. 1 is at the top, Config. 5 is at the bottom, and the others are ordered in between them systematically.

5.3 Simulation 3: 32 Actions

In Tables 5 and 6, we present the results obtained for Simulation 3, which involved 32 actions. The configurations in these experiments followed the same concept as depicted in Fig. 2 for Simulation 2. We can observe that the HDPA without the ADE required considerably more iterations for the case when the optimal and sub-optimal actions were in opposite parts of the tree (Row No. 1), i.e., compared with having the actions ordered in a descending order (Row No. 5). The numbers of iterations were approximately 17, 700 and 15, 800, respectively. Comparing the results in Table 5 with those in Table 6, we observe that the ADE HDPA had a more consistent performance in terms of the number of iterations and the standard deviations. For Row No. 1 in the tables, we see that the HDPA with the ADE required approximately 16, 350, while the HDPA without the ADE required 17, 700, which is approximately 8.26% worse.

Table 3. Experimental results for different action distributions without the ADE for Simulation 2, with 16 actions and 0.995 as the convergence criterion. The results were averaged over 100 experiments, with $\Delta = 6.75e-5$. The different rows represent different action configurations as described in the second column.

Row No.	Configuration characteristics	Avg	Std	Acc.
1	Config 1 in Fig. 2	12,691.92	509.64	100%
2	Config 2 in Fig. 2	12,362.99	413.44	100%
3	Config 3 in Fig. 2	11,674.32	100.95	100%
4	Ascending	11,285.18	83.24	100%
5	Descending	11,291.90	86.09	100%

Table 4. Experimental results for different action distributions with the ADE for Simulation 2, with 16 actions and 0.995 as the convergence criterion. The results were averaged over 100 experiments, with $\Delta = 6.75e-5$. The different rows represent different action configurations as described in the second column.

Row No.	Configuration characteristics	Avg	Std	Acc.
1	Config 1 in Fig. 2	11,556.77	111.29	100%
2	Config 2 in Fig. 2	11,540.57	106.94	100%
3	Config 3 in Fig. 2	11,543.75	94.47	100%
4	Ascending	11,573.89	119.07	100%
5	Descending	11,520.16	91.70	100%

Table 5. Experimental results for different action distributions without the ADE for Environments with 32 actions where 0.995 is the convergence criterion. The results were averaged over 100 experiments, with $\Delta = 4.5e-5$. The different rows represent different configurations as described in the second column.

Row No.	Configuration characteristics	Avg	Std	Acc.
1	As in Config 1 in Fig. 2	17,690.81	849.65	100%
2	As in Config 2 in Fig. 2	17,980.64	742.41	100%
3	As in Config 3 in Fig. 2	16,911.9	600.69	100%
4	Ascending	15,807.09	89.47	100%
5	Descending	15,806.49	90.23	100%

Table 6. Experimental results for different action distributions with the ADE HDPA for 32 actions and 0.995 as the convergence criterion. The results were averaged over 100 experiments, with $\Delta = 4.5e-5$ and $\theta = 12$. The different rows represent different configurations as described in the second column.

Row No.	Configuration characteristics	Avg	Std	Acc.
1	As in Config 1 in Fig. 2	16,338.12	114.54	100%
2	As in Config 2 in Fig. 2	16,302.25	121.44	100%
3	As in Config 3 in Fig. 2	16,327.96	121.10	100%
4	Ascending	16,371.00	134.76	100%
5	Descending	16,256.67	107.27	100%

6   Conclusion

In this paper we have proposed the novel Action Distribution Enhancing (ADE) approach for optimally configuring the underlying hierarchical tree representing the distribution of the actions in the HCPA/HDPA. The ADE involves two phases, the first of which estimates the action probabilities very crudely, and subsequently assigns the actions at the leaves of the tree. The corresponding LA then operate in a hierarchical manner, each of them involving two actions. Our hypothesis was that if the leaves were arranged in an ascending/descending order, the collection of hierarchical automata would perform in their most optimized manner, and we confirmed this hypothesis by both a formal theoretical analysis and experimentally [16].

We have then proceeded to verify the power of incorporating the ADE into problems involving different numbers of automata, and various Environments with corresponding reward probabilities. Quite briefly stated, our simulation results uniformly confirm that the inclusion of the ADE significantly stabilizes and increases[5] the convergence speed of the hierarchical machine.

If we consider the chronology of LA from its infancy in FSSA through VSSA, the Estimator approaches, and the more-recent hierarchical schemes, we modestly believe that the inclusion of the ADE represents the state-of-the-art which will not be able to be surpassed too easily.

References

1. Ng, D.T.H., Oommen, B.J., Hansen, E.R.: Adaptive learning mechanisms for ordering actions using random races. IEEE Trans. Syst. Man Cybern. **23**(5), 1450–1465 (1993)
2. Zhang, X., Granmo, O.-C., Oommen, B.J.: On incorporating the paradigms of discretization and Bayesian estimation to create a new family of pursuit learning automata. Appl. Intell. **39**(4), 782–792 (2013)

---

[5] The speed of HDPA with ADE, compared with vanilla HDPA, indeed decreases a bit for the ascending/descending case. Nevertheless, considering the significant speed gain for other cases, the average speed increases.

3. Tsetlin, M.L.: Automaton Theory and the Modeling of Biological Systems. Academic Press, New York (1973)
4. Lakshmivarahan, S.: Learning Algorithms Theory and Applications, ed. 1. Springer, New York (1981). https://doi.org/10.1007/978-1-4612-5975-6
5. Narendra, K.S., Thathachar, M.A.L.: Learning Automata: An Introduction. Dover Books on Electrical Engineering Series, Dover Publications, Courier Corporation (2013)
6. Tsetlin, M.L.: Finite automata and modeling the simplest forms of behavior. Uspekhi Matem Nauk **8**(4), 1–26 (1963)
7. Lakshmivarahan, S., Thathachar, M.A.L.: Absolutely expedient learning algorithms for stochastic automata. IEEE Trans. Syst. Man Cybern. **SMC–3**(3), 281–286 (1973)
8. Oommen, B.J.: Absorbing and ergodic Discretion two-action learning automata. IEEE Trans. Syst. Man Cybern. **16**(2), 282–293 (1986)
9. Oommen, B.J., Agache, M.: Continuous and Discretion pursuit learning schemes: various algorithms and their comparison. IEEE Trans. Syst. Man Cybernet. Part B (Cybernetics) **31**(3), 277–287 (2001)
10. Zhang, X., Granmo, O.-C., Oommen, B.J.: Discretized Bayesian Pursuit – a new scheme for reinforcement learning. In: Jiang, H., Ding, W., Ali, M., Wu, X. (eds.) IEA/AIE 2012. LNCS (LNAI), vol. 7345, pp. 784–793. Springer, Heidelberg (2012). https://doi.org/10.1007/978-3-642-31087-4_79
11. Thathachar, M.A.L., Sastry, P.S.: Estimator algorithms for learning automata. In: Proceedings of the Platinum Jubilee Conference on Systems and Signal Processing, Department of Electrical Engineering, Indian Institute of Science (1986)
12. Zhang, X., Oommen, B.J., Granmo, O.-C.: The design of absorbing Bayesian pursuit algorithms and the formal analyses of their ϵ-optimality. Pattern Anal. Appl. **20**, 797–808 (2017)
13. Lanctot, J.K., Oommen, B.J.: Discretized estimator learning automata. IEEE Trans. Syst. Man Cybern. **22**(6), 1473–1483 (1992)
14. Yazidi, A., Zhang, X., Jiao, L., Oommen, B.J.: The hierarchical continuous pursuit learning automation: a novel scheme for environments with large numbers of actions. IEEE Trans. Neural Networks Learn. Syst. **31**(2), 512–526 (2020)
15. Omslandseter, R.O., Jiao, L., Zhang, X., Yazidi, A., Oommen, B.J.: The hierarchical discrete learning automaton suitable for environments with *Many* actions and *High* accuracy requirements. In: Long, G., Yu, X., Wang, S. (eds.) AI 2022. LNCS (LNAI), vol. 13151, pp. 507–518. Springer, Cham (2022). https://doi.org/10.1007/978-3-030-97546-3_41
16. Omslandseter, R.O., Jiao, L., Oommen, B.J.: Pioneering Approaches for Enhancing the Speed of Hierarchical LA by Ordering the Actions". Unabridged version of this paper. To be submitted for publication

A Robust Exploration Strategy in Reinforcement Learning Based on Temporal Difference Error

Muhammad Shadi Hajar[1]([✉]) [iD], Harsha Kalutarage[1] [iD],
and M. Omar Al-Kadri[2] [iD]

[1] Robert Gordon University, Aberdeen AB10 7GJ, UK
{m.hajar,h.kalutarage}@rgu.ac.uk
[2] Birmingham City University, Birmingham B4 7XG, UK
omar.alkadri@bcu.ac.uk

Abstract. Exploration is a critical component in reinforcement learning algorithms. Exploration exploitation trade-off is still a fundamental dilemma in reinforcement learning. The learning agent needs to learn how to deal with a stochastic environment in order to maximize the accumulated long-term reward. This paper proposes a robust exploration strategy (RES) based on the temporal difference error. In RES, the exploration problem is modeled using Beta probability distribution to control the exploration rate. Moreover, the most promising action is selected during the exploration with a view to maximizing the accumulated reward and avoiding un-rewardable wrong actions. RES has been evaluated on the k-armed bandit problem. The simulation results show superior performance without the need to tune parameters.

Keywords: Reinforcement learning · Exploration · Exploitation · Q-learning · k-armed bandit · ε-greedy · Softmax

1 Introduction

Reinforcement learning (RL) is a branch of machine learning where a learning agent tries to map situations to actions with a view to maximizing the long-term reward. Without prior knowledge, the learning agent must discover which actions are more rewardable. Taking action at any state affects not only the immediately received reward but all the subsequent rewards. Hence, it may prevent the learning agent from converging to the global optimum. This reliance on the return feedback from the environment necessitates the learning agent to explore the whole environment to take optimal actions and maximize the long-term rewards. Therefore, RL algorithms count on exploration to obtain sufficient informative feedback from the environment to exploit the most rewardable actions.

The exploration-exploitation trade-off is still a fundamental problem. When learning agents over-explore the environment, they cannot maximize the accumulated reward as exploratory actions may return minimum rewards. Moreover,

H. Aziz et al. (Eds.): AI 2022, LNAI 13728, pp. 789–799, 2022.
https://doi.org/10.1007/978-3-031-22695-3_55

exploiting uncertain action-value functions may yield less reward and make the convergence suboptimal. Thus, this problem is known as the exploration-exploitation dilemma, which has been widely investigated by mathematicians and is still unresolved [1]. ε-greedy and softmax exploration methods are widely used in the literature to balance exploration-exploitation. These two straight-forward algorithms perform well in some cases and thus are hard to beat [2]. However, they require rigorous parameter tuning and may not hold in dynamic environments.

The main contribution of this paper is proposing a robust exploration strategy (RES) based on Temporal Difference (TD) error to effectively balance exploration-exploitation with a view to maximizing the accumulated long-term reward. The well-known k-armed bandit problem has been used to demonstrate the robustness of the proposed method.

The remainder of this paper is organized into five sections as follows. Related work is given in Sect. 2. The used methodology is comprehensively discussed in Sect. 3, followed by the evaluation results in Sect. 4. Finally, Sect. 5 concludes the paper and highlights future work.

2   Related Work

Several approaches have been proposed to produce an efficient exploration algorithm in order to balance the exploration-exploitation trade-off. However, the main two approaches are the blind exploration approach, such as ε-greedy [1], and the value-based approach, such as softmax [3].

In blind exploration approach [3], the learning agent solely explores the environment based on randomly taken actions. It is a reward-free method to explore the environment without any kind of information. Thus the action selection process during exploration is uniform. In real life, ε-greedy is always the first choice for developers due to its simplicity and near-optimal results despite the time-consuming process of tuning its single parameter [4]. In this method, the parameter $\varepsilon \in [0, 1]$ controls the exploration rate as shown in Eq. 1. Although this method is widely adopted in RL, it has some drawbacks. First is the afore-mentioned issue of tuning the parameter ε. Second, choosing random action during the exploration may significantly degrade the learning agent performance as taking random action at time t could affect all future rewards negatively.

$$a_t = \begin{cases} \underset{a_t \in A_t}{argmax} \ Q(s, a) & with \ probability \ (1 - \varepsilon) \\ a \ random \ action & with \ probability \ \ \varepsilon \end{cases} \tag{1}$$

In a value-based approach, the learning agent takes informative action based on its estimation of the action-value functions. The exploration-exploitation trade-off is achieved by assigning probabilities to the available actions at time step t based on the current action-value functions. The predominant algorithm in this approach is the softmax action selection algorithm [3]. Softmax method is usually modeled using Boltzmann distribution as shown in Eq. 2, where the greedy action is always chosen with the highest probability. In contrast, other possible actions are weighted according to their corresponding action-value functions.

$$\pi(a|s) = Pr\{a_t = a|s_t = s\} = \frac{e^{\frac{Q(s,a)}{\tau}}}{\sum_1^n e^{\frac{Q(s,a)}{\tau}}} \qquad (2)$$

where τ is the temperature parameter, and it is used to control choosing the greedy action. When τ is decreased, the greedy action probability increases, and when $\tau \to \infty$, all possible actions will have the same probability, and hence the action will be selected randomly, as in ε-greedy. Tuning the temperature parameter is not straightforward [3]. Moreover, as the probabilities are calculated based on the actual estimated values, the action selection process is highly influenced by these values, which necessities re-tuning the temperature parameter whenever the reward function has changed, even for the same problem.

In [5] and [2], the authors proposed two methods based on ε-greedy and softmax algorithms, respectively. In both works, the authors used a method called Value-Difference Based Exploration (VDBE), which is a state-dependent exploration probability method to control the exploration-exploitation trade-off. Although the simulation results show promising results, there is still a need to tune the sensitivity parameter σ, which is used in both methods.

3   Methodology

In this section, the exploration-exploitation dilemma has been introduced, and the design requirement has been presented. Moreover, the proposed method has been discussed comprehensively.

3.1   Overview

In RL, the intelligent learning agent interacts with the environment E through a series of state-action pairs with a view to maximizing the accumulative long-term reward. The agent is modeled using the tuple (S, A, R), where S represents a set of states, A is the action set, and R is the reward function, as illustrated in Fig. 1. In value-based RL algorithms, the learning agent uses value functions to estimate the future reward of taking action a in a given state s. Therefore, the action-value function for taking action a given a state (s) and following policy π could be defined as in Eq. 3. At each time step t, the learning agent in state $s \in S$ takes an action $a \in A$, receives a reward $r_{t+1} \in R$, and moves from state s to state s'

$$q_\pi(s, a) \doteq \mathbf{E}_\pi[G_t|S_t = s, A_t = a] = \mathbf{E}_\pi[\Sigma_{n=0}^\infty \gamma^n R_{t+n+1}|S_t = s, A_t = a], \quad \forall s \in S \qquad (3)$$

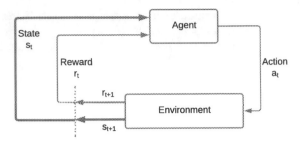

Fig. 1. RL Model

where $\mathbf{E}_\pi[G_t|S_t = s]$ is the expected discounted reward when the agent starts at state s and follows the policy π and $\gamma \in [0,1]$ is the discount factor for the future reward R.

3.2   Q-Learning

Temporal difference (TD) algorithms are a combination of Monte Carlo methods and Dynamic programming (DP) [6]. In TD learning, the agent learns directly from the environment without any prior knowledge, and thus it is called a model-free method. Q-learning is a temporal difference algorithm widely used to predict the action-value function $q_\pi(s,a)$ without any prior knowledge using trials and errors [7]. In Q-learning, the action-value function $Q(s_t,a_t)$ is updated independently of the policy followed by the agent to approximate the optimal action-value function q_*, which makes it applicable for many problems. Q-learning is proven to converge to q_*; however, there is still a requirement that all states still need to be visited and updated.

$$a_t^{*(i)} = \underset{a_t \in A}{argmax}Q_t(s_{t+1}, a_t) \tag{4}$$

$$Q_{t+1}(s_t, a_t) \leftarrow (1-\alpha)Q_t(s_t, a_t) + \alpha[r_{t+1}(s_{t+1}) + \gamma \underset{a \in A}{max}Q_t(s_{t+1}, a_t^*)] \tag{5}$$

where $\alpha \in [0,1]$ is the learning step size where small values slow the learning, and higher values could cause oscillations, and $\gamma \in [0,1]$ is the discount factor where small values make the learning agent nearsighted by ignoring the future rewards.

3.3   The Proposed Method

In RL, the environment can be modeled as a Markov decision process (MDP), as illustrated in Fig. 1. The learning agent interacts with the environment at a series of discrete-time steps $t = 0, 1, 2, ..$, and receives a reward after each interaction. In order to estimate the optimal policy π_*, the learning agent must balance between exploration and exploitation. In TD algorithms, the value functions are updated using an error function representing the difference between the predicted reward at a given state and the actually received reward, as shown in Eq. 6, which represents the TD error in Q-learning.

$$\delta_t \doteq r_{t+1} + \gamma Q(s_{t+1}, a_{t+1}) - Q(s_t, a_t) \tag{6}$$

Thus, the Q-learning updating algorithm could be described as in Eq. 7.

$$Q_{t+1}(s_t, a_t) \leftarrow Q_t(s_t, a_t) + \alpha \delta_t \tag{7}$$

When the Q function converges to the optimal value $Q \rightarrow Q_*$, the TD error converges to zero $\delta_t \rightarrow 0$. Therefore, the TD error could be used to estimate the exploration ratio. The learning agent should explore more when the TD error is high and less when it is small. This approach seems reasonable. However, there are some limitations. The TD error suffers from oscillations, which means that the learning agent could over-explore the environment. Moreover, there is a need to pick the most promising action to perform during the exploration phase. Therefore, the TD error could be used to evaluate the Q values differences trend, which then will be used to choose the most promising action. The TD error trend is exponentially smoothed using Eq. 8.

$$\Delta_{t+1} = \omega \Delta_t + (1 - \omega)\delta_t \tag{8}$$

RES uses a dynamic exploration rate θ. It has been evaluated using Beta probability distribution as in Eq. 9.

$$\theta_t = \frac{\alpha_t}{\alpha_t + \beta_t} \tag{9}$$

where α and β are the Beta distribution levels. After each action, the TD error δ_t and the exponentially smoothed trend Δ_t are evaluated as described in Algorithm 1. This process is modeled using Beta probability distribution to explore more when there is a promising action to explore. After each action, the learning agent compare δ_t and Δ_t. If the TD error is less than or equal to the smoothed trend, this action is considered a successful action, and the beta levels are updated using Eq. 10. Otherwise, it will be unsuccessful, and the levels will be updated using Eq. 11

$$\begin{aligned} \alpha_t &= \lambda\alpha_{t-1} + 1 \\ \beta_t &= \lambda\beta_{t-1} \end{aligned} \tag{10}$$

$$\begin{aligned} \alpha_t &= \lambda\alpha_{t-1} \\ \beta_t &= \lambda\beta_{t-1} + 1 \end{aligned} \tag{11}$$

where $\lambda \in [0, 1]$ is the longevity factor, which used to give more weight to recent observations. Adopting this method has two advantages. First, the amount of exploration will be decreased gradually for a static environment, which means over time, the learning agent will exploit the greedy action more to maximize the overall reward. Second, in a dynamic environment, this method is able to reflect the environment dynamicity into more exploration to help the algorithm re-converge to the global optimum. Therefore, higher values of λ give more weight to previous observations and fit dynamic environments, while smaller values give more weight to current observations and fit static environments.

The second important thing is choosing a promising action during the exploration instead of a random one. Randomly chosen action could be catastrophic in some applications as the chosen action at time t may affect all the future rewards. Therefore, in RES, more weight is given to actions with a high smoothed difference as they are regarded as promising actions to discover. To achieve that, RES uses the Boltzmann distribution [8] of the smoothed differences to calculate the weighted probabilities of the available actions, as shown in Eq. 12.

$$P_{\Delta_t^i} = \frac{e^{\Delta_t^i}}{\Sigma e^{\Delta_t}} : \forall i \in A \tag{12}$$

Algorithm 1: Updating Algorithm

Input: The exponentially smoothed trend: Δ_{t-1}^i
Input: The exploration threshold: θ_{t-1}
Input: The beta distribution levels: α and β
Output: The updated values $\Delta_t^i, P_{\Delta_t^a}, \theta_t$
$\delta_t^i = [r_{t+1}^i + \gamma Q^i(s_{t+1}, a_{t+1}) - Q^i(s_t, a_t)]$
$\Delta_t^i = \omega.\Delta_{t-1}^i + (1-\omega).\delta_t^i$
if $|\delta_t^i| \leq |\Delta_t^i|$ **then**
 $\quad | \quad \alpha = \lambda\alpha + 1$
 $\quad | \quad \beta = \lambda\beta$
else
 $\quad | \quad \alpha = \lambda\alpha$
 $\quad | \quad \beta = \lambda\beta + 1$
$P_{\Delta_t^i} = \frac{e^{\Delta_t^i}}{\Sigma e^{\Delta_t}} : \forall i \in A$
$\theta_t = \frac{\alpha}{\alpha+\beta}$

The aforementioned parameters are updated whether it is an exploration or exploitation cycle as shown in Algorithm 2. The uniform random number $\rho \in [0, 1]$ is drawn for each time step to choosing whether to explore or exploit. If it is an exploitation cycle, then the action will be chosen greedily. Otherwise, it will be weighted randomly chosen based on the Boltzmann distribution of the exponentially weighted TD errors.

4   Evaluation

The proposed method has been evaluated on the k-armed bandit problem, which is one of the common RL problems to evaluate the explorations exploitation methods [5]. It has been widely used to model different problems, such as economic [9] and routing problems [10,11]. The problem represents a bandit machine with multiple arms, pulling one of which gives the player a variable reward. The reward is usually stochastic and drawn from a pre-defined probability distribution. Hence, the learning agent (player) with no prior knowledge has to learn from the pulled uncertain rewards the most rewardable arm by exploring them, which leads to a loss in the gained rewards over time.

4.1   Experimental Setup

The proposed method has been evaluated over long run to prove that it converges to the optimal solution and achieves the maximum reward. The bandit machine consists of 10 levers as described in [1]. At each lever pull, the learning agent gets a stochastic reward drawn from a randomly defined Gaussian distribution $\mathcal{N}(10, 1)$, with mean $Q_*(s, a) = 10$ and standard deviation $\sigma = 1$. The learning agent can improve its selection policy within 2000 trials. Each experiment has been repeated 1000 times, and then the results have averaged out and reported.

RES has been contrasted with ε-greedy [1], softmax action selection [3], and VDBE-Softmax [2]. To ensure a fair comparison, the exploration rate of ε-greedy and softmax have been optimized. At the same time, the optimized value of the VDBE method for the $k-$armed bandit problem is adopted as reported in [5]. In

---

Algorithm 2: Exploration algorithm

Initialize Q values.
Initialize the differences probabilities:
$P_{\Delta_0^a} = \frac{1}{|A_0|} : \forall a \in A$
Initialize the exploration parameters:
$\alpha = 1, \beta = 1$
$\theta_0 = \frac{\alpha}{\alpha + \beta}$
while $TRUE$ **do**
    $\rho \leftarrow rand(0..1)$
    if $\rho \geq \theta$ **then**
        $a_{t+1} \leftarrow \underset{a_t \in A_t}{argmax}\, Q_t(s_t, a_t)$
        Parameters updating using Algorithm 1
        $Q_{t+1}(s_t, a_t) \leftarrow (1 - \alpha)Q_t(s_t, a_t) + \alpha[r_{t+1}(s_{t+1}) + \gamma \underset{a \in A}{max} Q_t(s_{t+1}, a_t)]$
    else
        $a_{t+1} \leftarrow weighted\_rand(P_{\Delta_t})$
        Parameters updating using Algorithm 1
        $Q_{t+1}(s_t, a_t) \leftarrow (1 - \alpha)Q_t(s_t, a_t) + \alpha[r_{t+1}(s_{t+1}) + \gamma \underset{a \in A}{max} Q_t(s_{t+1}, a_t)]$

RES, the longevity factor is a decay factor for the single exponential smoothing of the Beta distribution levels. Its value is set to 0.9 to slowly decrease the weight of old observations over time, which is widely used in the literature [12].

4.2   Simulation Results

In the first experiment, the simulation was run for a high number of iterations to ensure that all methods are converged as some methods show high performance at the beginning; however, over time, they perform poorly. Figure 2 shows the average reward of the four exploration methods. Both softmax and VDBE-Softmax perform well at the beginning as both use the Q function to choose actions during exploration. ε-greedy performs poorly at the beginning but shows good performance over time. The reason behind this behavior is that randomly chosen actions during exploration need more time to build the Q table, which makes even the exploitation actions inefficient. However, the performance enhances significantly when the learning agent gets more evidence from the environment over time.

On the other hand, RES shows superior performance over the long run. As it depends on the exponentially smoothed trend of the TD error, it needs a few exploration steps to estimate Δ_t. Once it gets a few exploration outcomes from the environment, it tends to explore the most promising actions based on its observations. Unlike other exploration methods, the convergence in RES is achieved separately. The most promising Q function converges faster than others as it will be excessively chosen, as shown clearly in Fig. 3, which illustrates the smoothed trend of the TD error over time. Evidently, Δ^1 converges faster to zero than others, although all other values converge over time but considerably slower than Δ^1. The reason behind this behavior is that Δ^1 is the most promising action to take. This could be obviously seen in Fig. 4, which shows the Q values for all actions over time. The most rewardable action is action one; hence, RES

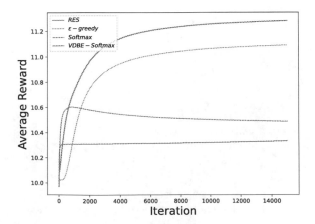

Fig. 2. The reward over time

explored this action more at the beginning, making it converge faster than other actions. Moreover, the high oscillations of this action value confirm that RES always exploits this action to gain maximum rewards.

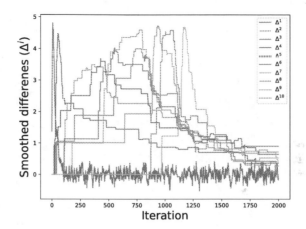

Fig. 3. The exponential smoothed differences over time

In the second experiment, the adaptability of RES is evaluated for a dynamic environment, where the action-value functions may get changed during the simulation. When the environment changes, the optimal action given a state s may become the worst. In light of that, the learning agent should act intelligently and explore the environment to re-converge to the global optimum again. After 50% of the iterations, the Gaussian distribution of the reward function has been changed randomly using the distribution $\mathcal{N}(rand(1, 15), rand(1, 5))$. Figure 5 and 6 show the average static and dynamic environment rewards, respectively.

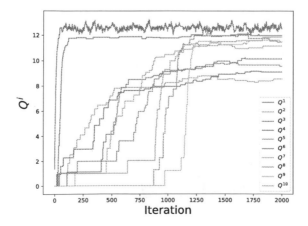

Fig. 4. Q values over time

For a static environment, as illustrated in Fig. 5, RES performs steadily and achieves the highest average reward over time. Moreover, when the environment has changed, as shown in Fig. 6, RES shows high adaptability to re-converge to the new global optimum. Once the change happens, it will be reflected by a change in the TD error and the exponential smoothed trend of the TD error. The former increase the exploration rate, while the latter makes the algorithm chooses the newest promising action.

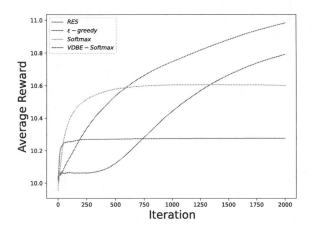

Fig. 5. The reward over time for static environment

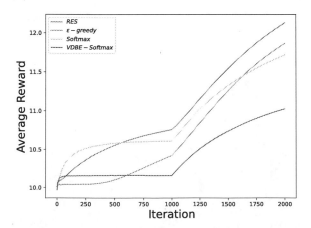

Fig. 6. The reward over time for dynamic environment

5   Conclusion and Future Work

Balancing the exploration-exploitation trade-off in reinforcement learning is still an open area of research. The dilemma lies between exploring the environment to obtain more informative data, which could be un-rewardable, or selecting actions that the learning agent can expect their reward. Our proposed action selection strategy, RES, presents a robust promising solution without the parameter tuning overhead. RES is an adaptive exploration strategy based on the exponentially smoothed trend of the temporal difference error and Beta probability distribution. This novel approach demonstrates outstanding performance on the well-known k-armed bandit problem. In the future, it will be evaluated on other temporal difference algorithms, such as SARSA. Moreover, it will be deployed on a more complicated problem to study its behavior, such as reinforcement learning-based routing protocols.

References

1. Sutton, R.S., Barto, A.G.: Reinforcement Learning: An Introduction. MIT Press (2018)
2. Tokic, M., Palm, G.: Value-difference based exploration: adaptive control between epsilon-greedy and softmax. In: Annual Conference on Artificial Intelligence, pp. 335–346 (2011)
3. Amin, S., Gomrokchi, M., Satija, H., van Hoof, H., Precup, D.: A survey of exploration methods in reinforcement learning. arXiv preprint arXiv:2109.00157 (2021)
4. Vermorel, J., Mohri, M.: Multi-armed bandit algorithms and empirical evaluation. In: Gama, J., Camacho, R., Brazdil, P.B., Jorge, A.M., Torgo, L. (eds.) ECML 2005. LNCS (LNAI), vol. 3720, pp. 437–448. Springer, Heidelberg (2005). https://doi.org/10.1007/11564096_42
5. Tokic, M.: Adaptive ϵ-greedy exploration in reinforcement learning based on value differences. In: Annual Conference on Artificial Intelligence, pp. 203–210 (2010)
6. Apostol, K.: Temporal Difference Learning, p. 96. SaluPress (2012)
7. Jang, B., Kim, M., Harerimana, G., Kim, J.W.: Q-learning algorithms: a comprehensive classification and applications. IEEE Access 7, 133653–133667 (2019)
8. Tijsma, A.D., Drugan, M.M., Wiering, M.A.: Comparing exploration strategies for Q-learning in random stochastic mazes. In: 2016 IEEE Symposium Series on Computational Intelligence (SSCI), pp. 1–8 (2016)
9. Li, J.: The k-armed bandit problem with multiple priors. J. Math. Econ. 80, 22–38 (2019)
10. De Oliveira, T.B., Bazzan, A.L., da Silva, B.C., Grunitzki, R.: Comparing multi-armed bandit algorithms and Q-learning for multiagent action selection: a case study in route choice. In: 2018 International Joint Conference on Neural Networks (IJCNN), pp. 1–8 (2018)
11. Hajar, M.S., Kalutarage, H., Al-Kadri, M.O.: DQR: a double Q learning multi agent routing protocol for wireless medical sensor network. In: 18th EAI International Conference on Security and Privacy in Communication Networks (SecureComm) (2022)
12. Hajar, M. S., Al-Kadri, M. O. & Kalutarage, H.: LTMS: a lightweight trust management system for wireless medical sensor networks. In: 2020 IEEE 19th International Conference on Trust, Security and Privacy in Computing and Communications (TrustCom), pp. 1783–1790 (2020)

Examining Average and Discounted Reward Optimality Criteria in Reinforcement Learning

Vektor Dewanto[✉] and Marcus Gallagher

School of Information Technology and Electrical Engineering,
University of Queensland, Brisbane, Australia
v.dewanto@uqconnect.edu.au, marcusg@uq.edu.au

Abstract. In reinforcement learning (RL), the goal is to obtain an optimal policy, for which the optimality criterion is fundamentally important. Two major optimality criteria are average and discounted rewards. While the latter is more popular, it is problematic to apply in environments without an inherent notion of discounting. This motivates us to revisit a) the progression of optimality criteria in dynamic programming, b) justification for and complication of an artificial discount factor, and c) benefits of directly maximizing the average reward criterion, which is discounting-free. Our contributions include a thorough examination of the relationship between average and discounted rewards, as well as a discussion of their pros and cons in RL. We emphasize that average-reward RL methods possess the ingredient and mechanism for applying a family of discounting-free optimality criteria to RL.

Keywords: Reinforcement learning · Optimality criteria · Average rewards · Discounted rewards · Discounting-free criteria

1 Introduction

Reinforcement learning (RL) is concerned with sequential decision making, where a decision maker has to choose an action based on its current state. Determining the best actions amounts to finding an *optimal* mapping from every state to a probability distribution over actions available at that state. Thus, one fundamental component of any RL method is the optimality criterion, by which we define what we mean by such optimal mapping.

The most popular optimality criterion in RL is the discounted reward [12,17]. On the other hand, there is growing interest in the average reward optimality, as surveyed by [10,19]. In this paper, we discuss both criteria in order to obtain a comprehensive understanding of their properties, relationships, and differences. This is important because the choice of optimality criteria affects almost every aspect of RL methods, including the policy evaluation function, the policy gradient formulation, and the resulting optimal policy, where the term policy refers

H. Aziz et al. (Eds.): AI 2022, LNAI 13728, pp. 800–813, 2022.
https://doi.org/10.1007/978-3-031-22695-3_56

to the above-mentioned mapping. Thus, the choice of optimality criteria eventually impacts the performance of an RL system (and the choices made within, e.g. approximation techniques and hyperparameters).

This paper presents a thorough examination of the connection between average and discounted rewards, as well as a discussion of their pros and cons in RL. The long version of this paper is available at https://arxiv.org/abs/2107.01348.

2   Preliminaries

Sequential decision making is often formulated as a Markov decision process (MDP) with a state set $\mathcal{S}$, an action set $\mathcal{A}$, a reward set $\mathcal{R}$, and a decision-epoch set $\mathcal{T}$. Here, all but $\mathcal{T}$ are finite sets, yielding an infinite-horizon finite MDP. At each decision-epoch (discrete timestep) $t \in \mathcal{T}$, a decision maker (henceforth, an agent) is in a state $s_t \in \mathcal{S}$, and chooses to then execute an action $a_t \in \mathcal{A}$. Consequently, it arrives in the next state $s_{t+1} \in \mathcal{S}$ and earns an (immediate) scalar reward $r_{t+1} \in \mathbb{R}$. For $t = 0, 1, \ldots, t_{\max}$ with $t_{\max} = \infty$, it experiences a sequence (trajectory) of $s_0, a_0, r_1, s_1, a_1, r_2, \ldots, s_{t_{\max}}$. The initial state, the next state, and the next reward are governed by the environment dynamics that is fully specified by three time-homogenous (time-invariant) entities, namely the initial state distribution $\mathring{p}$, the one-(time)step state transition distribution $p(\cdot|s_t, a_t)$, and the reward function $r(s_t, a_t) = \mathbb{E}_{p(\cdot|s_t,a_t)} \left[\sum_{r \in \mathcal{R}} \Pr\{r|s_t, a_t, S_{t+1}\} \cdot r \right]$. Therefore, $S_{t=0} \sim \mathring{p}$, $S_{t+1}|S_t = s_t, A_t = a_t \sim p(\cdot|s_t, a_t)$, and $r_{t+1}:=r(s_t, a_t)$, where S_t and A_t denote the state and action random variables, respectively. We assume that the rewards are bounded, i.e. $|r(s,a)| \leq r_{\max} < \infty, \forall s \in \mathcal{S}, a \in \mathcal{A}$, and the MDP is unichain and aperiodic.

The solution to a sequential decision making problem is an *optimal* mapping from every state to a probability distribution over the (available) action set $\mathcal{A}$ in that state. It is optimal with respect to some optimality criterion, as discussed later in Sect. 2.1. Any of such a mapping (regardless whether it is optimal) is called a *policy* and generally depends on timestep t. It is denoted as $\pi_t : \mathcal{S} \mapsto [0,1]^{|\mathcal{A}|}$, or alternatively $\pi_t : \mathcal{S} \times \mathcal{A} \mapsto [0,1]$, where $\pi_t(a_t|s_t):=\Pr\{A_t = a_t|S_t = s_t\}$ indicating the probability of selecting action $a_t \in \mathcal{A}$ given the state $s_t \in \mathcal{S}$ at timestep $t \in \mathcal{T}$. Thus, each action is sampled from a conditional action distribution, i.e. $A_t|S_t = s_t \sim \pi_t(\cdot|s_t)$.

The most specific solution space is the stationary and deterministic policy set Π_{SD} whose policies $\pi \in \Pi_{\mathrm{SD}}$ are *stationary* (time-invariant), i.e. $\pi:=\pi_0 = \pi_1 = \ldots = \pi_{t_{\max}-1}$, as well as *deterministic*, i.e. $\pi(\cdot|s_t)$ has a single action support (hence the mapping is reduced to $\pi : \mathcal{S} \mapsto \mathcal{A}$). In this work, we consider a more general policy set, that is the stationary policy set Π_{S}. It includes the stationary randomized (stochastic) set Π_{SR} and its degenerate counterpart: the stationary deterministic set Π_{SD}.

2.1   Optimality Criteria

In a basic notion of optimality, a policy with the largest *value* is optimal, i.e.

$$v_x(\pi_x^*) \geq v_x(\pi), \quad \forall \pi \in \Pi_{\mathrm{S}}. \tag{1}$$

Here, the function v_x measures the value (utility) of a policy π based on the *infinite* reward sequence that is earned by an agent following π. The subscript x indicates the specific type of value functions, which induces a specific x-optimality criterion, hence the x-optimal policy, denoted as π_x^*.

One intuitive value function is the *expected* total reward. That is,

$$v_{\text{tot}}(\pi, s) := \lim_{t_{\max} \to \infty} \mathbb{E}_{A_t \sim \pi, S_{t+1} \sim p} \left[\sum_{t=0}^{t_{\max}-1} r(S_t, A_t) \Big| S_0 = s, \pi \right], \quad \forall \pi \in \Pi_S, \forall s \in \mathcal{S}. \tag{2}$$

However, v_{tot} may be infinite (unbounded, divergent, non-summable). Howard [14] therefore, examined the *expected* average reward (also called the *gain*) defined as

$$v_g(\pi, s) := \lim_{t_{\max} \to \infty} \frac{1}{t_{\max}} \mathbb{E}_{A_t \sim \pi, S_{t+1} \sim p} \left[\sum_{t=0}^{t_{\max}-1} r(S_t, A_t) \Big| S_0 = s, \pi \right], \tag{3}$$

which is finite for all $\pi \in \Pi_S$ and $s \in \mathcal{S}$. For more details, including interpretation about the gain, we refer the reader to [10].

Alternatively, Blackwell [7, Sec 4] attempted tackling the infiniteness of (2) through the *expected* total discounted reward, which was studied before by [6]. That is,

$$v_\gamma(\pi, s) := \lim_{t_{\max} \to \infty} \mathbb{E}_{A_t \sim \pi, S_{t+1} \sim p} \left[\sum_{t=0}^{t_{\max}-1} \gamma^t r(S_t, A_t) \Big| S_0 = s, \pi \right], \quad \forall \pi \in \Pi_S, \forall s \in \mathcal{S}, \tag{4}$$

with a discount factor $\gamma \in [0, 1)$. In particular, according to what is later known as the truncated Laurent series expansion (7), Blackwell suggested finding policies that are γ-discounted optimal for all discount factors γ *sufficiently* close to 1. He also established their existence in finite MDPs. Subsequently, Smallwood [28] identified that the discount factor interval can be divided into a *finite* number of intervals, i.e. $[0 = \gamma_m, \gamma_{m-1}), [\gamma_{m-1}, \gamma_{m-2}), \ldots, [\gamma_0, \gamma_{-1} = 1)$, in such a way that there exist policies $\pi_{\gamma_i}^*$ for $0 \leq i \leq m$, that are γ-discounted optimal for all $\gamma \in [\gamma_i, \gamma_{i-1})$. This leads to the concept of Blackwell optimality. A policy π_{Bw}^* is Blackwell optimal if there exists a critical[1] discount factor $\gamma_{\text{Bw}} \in [0, 1)$ such that

$$v_\gamma(\pi_{\text{Bw}}^*, s) \geq v_\gamma(\pi, s), \quad \text{for } \gamma_{\text{Bw}} \leq \gamma < 1, \text{ and } \forall s \in \mathcal{S}, \forall \pi \in \Pi_S. \tag{5}$$

Note that whenever the policy value function v depends not only on policies π but also on states s from which the value is measured, the basic optimality (1) requires that the optimal policy has a value greater than or equal to the other policies' values in all state $s \in \mathcal{S}$.

In order to obtain Blackwell optimal policies, Veinott [33, Eqn 27] introduced a family of new optimality criteria. That is, a policy π_n^* is n-discount optimal for $n = -1, 0, \ldots,$ if

$$\lim_{\gamma \to 1} \frac{1}{(1-\gamma)^n} \left(v_\gamma(\pi_n^*, s) - v_\gamma(\pi, s) \right) \geq 0, \quad \forall s \in \mathcal{S}, \forall \pi \in \Pi_S.$$

---

[1] It is critical in that it specifies the sufficiency of being close to 1 for attaining Blackwell optimal policies.

Veinott showed that $(n = |\mathcal{S}|)$-discount optimality is equivalent to Blackwell optimality because the selectivity increases with n (hence, n_i-discount optimality implies n_j-discount optimalities for all $i > j$). Then, he developed a policy-iteration algorithm (for finding n-discount optimal policies) that utilizes the full Laurent series expansion of $\boldsymbol{v}_\gamma \in \mathbb{R}^{|\mathcal{S}|}$ as follows,

$$\boldsymbol{v}_\gamma(\pi) = \frac{1}{\gamma}\left(\frac{\gamma}{1-\gamma}\boldsymbol{v}_{-1}(\pi) + \boldsymbol{v}_0(\pi) + \sum_{n=1}^{\infty}\left(\frac{1-\gamma}{\gamma}\right)^n \boldsymbol{v}_n(\pi)\right) \quad \text{(Full expansion)}$$

(6)

$$= \frac{1}{1-\gamma}\boldsymbol{v}_{-1}(\pi) + \boldsymbol{v}_0(\pi) + \left\{\boldsymbol{e}(\pi,\gamma)\right\}, \quad \text{(Truncated expansion)} \quad (7)$$

where $\boldsymbol{v}_n \in \mathbb{R}^{|\mathcal{S}|}$ for $n = -1, 0, \ldots$ denotes the expansion coefficients. The truncated form first appeared (before the full one) in [7, Thm 4a], where $\boldsymbol{v}_{-1}$ is equivalent to the gain from all states $\boldsymbol{v}_g \in \mathbb{R}^{|\mathcal{S}|}$, $\boldsymbol{v}_0$ is equivalent to the so-called bias, and $\boldsymbol{e}(\pi,\gamma)$ converges to $\boldsymbol{0}$ as γ approaches 1, that is $\lim_{\gamma \to 1} \mathcal{O}(1-\gamma) = \boldsymbol{0}$.

2.2   Discounting with an Inherent Notion of Discounting

A environment with an inherent notion of discounting has a discount factor γ that encodes one of the following entities. Thus, γ is part of the environment specification (definition, description).

   Firstly, *the time value of rewards, i.e. the value of a unit reward t timesteps in the future is γ^t:* This is related to psychological concepts. For example, some people prefer rewards now rather than latter, hence they assign greater values to early rewards through a small γ (shortsighted). It is also natural to believe that there is more certainty about near- than far-future, because immediate rewards are (exponentially more) likely due to recent actions. The time preference is also well-motivated in economics. This includes γ for taking account of the decreasing value of money (because of inflation), as well as the interpretation of $(1-\gamma)/\gamma$ as a positive interest rate. Moreover, commercial activities have failure (abandonment) risk due to changing government regulation and consumer preferences over time.

   Secondly, *the uncertainty about random termination independent of the agent's actions:* Such termination comes from external control beyond the agent, e.g. someone shutting down a robot, engine failure (due to weather/natural disaster), and death of any living organisms.

   In particular, whenever the random termination time $T_{\max}$ follows a geometric distribution $Geo(p = 1-\gamma)$, we have the following identity between the total and discounted rewards for all $s \in \mathcal{S}$ and all $\pi \in \Pi_S$,

$$v_{T_{\max}}(\pi, s) := \mathbb{E}_{S_t, A_t}\left[\mathbb{E}_{T_{\max}}\left[\sum_{t=0}^{T_{\max}-1} r(S_t, A_t)\Big| S_0 = s, \pi\right]\right] = \underbrace{v_\gamma(\pi, s)}_{\text{See (4)}}, \quad (8)$$

where the discount factor γ plays the role of the geometric distribution parameter [25, Prop 5.3.1]. This discounting implies that at every timestep (for any state-action pair), the agent has a probability of $(1 - \gamma)$ for entering the 0-reward absorbing terminal state. Note that because γ is invariant to states and actions (as well as time), this basic way of capturing the randomness of $T_{\max}$ may be inaccurate in cases where termination depends on states, actions, or both.

3   Discounting Without an Inherent Notion of Discounting

From now on, we focus on environments without *inherent* notion of discounting, where γ is not part of the environment specification (cf. Sec 2.2). We emphasize the qualification "inherent" since any MDP can always be thought of having some notion of discounting from the Blackwell optimality point of view (5). This is because a Blackwell optimal policy is guaranteed to exist in finite MDPs [25, Thm 10.1.4]; implying the existence of its discount factor $\gamma_{\mathrm{Bw}} \in (0, 1]$, and of a (potentially very long) finite-horizon MDP model that gives exactly the same Blackwell-optimal policy as its infinite-horizon counterpart. See also [8, Ch 1.3].

When there is no inherent notion of discounting, the discount factor γ is imposed for bounded sums (Sect. 2.1) and becomes part of the solution method (algorithm). This is what we refer to as *artificial* discounting, which induces *artificial* interpretation as, for instance, those described in Sect. 2.2. The γ_{Bw} (mentioned in the previous paragraph) is one of such artificial discount factors. In addition to bounding the sum, we observe the following justifications that have been made for introducing an artificial γ.

3.1   Approximation to the Average Reward as γ Approaches 1

For recurrent MDPs, the gain optimality is the most selective because there are no transient states [11, Sec 3.1]. This implies that a gain optimal policy is also Blackwell optimal in recurrent MDPs, for which one should target the gain optimality criterion.

Nonetheless, the following relationships exist between the average reward v_g and discounted reward v_γ value functions of a policy $\pi \in \Pi_{\mathrm{S}}$. **Firstly**, for every state $s_0 \in \mathcal{S}$,

$$v_g(\pi, s_0) = \lim_{\gamma \to 1} (1 - \gamma)\, v_\gamma(\pi, s_0), \qquad\qquad \text{[25, Corollary 8.2.5]} \quad (9)$$

$$= (1 - \gamma) \sum_{s \in \mathcal{S}} p_\pi^\star(s|s_0)\, v_\gamma(\pi, s), \quad \forall \gamma \in [0, 1), \qquad \text{[27, Sec 5.3]} \quad (10)$$

where $p_\pi^\star(s|s_0)$ denotes the stationary probability of a state s, i.e. the long-run (steady-state) probability of being in state s when the MC begins in s_0. Here, (9) is obtained by multiplying the left and right hand sides of (7) by $(1 - \gamma)$, then taking the limit of both sides as $\gamma \to 1$. It is interesting that any discount factor $\gamma \in [0, 1)$ maintains the equality in (10), which was also proved by [29, p254].

The second relationship pertains to the gradient of the gain when a parameterized policy $\pi(\boldsymbol{\theta})$ is used. By notationally suppressing the policy parameterization $\boldsymbol{\theta}$ and the dependency on s_0, as well as using $\nabla := \partial/\partial\boldsymbol{\theta}$, this relation can be expressed as

$$\nabla v_g(\pi) = \lim_{\gamma \to 1} \left\{ \sum_{s \in \mathcal{S}} \sum_{s' \in \mathcal{S}} p_\pi^\star(s) \underbrace{\sum_{a \in \mathcal{A}} p(s'|s,a)\pi(a|s)\nabla \log \pi(a|s)\, v_\gamma(\pi, s')}_{\nabla p_\pi(s'|s)} \right\} \tag{11}$$

$$= \underbrace{\sum_{s \in \mathcal{S}} \sum_{a \in \mathcal{A}} p_\pi^\star(s)\pi(a|s)\Big[q_\gamma(\pi, s, a)\nabla \log \pi(a|s)\Big]}_{\text{involving the } \textit{discounted} \text{ state-action value } q_\gamma} + \underbrace{(1-\gamma)\sum_{s \in \mathcal{S}} p_\pi^\star(s)\Big[v_\gamma(\pi, s)\nabla \log p_\pi^\star(s)\Big]}_{\text{involving the } \textit{discounted} \text{ state value } v_\gamma},$$

$$\tag{12}$$

for all $\gamma \in [0, 1)$. Notice that the right hand sides (RHS's) of (11) and (12) involve the discounted reward value function, i.e. the state value function $v_\gamma(\pi, s), \forall s \in \mathcal{S}$ in (4) and the corresponding (state-)action value function $q_\gamma(\pi, s, a), \forall (s, a) \in \mathcal{S} \times \mathcal{A}$, where $v_\gamma^\pi(s) = \mathbb{E}_{A \sim \pi}\big[q_\gamma^\pi(s, A)\big]$. The identity (11) was shown by [4, Thm 2], whereas (12) was derived from (10) by [22, Appendix A].

Thus for attaining average-reward optimality, one can maximize v_γ but merely as an approximation to v_g because setting γ exactly to 1 in (9, 11) is prohibited by definition (4). It is an approximation in (10) whenever v_γ is weighted by some initial state distribution $\mathring{p}$ (such as in (15)) or *transient* state-distributions p_π^t, which generally differs from $p_\pi^\star$. In (12), the second RHS term is typically ignored since estimating $\nabla \log p_\pi^\star(s)$ is difficult in practice; such a difficulty motivates the development of the policy gradient theorem [29, Ch 13.2]. Consequently, $\nabla v_g(\pi)$ is approximated (although closely whenever γ is close to 1) by the first RHS term of (12), then by sampling the state $S \sim p_\pi^\star$ (for example, after the agent interacts long "enough" with its environment).

Moreover, approximately maximizing the average reward via discounting is favourable because discounting formulation has several mathematical virtues, as described in Sect. 3.3.

3.2   A Technique for the Most Selective Optimality with $\gamma \in [\gamma_{\mathrm{Bw}}, 1)$

As discussed in the previous Sec 3.1, γ-discounted optimality approximates the gain optimality as $\gamma \to 1$. This is desirable since the gain optimality is the most selective in recurrent MDPs. In unichain MDPs however, the gain optimality is generally underselective since the gain ignores the rewards earned in transient states. Consequently, multiple gain-optimal policies prescribe different action selections (earning different rewards) in transient states. The underselectiveness of gain optimality (equivalent to $(n = -1)$-discount optimality) can be refined up to the most selective optimality by increasing the value of n from -1 to 0 (or higher if needed up to $n = (|\mathcal{S}| - 2)$ in unichain MDPs) in the family of n-discount optimality.

Interestingly, such a remedy towards the most selective criterion can also be achieved by specifying a discount factor γ that lies in the Blackwell's interval, i.e. $\gamma \in [\gamma_{\text{Bw}}, 1)$, for the γ-discounted optimality (5). This is because the resulting $\pi^*_{\gamma \in [\gamma_{\text{Bw}}, 1)}$, which is also called a Blackwell optimal policy, is also optimal for all $n = -1, 0, \ldots$ in n-discount optimality [25, Thm 10.1.5]. Moreover, Blackwell optimality is always the most selective regardless of the MDP classification since γ-discounted criterion is model-classification-invariant. Thus, artificial discounting can be interpreted as a technique to attain the most selective criterion (i.e. the Blackwell optimality) whenever $\gamma \in [\gamma_{\text{Bw}}, 1)$ not only in recurrent but also unichain as well as the most general multichain MDPs.

Targetting the Blackwell optimality (instead of, gain optimality) is imperative, especially for episodic environments[2] that are commonly modelled as infinite-horizon MDPs (so that the stationary policy set Π_S is a sufficient space to look at for optimal policies). Such modelling is carried out by augmenting the state set with a 0-reward absorbing terminal state (denoted by s_{zrat}). This yields a unichain MDP with a non-empty set of transient states. For such s_{zrat}-models, the gain is trivially 0 for all stationary policies so that gain optimality is underselective. The ($n = 0$)-discount optimality improves the selectivity. It may be the most selective (equivalent to Blackwell optimality) in some cases. Otherwise, it is underselective as well, hence some higher n-discount optimality criterion should be used. It is also worth noting that in s_{zrat}-models, ($n = 0$)-discount optimality is equivalent to the total reward optimality whose v_{tot} (2) is finite [25, Prop 10.4.2]. The latter therefore may also be underselective.

Towards obtaining Blackwell optimal policies in unichain MDPs, the relationship between maximizing γ-discounted and n-discount criteria can be summarized:

$$\underbrace{\operatorname*{argmax}_{\pi \in \Pi_S} v_{\gamma \in [\gamma_{\text{Bw}}, 1)}(\pi, s)}_{\text{Blackwell-optimal policies}}$$

$$= \begin{cases} \operatorname*{argmax}_{\pi \in \Pi_S} \left[\underbrace{v_{-1}(\pi, s) = \lim_{\gamma \to 1} (1 - \gamma) v_\gamma(\pi, s)}_{\text{Based on (9)}} \right] & \text{if recurrent MDPs (see Sec 3.1),} \\[3ex] \operatorname*{argmax}_{\pi \in \Pi^*_{n=-1}} \left[\underbrace{v_0(\pi, s) = \lim_{\gamma \to 1} \frac{(1 - \gamma) v_\gamma(\pi, s) - v_{-1}(\pi, s)}{1 - \gamma}}_{\text{Based on (7)}} \right] & \begin{array}{l}\text{if unichain MDPs and} \\ (n = 0)\text{-discount is the} \\ \text{most selective,}\end{array} \\[3ex] \operatorname*{argmax}_{\pi \in \Pi^*_{n-1}} \left[v_n(\pi, s) \right] \text{ for } n = 1, 2, \ldots, (|\mathcal{S}| - 3) & \begin{array}{l}\text{if unichain MDPs and } n\text{-} \\ \text{discount is the most selective,}\end{array} \\[3ex] \operatorname*{argmax}_{\pi \in \Pi^*_{n=(|\mathcal{S}|-3)}} \left[v_{n=(|\mathcal{S}|-2)}(\pi, s) \right] & \text{if unichain MDPs,} \end{cases}$$

$$(13)$$

for all $s \in \mathcal{S}$, where Π^*_n denotes the n-discount optimal policy set.

From the second case in the RHS of (13), we know that v_0 can be computed by taking the limit of a function involving v_γ and $v_{-1}(= v_g)$ as γ approaches 1. This means that for unichain MDPs, the Blackwell optimal policies may be

---

[2] Episodic environments are those with at least one terminal state. Once the agent enters the terminal state, the agent-environment interaction terminates.

obtained by setting γ very close to 1, similar to that for recurrent MDPs (the first case) but not the same since the function of which the limit is taken differs.

In practice, the limits in the first and second cases in (13) are computed approximately using γ close to unity, which is likely in the Blackwell's interval: $\gamma_{\mathrm{Bw}} \leq (\gamma \approx 1) < 1$. Paradoxically however, this does not necessarily attain the Blackwell optimality because the finite-precision computation involving $\gamma \approx 1$ yields quite accurate estimation to the limit values: maximizing the first and second cases in the RHS of (13) with $\gamma \approx 1$ attains gain ($n = -1$) and bias ($n = 0$) optimality respectively, which may be underselective in unichain MDPs. Consequently in practice, the most selective Blackwell optimality can always be achieved using γ that is at least as high as γ_{Bw} but not too close to 1.

3.3   Several Mathematical Virtues of Discounting

Discounted reward optimality is easier to deal with than its average reward counterpart. This can be attributed to three main factors as follows.

Firstly, the discounted-reward theory holds regardless of the classification of the induced MCs, whereas that of the average reward involves such classification. Because in RL settings, the transition probability $p(s'|s, a)$ is unknown, average reward algorithms require estimation or assumption about the chain classification, specifically whether unichain or multichain. Nevertheless, note that such (assumed) classification is needed in order to apply a specific (simpler) class of average-reward algorithms: leveraging the fact that a unichain MDP has a single scalar gain (associated with its single chain) that is constant across all states, whereas a multichain MDP generally has different gain values associated with its multiple chains.

Secondly, the discounted Bellman optimality operator $\mathbb{B}_{\gamma}^{*}$ is contractive, where the discount factor γ serves as the contraction modulus, i.e. any real number in $[0, 1)$. That is,

$$\|\mathbb{B}_{\gamma}^{*}[\boldsymbol{v}] - \mathbb{B}_{\gamma}^{*}[\boldsymbol{v}']\|_{\infty} \leq \gamma \|\boldsymbol{v} - \boldsymbol{v}'\|_{\infty}, \quad \text{for any vectors } \boldsymbol{v}, \boldsymbol{v}' \in \mathbb{R}^{|\mathcal{S}|},$$

where in state-wise form, $\mathbb{B}_{\gamma}^{*}[\boldsymbol{v}](s) := \max_{a \in \mathcal{A}}\{r(s, a) + \gamma \sum_{s' \in \mathcal{S}} p(s'|s, a)v(s')\}$ for all $s \in \mathcal{S}$, and $\|\boldsymbol{v}\|_{\infty} := \max_{s \in \mathcal{S}} |v(s)|$ denotes the maximum norm. This means that $\mathbb{B}_{\gamma}^{*}$ makes $\boldsymbol{v}$ and $\boldsymbol{v}'$ closer by at least γ such that the sequence of iterates $\boldsymbol{v}^{k+1} \leftarrow \mathbb{B}_{\gamma}^{*}[\boldsymbol{v}^{k}]$ converges to the fixed point of $\mathbb{B}_{\gamma}^{*}$, which is the optimal discounted value $\boldsymbol{v}_{\gamma}^{*}$, as $k \to \infty$. That is, $\lim_{k \to \infty} \|\boldsymbol{v}^{k} - \boldsymbol{v}_{\gamma}^{*}\|_{\infty} = \boldsymbol{0}$. In the absense of γ, the contraction no longer holds. This is the case for the average reward Bellman optimality operator $\mathbb{B}_{g}^{*}[\boldsymbol{v}](s) := \max_{a \in \mathcal{A}}\{r(s, a) + \sum_{s' \in \mathcal{S}} p(s'|s, a)v(s')\}, \forall s \in \mathcal{S}$. As a result, the basic value iteration based on $\mathbb{B}_{g}^{*}$ is not guaranteed to converge [19, Sec 2.4.3].

The aforementioned contraction property also applies to the discounted Bellman *expectation* operator, i.e. $\mathbb{B}_{\gamma}^{\pi}[\boldsymbol{v}](s) := \mathbb{E}_{A \sim \pi}\left[r(s, A) + \gamma \sum_{s' \in \mathcal{S}} p(s'|s, A)v(s')\right]$ for all $s \in \mathcal{S}$. Consequently, we have $\lim_{k \to \infty} \|\boldsymbol{v}^{k} - \boldsymbol{v}_{\gamma}^{\pi}\|_{\infty} = \boldsymbol{0}$. Here, the discounted policy value $\boldsymbol{v}_{\gamma}^{\pi}$ is the fixed point of $\mathbb{B}_{\gamma}^{\pi}$ such that $\boldsymbol{v}_{\gamma}^{\pi} = \mathbb{B}_{\gamma}^{\pi}[\boldsymbol{v}_{\gamma}^{\pi}]$, which is known as the discounted Bellman evaluation equation (BEE). Related to this

BEE, the discount factor $\gamma \in [0, 1)$ also plays an important role in the convergence of TD-based parametric value aproximators in that the inverse $(\boldsymbol{I} - \gamma \boldsymbol{P}_\pi)^{-1}$, which involves in the approximator's minimizer formula, exists [29, p206]. This is in contrast to the matrix $(\boldsymbol{I} - \boldsymbol{P}_\pi)$ whose inverse does not exist [25, p596].

Third, discounting can be used to reduce the variance of, for example policy gradient estimates, at the cost of bias-errors [1,16]. In particular, the variance (*the bias-error*) increases (*decreases*) as a function of $1/(1 - \gamma) = \sum_{t=0}^{\infty} \gamma^t$. This is because the effective number of timesteps (horizon) can be controlled by γ. This is also related to the fact that the infinite-horizon discounted reward v_γ (4) can be ϵ-approximated by a finite horizon τ proportional to $\log_\gamma(1 - \gamma)$. That is, $\tau = \left\lceil \log_\gamma \frac{(1-\gamma)\epsilon}{r_{\max}} \right\rceil$, where $\lceil x \rceil$ indicates the smallest integer greater than or equal to $x \in \mathbb{R}$.

4   Artificial Discount Factors Are Sensitive and Troublesome

The artificial discount factor γ (which is part of the solution method) is said to be *sensitive* because the performance of RL methods often depends largely on γ. Fig 1a illustrates this phenomenon using Q_γ-learning with various γ values. As can be seen, higher γ leads to slower convergence, whereas lower γ leads to suboptimal policies (with respect to the most selective criterion, which in this case, is the gain optimality since the MDP is recurrent). This trade-off is empirically balanced around $\gamma_{\mathrm{Bw}} \approx 0.83$. Such trade-off is elaborated more in Sects. 4.1 and 4.2.

The artificial γ is *troublesome* because its critical value, i.e. γ_{Bw}, is difficult to determine, even in DP where the transition and reward functions are known [13]. This is exacerbated by the fact that γ_{Bw} is specific to each environment instance (even from the same environment family, as shown in Fig. 1a). Nevertheless, knowing this critical value γ_{Bw} is always desirable. For example, despite the gain optimality can be attained by having γ very close to 1, setting $\gamma \leftarrow \gamma_{\mathrm{Bw}}$ (or some value around it) leads to not only convergence to the optimal gain (or close to it) but also faster convergence in recurrent MDPs, as demonstrated by Q_γ-learning (Fig. 1b). We can also observe that the discounted $v_{\gamma \approx \gamma_{\mathrm{Bw}}}$-landscape already resembles the gain v_g-landscape. Thus, for obtaining the gain-optimal policy in recurrent MDPs, γ does not need to be too close to 1 (as long as it is larger than or equal to γ_{Bw}); see also (13).

Apart from that, γ is troublesome because some derivation involving it demands extra care, e.g. for handling the discounted state distribution in discounted-reward policy gradient algorithms [23,30].

4.1   Higher Discount Factors Lead to Slower Convergence

According to (9), increasing the discount factor γ closer to 1 makes the scaled discounted reward $(1 - \gamma)v_\gamma$ approximate the average reward v_g more closely.

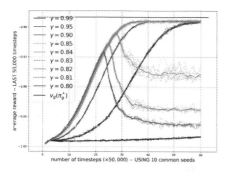

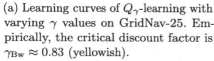

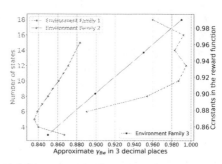

(a) Learning curves of Q_γ-learning with varying γ values on GridNav-25. Empirically, the critical discount factor is $\gamma_{Bw} \approx 0.83$ (yellowish).

(b) The critical γ_{Bw} as a (non-trivial) function of number of states, and of some constant in the reward function on three environment families.

Fig. 1. Empirical results illustrating the sensitivity and troublesomeness of artificial γ. In (a), the black horizontal line on the top is the optimal gain $v_g(\pi_g^*)$. (Color figure online)

This means that a discounted-reward method with such a setting obtains more accurate estimates of gain-optimal policies. However, it suffers from a lower rate of convergence (to the approximate gain optimality), as well as from some numerical issue (since for example, it involves the term $1/(1-\gamma)$ that explodes as $\gamma \to 1$). This becomes unavoidable whenever γ_{Bw} is indeed very close to unity because the most selective Blackwell optimality (equivalent to gain optimality in recurrent MDPs) requires that $\gamma \geq \gamma_{Bw}$.

The slow convergence can be explained by examining the effect of the effective horizon induced by γ. That is, as γ approaches 1, the reward information is propagated to more states [5, Fig 1]. From discounted policy gradient methods, we also know that an i.i.d state sample from the discounted state distribution is the last state of a trajectory whose length is drawn from a geometric distribution $Geo(p = 1 - \gamma)$. Evidently, the closer γ to 1, the longer the required trajectory. Also recall that such a geometric distribution has a mean of $1/(1-\gamma)$ and a variance of $\gamma/(1-\gamma)^2$, which blow up as $\gamma \to 1$.

There are numerous works that prove and demonstrate slow convergence due to higher γ. From them, we understand that the error (hence, iteration/sample complexity) essentially grows as a function of $1/(1-\gamma)$. Those works include [21] for Q-learning with function approximators, [29] for TD learning, and [1] for policy gradient methods. We note that for a specific environment type and with some additional hyperparameter, [9] proposed a variant of Q-learning whose sample complexity is independent of γ.

4.2   Lower Discount Factors Likely Lead to Suboptimal Policies

Setting γ further from 1 such that $\gamma < \gamma_{Bw}$ yields γ-discounted optimal policies that are suboptimal with respect to the most selective criterion (see Fig. 1a).

From the gain optimality standpoint, lower γ makes $(1-\gamma)v_\gamma$ deviate from v_g in the order of $\mathcal{O}(1-\gamma)$ as shown by [4]. More generally based on (13), $\gamma < \gamma_{\mathrm{Bw}}$ induces an optimal policy $\pi^*_{\gamma<\gamma_{\mathrm{Bw}}}$ that is not Blackwell optimal (hence, $\pi^*_{\gamma<\gamma_{\mathrm{Bw}}}$ is also not gain optimal in recurrent MDPs). This begs the question: is it ethical to run a suboptimal policy (due to misspecifying the optimality criterion) in perpetuity?

For a parameterized policy in recurrent MDPs, $\gamma < \gamma_{\mathrm{Bw}}$ induces a discounted v_γ-landscape that is different form the gain v_g-landscape. Such a discrepancy becomes more significant as γ is set further below 1. Therefore, the maximizing parameters do not coincide, i.e. $\mathrm{argmax}_{\boldsymbol{\theta}}\, v_{\gamma<\gamma_{\mathrm{Bw}}}(\pi(\boldsymbol{\theta})) \neq \mathrm{argmax}_{\boldsymbol{\theta}}\, v_g(\pi(\boldsymbol{\theta}))=:\boldsymbol{\theta}^*_g$, where $\boldsymbol{\theta} \in \Theta$ denotes the policy parameter in some parameter set. Interestingly, $v_{\gamma<\gamma_{\mathrm{Bw}}}(\pi(\boldsymbol{\theta}^*_g))$ is a local maximum in v_γ-landscape so that the $(\gamma < \gamma_{\mathrm{Bw}})$-discounted-reward optimization is ill-posed in that the (global) maximum is not what we desire in terms of attaining the most selective criterion.

An error bound is established in [24, Thm 2] due to misspecifying a discount factor $\gamma < \gamma_{\mathrm{Bw}}$. Subsequently, Jiang [15] refined the aforementioned error bound by taking into account the transition and reward functions. They also highlighted that such an error as a function of γ is *not* monotonically decreasing (with increasing γ). This is consistent with what Smallwood [28] observed, i.e. multiple disconnected γ-intervals that share the same γ-discounted optimal policy. We note that specifically for sparse-reward environments (where non-zero rewards are not received in every timestep), a lower discount factor $\gamma < \gamma_{\mathrm{Bw}}$ is likely to improve the performance of RL algorithms [24, Thm 10]. They argue that lowering γ decreases the value approximation error $\|v^*_\gamma - \hat{v}^*_\gamma\|_\infty$ more significantly than it increases the γ-misspecification error $\|v^*_{\gamma_{\mathrm{Bw}}} - v^*_\gamma\|_\infty$, where $\hat{v}^*_\gamma$ denotes an estimate for v^*_γ.

5   Benefits of the Average Reward in Recurrent MDPs

In this Section, we enumerate the benefits of directly maximizing the average reward (gain) optimality criterion in recurrent MDPs. Loosely speaking, such a recurrent structure is found in continuing environments[3] with cyclical events across all states. For an episodic environment, we can obtain a recurrent MDP by explicitly modelling the episode repetition. For a non-exhaustive list of continuing and episodic environments, refer to [10].

The combination of recurrent MDPs and gain optimality is advantageous. There is no discount factor γ involved (as a by-product), removing the difficulties due to artificial discounting (Sect. 4). Other benefits are described next.

5.1   Unconditionally the Most Selective Criterion

Gain optimality is the most selective criterion for recurrent MDPs. This is because all states are recurrent so that the gain is all that is needed to quantitatively measure the quality of any stationary policy from those states (such a gain

---

[3] Continuing environments are those with no terminal state; cf. episodic environments.

quantity turns out to be constant for all states in recurrent or more generally unichain MDPs). Recall that the gain is concerned with the long-run rewards (3), and recurrent states are states that are re-visited infinitely many times in the long-run.

Gain optimality therefore is equivalent to Blackwell optimality *uncondition-ally*, as well as *instantaneously* in that there is no need for hyperparameter tuning for the optimization objective function itself. This is in contrast to γ-discounted optimality, where it is equivalent to Blackwell optimality if $\gamma \geq \gamma_{\mathrm{Bw}}$ as in (5), and since γ_{Bw} is unknown, tuning γ is necessary (Sect. 3.2).

5.2   Uniformly Optimal Policies

Since recurrent (up to unichain) MDPs have only a single recurrent class (chain), the gain of any policy is constant across all states. As a result, average-reward policy gradient methods maximize an objective (14) that is independent of initial states, or generally of initial state distributions. The resulting gain-optimal policies are said to be *uniformly optimal* because they are optimal for all initial states or all initial state distributions [2, Def 2.1].

In contrast, the discounted-reward counterpart maximizes an objective (15) that is defined with respect to some initial state distribution $\mathring{p}$ (since the discounted-reward value v_γ depends on the state from which the value is measured, as in (4)). Consequently, the resulting γ-discounted optimal policies may not be optimal for all initial states; they are said to be *non-uniformly optimal*, as noted by [3, p41]. This non-uniform optimality can be interpreted as a relaxation of the uniform optimality in DP, which requires that the superiority of optimal policies π_γ^* holds in every state, i.e. $v_\gamma(\pi_\gamma^*, s_0) \geq v_\gamma(\pi, s_0)$, for all states $s_0 \in \mathcal{S}$ and all policies $\pi \in \Pi_{\mathrm{S}}$, see Sect. 2.1.

The objectives of average- and discounted-reward policy gradient methods:

$$\text{Average-reward policy gradient objective:} \quad \underset{\theta \in \Theta}{\operatorname{argmax}}\, v_g(\pi(\boldsymbol{\theta})), \tag{14}$$

$$\text{Discounted-reward policy gradient objective:} \quad \underset{\theta \in \Theta}{\operatorname{argmax}}\, \mathbb{E}_{\mathring{p}}\left[v_\gamma(\pi(\boldsymbol{\theta}), S_0)\right], \tag{15}$$

where $\boldsymbol{\theta} \in \Theta$ is the policy parameter and $S_0 \sim \mathring{p}$ is the initial state.

5.3   Potentially Higher Convergence Rate

Without delving into specific algorithms, there are at least two reasons for faster convergence of average-reward methods (compared to their discounted-reward counterparts), as hinted by [26, Sec 6] and [32, p114]. **First** is the common gain across states. Such commonality eases the gain approximation in that no generalization is needed, i.e. a single gain estimation is all that is needed for one true gain of all states in unichain MDPs. **Second**, average-reward methods optimize solely the gain term in the Laurent series expansion of v_γ in (6). On the other hand, their discounted-reward counterparts optimize the gain, bias, and higher-order terms altogether simultaneously, whose implication becomes more substantial as γ is set further below 1, as can be observed in (7).

6   Conclusions

We contribute a thorough examination of the relationship between average and discounted rewards, as well as a discussion of their pros and cons in RL. Our examination here is devised through broader lens of refined optimality criteria (which generalize average and discounted rewards), inspired by the seminal work of Mahadevan [20]. It is also broader in the sense of algorithmic styles: value- and policy-iteration, as well as of tabular and function approximation settings in RL. We provide updates and extensions to the existing comparison works, such as [18,31], in order to obtain a comprehensive view on discounting and discounting-free RL.

Maximizing the average reward has a number of benefits that make it worthwhile to use and to investigate further. It is the root for approaching Blackwell optimality through Veinott's criteria, which are discounting-free (eliminating any complication due to artificial discounting). Future works include examination about exploration strategies: to what extent strategies developed for the discounted rewards applies to RL aiming at discounting-free criteria.

References

1. Agarwal, A., Kakade, S.M., Lee, J.D., Mahajan, G.: On the theory of policy gradient methods: optimality, approximation, and distribution shift. arXiv: 1908.00261 (2019)
2. Altman, E.: Constrained Markov Decision Processes. Taylor & Francis, Boca Raton (1999)
3. Bacon, P.L.: Temporal Representation Learning. Ph.D. thesis, School of Computer Science, McGill University, June 2018
4. Baxter, J., Bartlett, P.L.: Infinite-horizon policy-gradient estimation. J. Artif. Intell. Res. **15**(1), 319–350 (2001)
5. Beleznay, F., Grobler, T., Szepesvari, C.: Comparing value-function estimation algorithms in undiscounted problems. Technical report (1999)
6. Bellman, R.: Dynamic Programming. Princeton University Press, New York (1957)
7. Blackwell, D.: Discrete dynamic programming. Ann. Math. Stat. **33**(2), 719–726 (1962)
8. Chang, H.S., Hu, J., Fu, M.C., Marcus, S.I.: Simulation-Based Algorithms for Markov Decision Processes, 2nd edn. Springer, London (2013). https://doi.org/10.1007/978-1-4471-5022-0
9. Devraj, A.M., Meyn, S.P.: Q-learning with uniformly bounded variance: Large discounting is not a barrier to fast learning. arXiv: 2002.10301 (2020)
10. Dewanto, V., Dunn, G., Eshragh, A., Gallagher, M., Roosta, F.: Average-reward model-free reinforcement learning: a systematic review and literature mapping. arXiv: 2010.08920 (2020)
11. Feinberg, E.A., Shwartz, A.: Handbook of Markov Decision Processes: Methods and Applications, vol. 40. Springer, New York (2002). https://doi.org/10.1007/978-1-4615-0805-2
12. Henderson, P., Islam, R., Bachman, P., Pineau, J., Precup, D., Meger, D.: Deep reinforcement learning that matters. In: Proceedings of the AAAI Conference on Artificial Intelligence (2018)

13. Hordijk, A., Dekker, R., Kallenberg, L.C.M.: Sensitivity analysis in discounted Markovian decision problems. Oper. Res. Spektrum **7**, 143–151 (1985)
14. Howard, R.A.: Dynamic Programming and Markov Processes. Technology Press of the Massachusetts Institute of Technology (1960)
15. Jiang, N., Singh, S., Tewari, A.: On structural properties of MDPs that bound loss due to shallow planning. In: Proceedings of the Twenty-Fifth International Joint Conference on Artificial Intelligence (2016)
16. Kakade, S.: Optimizing average reward using discounted rewards. In: Proceedings of the 14th Annual Conference on Computational Learning Theory (2001)
17. Machado, M.C., Bellemare, M.G., Talvitie, E., Veness, J., Hausknecht, M.J., Bowling, M.: Revisiting the arcade learning environment: evaluation protocols and open problems for general agents. Journal of Artificial Intelligence Research (2018)
18. Mahadevan, S.: To discount or not to discount in reinforcement learning: A case study comparing R-learning and Q-learning. In: Proceedings of the 11th International Conference on Machine Learning (1994)
19. Mahadevan, S.: Average reward reinforcement learning: foundations, algorithms, and empirical results. Mach. Learn. **22**, 159–195 (1996)
20. Mahadevan, S.: Sensitive discount optimality: Unifying discounted and average reward reinforcement learning. In: Proceedings of the 13th International Conference on Machine Learning (1996)
21. Melo, F.S., Ribeiro, M.I.: Q-learning with linear function approximation. In: Computational Learning Theory (2007)
22. Morimura, T., Uchibe, E., Yoshimoto, J., Peters, J., Doya, K.: Derivatives of logarithmic stationary distributions for policy gradient reinforcement learning. Neural Comput. **22**(2), 342–376 (2010)
23. Nota, C., Thomas, P.S.: Is the policy gradient a gradient? In: Proceedings of the 19th International Conference on Autonomous Agents and MultiAgent Systems (2020)
24. Petrik, M., Scherrer, B.: Biasing approximate dynamic programming with a lower discount factor. In: Advances in Neural Information Processing Systems, vol. 21 (2008)
25. Puterman, M.L.: Markov Decision Processes: Discrete Stochastic Dynamic Programming, 1st edn. Wiley, New York (1994)
26. Schwartz, A.: A reinforcement learning method for maximizing undiscounted rewards. In: Proceedings of the 10th International Conference on Machine Learning (1993)
27. Singh, S.P., Jaakkola, T.S., Jordan, M.I.: Learning without state-estimation in partially observable Markovian decision processes. In: Proceedings of the 11th International Conference on Machine Learning (1994)
28. Smallwood, R.D.: Optimum policy regions for Markov processes with discounting. Oper. Res. **14**, 658–669 (1966)
29. Sutton, R.S., Barto, A.G.: Introduction to Reinforcement Learning. MIT Press, Cambridge (2018)
30. Thomas, P.: Bias in natural actor-critic algorithms. In: Proceedings of the 31st International Conference on Machine Learning (2014)
31. Tsitsiklis, J.N., Van Roy, B.: On average versus discounted reward temporal-difference learning. Mach. Learn. **49**, 179–191 (2002)
32. Van Roy, B.: Learning and Value Function Approximation in Complex Decision Processes. Ph.D. thesis, MIT (1998)
33. Veinott, A.F.: Discrete dynamic programming with sensitive discount optimality criteria. Ann. Math. Stat. **40**(5), 1635–1660 (1969)

Correction to: Efficiency and Truthfulness in Dial-a-Ride Problems with Customers Location Preferences

Martin Aleksandrov and Philip Kilby

Correction to:
Chapter "Efficiency and Truthfulness in Dial-a-Ride Problems with Customers Location Preferences" in: H. Aziz et al. (Eds.):
AI 2022: Advances in Artificial Intelligence, **LNAI 13728,**
https://doi.org/10.1007/978-3-031-22695-3_48

In the originally published version of chapter 48, Table 1 and the equation on page 696 contained errors. This has been corrected.

The updated original version of this chapter can be found at
https://doi.org/10.1007/978-3-031-22695-3_48

Author Index